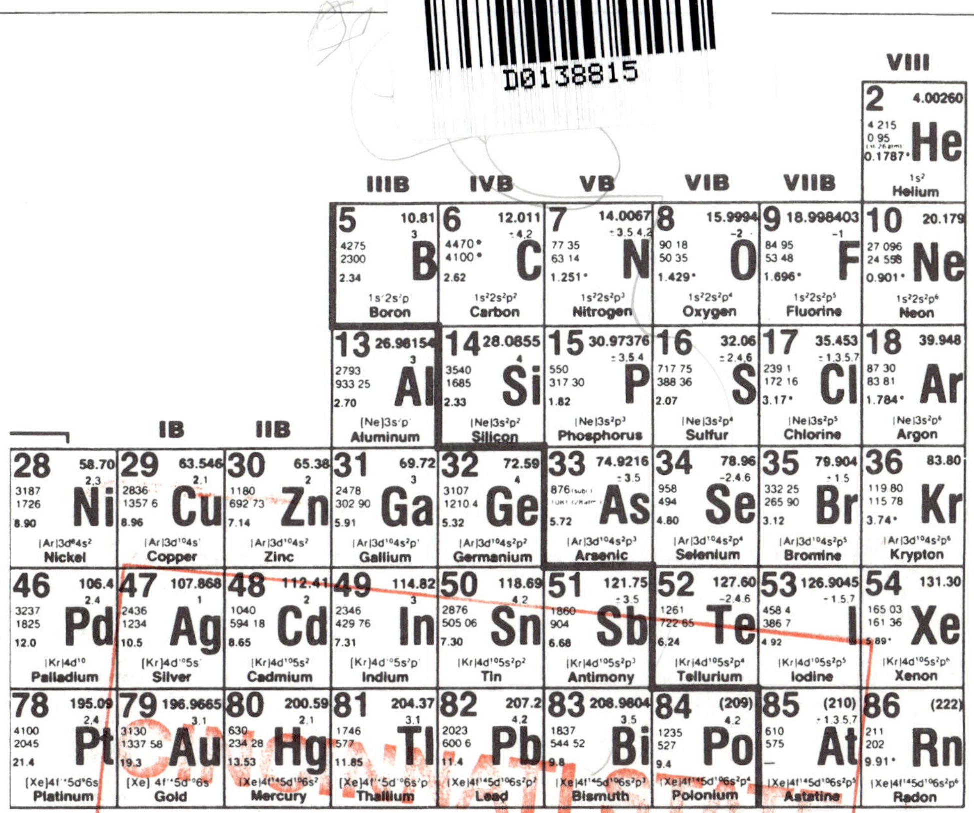

† The names and symbols of elements 104 - 106 are those recommended by IUPAC as systematic alternatives to those suggested by the purported discoverers. Berkeley (USA) researchers have proposed Rutherfordium, Rf, for element 104 and Hahnium, Ha, for element 105. Dubna (USSR) researchers, who also claim the discovery of these elements have proposed different names (and symbols).

The A & B subgroup designations, applicable to elements in rows 4, 5, 6, and 7, are those recommended by the International Union of Pure and Applied Chemistry. It should be noted that some authors and organizations use the opposite convention in distinguishing these subgroups.

No.	Atomic weight	Oxidation states	Boiling pt.	Melting pt.	Density	Symbol	Configuration	Name
64	157.25	3	3539	1585	7.89	Gd	$[Xe]4f^75d^16s^2$	Gadolinium
65	158.9254	3,4	3496	1630	8.27	Tb	$[Xe]4f^96s^2$	Terbium
66	162.50	3	2835	1682	8.54	Dy	$[Xe]4f^{10}6s^2$	Dysprosium
67	164.9304	3	2968	1743	8.80	Ho	$[Xe]4f^{11}6s^2$	Holmium
68	167.26	3	3136	1795	9.05	Er	$[Xe]4f^{12}6s^2$	Erbium
69	168.9342	3,2	2220	1818	9.33	Tm	$[Xe]4f^{13}6s^2$	Thulium
70	173.04	3,2	1467	1097	6.98	Yb	$[Xe]4f^{14}6s^2$	Ytterbium
71	174.967	3	3668	1936	9.84	Lu	$[Xe]4f^{14}5d^16s^2$	Lutetium
96	(247)	3	—	1340	13.511	Cm	$[Rn]5f^76d^17s^2$	Curium
97	(247)	4,3	—	—	—	Bk	$[Rn]5f^97s^2$	Berkelium
98	(251)	3	—	900	—	Cf	$[Rn]5f^{10}7s^2$	Californium
99	(252)	—	—	—	—	Es	$[Rn]5f^{11}7s^2$	Einsteinium
100	(257)	—	—	—	—	Fm	$[Rn]5f^{12}7s^2$	Fermium
101	(258)	—	—	—	—	Md	$[Rn]5f^{13}7s^2$	Mendelevium
102	(259)	—	—	—	—	No	$[Rn]5f^{14}7s^2$	Nobelium
103	(260)	—	—	—	—	Lr	$[Rn]5f^{14}6d^17s^2$	Lawrencium

SARGENT-WELCH SCIENTIFIC COMPANY
7300 NORTH LINDER AVENUE, SKOKIE, ILLINOIS 60077

Catalog Number S-18806

ENGINEERING MATERIALS TECHNOLOGY

Structures, Processing, Properties & Selection

THIRD EDITION

JAMES A. JACOBS
Professor of Technology
Norfolk State University

THOMAS F. KILDUFF
Professor Emeritus
Thomas Nelson Community College

Prentice Hall
Upper Saddle River, New Jersey *Columbus, Ohio*

Library of Congress Cataloging-in-Publication Data
Jacobs, James A.
Engineering materials technology: structures, processing, properties & selection/James A. Jacobs, Thomas F. Kilduff.—3rd ed.
p. cm.
Includes bibliographical references and index.
ISBN 0-13-398793-0
1. Materials I. Kilduff, Thomas F.
TA406.J26 1997 96-1768
620.1'1—dc20 CIP

Cover photos: Lanxide Corporation; MacSteel–Fort Smith, Arkansas; Lanxide Corporation; Ernest J. Garcia, Sandia National Laboratories
Editor: Stephen Helba
Production Editor: Patricia S. Kelly
Design Coordinator and Cover Designer: Jill E. Bonar
Marketing Manager: Danny Hoyt
Production Manager: Pamela D. Bennett

This book was set in Times Roman and Univers by Graphic World and was printed and bound by Quebecor Printing/Book Press. The cover was printed by Phoenix Color Corp.

Printed in the United States of America
10 9 8 7 6 5 4 3 2

ISBN: 0-13-398793-0

Prentice-Hall International (UK) Limited, *London*
Prentice-Hall of Australia Pty. Limited, *Sydney*
Prentice-Hall of Canada, Inc., *Toronto*
Prentice-Hall Hispanoamericana, S. A., *Mexico*
Prentice-Hall of India Private Limited, *New Delhi*
Prentice-Hall of Japan, Inc., *Tokyo*
Simon & Schuster Asia Pte. Ltd., *Singapore*
Editora Prentice-Hall do Brasil, Ltda., *Rio de Janeiro*

As authors we are indebted to the many people whose constructive suggestions and cooperation helped bring this book to completion. However, most of all, we dedicate this book to our wives, Martha and Virginia, and daughters, Sherri, Tammi, Jeanene, Liz, and Suzy, to all of whom we owe a major debt for their understanding, unfailing support, optimism, and tangible support as we toiled for many years.

PREFACE

Because of the impact of materials science and engineering on technology and society, many people can benefit from a study of engineering materials technology. We wrote the first edition of this book for students in engineering and industrial technology programs. After it was published, we learned that many other people found our presentation and pedagogy refreshing and useful for them. Included in this group were professors and secondary school teachers who taught materials science and technology courses for non-technical students, industrial trainers who instructed managers and apprentices on new developments in materials and processing technology, teacher education professors who found the need to include more about the nature of materials in their courses, and many others who needed a very readable text/ready-reference book on materials technology.

Our academic colleagues, students, and industrial advisors provided extremely valuable input in the form of thorough reviews and suggestions for this third revision. We hope that you find a good balance in the manner in which we pared, rearranged, and added to this edition.

Building on experience from past editions and looking at formats and pedagogies from other books that we found effective as tools of learning, regardless of the subject matter, this edition should provide more synergism in the way we integrated common themes throughout. Those themes include (1) relationships among processing, structure, properties, and applications; (2) the importance of considering the total Materials Cycle in materials synthesis, processing, selection, and economics; (3) the need for environmentally-friendly design and manufacturing of systems and products; (4) new and/or improved technologies that influence many aspects of engineering materials technology such as micro/nanotechnology, recycling, surface engineering, smart materials and intelligent structures, and biomimetics; (5) encouragement for the reader to explore the many resources and databases outside this book; and (6) the role of standard practices in all aspects of design, testing, processing, manufacturing, selection, and applications of materials.

Also, students are continually prompted to observe materials all around them. Looking at applications for proper and improper materials usage reinforces concepts studied and rationale for their study. The study of materials science, engineering, and technology provides a natural vehicle for comprehension of concepts in math and science.

This book engages the reader in thought and activities that begin to develop problem-solving abilities in such areas as materials evaluation and selection, materials processes selection, failure analysis, and materials testing.

It seeks to develop competencies in applications of current engineering materials and to prepare the user to work with future materials (advanced materials, smart materials, high performance materials, designer materials, etc.) and advanced materials processes (net- and near-net shape, rapid solidification, rapid prototyping, etc.). Keeping in mind the limited time available to teach such a broad course, we provide some sample listings of materials data to reflect the nature of such information, but encourage readers to consult handbooks, standards, and the other reference tools of the field to appreciate the wide range of available data.

We emphasize materials in the systems context and integrate information about many related developments that have thrust materials engineering and materials science into the limelight. Our intent is to place materials in their proper context to help you see their relevance. Among the issues discussed are global competitiveness, concurrent engineering, use of a "green" approach to technology, alternative powered vehicles, use and disposal of hazardous materials, new instruments for characterizing, testing, synthesizing and processing, computerized databases, AI, CIM, total quality emphasis, and recycling.

ORGANIZATION

The book is organized into self-containing modules that consist of module outcomes that present what the reader should be able to do after studying the module; Pause & Ponders that present case studies or applications of materials developments, aimed at sparking the reader's interest and providing a reason to learn the content; concepts on the module's topic supported by solved problems and numerous illustrations and examples representing a full range of products, systems, and technologies; Applications & Alternatives that reinforce the concepts learned by illustrating current use of the information; Self-Assessments that provide open-ended essay questions, problems to solve, and objective questions; References & Related Readings that provide information on our sources of information, plus useful handbooks, journals, periodicals, and similar data; listings of Experiments and Updates from past National Educators Workshops, which can be obtained by using information in the Instructor's Manual. The experiments and demonstrations listed in each module, which have been tested and thoroughly peer reviewed, offer some excellent activities to supplement the text. The answers to all the problems and questions for the Self-Assessments are available in the Instructor's Manual.

Please become familiar with the Contents and Index sections. We have taken great care to fully develop these features in order to help you make the many cross references required of this wide-ranging discipline of engineering materials technology. With the Contents you can locate many useful aids in the Appendix such as Greek symbols, SI/US Customary conventions and conversions, tables of properties, ASTM abbreviations, hardness to strength conversions, trade names, material selection guides, and more. Within modules you will find useful visual aids such as our hardness scale comparison figure. All of these features are extensively cross referenced in the Index. This book should be a useful guide or ready reference, much like *Machinery's Handbook,* which you should keep at your side as you engage in engineering/industrial problem solving.

We would like feedback from readers of our work. Please write to us through Prentice Hall.

NEW:UPDATES AND OTHER RESOURCES

As creator in 1986 of the National Educators Workshop series (NEW:Update), Jim Jacobs has co-directed these annual workshops which aimed at providing emerging concepts on engineering materials, science, and technology with emphasis on laboratory experiments. Major sponsors of these workshops and related research were the National Aeronautics and Space Administration (NASA), National Institute of Standards and Technology (NIST), and Department of Energy (DOE). Faculty from community colleges, four-year colleges, and major universities from across the United States, Canada, and other nations gathered at these sessions to share ideas on more effective materials education. Many of the leading corporate and governmental agencies with materials science and engineering laboratories provided updates at the workshops on the full range of materials developments. The seminars presented at these workshops provided us with an opportunity to learn of the latest improvements.

Among those who provided updates were the NASA Langley Research Center, Materials Science and Engineering Laboratory of NIST, Oak Ridge National Laboratory (ORNL), Ford Motor Company, Pacific Northwest National Laboratory (PNNL), IBM Almaden and Yorktown Research Laboratories, DuPont Corporation, Battelle, Sandia Laboratory, Bell Core, General Electric, Materials Research Laboratory of The Pennsylvania State University, Corning Glass Works, American Society for Testing and Materials, National Center for Excellence in Metalworking Technology, and Naval Research Laboratory. The experience gained from the years of NEW:Updates and working with professors, engineers, and scientists in government and industry has provided valuable ideas for this revision. Experiments from the NEW:Updates are listed in each module, and copies are available by writing Jim Jacobs.

We have benefitted from the explosion of new reference materials related to engineering materials technology. For example, *Advanced Materials & Processes* (AM&P), ASM International's periodical, and the *Journal of Materials Education* (JME), published by the Materials Education Council of the U.S., provide excellent coverage for current and future materials developments. We recommend AM&P and JME as useful supplements to our book along with other periodicals such as SME's *Manufacturing Engineering* and those listed in the References and Related Readings section of each module.

ACKNOWLEDGMENTS

We have been most fortunate to receive so much input from so many sources. From those hundreds of educators, researchers, engineers, scientists, and students who participated for more than eleven years in the National Educators' Workshops series, we gained valuable insights and resources that helped us to keep our material current and appropriate.

We wish to thank our colleagues in community colleges and four-year institutions who served as reviewers for this edition, and list their names below. We appreciate the detailed reviews from Patricia L. Olesak of Purdue University, Alden P. Gandreau of Hudson Valley Community College, and William Wait of SUNY-Oswego.

Illustrating a text of this nature is a very difficult job. We appreciate the assistance of the numerous colleagues, societies, companies, and governmental laboratories, most of whom are listed with each figure's caption, for providing the valuable information and illustrations.

We especially wish to acknowledge invaluable assistance from the following people for all editions for their various inputs:

Ed Widener—Purdue University
Clint Bertrand—Texas A&M University
Jim Nagy—Erie Community College
Michael Wehrein—Lethbridge CC, Alberta, Canada
Rhonda Housley—Spartanburg Technical College
Carl Metzloff—Erie Community College
Larry Helsel—Eastern Illinois University
Sarah Joy Nichols—Cape Fear Community College
Richard Cowan—Ohio Northern University
Seth Bates—San Jose State University
James A. Clum—SUNY—Binghamton
Joseph Neville—Wentworth Institute of Technology
Kathleen Kitto—Western Washington University
David A. Smith—Lehigh University
James DeLaura—Central Connecticut State U.
Wayne Elban—Loyola College
Steve Piippo—Richland High School
Roy Bunnell, Mike Schweiger, Irene Hays, Denis Stracher, and Eugene Eschbach,—Battelle, Pacific Northwest National Laboratories
James E. Gardner & Charles E. Harris—NASA Langley Research Center
Laurie A. George—American Dental Association, Health Foundation
Robert Berrettini—Editor of JME
Hugh Baker—ASM International
Irvin Poston—General Motors Technical Center
Thomas C. Holka & Bennie S. Bailey—Ford Motor Company
Leonardo E. Boulden—Howmet Corporation
Douglas Craig—Director of Metals and Ceramics at Oak Ridge National Labs
Jonice Harris and Anna Fraker—National Institute for Standards & Technology
Louis Nenninger, II,—Ceramic Engineer for Owens-Illinois
Sankar Sastri—New York Technical College
Peter Beardmore—Director of Ford Motors' Materials Research Staff
George G. Marra—Deputy Director, U.S. Forest Products Laboratory, Madison, Wisconsin
Charles V. White—GMI Engineering and Management Institute
Cindy Neilson—Ceramic Engineer, IBM
Walter E. Thomas—Western Carolina University
Frank J. Rubino—Middlesex County College
Howard Hull—New York City Technical College
Charles Flanders—Texas A & M University
Mario J. Restive—Mohawk Valley Community College
Richard I. Phillips—Southwest Missouri State University
Peter Route—Wentworth Institute
Lynda L. Anderson—SCCC
Pravin Raut—Savannah State College
and many others.

Contents

Module 1
Engineering Materials Technology

After studying this module, you should be able to do the following:

1.1. Recall the nature of the broad field of engineering materials technology and the contributions of materials science and engineering to materials developments, processing, and manufacturing.

1.2. Explain the significance and consequences of the five stages in the Materials Cycle.

1.3. Describe the characteristics of the ideal engineering material and discuss obstacles to changing from one material to another.

1.4. Cite examples of recent developments in materials science and engineering that have affected technologies such as information, transportation, production, medical, and entertainment.

1.5. Discuss compromise and trade-off in materials selection and other materials selection factors, including tools used for selection.

1.6. Define algorithm and describe examples of using algorithms to solve an engineering problem.

1.7. Synthesize the personal and societal values gained from the study of engineering materials technology in terms of career preparation, consumerism, and citizenship.

Figure 1-1 Advances in engineering materials technology allow for improvements in many products. Advanced composites developed by the aerospace industry are now found in many types of sporting equipment. The OCLV frame for the performance competition racer, seen here, uses carbon fiber composite technology similar to that used for stealth fighters and the space shuttle. (Trek Bicycle Corporation.)

PAUSE & PONDER

We live in a ***materials world!*** Many of the advances in society, with the aid of technology, resulted from discoveries and developments in engineering materials. You exist in a world made of an infinite variety of materials. Materials such as the paper these words are printed on and the wooden, metal, or fabric seat that you are sitting on are often taken for granted by people. Recent decades have seen an explosion of interest in and knowledge about making our lives better by improving materials and materials processing for finer engineered products. Some people call this the ***Materials Age*** because of all these developments.

We want you to PAUSE for a moment and PONDER some of the products around you. Think about what materials went into their construction and the types of processes used to fashion these products. Can you name five different materials on products around you?

This book will open your eyes to the fantastic ***materials world.*** Through this study you will expand your knowledge about and competencies in structure, processing, properties, and processing of engineering materials. Such information can serve you well as a consumer and citizen, and in your career, because the one constant about materials technology is its constant change. As consumers demand better products, the technological team of technicians, scientists, engineers, crafters, and technologists will improve the materials and processes for better design and higher quality. One example of applying new materials and processing technology is seen in the Optimum Compaction Low Void (OCLV) bicycle in Figure 1-1.

To produce a world-class competition bike, the U.S. company Trek Bicycle Corporation set out to design an agile, lighter, and more durable frame. The uniquely constructed one-piece main frame consists of 65% carbon fiber strands that have been compacted in an epoxy resin matrix with a process that yields less than 1% void (air pockets) in the finished composite frame. Trek advertises the OCLV carbon frames as the world's lightest, stiffest, and strongest. Of their frames made of ZX aluminum alloy, Trek states that the alloy achieves desirable properties by combining aluminum, with its low density, with magnesium for tensile strength and zinc to improve corrosion resistance.

Attached to the rear of the OCLV frame is a 6061 T aluminum alloy rear triangle that is Tungsten Inert Gas (TIG) welded for a light and stiff structure. Trek maintains that human crafters provide the precision and passion for bike assembly that robots don't possess. It's easy to accept the passion of humans, but many will question whether humans perform with greater precision than robots.

As you see in Figure 1-1, the OCLV model bike uses various materials, including parts made of rubber, aluminum, steel, and plastic, that complement the carbon composite frame. As with many newer product designs, adhesive bonding plays a role in joining dissimilar materials. Advanced Bonding Technology (ABT) relies on epoxy resin to glue components into alignment to achieve the designed geometry without the need for post-fabrication adjustments, as required in other fabrication techniques. Teflon-impregnated composite bearings are also bonded into the rear aluminum triangle and main frame to eliminate maintenance and lubrication.

This Pause & Ponder sets the tone for this book. Your study should help you to (1) develop an awareness of the importance of engineering materials in everyday life; (2) recognize society's dependence on materials; and (3) appreciate the value of a knowledge of engineering materials technology for you as a consumer, citizen, and member of the technological workforce.

The Trek bicycle and its innovations portray the aggressive nature with which forward-thinking businesses seize on advances in materials technology to propel their corporations beyond their competitors. Their engineers can select from ***designer materials*** rather than settling for *off-the-shelf materials* (materials with limited structure and properties). Throughout this module and the text, we will cite many examples of unique applications of materials and processing technology that set the pace for superior products and systems in fields such as transportation, construction, communication, recreation, health, and entertainment. The examples will help you to develop important competencies in engineering materials technology that will serve you well.

Once more, we ask you to PAUSE for a moment and PONDER this brief discussion of the materials and processes used in a high-tech bike. As you read this module, reflect on the discussion to help in comprehending the new concepts presented. We have designed this book to provide you with an understanding of how materials science and engineering fit into the many aspects of technology and society. Industries such as automotive, aircraft, sports, computer, and entertainment rely heavily on materials technology. The illustrations, case studies, problems, and other topics were selected to give you a feel for the significance of both engineering materials and technical problem-solving required in modern industry and society. Enjoy!

1.1 ENGINEERING MATERIALS TECHNOLOGY

Materials are the matter of the universe. These substances have properties that make them useful in structures, machines, devices, products, and systems. The term ***properties*** describes behavior of materials when subjected to some external force or condition. For example, the tensile strength of a metal is a measure of the material's resistance to a pulling force. The ***Family of Materials*** (which will be explained in more detail in Module 2) consists of four main groups of materials: ***Metals*** (e.g., steel), ***Polymers*** (e.g., plastics), ***Ceramics*** (e.g., porcelain), and ***Composites*** (e.g., glass-reinforced plastics). The materials in each group have similar properties and/or structures, as will be described later.

Engineering materials is a term often used loosely to define most materials that go into ***products*** and ***systems.*** A telephone is a product that would be part of a telephone system composed of many telephones, wires, fiber optics, switches, computers, and so on. Engineering materials can also have a more specific meaning that refers to materials whose structure has been designed to develop specific properties for a given application. For example, *engineering plastics* are

> those plastics and polymeric compositions for which well-defined properties are available such that engineering rather than empirical [trial and error] methods can be used for the de-

sign and manufacture of products that require definite and predictable performance in structural applications over a substantial temperature range. (ASTM, 1990, p. 167)

In other words, engineering plastics such as polycarbonates and acetals could replace more ***conventional engineering materials*** such as steel and wood because their properties are competitive for structural components such as piping, cams, and gears. On the other hand, general-purpose plastics, such as polystyrene and vinyls, do not possess the properties to carry heavy loads but serve as packaging, upholstery, and so on.

The field of ***materials engineering*** "deals with the synthesis* and use of knowledge [structure, properties, processing, and behavior] in order to develop, prepare, modify, and apply materials to specific needs" (National Research Council, 1989, p. 20). Materials engineers have become very much in demand as we seek to improve the efficiency of products.

Materials science and engineering (MSE) has become a major field of study, one critical to many other fields. As defined by a National Academy of Sciences study, MSE involves the generation and application of knowledge relating the *composition, structure,* and *processing* of materials to their *properties* and *uses.* The "science" focuses on discovering the nature of materials, which in turn leads to theories or descriptions that explain how structure relates to composition, properties, and behavior. The "engineering," on the other hand, deals with use of the science in order to develop, prepare, modify, and apply materials to meet specific needs. The field is often considered an engineering science because of its applied nature. Materials science and engineering is interdisciplinary or multidisciplinary, embracing areas such as metallurgy, ceramics, solid-state physics, and polymer chemistry.

Engineering materials technology covers fields of applied science related to materials, materials processing, and the many engineering specialties dealing with materials such as research and development, design, manufacturing, construction, and maintenance.

Many new processes have evolved for the manufacture of engineering materials. Through these processes, coupled with the design of engineering materials, we now enjoy the benefits of superior engineering materials.

1.2 THE MATERIALS CYCLE

To better understand engineering materials technology, it is useful to view the materials cycle. The ***materials cycle*** can be broken into phases as shown in Figure 1-2. An explanation of each of the stages listed in the figure will provide you with insight into the importance of materials, how they affect our lives, and their total impact on society.

1.2.1 Extracting Raw Materials

The earth has provided us with the basic ingredients for producing an unlimited variety of materials. The basic building blocks of these materials are the 106 known chemical elements that comprise the *periodic table of elements,* shown in Table 10.1 of the appendix.

*Several terms, such as *synthesis* and *process,* introduced in this module will be developed later in the book. Our thorough index allows easy cross-reference of most terms.

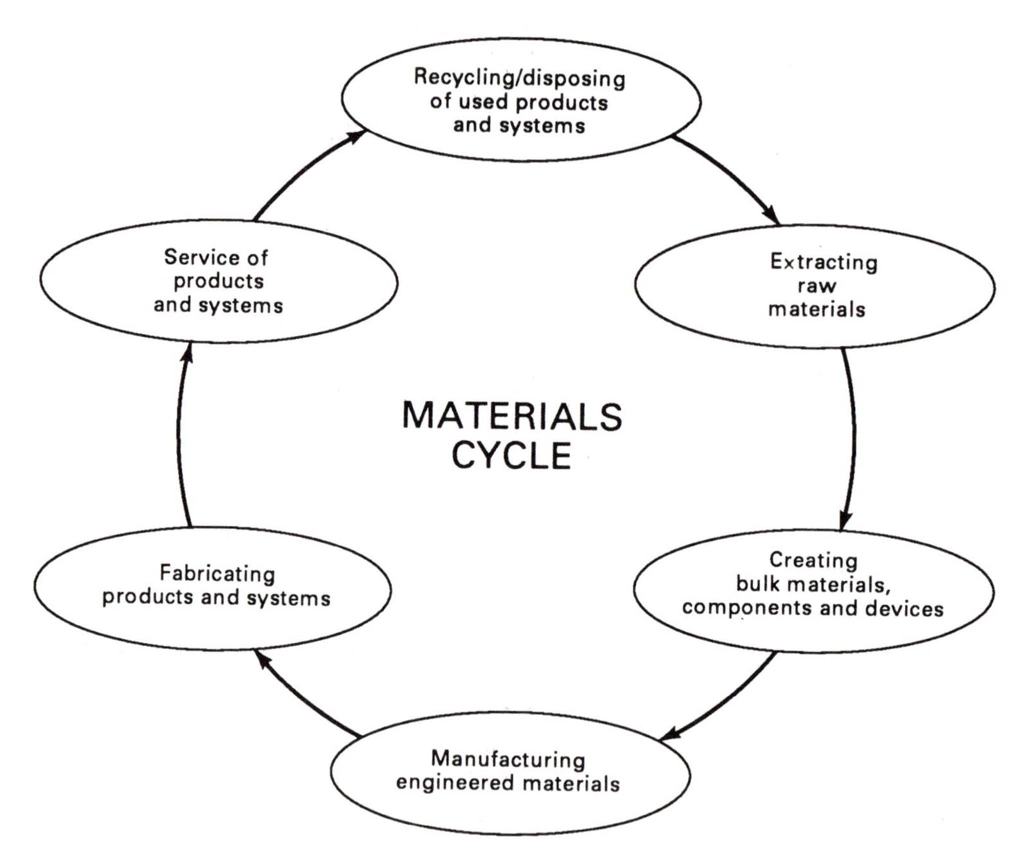

Figure 1-2 Materials Cycle.

Some of these elements, in the form of *solid materials,* are readily available to use, such as gold. Pure carbon in the crystalline form of diamond, the hardest known material, also takes little processing. But other elements are locked in the earth, ocean, or atmosphere.

It is often necessary to mine ore to obtain raw materials. Aluminum (Al), for example, is the most abundant metal; it comprises more than 15% of the earth's crust. However, when it is extracted from the earth, Al is combined with oxygen (O) to form alumina (Al_2O_3). It takes large amounts of electricity to *extract* Al from bauxite (Al_2O_3) ore. It takes 95% less energy to produce aluminum by *recycling* scrap aluminum products.

Synthesis involves transforming gases, liquids, and solid elements by chemical and physical means, where atoms and molecules are combined to form solid materials. An example of synthesis is production of a bulk material, such as nylon. The synthesis of polymers from ***raw materials,*** including coal, petroleum, water, and air, yields substances for producing *bulk materials* such as bulk nylon. Synthesis requires constant research by scientists to improve synthesis techniques.

While also dealing with atoms and molecules, ***materials processing*** "includes control of structure at higher levels of aggregation and may sometimes have an engineering aspect" (National Research Council, 1989, p. 224). Materials processing will yield bulk materials such as nylon tubing, individual components such as ceramic jet nozzles, devices such as semiconductors for computers, structures such as automobile frames, and systems such as fiber-optic communications systems. All the products and systems seen in the photos of this module resulted from materials processing. Lyle Swartz (1996) coined the term ***scitech*** to describe the complex intertwining of science with technological applications such as materials processing and manufacturing.

Materials processing has historically been considered technology rather than science. Generally, trial and error was used to develop the technology rather than scientific inquiry and theoretical analysis. But more recently, applications in fields such as applied chemistry, solid-state physics, and fracture mechanics have caused a merging of science with engineering and the development of the discipline of materials science and engineering.

The syntheses phases of materials technology have been considered more basic science involving scientists such as metallurgists, polymer chemists, and nuclear physicists. However, the development of advanced ceramics and advanced composites as well as that of electronic materials has caused a melding of disciplines. Many universities now offer degrees in materials science and engineering because of the need to ensure that the synthesis phase incorporates considerations for the processing of raw materials. These degrees might be considered *scitech* degrees.

One example of emerging materials technology is ***shape-limited synthesis,*** a new method developed to produce materials that blend synthesis with processing by beginning with one of the chemical agents already in the form of the final shape (see Module 8). This technology contrasts with the conventional procedure of first synthesizing liquids, particles, powders, or pellets, which are then processed by casting or molding into the final shape. These techniques are used in the production of advanced ceramics and composites.

During the raw materials extraction phase, there must be a concern for the by-products of refining and synthesis. Are gases released to the atmosphere that may have adverse effects on people, animals, vegetation, and even on other materials? Is the water used in separation processes returned safely to rivers and streams, or into the ground? Does the waste contain heavy metals or toxic chemicals that would be dangerous to the animal food chain (insects, birds, fish, cattle, then humans)?

1.2.2 Creating Bulk Materials, Components, and Devices

Bulk materials are the products of synthesis, materials extraction, refinement, and processing. There are many bulk materials with which you are familiar, such as fir plywood, sheet steel, acrylic tubing, window glass, copper wire, and concrete. Bulk materials are usually made in large quantities through continuous processing and then supplied to manufacturers of components and devices.

Components include gears, electrical wires, screws, nuts, jet engine turbine blades, brackets, levers, and the thousands of constituent parts that go into many products and systems. ***Devices,*** which include microprocessors, resistors, switches, and heating elements, are usually more complex than components and are designed to serve a specific purpose. ***Products*** are individual units, such as roller-blade skate sets, television sets, chairs, and telephones. ***Systems*** are an aggregate of products, components, and devices. For example, a telephone system is made up of millions of products (e.g., telephones, microwave transmitters, and computers), components (e.g., optical fiber, copper wire, and lasers), and devices (e.g., switches, relays, and microprocessors).

The source of bulk materials varies by material type. Some materials go through all stages from extraction of raw materials to production of finished parts by the same producers, but this would be an exception. Normally, specialists are involved in the various stages. For example, production of bulk polymers (plastic, rubber, paint, adhesives) is usually done by materials makers known as *polymer manufacturers.* These manufacturers are often located near oil refineries because most plastics are petroleum based (see Figure 6-5). The flow diagram for polymer production shows that the manufacturer begins with raw materials such as crude oil, natural gas, trees, and cotton to make chemical compounds such as ethane, trichloromethane, hydrogen fluoride, and ethylene chloride. From these raw materials, monomers (single molecules) are produced and include ethylene, methylmethacrylate, and vinyl chloride. Through heat, pressure, and the addition of chemicals (catalysts), the monomers are polymerized to form long-chain polymers. The final polymer (resin) is a bulk "virgin material" ready to go to the next stage of production, which is normally handled by ***fabricators.*** Polymer manufacturers may sell monomers to ***processors***; they polymerize the monomers into bulk plastics such as acrylic sheet or PVC [poly (vinyl chloride)] tubing. The processors then sell their bulk plastics to fabricators, who then produce engineered materials such as acrylic lenses for automobile brakelights.

Figure 1-3a depicts how advances in engineering materials technology have helped to improve properties of engineering materials. As the curve shows, traditional materials such as stone and cast iron had very low strength-to-density ratios *(specific strength).* In the latter half of the twentieth century, specific strength has increased in engineered materials to over 10×10^6 in. for advanced composites, compared to less than 0.5×10^6 in. with traditional, "off-the-shelf," engineering materials of wood, stone, and metals (National Research Council, 1989, p. 20). Figure 1-3b shows some applications of advanced ceramic and advanced composites, new breeds of materials, for engines.

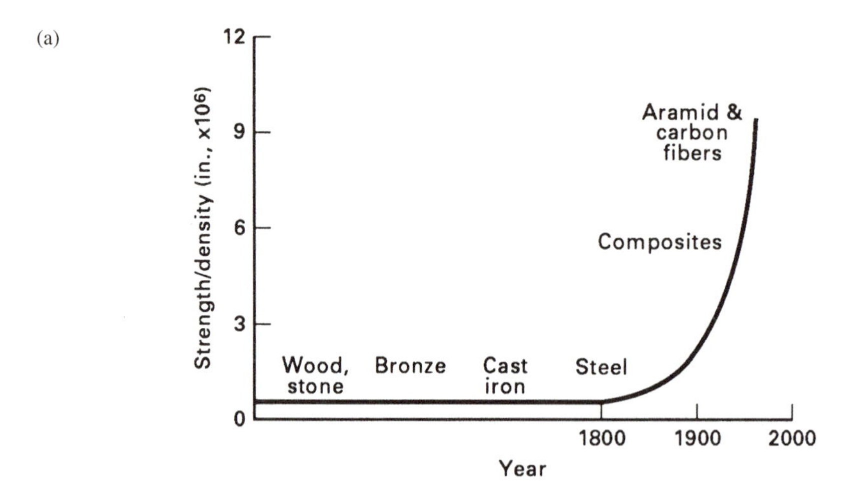

Figure 1-3 (a) Creating bulk materials and components using innovative techniques yields large gains. Modern materials have 50 times higher strength-to-weight ratios than cast iron and other early engineering materials. (Oak Ridge National Laboratory.)

(b)

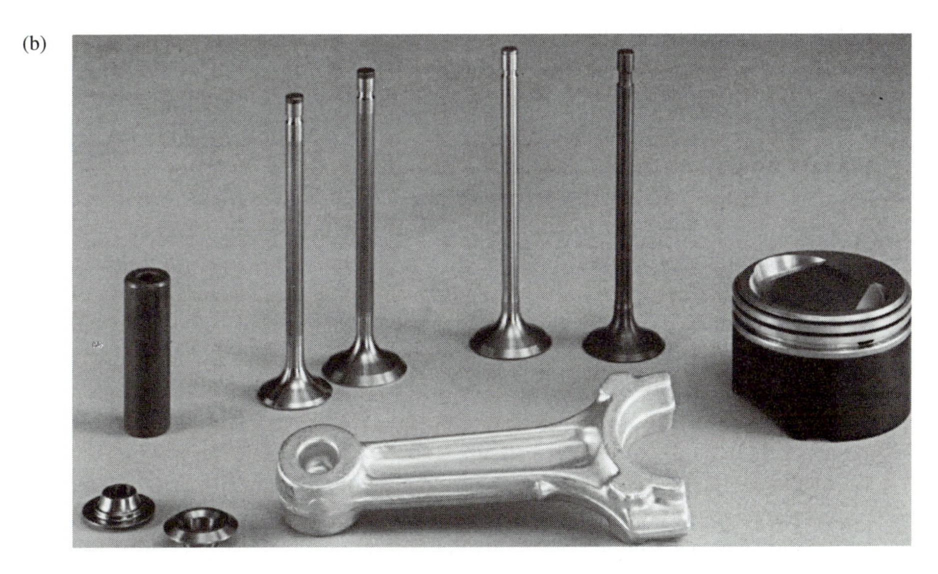

Figure 1-3 (b) More efficient engine components made of *advanced ceramics* and *advanced composites* replace *traditional materials.* Clockwise from upper left: silicon-nitride piston pin, two titanium valves, rapid solidification (RS) aluminum valve, silicon-nitride valve, metal-matrix composite (MMC) piston, RS aluminum connecting rod, and two titanium valve spring retainers. *(Advanced Materials & Processes.)*

Pre–industrial-revolution production techniques meant that production was done manually and individual components were cast or cut and *wrought* (shaped with heat and force) from bulk materials. Next, the bulk materials were turned into finished "customized" products; this required considerable skill and labor.

Mass production, which evolved through the industrial revolution, made it possible to produce large numbers of parts using machines that reduced the amount of human labor required. More jobs were created as demand for affordable products grew in developed countries. The materials science and engineering (MSE) field was in a formative stage as metallurgy began to evolve as a science aimed at better understanding the nature, properties, and processibility of metals and their alloys. Then, during the middle 1900s, MSE began to grow in importance as society began to demand more from technology; conventional materials could no longer meet the needs for such materials properties as higher temperatures, greater strength-to-weight ratios, and directional properties. It became necessary to engineer materials as well as products and systems.

1.2.3 Manufacturing Engineered Materials

The new generation of engineering materials are often ***designed materials***; they have been *engineered* to provide designated properties. In other words, instead of designers selecting from a list of available materials, they may specify the desired properties for their needs, and then rely on materials engineers and technologists to create materials to suit the need. ***Advanced composites*** are examples of engineered materials because engineers and technicians determine how reinforcing fibers should be aligned to withstand the stresses that a product will encounter under service conditions. These composites may be made of plastic resins that can withstand higher temperatures than general-purpose plastics. ***Advanced ceramics,*** one class of engineered materials, may be tougher than the normally brittle ceramics. Figure 7-22 shows the technique for manufacturing an engineered ceramic: chemical vapors of silicon carbide form a matrix as they are infiltrated into reinforcing fibers of silicon carbide for net-shape advanced ceramic composite parts.

Figures 1-4 and 1-5 depict the nature of manufacturing technology for engineered materials. For example, advancement of composite science and technology involves an integrated program, as seen in Figure 1-4, where materials science couples with the studies of engineering mechanics (study of forces in a materials system), durability, and engineering design; these studies then converge with issues of product life-cycle, fundamental laws (e.g., physics and chemistry), how material components interact, the quality and reliability of the materials system, and finally, how is cost affected by the system. Figure 1-5 shows an example of an advanced composite product that has resulted from the systems approach to manufacturing. The tennis racquet on the left has a pointer indicating the sweet spot of the racquet. The graphite fibers within the epoxy matrix, racquet frame were "engineered" to provide the proper directional properties to withstand impact from a tennis ball in the sweet spot region. An equivalent amount of force improperly applied to the racquet in another region, as seen by the pointer in the upper right, resulted in racquet failure. The directional properties of the materials in the frame were engineered only for normal impact from proper use of the racquet.

Many new processes have evolved for the manufacture of engineering materials, as shown in Figure 1-6. At each stage of the materials cycle, we must look at various issues,

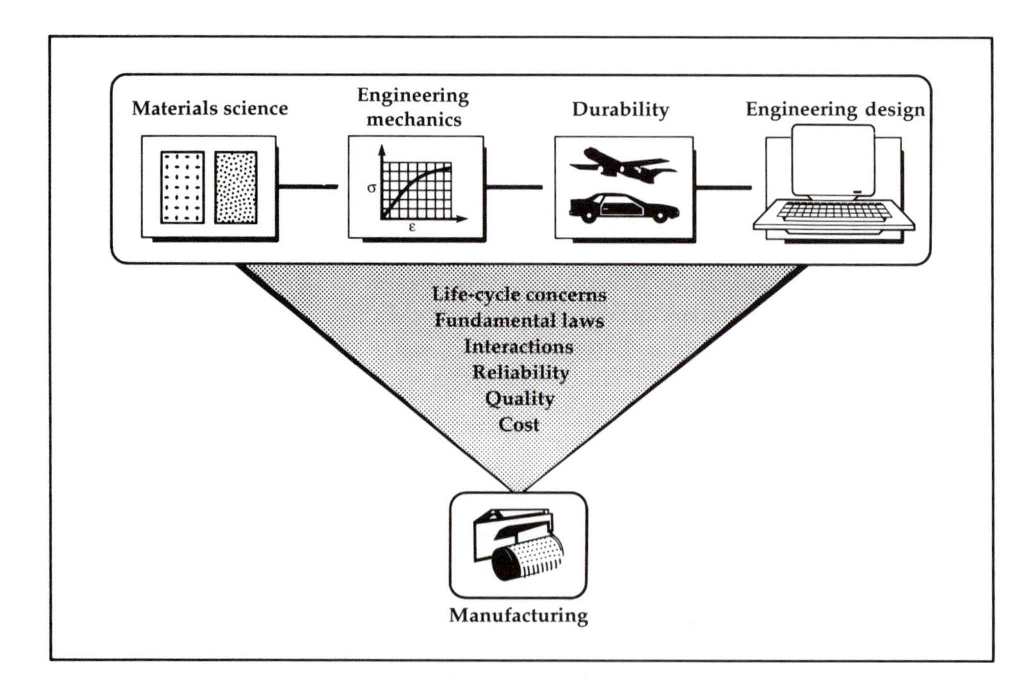

Figure 1-4 Manufacturing Engineering Materials—An integrated program combines traditional strengths with work in manufacturing processes to yield an expanding science base, new engineering methodologies and practice, and a unified-life-cycle approach to manufacturing. *(Advanced Materials & Processes.)*

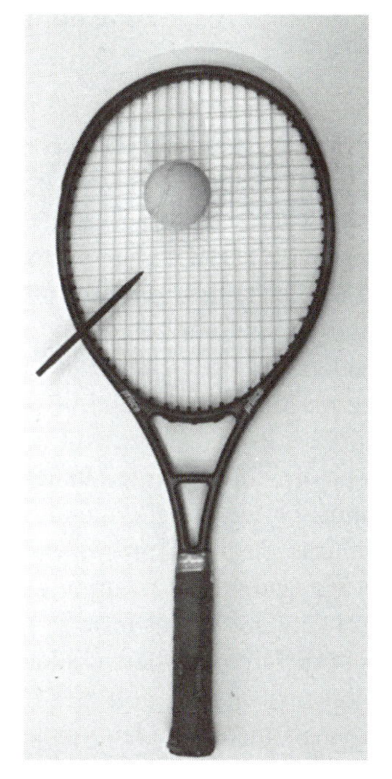

Figure 1-5 Left: Engineered graphite epoxy framed tennis racquet provides directional properties for a large sweet spot (at arrow) and considerable power. Right: Force (at arrow) improperly exerted on the frame, instead of the strings, caused the racquet to fail catastrophically.

Figure 1-6 Manufacturing Engineering Materials with Improvements—An engineering thermoplastic ABS resin, used in these interior door panels, elevates this grade of Cycolac® so that it's a competitor with the popular polypropylene plastic. Advantages of the ABS include reduced glare, for customer acceptance, and molded-in color, which is preferred by processors because of the potential to eliminate painting, lower component cost, reduced emission of VOCs (volatile organic compounds), and enhanced recycling. (G.E. Plastic.)

ranging from materials and production costs, to environmental impacts, to consumer acceptance. Note from Figure 1-6 how producers of engineering materials strive to address issues such as lowering cost, reducing processing steps, recycling materials, and improving air quality (reducing VOCs [volatile organic compounds] improves air quality). ***Intelligent processing of materials (IPM)*** is an evolving technology that uses computer modeling of processes and sensors put in place to monitor and permit control of processing. IPM yields better quality and more reliable products.

Another group of engineered materials is ***smart materials,*** a term referring to a variety of liquids and solids that have the ability at a predetermined condition to sense stresses and respond to alter their properties. For example, smart glass will darken when an electrical current is passed through a laminated grid in the glass. Electrical current can also thicken certain fluids. Optical fibers, metal fibers, and electrorheological (ER) fluids inserted into solids such as plastic composites, aluminum sheet, or concrete can sense stresses and cracking to provide early warning of probable failure. These fibers might be used in engineering applications such as insertion into aircraft wings and highway bridges to provide early warning of failure. Fibers could also serve to change the stiffness of automotive springs, helicopter blades, or golf clubs.

In the search for smart materials, ***biomimicking*** has been employed to study nature and attempt to mimic its wonders. Examples of biomimickry include efforts to reproduce the way that spiders produce very strong fibers and the procedures used by mollusks in building their shells, which may lead to better techniques for fabricating integrated circuits for computers and microprocessors.

1.2.4 Fabricating Products and Systems

Once engineering materials have been manufactured, they are assembled into many useful products and systems. New fabricating techniques are evolving through improvements in manufacturing engineering. ***Manufacturing engineering*** is the study of techniques to turn bulk materials into finished products and systems. Once a field of empirical methods in which processes were handed down from generation to generation, with improvements coming gradually as a result of observation and analysis, manufacturing has recently become more involved, with detailed study and application of theoretical principles. College programs leading to degrees have sprung up for manufacturing technology and manufacturing engineering at levels from associate degrees through doctorates in engineering.

Automation is the common element today in manufacturing, with less manual labor involved. Computers, sensors, robotics, machine vision, adaptive control, and artificial intelligence are applied by manufacturing engineers to perform the manufacturing processes once carried out by humans. Improved quality, smaller lot sizes, more product options, and reduction in price have been the benefits of the improved technology. Fabrication of products and systems is done in a safer work environment, where much of the handling of parts is done by machines. Just-in-time (JIT) techniques that rely on computer assistance keep raw materials and parts moving with a minimum of warehousing. People who wish to work in manufacturing are expected to be well educated because of their newer role as *problem solvers* rather than laborers.

A clear understanding of the total materials cycle is now necessary by the entire technological team of crafters, technicians, technologists, engineers, and scientists. Computer-integrated manufacturing (CIM) is a technological team approach to manufacturing that places the materials engineers and technologists together with the design engineers, technicians, and manufacturing engineers, plus the environmental engineers and even marketing personnel for a ***systems approach*** to products. New laws regarding the environment and natural resources are intended to ensure that the technological team considers the effect of all stages of the materials/product cycle to have appropriate safeguards that protect the environment as well as present and future generations. See Figure 1-6 for an example.

The change in the nature of fabricating products has meant numerous changes in the workplace. Manufacturing jobs for crafters and production workers (skilled, semiskilled, and unskilled labor) have decreased drastically. Jobs call for educated people who can design products and manufacturing systems; set up, monitor, troubleshoot, and service manufacturing equipment; and serve as liaison (go-between) among various stages of the total materials/product cycle to ensure that quality is always improving and customer needs are met.

1.2.5 Service of Products and Systems

Shifts in manufacturing have resulted in a larger service work force and a smaller manufacturing work force. The complexity of products makes it harder for the average person to make repairs on his or her own products. Special diagnostic equipment is used to analyze everything from automobiles to robots to appliances.

The demand for better quality in products and systems has resulted in improved, long-term warranties. Manufacturers are very interested in analyzing materials that fail so that they can improve materials engineering and product design. Proposed laws, and some laws already on the books, place responsibility for disposal of materials/products on the manufacturer once the service life is over. ***Design for assembly*** places emphasis on designed products that lend themselves to easy assembly by robots and other automated equipment. ***Design for disassembly*** is a concept that places recycling at the beginning or design stage of the materials cycle to ensure that waste going into municipal landfills will be minimized.

1.2.6 Recycling/Disposal

The last stage of the materials cycle can become the first stage through the resurrection of material when recycling is employed. Most materials can be recycled. However, it is very difficult for manufacturers to develop a full materials cycle that will ensure recycling. Examples of industries with successful recycling programs are steel and aluminum. These materials have been recycled successfully because of the tremendous savings gained through the production of bulk materials using recycled materials. It has taken legislation to force recycling, even of aluminum, as well as of glass and paper. This recycling has been accomplished by requiring deposits on beverage containers to provide financial incentives for people to return cans and bottles.

Federal, state, and local laws have put mandates on recycling by restricting the amount of municipal solid waste that can be placed in landfills. Clean air and water regulations have restricted the amount and type of waste that can be incinerated or dumped into the ocean. But much remains to be accomplished to develop the proper attitudes and habits among our citizens if we are to make the total materials cycle efficient and thus protect the environment and natural resources for future generations.

One example of the problems involved relates to materials for packaging. Manufacturers of polystyrene packaging (often improperly called styrofoam) have been working with McDonald's and other fast-food restaurants to head off complaints based on the fact that most plastics are not biodegradable. Citizen groups have called for elimination of the polystyrene clamshells used to package hot dogs, hamburgers, and other sandwiches, as well as for drink cups. Manufacturers set up a recycling system to reclaim the polystyrene. Because of the pressure and fear of losing business, most fast-food chains moved back to paper packaging before the recycling system could be fully implemented. Was this the best action?

First, we should recognize that paper will *not* biodegrade in the modern landfill. Modern landfills are designed to stop materials from breaking down, because they can then leach into the groundwater. So going to paper packaging of foods might actually create more solid waste because this type of paper is difficult to recycle. Complex issues are involved in decisions regarding the use of plastics versus paper or glass. Most of our local, state, and federal legislators lack technical backgrounds and therefore are not qualified to decide some of these complex issues. It is therefore vital that citizens prepare themselves with a knowledge of materials technology to provide input to the legal issues. Recycling and disposal of materials is one area that will continue to be a major issue.

1.2.7 Consequences of the Stages of the Materials Cycle

Mistakes costly to the environment and society have been made with mining, well drilling, and tree harvesting. Mine tailings left behind when mining ore are washed by rain, causing the leaching of concentrated minerals that might be harmful to plants and animals in the streams, rivers, and lakes. Oil drilling can result in harmful spillage. Clearcutting of trees can result in severe soil erosion and can upset wildlife.

On the other hand, if we did not extract these raw materials from the earth, civilization would not benefit from the valuable resources harvested, people would not have employment in the related fields, and we would not have a thriving commerce that offers a good quality of life. In other words, we have made trade-offs, some conscious, some not. For example, the plastics industry is a comparatively young industry that evolved during a time when oil and gas seemed limitless. There was not much knowledge or concern about pollution, global warming, the ozone layer, or the scarcity of landfill space for municipal solid waste. Plastics offered us sanitary packaging, great strength-to-weight ratios in materials, improved design flexibility, lower-cost products, and other advantages over the traditional materials such as metal, wood, paper, and ceramics.

Now that we recognize such problems as limited resources, energy waste, harmful by-products of materials processing, and materials disposal, the expectation is that everyone will apply knowledge of materials to ensure that our natural resources are best utilized and that engineering materials technology safeguards the environment.

1.3 MATERIALS SELECTION

We are surrounded by materials and we rarely think about how these materials are selected. Why was your desk made of solid wood, plywood, or plastic-laminated particle board? Why have so many plastics replaced steel and zinc in automobiles? What is the controversy about using foamed polystyrene plastic to package fast-food?

While you might take for granted the materials that make up your products, you can be sure that the designers did not. People who design homes, cars, aircraft, clothing, furniture, and other products or systems devote a lot of attention to the selection of the materials they use. Material selection might make or break a company. But how do the designers make that selection to arrive at the best material?

Did you realize that there are hundreds of different types of plastics? They range from very soft to quite hard, inexpensive to very expensive, and transparent to opaque. Wood, too, is available in numerous varieties, ranging from the very soft, lightweight balsa wood used for model airplanes to a group of birch and beech trees that are so heavy and hard that they are sometimes called ironwood. Lignum vitae, one of the hardest and heaviest woods, has been used for bearings and pulley sheaves on ships since ancient times.

Metals in combination with other metals or nonmetal elements, known as ***alloys,*** include many varieties of steels (iron and carbon), aluminum alloys, brass (copper and zinc), and hundreds of others. Steel, the most common production metal, is found in the bodies of cars, as railroad-train wheels, and for piano wire.

Some glasses, such as most drinking glasses, are very delicate and cannot stand much temperature change, but thermally hard glasses are used as peep holes on furnaces. Cement is one type of ceramic, as is glass. Clays are ceramic raw material; they are used for making dishes, toilets, and spark plug insulators.

With the nearly limitless range of materials available to the designer or architect, how do they make a materials selection for products and buildings? What selection criteria are most important?

1.3.1 The Ideal Material

The engineer, technologist, technician, or architect searches for the ideal materials to suit the designated need. What is an ***ideal material***? Among other characteristics, we can list the following for the ideal material.

1. Endless and readily available source of supply
2. Cheap to refine and produce
3. Energy efficient
4. Strong, stiff, and dimensionally stable at all temperatures
5. Lightweight
6. Corrosion resistant
7. No harmful effects on the environment or people
8. Biodegradable
9. Numerous secondary uses

It is a very complex process for the designer to find the ideal material for a specific product.

1.3.2 Selection and Compromise

Compromise is the rule, not the exception, in materials selection. For example, ***space-age materials*** such as graphite/epoxy composites and Kevlar (aramid fiber) epoxy composites possess strength-to-weight ratios three to five times greater than those of steel. So why don't automakers use graphite epoxy and aramid epoxy composites to make car bodies, axles, and drive shafts? Wouldn't use of these composites allow for much more fuel-efficient vehicles? Yes, the space-age materials would be much more fuel efficient—but because these materials can cost 15 to 40 times as much as steel, an $11,000 Ford might cost $300,000. Who could afford to buy it?

Ford Motor Company experimented with their popular Taurus model by building a prototype chassis that reflects designers' keen interest in using reinforced-plastic composite technology for improved vehicles. Instead of using the more exotic fiber-reinforced plastic (FRP) composites such as graphite epoxy, they used glass-reinforced plastics (GRP). In the Taurus prototype, they were able to use only 8 or 10 GRP parts to replace over 400 stamped steel parts. A GRP chassis would save about 30% of the car weight because a lighter chassis means downsizing engines, brakes, and other components. Additional benefits come through more efficient manufacturing techniques. The Partnership for a New Generation Vehicle (PNGV), discussed later in this book, was formed among the big three U.S. automakers and the federal government to explore alternative materials and designs.

Throughout this book, we provide numerous examples of alternative designs, materials, and processes from many fields of technology. These examples help to illustrate the complexity of materials selection. As pointed out before, materials selection is not simply a matter of making decisions about cost and materials properties.

1.3.3 Obstacles to Change

Switching from traditional materials such as steel and concrete to newer materials such as plastic-based composites seems a simple, straightforward approach for the contemporary designer. The newer materials are often superior, but sometimes there are complications. Often, lack of experience with a new material causes hesitation by designers. Departures from tried-and-true materials may be costly, as was the case with plumbing pipes. With a shortage of copper and the health threat of lead–tin alloyed soldered water pipes, there was a quest to find substitute materials. The answer seemed to lie in polybutylene oxide plastic pipe, which relied on metal/plastic compression fittings. The compression fittings seemed to be an excellent substitute for the highly skill-oriented sweat soldering required of copper pipes. Unfortunately, builders and municipal water departments that switched to the polybutylene soon learned that this new water piping system had its problems.

With time, many of the joints began to leak. The designers who selected the new pipe were led to believe that fabricating with compression joints was a rather simple task. The makers of the polybutylene system claimed that failures were due to improper installation of the compression joints. The users of the new piping blamed the problem on oxidizing polymers (plastics) and faulty aluminum compression rings.

This controversy focuses attention on a leading problem of all new materials: It requires time before both designers and fabricators gain sufficient experience to become comfortable with them and the associated processes required to make products or systems. This problem is exacerbated when human life might be in jeopardy, such as when designing for aircraft. As a consequence, new materials and processes are usually slower to enter the marketplace than might be expected. These issues are all a part of engineering problem solving. Materials selection is a problem-solving issue that requires an algorithm for its solution.

1.3.4 Algorithm for Materials Selection

Engineering requires clearly stated, unambiguous steps for problem-solving. ***Algorithms*** are well-defined methods for solving specific problems. Computer programs are written after an algorithm has been developed to lay out clearly the steps that the program is to solve. For example, you could write a simple algorithm to calculate the strength required of a light pole to withstand the pushing forces (compression) from a light fixture. A much more complex algorithm would be required to select a piston connecting rod for an internal-combustion engine. The first problem requires only the selection of a material of suitable size/strength to hold up the light fixture, and almost any material would suffice as long as it was sufficiently strong and pleasing to the user. On the other hand, a connecting rod will undergo many types of mechanical stress, ranging from compressive to tensile to torsional to gravity forces, in addition to thermal stress from the combustion chamber. How does the designer match component requirements with available materials?

1.3.4.1 Selection tools. To aid in the creation of materials selection algorithms, databases must be available to answer questions on material suitability. A materials ***database*** involves tables listing properties of materials, such as tensile strength, hardness, corrosion resistance, and the ability to withstand heat. Thousands of reference books are available with such data. Much of these data are computerized to allow easier access.

Certain graphical techniques aid the designer in materials selection. In Figure 1-7 a graphical plot of two important properties shows a relationship of stiffness (Young's modulus) to weight (density) of major groups of materials. For example, ellipses encircle the stiffness-to-density ratio, which represents those of most engineering ceramics and engineering alloys. By encircling major materials groups, a designer can see that both groups of materials possess high stiffness but at the expense of high density. On the other hand, engineering composites, as a group, possess stiffness that approaches or may equal that of engineering ceramics and alloys; however, composites have the advantage of much lower density. Similar graphical plots are possible for a variety of properties ratios that allow the narrowing of materials as design requirements narrow the choices.

1.3.4.2 Properties of materials. Periodicals can provide current data and performance criteria that involve structural (load-bearing) materials:

1. Strength (tensile, compressive, flexural, shear, and torsional)
2. Resistance to elevated temperatures
3. Fatigue resistance (repeated loading and unloading)
4. Toughness (resistance to impact)

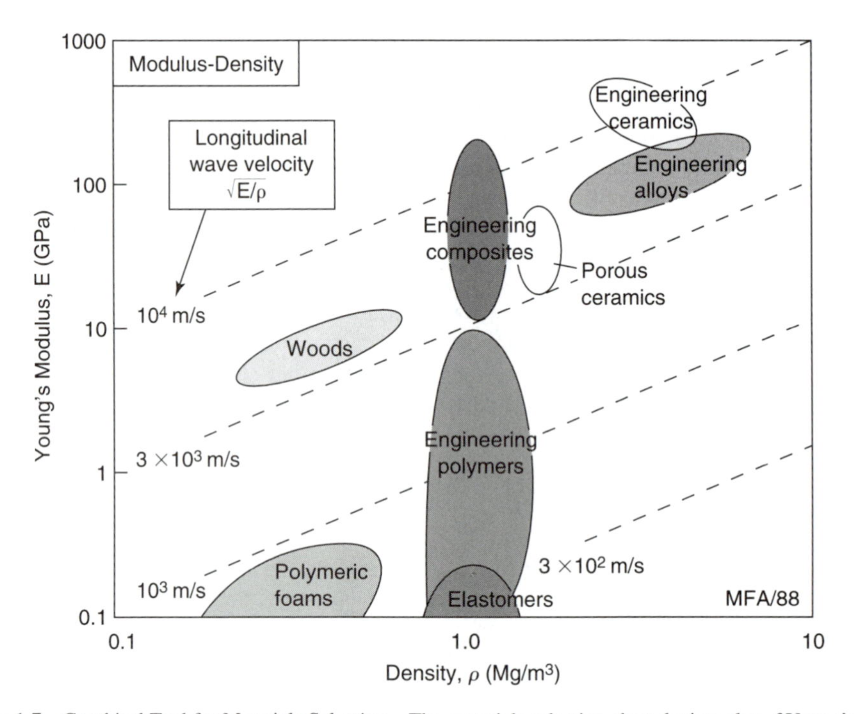

Figure 1-7 Graphical Tool for Materials Selection—The materials selection chart depicts plot of Young's modulus and material density. Boundaries encircle the clustering plots of these two properties to allow comparisons of groups of materials. (Ashby, Michael F., "Materials in Mechanical Design," *Journal of Materials Education*. Vol. 15, pp. 143–166, 1993.)

5. Wear resistance (hardness)
6. Corrosion resistance

Such publications present values for the performance criteria (properties) for metals, polymers, and ceramics, with updates on newer materials such as aramid fibers, zinc aluminum alloys, and superalloys.

Various periodicals, such as *Machine Design* and *Modern Plastics,* have annual materials selectors that provide general information on properties for a long list of materials. There are also many handbooks, such as the multiple-volume *Metals Handbook* and the series of *Engineered Materials Handbooks* that cover nonmetals; both are published by ASM International.

Tables as found in Module 10, covering representative materials, provide comparative data on selected ceramics, metals, and nylon. The designer must use such data to determine if a material has appropriate physical, mechanical, and chemical properties to withstand the service conditions to which a part will be subjected. For example, from Table 10-9 we learn that nylon is an engineering thermoplastic and is used for some bearing applications. By turning to Table 7-2, Typical Properties of Selected Engineering and Technical Ceramics, which compares selected plastics, metals, and ceramics, you will note that nylon has a melting point of 215°C and a maximum service temperature of 422 K (149°C). On the other hand, alumina (used for spark plugs and computer modules) has a melting point of 2050°C and a maximum service temperature of 2222 K. Alumina is also much harder than nylon: 9 Mohs versus 2. The modulus of elasticity [given in megapascal (MPa)] is a comparison of stiffness; alumina is very stiff (modulus of 34.5×10^4 MPa) and brittle, whereas nylon is flexible (modulus of 0.33×10^4 MPa). From the comparison of hardness and temperature resistance, it is clear why ceramics are selected for furnaces, ovens, and other applications where high abrasion and high heat are present. On the other hand, nylon is selected for rope, tubing, and gears and bearings in electrical appliances.

The two tables discussed above provide general data on properties for a simple comparison. Selection of specific materials requires many more detailed specifications. General databases from handbooks will provide much detail, but the final selection often requires that material manufacturers supply their own properties database for their product lines. While databases are imperative in the initial selection steps, there are other factors that complicate materials selection.

1.3.4.3 Materials systems. Materials rarely exist in isolation, without interacting with other materials. Rather, a combination of materials are selected to complement one another. In a successful ***materials system,*** each component is compatible with the others while contributing its distinctive properties to the overall characteristics of the system of which it is a part. A state-of-the-art telephone is a good example. The casing might be a tough ABS plastic, which houses a microchip (a solid-state ceramic device) that provides memory and sound-transmission capabilities. Copper leads join the circuitry together. There might be a battery and a ceramic light-emitting diode to show when the battery is low. The acid in the battery must be isolated to prevent corrosion, and the copper leads must be insulated so that they do not short out. Each component is made of materials that meet the demands of the physical and chemical environment normally encountered when using the system.

1.3.4.4 Additional selection criteria. *Existing specifications* have a lot of influence on the choice of material. These specifications or "standards" are used when redesigning an improved model of the product. When the materials-selection algorithm results in selection of a new material, it might not be covered by current specifications from such standardization agencies as the National Institute of Standards and Technology, Underwriters' Laboratory, fire departments, American Society for Testing and Materials, or the U.S. Food and Drug Administration. The Occupational Safety and Health Act (OSHA) of 1970 sets forth conditions of safety that must be met by those involved in the manufacture or use of goods and services in the United States. It might take considerable time for these agencies to alter their specifications to include the new material or they might not approve its use.

Availability is another concern of the designer. Will the material be easily available in the quantities and sizes required by the production demands? Also, will it be available in the shapes required? Aluminum extrusions, for example, are available in many varieties of standard shapes, such as round, oval, and square tubing. In the past, designers were limited to existing materials such as metal alloys, woods, or concrete. Now, it is possible to start from scratch at the synthesis stage to have materials engineers design a materials system to provide properties to meet the expected needs.

Processibility, the ease with which raw materials can be transformed into a finished product, is of paramount concern. Much of the current focus is on low-energy processing. Companies may have difficulty processing the new material on existing equipment. Can they afford to invest in new equipment? Today, the reverse question is usually asked: Can we afford not to use the new material and process? If we do not, the competition might make the change and run us out of business with their superior product. Many new technologies are now available.

Near-net-shape production involves incorporating numerous separate parts into a single, integrated assembly, thus saving overall production costs. This new technology is receiving considerable attention from various industries. Net-shape and near-net-shape processing aims to improve product reliability while reducing costs of materials and processing. One such example, the Super Plug™, a patented door hardware module (see Figure 1-8), replaced the 61 separate, stamped and formed steel, molded plastic, and rubber door parts with one plastic/glass composite system made of XENOY® polycarbonate/polyester blend with 30% glass fiber. The single system not only saves materials and processing costs but also provides manufacturing cost savings at assembly and yields overall weight reduction for fuel savings for the life of the vehicle. Another picture of the module is shown with the door in Figure 6-21.

The near-net-shape concept is a growing trend because it is consistent with the new thinking about design/manufacturing; that is, the concept design, engineering analysis, materials selection, and processing are now orchestrated in a team approach. In traditional practice, design was handled by an engineering group. It sent the drawings to manufacturing; that department was responsible for determining how to produce the part. Often, there was conflict in the demands that each group made on the other. The new CAD/CAM technologies force an integrated approach. Materials selection is central to the design process and has great bearing on manufacturing techniques. ***Expert systems*** as a part of computer-based ***artificial intelligence*** are becoming available to help make selections from the increasing number of variables and levels of optimization inherent in materials.

Plastics, ceramics, and composites have become increasingly competitive because engineering materials and near-net-shape products and parts are suitable for plastics such as polystyrene, nylon, polypropylene, polycarbonate, and glass-filled epoxy. The metal industry has also brought out new processes and refined older ones to simplify and integrate parts for near-net-shapes.

Quality and performance are two aspects that achieve consumer satisfaction. The high cost of most durable goods and the competition for customer acceptance has resulted in extended warranties. Materials selection must ensure that parts will not rust, break under repeated stress, or fail to perform in any other way for the predicted service life of the product.

Consumer acceptance includes many factors beyond excellent quality and high performance; there are also societal aspects. Society as a whole as well as governmental agencies are requiring a closer look at manufactured products. Any product has to be considered in terms of its total life cycle. What are the results of the processing methods? Are

(a)

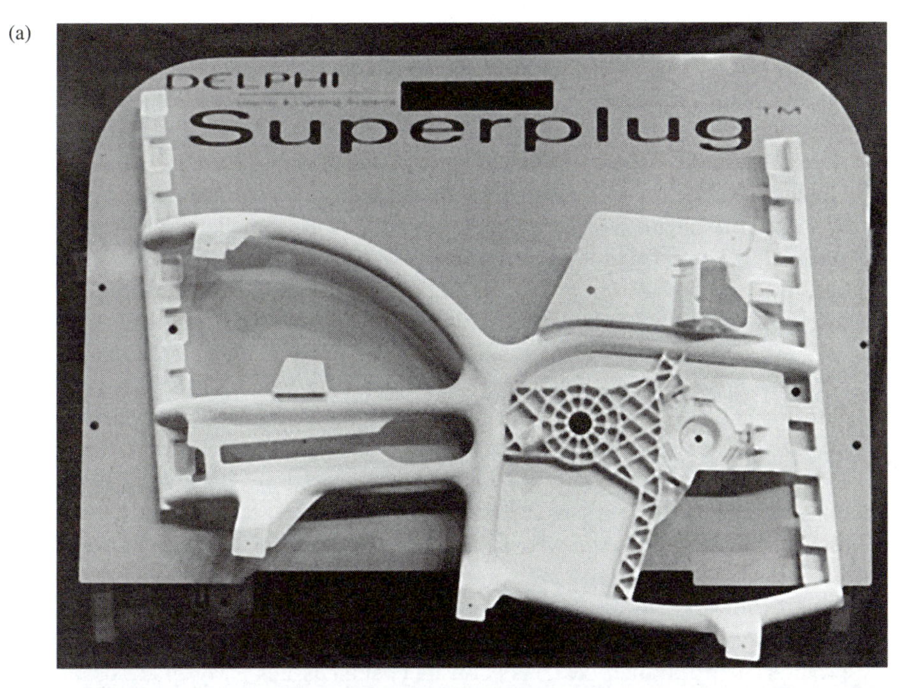

(b)

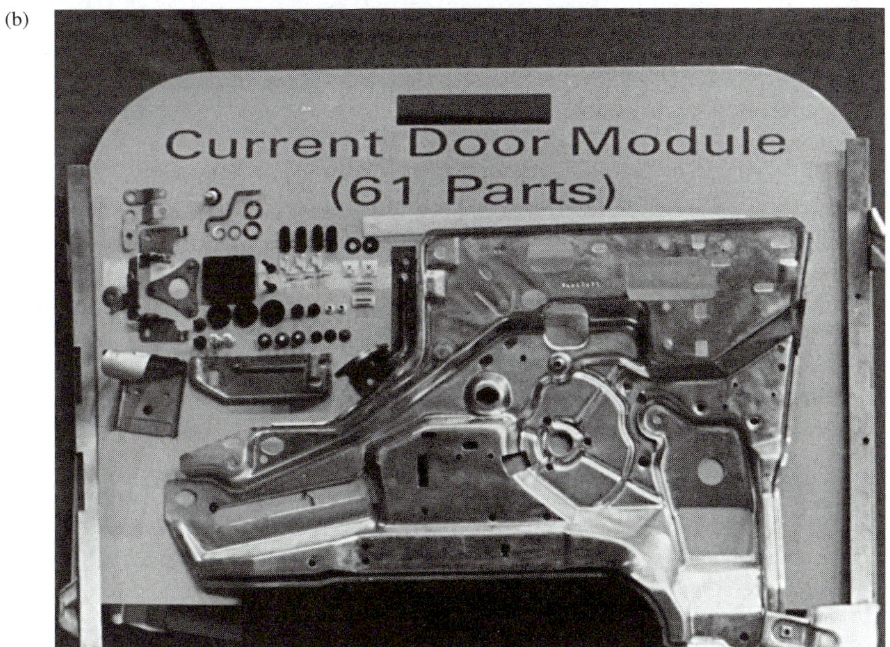

Figure 1-8 (a) G.E. Plastics Super Plug™ is an example of net-shape processing. The glass-reinforced plastic composite provides savings in both production and operating costs over the current door module (b), which is made of stamped steel and uses up to 61 parts. (G.E. Plastics.)

polluting gases being released into the environment, or are toxic metals and chemicals being flushed into our rivers and streams? During use, does the product safeguard our health? At the end of the product's useful life, how can it be disposed of safely? Municipal solid waste is a hidden product cost that we pay in the form of higher taxes and a poorer quality of life. In the example given before, fast-food restaurant chains moved away from polystyrene packages because the public felt that these plastic containers were more harmful to the environment than paper packaging. Soft-drink manufacturers are moving toward *reusable* plastic bottles. Currently, poly(ethylene terephthalate) (PET) is used for 2-liter soft-drink bottles; it is a good recycling plastic. But a more durable bottle of polycarbonate can be rewashed and refilled just as many glass bottles are now handled.

Design for disassembly has become a theme in much of product design by major corporations. Europe, which has a higher degree of ecological concern, has led the way. With the desire to facilitate recycling, manufacturers of small appliances and durable goods are establishing procedures to ensure that products can be broken into components for easy sorting prior to recycling. Among the procedures are reducing the variety of plastics, adding labels to plastics for easy identification of plastic type, and eliminating screws and adhesives so that parts will disassemble easily.

One of the latest software programs is designed specifically to make products easier to fix. Known as ***Design for Service,*** this program takes its place alongside previous software programs called ***Design for Assembly*** and ***Design for Manufacturability.*** This new program helps product designers consider repair issues early in the design stage. Objectives of the program include making repairs less costly and extending the functioning life of products. Environmental issues such as recycling are directly addressed by this new computer software, which may have customers fixing products rather than tossing them out. In addition, this software augments previous software that addresses the need for disassembly of a product for whatever reason.

More often than not, ***cost*** is the primary selection criterion that will determine the final choice of material. In other words, if several materials have the specified physical, mechanical, and chemical properties, and are suitable for the processing technique selected, the lower-cost materials would be the logical choice. Determining cost is not as simple as it may seem. For example, a variety of plastics, including PET [poly(ethylene terephthalate)] and HDPE (high-density polyethylene), have replaced glass as containers for soft drinks, milk, and juices. Although the initial cost of plastic may be greater, the plastic bottles provide savings due to their toughness (less breakage) and the savings in shipping (PET and HDPE are much lighter than glass bottles).

Figure 1-9(a) shows a model stock panel factory. The plant can fabricate composite wall systems [Figure 1-9(b)] that combine a corrugated engineered-wood by-product with a low-density reinforced PPO [poly(phenylene oxide)] foam and recycled polystyrene, which is covered with a fire-retarding glass-reinforced plastic. This type of wall panel can cut down on building costs and promote recycling while improving overall quality.

Product liability is civil (as opposed to criminal) liability of the manufacturer to an ultimate user for injury resulting from a defective product. ***Caveat emptor*** (let the buyer beware) was once the rule. Today, numerous liability laws are in effect. For those involved in materials, particularly materials selection in the design process, the trend is for courts and juries to identify members of the design team as being responsible for some fault in a

(a)

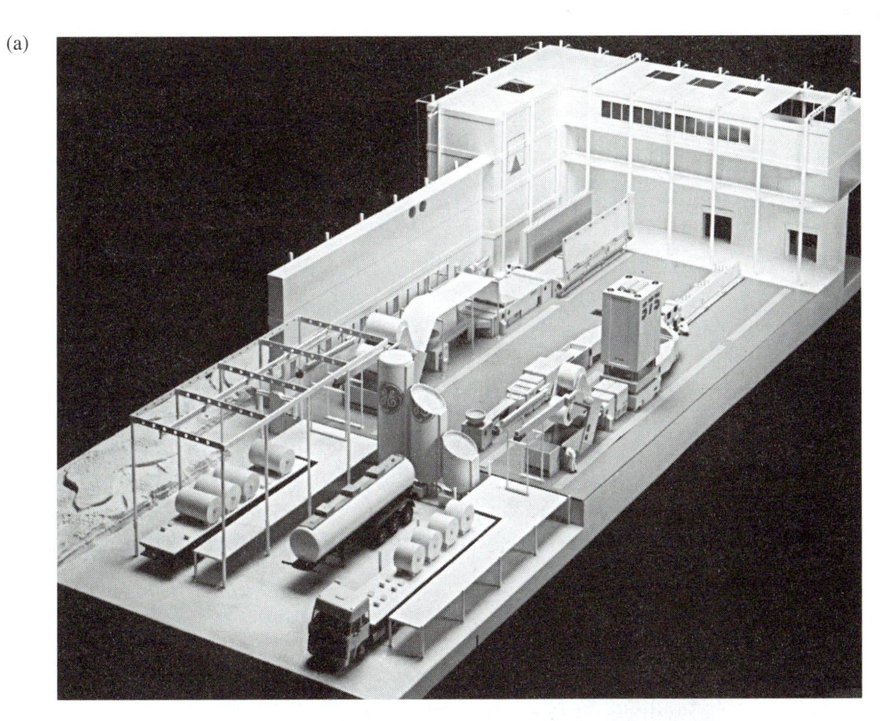

(b)

Figure 1-9 (a) Designers at G.E. Plastics have developed a scale model of a panel stock factory and finishing facility. After panels are manufactured in one plant, they will be customized for assembly with doors and windows in an automated finishing components factory. (b) A developmental wall system combines an experimental corrugated strand product with low-density polyphenylene oxide foam and reclaimed hamburger packages. The panel is covered with a glass-reinforced plastic that is an excellent fire retardant. (General Electric Plastics.)

product-liability action. Therefore, it is imperative to obtain and use the latest information about materials selection, particularly materials' long-term characteristics.

1.3.4.5 Materials selection in this book. Many of the illustrations and tables in this book provide examples of materials selection. The Self-Assessment at the end of each chapter will provide practice in materials selection as well as testing other knowledge gained from the module text. You are also encouraged to constantly observe materials applications in your everyday life. Keep a *materials journal* in which you write materials applications as "observations." Next to the observations section, place a section entitled "analysis"; there, record the positive or negative aspects of the observed use of materials. For some observations, state how you could make materials substitutions to improve on the designer's choice. Discuss these with others. As you progress through the book, your observations and analyses will become more sophisticated. This valuable knowledge will serve you well as a consumer, as an intelligent citizen, and in your career.

1.4 TECHNOLOGICAL LITERACY

Our society continues to become more complex. Corporate "downsizing," company restructuring, and similar trends that aim at fewer workers for greater profits place people in jeopardy of losing their jobs. To avoid being a victim of technological advancements, we must be prepared to change jobs several times during our working lifetime. To be agile in the job market, one must become ***technologically literate,*** that is, understand the language and concepts of technology to understand new technological advances. Technological literacy is not only important from a career standpoint, it is also required to function in our ever-changing technological society.

1.4.1 The Materials Consumer

Each year, an increasing number of newly developed materials are substituted for more familiar materials with limited properties. Today, 40% or more of a manufactured item's price represents materials cost. To be an informed and intelligent consumer requires a basic understanding of materials. The selection of a product can be improved by a greater understanding of the nature and properties of the materials used in the product. After a product has been purchased, the problem of failure, sometimes by the *abuse* or *misuse* of the product, can be lessened by such knowledge. Learning about the structure of materials, hence how materials behave, should permit an intelligent analysis of a failure and possibly pinpoint its source and cause. A knowledgeable consumer stands a much better chance of success in demanding remedial action from both manufacturers and retailers of faulty products than does one with less knowledge of the behavior of materials and a poor technical vocabulary with which to explain such behavior.

Wood is our oldest building material and possesses unique structure, warmth, and beauty, but perhaps more important, the ability to renew itself makes it equal in importance to the newer materials (Figure 1-10). In 1850, about 90% of U.S. energy came from wood, both in the raw state and refined into charcoal for fuel in glass and iron making. By 1900, wood accounted for only 20% of the nation's fuel and continued to decline. Wood use as

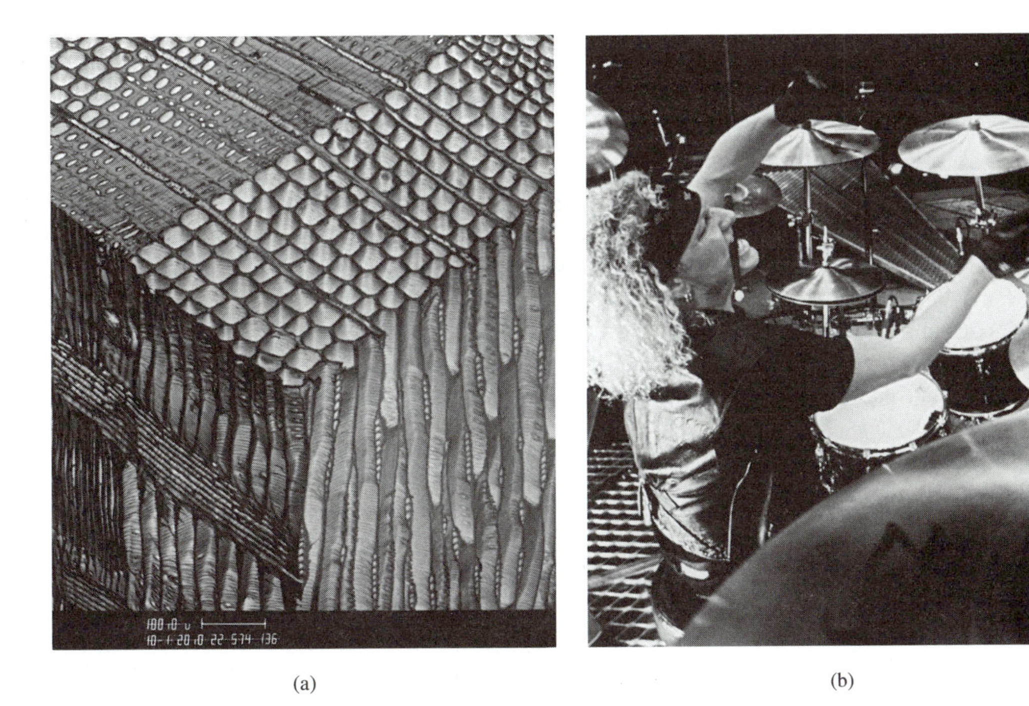

Figure 1-10 Wood's Complex Structure. (a) Electron microscope reveals microstructure. (Dr. Wilfred A. Côté, SUNY College of Environmental Science and Forestry.) (b) Improving on wood—a new composite material for entertainment technology. The drumstick combines the strength and vibration absorption efficiency of an aluminum core with the strong sound-making ability of the cover molded of a thermoplastic polyurethane resin. The resin is hard enough to produce strong sound and not wear out, but soft enough not to damage the drum sets. When the covers wear out, they can be replaced for less than half the cost of wooden drumsticks. The first major endorsement of the new advanced high efficiency alloy drumstick (AHEAD) was by a heavy metal rock star, drummer Matt Sorum of the band Guns 'N' Roses. *(Advanced Materials & Processes.)*

a fuel is now on the rise in both home and industry. Industries such as lumber and paper make extensive use of wood residues, including waste chemical by-products and bark. Wood does not figure as a major fuel for meeting the void left by depleted fossil fuels, but it holds promise as a substitute for more energy-intensive materials.

In construction materials for similar phases of building, the comparison of fuel required for the preparation of building products shows that wood requires the least energy, plastic nearly six times, and steel about eight times the amount of fuel needed for wood products. Concrete, glass, and other ceramics also show great promise as future materials for building and manufacture because the raw materials are as vast as the sands of a desert. Scarcity of old-growth trees and competition for high-grade wood to make lumber, furniture, and so on, has resulted in the development of new composites that may be superior to wood for many construction and nonbuilding applications. [See Figure 1-10(b).]

1.4.2 The Intelligent Citizen

The technology of materials also provides us with the necessary knowledge to make decisions based on personal values relating to political, social, and ecological issues. *A better informed citizen is a better citizen,* who is much needed in today's changing technological society in which the great issues over energy and materials resources are being debated and voted on.

Long-range industrial research and development of new alloys and nonmetals, intensification of programs to conserve and reclaim metals, enlargement of the search for and development of new domestic sources, and the utilization of ocean resources are all affected by issues that find a source in the political, social, economic, or ecological spheres within our society. The fact that the United States is dependent on foreign sources for most metals (with the exception of iron and copper), plus the negative effects of our high rate of consumption of the world's known reserves, places greater emphasis on a good working knowledge of the technology of materials so that citizens can guide their government representatives in making the correct decisions on matters that will have a lasting effect on our lives and living standard.

APPLICATIONS & ALTERNATIVES

Materials systems in biotechnology. Dramatic developments occur weekly in medicine due to advances of technology. The field of ***bioengineering*** combines biology and medicine with many specialties in engineering and materials science to foster technological developments to aid human beings and animals. Bioengineering produces artificial organs, limbs, and related anatomic structures. The materials aspect of these devices becomes a critical concern because of the need for physiological compatibility between the materials and the human or animal systems. The materials system described next introduces you to topics that you will study in following modules.

We can find many examples of materials joined into systems that use complementary properties of the various materials. Prosthetic devices used to replace defective joints in knees and hips provide an excellent case of such a materials system. These artificial joint implants are often necessary in patients suffering from accidents or disease. They consist of metal alloys, plastics, composites, and adhesives. Figure 1-11(a) shows a PCA (porous-coated anatomic) (PCA is a trademark of Howmedica, Inc.) knee replacement made of a cobalt alloy, sintered cobalt alloy spheres, ultrahigh-molecular-weight polyethylene (UHMWPE),* and a copolymer adhesive of poly(methyl methacrylate) and styrene (PMMA-PS). Figure 1-11(b) shows a view from the front of the human leg with the PCA in place and a lag screw rejoining the fibular bone that was cut as a part of the operation. Figure 1-11(c) shows an x-ray taken 6 months after the operation with the PCA in place.

Such orthopedic surgical operations, known as total joint orthoplasty, date back to 1940, when Vitallium first served as the metal alloy in hip-replacement devices. The alloy Vitallium consists of 65% cobalt with chromium and molybdenum and provides a hard, tough, noncorroding metal. The UHMWPE that bonds to the lower half of the PCA

*American Society for Testing and Materials (ASTM) and American National Standards Institute (ANSI) abbreviations, acronyms, and symbols are used throughout this book.

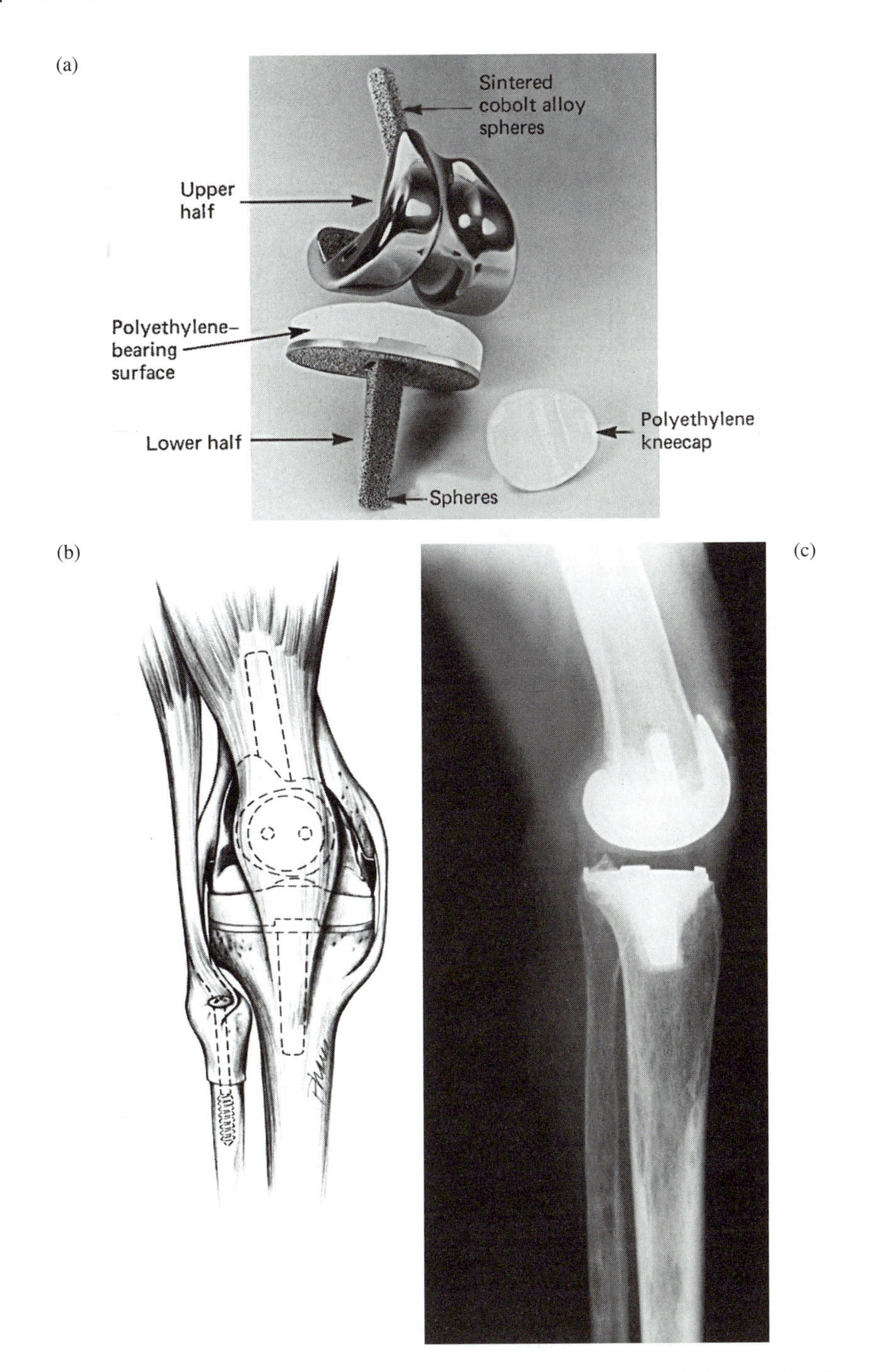

Figure 1-11 Human knee joint replacement. (a) PCA™ total knee joint. (b) Frontal view drawing of total knee joint in human leg. (c) Post-operative 6-month follow-up x-ray; note polyethylene bearing surface and kneecap do not show up on x-ray. (Howmedica, Inc.)

provides a self-lubricating bearing surface on which the highly polished Vitallium part hinges. You can imagine the type of environment that the PMMA-PS adhesive must endure because of body fluids present during implantation, thus bearing witness to the value of this adhesive to achieve a good bond, which most other adhesives could not accomplish. This copolymer adhesive cures rapidly while providing a free-radical reaction (the mixing of ingredients does not require close control for a proper bond). Since the operation requires cutting open the patient, the fast-setting nature of PMMA-PS is required to speed along the operation. Unfortunately, PMMA-PS cures into a brittle structure prone to fatigue fracture. The adhesive often deteriorates by loosening after years of service. The addition of a composite surface made of chrome–cobalt sintered spheres (see Figure 1-11) provides a porous surface designed to increase the surface area contacting the adhesive and improves the fatigue life of the PMMA. Tests revealed that cemented composite surfaces increase the ultimate tensile strength to 1,733,000 pounds per square inch (psi), compared to cemented surfaces of waffled Vitallium at 670,000 psi and cemented surfaces of UHMWPE at 175,000 psi. Dynamic testing revealed an improved fatigue life of the porous coating 2.7 times longer than the waffled metal surface with a tension–compression load at 625 lb.

Because of a concern for long-term failure with PMMA, some orthopedic surgeons implant the total joint prosthesis without an adhesive. Instead, they use either cobalt–chromium beads or titanium wire mesh that joins to the metal parts to increase metal porosity at the bone-to-prosthesis interface. Bone tissue then grows into the porous surface, creating a stable interface.

Problems remain because some people exhibit sensitivity to cobalt. Also, studies reveal cobalt to be a cancer-causing agent when injected into rats. In the constant search for the ideal material, titanium, ceramics, and composites offer possibilities for implant material alternatives.

What materials systems can you recall? Watch for uses of materials systems as you read newspapers and magazines or observe products because often an idea used in one field can be modified and employed in another. Can you see applications for materials systems in other fields that use the principles discussed here?

SELF-ASSESSMENT

1-1. Give two examples to describe where failure to consider the total materials cycle had harmful results.

1-2. How have the changes in use of materials and manufacturing techniques affected the work force in modern industry?

1-3. Name a major obstacle to changing from one material to another.

1-4. What must a designer do to make a material selection since there is rarely one ideal material for a given application?

1-5. After all other considerations are made, what is very often the major selection criterion that determines the final choice of material?

1-6. What is a materials database? Name examples of available databases?

1-7. What advantage does net-shape processing offer?

1-8. Because environmental and social factors place emphasis on ensuring that materials be recycled instead of becoming municipal solid waste, what is a new design technique that will aid in recycling?

1-9. Give three reasons why technological literacy is important to you.

1-10. Define the terms **(a)** *engineering materials technology*; **(b)** *materials science*; **(c)** *materials engineering.*

1-11. Give your own example of a materials system.

1-12. In an age of exotic materials, why is wood (one of the oldest materials) still valuable?

1-13. In the section Pause & Ponder, we asked you to begin to take note of materials. Name several examples of materials applications you can see now.

OBJECTIVE QUESTIONS

1-1. Recall the Pause & Ponder in light of what you have read in this module. Which statement is MOST TRUE about materials technology?

- **a.** Materials science and engineering usually follows the lead of developments in other technologies.
- **b.** Industry and the consumer are quick to accept new materials and processes innovations.
- **c.** Developments in materials science and engineering have allowed for innovations in most other technologies.
- **d.** Existing specifications and laws make it easy to introduce new materials in fields such as building construction and commercial aircraft.

Questions 1-2 through 1-5: choose the BEST TERM to match the definition.

a. engineering plastics, **b.** engineering materials, **c.** engineering materials technology, **d.** materials, **e.** materials science, **f.** materials engineering, **g.** conventional materials, **h.** materials science and engineering.

1-2. A term that loosely defines most materials in products and systems.

1-3. A field that deals with developing, preparing, modifying, and applying materials to a specific need.

1-4. A broad term that covers fields of applied science related to all aspects of materials, materials processing, and the many engineering specialties dealing with materials such as those used in manufacturing and construction.

1-5. A field that involves the generation and application of knowledge relating the composition, structure, and processing of materials to their properties and uses.

1-6. Which concept of the Materials Cycle has been LEAST AFFECTED by recent trends:

- **a.** Concern for the entire cycle that includes effects of harmful by-products and unrecycled old products
- **b.** Ability to design materials to meet ever-increasing demands of technology rather than use "off-the-shelf" materials
- **c.** Need for a more highly educated work force to support new manufacturing techniques
- **d.** Desire to achieve the maximum properties from any given material

1-7. Transforming gaseous, liquid, and solid elements by chemical and physical means, where atoms or molecules are combined to form solid materials.

a. Wroughting **b.** Synthesis **c.** Analysis **d.** Processing

1-8. Control of structure at higher levels of aggregation to yield bulk materials.

a. Wroughting **b.** Synthesis **c.** Analysis **d.** Processing

1-9. Composites produced with innovative processes that yield high strength and stiffness plus weight reduction.

a. Advanced composites **b.** Fiberglass **c.** Reinforced concrete **d.** Hydrocarbons

1-10. Manufacturing engineering has evolved to a sophisticated field of study. Which two aspects of fabricating products and systems contribute most to the new approaches of this field?

a. Systems approach **b.** Chemistry **c.** Automation **d.** Mathematics

1-11. What approach to design reflects a new concern for disposing of products after their useful life?

a. JIT **b.** Design for assembly **c.** Design for disassembly **d.** CIM

1-12. Which is *not* an obstacle to acceptance of new engineering materials and processes?

a. Lack of experience by designers and fabricators

b. Concern for safety with new techniques

c. Need for compromise between favorable properties and cost

d. Concern for keeping up with one's competition

1-13. What factor in materials selection usually dominates the final choice?

a. Cost **b.** Processibility **c.** Recyclability **d.** Weight

1-14. Which term describes well-defined methods for solving specific problems, such as those used to develop an approach to materials selection?

a. Algorithm **b.** JIT **c.** Design **d.** CIM

1-15. Which of the following best fits the category of a material system?

a. Gold **b.** Battery **c.** Plastic spoon **d.** Titanium

REFERENCES & RELATED READINGS

AMERICAN SOCIETY FOR TESTING AND MATERIALS. *Compilation of ASTM Standard Definitions,* 7th ed. Philadelphia: ASTM, 1990.

ASHBY, MICHAEL F. "Materials in Mechanical Design," *Journal of Materials Education.* Vol. 15, pp. 143–166, 1993.

ASM INTERNATIONAL. *ASM Engineered Materials Handbook Desk Edition,* 1995.

ASM INTERNATIONAL. *ASM Materials Engineering Dictionary,* 1992.

ASM INTERNATIONAL. "Guide to Selecting Engineering Materials," special issue, *Advanced Materials and Processes,* June 1995.

BATES, SETH. "Manual and Computer-Aided Materials Selection for Industrial Production: An Exercise in Decision Making," *Proceedings of National Educators' Workshop, NEW: Update 89,* NASA Conference Publication 3074, pp. 133–140.

CANBY, THOMAS Y. "Advanced Materials: Reshaping Our Lives," *National Geographic,* December 1989, pp. 746–781.

NATIONAL RESEARCH COUNCIL. *Materials Science and Engineering for the 1990s: Maintaining Competitiveness in the Materials Age.* Washington, D.C.: NRC, 1989.

ROGERS, CRAIG A. "Intelligent Materials Systems: The Dawn of a New Materials Age," Center for Intelligent Materials and Structures, Virginia Polytechnic Institute and State University, 1992.

SWARTZ, LYLE H. "Industry, Government Cooperation: New in a Changing World," *Advanced Materials & Processing,* February 1996, pp. 110–116.

WEISER, MARTIN W., and OLDEN L. BURCHETT. "Materials Selection Is a Major Component of Efficient Product Realization," *ASEE Annual Conference,* Anaheim, CA, June 25–28, 1995.

Periodicals

Advanced Materials and Processes
Automotive Engineering
Machine Design
Manufacturing Engineering
Popular Science

Module 2
Nature and Family of Materials

After studying this module, you should be able to do the following:

2.1. Using the materials and civilization timeline, give an example of how materials technology developments sparked new technology in the Scientific Age.

2.2. Use the periodic table of elements and illustrations to determine and/or calculate the atomic structure of elements, ionization potential, and electronegativity.

2.3. Apply the SI rules for prefixes in expressing multiple quantities.

2.4. Use sketches and explanations to describe the microstructure and macrostructure of solid-state materials plus bonding forces within materials.

2.5. Discuss the significance of fullerenes and the polymorphs of carbon.

2.6. Name the five main groups that make up the family of materials; cite examples of materials within each group; and sketch the typical microstructure for ceramics, metals, polymers, and composites.

2.7. Describe intelligent materials systems (smart materials) and discuss some goals of materials science and engineering (MSE) for this class of materials.

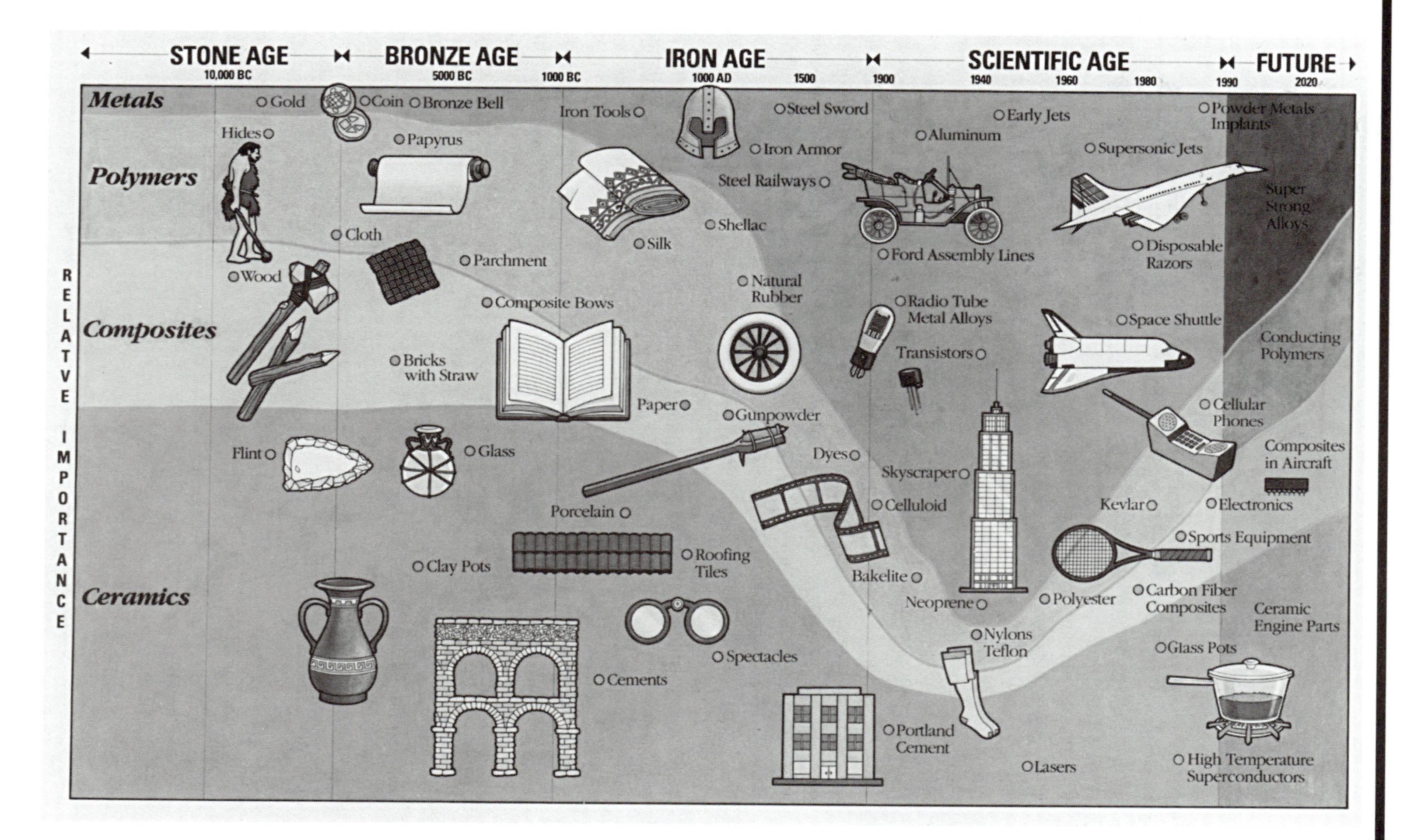

Figure 2-1 Material and civilization timeline. (From Future Materials exhibit, Franklin Institute Science Museum.)

PAUSE & PONDER

Recalling our study of history, materials technology played a major influence on societies. Periods of civilization were named for the materials that dominated each era: from the earliest time of primitive people in the Stone Age, through the Bronze Age and the Iron Age. The era we live in has been given many names, including the Scientific Age, the Information Age, the Space Age, and even the Materials Age. Figure 2-1 provides a historical perspective of the concurrent development of civilization and materials technology. This figure, provided by the Franklin Institute's "Future Materials" exhibit, allows a graphical comparison of the relative importance of materials through the ages. Note that the materials are placed into four groups: metals, polymers, composites, and ceramics. In this module, we will provide an in-depth look at these groupings as we look at the family of materials.

The Stone Age, which includes prehistory, was so named because the evolution of civilization began to make great strides with the development of tools. Naturally occurring flint allowed for shaping, and people discovered that flint and other stones could be joined with sticks to produce tools and weapons. Clay dug from the earth was formed and dried into pottery. People of this period (which might be more accurately called the ceramic, hide, and wood age) used the naturally occurring "engineering materials" for tools.

The Bronze Age represented another major leap forward as "materials technology" brought about the first alloying, when copper and tin were melted together to yield a superior metal useful for shields, knives, ornaments, cups, and urns. Note on Figure 2-1 that ceramics, still widely used, were also advancing during the Bronze Age with the development of glass and composite bricks (clay and straw). Once iron was developed, around 1100 B.C., it became of great importance to society. Iron's abundance and fine engineering properties proved superior to bronze. With the eventual alloying of iron with carbon and other elements, steel became the superior engineering material; to this day steel has been dominant, although often challenged. During the Iron Age and into the Scientific Age, thousands of iron and steel alloys were developed by metallurgists to meet the demands of the industrial revolution; then mass production received a great boost from the Ford assembly line.

Aluminum, the third most abundant element in the earth's crust, remained a rare and expensive metal until 1886, when an American, Charles Martin Hall, and a Frenchman, Paul Heroult, independently devised a practical process for electrolytic production of aluminum. Still used today, the Hall-Heroult process is the largest consumer of electricity in the United States, making the recycling of aluminum products most worthwhile. Now iron and steel had competition from a lightweight, corrosion-resistant metal. Plastics and rubbers, including Bakelite, Celluloid, Neoprene, nylon, and Teflon (Figure 2-1), were also being developed, which could substitute economically for natural materials such as ivory, tortoiseshell, wood, and silk. Experimentation with combinations of metals (alloys) aimed at meeting the needs of communications, transportation, and construction. Building tall skyscrapers with new types of steel supported the growth of large cities. Soon the radio vacuum tube was improved upon with transistors, and eventually, electronics benefited from the microchip as transistors were reduced in size and combined with other circuit components to produce integrated circuits made with semiconductor materials. It is exciting today to follow the application of new electronics materials to computer technology in areas such as artificial intelligence (AI), which mimics human reasoning, and virtual reality (VR), which provides vast new dimensions to visualization using computer graphics.

Now as we move into the future with results of research and development from the Scientific Age or Age of Materials, technology is beginning to shift away from the dominance of iron and steel. Note in Figure 2-1 that the lines separating the relative importance of metals, polymers, composites, and ceramics are altering again. Advanced materials such as the Fiber Kevlar used in aircraft and spacecraft, high-performance ceramics for engine parts, and high-temperature ceramic superconductors are causing the zones of relative importance for ceramics and composites to widen. Continued use of powder metals, superstrong alloys, and conducting polymers shows that the polymer and metal technologies will maintain their importance but not the dominance of recent eras.

Beyond the year 2000, improved materials—the advanced materials and the smart materials—will be called on for use in new products and systems. These materials will be produced through scitech as materials engineers work with manufacturing engineers and materials scientists to invent new materials processing. As you study the nature and family of materials, keep in mind the historical perspective of materials and recognize that technology can cause many shifts in the relative significance of materials. It is important for you to understand the similarities and differences in the main groups of materials and to keep abreast of new developments.

2.1 NATURE

Through an understanding of the nature and structure of materials, one can predict how materials should behave when exposed to certain forces and environments. The nature of materials results from their composition and structure, which determine their properties. In this module, we will present basic concepts of atomic and molecular structure that lay a foundation for the discussion throughout the book on the nature, properties, and uses of materials. The ability to grasp these fundamental materials technology concepts not only allows you to deal with today's materials and processes, but also prepares you for new developments in materials.

2.1.1 Internal Structure

The structure of materials follows the structure of a building because both develop through the joining of smaller units. Particles of earth form bricks, bricks are stacked together to form walls, walls form rooms, and rooms make a building. *Matter* is anything that has mass and occupies volume. It exists in one of four physical states: solid, liquid, gas, or plasma. ***Plasma*** exists only under drastic conditions. At very high temperatures, matter is shredded into positively charged particles. These charged particles move about in a cloud of negatively charged electrons. On the surface of the sun, hydrogen and helium gases are changed into plasma by the heat of the sun (6000°C). When two or more ***atoms*** are joined by covalent bonds, the unit of matter formed is called a ***molecule.*** An ***element*** is a substance that cannot be broken down any further by chemical reaction. Examples of elements are carbon, oxygen, or sulfur. Elements combine to form compounds. An example of a compound is NaCl. A ***compound*** is a substance that can be broken down into elements by a chemical reaction. Molecules combine through chemical reactions to form polymeric substances known as ***polymers.*** Elements, molecules, and compounds exist in many forms, such as a gas, liquid, or solid as described above.

2.1.1.1 Atomic structure. An ***atom*** is the smallest particle of an element that possesses the physical and chemical properties of that element. It is the smallest particle of an element that can enter into a chemical change. Atoms are the basic building blocks of matter. The average diameter of an atom is only about 10^{-10} meter. It takes more than 10^6 atoms edge to edge to make the thickness of this page.

What the atom really looks like or how the charged particles distribute themselves and move within the atom is not exactly known. Consequently, scientists resort to various models of an atom to help explain its characteristics. One model, the planetary model, appears to be the most popular in explaining the nature of an atom (see Figure 2-2). However, this model is far from being the most accurate. The atom is pictured as a sphere with a very dense center, the nucleus, containing the protons and neutrons around which negatively charged particles (the electrons) are moving in circular or elliptical orbits, much like the motion of the planets around the sun. The diameter of the nucleus is about 1/10,000 the size of an atom, but it contains more than 90% of the mass of the atom. If one represented the nucleus as a golf ball, the electrons would revolve in a sphere with a radius of about 1½ miles. Another model of the atom, sometimes sketched in three dimensions, attempts to show the spaces within which the electrons are confined as fuzzy volumes. The atom can be thought of as a sphere of empty space with 90% of its mass located in its center or nucleus.

For our purposes, the nucleus consists of protons and neutrons. A ***proton*** is a particle of matter that carries a positive electrical charge equivalent to the negative charge on an electron. ***Neutrons*** are uncharged particles in the nucleus with a mass nearly equal to the proton's mass. The ***atomic number*** of an element is equal to the number of protons in the nucleus. The number of protons in the nucleus equals the number of electrons in the

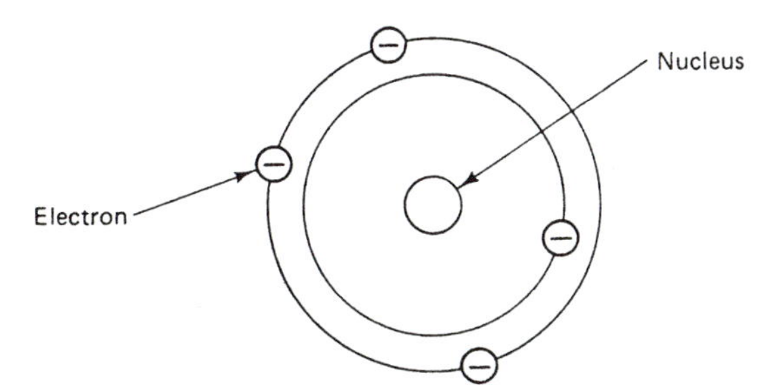

Figure 2-2 Planetary model of an atom.

atom in the balanced, neutral, or equilibrium state. Elements in the periodic table (see Table 10-1A in the appendix) are in order according to their atomic numbers.

All atoms of an element contain the same number of protons, but these atoms of the same element can contain different numbers of neutrons. In other words, these atoms are different, but their positive and negative charges are identical. Such atoms with different numbers of neutrons are ***isotopes.*** For example, hydrogen exists in three isotopic forms (see Figure 2-3). Tin has 10 naturally occurring isotopes. The total number of protons and neutrons in the nucleus is the ***mass number of the isotope.*** Protons and neutrons are referred to as ***nucleons.*** These nucleons, called ***quarks,*** are now believed to be complicated little systems of the basic building blocks of all materials. A standard notation for expressing the mass number is to write the symbol of the element, then write the mass number as a superscript (above the line) and its atomic number as a subscript (below the line). Figure 2-3 shows this notation for the isotopes of hydrogen.

In summary, an atom (see Figure 2-4) of an element is made up of a dense core, the ***nucleus,*** consisting of protons and neutrons surrounded by even smaller particles called ***electrons.*** The electrons carry a negative (−) electrical charge much like the charge at the

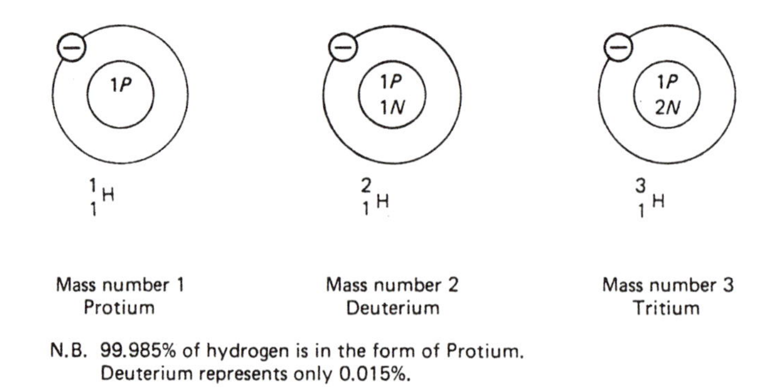

Figure 2-3 Isotopes of hydrogen.

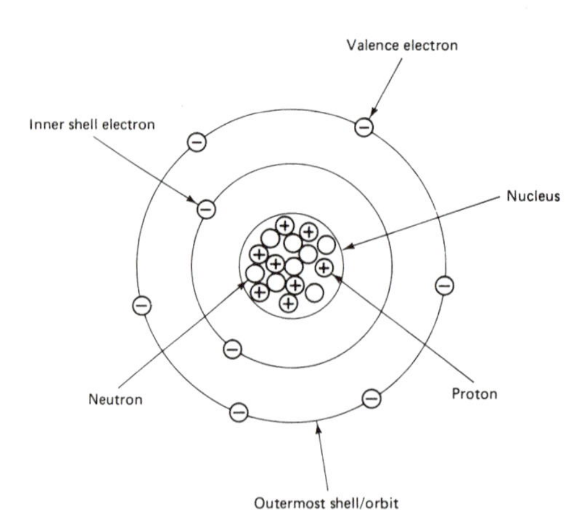

Figure 2-4 Subatomic parts of an atom. This atom is shown in an equilibrium, balanced, or electrically neutral state.

negative (−) terminal of a battery. Those electrons that occupy the outermost ring or shell from the nucleus are ***valence electrons.*** If the atom is in a balanced or equilibrium state (see Figure 2-4), it possesses the same number of protons as electrons, and the electrical charge carried by the atom is zero. The negative (−) charge of the electrons is balanced by the positive (+) charge of the protons in the nucleus.

The process of pulling away (removing) or adding valence electrons from a balanced or neutral atom is called ***ionization.*** The atoms with unbalanced electrical charge are called ***ions*** (see Figures 2-5 and 2-6). The tendency of atoms to lose their valence electrons and the energy required to remove these electrons will be discussed in the treatment of the periodic table and its underlying concepts.

Electronic Structure. Many new theories were developed between 1900 and 1930 that laid the basis for a new branch of physics called ***quantum mechanics.*** One theory was the idea that light is quantized, that it is made up of discrete amounts of energy. In a similar manner, we consider our present-day model of an atom with its electrons as having both wavelike and particlelike characteristics. The position of these electrons in relation to the nucleus of an atom, according to this new theory, must be described in terms of probability distribution (electron cloud) rather than the precisely determined values of classical theory. Four parameters called *quantum numbers* characterize an electron as to its size, shape, and spatial orientation. Shells or principal energy levels are specified by a principal quantum number n, where $n = 1$ through 7 or K, L, M, N, O, P, and Q [see Figure 2-7 (a)]. The second quantum number, l, signifies the sublevel, subshell, or orbital (s, p, d, f). The number of energy states for each subshell is determined by the third quantum number m_l. For the s subshell, there is a single energy state; for p, there are 3; for d, there are 5; and for f, there are 7. The fourth quantum number m_s relates to the electron spin moment, for which there are two values possible ($+\frac{1}{2}$ and $-\frac{1}{2}$), one for each of the spin orientations. Spin momentum controls the manner in which electrons fill their subshells and also the magnetic properties of materials (see Module 9).

Three basic rules define the order in which electrons fill their subshells or establish their electron configurations. One says that the electrons occupy their subshells or orbitals in such a way as to minimize the energy of the atom. A second rule states that only two electrons may exist in the same orbital, and these electrons must have opposite spins. The last rule says that when orbitals of equal energy are being filled, the electrons initially

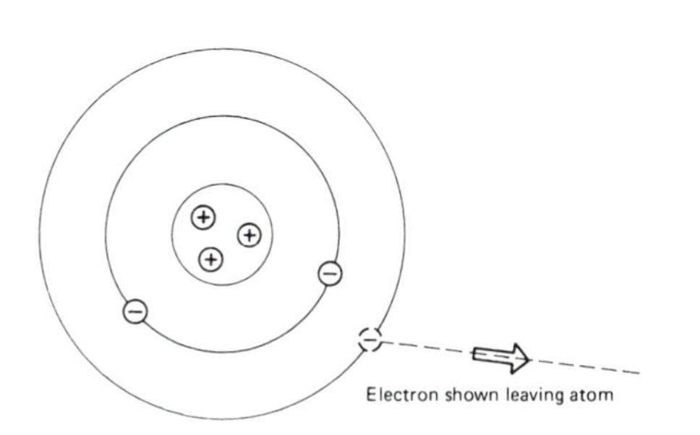

Figure 2-5 Positive ion or cation.

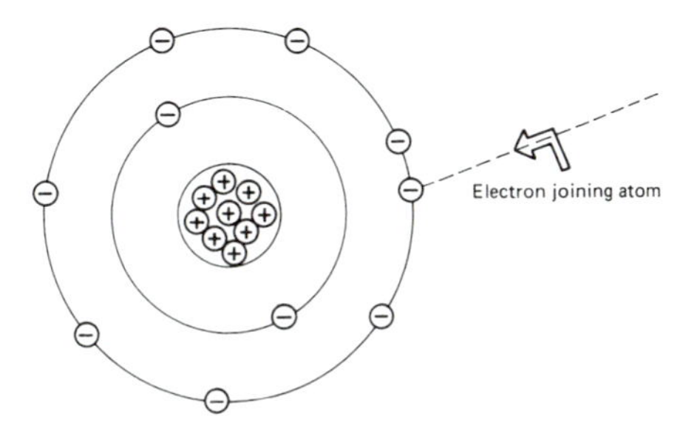

Figure 2-6 Negative ion or anion.

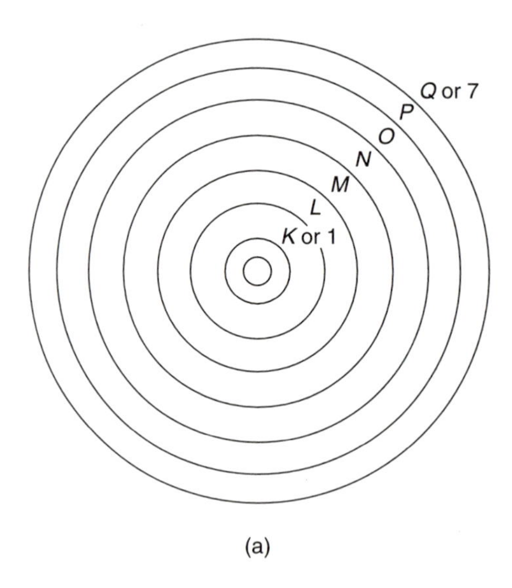

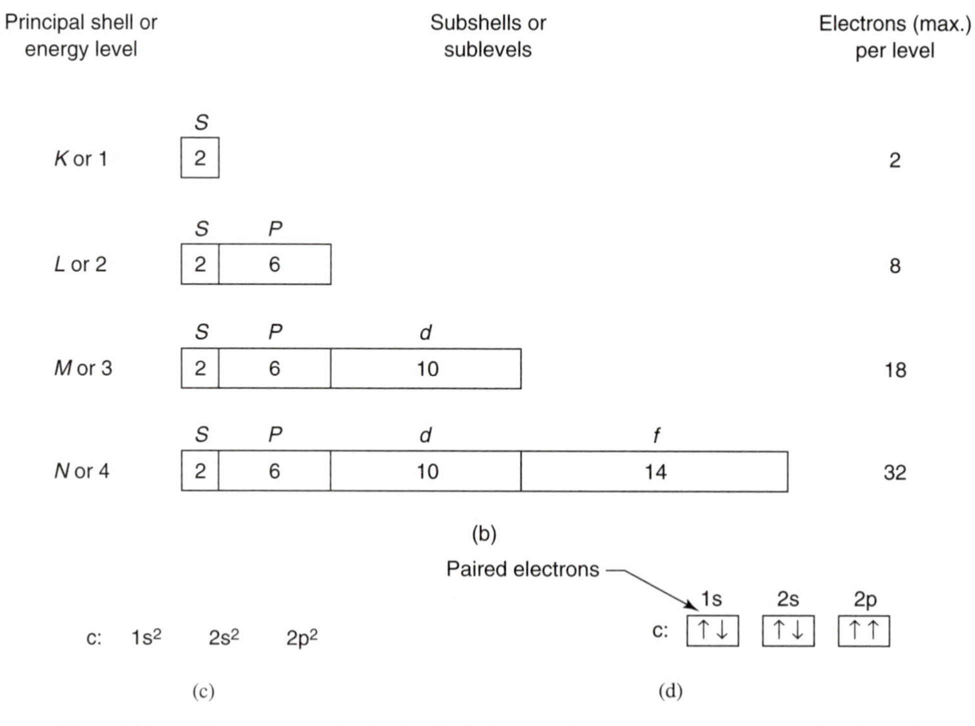

Figure 2-7 (a) Electron energy levels. (b) Block diagram of the first four energy levels showing their respective sublevels with their maximum number of electrons. (c) and (d) Electron configurations for the element carbon (C), atomic number 6 in period 2 of the periodic table.

occupy these orbitals singly. The result of this rule is that electrons try to get as far apart from each other as possible. To accomplish this, they favor occupying empty orbitals of equal energy as opposed to pairing up with other electrons in half-filled orbitals. As a result, atoms tend to have as many unpaired electrons as possible. The orbital diagram is used to demonstrate electrons filling orbitals. Figure 2-7(c) and (d) shows two ways to depict the electron configuration of the element carbon (C). This diagram is similar to Figure 2-7(a) and (b) because each subshell is broken down into individual orbitals of equal energy. Electrons are shown as arrows. An arrow pointing up corresponds to one type of spin ($\frac{1}{2}$) and an arrow pointing down to the other ($-\frac{1}{2}$). Electrons in the same orbital with opposing (opposite) spins are said to be paired ($\uparrow\downarrow$). The electrons in 1s and 2s orbitals are paired. Electrons in different, singly occupied orbitals of the same subshell have parallel spins (arrows pointing in the same direction). This orbital diagram represents the carbon atom in its most stable or ground state. The order in which all electrons fill subshells, and thus minimize the energy of the atom, as established by experiment, is (with but a few exceptions) as follows:

1s, 2s, 2p, 3s, 3p, 4s, 3d, 4p, 5s, 4d, 5p, 6s, 4f, 5d, 6p, 7s, 5f, 6d, 7p

There are other techniques or methods used to determine the order of filling; they are not discussed in this text.

The energy levels in atoms have been given letter names starting with K, the first and lowest energy level (closest to the nucleus), and proceeding outward to Q, the seventh level, where the electrons have the highest energy. Each level can hold only a certain number of electrons at any one time. To determine the maximum number of electrons at any energy level, the relationship $2n^2$ can be used. The number of the energy level is substituted for n. For example, the third or M energy level can contain a maximum (2×3^2) or 18 electrons. Figure 2-7(a) contains a simple sketch of all energy levels drawn concentric with the nucleus of an atom. Some energy levels are listed along with the maximum number of electrons permitted at each level in Figure 2-7(b).

The manner in which the electrons of an atom distribute themselves in the ground, or lowest energy, state is the ***electron configuration*** of the atom. Learning the electron configuration of atoms permits us to predict what the atoms will do in the presence of other atoms. In other words, the number of electrons in an atom's outermost energy level (the valence electrons) is the key that controls the chemical properties of an element. As an example, hydrogen with an atomic number of 1 in the ground state has its one electron in the K level. Carbon has six electrons; two fill the K level and the remaining four are in the L level. Atoms that have a completed energy level are chemically inert; that is, their atoms do not react with other atoms. These atoms do not attract other electrons, nor do they readily lose their outer electrons. Hence, these atoms possess atomic stability. Helium has a complete K energy level with its two electrons. Some atoms have the tendency to build up the number of electrons in their levels to eight. (This tendency is known as the octet rule, or the rule of eight.) Once they arrive at eight, the next electron goes into the next higher level. Those atoms that have only one electron in their outermost energy level are extremely reactive; they tend to lose their electron. Examples of such elements are lithium and sodium.

Probing deeper into the electron world, scientists have determined that the energy levels (shells) are divided into sublevels (orbitals) that also contain a maximum number of electrons. Table 2-1 contains an example of sublevels for selected elements, which are given lowercase-letter names.

The K level has only one sublevel, the s sublevel. Using a block diagram, the energy levels could be represented as shown in Figure 2-7(b). The electrons fill these sublevels in a particular way. The first 18 elements of the periodic table build up their permitted number in a routine manner. Once the K level reaches its maximum of two electrons, the electrons begin filling up the L level until a maximum of eight electrons is reached, as with neon (Ne). Then the third energy level from the nucleus (M level) with its three sublevels (s, p, d) begins filling to its maximum of 18. This point is reached with the element argon (Ar, atomic number 18). Note that both neon (Ne) and argon (Ar) have a maximum possible number of electrons in their outermost energy levels. These elements are classified as chemically inert gases; that is, they are very stable elements and do not react with other elements. Neon (Ne), with its 10 electrons, has an electron configuration: Ne: $1s^2\ 2s^2\ 2p^6$. Note that it has a full L, or 2, level of eight electrons (octet rule). This neon core of electrons can be represented as [Ne]. Argon, atomic number 18, can be represented in a similar way as [Ar]. Again, note that its outermost level (M, or 3) has eight electrons. To represent the electron configuration of titanium (Ti, atomic number 22) showing the order in which the electrons fill the orbitals and using the argon core symbol, we write: Ti: [Ar] $3d^2$ $4s^2$. The longer version of the Ti electron configuration follows in the next paragraph. Table 2-1 contains other elements' electron configurations.

Using a shorthand notation, the electronic configuration of oxygen is ^{8}O: $1s^2$, $2s^2$, $2p^4$. Note that the L energy level has only six electrons, but the p sublevel could contain a maximum of eight electrons. As a result, oxygen tends to accept two additional electrons to form a stable outer energy level. By definition, oxygen is an acceptor of electrons and hence is classified as a ***nonmetal.*** A ***metal*** such as aluminum, with its electronic configuration $^{13}A1$: $1s^2$, $2s^2$, $2p^6$, $3s^2$, $3p^1$, has a stable L energy level below the farthest energy level, the third or M shell. The three valence electrons in the M level are close to those in the L level and possess similar bonding energies, which results in these three electrons having a tendency to stay with their atom. In addition, to achieve stability, aluminum would need the addition of five more electrons, which is unlikely to occur. Hence, aluminum is classified as a metal. One final example is titanium (Ti), with the electronic no-

TABLE 2-1 ELECTRON CONFIGURATIONS OF SELECTED ELEMENTS

Element	Atomic number	K 1 s	L 2 s p	M 3 s p d	N 4 s p d f
Oxygen	8	2	2 4		
Sodium	11	2	2 6	1	
Aluminum	13	2	2 6	2 1	
Silicon	14	2	2 6	2 2	
Titanium	22	2	2 6	2 6 2	2
Arsenic	33	2	2 6	2 6 10	2 3

tation ^{22}Ti: $1s^2$, $2s^2$, $2p^6$, $3s^2$ $3p^6$, $3d^2$, $4s^2$. This element is chosen to show that the filling of electron levels is somewhat irregular. Note that two electrons have filled the 4s sublevel, whereas the 3d sublevel is far from being filled, with only two electrons [see Figure 2-7(b)]. For an example of the use of this notation, refer to Section 3.4.9.

On the other hand, let us look at some elements that have only one electron in their outermost energy level. Lithium (Li), atomic number 3, with two electrons in the K level and only one electron in the L level, is extremely reactive. The same can be said of sodium (Na) and potassium (K). For elements with atomic numbers greater than 18, starting with potassium (K), the routine followed by the next 18 elements in filling their electron levels is somewhat different.

Through an understanding of the electronic structure of atoms, the properties of elements can be ascertained and the bonding of atoms of elements together to form molecules, compounds, or alloys can be explained. Additionally, these electrons determine the size of atoms, the degree of electrical and thermal conductivity of materials, and the optical characteristics of materials.

When atoms take part in chemical reactions, there is an energy change or transfer, usually resulting in the release of energy in the form of heat. This energy change brings about a rearrangement of the electrons, and the end result is an electron configuration with lower energy than any of the reactant elements. Only the outermost (valence) electrons are rearranged. The electrons in the inner energy levels (or shells) are too stable, or shielded from this activity. More will be said about this phenomenon when chemical bonding of atoms is discussed later in this module.

Periodic Table. Elements are the basic building blocks of all macroscopic matter. Each has its own chemical symbol, a shorthand abbreviation of its chemical name, and *each has its own specific properties.* It is important to reiterate that elements with the same number of valence electrons have similar properties. Scientists have long recognized that some elements had similar properties while others, like sodium and chlorine, had very different properties. Attempts to classify the elements met with success in 1869, when Dimitri Mendeleev formulated the periodic table of the elements. He listed the elements in order of their atomic mass and discovered that elements with similar properties recur in a periodic manner. This recurrence of similar properties allowed Mendeleev to place these elements in tabular form, with similar elements in the same column. At that time, knowledge of the internal structure of the atom was unknown. All versions of the modern periodic table arrange the elements in order of increasing atomic number, which usually matches the order of increasing atomic mass.

Refer to Table 10-1A, Periodic Table of the Elements, for the following discussion. The horizontal rows, called ***periods,*** are numbered from 1 to 7. The vertical columns, called ***groups,*** are numbered in two different ways. If only the main groups (also referred to as representative elements) are numbered, the roman numerals I to VIII are used. If the transition element groups are also numbered, then the arabic numbers 1 to 18 are used. The 10 elements that fall between groups II and III in periods 4, 5, and 6 (a total of 18 elements) are known as ***transition elements.*** Their properties are attributable to the delayed filling of the next-to-last or outermost energy level. Period 6 contains 14 additional elements between lanthanum (La) and hafnium (Hf), which are called lanthanides. The total number of elements in this period is 32. The 14 elements in period 7 that form a similar series are called actinides. To give the table a convenient size and shape, these two series of transition elements are set off separately at the bottom of the table.

The elements can be classified broadly into three basic groupings (metals, metalloids, and nonmetals) because they have certain common properties. ***Metals*** have certain characteristic properties: they are solids at room temperature (with the exception of mercury), they are malleable and ductile, and they conduct electricity. ***Nonmetals*** (a total of 17 elements) are poor conductors of heat, they are mostly insulators, and they tend to be brittle and fracture easily. ***Metalloids,*** or ***semimetals,*** have some properties like the metals, i.e., they conduct electricity, and other properties like nonmetals. They consist of seven elements (B, Si, Ge, As, Sb, Te, and At) adjacent to the zigzag border between the metals and the nonmetals. Semimetals are brittle and are poor conductors of heat and electricity. Silicon and germanium, for example, are called semiconductors and they are playing a major role in our technological world (see Module 9).

The atomic radius of an atom (size of atom) decreases as one moves from left to right across a period. As one moves down a particular group of elements, the size of the atoms increases. Ionization, defined previously, involves the loss of a valence electron by an atom. The energy required to accomplish this is known as the ***ionization potential.*** The first ionization potential is a measure of the energy needed to remove the first electron from the atom in its ground or equilibrium state. In general, ionization potentials follow the periodic relationship and decrease as the atomic numbers of the elements increase in a given periodic group. The lower the ionization potential, the greater the metallic character. For all elements, the ionization potential is usually listed in the periodic table expressed in electron volts (eV), or other energy terms, such as calories per mole. For example, it takes 13.6 eV of energy (313 kcal/mol) to free the electron from the hydrogen atom.

This binding energy increases as the size of the atom decreases. In other words, the farther away an electron is from its nucleus, the smaller the force of attraction between it and its positively charged nucleus. Metallic elements on the far left of the table readily ionize, whereas atoms of nonmetals located on the right of the table more readily share or accept additional electrons and become negative ions. The observation to be made is that the tendency of atoms to ionize depends on their relative positions in the periodic table. ***Electronegativity*** is the degree to which an atom attracts electrons. It is measured on a scale from 0 to 4.1, as shown in Table 10-1B. The nonmetallics, which accept electrons to produce negative ions (anions), are the most electronegative elements, and groups VIB and VIIB contain the most electronegative elements, with a value of 3.98 for fluorine (F). Francium (Fr) has the lowest value, 0.7.

Electropositive elements are metallic in nature and give up electrons in chemical reactions to produce positive ions (cations). The most electropositive elements are found in group 1A. Another term that describes this phenomenon is ***electron affinity,*** which is a measure of the attraction between the electron and the nucleus. Figure 2-8 is a graph based on the Pauling electronegativity scale showing the percentage of ionic bonding (character) in a compound versus the difference in electronegativity between two elements in the compound. If the difference between the electronegativity of two elements is greater than 1.7, the bond is ionic.

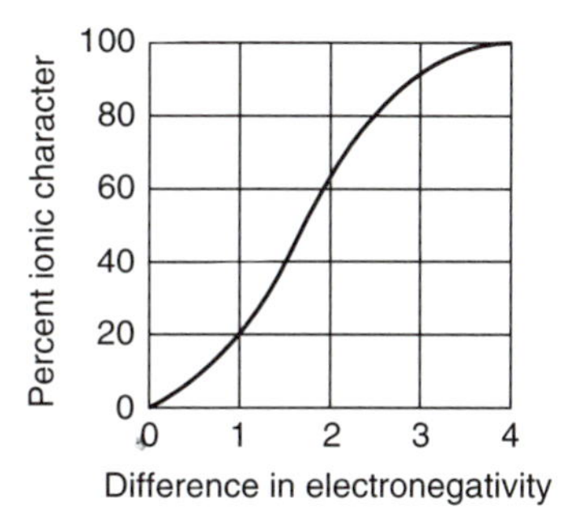

Figure 2-8 Percent ionic character of a single bond plotted as a function of the difference in electronegativities of the two bonded atoms.

Illustrative Problem

Determine the type of bonding in the compound MgO.

Solution: Using Table 10-1B, the electronegativity for oxygen is found to be 3.44 and for magnesium, it is 1.31. Entering a difference of 2.13 in the graph shown in Figure 2-8, the percent ionic bonding is about 68%, leaving 32% of the bonding covalent. A more accurate solution is obtainable using data in the small table at the top of Table 10-1B.

2.1.1.2 Molecular structure and bonding. We mentioned previously that atoms join together to form molecules and that molecules combine to form various compounds. A ***compound*** is a substance containing two or more elements combined in fixed proportions. Compounds lose the characteristics of the elements; for example, sodium Na combines with chlorine Cl to make the compound sodium chloride (NaCl), as described in the next section.

In some molecules, the atoms are of the same element. Fluorine (F) exists as a diatomic molecule under normal conditions because two fluorine atoms combine to form one fluorine molecule. All the group VIIB elements form diatomic molecules. In this section, we will gain further understanding about how the 100 or so elements combine or chemically bond to form not only more than 3 million organic compounds, but the various inorganic and bulk metals such as iron or copper.

Molecules and compounds are formed by chemical bonding. ***Chemical bonding*** can be explained simply as the end product of the interaction of the electrical forces of attraction and repulsion between oppositely charged or similarly charged particles of matter. Chemical bonds are formed by the electrons in the atom's outermost energy levels or regions. In our study of these electron energy levels, we have learned that, generally, atoms that have eight electrons in their outermost orbit are very stable (in equilibrium state). This tendency to start filling the next-higher level once an atom has eight electrons in an energy level is known as the *octet rule,* described previously. Atoms with fewer than eight outermost electrons attempt to seek this stable condition by bonding with each other so that each atom can attain this stable configuration of eight electrons (valence electrons) in its outermost energy level. One way that atoms can achieve this stability is by sharing electrons with other atoms that also need eight electrons in their outermost energy levels. For purposes of study, we can divide the various types of bonding into two groups, strong or ***primary bonding,*** and weak or ***secondary bonding.*** In the first group, we include covalent,

N ≡ N

N_2 molecule **Figure 2-9** Triple covalent bond.

ionic, and metallic bonding. The second group includes weak atomic bonds, electric dipoles, polar molecules, and the hydrogen bond. The forces that produce secondary bonds are known as ***van der Waals forces.*** It is emphasized that atoms of one material do not bond solely by one particular bonding mechanism. One type of bond can predominate in a material, but in many materials the various bonding types are represented and mixed-bond types of materials are produced.

Covalent Electron Sharing, or Electron-Pair Bonding. The sharing of electrons between two or more atoms is known as ***covalent*** or ***shared electron-pair bonding.*** As an example, fluorine (F) atoms have a total of nine electrons: two in the inner energy level and seven in the outermost level. Two fluorine atoms can share one of the electrons with the other to form a single covalent bond, depicted with a single dash representing the bond. Bonding of the elements is indicated with the small bar (—). Each bar indicates a covalent bond. Covalent bonding gets its name from the fact that a pair of valence electrons is part of the electron structure of each atom bonding the two atoms together. In general, organic compounds (complex compounds containing carbon and/or hydrogen) are formed by covalent bonding. A ***triple covalent*** bond is one in which two atoms *share three* of their electrons with each other. Figure 2-9 illustrates this with nitrogen (N). Nitrogen needs eight electrons to fill the outer energy level. It has only five. Hence, it shares three with another nitrogen atom, forming a triple bond.

Organic chemistry is the chemistry of carbon compounds. Of all the millions of known chemical compounds, more than 95% are compounds of carbon. The simplest organic compounds are the ***hydrocarbons*** (HCs), which contain only hydrogen and carbon. ***Alkanes,*** also known as saturated HCs, are a series of related compounds that contain only single bonds. Figure 2-10 shows the first three compounds in this series: methane, ethane, and propane. They all have a continuous, open-ended chain of carbon atoms and possess similar chemical properties. Each bar in Figure 2-11 represents the single bond of a hydrogen atom with a carbon atom to form the compound methane ethylene. Ethylene is the first member of a family of HCs called ***alkenes.*** The structure of ethylene as shown in Figures 2-10 and 2-12 is the basic compound for producing polyethylene. More than half of the production of ethylene goes into the manufacture of polyethylene. In the form of ethylene glycol, it is also the major component of most brands of antifreeze. In these figures, the C represents the carbon atom and the H represents the hydrogen atom. Note in Figure 2-10 that, as the molecular weight increases, the compounds change from a gaseous state to a liquid and finally to a solid as more and more molecules bond to their chains of molecules. Each alkene contains a carbon-to-carbon double bond. Acetylene is the first member of the ***alkyne*** family that contains a triple bond. Collectively, alkenes and alkynes are called unsaturated HCs because they can add more hydrogen atoms to form saturated HCs or ***alkanes.*** One unusual feature of unsaturated HCs is that they can also add to each other to form very large molecules called polymers (discussed in Module 6).

Benzene and similar compounds such as toluene are referred to as ***aromatic hydrocarbons*** because some of the first benzene-like compounds had strong aromas. Today, any

Compound name	Pictorial	Molecular weight	Chemical equation	Structural diagram
Methane	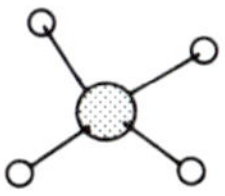	16 Gas	CH_4 (single bond)	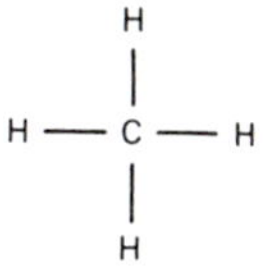
Ethane	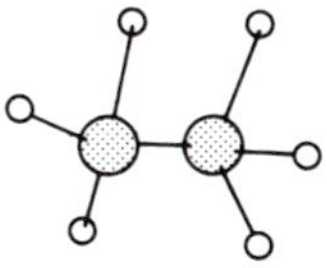	30 Gas	C_2H_6 (single bond)	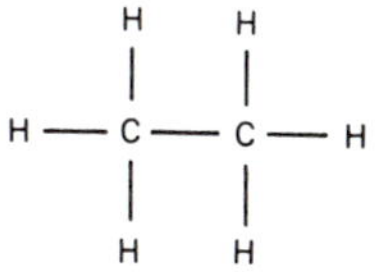
Propane	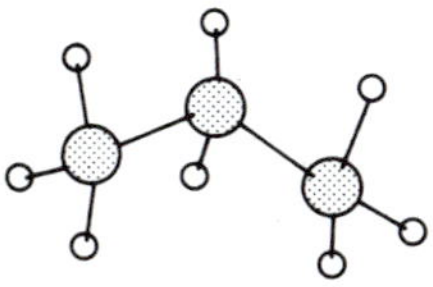	44 Gas	C_3H_8 (single bond)	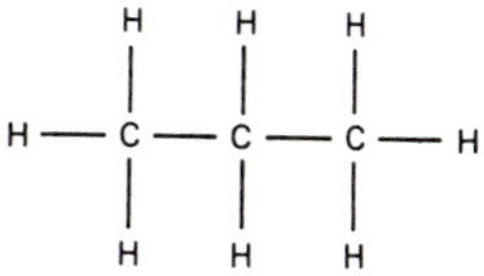
Ethylene	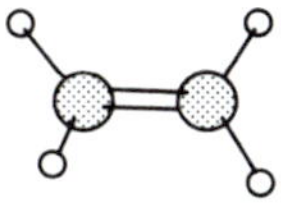	28 Gas	C_2H_4 (double bond)	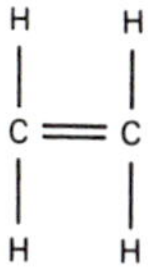
Acetylene	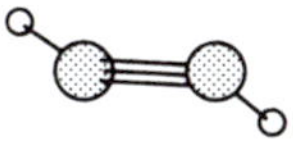	26 Gas	C_2H_2 (triple bond)	
Paraffin	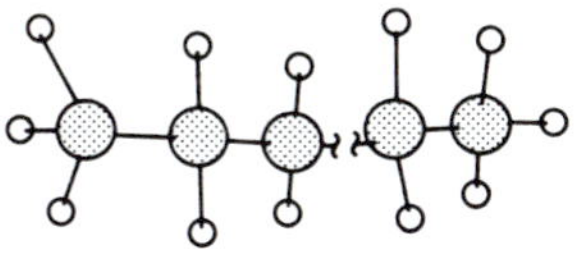	225 Waxy solid	$C_{16}H_{33}$ (single bond)	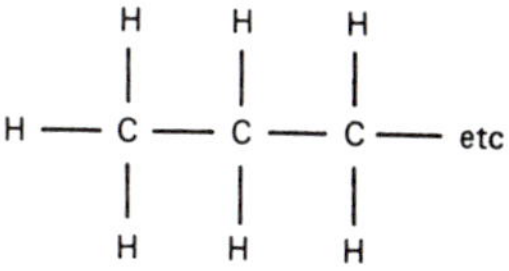
Polyethylene	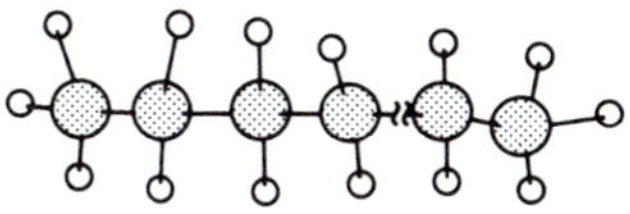	1402 Solid	$C_{100}H_{202}$ (single bond)	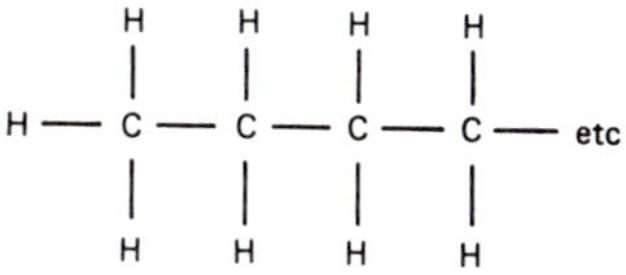

Carbon Hydrogen

Figure 2-10 Hydrocarbons.

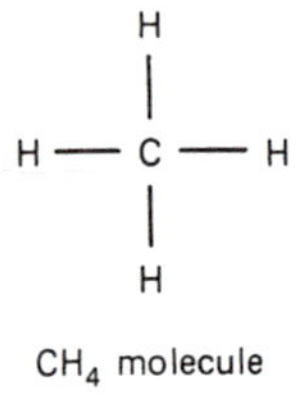

H — C — H (with H above and below)

CH_4 molecule

Figure 2-11 Covalent bonding of methane.

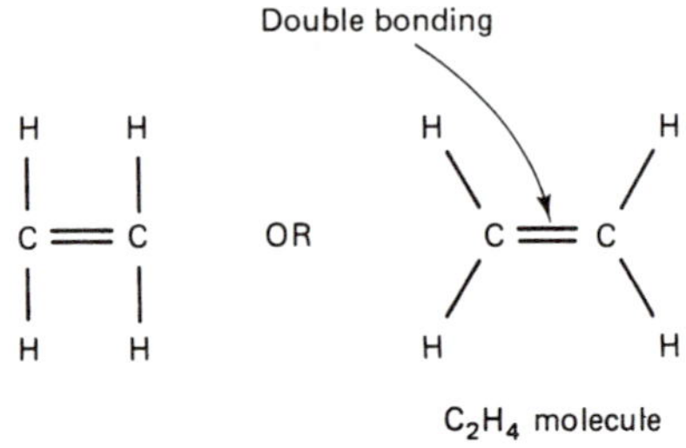

Double bonding

H H / C = C / H H OR H, H / C = C / H, H

C_2H_4 molecule

Figure 2-12 Covalent bonding of ethylene.

compound that contains a benzene ring or has similar properties to those of benzene is called *aromatic.* Figure 2-13 shows many ways of representing benzene. All possible structures have double bonds and some have triple bonds, but benzene doesn't react that way chemically. Because the three pairs of electrons that would form double bonds are not localized (they seem to lie above and below the plane of the carbon atoms), the popular way of representing benzene is with a circle within a hexagon. The circle represents the six unassigned electrons and the hexagon represents the ring of six carbon atoms. This ring of electrons resists being disrupted, providing the benzene molecule with a stable structure. Aromatic HCs are all liquids that are lighter than water. They are used mainly as solvents.

The chemical equation (Figure 2-13) may be considered as a one-dimensional description of a chemical compound or molecule, the structural description as two-dimensional, and the pictorial as a three-dimensional view. Covalent bonding occurs between atoms of nonmetallic elements such as hydrogen, fluorine, chlorine, and carbon, and in molecular compounds such as water, carbon dioxide, ammonia, and ethylene. By sharing electrons, the atoms achieve stability through the creation of interatomic electrical forces, which bond the atoms together to form molecules and compounds. Research has shown

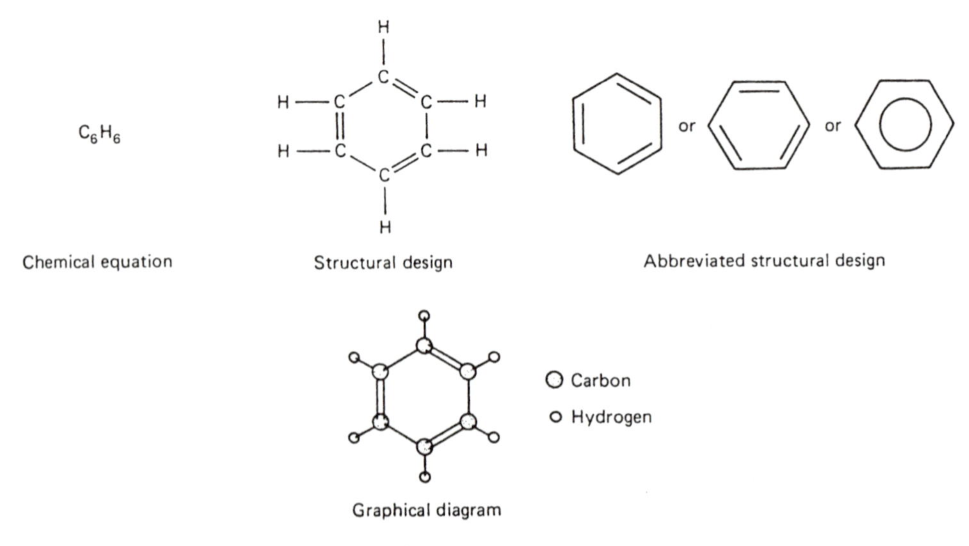

Figure 2-13 Benzene or benzene ring.

that double or triple bonds are stronger than single bonds between the same two atoms. Carbon forms four single covalent bonds.

The many forms ***(polymorphism)*** of carbon serve to illustrate the array of properties achieved by variations in bonding and location of atoms. Carbon exists as diamond, the hardest natural material known, and graphite, an effective lubricant. ***Diamond*** has a cubic symmetry [Figure 2-14(a)], with perfect covalent bonding that joins all atoms into a single molecule. ***Graphite*** develops into sheets with good covalent bonding, but between sheets van der Waals bonds are weak and shear easily, thus allowing the sheets to slide past each other [Figure 2-14(b)]. ***Fullerene*** and ***buckyball*** are the two common names for the

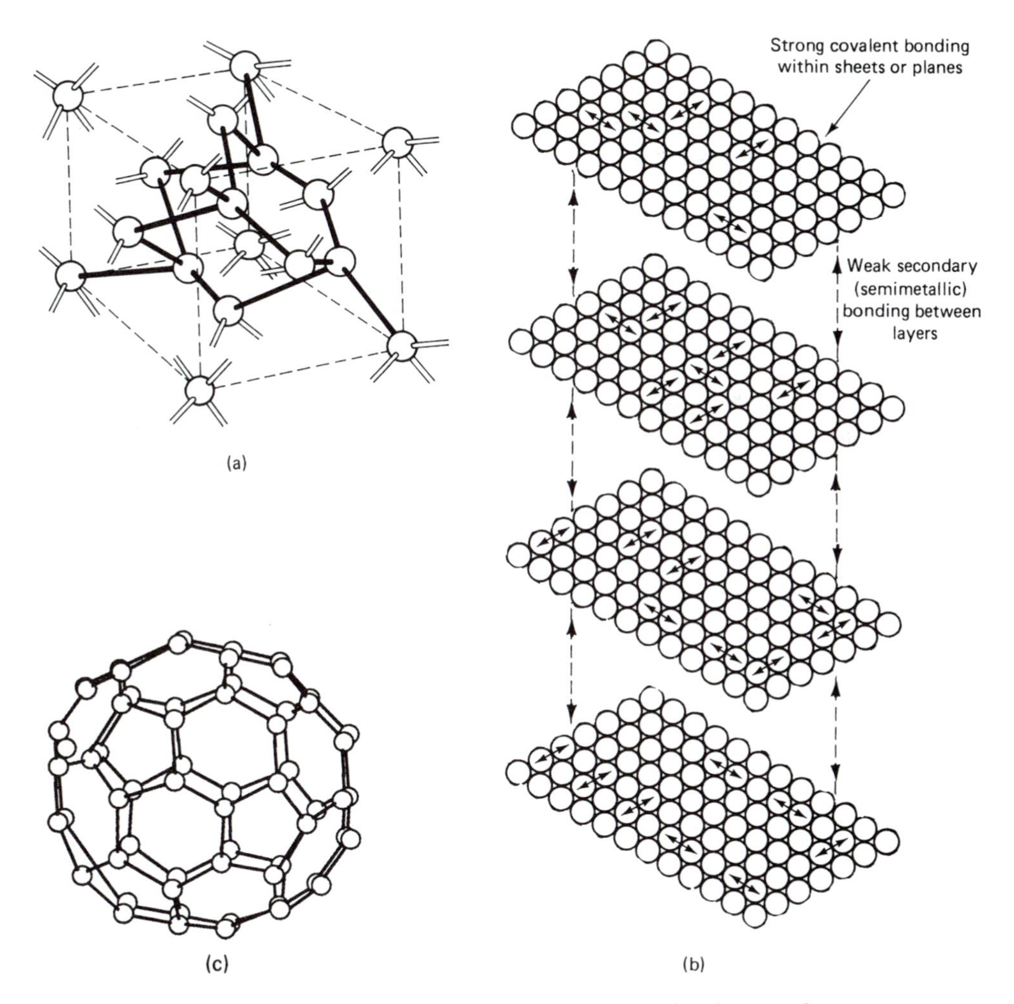

Figure 2-14 Polymorphism of carbon. (a) Diamond's cubic structure: each carbon atom forms strong covalent bonds with four other carbon atoms to develop a tetrahedron in the same manner as silicon and germanium; diamond cutters use a sharp instrument and a sharp blow to split crystal along cleavage planes (111) to produce perfect smaller jewels. (b) Layered structure in graphite: van der Waal bonding between layers allows easy cleavage into sheets, thus providing good lubrication properties. (c) C-60 or fullerene molecules have a spherical shape, like a soccer ball, composed of 60 carbon atoms covalently bonded together.

buckminsterfullerene molecule. Figure 2-14(c) depicts the recently discovered carbon form ^{60}C, which draws its name from Buckminster Fuller, inventor of geodesic domes, because this complex molecule has a soccer-ball shape, composed of 20 hexagons and 12 pentagons.

The discovery of fullerenes [^{60}C molecules; Figure 2-14(c)] was announced by Richard E. Smalley in 1985. The scanning tunneling microscope, discussed in Module 3, allowed the visual proof of the discovery. This new polymorph of carbon holds much promise as materials scientists and engineers unfold the techniques for processing it. The two other polymorphs of carbon are diamond and graphite [Figure 2-14(a) and (b)]. The sooty substances, which are generated at about 2500°C, can be produced as helixes much like DNA, the basic building block of all living cells. The helix form of fullerene is another candidate for making resistance-free electrical superconductors. On the other hand, adding sodium atoms to ^{60}C spheres will yield electrical insulators.

Among other applications of fullerenes are superstrong fibers for construction materials or thin diamond films for hard-coat cutting tools and electronic circuitry. When joined with ammonia atoms, buckyballs become magnetic. Steel is an alloy of iron, carbon, and other elements. When the ^{60}C tubular molecules were alloyed in steel, greater strength was achieved while using conventional steel-making techniques.

Fullerenes can be expected to play a key role in many materials applications because this unique substance provides alternatives in numerous phases of materials science and engineering. Rapid development of ^{60}C will mirror those of many other new materials.

Materials science and engineering have the benefit of powerful new tools, such as the scanning tunneling microscope to aid in unlocking the secret of structure. Buckyballs are much like charm bracelets that allow one to attach many charms of different types. ^{60}C permits the combination of numerous elements and molecules that will keep chemists and materials scientists at work for many decades producing new materials.

Ionic Bonding (Electron Swapping). Instead of sharing electrons, some elements actually swap or ***transfer*** electrons to other elements. This type of bonding is called ***ionic.*** In general, inorganic compounds bond ionically. Elements having outer electron levels that are almost full or almost empty tend to gain or lose electrons to complete their outer energy levels. A classic example of ionic bonding of two elements is sodium (Na) and chlorine (Cl). From the periodic table, we learn that neutral sodium has 11 electrons, one of which is in the outer energy level. This single valence electron is weakly attracted to the positive nucleus. Chlorine (Cl) has 17 electrons, leaving seven electrons in the outer or third energy level. As you recall, chlorine needs eight electrons in its third energy level to attain stability. Since each atom attempts to attain equilibrium (lowest energy state) by completing its required number of electrons for each energy level, the sodium atom gives up its one electron through the action of a strong driving force of the chlorine atom to form a stable outer shell (Figure 2-15). This results in a temporary unstable condition that is soon changed by the sodium atom, which has now become a positive ion (cation) due to the loss of one of its electrons. The positively charged sodium atom (cation) interacts with the chlorine atom, which is now a negative ion (anion) due to the gain of the electron from the sodium atom. This interaction is a strong electrical (electrostatic) force between oppositely charged ions forming the ionic bond, which produces the molecule of sodium chloride (NaCl), common salt. Each of these elements has its own characteristic properties. Sodium is an active metal and chlorine is a poisonous gas. Yet combined as an ionic

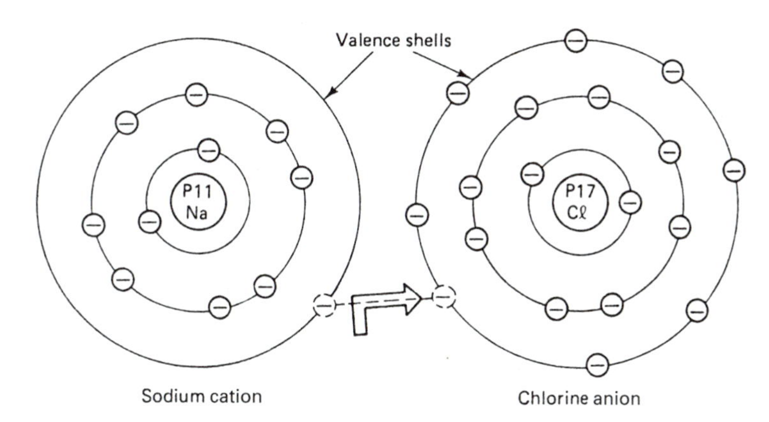

Figure 2-15 Ionic bonding (electron swapping) of a sodium chloride molecule. Upon bonding both atoms have the equivalent of eight electrons in their outer shells.

molecule, they have the well-known properties of common salt. Sodium and chloride ions are also different from their respective neutral substances (atoms). The greenish-yellow element chlorine, whose atoms have seven valence electrons, is very active. Chloride ions, in the formation of NaCl, have satisfied the octet rule with eight electrons in their outermost shells, and thus are stable. Otherwise, Cl^{-1} ions are extremely active, reacting with metals to produce corrosion. All these bonds are electrostatic in origin, so that the chief distinction among the various ways in which atoms bond to form molecules is in the distribution of the electrons around the atoms and molecules. ***Metals,*** defined later, can be characterized by the tendency of their atoms to lose their valence electrons. An ***active metal*** is one that loses its electrons quite readily. Similarly, an ***active nonmetal*** is one whose atoms readily accept electrons. We also know that nonmetallic elements have a larger number of valence electrons than do metallic elements.

Metallic Bonding (Electron Swarming). Metals contain one, two, or three valence electrons. These are shielded from the strong attractive forces of the positive nucleus by the inner electrons and thus they bond to the nucleus relatively weakly. Consequently, when in the company of other metal atoms, these metal atoms will lose their weakly held valence electrons, which, in turn, enter a common free-electron cloud, swarm, sea, or gas. These free or delocalized electrons can then move in three dimensions. Once these electrons leave their atoms, the atoms become positive ions (cations). An electrostatic balance is maintained between the cation and the ***electron cloud,*** which results in the cations arranging themselves in a three-dimensional pattern as shown in Figure 2-16. This electrostatic balance is the glue that bonds the metallic structure. The ever-moving free-electron cloud acts like a matrix surrounding the positive metallic ions and provides rigidity. If these ions were represented as spheres, their space arrangement in bulk metal could be likened to a box full of table tennis balls.

Secondary Bonding. The van der Waals bonds, which are much weaker forces of attraction, produce bonding between atoms and between molecules. Group VIIIA elements (inert gases) have a full complement of valence electrons in their outer levels. Thus, they have no inclination to lose, gain, or share electrons. At ordinary temperatures, they remain as single atoms (monatomic). Only at very low temperatures will these gases condense. It

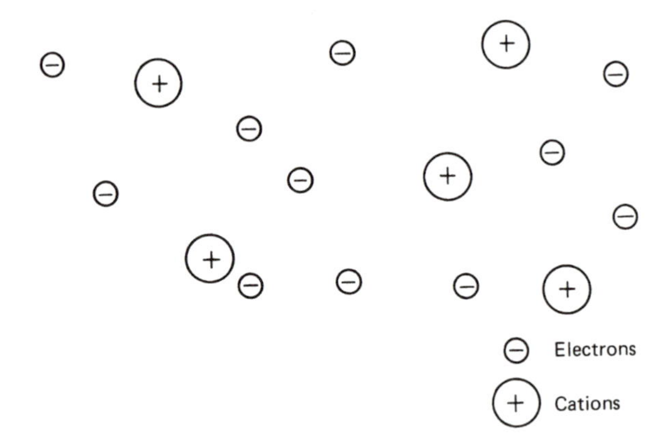

Figure 2-16 Metallic bonding or electron swarming.

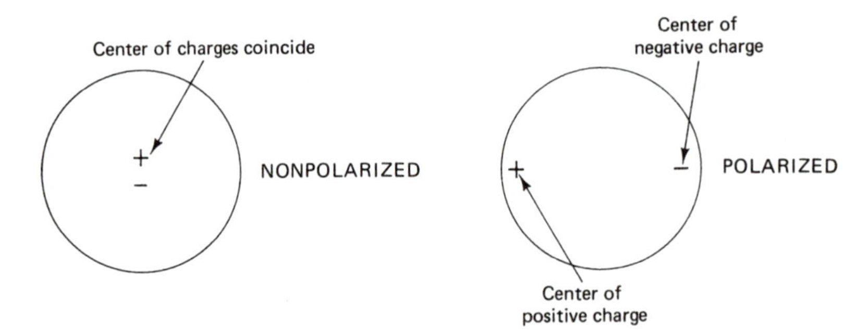

Figure 2-17 Schematic of a nonpolarized and a polarized molecule.

is the presence of van der Waals forces that permits this condensation. A similar situation occurs with covalently bonded molecules that have achieved equilibrium by sharing electrons. The van der Waals forces also permit these molecules to condense at low temperatures. All molecules consist of distributions of electrical charge. When two molecules are in close proximity, these electrical charge distributions interact and create intermolecular forces. The van der Waals forces of attraction between molecules have their origin in the forces of attraction of the nucleus (positive) of one molecule for the electron charge (negative) of a neighboring molecule, which form fluctuating dipoles in the two molecules. Such forces increase with an increase in the number of electrons per molecule because the electrons in the large molecule are more readily polarized by these forces.

In a ***polar molecule*** or ***electric dipole,*** the charges are polarized; that is, both the positive and negative charges are localized within the molecule. Figure 2-17 illustrates these two situations. A hydrogen molecule is a simple example of a ***nonpolar molecule.*** This diatomic molecule, in which both atoms are of the same element and are covalently bonded together, is sketched in Figure 2-18. Each atom shares the bonding electrons equally, producing an electrical charge distribution that is symmetrical about a line joining the two nuclei.

Any diatomic molecule, such as hydrogen chloride (HCl), in which the atoms are different from each other is polar [Figure 2-19(a)]. The negative pole is located at the atom that is more electronegative (attracts electrons more strongly). In a polar molecule, the

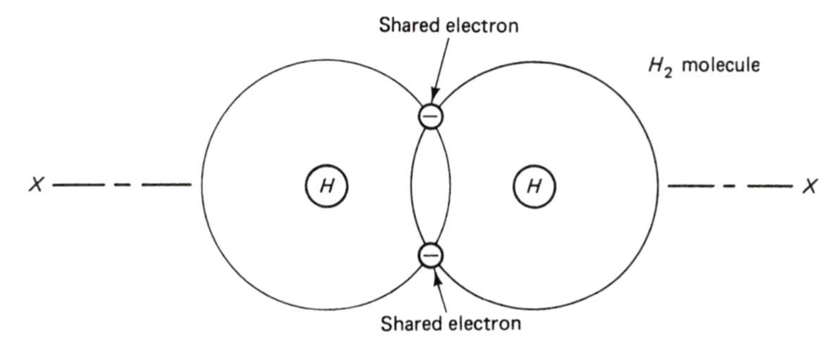

(a) Both atoms share two electrons equally forming a single covalent bond to complete each atom's outer energy level (K level).

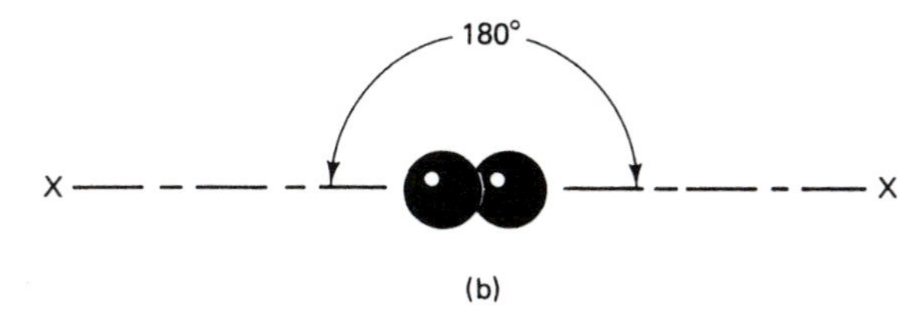

(b)

Figure 2-18 (a) A nonpolar diatomic hydrogen molecule. (b) A linear hydrogen molecule.

electrons are not shared equally. The chlorine atom, being more electronegative, attracts more electrons than the hydrogen atom does, which causes the shared electrons to be associated more closely with the chlorine nucleus. Figure 2-19(b) is a three-dimensional sketch of this molecule showing that, like the hydrogen molecule, it has an electrical charge distribution with a symmetry around the line *x–x*.

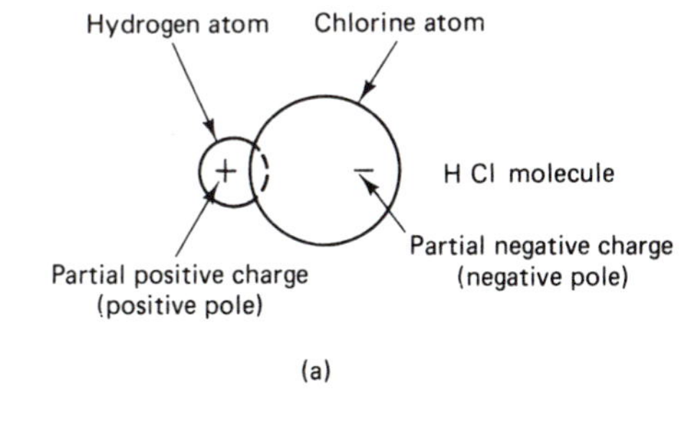

(a)

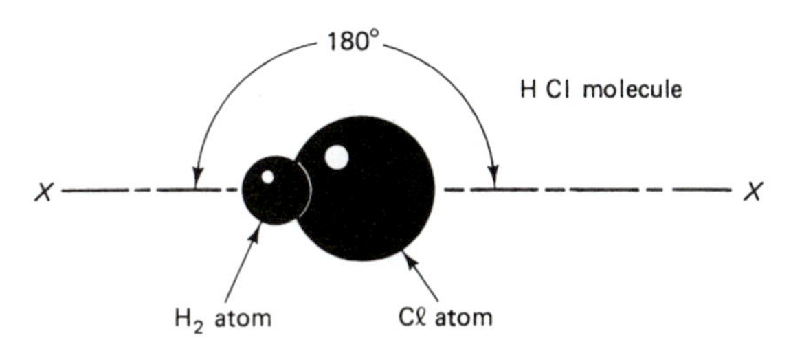

(b)

Figure 2-19 (a) A polar diatomic molecule of hydrogen chloride (HCl). (b) A linear hydrogen chloride molecule.

A good example of a polar molecule containing more than two atoms is water (H_2O), sketched in Figure 2-20. Oxygen has a greater attraction for electrons than does hydrogen. As a result, the four electrons being shared in the two covalent bonds remain closer to the oxygen atom. This unequal sharing of electrons has the effect of creating a partial positive charge between the hydrogen atoms and a partial negative charge on the oxygen, which produces a *polar covalent bond.* The three-dimensional sketch [Figure 2-20(a)] of the water molecule with the oxygen atom in the center of a cube and with two hydrogen atoms occupying two of the eight corners shows a bent molecule (not symmetrical), with an angle between the two hydrogen atoms of about 105° [Figure 2-20(b)]. Most covalent bonds are polar. The asymmetrical charge distribution on the water molecule results in the formation of regions of positive (+) and negative (−) charges located at a maximum distance from each other [Figure 2-20(c)]. If the charge distribution were symmetrical, the positive (+) and negative (−) charges would cancel each other, as both centers would be at the center of the molecule. Thus, each center of charge can exert an attractive force on an adjacent charge of opposite sign. This statement explains, in part, how water molecules bond together to form liquid water. As pictured, the hydrogen atoms bond on one side of the oxygen, leaving the protons of the hydrogen atom farther from the nucleus of the oxygen atom.

The covalent bonds between the hydrogen and oxygen atoms in a water molecule, being polar, result in an asymmetrical charge distribution. The oxygen atom, having a slight negative charge, attracts a positive hydrogen atom belonging to an adjacent water

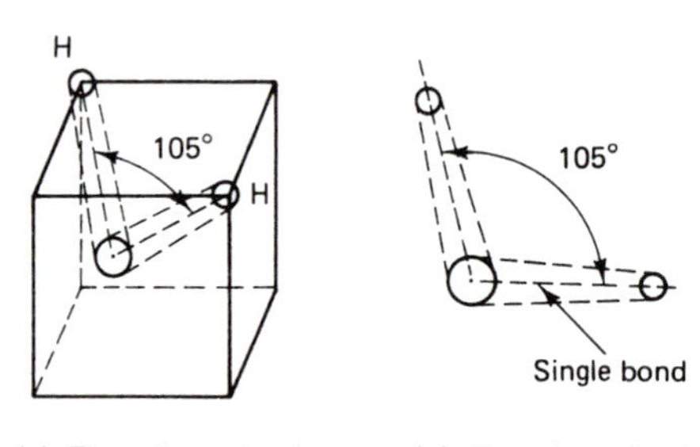

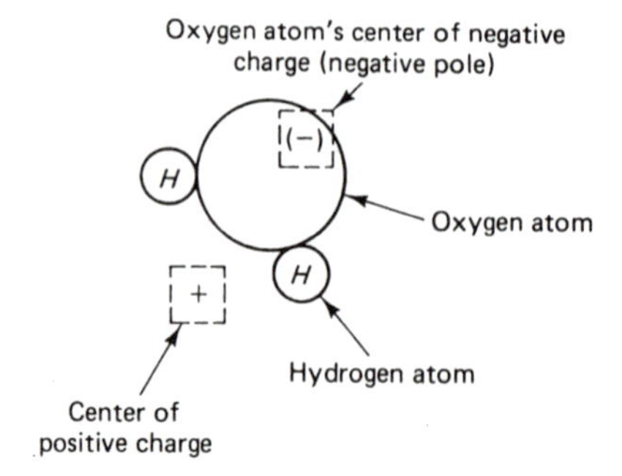

Figure 2-20 Various sketches of a water molecule (H_2O).

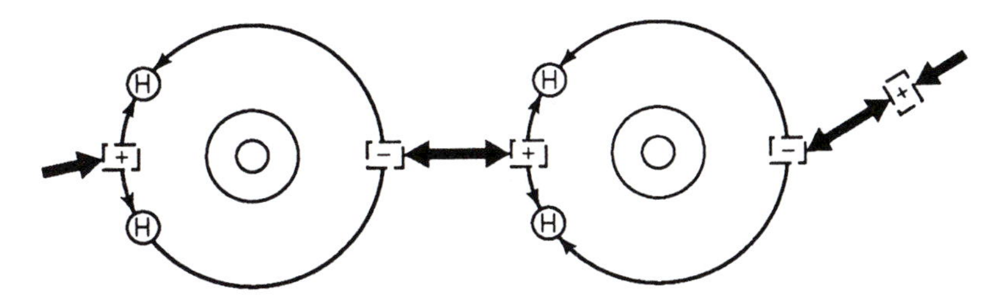

Figure 2-21 Hydrogen bond (bridge) in liquid water.

molecule and forms a ***hydrogen bond,*** sometimes called a ***hydrogen bridge.*** This bond is the strongest of the secondary bonds. Figure 2-21 shows one schematic for representing the hydrogen bond between two water molecules in the liquid phase. The relatively high boiling temperature as well as other extraordinary properties of water are attributable to the bonding action of the hydrogen bridging.

As do all bonds, the hydrogen bond affects the properties and behavior of materials, particularly polymeric materials. In thermoplastics, the hydrogen bond joins long chain-like molecules to each other. These relatively weak bonds can be easily loosened or broken by heating, permitting flow to take place. This explains why thermoplastics can be converted by heating to a soft, flexible state and then back to rigid, solid plastic material upon cooling.

Our discussion of chemical bonding is well summarized in Table 2-2, which shows one secondary bond and the three primary bonds. Further discussion of these chemical

TABLE 2-2 SUMMARY OF ATOMIC BONDING

Type of bond	Number of electrons "shared"	Types of atoms involved	Remarks
van der Waals (molecular)	0	Same	Weak electrostatic attraction due to asymmetrical electrical charges in electrically neutral (as a whole) atoms or molecules
Ionic	1 (or more transferred)	Different	Strong electrostatic attraction
Covalent		Same or different	Electron pair "revolves" in common orbit about both nuclei, one atom supplying one electron
Normal	2		
Coordinate	2	Different	Electron pair "revolves" in common orbit about both nuclei, one atom supplying both electrons, the other atom supplying none (quite rare)
Metallic	∞	Same	General attraction of a very large number of positive (metallic) ions for a dispersed cloud of electrons

Source: C.O. Smith, *The Science of Engineering Materials,* 2nd ed., Prentice Hall, Englewood Cliffs, N.J., 1977.

bonding mechanisms comes up in dealing with the various groups within the family of materials, seen in Table 2-3.

2.1.2 Solid State

A ***solid*** is a sample of matter that has a fixed volume or size and a fixed shape that it retains indefinitely without any need to confine it. Picturing the atom as a sphere of microscopic size, atoms that make up a solid are packed closer together than the same atoms making up a liquid or gas. We can say that solids do not flow like liquids nor do they expand like gases. The atoms of a solid do not possess the kinetic energy or the motion they would possess if they were in a liquid or gas. These atoms are quite subdued and do not act as independent units, but move with their neighboring atoms. As we will see, solids can be grouped into several categories, such as metals, plastics, ceramics, and composites. The attractive forces between atoms in a solid are much greater than between atoms in a liquid or gas. As such, these forces may position the atoms in some orderly geometric arrangement to form a ***crystalline structure*** (as in most metals) in which the atoms vibrate, rotate, and oscillate around fixed locations, maintaining a minimum dynamic equilibrium between adjacent atoms. From this, we can conclude that the geometric arrangement of atoms is related to its energy content. If the arrangement is orderly, this implies a minimum energy content. On the other hand, a disorderly, random distribution of particles (producing an ***amorphous structure,*** as in many plastics) is an indicator of motion. Hence, unbalanced forces are acting on the atoms, and thus an increase in internal energy by the system of particles is necessary to sustain such motion of particles from a small energy demand for vibration to a larger energy requirement for translational motion. Finally, another indicator of energy is temperature. The higher the temperature of the system of particles, the higher the disorder of the particles will be and the more energy that will be possessed by the system. With solids, we know that most of the motion of the atoms is confined to a vibrational type.

2.1.3 Processing and Structure

Throughout this book, you will learn how internal structure is altered as a result of processing. If materials are allowed to cool slowly as they transform from the liquid to the solid form, they will achieve equilibrium, which results in their natural internal structure and a given set of properties. However, if the cooling process is accelerated or interrupted, an entirely different structure and set of properties often results. For example, in casting molten metals to produce a desired shape, the rate of cooling is normally around 1 kelvin per second. But with use of a rapid-solidification processing technique, the molten metal may be cooled at a rate of 10^8 kelvin per second, which results in disorderly structure and unique magnetic properties not available in the normally cooled metal. Module 3 covers the rapid solidification process.

Structure can also be changed when materials are in the solid state. If the proper ingredients have been mixed together during material formulation, it becomes possible to reheat the solid, then vary the rate of cooling to obtain a variation in properties such as hardness or electrical conductivity. The process of bending or compressing a solid material will alter the internal structure in a manner that will possibly increase the hardness and improve strength.

TABLE 2-3 FAMILY OF MATERIALS

Group	Subgroup	Examples
Metallics (metals and alloys)	Ferrous	Iron
		Steel
		Cast iron
	Nonferrous	Aluminum
		Tin
		Zinc
		Magnesium
		Copper
		Gold
	Powdered metal	Sintered steel
		Sintered brass
Polymerics	Human-made	Plastics
		Elastomers
		Adhesives
		Paper
	Natural	Wood
		Rubber
	Animal	Bone
		Skin
Ceramics	Crystalline compounds	Porcelain
		Structural clay
		Abrasives
	Glass	Glassware
		Annealed glass
Composites	Polymer-based	Plywood
		Laminated timber
		Impregnated wood
		Fiberglass
		Graphite epoxy
		Plastic laminates
	Metallic-based	Boron aluminum
		Primex
	Ceramic-based	Reinforced concrete
		CFCC
	Cermets	Tungsten carbide
		Chromium alumina
	Other	Reinforced glass
Others	Electronic materials	Semiconductors
		Superconductors
	Lubricants	Graphite
	Fuels	Coal
		Oil
	Protective coatings	Anodized aluminum
	Biomaterials	Carbon implants
	Smart materials	Shape memory alloys
		Shape memory polymers

2.2 FAMILY OF MATERIALS

There is nearly a limitless variety of materials. With so many materials, how can one be expected to understand them? The method used in this book is to consider all materials as members of a big family. Materials that possess common characteristics are then placed into their own group within the family. Even though overlaps exist in the grouping system, it is easier to understand materials when relationships are identified.

As seen in Table 2-3, four groups, *metallics, polymers, ceramics,* and *composites,* comprise the main groups; a fifth group of *other materials* is used for materials that do not fit well into the four groups. Each group then divides into subgroups. Some examples of each subgroup are listed to illustrate common materials in the subgroup. Figure 2-22 illustrates the variety of materials used in one of the communications systems: telecommunications. Within any system such as transportation, communications, or construction, numerous *materials systems* exist. These systems must pull materials from the groups within the material's family to meet their special needs. The wise materials user recognizes the

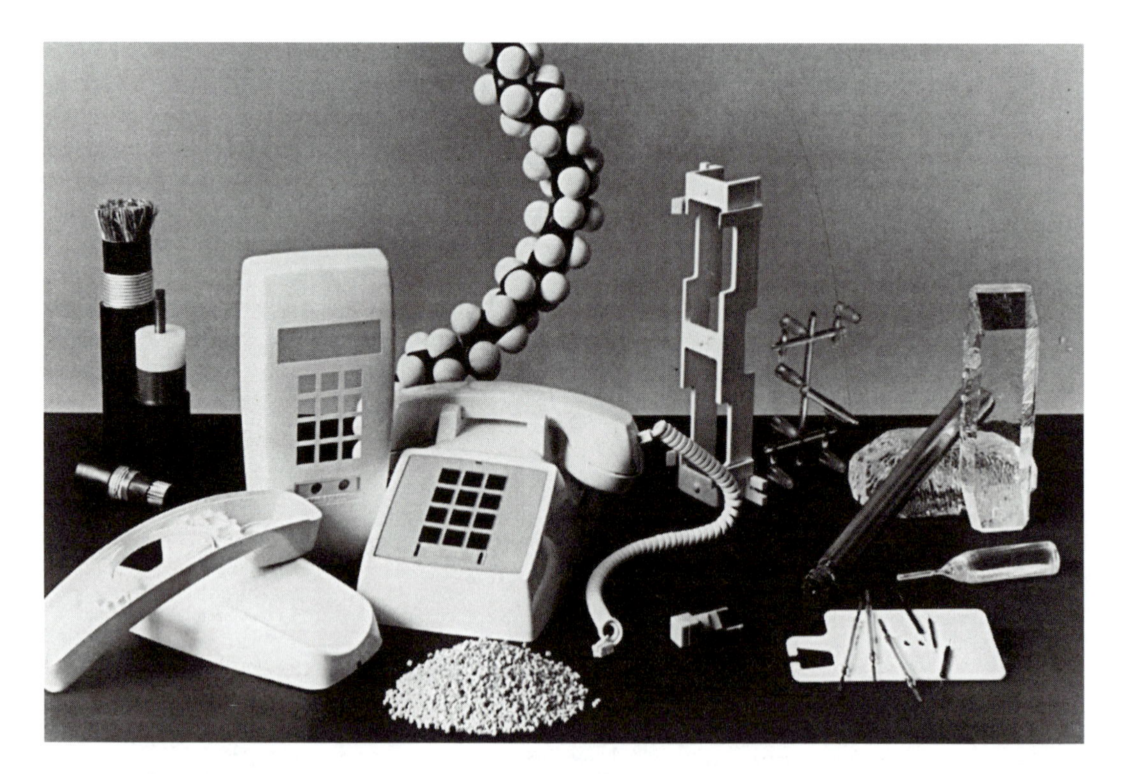

Figure 2-22 Family of materials in communication. One of the major factors influencing performance of telecommunications equipment is performance of materials used in its construction. Plastics (polymers) for outdoor, undersea, and underground cables; magnetic alloys (metals); fiber optics; synthetic single crystals (ceramics); and fiber-reinforced rubber (composites) are five principal areas in which materials research and development discoveries greatly influence telecommunications. (Bell Laboratories.)

many options available because of the compatibility between groups and then proceeds to develop a materials system that meets a specific need.

2.2.1 Other Systems of Grouping

The grouping system used in this book follows closely grouping systems common to engineering. Slight differences exist in other grouping systems. For example, one system might divide materials into metals, polymers, ceramics, glass, wood, and concrete. Some systems divide all materials into three groups: organics, metals, and ceramics. Then they distinguish between natural and synthetic materials. Others group materials as crystalline or amorphous. Regardless of the system, some inconsistencies develop because of the nearly limitless variety of materials. Still, the advantages of grouping outweigh the limitations.

In this section, we will present the system of grouping for this book and study briefly the general nature of each group and subgroup. Detailed study of each group follows in subsequent modules.

2.2.2 Metallics

Metallics, or metallic materials, include metal alloys. In a strict definition, *metal* refers only to an element such as iron, gold, aluminum, and lead. (See the discussion of metals and nonmetals in Section 2.1.1.1.)

2.2.2.1 *Metallic* defined. The definition used for a metal will differ depending on the field of study. Chemists might use a different definition for metals than that used by physicists.

- ***Metals*** are elements that can be defined by their properties, such as ductility, toughness, malleability, electrical and heat conductivity, and thermal expansion.
- ***Metals*** are also large aggregations (collections of millions of crystals composed of different types of atoms held together by metallic bonds; see Figure 2-16). Metals usually have fewer than four valence electrons (electrons in the outer orbits of their atoms), as opposed to nonmetals, which generally have four to seven. The metal atom is generally much larger than the atom of the nonmetal.
- ***Alloys*** consist of metal elements combined with other elements. Steel is an iron alloy made by combining iron, carbon, and some other elements. Aluminum–lithium alloys provide a 10% saving in weight over conventional aluminum alloy.

2.2.2.2 Types of metallics. While metals comprise about three-fourths of the elements that we use (see metal groups in the periodic table), few find service in their pure form. There are several reasons for not using pure metals; pure metals may be too hard or too soft, or they may be too costly because of their scarcity, but the key factor normally is that the desired property sought in engineering requires a blending of metals and elements. Thus, the combination forms (alloys) find greatest use. Therefore, *metals* and *metallics* become interchangeable terms. Metallics are broken into subgroups of ferrous and nonferrous metals.

Ferrous. ***Ferrous*** is a Latin-based word meaning "iron." Ferrous metals include iron and alloys of at least 50% iron, such as cast iron, wrought iron, steel, and stainless steel. Each of these alloys is highly dependent on possessing the element carbon.

Steel is our most widely used alloy. Sheet steel forms car bodies, desk bodies, cabinets for refrigerators, stoves, and washing machines; it is used in doors, "tin" cans, shelving, and thousands of other products. Heavier steel products, such as plate, I beams, angle iron, pipe, and bar, form the structural frames of buildings, bridges, ships, automobiles, roadways, and many other structures.

Nonferrous. Metal elements other than iron are called ***nonferrous*** metals. The nonferrous subgroup includes common lightweight metals such as titanium and beryllium and common heavier metals such as copper, lead, tin, and zinc. Among the heavier metals is a group of white metals, including tin, lead, and cadmium; they have lower melting points, about 230° to 330°C. Among the high-temperature (refractory) nonferrous metals are molybdenum, niobium, tantalum, and tungsten. Tungsten has the highest melting point of all metals: 3400°C. Metal alloys other than iron are called nonferrous alloys. The possible combinations of nonferrous alloys are practically endless.

Powdered Metals. Alloying of metals involves melting the main ingredients together so that on cooling, the metal alloy is generally a nonporous solid. Powder metal is often used instead because it is undesirable or impractical to join the elements through alloying or to produce parts by casting or other forming processes. ***Powdered metal*** is sometimes called ***sintered*** metal. As shown in Figure 2-23, this process consists of producing small particles, compacting, and *sintering* (applying heat below the melting point of the main component). The squeezing pressure with added heat bonds the metal powder into a strong (sometimes porous) solid. Powdered metals can be ferrous, nonferrous, or a combination of ferrous and nonferrous elements with nonmetallic elements.

2.2.3 Polymerics

Polymeric materials basically are materials that contain many parts. ***Poly*** means "many" and *mer* comes from *meros,* which means "parts." A ***polymer*** is a chainlike molecule made up of smaller molecular units (monomers). The ***monomers,*** made up of atoms, bond together covalently to form a polymer that usually has a carbon backbone.

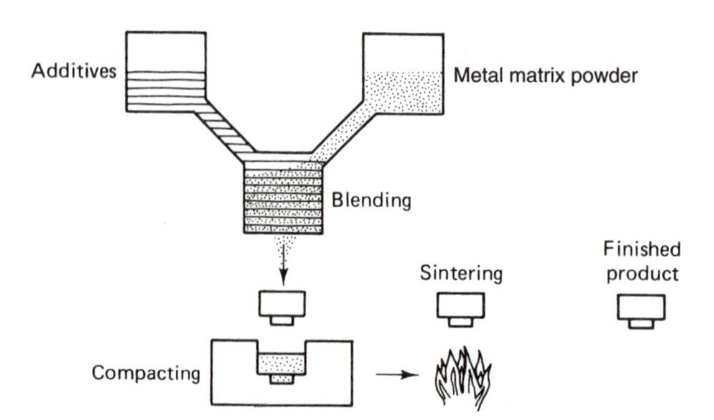

Figure 2-23 Powdered metal.

Figure 2-10 shows a simple sketch of an ethylene monomer. The atoms of carbon (C) and hydrogen (H) form covalent bonds. In Figure 2-10, many of the ethylene monomers have joined into a polyethylene polymer. Thousands of polyethylene polymers join together to form polyethylene plastic. The same process of polymerization is responsible for the formation of human-made rubbers and plastics; natural fibers; wood; rubber; animal bone; skin; and the tissues of humans, animals, and insects.

2.2.3.1 Plastics. The term ***plastic*** is used to define human-made, polymer resins containing carbon atoms covalently bonded with other elements, along with organic and inorganic substances. The word *plastic* also means *moldable* or *workable,* such as with dough or wet clay. Plastic materials are either liquid or moldable during the processing state, after which they turn to a solid. After processing, some plastics cannot be returned to the plastic or moldable state; they are ***thermosetting plastics*** or ***thermosets.*** *Thermo* means "heat," and *set* means "permanent." Common thermosetting plastics include epoxy, phenolic, and polyurethane. Other plastics can be repeatedly reheated to return to the plastic state; they are ***thermoplastics.*** Examples of thermoplastic plastics are acrylics (e.g., Plexiglas or Lucite), nylon, and polyethylene. Although at present most plastics are produced from oil, they can also be made from other organic (carbon) materials, such as coal or agricultural crops, including wood and soybean. As our limited supply of oil is depleted, the major sources of polymerics will change.

2.2.3.2 Wood. Of all the materials used in industry, *wood* is the most familiar and most used. Wood is a natural polymer. In the same manner that polymers of ethylene are joined to form polyethylene, *glucose* monomers polymerize in wood to form ***cellulose*** polymers ($C_6H_{10}O_5$). Glucose is a sugar made up of carbon (C), hydrogen (H), and oxygen (O). Cellulose polymers join in layers with the gluelike substance ***lignin,*** which is another polymer.

2.2.3.3 Elastomers. Prior to World War II, most rubbers *(elastomers)* consisted of natural rubbers. Today, synthetic or human-made rubbers far exceed the use of natural rubbers. An ***elastomer*** is defined as any polymeric material that can be stretched at room temperature to at least twice its original length and return to its original length after the stretching force has been removed. Some stretch to over 10 times their original length. Elastomers are able to store energy, so they can return to their original length and/or shape repeatedly.

Elastomers have a molecular, amorphous structure similar to that of other polymeric materials. This *amorphous,* or shapeless, structure consists of long coiled-up chains of giant molecules (polymers) that are entangled with each other. Adjacent polymers are not strongly bonded together. As seen in Figure 2-24(a), when a tensile force (pulling force) is applied, these coils straighten out (***bond straightening***) and snap back like springs to their original coiled condition on removal of the force. When this same type of force is strong enough, not only do the polymers increase in length as a result of the bonds lengthening between individual atoms, they also lengthen [***bond lengthening,*** Figure 2-24(b)] due to

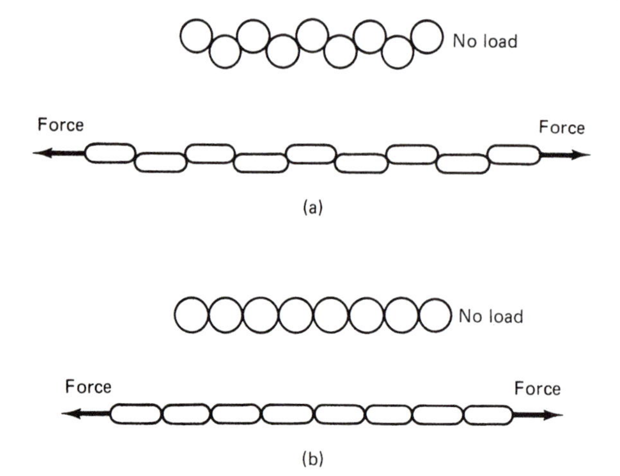

Figure 2-24 Elastomer structure. (a) Bond straightening. (b) Bond lengthening.

the unwinding coils, resulting in a temporary structure that approaches crystallinity, or a more orderly structure. Remember that with crystallinity comes strength.

To further increase the strength of the elastomers, the process of ***vulcanization*** is used to form the necessary cross-links (strong bonds) between adjacent polymers (to be discussed later in Section 6.4.2.2). Vulcanization is a chemical process (invented by Charles Goodyear in 1839) that produces covalent bonding between adjacent polymers with the help of a small amount of sulfur. Cross-links tie the polymers together to produce a tough, strong, and hard rubber for many uses in industry, such as in automobile tires.

2.2.3.4 Other natural polymers. A most amazing natural polymer is human skin, which has no equal substitute. Animal skin or hide in the form of fur and leather has limited industrial use because synthetic materials have been developed that offer greater advantages to the designer than those of the natural polymers. Medical science continues to study such natural polymers as bones, nails, and tissues of human beings and animals to synthesize these materials for replacement when they are damaged due to injury or illness. Bioengineering and biomechanics are newer fields that integrate engineering and medicine to solve material problems in the treatment of humans.

2.2.4 Ceramics

Ceramics are crystalline compounds combining metallic and nonmetallic elements. ***Glass*** is grouped with ceramics because it has similar properties, but most glass is *amorphous.* Included in ceramics are porcelain such as pottery; abrasives such as emery used on sandpaper; refractories (materials with good resistance to heat) such as tantalum carbide, with a melting temperature of about 3870°C; and structural clay such as brick. Ceramics, including glass, are hard, brittle (no internal slip), stiff, and have high melting points. Ceramics primarily have ionic bonds, but covalent bonding is also present. Silica is a basic

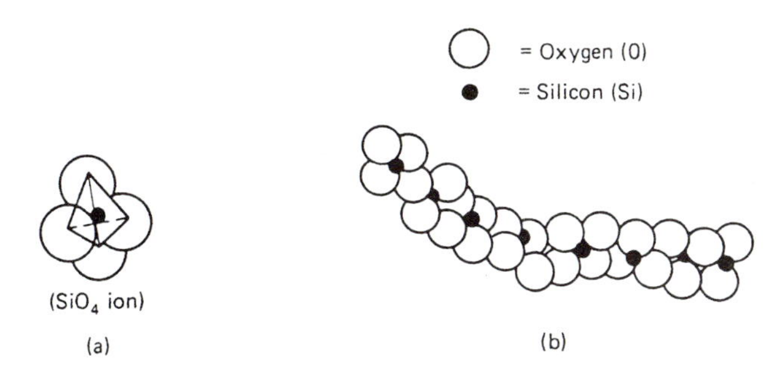

Figure 2-25 Ceramic structure. (a) Silicate tetrahedron. (b) Single chain of silicate.

unit in many ceramics. The internal structure of silica has a pyramid (tetrahedron) unit as diagrammed in Figure 2-25(a), which shows both the graphical diagram and the chemical equation. These silicate tetrahedrons join into chains, with the graphical diagram shown in Figure 2-25(a). In Figure 2-25(a) and (b), note how the larger oxygen (O) atoms surround the small silicon (Si) atom. The silicon atom occupies the space opening (interstice) between the oxygen atoms and shares four valence electrons with the four oxygen atoms. Chains such as shown in Figure 2-25(b) are extremely long and join together in three dimensions. The chains are held together by ionic bonds, whereas individual silica tetrahedrals bond together covalently. Silica is combined with metals such as aluminum, magnesium, and other elements to form a wide variety of ceramic materials.

2.2.5 Composites

By strict definition, a ***composite*** is a material containing two or more integrated materials (constituents), with each material keeping its own identity. Normally, combining of the materials serves to rectify weaknesses possessed by each constituent when it exists alone. By this strict definition, many natural materials exist as composites; for example, wood is a combination of cellulose and lignin, but wood and natural materials are not classified as composites. Some familiar composites include plywood, laminated dimes and quarters, and shoe soles.

While most of the groups of the family of materials could be classed as composites because of the way they are placed in service (such as painted steel or case-hardened steel), the composite classification commonly refers to materials developed to meet the demands in building, electronic, aerospace, and auto industries. With an ever-increasing use of composites, they are truly the material of today and the near future because composites can be designed to be stronger, lighter, stiffer, and more heat resistant than natural materials or to possess properties required by technology that are not available in a single material. Composites allow the designer to select the right combination of materials to perform safely at the lowest cost.

The subgroups of composites shown in Table 2-3 include polymer based, metallic based, ceramic based, cermets, and others. It is also possible to classify composites by their structure. Composite structures include layers, fibers, particles, and any combination of the three. ***Layered composites***, shown in Figure 2-26, consist of ***laminations*** like a

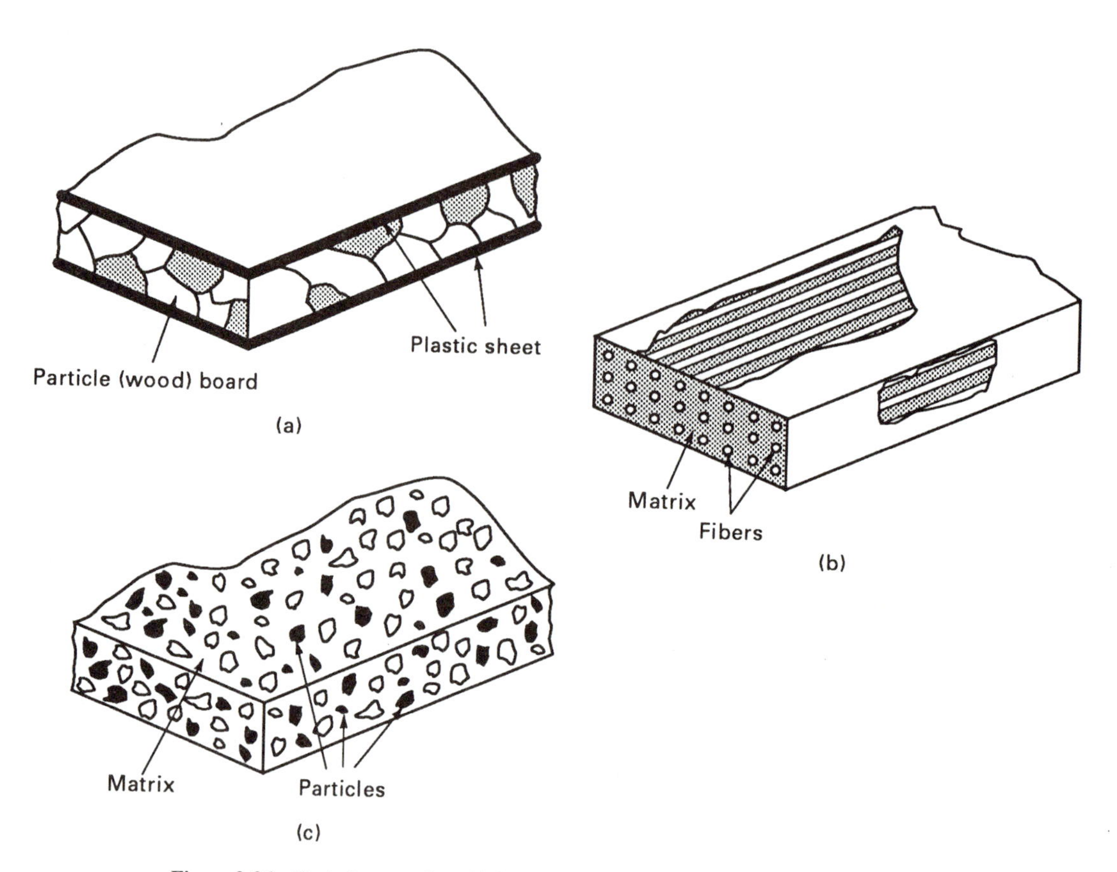

Figure 2-26 Typical composites. (a) Layered composite. (b) Fiber composite. (c) Particle composite.

sandwich. The laminations are usually bonded together by adhesives, but other forces could be used, such as those provided by welding.

Fibers and particles are integrated into composites by suspending them in a ***matrix*** or by the use of *cohesive* forces. The matrix is the material component, such as plastic, epoxy cement, rubber, or metal, that surrounds the fibers or particles. ***Cohesive*** forces involve the molecular attraction of one constituent to the other.

Examples of layered composites [Figure 2-26(a)] include plywood, laminated boards such as particleboard (wood chips) covered with plastic sheets, thermocouples (different metals welded together), safety glass (plastic sandwiched between sheets of glass), cardboard, and Alclad (copper alloy core that gains corrosion resistance from outside covers of aluminum). ***Particle composites*** include concrete (gravel, rock, and sand in Portland cement matrix), and particleboard (wood chips in resin matrix), as shown in Figure 2-26(c). ***Fiber composites,*** such as reinforced plastics (FRPs), include fiberglass (glass fibers in a polyester plastic matrix), hardboard (vegetable fibers in an adhesive matrix), boron fibers in an aluminum matrix, reinforced glass (wire mesh in a glass matrix), and graphite epoxy (graphite fibers in an epoxy plastic matrix), as shown in Figure 2-26(b).

2.2.6 Other Materials

This group is used to include materials that do not fit well into the groups discussed previously. Comparing the materials in the "other" group with the rest of the family of materials, Table 2-3 reveals that most come from the metallics, polymers, ceramics, and composite groups, but their applications justify a separate study. Some are discussed in Module 9.

Materials systems as described in Module 1 open up a whole new opportunity for grouping. Each major group of our family of materials has some materials from a newly evolving subgroup known as intelligent material systems.

Intelligent material systems, smart materials, or ***smart structures*** are materials and material systems designed to mimic biological organisms and offer the ultimate material system that can place control and feedback into a material structure. These materials take their cue from biological elements such as muscles, nerves, and bodily control systems that adapt to environmental changes. Just as alchemists long yearned to turn lead into gold, so too, materials scientists and engineers seek to endow materials with abilities such as 1) making controlled adaptations to changes in stress or heat, 2) making self repairs, and 3) providing feedback information on conditions that may have caused a material failure.

Science fiction such as the movie *Terminator II* has long foretold of "morphing" (changing shape and structure) or ***biomimetics*** (materials mimicking living tissue). As foretold in science fiction, materials designers continue to evolve material structures that do allow for modification to unpredictable environments. Also, feedback provides information to designers and users throughout the life of the material or structure. An example seen later in Module 6, a smart house, uses acrylic windows that change from clear to translucent white with the changing of outside light conditions. The transformation results from liquid crystals embedded in the glazing. Another feature available for smart houses includes new exterior walls that change from light colors that reflect heat in the summer to dark colors that absorb heat in the winter.

Monitoring systems using piezoelectric devices (explained in Module 9) are embedded into or adhesively bonded onto structures such as bridges, aircraft skin, building walls, and similar load-bearing structures. Connected to recording instruments, these intelligent structures provide information to help predict material/structural failures caused by aging, earthquakes, storms, or other stresses.

Nitinol, an alloy of nickel and titanium, illustrates how "smart materials" can be tailored to provide unique properties. The alloy is a ***shape memory alloy (SMA).*** SMAs, through heat treating, can be trained to hold a shape that they will always "remember," even if drastically bent out of shape. As seen in Figure 9-53 in Module 9, an eyeglass frame temple made of an SMA can be wrapped around one's fingers and then snapped back into its original shape. Through training, it is possible to set the shape memory so that it will change from a new shape back to its designed shape with application of a desired temperature, with $\pm$ 1°C accuracy. Nitinol is available in a variety of forms, including wire, sheet, ribbon, and springs. It has high operating temperature, abrasion and corrosion resistance, and high tensile strength. The Nitinol alloy offers a variety of potential uses, ranging from moving robotic fingers by application of electrical currents to stress/strain sensors imbedded in bridges and other structures.

Biological structures have evolved over the millennia, and we can learn from their traits and have begun to imitate nature with designs of synthetic materials. Some of the design goals for smart materials and intelligent structures follow:*

1. Cost-efficient, durable structures whose performances match demands on the structure.
2. Change properties, color, shape, and manner to handle external physical loads to repair damage or make repairs from damage.
3. Possess the five senses of smell, taste, hearing, sight, and touch.
4. Allow structures to learn, grow, survive, and age with grace and simplicity.
5. Transmit information back to designers and user.
6. Incorporate adaptive features and intelligence to reduce mass and energy needs.
7. Allow for specification of materials and structural requirements to arrive at designs that are affordable while fulfilling design objectives.

The few examples of intelligent material systems cited here and in other parts of this book reveal "the dawn of a new materials age," as explained by Rogers (1992). You should watch for this evolution of new materials. The concepts learned here will allow you to comprehend the innovations.

APPLICATIONS & ALTERNATIVES

This module introduced the family of materials grouping system as a means to help organize your thinking about the infinite variety of engineering materials. The Boeing 777 in Figure 2-27 represents a new breed of airplane, the latest in the application of engineering materials technology. Several illustrated discussions throughout this book will focus on the 777 and the innovative application of design, materials, and manufacturing techniques used on this new generation of airplane. With an estimated development cost of $5 billion and an average per plane cost of $100 million, the 777 relied on newer approaches to product design, development, and manufacture.

An approach to design/manufacturing, ***concurrent engineering,*** is a highly coordinated team approach that commences when a project begins. All parties (designers, manufacturers, suppliers, and customers) provide input throughout the project. Another useful tool is ***simulation,*** which uses computers and testing equipment to replace prototypes, live test beds, and other traditional approaches to development and saves time and money. Another approach involved *extensive component testing* to ensure reliability of each component, such as engines or landing gears, often done prior to assembly on the aircraft.

The 777 was designed to meet projected twenty-first century air travel patterns that use planes smaller than the Boeing 747 but larger than the 767. Designed with more reliable and very large engines, the 777 twin engine is built for transcontinental flight. Figure 2-27 shows the first Boeing 777 lifting off for its maiden test flight.

*List adapted from Craig Rogers, 1992. Two excellent case study sources on the 777 are the PBS video series *21st Century Jet* (1 800 828-4PBS) and *21st Century Jet: The Making and Marketing of the Boeing 777* (Karl Sabbagh, Scribner).

Figure 2-27 The first Boeing 777 lifting off for its maiden test flight. (The Boeing Company.)

Pause for a moment and think about each of the main groups of the family of materials. How many applications can you think of for each main group that was used on this aircraft? Later in the text, we will show you some innovative applications of materials on the 777.

SELF-ASSESSMENT

2-1. From the timeline (Figure 2-1) discussed in the section Pause & Ponder, what principal group of our family of materials is regaining prominence as we move into the future? Explain why.

2-2. Using the timeline in Figure 2-1, give an example of how a material development sparked a new technology in the Scientific Age.

2-3. How many elements have been discovered to date?

2-4. Why do we need models of atoms?

2-5. What is an isotope?

2-6. Using a periodic table, determine the number of protons in the element rubidium.

2-7. Fill in Table 2-1 for the element gallium. Write the shorthand notation for the electronic configuration of iron.

2-8. Verify that the size of an atom of lithium is larger than an atom of fluorine.

2-9. The formula $2n^2$ is an easy way to determine the maximum possible number of electrons in any energy level of an atom. Using the formula, calculate the maximum number of electrons in the P energy level.

2-10. The key word in describing ionic bonding is *transfer.* What key word can be used to describe covalent bonding?

2-11. What is the name of the recently discovered carbon structure that resembles a soccer ball?

2-12. Give examples of ionic, covalent, and metallic bonding.

2-13. Express the average interatomic distance in units of nanometers and inches.

2-14. What is the relationship between processing and structure?

2-15. In which family group belong the materials with large chainlike molecules that consist of smaller molecules or monomers? Are they generally amorphous or crystalline?

2-16. When a metallic contains at least 50% iron, what is its classification?

2-17. What is the name for a metallic consisting of metal elements and other elements? Give three examples.

2-18. Define *composites.* List three composite structures.

2-19. In the ceramic grouping, most materials are crystalline, but which one has an amorphous structure?

2-20. List three examples of lightweight nonferrous metals or alloys, three examples of heavier nonferrous metals, and one example of a refractory metal.

2-21. What is the subgroup of metallics in which metal particles are joined by pressure and sintering heat?

2-22. What subgroup covers human-made polymers containing carbon atoms covalently bonded together with other elements? List the two divisions of this polymeric material and give an example of each.

2-23. Name three natural polymeric materials.

2-24. How do elastomers differ from other polymeric materials? How can the strength of rubber be improved?

2-25. How does the energy shortage threaten the supply of plastics and rubber?

2-26. What is the prime purpose of selecting a composite material over material from the other family groups?

2-27. List three materials that were grouped in the "other" family group of materials.

OBJECTIVE QUESTIONS

2-1. The largest consumer of electricity in the United States is the refining of the metal

a. Steel **b.** Copper **c.** Aluminum

2-2. A _________ is a substance that cannot be broken down any further by a chemical reaction.

a. Molecule **b.** Compound **c.** Element **d.** Atom

2-3. A compound is a substance that can be broken down by chemical reactions into

a. Elements **b.** Molecules **c.** Atoms

2-4. Protons and neutrons are known as

a. Quarks **b.** Electrons **c.** Neutrinos **d.** Nucleons

2-5. Another name for an electron shell is

a. Quantum **b.** Principal energy level **c.** Energy state **d.** Ion

2-6. Two electrons can exist in the same orbital if they have

a. The same energy **b.** The same polarity **c.** Opposite spins

2-7. The maximum number of electrons in the L, or second, energy level is

a. 2 **b.** 18 **c.** 8 **d.** 0

2-8. Inert gases can also be called

a. Stable **b.** Group 5 elements **c.** Noble gases **d.** Active

2-9. The core electrons in neon (Ne) number

a. 2 **b.** 10 **c.** 18 **d.** 6

2-10. Horizontal rows in the periodic table are called

a. Groups **b.** Transition elements **c.** Periods **d.** Metals

2-11. Metalloids consist of how many elements?

a. 6 **b.** 10 **c.** 7

2-12. The electronegativity of iron (Fe) is

a. 1.83 **b.** 3.98 **c.** 0.93

2-13. A distinction that sets intelligent structures apart from other advanced materials is:

a. Cost per unit volume is often higher than traditional materials.

b. Ability to adapt to changes in environment.

c. Requires more exotic processing.

d. Benefits from superior properties offset material cost.

ACTIVITIES

2-1. In this module, we have focused on the variety of materials in use today. To aid in your understanding of the application of various materials and assist in your becoming a wise "materials consumer," you should:

(a) Observe the materials used in everyday products such as tables and chairs, buildings, roads, street lights, home appliances, shop tools, laboratory instruments, bicycles, automobiles, electronic equipment, clothing, and any other products you encounter.

(b) Ask yourself, your friends, your instructor, or knowledgeable people: Why did the designer select that particular material or combination of materials for this product? Would another material be less costly, be lighter, resist corrosion, save energy, recycle rather than pollute, and so on?

(c) Gather a variety of materials and compare their features, or construct a mobile of various materials with labels on each material. (Do a neat job and you will have a nice art object.)

2-2. Gather the following: soup can, seamless soda can, penny, piece of solder, lead sinker, screw, and washer. Use a magnet on each to see which attracts the magnet.

(a) Of the metals that attract the magnet, what properties do they possess?

(b) The soda cans and the soup cans are also referred to as *tin cans.* Are they made of tin? Explain.

2-3. Refer to periodicals such as *Popular Science, Machine Design, Plastic World,* and *Smithsonian,* or other science and technology magazines, to read about new uses of the family of materials.

2-4. Select appropriate materials to construct a container to hold a single uncooked egg. To test the success of your container, drop it from 10 meters. If you want to make a contest of various designs by other students, use the weight of the containers to determine the best design. The design of lightest weight that survives the drop with no harm to the egg would be best. If no eggs survive, the instructor would choose the most creative design employing the widest variety of materials.

REFERENCES & RELATED READINGS

ASHBY, M. F. "Technology of the 1990s: Advanced Materials and Predictive Design," *Philosophical Transactions of the Royal Society of London,* Vol. A322, 1987, pp. 393–407.

ASHBY, M. F., and D. R. H. JONES. *Engineering Materials: An Introduction to Their Properties and Applications.* Elmsford, N.Y.: Pergamon, 1980.

ASIMOV, ISAAC. *Building Blocks of the Universe.* New York: Abelard, 1957.

BELLMEYER, FRED W. *Synthetic Polymers.* Garden City, N.Y.: Doubleday, 1972.

BETTS, JOHN E. *Physics for Technology.* Reston, Va.: Reston, 1964.

BROWN, THEODORE L., and H. EUGENE LEMAY, JR. *Chemistry: The Central Science,* 2nd ed. Englewood Cliffs, N.J.: Prentice Hall, 1981.

CHAUDHRY, A., T. JOSEPH, F. SUN, and C. ROGERS. "Local-Area Health Monitoring of Aircraft via Piezoelectric Actuator/Sensor Patches." North American Conference on Smart Structures and Materials, San Diego, Calif., 26 Feb.–3 March, 1995.

CHRISTIANSEN, G. S., and PAUL H. GARRETT. *Structure and Change: An Introduction to the Science of Matter.* San Francisco: W.H. Freeman, 1960.

COTTERILL, RODNEY. *The Cambridge Guide to the Material World.* Cambridge: Cambridge University Press, 1989.

CRAIG, DOUGLAS F. *"Structural Ceramics," National Educators' Workshop, NEW: Update 91,* Oak Ridge National Laboratory, Oak Ridge, Tenn., November 12, 1991.

DRAGO, R. S. *Prerequisites for College Chemistry.* New York: Harcourt Brace Jovanovich, 1966.

EASTERLING, KEN. *Tomorrow's Materials.* London: The Institute of Metals, 1988.

EWEN, DALE, AND LERAY HEATON. *Physics for Technology Education.* Englewood Cliffs, N.J.: Prentice Hall, 1981.

GAMOW, GEORGE A. *Mr. Tomkins Explores the Atom.* New York: Macmillan, 1945.

GIANCOLI, DOUGLAS C. *Physics.* Englewood Cliffs, N.J.: Prentice Hall, 1980.

NATIONAL RESEARCH COUNCIL. *Materials Science and Engineering for the 1990s: Maintaining Competitiveness in the Age of Materials.* Washington, D.C.: NRC, 1989.

PETERSON, I. "Buckyballs for Scanning Surfaces," *Science News,* March 30, 1991.

ROGERS, CRAIG A. "Intelligent Materials Systems: The Dawn of a New Materials Age." Center for Intelligent Materials and Structures, Virginia Polytechnic Institute and State University, 1992.

SHACKELFORD, JAMES, AND WILLIAM ALEXANDER. *The CRC Materials Science and Engineering Handbook.* Boca Raton, Fla.: CRC Press, 1991.

SMITH, CHARLES O. *Science of Engineering Materials.* Englewood Cliffs, N.J.: Prentice Hall, 1977.

Periodicals

Adhesive Age
Advanced Materials and Processes
Ceramics Monthly
Chemical and Engineering News
Chemical Engineering
Discover
Journal of Materials Education
Machine Design
Manufacturing Engineer
NASA Technology Briefs
Plastic Technology
Plastic World
Popular Science
Scientific American

Module 3

Processing and Structure of Solid Materials

After studying this module, you should be able to do the following:

3.1. Discuss the relationships between processing and structure of materials.

3.2. Recall the synthesis and processing capabilities of contemporary materials technology at the various scales (macro, micro, nano, and atomic).

3.3. Recall techniques and instruments used to determine material structures, including the capabilities of the optical, electron, and scanning tunneling microscopes.

3.4. Identify characterization components of substrate and coating on a material specimen, including oxide inclusion, substrate roughness, void, pore, crack, cohesive strength between particles, particles, phases, and grains. Explain qualitative and quantitative material characterization.

3.5. Use sketches, unit cell structures, and standard notations to explain and solve problems related to crystallography. Determine the unit cell structure of given elements from a periodic table of elements for metallurgists.

3.6. Explain the alloying and doping process of materials in terms of solutions, alloying elements, packing factors, crystal imperfections, and impurities, and use appropriate sketches. Cite specific practical applications for alloying and doping.

3.7. Determine basic information from phase diagrams and tables on engineering materials.

3.8. Recognize Greek symbols, and determine their meaning and proper pronunciation.

3.9. Apply SI rules of usage, make conversions of U.S. customary units to SI (and vice versa), and use both for computations.

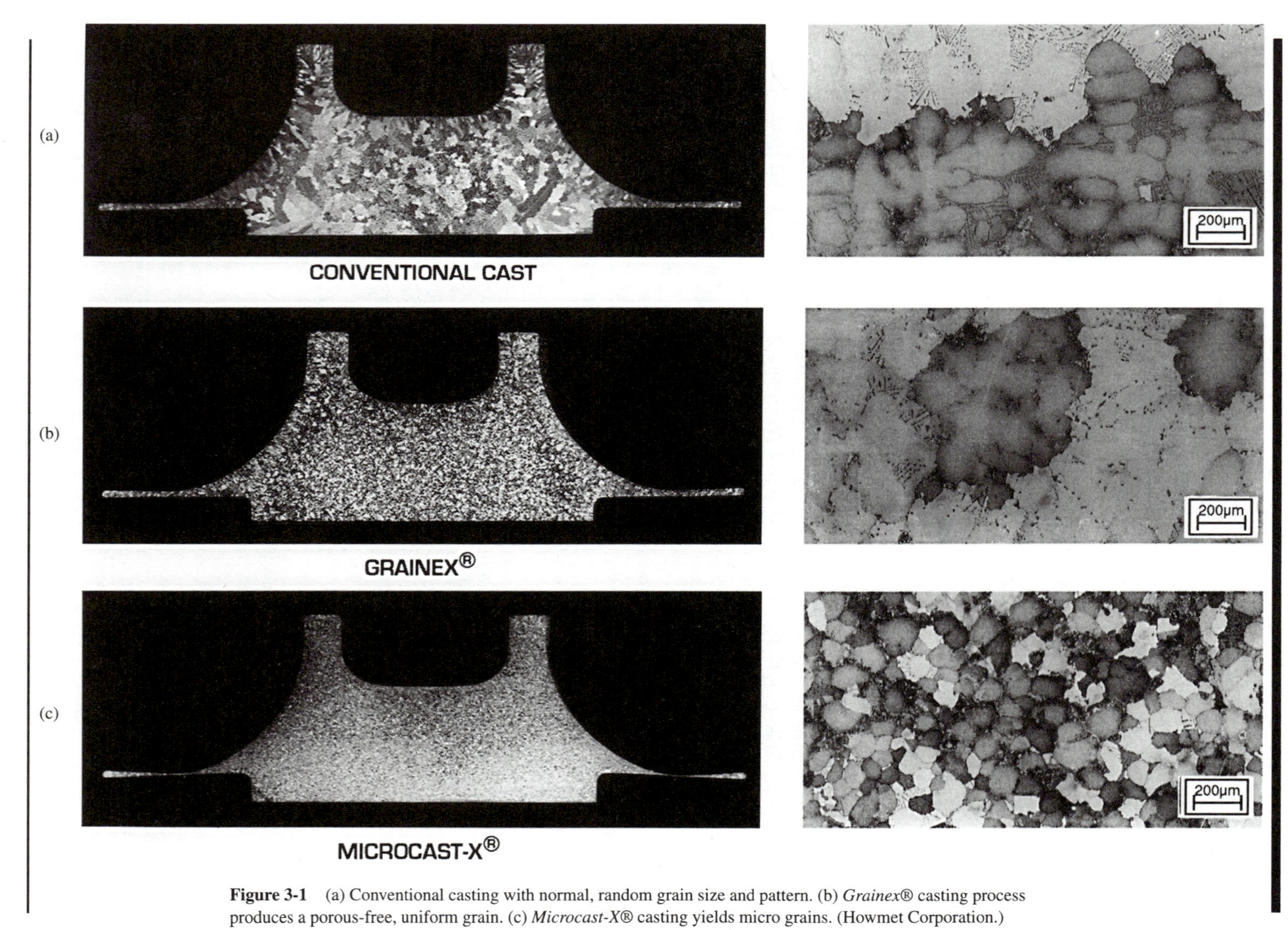

Figure 3-1 (a) Conventional casting with normal, random grain size and pattern. (b) *Grainex®* casting process produces a porous-free, uniform grain. (c) *Microcast-X®* casting yields micro grains. (Howmet Corporation.)

PAUSE & PONDER

Understanding the natural structure of materials allows for manipulation of structures to achieve results different from those that would normally occur. An example of such processing to achieve a desired structure resulted from an engineering need to meet demands for higher operating temperatures in the metal alloys for aircraft gas turbine engines. The industry sought to improve grain structure to avoid fatigue failures. Failures occurred because conventional casting [Figure 3-1(a)] yields random grain size.

Many years of process evolution were necessary to achieve grains that are uniform in all directions, or ***equiaxed grains,*** as produced by the investment casting processes. A further improvement in uniformity of grain size in those parts with unequal cross-sections was achieved by agitating the mold while the alloy was still molten, much like a clothes washing machine agitates its load (Figure 3-2). Coupled with agitation, hot isostatic pressure (HIPS) in the ***Grainex®*** process produces a porous-free, uniform grain [Figure 3-1(b)].

As research continued for superior cast alloys with refined grain, Howmet Corporation developed ***Microcast-X®***. This process was achieved by using a technique from the earlier equiaxed casting that involved reducing super heating of the molten alloy. The micro grains that result from Microcast-X [Figure 3-1(c)] are good for fatigue behavior in large, highly stressed ***structural castings***. However, the very small grains lower high temperature strength properties above 1500°F.

Controlling the cooling rate of a bulk material can also be done for lower operating temperature metals. Normally, metals must be heated to the liquid state before they can be formed into complex shapes. Less complex shapes can be achieved in metals by heating them to a point in which they become ***plastic,*** or formable. These processes include extrusion, drawing, bending, and so on. Some softer metals and thinner, harder metals can be cold formed at room temperature.

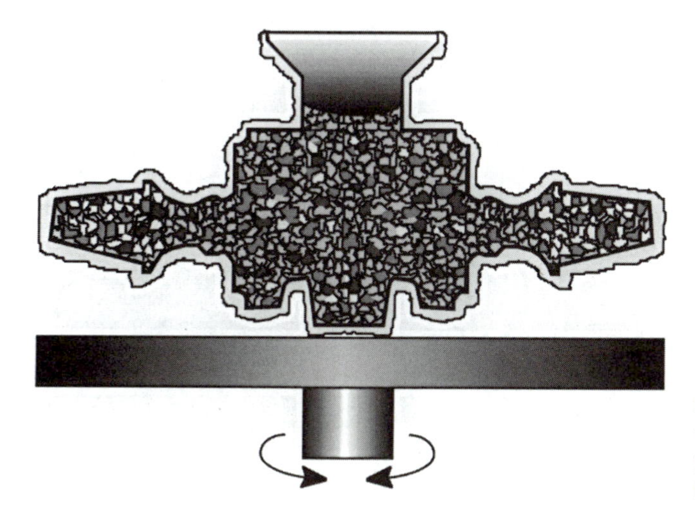

Figure 3-2 *Grainex®* casting process involves agitating the mold to produce a porous-free, uniform grain. (Howmet Corporation.)

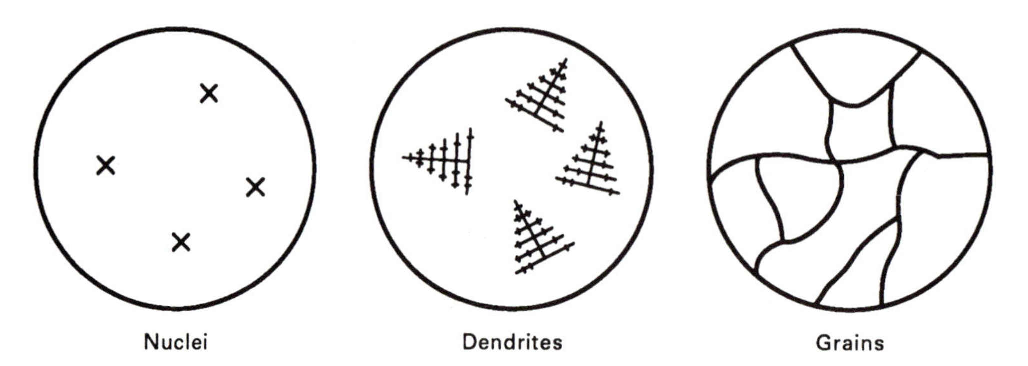

Figure 3-3 Stages of a metal cooling through equilibrium: first nucleates, then forms dendrites, and finally cools into a crystalline solid of many metal grains.

Thixotropic (read "thicks-oh-tropic") ***metals*** exist at a semisolid state; they liquefy while under shearing stress and then solidify when left standing. Jello is semisolid but, of course, much softer than a thixotropic metal. It is possible to turn metal alloys into the thixotropic state by stirring metal vigorously as it cools from the liquid state en route to the solid state. When metal cools under ***equilibrium*** conditions (slow, natural change from liquid to solid) as shown in Figure 3-3, crystals form: first, by ***nucleating*** (seed crystals

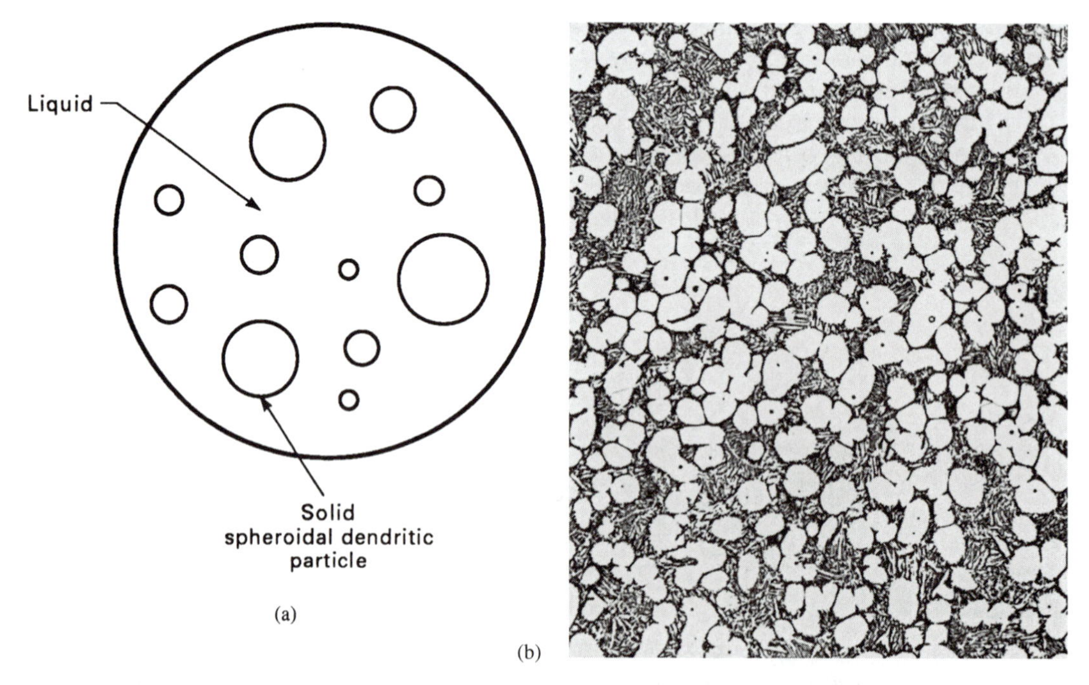

Figure 3-4 Semisolid, thixotropic alloys form nearly spheroidal dendrites. (a) Diagram of SSM structure. (b) Photomicrograph reveals how continuous stirring during solidification of semisolid material (aluminum casting alloy) results in a microstructure of large equiaxed solid particles within the formerly liquid eutectic mixture. The solid particles agglomerate during the SSM processing, affecting flow behavior. (ASM International.)

begin to form), then with further cooling growing to form ***dendrites*** (tree-shaped structures), and finally growing into ***grains*** or ***crystals.*** If the metal is stirred vigorously while in a state between liquid and solid (much like the mush of melting snow, which is part ice crystals and part water), the dendrites will form spherelike shapes (Figure 3-4) rather than the normal treelike dendrites to produce a ***semisolid material (SSM).***

Injection molding, shown in Figure 3-5, utilizes a screwing action that forces thixotropic magnesium through a nozzle into a mold or die made of very hard tool steel or similar die stock. The mold has had a cavity formed into it in the reverse shape (female) of the part to be produced. Injection molding is a process often used for molding plastic and ceramic parts, but not metals. Dow Chemical has developed a process for injection molding of magnesium parts that is possible through the use of thixotropic alloys. ***Casting*** and ***molding*** are shaping operations in which the starting material is a heated liquid or semifluid. When referring to metals, the term *casting* is normally used, whereas *molding* is the common term used for plastics. Both processes are used in glassworking. Figure 5-61(a) shows the sequence of operations in the investment casting of titanium alloys.

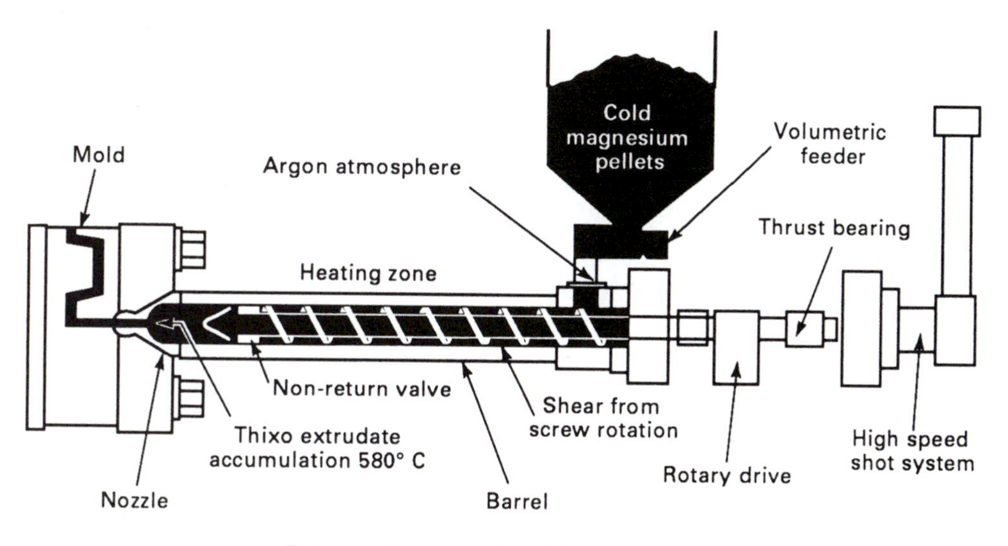

(a)

(b)

Figure 3-5 (a) Dow Chemical's thixotropic magnesium injection molding process. (b) SSM magnesium alloy parts formed by thixotropic injection molding. (ASM International and Thixomat Corp.)

In Figure 3-5, you will notice that cold magnesium pellets (in a thixotropic state) are fed through a hopper into an argon atmosphere, then into the heated barrel of the injection-molding machine. Argon is an ***inert gas*** and, as such, protects the magnesium from oxygen in the factory atmosphere. The shearing action of the screw, acting on the magnesium pellets, transforms them into a thixo-extrudate at 580°C. When enough of the extruded magnesium slurry (solid and liquid) collects in the heating zone chamber, it is injected by the force of the high-speed shot system, which pushes against the thrust bearing, causing the reciprocating screw to inject slurry through the nonreturn valve past the nozzle into the mold cavity. The molded SSM part is the equivalent to a part produced through melting and casting in the liquid state. The injection process offers advantages over die casting: improvements in mold cycle time, less downtime, and improved part quality.

The processing techniques explained above illustrate the desire to apply technology that provides an end product of the highest quality with as few steps as possible. In this module, we provide information on the basic structure of materials and how structure is altered by processes and synthesis. There is a never-ending quest for improved techniques that can form and treat materials so that the ideal material can be produced for the perfect product or system. The quest requires constant study to keep up with new developments. This module can prepare you to deal with today's materials and processes and to adapt to coming changes.

3.1 SYNTHESIS AND PROCESSING

Solid materials are products of the conversion of raw materials into bulk materials. With many polymers, synthesis is a part of raw materials production. Newer materials, such as artificially structured materials, are also synthesized. The result of raw materials processing or synthesis with conventional means provides us with familiar structures such as polycrystalline metals, amorphous plastics, and crystalline ceramics. ***Artificially structured materials,*** which are synthesized on micro, nano, and atomic scales, have engineered unique structures that may be ultrapure and/or possess electronic circuitry and optical properties based on a special design. In the modules on metals, polymers, ceramics, and composites, you will learn that a newer process, such as rapid solidification of metal, provides an amorphous structure rather than the natural polycrystalline structure. Chemical vapor deposition (CVD) and molecular beam epitaxy (MBE) are recent synthesis techniques that allow for the production of "designer materials." Many ***"beam" processing*** and synthesis techniques have evolved with the use of laser, ion, and electron beams to produce micro- and nano-scale structures, as shown in Table 3-1.

Manufacturing processes are traditionally used to convert the bulk synthesized or processed materials into structural shapes and final subassemblies of products. Manufacturing processes will usually change the structure of the bulk material, thereby requiring a variety of "secondary processes" to ensure that the final product has the necessary structure. ***Near-net-shape*** and ***net-shape processing*** aim to achieve a final product or specified

TABLE 3-1 COMPARISON OF OUR CONTEMPORARY CAPACITY TO PREPARE NEW MATERIALS OR COMPOSITES ARRANGED AS A HIERARCHY OF SIZE

Scale	Capacity of the technology	Type of materials synthesized or processed
Macro (cm–m)	Excellent	Materials for very large structures have been optimized (e.g., reinforced concrete, airframe alloys)
Micro (mm–μm)	Very good	Lab scale and small objects (GRP to graphite–epoxy composites, transparent Lucalox, composite transducers, etc.)
Nano (0.1–10 nm)	Primitive but new	New materials designed and made (intercalates, glass-ceramics, toughened zirconia, nanocomposite desiccants, etc.)
Atomic (0.1 nm)	Poor	New materials found by serendipity (penicillin, Teflon, magnetic garnets, 1:2:3 superconductors, etc.)

Source: Adapted from Roy (1992).

structure through the initial synthesis of raw material processing, thereby avoiding secondary processes.

Aluminum serves as an ideal example. Although aluminum is the most plentiful metal in our earth's crust, it requires enormous amounts of electrical and chemical energy to extract it from bauxite ore. Great savings are realized by recycling, which makes it practical and desirable for all concerned. Often, however, bulk aluminum must go through many ***secondary processes,*** such as machine cutting, welding, grinding, and heat treating, to achieve a finished product. Using near-net-shape processing to go from the raw aluminum to the final product, while achieving the desired structure and properties, would save even more energy. In Module 8, we will explain how pressureless, infiltration processes produce near-net-shape metal-matrix composite (MMC) parts.

Table 3-1 provides a comparison of contemporary synthesis/processes methods based on the scale (macro, micro, nano, or atomic) of the technique employed. Note that as the scale gets smaller, the capacity of the technology diminishes. From Figure 2-1, showing the timeline of materials development, the immediate future of materials technology is classified as the nanocomposite era (defined later in this module). The ability to develop improved instrumentation to explore and manipulate matter at the nano and atomic scales will determine the winners of the race to place products in the marketplace. All developed countries and many developing countries are investing heavily to improve or maintain their competitiveness in the races for materials such as those listed in the nano and atomic categories for use in products such as high-definition television (HDTV), superior computer technology (neural networks, multimedia, artificial intelligence, virtual reality), improved transportation (magnetic-levitation trains and surface seacraft), better communications and entertainment technology, and smart homes and smart highways.

With instruments such as the scanning tunneling microscope (described later), materials scientists may view and manipulate the nano and atomic levels for better understanding and control of structure and improved synthesis and processing. At the nano and atomic scales, it is becoming possible to produce new structures such as ***nanocomposites,*** materials with atoms and molecules arranged into unique phases to provide designed properties. We will see later in Figure 3-16 that ***nanoscale mixing*** will provide properties such as the following (Roy 1992):

1. *Chemical/structural:* control of phases formed, lower reaction and sintering temperatures, control of microstructure, control of morphology via solid-state epitaxy
2. *Mechanical:* fivefold increase in strength as size goes from 100 to 20 nm; superhard (> diamond) materials
3. *Optical:* tenfold increase in luminescence
4. *Electrical:* change in fundamental conduction
5. *Magnetic:* newer super-, para-, and ferrimagnets

3.1.1 Nanotechnology

Nanotechnology provides the potential to design and manipulate material structures at the atomic level. Electronic components such as transistors, capacitors, and diodes with dimensions less than 0.25 micrometers are emerging from the nation's laboratories. Units such as micrometers are less useful as units of measure. The next smallest unit of measure is the ***nanometer*** *(nm),* defined as one billionth of a meter. (To emphasize its size, a human hair has a diameter that is 10 times the size of a nanometer.) The diameter of most atoms is 0.1 to 0.4 nanometers (see Figure 3-15). This new technology is being investigated for use in manufacturing. Starting at the molecular level and assembling any number of molecules of any size, material can be constructed with perfect precision. This attempt has been given the name of *molecular manufacturing, nanofabrication manufacturing,* or *molecular nanotechnology.*

Nanotechnology is defined as the processing of materials into microelectromechanical structures and devices whose sizes vary from hundreds to hundredths of a micrometer (mm). Combining science and technology, nanotechnology builds at the atomic and molecular levels. The materials involved can include diamond films, organic films, semiconductors, metal films, dielectrics, ferroelectric films, and piezoelectric films. Processing techniques involved include ion implantation, optical and e-beam lithography, plasma and wet etchings, and physical and chemical vapor deposition. These processes are producing flat panel displays, microelectronics, optoelectronics, and solid-state sensors. Many of these topics are covered throughout this book.

3.2 DETERMINING MATERIAL STRUCTURE

The kinetic theory of matter describes atoms in a solid as being at their lowest energy state. In this state, their motions have slowed considerably. These atoms of solids are so close to their neighboring atoms that they can no longer act as independent particles. They vibrate about their equilibrium positions, depending on the degree of remaining kinetic energy (thermal energy) they may possess. They can also move in conjunction with their neighboring atoms. But they can and do move. In the discussion about bonding in Module 2, we learned that two or more atoms can chemically join each other, depending on their electronic structures. We also learned that atoms, ions, and molecules are of different sizes.

The internal structure of solids determines to a great extent the properties of the solid or how a particular solid material will perform or behave in a given application. The study

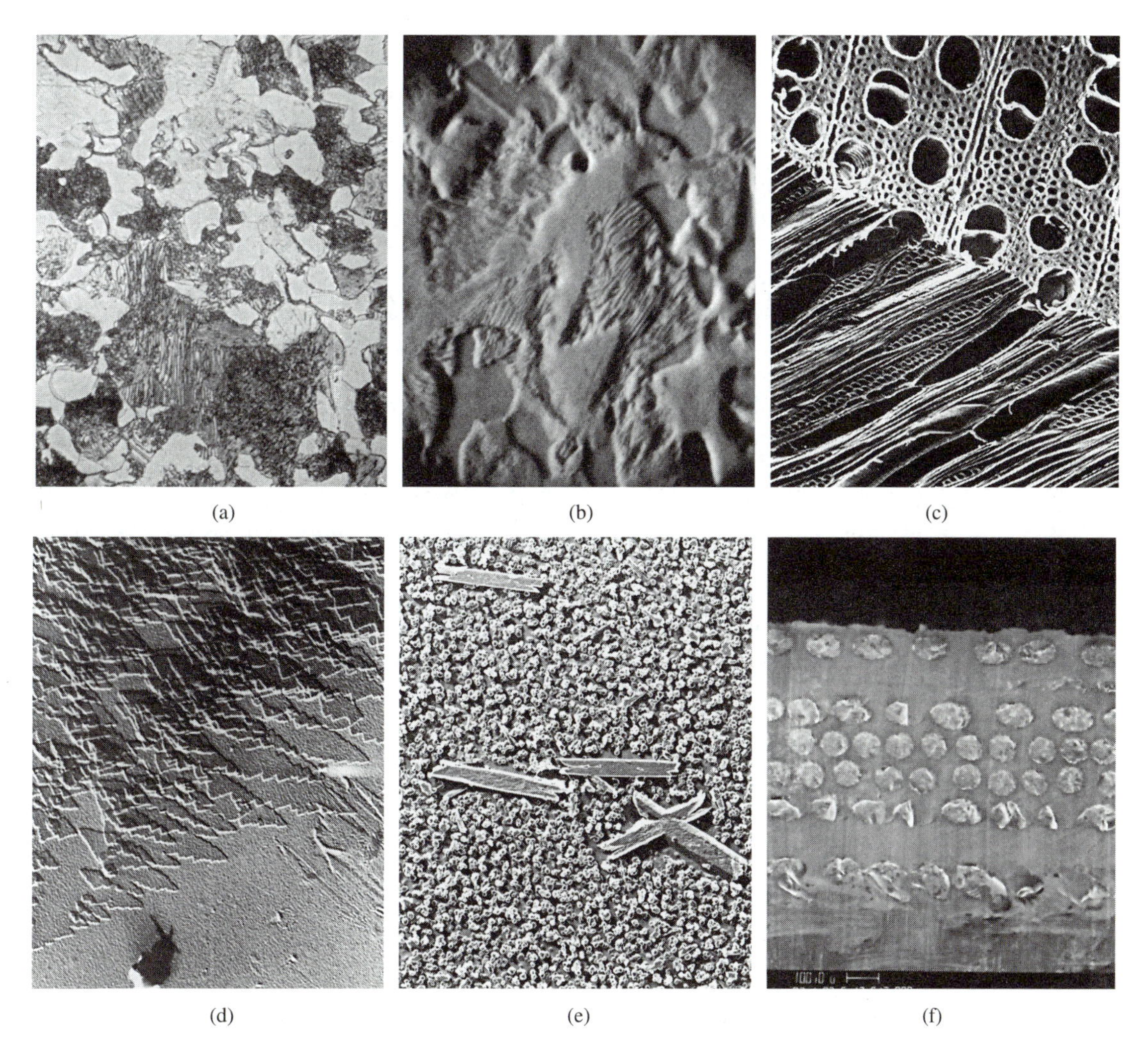

Figure 3-6 Photomicrographs of structural features of some representative materials. (a) Steel (SAE/AISI 1045) 1000 ×. (NASA). (b) Steel (SAE/AISI 1045) SEM 2550 ×. (NASA). (c) Wood (maple) 415 ×. (U.S. Forest Products Lab.) (d) Plastic (polyethylene) dendrite crystals. (Dr. Philip H. Geil—U. of Illinois.) (e) Glass ceramic: crystals growing in amorphous glass. 17,900 ×. (Corning Glass Works.) (f) Fiber metal composite. (NASA.)

of the internal structure is pursued in various ways using x-rays as well as electron, proton, or neutron beams to disclose crystal size, crystal structures, crystal imperfections, bonding types, spacing of adjacent atoms and adjacent planes, and the different atoms in a solid. A brief knowledge of some of the more important tools in the field of spectroscopy and microscopy that permit a still deeper probe into the internal structure of materials is vital to a greater understanding of the effects of structure on the behavior of natural and human-made materials. Figure 3-6 shows photomicrographs of varying magnification produced with the aid of various instruments used in materials science.

3.2.1 Spectroscopy

White light, when refracted (light rays undergo a change in direction when passing from one medium to another), is broken up into a number of different colors and forms a spectrum. Each spectral color corresponds to a particular wavelength. Every chemical element (as well as its atoms) produces a characteristic spectrum, which may be detected and measured by a spectroscope. Spectroscopy using such an instrument studies substances through an analysis of their spectra. In other words, ***spectroscopy*** permits the precise measurement of the wavelengths of atoms' radiations, which are discrete and distinct for each atom. When atoms are exposed to high energy from an outside source, such as infrared, atomic absorption, arc-spark, or glow discharge, they become excited and their electrons move to higher energy levels. Becoming unstable, the electrons tend to return to their lower energy levels. The interchanges of energy between radiation and matter take place in discrete units called ***quanta*** (discussed in Modules 4 and 9). Under excitation an atomic substance will reach a state of dynamic equilibrium in which some atoms are releasing energy at the same rate that it is being absorbed by others. An atom loses energy when its excited electrons transit to a lower energy level, giving off a ***photon*** of light. The wavelength of this photon is related to the magnitude of the change of energy. Such transitions, when observed by the spectroscope, produce atomic spectra (Figure 3-7) with discrete and distinguishing wavelengths corresponding to the various electron transitions between energy levels. These studies permit scientists to increase their knowledge of atoms, electronic structures, and various bonding mechanisms.

3.2.2 X-Ray Diffraction

One of the most useful tools in the study of crystal structures of solids is ***x-ray diffraction.*** An x-ray diffractometer is shown in Figure 3-8. X-rays have wavelengths about equal to the diameters of atoms (10^{-10} meters) or about the same length as the spacing between atoms (or ions) in solids. When these x-rays are directed at a solid with a crystalline structure, the waves are refracted. The equation, known as *Bragg's law,* $n\lambda = 2d \sin \theta$, relates the wavelength, λ, to the distance between planes of atoms, d, and the glancing angle between the incident beam and the plane of atoms, θ. Figure 3-9(a) shows a sketch of this relationship. If the incident beam strikes the planes of atoms at some arbitrary angle, the reflected beam may be nonexistent, because the refracted rays from the atomic planes will be out of phase and produce destructive interference [Figure 3-9(b)] and cancellation of the reflected beam. It has been determined that, at a particular angle θ, the reflected beam will be in phase and produce constructive interference [Figure 3-9(b)]. Mathematically, this can be expressed as $\lambda_1 = n\lambda_2$, where $n = 1, 2, 3, \ldots$; the distance traveled from the different parallel planes of atoms represents an integral number *(n)* of wavelengths. Referring to Figure 3-9(a), this statement may be written as

$$ACA' = BEB' - n\lambda \quad \text{where } n = 1, 2, 3, \ldots$$

Knowing the wavelength of the incident beam of x-rays, the glancing angle can be measured experimentally, and solving Bragg's law for $d = n\lambda/2 \sin \theta$, the interplanar distance can be calculated. Figure 3-10 shows a photographic plate on which the constructive waves produce a series of dots, indicating that the x-rays are scattered from crystals at only

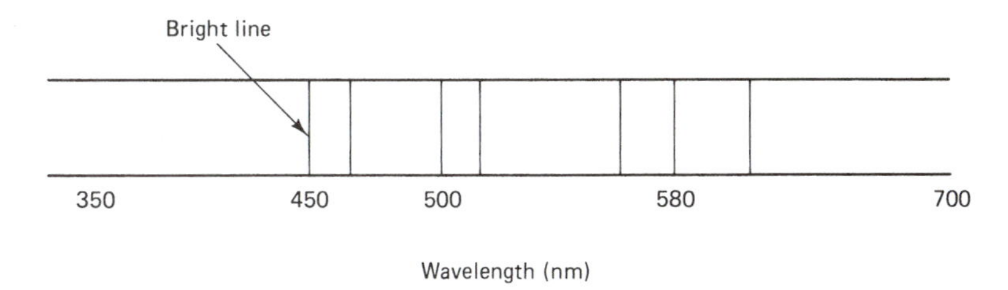

Figure 3-7 Atomic spectrum of sodium in the visible region. (Adapted from D. C. Giancoli, *Physics,* Prentice-Hall, Inc., Englewood Cliffs, N.J., 1980.)

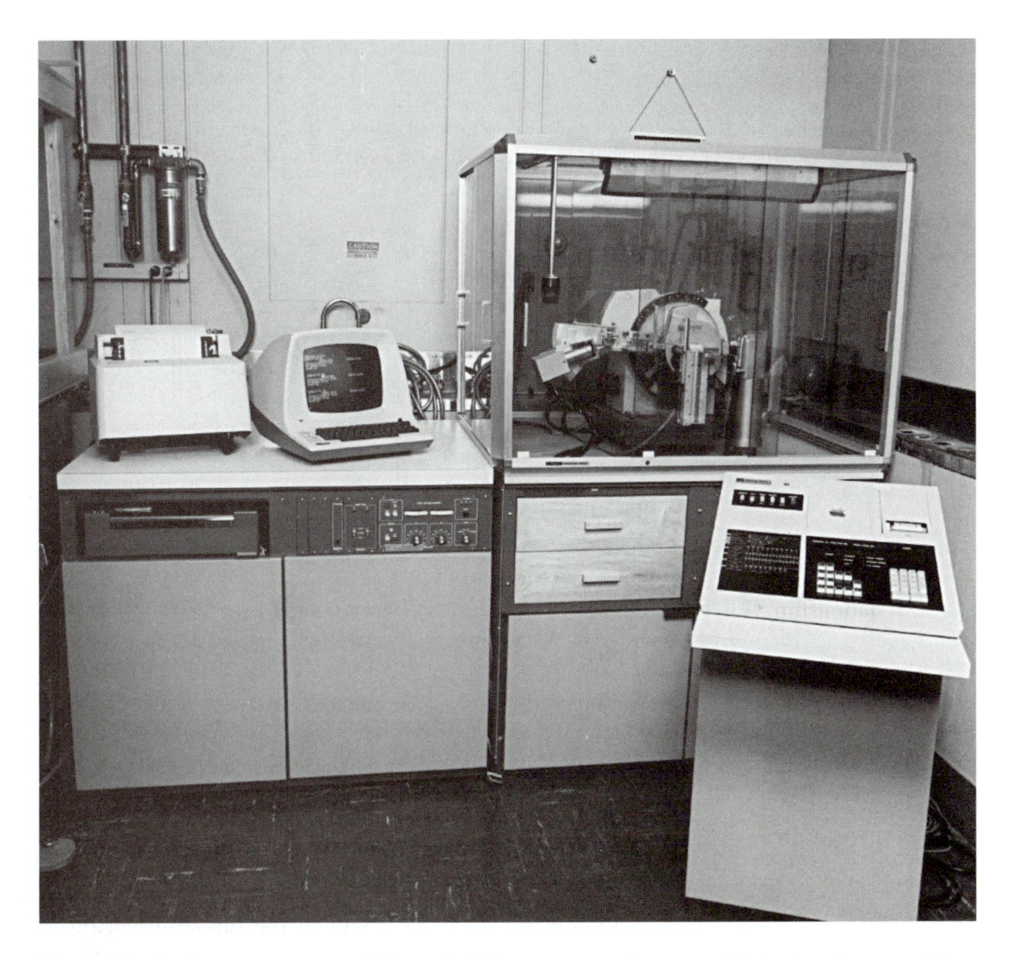

Figure 3-8 A microcomputer-controlled powder diffractometer goniometer used to analyze the composition of various substances. (General Electric.)

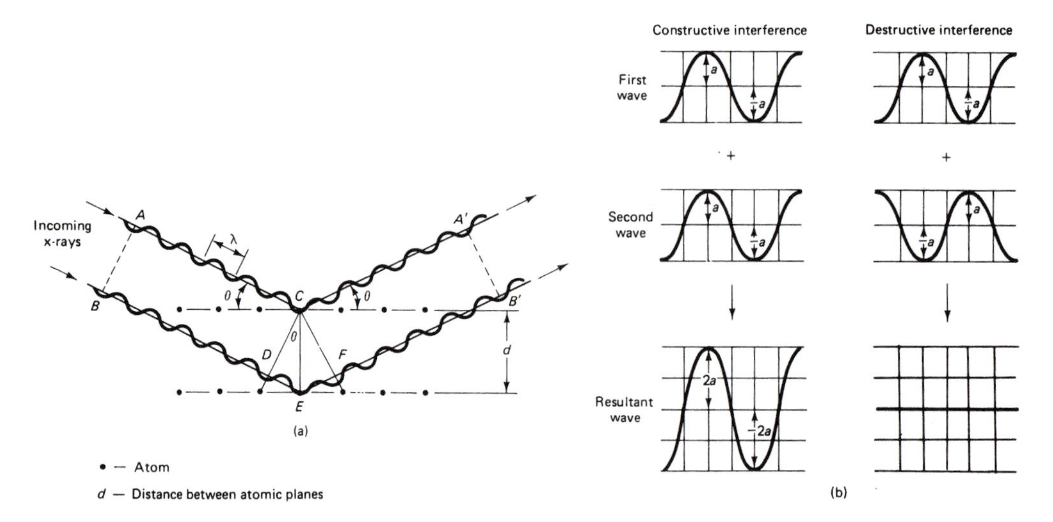

Figure 3-9 (a) Scattering of x-rays by atoms in parallel planes. (b) Constructive and destructive interference of waves. (Theodore L. Brown and H. Eugene LeMay, Jr., *Chemistry—The Central Science,* 2nd ed., Prentice-Hall, Inc., Englewood Cliffs, N.J., 1981.)

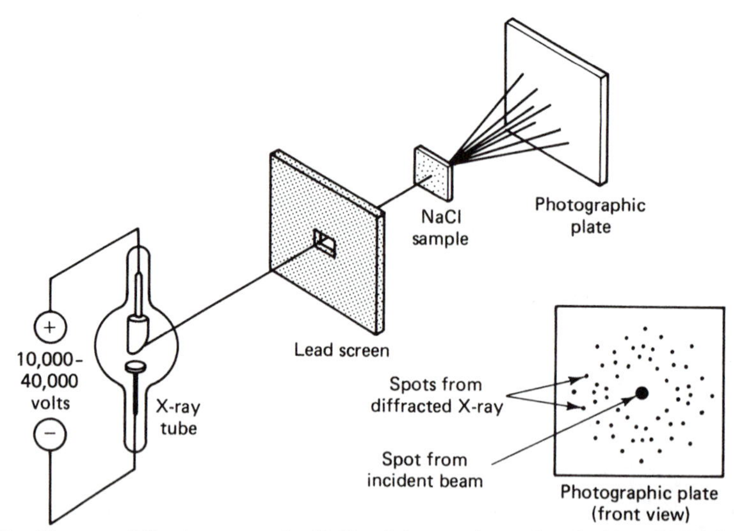

Figure 3-10 The x-ray diffraction pattern for NaCl and the experimental method by which it is obtained. (T. L. Brown and H. E. LeMay, Jr., *Chemistry—The Central Science,* 2nd ed., Prentice-Hall, Inc., Englewood Cliffs, N.J., 1981.)

certain angles. All the various angles at which diffraction occurs are determined by measurements on the photographic film. By studying the directions of diffracted x-ray beams, as well as the intensities produced by this powerful tool, much can be learned about the crystal structure of solids.

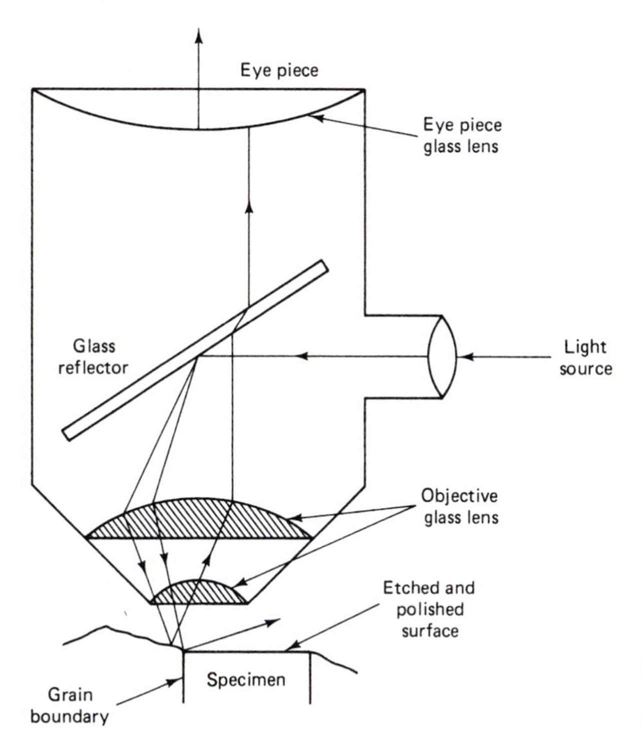

Figure 3-11 Schematic of an optical system of a metallographic microscope detecting a grain boundary in a crystal.

3.2.3 Optical Microscopy

The human eye can distinguish between two points in space only if the two points are approximately 0.1 mm apart at a distance of 25 cm. Microscopic objects perhaps only 0.5 μm apart with the eye focused at the normal reading distance of 25 cm can be seen as an image magnified by a factor of 200 or more. The compound microscope fulfills this requirement by having two lenses, the objective and the eyepiece, to give two stages of magnification. The quality of the microscope image is determined by the magnification, degree of diffraction, distortion due to aberrations of the lenses, and contrast. ***Metallographs*** are optical, metallurgy microscopes capable of photographing the images of solid specimens (see Figures 3-11 and 3-12). Opaque specimens, such as metallurgical samples, are typically observed with a reflection microscope in which light is reflected from the sample. Specimens often show very poor contrast between adjacent areas, even though these areas differ in chemical composition or morphology. As a result, most metallographic specimens are acid-etched to make specimen details visible. The ***etching*** reacts with the grain boundary areas at a different rate than the grains themselves. When incident light rays strike these areas with their different crystal orientations, as well as their difference in rate of etching, the reflected light travels back to the eye in different amounts (see Figure 3-13). If most of the incident light is reflected back to the eye, the grain will appear bright. Other crystals reflecting less light will appear darker. A ***photomicrograph*** of a specimen taken

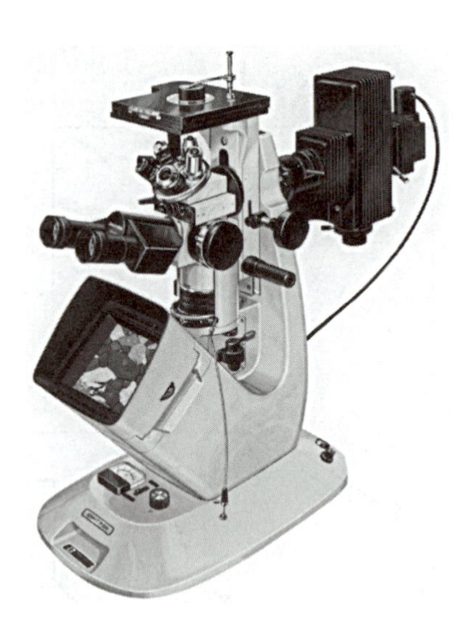

Figure 3-12 Photograph of a metallograph. (Buehler Ltd.)

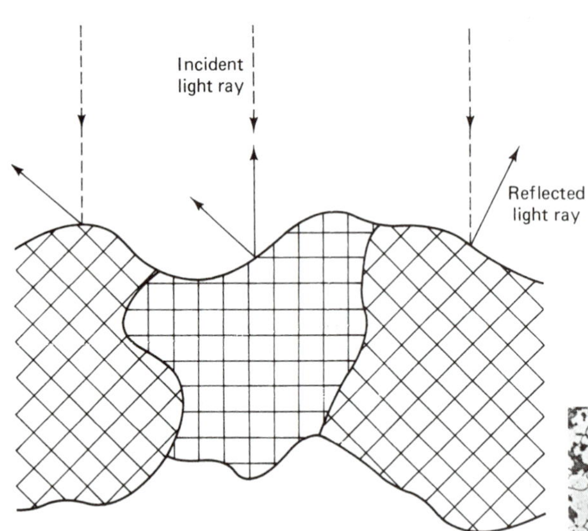

Figure 3-13 Sketch showing identification of individual crystals by reflected light rays.

Figure 3-14 A photomicrograph of low carbon steel. (Buehler Ltd.)

through a metallograph (Figure 3-14) shows the details of the microstructure, including the grain boundaries. Photomicrographs of various materials are shown throughout the book.

3.2.4 Electron Microscopy

The maximum magnification produced by optical microscopes is about 2000× . This limitation is imposed by the wavelengths of visible light, which limit the resolution of minute details in the specimen being observed. *Resolving power* is a term used to describe the ability of a lens to reveal detail in an image. Optical microscopes (discussed previously) do not have the ability to discern details smaller than 10^{-6} m. To observe objects in the 10^{-10} m range, high-energy electrons with wavelengths around 4×10^{-11} are used. These electrons are similar to visible light because they both are forms of electromagnetic radiation. By using magnetic lenses to focus the electrons [see Figure 3-15(a)], their resolving power can be increased greatly to give magnifications over 200,000 ×. It must be remembered that metallic specimens are opaque to electrons (i.e., electrons are absorbed by the metallic atoms).

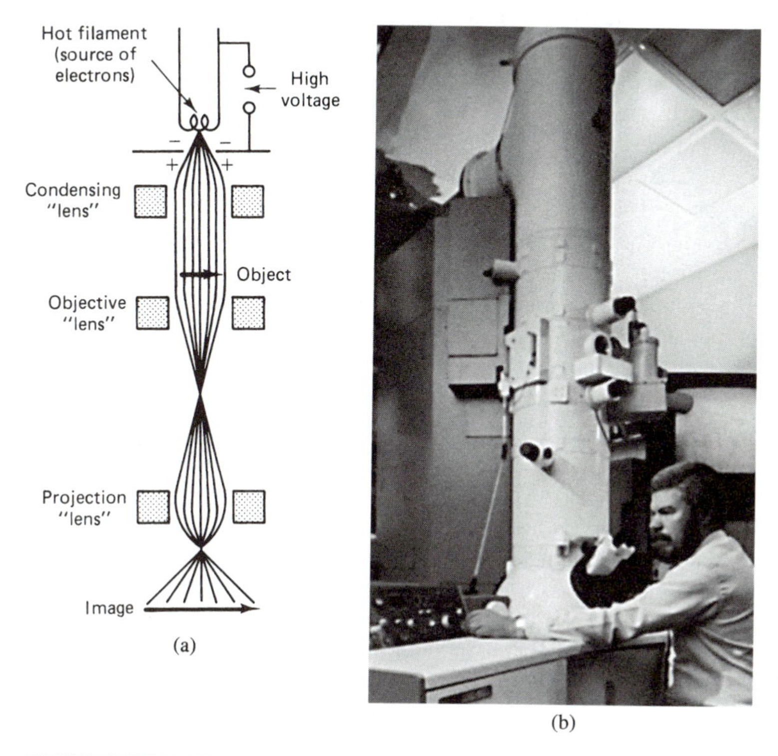

(c)

Figure 3-15 Transmission. (a) Electron microscope. The squares represent magnetic field coils for the "magnetic lenses." (Douglas C. Giancoli, *Physics—Principles with Applications.* © 1980, reprinted with permission of Prentice-Hall, Inc., Englewood Cliffs, N.J.) (b) Photograph of a transmission electron microscope (TEM). (c) TEM micrograph comparing whiskers of ceramic (SiC) to a human hair (ORNL).

The ***transmission electron microscope*** (TEM) (see Figure 3-15) has been a driving force for the development in electron optics earlier in this century. In brief, an unfocused electron beam is incident on a very thin sample and, by using the transmitted electrons, magnifying projection optics form an image of the sample on photographic film with a resolution of a few billionths of a meter. Because the electron beam passes through the specimen, details of the internal microstructure are made observable. However, solid materials are highly absorptive to electrons and specimens must be prepared in the form of a very thin foil. The transmitted beam is projected onto a fluorescent screen or photographic film so that the image may be viewed.

The ***scanning electron microscope*** (SEM) scans the surface of a specimen with an electron beam and the reflected (back scattered) beam of electrons is collected and displayed at the same scanning rate on a cathode ray tube. The image on the screen contains the surface features (with good depth of field) of the specimen, which may be photographed. The major requirement for the specimen is that it be electrically conductive. Accessory equipment allows analysis of the chemical composition of very localized areas of the sample.

An example of the use of electron microscopy can be found in the research and development of synthetic crystalline materials produced by ***molecular beam epitaxy*** (MBE). These materials not found in nature have extremely thin layers of atoms deposited alternately on a semiconducting substrate or base. By controlling the chemical composition and thickness of the layers, various crystalline structures are formed with varying electrical, mechanical, and optical properties. The MBE method (described in Module 9) would not be possible without the prior development of electron microscopes with small wavelengths (0.2 nm) and large magnifications, which permitted scientists to verify that their efforts were indeed producing a crystal with extremely thin atomic layers of alternating composition. Atomic layers as thin as one ten-millionth of a centimeter were determined to exist in these new human-made materials.

Knowing that (1) changes in the internal structure of solids are accompanied by a change in energy, and (2) the release of this energy in the form of heat ***(exothermic)*** or the absorption of this heat ***(endothermic)*** can be detected and measured leads to another technique that increases our knowledge about the internal structure of solids. ***Differential thermal analysis*** (DTA) techniques produce temperature patterns that can be interpreted to obtain information about the various structural changes that solids undergo as a result of the application of different external forces. Finally, even the vibration of atoms in a molecule can be detected using infrared spectroscopy. Such techniques, for example, are used to identify unknown polymeric materials.

3.2.5 Other Instruments

The ***scanning tunneling microscope*** (STM) is a newer form of electron microscope invented in the 1980s by two Swiss scientists, Binnig and Rohrer, at IBM that uses a very fine metal tip to scan across a sample at a height of a few atomic spacings. At such spacings, a tunneling current occurs when only a few volts are applied between the tip and sample. This tunneling current is sensitive to the tip-to-sample spacing and is used to measure and control the height of the tip. In this way, scanning micrographs of the surface with a height resolution of about one atom and lateral resolution of about one to three atomic spacings can be produced. Also, the electron energy distribution of the tunneling current depends on the electronic properties of the sample and is used to select or view different types of surface atoms. This instrument made it possible to see the new polymorph of carbon known as *fullerenes.*

Scanning probe microscopy (SPM) involves many related technologies, all of which operate by scanning a fine probe tip over a sample surface (Figure 3-16 shows a schematic of the probe and surface) to generate a high-magnification ($10^9\times$), three-dimensional image. Piezoelectric ceramics, discussed in Module 9, control the probe motions in the x, y, and z directions with nm resolution [see Figure 3-16(c)]. SPM can be used under air, liquid, or vacuum conditions. Depending on the type of SPM and how many additional capabilities it has, its price ranges from $60,000 to $500,000. What is noteworthy is that the sizes of these pieces of equipment range from palm- to desk-size. Their importance is so momentous because SPMs are the connecting bridge between our world and the world of the atom and the molecule. Scientists who are now crossing this bridge are engaged in developing a new technology (discussed in various sections of this book) called by various names: *nanotechnology, molecular manufacturing,* or *molecular nanotechnology.* Regardless of the name, this new technological revolution will someday bring us the capability to manufacture materials as nature does by assembling individual atoms [see Figure 3-16(f)] and/or molecules to construct whatever material is needed. A few examples of available techniques using SPM are listed below:

Atomic force microscopy (AFM)—a cantilever probe senses electrostatic interactions with the specimen (see Figure 3-16). It can work with samples in their natural state.

Lateral force microscopy (LFM)—frictional response between the cantilever probe and the surface is plotted.

Figure 3-16 (a) Atomic force microscope has a resolution of 10×10^{-10} m and a scan range of 100^{-6} m. (Park Scientific Instruments)

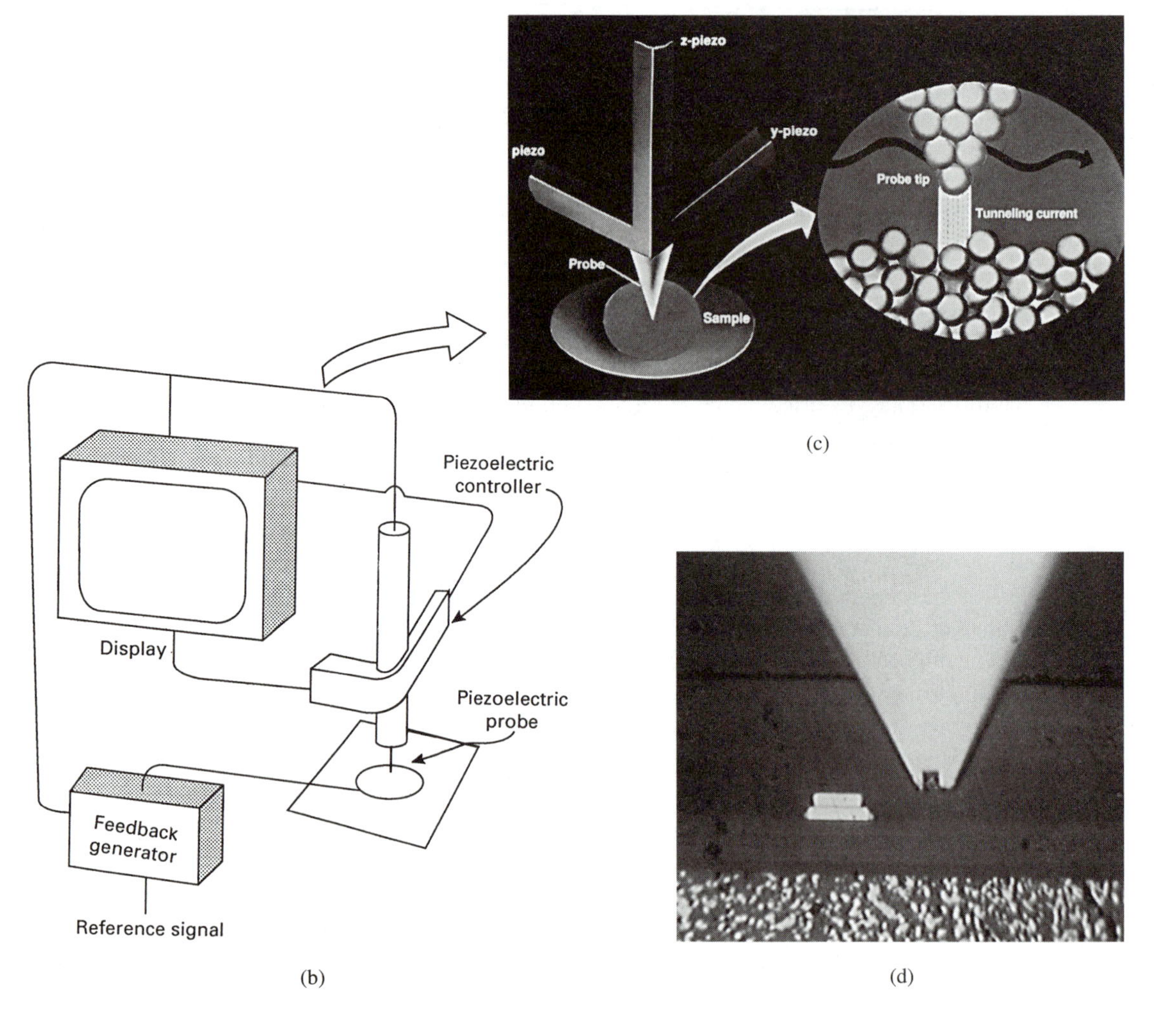

(c)

(b)

(d)

Figure 3-16 (b) Scanning tunneling microscope (STM) schematic diagram. (c) Enlarged diagram of probe. (d) Enlarged photo shows tungsten probe tip 0.2 nanometers wide to probe conductor and semiconductor materials.

Modulated force microscopy—maps the variation in sample compliance (hardness) due to changes in cantilever modulation.

Electric force microscopy (EFM)—measures the electrostatic force of attraction or repulsion between the tip and the sample.

Magnetic force microscopy (MFM)—measures the magnetic force of attraction or repulsion between the tip and sample.

Scanning thermal microscopy (SThM)—uses an electrically resistive probe to map thermal conductivity or temperature variations between the tip and the sample.

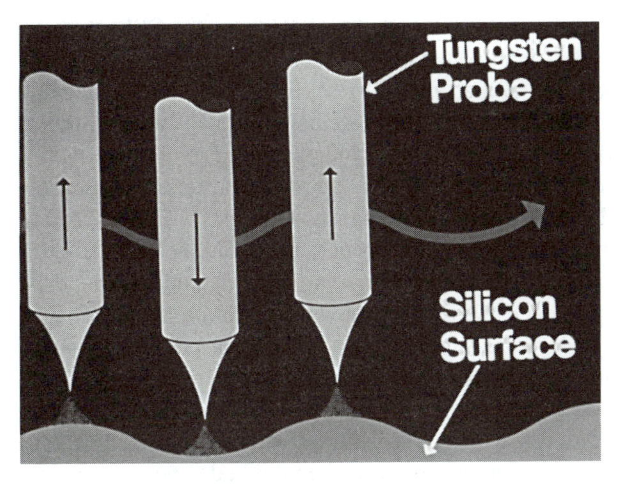

(e)

Figure 3-16) (e) Schematic showing how tungsten probe moves along and scans silicon surface. (f) Image from STM of gallium arsenide (GaAs) atoms. (g) Nanoscale composite mixing—top: STM image of silicon surface; middle: a voltage pulse of +3 volts applied between probe tip and surface led to formation of a mound of silicon atoms (large white spot) surrounded by a black matlike area; bottom: a second voltage pulse of +3 volts leads to removal of the mound, then with the probe in a different position, application of pulse −3 volts led to a cluster (white spot) being deposited on the surface, leaving a large empty area (large black spot). (IBM Research.)

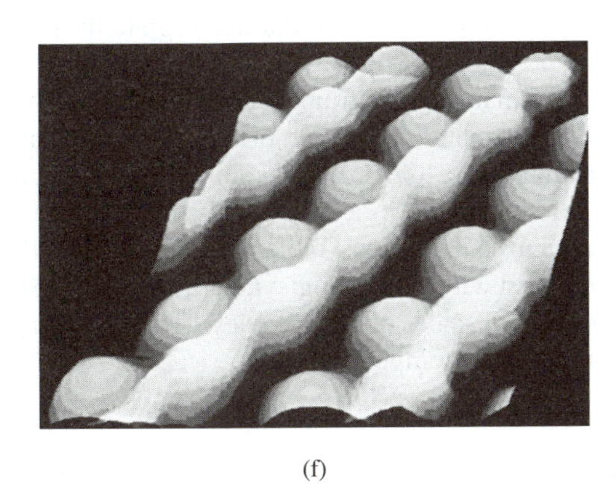

(f)

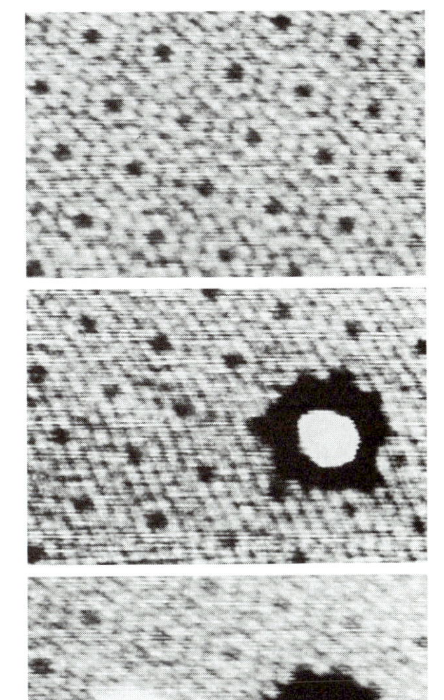

(g)

New techniques are continually being developed. An example is the SPM with an SEM. This technique combines the best features of both microscopes into a powerful instrument in the study and characterization of materials. The scanning ion-conductance microscopes (SICM) and the laser force microscope (LFM) are helping to advance synthesis of high-temperature superconductors and to improve the understanding of all materials.

In 1951, the ***field-ion microscope*** was introduced. It can provide direct images of individual atoms in a solid surface. It achieves its high magnification from a radial projection of ions from the tip of a sharp needle to a fluorescent screen. The needle, called a "field emitter," or more commonly, a tip, has a radius of curvature at the end of around 10 nm. Ions are generated above the tip surface in a process known as *field ionization.* The tip is placed in a vacuum chamber and a low-pressure "imaging gas" is introduced. An electric field is also applied. The imaging gas atoms in the vicinity of the tip are ionized by an electron tunneling from the tip surface to the specimen surface. The positive field ions formed follow the electric field lines away from the tip surface to the fluorescent screen, where they produce image spots that are representations of individual atoms. From the symmetry of the pattern of spots, the Miller indices (Section 3.4.3) of the observed planes of atoms can be determined. The field-ion microscope is similar in design to the *field-emission microscope* invented in 1936.

These are but a few of the many sophisticated instruments and techniques that allow scientists today to measure the features of materials, define their properties, manipulate their structure at the atomic and molecular levels to increase scientific knowledge, assist in the manufacture of products, and contribute to the process of analyzing applications. As stated earlier regarding synthesis and nanoscale mixing (Table 3-1, and illustrated in Figure 3-16), nanoscale technology is achieved with processing/syntheses using probe microscopes, beam processing (MBE, laser, etc.) at scales of 50 nm or less. These techniques are carried out on advanced metals, ceramics, and composites.

3.3 MATERIALS CHARACTERIZATION

Just as humans have fingerprints unique to each person, materials have unique characteristics. ***Characterization*** allows the identification and analysis of unique properties of a material's microstructure, revealed as defects, crystal structure, and impurities, as well as macrostructural properties such as surface roughness. Characterization may use some of the same tools and techniques as those used for destructive or nondestructive testing and evaluation (see Module 4), but characterization focuses more on structure and composition. The analysis can be qualitative or quantitative; it can be used to provide chemical or physical characteristics of the material. ***Qualitative characterization*** has been practiced since early times as crafters examined wood, metal, or stone to determine the suitability of that material for the tool or product they would craft. Figure 3-17 shows some of the characteristics for analysis of a material's substrate and coating.

Quantitative characterization results from the ability to link computers with analysis instruments and to capture large amounts of data to make a comparison against a standard. Quantitative characterization is a recent technology that has evolved as a result of the

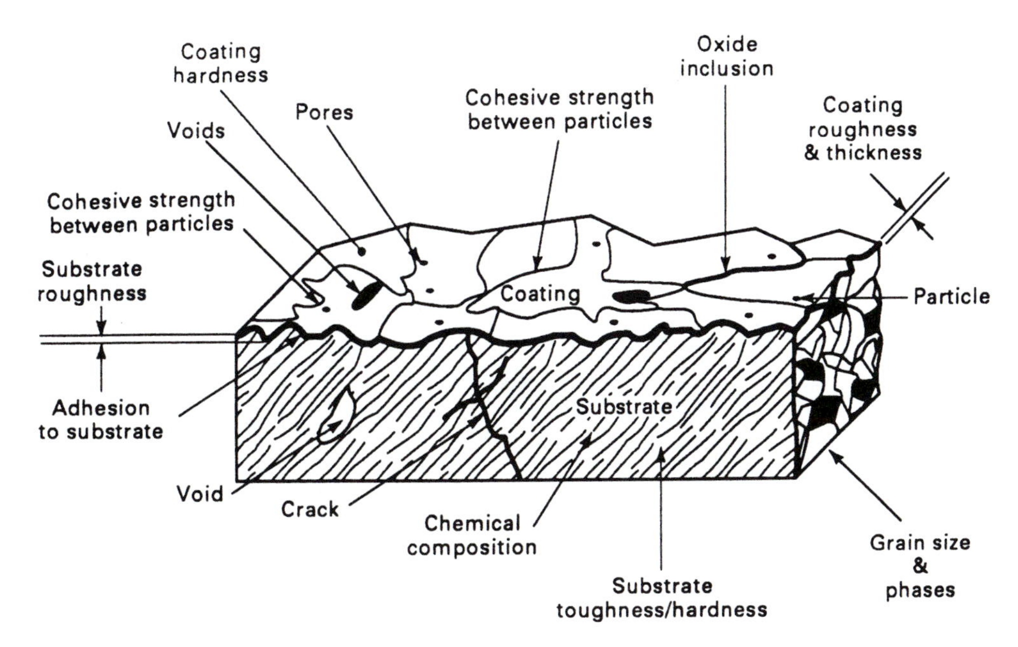

Figure 3-17 Characterization components of a substrate material and coating.

need for improvement over older evaluation techniques that could not provide the precision required by advanced materials. Very hard materials or materials such as ceramic fibers in softer polymer composite matrices are often difficult to prepare for analysis by conventional metallographic techniques. Figure 3-18 presents three micrographs of a steel substrate spray-coated with tungsten carbide (WC) using high-velocity oxyfuel (HVOF) on steel. The results demonstrate that different means of preparing the same specimen for analysis can yield different results. For characterization techniques to be useful, they must provide true micro/macrostructural information. On the specimen shown, polishing with silicon carbide paper or fixed diamond platens seems to indicate voids in the coating. However, examination of the same specimen after diamond lapping revealed that it was the polishing medium that caused pullout of the coating.

To overcome problems of inaccurate characterization, both improved specimen preparation and analysis may be required (see Figure 3-18). For example, digital analysis is now possible for images obtained with microscopes and x-rays, or generated from acoustic (sound) devices. Digital analysis provides data on features such as porosity, grain size, and surface flaws. ***Artificial intelligence*** (AI), the use of computers to solve problems using methods similar to those used by humans, has been employed to make quantitative characterization and statistical analysis of structures that would be too tedious for humans. Advances in digital analysis have come on the heels of methods to enhance optical and other imaging techniques with technology such as machine vision, which feeds special data into image analyzers. Improved methods of spectrometry, such as flame atomic absorption, allow chemical characterization of trace elements in high-performance alloys

Figure 3-18 Using three methods to prepare samples of steel coated with tungsten carbide yielded three different results, making analysis difficult. This demonstrates the importance of proper specimen preparation before analysis.

down to a few parts per million (ppm). These analyses permit the use of chemically characterized alloyed parts to be qualified for use in very critical applications, such as aircraft gas turbines.

3.4 STRUCTURE OF SOLID MATERIALS

In simplified form, structure is the "cause" that produces the properties ("effect") possessed by a material. In fact, there are many levels of structure, from the atomic to the macroscopic, each with its own chemical composition and distribution of its structural components (i.e., atoms, molecules, crystals, and phases, or what can be called the basic building blocks of materials). All these material structures give rise to an infinite array of materials, each possessing its own unique set of properties. All who work with materials or use them must constantly be aware that subjecting a material to external forces and/or conditions (impact, changes in temperature, corrosive conditions, radiation) will, in the majority of instances, have an effect on its structure with attendant change in its properties. Some of these changes can be handled quite satisfactorily by the material if the design process was correct in every respect. A final observation of the awareness of the many structures in a material helps us understand the role of material scientists, who are constantly working to develop materials with different structures or with different combinations of structures with the right mix of properties to help solve our technological problems.

Definitions of solids most always contain the phrase "definite volume and shape," together with the fact that solids maintain their shape to a varying extent when subjected to

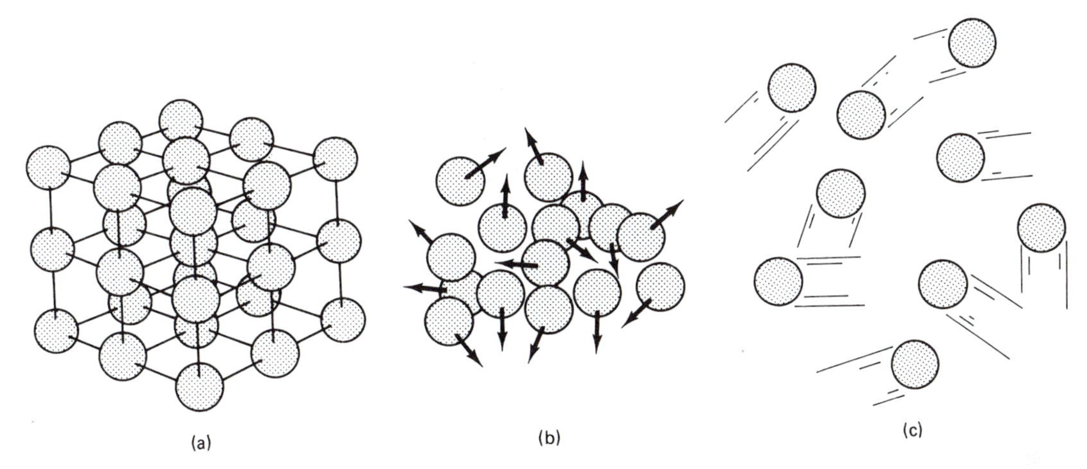

Figure 3-19 How atoms are arranged in (a) a crystalline solid, (b) a liquid, (c) a gas. (D.C. Giancoli, *Physics,* Prentice-Hall, Inc., Englewood Cliffs, N.J., 1980.)

external mechanical forces. Solids occur in two *basic* forms: crystalline and amorphous. A true ***crystalline solid*** possesses an ordered, three-dimensional, geometric arrangement that repeats itself. A metal would be representative of a crystalline solid (though not perfectly crystalline). An ***amorphous solid,*** on the contrary, contains no repetitious pattern of atom locations to any extent. The classic example of an amorphous solid, glass, is sometimes referred to as a supercooled liquid because of the random nature of its atomic arrangement (see Figure 7-40). Figure 3-19 illustrates the different atomic arrangements of three states of matter, emphasizing the different degrees of disorder and closeness of their positions in relation to each other.

In our treatment of ***crystallography*** (the study of crystalline structures), we will limit ourselves to the orderly arrangement of the atoms in their microscopic world. In so doing, we represent atoms, ions, or molecules essentially as spheres of varying sizes occupying points at various distances from each other in space—hence, the need for an ***axis system.*** Such a system is shown in Figure 3-20.

The arrowhead on each axis points in the positive direction. The negative directions, therefore, are opposite to the arrowheads. The *x, y,* and *z* axes are, in this case, perpendicular to each other, with their origins coinciding at any convenient point. A box with dimensions *a, b,* and *c* is sketched in Figure 3-21, using this axis system. If this solid figure were a cube, these dimensions would be of the same magnitude. However, in one important crystal system they are not equal in magnitude. Another interpretation of these letters (or *lattice parameters*) is that they represent distances from the origin to points on the axes. The letter *b* would, for example, represent the perpendicular distance from the origin to a point on the *y* axis.

The three axes in Figure 3-22, which are infinite in length, form three mutually perpendicular planes that are also infinite in size. A box is drawn using these axes as sides to help illustrate some basics concerning crystal systems. The *x–y* horizontal plane forms the base of the box. The top of the box is contained in a plane parallel to the *x–y* plane. A

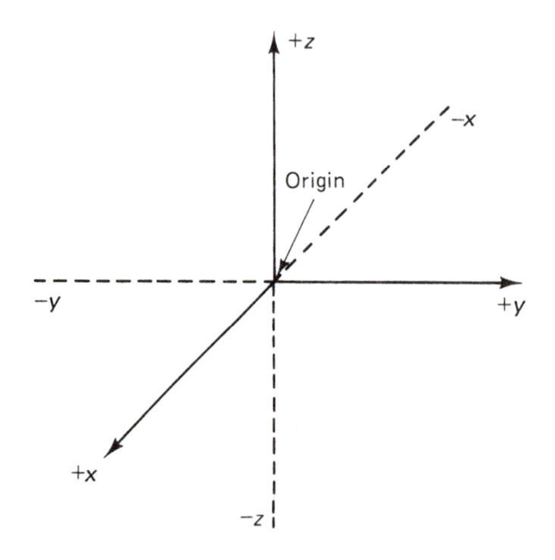

Figure 3-20 A standard axis system.

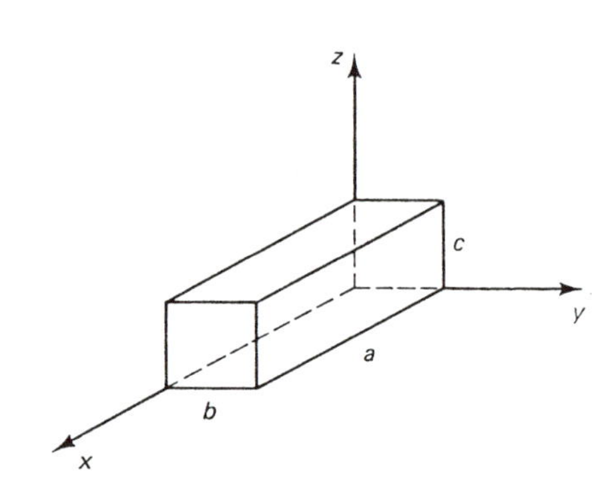

Figure 3-21 A box aligned on an axis system.

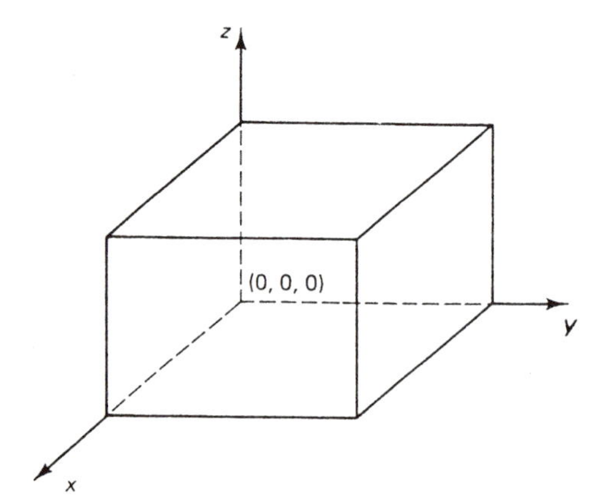

Figure 3-22 A three-dimensional box aligned on an axis system with an origin at the lower-left rear corner.

portion of the x–y plane contains the rear (as looked at by the viewer) end of the box, and the front end would be contained in a plane parallel to the y–z plane. There are countless planes parallel to these three primary (x–y, y–z, and x–z) planes, which can be referred to using a standard technique called ***Miller indices*** (see Section 3.4.3).

3.4.1 Unit Cells and Space Lattices

In our study of crystals, the axes system just described (called a ***simple cubic crystal system***) is modified slightly for the sake of uniformity. The standardized axes system shown in Figure 3-23 retains the x, y, and z axes. The box is now called a *unit cell* (defined later). The angles between the principal planes are named α (Greek letter alpha), β (Greek letter beta), and γ (Greek letter gamma). For example, in Figure 3-24, α is the angle measured in degrees between the x–y and x–z planes; angle β, between x–y and y–z planes; and angle γ, between the x–z and y–z planes. The sides of the box (unit cell), labeled a, b, and c, are the lattice parameters in the x, y, and z directions, respectively. These distances are also

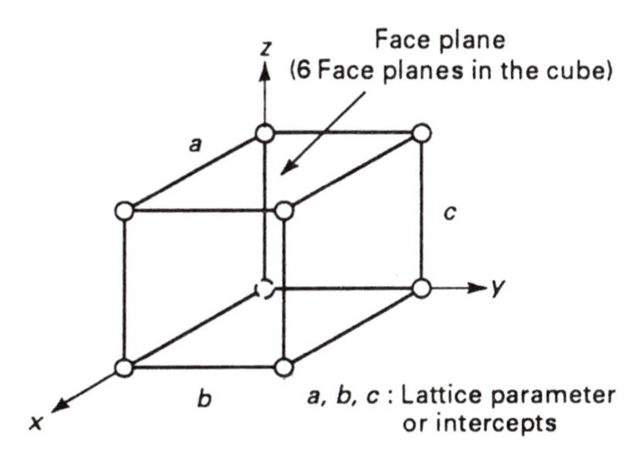

Figure 3-23 A simple crystal lattice unit cell with equal intercepts a, b, and c.

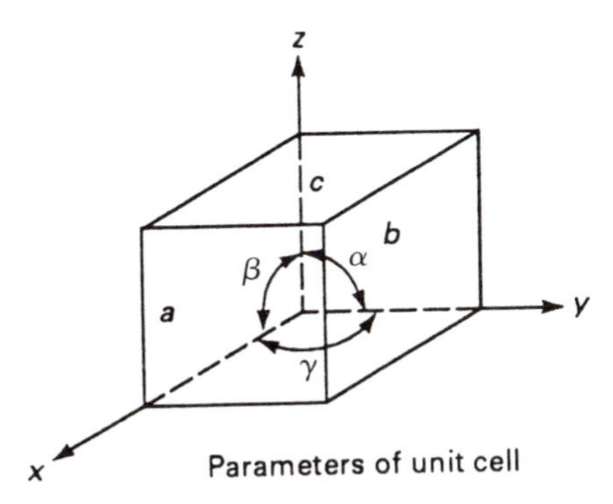

Figure 3-24 Angles between planes and intercepts in a unit cell.

known as ***intercepts.*** To describe a particular axes or crystal system adequately, all six of the preceding dimensions are needed.

The term ***unit cell*** is used to describe the basic building block or basic geometric arrangement of atoms in a crystal. You can compare a unit cell to a single brick in a brick wall. Knowing that atoms are located at each corner of the single brick, it is easy to picture the atomic structure of a crystal, with such a unit cell repeating itself in three-dimensional space.

If you repeat the unit cell in all three dimensions, you create a crystalline structure with a definite pattern. This larger pattern of atoms in a single crystal is known as a ***space lattice*** or ***crystal lattice.*** A space lattice is three sets of straight lines at angles to each other, constructed to divide space into small volumes of equal size, with atoms (ions or molecules) located at the intersections of these lines or between the various lines. We must remember that the lines and points in a space lattice are only imaginary. The lattice concept is used to show the positions of atoms, molecules, or ions in relation to each other. Atoms may be represented by circles, spheres, table tennis balls, or tennis balls located either at the intersection of these lines or between these lines. A part of a space lattice is sketched in Figure 3-25, with small spheres representing atoms. We must also remember that the

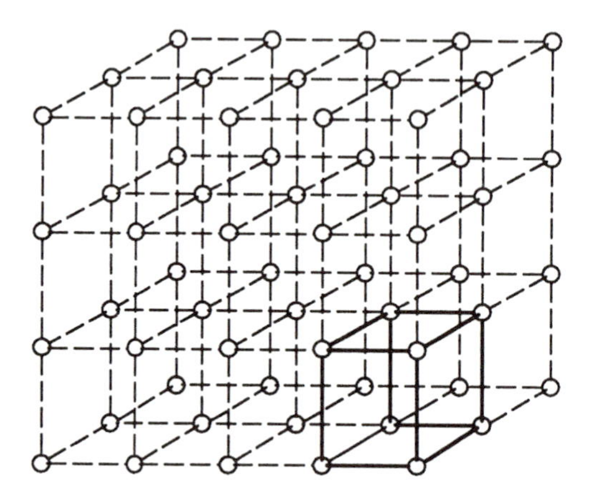

Figure 3-25 A space or crystal lattice with a simple cubic unit cell outlined.

actual atoms in solids are located as close to each other as possible, thus attaining the lowest possible energy level. Two atoms closest to each other would be represented by two spheres touching each other. The closer the atoms are, the denser the solid.

3.4.2 Crystal Systems

Mostly pure substances form crystals in the solid state. In this section, we emphasize crystal systems composed of atoms and ions, but the structural particles of crystalline solids can be atoms, ions, or molecules. Solid methane, CH_4, a molecular solid, has a face-centered cubic (fcc) structure, which means that there is a CH_4 molecule at each corner and at the center of each face in its unit cell. The forces acting among these structural particles may be metallic bonds, interionic attractions, van der Waals forces, or covalent bonds. Most metals, a good many ceramic materials, and certain polymers crystallize when they solidify. Most metals have a cubic close-packed, hexagonal close-packed, or body-centered cubic lattice structure. There are some exceptions. Tin (Sn) and indium (Im) form body-centered tetragonal structures. (Figure 3-29 shows a bcc tetragonal unit cell.) Table 3-2 indicates this tetragonal structure as XX showing, among other things, that this structure is not a common one for pure elements in the periodic table. Zirconia (ZrO_2), most important in improving toughness (see Module 7), forms a tetragonal unit cell structure. Martensite, a form of tough steel, gains its hardness and strength from its hard, brittle, tetragonal phase that is supersaturated with carbon.

The cubic system's basic structure has been demonstrated in Figures 3-20 to 3-24, assuming that the sides of the boxes described are all equal in length ($a = b = c$). It is possible for atoms to form three different patterns or arrangements with this cube. These patterns, of which there are a total of 14, are known as ***Brevais lattices.*** Each pattern can be described in detail by using a unit cell as a reference. The *cubic axis system,* with its mutually perpendicular axes and equal lattice parameters, has already been described. The cubic coordinate axis system allows us to form three basic unit cells by placing atoms, ions, or molecules at various lattice positions.

The simple ***cubic unit cell*** (Figure 3-26) consists of eight atoms located at each corner of the cube. This structure does not exist in nature because the space in the center would need to be filled by other atoms. The point to be made is that this representation of

PERIODIC SYSTEM OF THE ELEMENTS

H-1 ▲ (−58) ⊗ XX ◀

Adapted Primarily for Ferrous Metallurgists

Atomic size factors (in parentheses) are % smaller (−) or larger (+) than gamma (FCC) iron at 75 F. Lattice environment (Coordination No.) is taken into account; CN is 12 except 6 for interstitials H, B, C, N & O. Groups VI, VIb, VII & VIIb form ionic compounds with the metals. Atomic size is based largely on work of W. Hume-Rothery and associates and L. Pauling (Some values, such as those for H & O, are approximate). Alloying valences are those of Pauling.

0	I	II	III	IV	V	VI	VII
He-2 FCC (Others)	Li-3 ⊗(+23) BCC* HCP†	Be-4 ●(−11) HCP* BCC ◗	B-5 ◬(−29) ⊗XX ▮	C-6 ▲(−34) ⊗XX ◆	N-7 ▲(−36) ⊗XX ◆	O-8 ▲(−33) ⊗ XX	F-9
Ne-10 FCC	Na-11 ⊗(+50) BCC* HCP	Mg-12 ⊗(+27) HCP	Al-13 ◑(+14) FCC ◗	Si-14 ●(+7) XX ◗	P-15 ●(+2) XX ◗	S-16 ●(+1) XX ▮	Cl-17 XX

0	Ia	IIa	IIIa	IVa	Va	VIa	VIIa	VIII	VIII	VIII	Ib	IIb	IIIb	IVb	Vb	VIb	VIIb
Ar-18 FCC	K-19 ⊗(+86) BCC	Ca-20 ⊗(+56) FCC* BCC	Sc-21 ⊗(+29) HCP* BCC	Ti-22 ◑(+16) HCP* BCC ◗	V-23 ●(+6) BCC ◗	Cr-24 ●(+1) BCC ◗	Mn-25 ●(+1) XX* FCC‡ ◀	Fe-26 ●(0) BCC* FCC	Co-27 ●(−1) HCP* FCC ◀	Ni-28 ●(−1) FCC ◀	Cu-29 ●(+1) FCC ◆	Zn-30 ●(+6) HCP ◗	Ga-31 ●(+12) XX	Ge-32 ●(+9) XX ◗	As-33 ●(+11) XX ◗	Se-34 ●(+11) XX	Br-35 XX
Kr-36 FCC	Rb-37 ⊗(+97) BCC	Sr-38 ⊗(+71) FCC* HCP‡	Y-39 ⊗(+42) HCP* BCC	Zr-40 ⊗(+27) HCP* BCC ▮	Cb-41 ◑(+15) BCC ▮	Mo-42 ●(+10) BCC ◗	Tc-43 ●(+8) HCP	Ru-44 ●(+6) HCP ◆	Rh-45 ●(+6) FCC ◀	Pd-46 ●(+9) FCC ◀	Ag-47 ◑(+14) FCC	Cd-48 ⊗(+20) HCP	In-49 ⊗(+25) XX	Sn-50 ⊗(+23) XX ◗	Sb-51 ⊗(+27) XX ◗	Te-52 ⊗(+27) XX	I-53 XX
Xe-54 FCC	Cs-55 ⊗(+112) BCC	Ba-56 ⊗(+76) BCC	La-57 ⊗(+48) HCP* FCC‡	Hf-72 ⊗(+26) HCP* BCC ▮	Ta-73 ◑(+16) BCC ▮	W-74 ●(+11) BCC ◗	Re-75 ●(+9) HCP ◆	Os-76 ●(+7) HCP ◀	Ir-77 ●(+8) FCC ◀	Pt-78 ●(+10) FCC ◀	Au-79 ◑(+14) FCC ◆	Hg-80 ⊗(+25) XX	Tl-81 ⊗(+36) HCP* BCC	Pb-82 ⊗(+39) FCC	Bi-83 ⊗(+35) XX	Po-84 ⊗(+40) XX	At-85
Rn-86	Fr-87	Ra-88	Ac-89 ⊗(+49) FCC					Note 1: The rare-earth (lanthanide, 58-71) and actinide (90-103) series are omitted. Note 2: Valence is 4 for C; 3 for N and P. Note 3: (1) and (2) are not alloying valences.									
Alloying Valence	1	2	3	4	5	6	6	6	6	6	5.56	4.56	3.56	2.56 Note 2	1.56 Note 2	(2) Note 3	(1) Note 3

SUBSTITUTIONAL SOLID SOLUTIONS

- ● FAVORABLE SIZE FACTOR: 0 TO ± 13%
- ◑ BORDERLINE SIZE FACTOR: ± 14 TO ± 16%
- ⊗ UNFAVORABLE SIZE FACTOR: > ± 16%

INTERSTITIAL SOLID SOLUTIONS

- ▲ FAVORABLE SIZE FACTOR: > (−40%)
- ◭ BORDERLINE SIZE FACTOR: (−30) TO (−40) %
- ◬ UNFAVORABLE SIZE FACTOR: < (−30%)

STRUCTURE

- BCC - BODY CENTERED CUBIC
- FCC - FACE CENTERED CUBIC
- HCP - HEXAGONAL CLOSE PACKED
- XX - NOT BCC, FCC OR HCP USUALLY MORE COMPLEX
- * - STRUCTURE AT 75 F
- † - ALSO FCC ‡ - ALSO BCC

TYPE OF GAMMA IRON (FCC) FIELD IF ALLOYED WITH IRON

- ◗ GAMMA LOOP, LIKE Cr
- ▮ LIMITED GAMMA LOOP, LIKE B
- ◀ OPEN GAMMA REGION, LIKE Ni
- ◆ LIMITED GAMMA REGION, LIKE C

Prepared by O. O. Miller

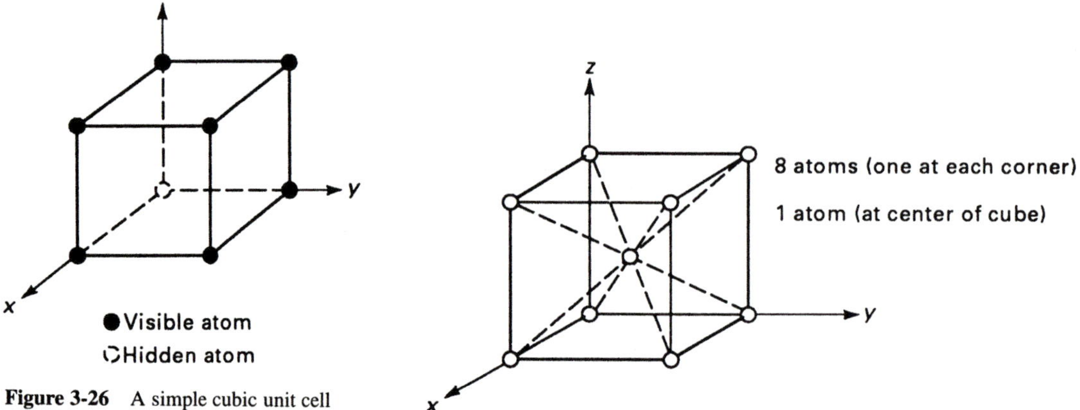

Figure 3-26 A simple cubic unit cell showing atoms only at the corners of the cube.

Figure 3-27 A body-centered cubic unit cell.

atoms in a solid shows only the location of atoms. It must be remembered that if you represented these eight atoms with hard rubber balls and arranged them in accordance with this simple cubic unit cell, all eight atoms would be touching each other.

Another unit cell (Figure 3-27) is known as the ***body-centered cubic*** (bcc). It is similar to the simple cubic unit cell, but contains an additional atom located in the center of the cube. The third type of unit cell formed from the cubic axis system is the ***face-centered cubic*** (fcc). One atom at each corner and one in the center of each of the cube faces make up the complement of atoms. There is no atom at the center of the cube (Figure 3-28).

For both the bcc and fcc structures, the same type of atoms must fill both the center/face-centered locations, as do the cell corners. If, in fact, a different type of atom fills the center position than the cell corners in a body-centered cubic structure, the structure is

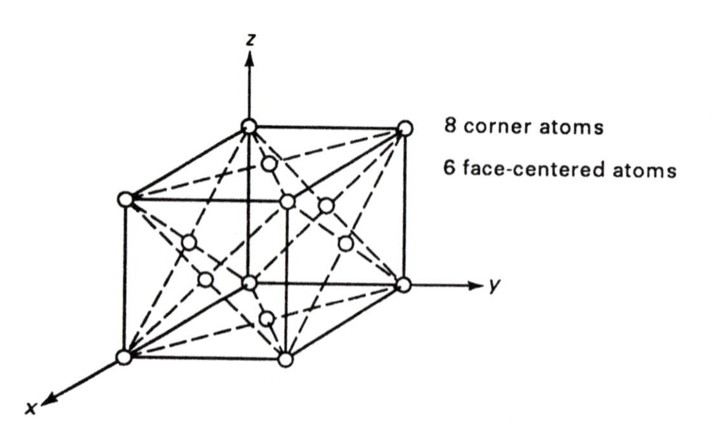

Figure 3-28 A face-centered cubic unit cell.

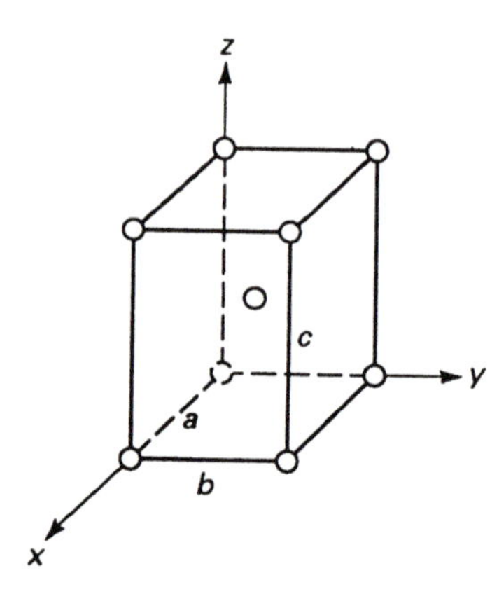

Figure 3-29 A body-centered tetragonal crystal lattice unit cell.

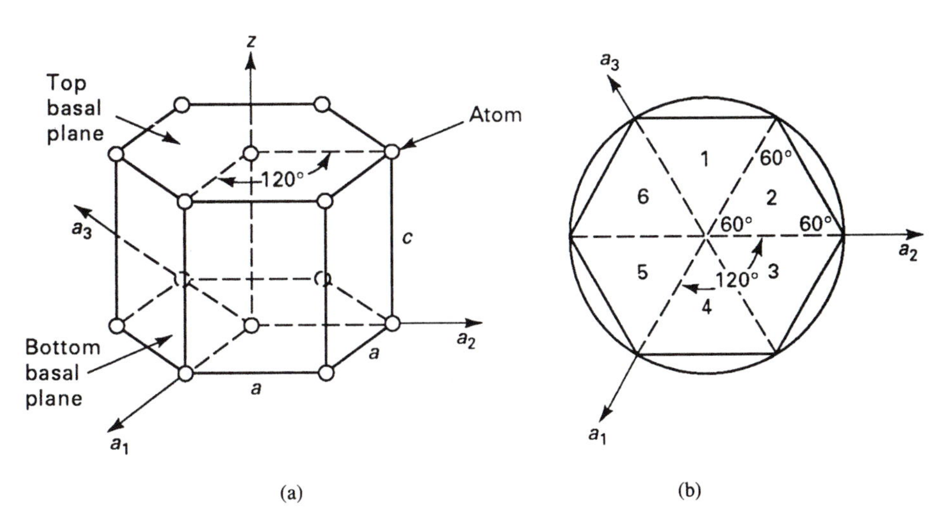

Figure 3-30 (a) A hexagonal crystal lattice unit cell. (b) Top plane of unit cell showing six equilateral triangles.

correctly called a simple cubic (sc) structure. An alloy made up of about 50% copper and 50% zinc forms such a structure.

The ***tetragonal crystal system*** has similar unit cells to the cubic, but the sides are not equal. As an example, the ***body-centered tetragonal*** (bct) crystal lattice unit cell is shown in Figure 3-29. Tin forms a tetragonal unit cell. The tetragonal is similar because the axes are all normal to each other. The difference lies in the length of the intercepts. The x and y intercepts have the same magnitude. The z intercept is larger than the x or y intercept. Martensite, a combination of iron and carbon that is contained in a hard steel, has its atoms of iron and carbon in a tetragonal lattice structure (Module 5).

The ***hexagonal crystal system*** [Figure 3-30(a)] can best be described using three axes (a_1, a_2, and a_3) in the x–y plane 120° apart and a fourth axis (z) at 90° to the x–y plane. The intercepts along the three axes in the horizontal plane are equal in length ($a = a = a$), but the fourth intercept, labeled c, is of a different length. This unit cell is made up basically of two parallel planes (top and bottom basal) separated by a distance equal to the dimension c. The atoms shown in the figure trace out a right hexagonal prism. Each of these two places can be divided into six equilateral triangles, with each side equal to the intercept a [Figure 3-30(b)].

Atoms of solid materials do not form the purely hexagonal unit cell as in Figure 3-30 because they cannot satisfy equilibrium conditions by being so far apart. In other words, they would be unstable. Consequently, they form the hexagonal unit cell, called ***close-packed hexagonal*** (cph), as shown in Figure 3-31, with its three mid-plane atoms at a distance of $c/2$. Zinc, titanium, and magnesium form cph unit cells.

In summary, there are a total of seven crystal systems, of which three were mentioned above. Two examples of some of the unit cells from the orthorhombic and monoclinic crystal systems are shown in Figure 3-32. Not discussed are the triclinic and rhombohedral systems. Using these seven axes systems, atoms can form 14 patterns in space

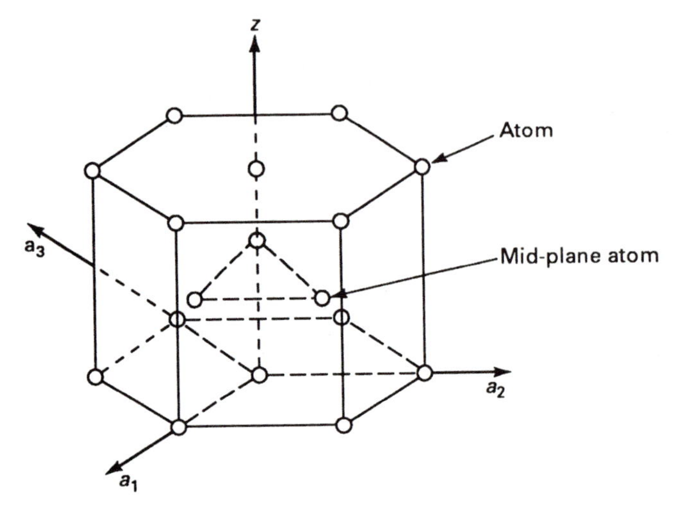

Figure 3-31 A close-packed hexagonal crystal lattice unit cell.

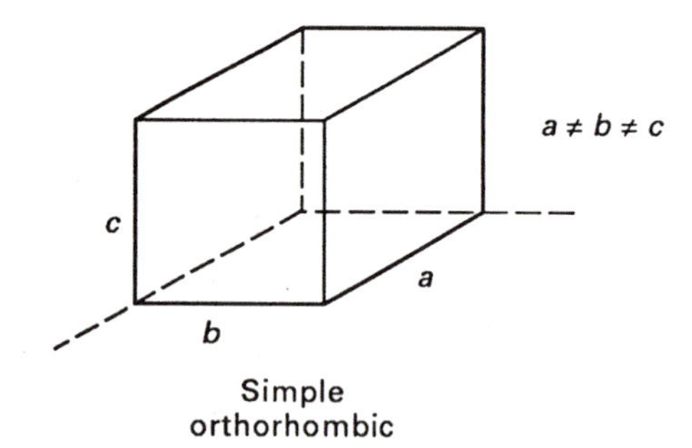

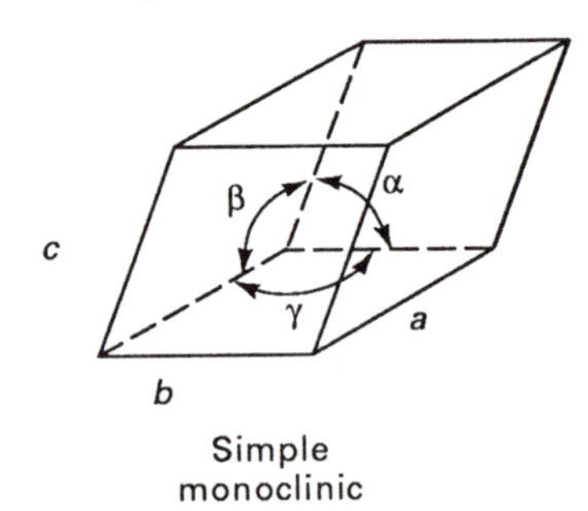

Figure 3-32 Orthorhombic and monoclinic unit cells.

(Brevais lattices), of which five have been mentioned and described. As an example, using unit cells for the cubic crystal system, atoms can form three different patterns: simple cubic (Figure 3-26), bcc cubic (Figure 3-27), and fcc cubic (Figure 3-28). Table 3-3 lists the intercepts and the angles between axes for five of seven crystal systems with which the earth's elements form their particular atomic structures.

3.4.3 Crystallographic Planes, Miller Indices, and Crystal Directions

The amazing thing about crystallographic structures is that any variation in the arrangement of atoms or the stacking of the planes formed by the atoms has a pronounced effect on the properties of the solid material. Atoms within the solids tend to move in coordination with each other. Solids tend to deform in the directions along the planes that are the most closely packed because it is easier for an atom (or ion) to move in these close-packed planes. Planes of atoms move, or slip, in parallel directions much like two adjacent cards in a deck of playing cards. This movement of atoms in a solid results from the application of some type of external force, such as a compressive or tensile load. The degree of movement of planes of atoms gives an indication of the ductility and strength of the particular

TABLE 3-3 CRYSTAL SYSTEMS SUMMARY

System	Intercepts	Angles between axes
Cubic	$a = b = c$	$\alpha = \beta = \gamma = 90°$
Tetragonal	$a = b \neq c$	$\alpha = \beta = \gamma = 90°$
Hexagonal	$a = a \neq c$	$\alpha = \beta = 90°; \gamma = 120°$
Monoclinic	$a \neq b \neq c$	$\alpha = \gamma = 90° \neq \beta$
Orthorhombic	$a \neq b \neq c$	$\alpha = \beta = \gamma = 90°$

material. Therefore, it is important to know something about how the principal crystal structures differ in the location of the atoms in their space lattices/unit cells and something about the planes in these structures.

First, let us consider how atoms (or ions) move in a plane. A simple experiment can show this phenomenon. Take a number of pennies and line them up in two rows, all touching, with an equal number in each row. If you push one row with your finger in one direction parallel to the row and do the same with the other row, but in the other direction, each row will move easily. Next, if you line up the objects in less than a perfect arrangement with maybe only one of the objects projecting into the other row and then push the rows as before, the result is very little movement in the direction of the pushing force and considerable movement in many directions. Later you will learn that metals and metal alloys that arrange themselves with fcc unit cells exhibit the most ductility mainly because they possess the greatest number of close-packed planes in the greatest number of directions. Also, if not parallel, these planes will intersect and movement of atoms may be hindered or stopped completely at some stage.

The number of atoms in any atom plane depends on the stacking sequence of the planes. The atom planes in these crystal lattices that have the largest number of atoms are called ***close-packed planes.*** Representing atoms (or ions) as spheres, the closest packing of atoms occurs when each atom has twelve other atoms of equal size around it.

Of all the crystalline materials found in nature, the most important fall into only three stacking sequences. The stacking of four atom planes in a particular order creates the fcc arrangement of atoms. Of course, in a solid material there would be an infinite number of planes stacked in this particular way. Fcc metals such as Al, Cu, and Si therefore have four sets of nonparallel, close-packed planes in their unit cells, as discussed later. Bcc unit cells for elements such as Cr, Li, or Mo contain no close-packed planes. Cph unit cells are formed using a different sequence of stacking. As a result, they have two hexagonal bases (basal planes), as shown in Figure 3-30.

Numerous exercises can be developed to help gain better visualization of these structures. In so doing, remember that no adjacent planes can be identical. Any deviation from the stacking sequences mentioned above are considered *stacking faults*.

In the bcc crystal system, we note in observing the bcc unit cell in Figures 3-27 and 3-33 that, by use of a cutting plane through this cell, the plane containing the most atoms would be the one that passed through four of the eight corner atoms and the one in the center of the cube. All figures are expanded views. Actually, the atoms are much closer. This diagonal plane could be a (110) plane, as sketched in Figure 3-33. It should be noted that, although these planes contain the most atoms, the atoms are not as closely packed as they

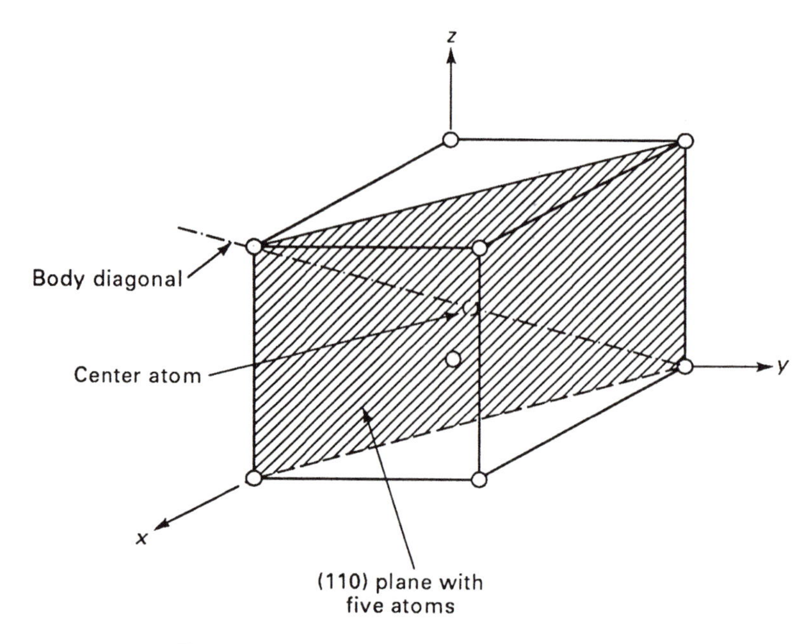

Figure 3-33 A bcc unit cell showing (110) plane.

could be. A view of this diagonal plane from a position at right angles to it, as in Figure 3-34, shows that the spaces are not occupied by atoms, the diagonal has a length of $4r$ (r is the radius of each atom), and the atoms touch each other along the bcc unit cell body diagonal.

In the *fcc unit cell,* the planes with the most atoms can be represented by plane (111) as sketched in Figure 3-35 or seven other planes with the same number of atoms. Observe that the plane contains three face-centered atoms and three corner atoms. Figure 3-36 shows these atoms all in contact with each other with a minimum of unoccupied space between them. With atoms of the same size, six atoms are touching each atom in such a plane.

The *cph unit cell* readily tells us the most closely packed planes of atoms are the basal planes and all planes parallel to them. This means that a metal with a cph crystal lattice can slip in only one direction. In other words, this metal would be brittle. The bcc or fcc metals, having a greater number of slip planes, can slip in many directions and are thereby classified as ductile metals. It must be remembered that, at this stage, we are talking about the atomic structure of millions of atoms in one crystal. The vast majority of solid materials are polycrystalline (i.e., composed of millions and millions of individual crystals whose axes are oriented in random directions). This random orientation could result in a blocking effect; that is, a crystal would stop the slippage of atoms in any adjacent crystal. More details of the movement of atoms will be discussed later.

3.4.3.1 Miller indices for atomic planes and directions. Because atoms move, it is appropriate to be able to describe the location of atoms in a unit cell as well as

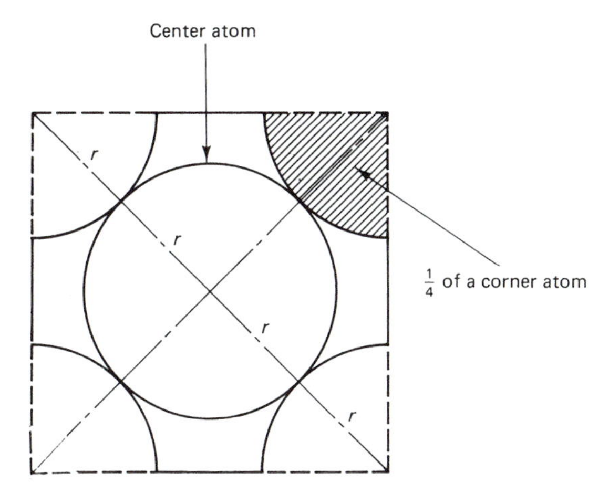

Figure 3-34 A bcc unit cell showing (110) plane.

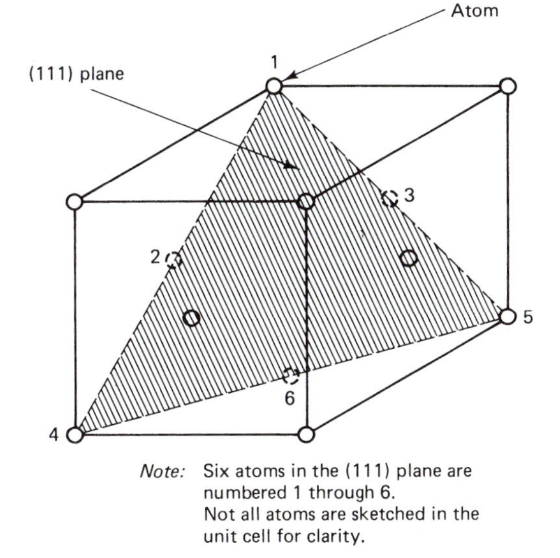

Note: Six atoms in the (111) plane are numbered 1 through 6. Not all atoms are sketched in the unit cell for clarity.

Figure 3-35 (111) plane in fcc unit cell.

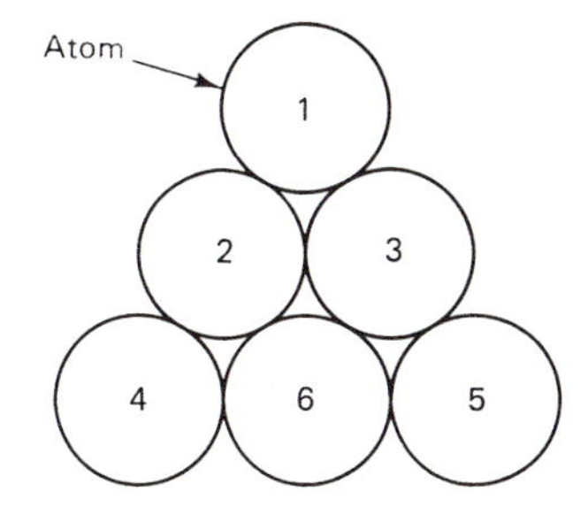

Figure 3-36 Normal view of diagonal plane (111) in fcc unit cell showing six numbered atoms.

the direction of their movement. The sites or locations of atoms and/or points in a unit cell are described by their axial coefficients expressed in unit-cell dimensions. These coefficients (also called coordinates or intercepts) are the respective distances along the *x, y,* and *z* axes from the origin of the axis system. By convention, the origin is located at the lower-left rear corner of the unit cell (Figure 3-22). However, it may be moved temporarily to another position if it is so indicated. In Figure 3-22, for example, the location of the atom shown at the origin of the axis system would have the coordinates 0,0,0, as indicated. The atom above the origin on the *z* axis in Figure 3-23 at a distance of 1 unit length would have the coefficients 0,0,*c* or 0,0,1. In Figure 3-28, the face-centered atom in the *x–y* plane is located by the coefficients ½, ½,0. Note that the distances making up the coordinates are separated by commas. The planes in a unit cell are designated by their *Miller indices.*

The procedure for determining the Miller indices of a plane is illustrated in the following problems.

Illustrative Problem

In Figure 3-37, two planes *A* and *B* are shown with their intercepts. The four-step procedure for determining the Miller indices of plane *A* is as follows:

Step 1. Record the intercepts in order of the *x, y,* and *z* axes.

x axis	intercept 1
y axis	intercept ½
z axis	intercept ⅓

Step 2. Clear any fractions by taking the reciprocals of the intercepts.

x axis	$1/1 = 1$
y axis	$1/\frac{1}{2} = 2$
z axis	$1/\frac{1}{3} = 3$

Step 3. Clear any fractions arrived at in step 2 by multiplying by a common multiplier.

no fractions

Step 4. Record the numbers (indices) in parentheses with no commas separating the numbers.

(123)

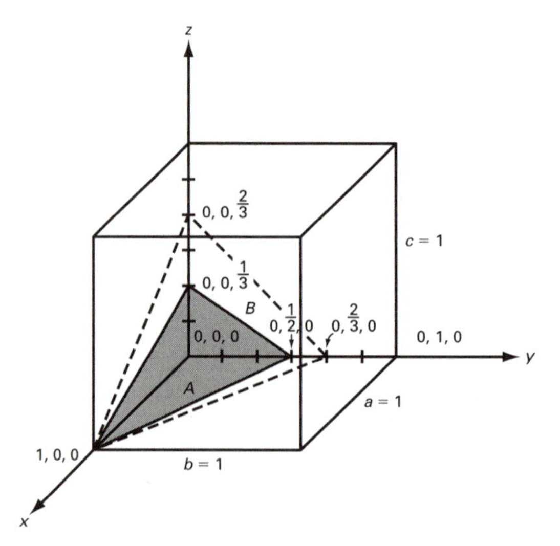

Figure 3-37 Miller indices for atomic planes for a cubic unit cell. Plane *A* is shaded; plane *B* is drawn with dashed lines.

Illustrative Problem

Determine the Miller indices for plane *B*.

Step 1. Intercepts are:

x axis	1
y axis	$\frac{2}{3}$
z axis	$\frac{2}{3}$

Step 2. Taking reciprocals:

x axis	$1/1 = 1$
y axis	$1/\frac{2}{3} = \frac{3}{2}$
z axis	$1/\frac{2}{3} = \frac{3}{2}$

Step 3. Clearing fractions:

common multiplier = 2

$$1 \times 2 = 2 \qquad \frac{3}{2} \times 2 = 3 \qquad \frac{3}{2} \times 2 = 3$$

Step 4. Record numbers (indices):

(233)

Illustrative Problem

Determine the Miller indices for the face plane in Figure 3-23 that parallels the x–z plane at a distance, b = 1.

Step 1. Intercepts:

x axis	∞	(plane intercepts x axis at infinity)
y axis	1	
z axis	∞	(plane intercepts z axis at infinity)

Step 2. Reciprocals:

x axis	$1/\infty = 0$
y axis	$1/1 = 1$
z axis	$1/\infty = 0$

Step 3. No fractions.

Step 4. (010) indices.

Figure 3-33 shows the (110) plane in a bcc unit cell, and Figure 3-35 shows a sketch of the (111) in an fcc unit cell.

3.4.3.2 Crystal directions. Many materials are anisotropic; consequently, their properties are not the same in all directions. To specify directions in a crystal structure, Miller indices are used. As with Miller indices for specifying points and planes in a crystal-lattice system, there is a procedure for determining the Miller indices for directions. The first step is to determine the coordinates of two points that lie in the particular line of direction. The first point, sometimes called the "head point," is farthest from the origin. Using the origin as the second point simplifies the procedure. The second step is to subtract the second point from the first point. The third step calls for clearing of any fractions to obtain indices with the lowest integer values. The fourth step is the writing of the indices in square brackets without commas. Negative integer values are indicated by the use of a bar placed over the integer. The procedure is illustrated in the following problem. (Refer to Figure 3-38.)

Illustrative Problem

In Figure 3-38, find the direction for a line passing through point A with coordinates ½,1,0 and the origin of the unit cell.

Step 1. Axis system, defined in Figure 3-38

Step 2. Head point coordinates ½,1,0

Tail point coordinates − 0,0,0

½,1,0

Step 3. Clear fractions (multiply by 2) and reduce to lowest integers: 1 2 0

Step 4. Enclose in brackets without commas: [120]

In more complex crystal systems, the determination of directions requires temporary relocation of the origin to another point in the unit cell to simplify the procedure. If the

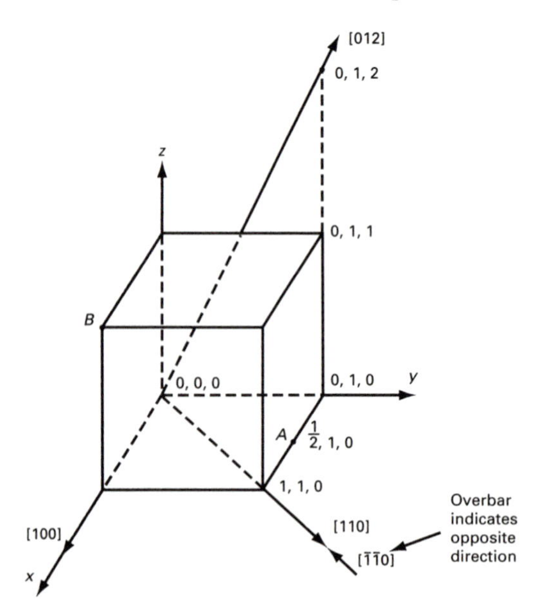

Figure 3-38 A cubic unit cell with some intercepts and crystal directions indicated. Note use of square brackets and the omission of commas to express Miller indices of direction.

properties of a crystal measured along two different directions are identical, the two directions are termed *equivalent.* Examples of equivalent directions are the face diagonals or the body diagonals in a cubic crystal. Equivalent directions are referred to as a *family of directions.* For example, the directions for the face diagonals (a family of directions) is specified by using spread or angle brackets: $\langle 1\ 1\ 0 \rangle$. This family consists of 12 directions, which can be arrived at by taking all the various ways (permutations) of expressing the three digits, 1, $\bar{1}$, and 0.

3.4.4 Coordination Number

To describe how many atoms are touching each other in a group of coordinated atoms, the term ***coordination number*** (CN) is used. The CN is the number of neighboring atoms that each atom has immediately surrounding it. Note in Figure 3-31 that each upper and lower basal plane of a cph unit cell contains an atom at its center. Each atom touches six atoms in its own plane, plus three atoms above and below in adjacent planes. Consequently, the CN for these atoms would be 12. The number of nearest atoms is dependent on two factors: (1) the type of bonding, and (2) the relative size of the atoms or ions involved. In our discussion of bonding, for example, we learned that valence electrons determine the type of bonding as well as the number of bonds an atom or ion can have. Carbon (C) in group IV has four covalent bonds and therefore a CN of 4. The group VII elements, such as chlorine (Cl), form only one bond (CN is 1). The relative size of the atoms determines how many neighboring atoms will touch another atom. Ionic bonding involves ions of different charges, hence, different sizes. The limiting factor in this case is the ratio of the size (radii) of the combined atoms of ions. The minimum ratios of atomic (ionic) radii produce various CNs. During ionization, atoms decrease in size when they change to cations and increase in size as they form anions.

Figure 3-39 represents the five ions occupying one of the six faces of the fcc unit cell for NaCl. The Na ion is just the right size to fit between the Cl ions at the corners of the unit cell. Thus, the ions are closely packed, with each cation separated from other cations by a layer of anions. Each cation and each anion are shared equally by six oppositely

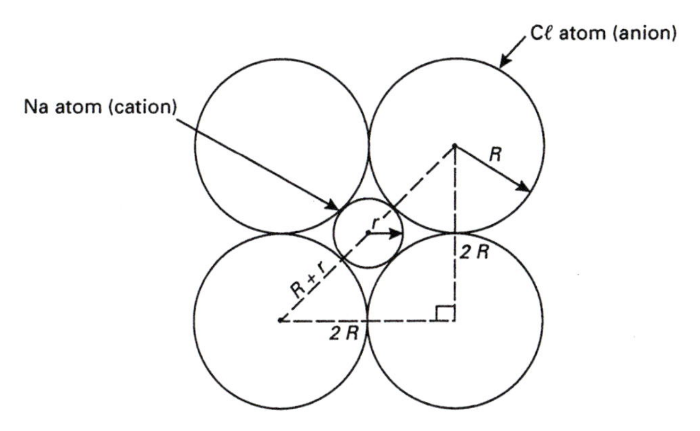

Figure 3-39 A face plane in NaCl crystal.

charged ions. Therefore, a CN of 6 describes this geometric arrangement. The center-to-center distance between the Na ion (radius = r) and its six neighboring Cl ions (radius = R) is determined to be the hypoteneuse of an isosceles right triangle (two legs or sides of equal length). This triangle is outlined in Figure 3-39 with its sides labeled 2R. Using Pythagorean theorem $x^2 + y^2 = z^2$ produces

$$(2R)^2 + (2R)^2 = (2r + 2R)^2$$

Taking the square root of both sides gives the expression

$$\sqrt{2}\,(2R) = (2r + 2R)$$

Solving for the radius (r) of the smaller ion in terms of the larger ions:

$$r = (\sqrt{2} - 1)\,R = 0.414R$$

These data show that the smallest ion that will just fill the spaces between six larger ions or that just touches six ions must have a radius that is at the minimum 0.414 times the size of the larger ions. This same result can be obtained by using a proportion between the isosceles triangle in Figure 3-39 and the standard 45–45–90 degree right triangle.

$$\frac{1}{2R} = \frac{\sqrt{2}}{(2r + 2R)}$$

Cross multiplying:

$$r + R = \sqrt{2}R$$

Solving for r:

$$r = \sqrt{2}R - R = R(0.414)$$

Table 3-4 reveals that, to have a CN of 6, the ratio of r/R must be at least 0.414.

In the case of sodium chloride (NaCl), r/R is 0.95/1.81 = 0.520. This being greater than 0.414, NaCl qualifies for a CN of 6. Note that, as the difference between r and R decreases, higher CNs are possible. A CN of 12 is the maximum, which occurs when the atoms (ions) have the same radius and the ratio becomes 1. In other words, as the r gets smaller than R (radius of surrounding atoms), the fewer neighboring atoms can make contact or touch the smaller atoms. Table 3-4 lists the minimum radii ratios for some common CNs.

TABLE 3-4 MINIMUM RADII RATIOS FOR CNs

CN	$\frac{r}{R}$
3	≥ 0.155
4	≥ 0.225
6	≥ 0.414
8(bcc)	≥ 0.732
12 (cph or fcc)	1.0

r, radius of smaller atom; R, radius of larger atom; $\geq$, greater than or equal to.

TABLE 3-5 CHARACTERISTICS OF SELECTED METALS

Element (symbol)	Crystal structure (20°C)	Atomic radius (10^{-10} m)	Melting point (K)
Beryllium (Be)	cph	1.14	1562
Aluminum (Al)	fcc	1.43	933
Titanium (Ti)	cph	1.46	1945
	bcc 877°C		
Iron (Fe)	bcc	1.24	1810
	fcc 910–1399°C	1.27	
	bcc 1399–1538°C		
Cesium (Cs)	bcc	2.62	300
Tungsten (W)	bcc	1.37	3680

With a CN of 12, each atom has contact with 12 other atoms. Each atom in an fcc or cph unit cell meets this description, provided that their radii are of similar size.

3.4.5 Allotropy/Polymorphism

It is of interest to list some common materials and show their crystal structure and melting point to illustrate the concept of the movement of atomic planes within a solid. Table 3-5 contains several elements that exist in more than one crystal structure depending primarily on the temperature. This phenomenon is known as ***allotropy*** or ***polymorphism*** (*poly,* "many"; *morph,* "shape"). Actually, an allotropic material, after changing to one structure, can reverse the phenomenon and return to its previous structure; a polymorphic material does not possess this reverse phenomenon. Over one-fourth of the elements are allotropic. Steel owes its existence to this property. Because of it, steel can be produced from iron. When iron is heated to above 910°C, its structure changes from bcc to fcc, allowing for a much greater absorption of carbon atoms (2% maximum). Figure 3-40 shows one carbon atom in an fcc unit cell of iron.

In general, we now know that any allotropic change in an element means different properties. All the properties of steel change almost instantaneously at 910°C. The best example of a change in properties with crystal structure is that of carbon (C). One polymorph of pure carbon is graphite, a black, greasy, low-strength material. A second polymorph is

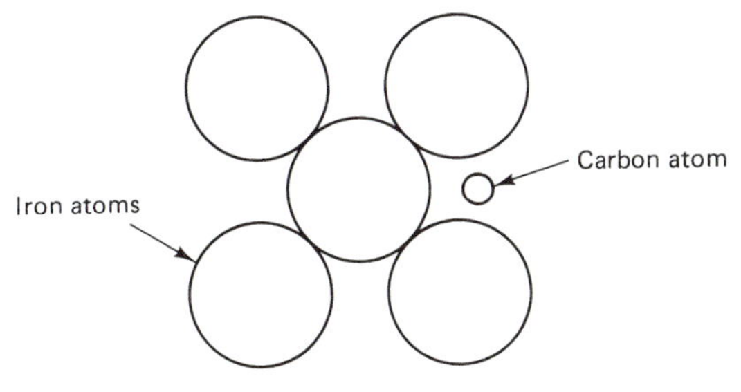

Figure 3-40 A face plane of an fcc unit cell for iron showing a carbon atom located in an interstice.

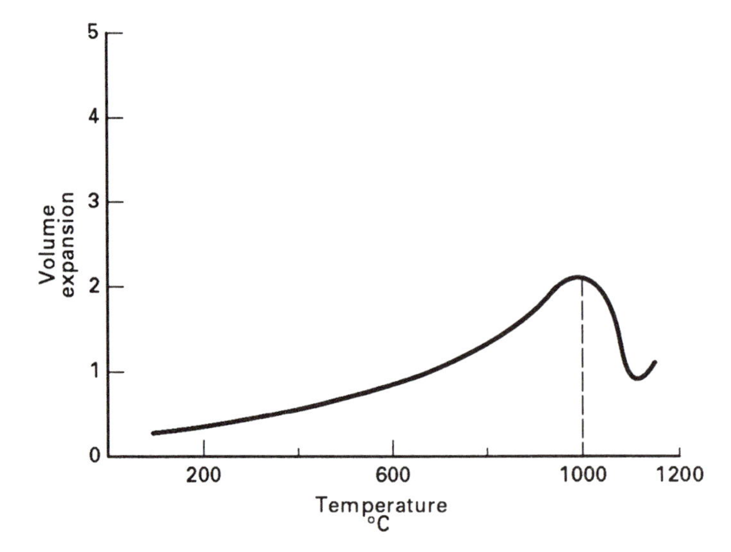

Figure 3-41 Volume expansion curve for ZrO_2 emphasizing the effects of the polymorphic transformation on the rates of expansion and the abrupt changes in volume.

diamond, the hardest naturally occurring substance. A third naturally occurring polymorph is the recently discovered fullerenes (see Figure 2-14). A final example is zirconia (ZrO_2), a ceramic used to toughen alumina (Al_2O_3) and described in detail in Module 7 (see Section 7.4). ZrO_2 is polymorphic with an fcc structure at high temperature. As it cools, it changes first to a simple tetragonal structure, followed by a further change to a monoclinic polymorph. Figure 3-41 is a sketch of the change in volume of ZrO_2 as it goes through these transformations on being cooled. These changes require the consumption of energy that would otherwise be used to help a crack propagate through the ceramic alumina. The end result is alumina toughened by this addition of zirconia.

3.4.6 Volume Changes and Packing Factor

In discussing crystal structure changes, we mentioned that every change in atomic structure brings changes in properties of the solid. One of these changes is volume. When steel transforms from a bcc to fcc structure, it decreases in volume. The explanation for this phenomenon involves the density of atoms in the various unit cells. ***Density*** is the ratio of the mass to the volume of a substance, which stays constant provided it is nonallotropic. With mass measured in kilograms (kg) and volume in cubic meters (m^3), density has the units of kilograms per meter cubed (kg/m^3). Other typical units are g/cm^3.

The ***atomic packing factor*** (APF), or *packing factor* (PF), is the ratio of the volume of atoms present in a crystal (unit cell) to the volume of the unit cell. In calculating the volume of an atom, we assume the atom is spherical. The difference between the PF and unity (1) is known as the *void fraction,* that is, the fraction of void (unoccupied or empty) space in the unit cell. Using the *simple cubic crystal unit cell* sketched in Figure 3-42, with atoms of equal radius *(R)* located at each corner, the volume of the cell occupied by the eight atoms is equivalent to one atom. Each corner atom contributes one-eighth of its volume. Therefore, the volume of one atom is $\frac{4}{3}\pi R^3$. The (100) face plane of this simple cubic unit cell shows four corner atoms touching each other (see Figure 3-43). The relationship of the

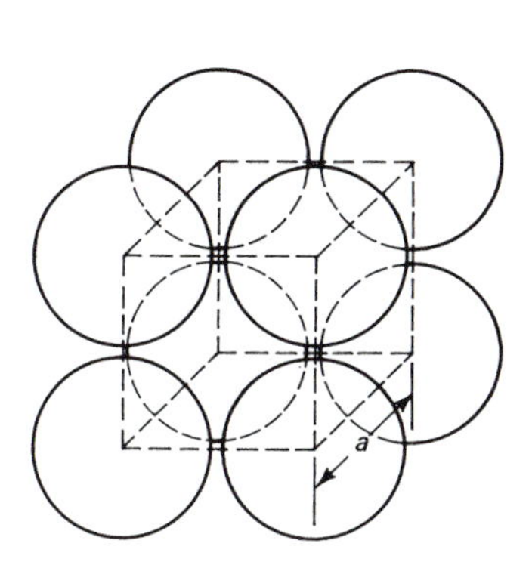

Figure 3-42 A simple cubic unit cell.

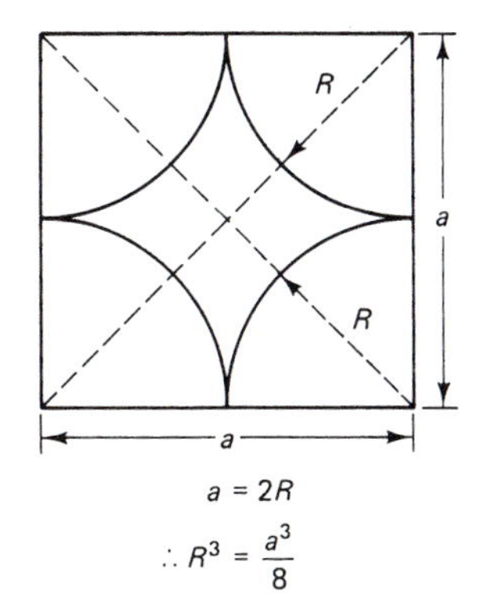

Figure 3-43 (100) face plane of simple cubic unit cell.

radius R of an atom to the lattice parameter or edge length a of the unit cell is $2R$. The volume of the unit cell is a^3. The volume of atoms is $(\pi/6)a^3$. Thus,

$$\text{PF} = \frac{\text{volume of atoms}}{\text{volume of unit cell}} = \frac{(\pi/6)a^3}{a^3} = \frac{\pi}{6} \cong 0.52$$

Solving for the PF in terms of the radius R produces the same results. The void factor is therefore $1 - 0.52 = 0.48$. What the calculation tells us is that only about half (52%) of the space in the simple cubic unit cell is occupied by the atoms. This is too inefficient, so atoms of metals do not crystallize in this structure. Remember, the closer the atoms come to each other, the less energy they have and the more stable is their structure.

The *bcc unit cell,* as you recall, is quite similar to the simple cubic unit cell with the addition of one atom in the very center of the unit cell. Therefore, the bcc unit cell contains the equivalent of two atoms.

For the *fcc unit cell,* there are four net atoms. Each of the eight corner atoms contributes one-eighth of an atom. Each of the six face atoms contributes one-half. The total is $1 + 3$, or 4, atoms. Notice that the atomic radius *(R)*, or a as used in the formula, cancels out in all these calculations, which tells us that the PF is not dependent on the radius of the spheres being packed if all the atoms are of the same size.

The fcc structure has the maximum PF for a pure metal. The cph structure also has a PF of 0.74. Finally, we should note that the coordination number varies directly with the PF. As an example, the $CN_{bcc} = 8$ and $PF_{bcc} = 0.68$; the $CN_{fcc} = 12$ and $PF_{fcc} = 0.74$.

In our study of metals, we will see that pure iron will change its structure on heating from bcc to fcc at 910°C. Knowing the PF for both these structures will lead us to the conclusion that iron will contract in volume as it is heated above 910°C. This change in structure forms the basis for the production of steel as well as the heat treatment of steel.

3.4.7 Crystal Imperfections

So far in our study of crystal structures we have dealt with atom arrangements that are perfect in every way. These structures have ***long-range order*** because the orderly arrangement of atoms extends throughout the entire material, forming a regular gridlike lattice or pattern. To grow a perfect crystal requires laboratory conditions, and even with such a controlled environment we have achieved limited success. Of course, this lack of perfection in the microstructure of materials used by people is far from being all bad. If it were not for imperfections of many kinds in solid materials, these solids would not possess the properties that we desire them to have. An example would be the heat-treating process used with high-carbon steel to change the properties of the steel to suit certain conditions demanded by our present technological age. Without imperfections in the crystalline arrangement of atoms, these processes would be severely limited in, if not incapable of, changing the structure and hence the properties of steel. The whole semiconductor industry owes much of its existence to the imperfections in bonding arrangements of the atoms' outer shell electrons. Therefore, it is now essential to delve into the several imperfections of a solid's atomic structure so that we can learn how to take advantage of such disorder in the atomic structure.

It is important to point out here that we are still talking about crystalline materials and not amorphous (noncrystalline) materials. We know that ***amorphous*** materials have no regular atomic structure. Known as ***short-range-order*** materials, their order is limited to an atom's nearest neighboring atoms. For example, water has short-range order due to its covalent bonding between the hydrogen and oxygen atoms. Even with its secondary bonding (hydrogen bridging), the water molecules are randomly joined together. Glass and polymers also have short-range order. This is not to say that amorphous materials are of little use. Many materials are being produced by processes such as powder metallurgy and rapid solidification that are amorphous in structure and that possess remarkable properties needed for some specific application.

Crystal imperfections fall into two categories: those involving impurity atoms and those in which there is some disorder in the atomic structure brought about by something other than impurity atoms.

3.4.8 Crystal Impurities

The word ***impurity*** comes from the use of the word when referring to the small percentage of copper in sterling silver that distinguishes sterling from pure silver. One talks about the impurities in copper that reduce its conductivity. In many instances, these impurities are purposely added to improve a material's properties and/or reduce its cost. Such a solid, called an *alloy,* is a material composed of two or more elements, at least one of which is an element that possesses metallic properties. Wrought alloys are those that can be mechanically deformed. Cast alloys are those alloys whose brittleness is such that they cannot be cold worked.

Brass is an alloy consisting of copper to which has been added some zinc. The addition of zinc has a great effect on the hardness, strength, ductility, and conductivity of the pure copper. Our objective in the following discussion is to explain in simple terms why adding these impurities (the zinc atoms) to the copper atom produces such differences in the properties between the pure metal and the alloy.

Before proceeding, definitions of the words *solution, solvent, solute, mixture, alloy,* and *diffusion* are required. A ***solution*** is a homogeneous mixture of chemically distinct substances that forms a phase. A ***phase*** is defined briefly as a physically distinct material that has its own structure, composition, or both. Uses of phases in solids are discussed in this module. The atoms or molecules of one substance are uniformly distributed throughout the other on a random basis. The substance present in the greatest proportion is the ***solvent.*** The other substance or substances present are the ***solutes.***

A ***mixture,*** on the other hand, is a material that has no fixed composition and contains more than one phase. The components (substances) can be identified and separated by physical means. Thus, a mixture of sugar and salt crystals is not homogeneous nor is it random. One can readily see the two distinct crystalline phases and, with some patience, segregate one crystal phase from the other. However, if you dissolve sugar and salt in water in a dilute concentration, they form a liquid solution. In this situation, the salt and sugar lose their individual identities by dissolving in water.

Air is an example of a solution of many gases dissolved in another gas. A similar situation occurs in solids, producing *solid solutions.* A ***solid solution*** is simply a solution in the solid state that consists of two kinds of atoms combined in one type of space lattice. In other words, when two elements are soluble in each other in the solid state, they produce a solid solution. ***Alloys,*** then, are a combination of a metal and one or more other elements forming either a mixture or a solid solution. Steel is a mixture of iron with a bcc structure and cementite. Brass is a solid solution because its single-phase structure is all fcc; it is a solid solution of copper (solvent) to which some zinc (solute) has been purposely added. The zinc atoms are diffused into the atomic structure of the copper on a random, uniform basis. ***Plastic alloys*** are blends of polymers or copolymers with other polymers or elastomers; they are also called *polymer blends.* Styrene-acrylonitrile is a plastic alloy.

Diffusion comes from the Latin verb meaning "to pour out" or "to spread out." In material science, it means the intermingling in solid materials of atoms (in metals), ions (in ceramics), or molecules (in polymers). This active movement of particles is fairly well understood through our experiences with gases and liquids. A bottle of perfume uncapped in one part of a room will soon disclose its presence to people in a distant corner of the room by the diffusing of perfume atoms through the air in the room. Salt or sugar dissolved in water will diffuse throughout the water and form a solution in which the salt or sugar atoms will be evenly distributed throughout the water.

Our main preoccupation in our present study of materials is with solids. Diffusion takes place in solids, too. We know that the individual atoms and molecules in a gaseous or liquid state are relatively far apart, offering little opposition to other atoms migrating through them. In solids the atoms are held tightly and close together. If a metallic crystalline solid were formed with a perfect crystalline structure, atoms would find it impossible to move about. But in our study of imperfect crystal structures, we will learn that point defects are the rule and not the exception. These defects are one main reason why atoms of a solid can actively move about within the atomic structure of the solid. Combine the presence of vacancies (the absence of an atom or atoms in a lattice site) with the fact that each atom possesses sufficient energy to cause it to vibrate about its position in the lattice structure and it is fairly easy to visualize why certain atoms possessing higher average energy in the crystal structure can break their bonds and "jump" from one lattice site to one that is not occupied (a vacancy) in the lattice structure. Once the atom moves to a new site, it leaves behind another vacancy. In other words, the atom exchanges positions with a vacancy.

The phenomenon of diffusion in solid materials is especially important in understanding the manufacturing and functioning of semiconductor materials; the carburizing of steel in surface or case hardening; the production of metal alloys including steel, the primary alloy of iron; and the heat treating of aluminum alloys, called precipitation hardening. ***Precipitation*** is a change in the solid state (change in phase) brought about by diffusion. How this precipitation is controlled to produce strong aluminum alloys is discussed in Module 5. These metallurgical processes are discussed elsewhere in this book.

In polymeric materials, diffusion is aided by defects in the molecular structure similar to those in metals. The diffusion of a penetrant into a solid polymer is of great importance. If the penetrant is a gas, the gas may permeate through the solid (even glass). Polymer films are designed to prevent the diffusion of gases and water vapor into foods that have been wrapped in the film. Liquid solvents may permeate a polymer and produce a softening of the polymer, color changes, and odors.

The ***doping*** (alloying with a very small amount of alloy addition) of silicon with phosphorus in the alloying of silicon to increase its electrical conductivity is another diffusion process that depends on point defects in the silicone structure. Many sintering processes, such as powder metallurgy and the large number of welding and brazing processes for the joining of two metals by local ***coalescence*** (to grow, unite into a single body), depend on the transport of atoms by diffusion.

Because the internal energy possessed by atoms is related to temperature, an increase in temperature will increase the rate of diffusion. Of course, if there are no vacancies in the structure, very little, if any, diffusion will occur.

Another factor that affects the diffusion of atoms, molecules, or ions is the type of bonding of the matrix atoms (i.e., strong bonding takes more energy to break the bonds). An example is high-melting-point solids. Smaller permeating atoms stand a better chance of diffusion through a structure of larger atoms, as in the case of carbon atoms diffusing through a structure of iron atoms. Our study of the microstructure of solids tells us that a lattice structure that contains loosely packed atoms (less dense) will offer less resistance to diffusing atoms than one whose structure contains tightly packed atoms. Finally, diffusion depends on time. This translates usually into allowing sufficient time for diffusion to take place at some higher temperature, as in an oven or furnace.

In summary, diffusion is the process depended on by material scientists to change the microstructure of solids and thus vary the properties of the many solid-state materials in use by our society. Having completed a brief background discussion of solvents, solutes, solid solutions, alloys, and the very important phenomenon of diffusion, we will return to the subject of crystal imperfections. The first of two categories of such imperfections in a crystal structure involves, as mentioned, impurity atoms, which produce two types of solid solutions, substitutional and interstitial solid solutions.

3.4.9 Substitutional Solid Solutions

In a substitutional solid solution (Figure 3-44), the solute atoms replace some of the solvent atoms in a crystal structure of the solvent. Using brass as our solid solution, up to about 40% of the copper atoms can be replaced by zinc atoms. This is possible because the atoms of copper and zinc are much alike. Their atoms are about the same size (the atomic

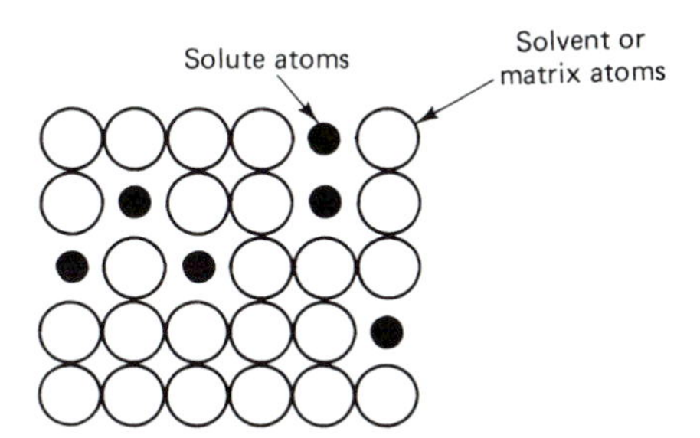

Figure 3-44 Substitutional solid solution.

radius of the copper atom is 1.278×10^{-10} m and of zinc, it is 1.39×10^{-10} m). Their electron structures are comparable:

$$^{29}\text{Cu}: 1s^2, 2s^2, 2p^6, 3s^2, 3p^6, 4s^1, 3d^{10}$$

$$^{30}\text{Zn}: 1s^2, 2s^2, 2p^6, 3s^2, 3p^6, 4s^2, 3d^{10}$$

Their crystalline structures are both fcc with a CN of 12. Another good example of a solid solution is Monel. Monel is a solution of copper in nickel (about 70% Ni and 30% Cu). The range of solubility goes from practically no nickel to almost 100% nickel. Again, these two elements have a common crystalline structure (fcc), and the atomic radius of nickel is 0.1246 nm. Therefore, Monel is a substitutional solid solution. Another interesting fact about this type of solid solution is that atoms may fill only one type of site in the lattice structure of the solvent atoms. For example, in the alloy of copper and gold, the majority of copper atoms occupy the face-centered sites and the gold atoms, the corner sites of the face-centered cubic unit cell. An ***ordered substitutional solid solution*** is formed. As a rule, two distinct elements may form a substitutional solid solution if the sizes of their atoms do not differ by more than 15%. Of course, there are further restrictions on the degree of solubility brought about by any differences in their crystal or electron structure. The solute or impurity atoms, although of similar size, may be larger or smaller than the solvent atom, which will produce only a slight distortion in the lattice structure.

3.4.10 Interstitial Solid Solutions

The second type of solid solution formed by impurity atoms is the ***interstitial*** (Figure 3-45). If the impurity atoms take up sites in the lattice structure that are normally unfilled or unoccupied by the pure (solvent) atoms, they form an interstitial solid solution. These

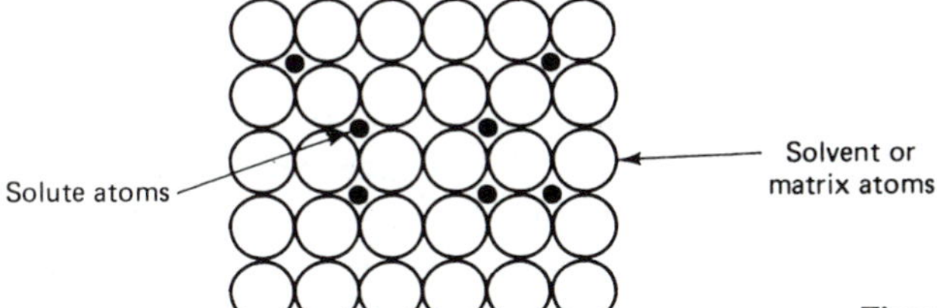

Figure 3-45 Interstitial solid solution.

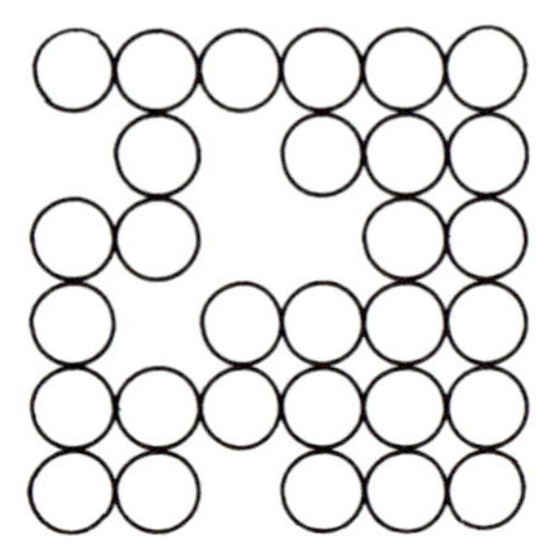

Figure 3-46 Point defects (vacancies).

normally unfilled voids or vacant spaces are called ***interstices.*** In the fcc unit cell, we know there is a relatively large interstice in the center and smaller interstices near each corner atom. It is worthwhile to point out that steel making is made possible because of the formation of an interstitial solid solution. First, we know that iron is allotropic. At temperatures below 910°C, iron is in the bcc form. Above that temperature, the bcc structure changes to fcc to accommodate a higher energy level of the atoms. In the fcc structures, carbon atoms can form in the interstices of the iron unit cell. At temperatures below 910°C, the bcc structure contains no room for the carbon atoms to fit between the iron atoms. This fact forms the basis for many of the heat-treating procedures used to produce a multitude of steels with the many different properties required by our technological society. We will discuss heat treating in detail later in the book.

3.4.11 Crystal Defects

The second category of crystal imperfections or lattice defects, a disorder of the crystal structure, is brought about by some mechanism such as thermal agitation of the crystal during its formation, the effects of gravity, or the result of high-energy radiation. Such deviations from the perfect crystal are classified for purposes of explanation as *point defects, line defects,* and *area defects.* Actually, defects occur in all combinations.

Point defects (1 or 2 atom positions), the simplest and best understood, affect only the small volume of the crystal surrounding a single lattice site. One such point defect is a ***vacancy,*** that is, the absence of an atom at a lattice site in the otherwise regular crystal (see Figure 3-46). As a result, the electronic bonding of the adjacent atoms is disrupted, which changes the effective radii of these atoms. This weakens the crystal. If sufficient vacancies were produced by heating of a crystalline solid, the crystal structure would lose its long-range symmetry and order, resulting ultimately in porosity or a change to a fluid. It is important to note that these local imperfections in the crystal structure produce a disequilibrium that has a great effect on the important properties of crystalline solids such as density, mechanical strength, diffusion, and electrical conductivity. By themselves, point defects do not affect strength as much as they affect diffusion—the migration of atoms.

Another point defect is called an ***interstitial defect*** or ***interstitialcy.*** This is produced by the presence of an extra atom in a void, the space between normal lattice positions. This interstitial atom may be added specifically as an alloying element or it might be an impurity atom indigenous to the solid. Other impurity atoms may deposit themselves in a lattice position reserved for atoms of the solid. Regardless of how they are formed, these point defects produce local aberrations in the atomic arrangement in the crystal, which produce varying degrees of local disorder in the bond structure and energy distributions in the solid. It is worthwhile to mention that disorder can occur below the atomic level (i.e., at the subatomic level). The imperfections in the electronic structure of atoms exist also. This fact has been capitalized on by the semiconductor industry, which produces materials with varying electrical properties. Further treatment of point defects in the electronic structure of atoms will be included under the topic of electronic bond theory of nonmetallic crystalline solids. It should be evident now that imperfections in the atomic, ionic, and electronic structure of solids do not all have a negative effect on a solid's properties. Some imperfections can improve certain properties; others may degrade some properties. All imperfections do not affect the strength of solids.

So far we have discussed point defects in terms of metallic crystals, with the exception of the brief mention of nonmetallic crystals with the "imperfections" in their electronic structures. As we know, all metals crystallize when they solidify, provided that they are given enough time to form their inherent structures. So do most ceramics and some polymeric (plastic) materials. In the case of ceramic materials, the crystal structures are more complex than those of metals, and the bonding is ionic, covalent, or a combination of metallic and covalent. With ionic bonding the atoms, of course, become ions. In the ionic solid sodium chloride (NaCl), each Na^+ cation finds itself surrounded by six equally spaced Cl^- anions, and each Cl^- anion is surrounded by six Na^+ cations (CN = 6). Figure 3-47 is a sketch of a model of the unit cell for NaCl, upon which the crystalline structure is built. Each unit cell has four sodium ions and four chlorine ions associated with it. In this model the very center of the cube is occupied by a sodium ion. Each of the other 12

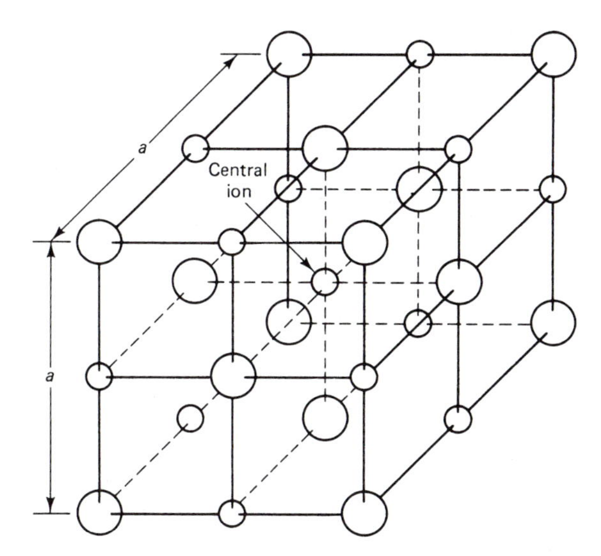

Figure 3-47 A unit cell of NaCl.

sodium ions is shared by three adjacent unit cells so that each contributes one-fourth of an ion to the unit cell sketched. The total number of sodium ions, therefore, is $1 + \frac{1}{4}(12) = 4$. The 14 chlorine ions distribute themselves in a similar fashion and contribute a total of 4 chlorine ions to the unit cell under discussion. Each set of ions forms an fcc structure with equal numbers of sodium and chlorine ions. The radii ratio of sodium to chlorine is 0.54, with the Na ion being almost half the size of the Cl ion. This difference in size of the ions places restrictions on other ions that could replace them. Furthermore, the replacement ions must have the same number of exterior (valence) electrons. With these restrictions in mind, point defects in ionic or ceramic crystals are the rule rather than the exception. A vacancy may consist of pairs of ions of opposite charge. An interstitialcy would consist of a displaced ion located at an interstitial site in the lattice structure.

The second type of imperfection, the (one-dimensional) *line defect,* is also known as a dislocation (see Figure 3-48). A ***dislocation*** is a linear array of atoms along which there is some imperfection in the bonding of the atoms. An undeformed crystal lattice is represented by Figure 3-49. Figure 3-50(a) and (b) shows what appears as an extra or incomplete plane of atoms (*A–B*) that causes distortion of the crystal structure. The two-dimensional representation also has atoms behind each atom, shown as open dots. The atoms in the area circled by a dashed line represent a center or core of poorly bonded atoms that extend back into the material along a line normal to the paper. This line defect is known as an ***edge dislocation,*** whose symbol is ($\perp$). A force *P,* acting as shown, would cause the rows of atoms to move in the direction of the force, each row moving one at a time much like dominoes striking each other. As each row of atoms moves, the next follows in turn until all planes of atoms in the area have been displaced sufficiently to make all planes continuous, as shown in Figure 3-50(d).

This line of local disturbance represented by the five atoms circled in Figure 3-50(a), which may extend to the boundaries of the crystal, is a region of higher energy. As stated previously, this region contains the line defect or dislocation. The dislocation plane contains this line. Above this line the atoms are under a compressive stress, while those atoms below the line are experiencing a tensile stress. These bonding forces are not as strong as in a perfect crystal lattice, which permits a relatively small shear force to break the bonds, allowing the dislocation to move. The bonds re-form after the dislocation passes.

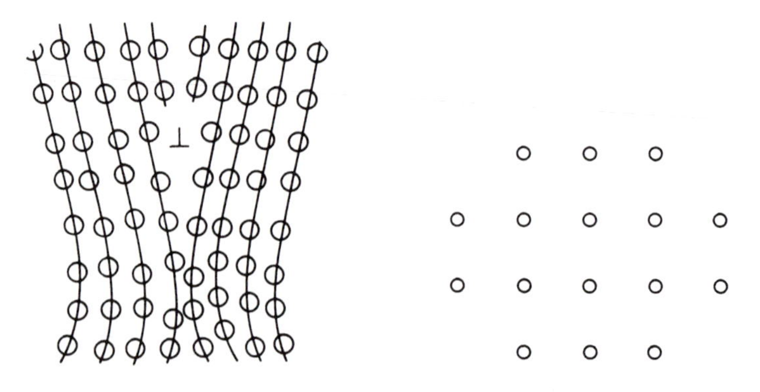

Figure 3-48 Line defects (dislocation). **Figure 3-49** Undeformed crystal.

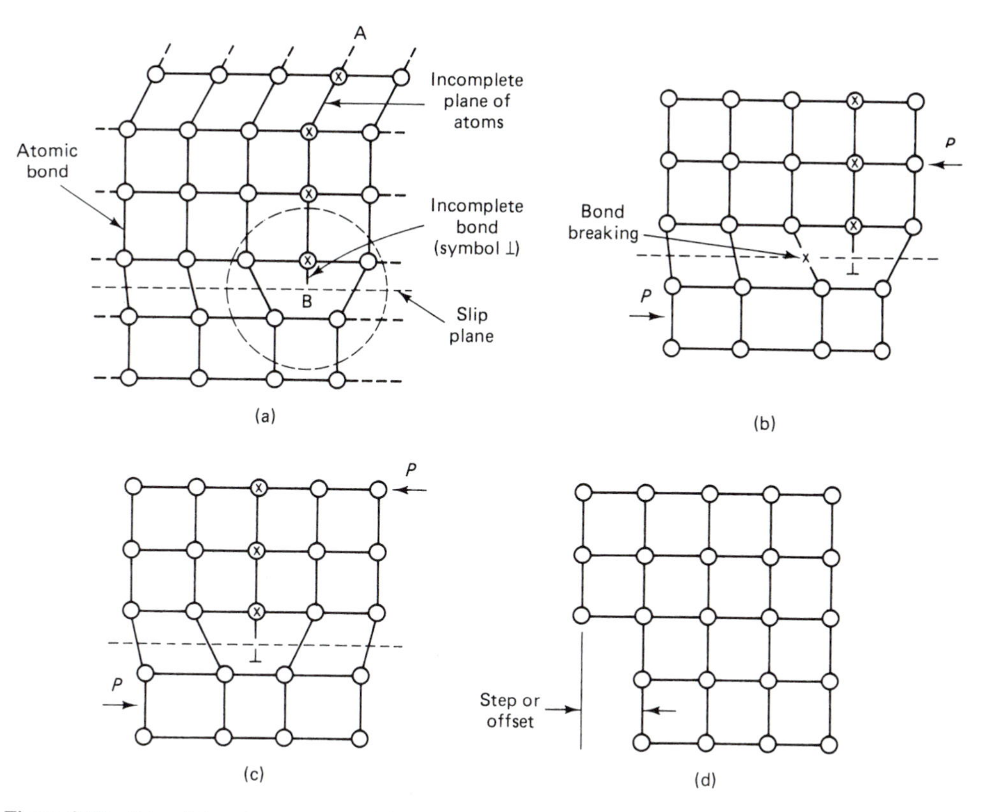

Figure 3-50 Edge dislocation movement (slip). (a) Dislocation—incomplete row of atoms above slip plane. (b) Shear force *(P)*, causing dislocation to move. (c) Dislocation moved one row to the left. (d) Dislocation reaches surface of crystal, producing plastic strain (deformation).

The successive passage or slipping of planes of atoms has been likened to the sliding of a large, heavy rug. To move the rug requires a large force. However, if you make a wrinkle in the rug and push the wrinkle a little at a time, the rug can be moved. The small movement of the rug to make the wrinkle can be thought of as the slipped portion of the rug; the other portion of the rug, the unslipped region. The wrinkle is the dislocation that separates these two regions.

The displacement of atoms ***(slip)*** is in a direction perpendicular to the dislocation line and/or plane. Where the direction of slip is parallel to the dislocation line, the line defect is called a ***screw dislocation,*** denoted by the symbol ↻. This is depicted in Figure 3-51. As with point defects, many line defects are actually combinations of edge and screw dislocations, producing curved dislocation lines or loops that start and end within the crystal. Dislocations originate during crystallization or plastic (inelastic) deformation. More will be said about dislocations when the topic of plastic deformation is treated later in this book.

However, before leaving dislocations, the following observations are made. The ideal crystal structure would contain no deformations. Experience with human-made,

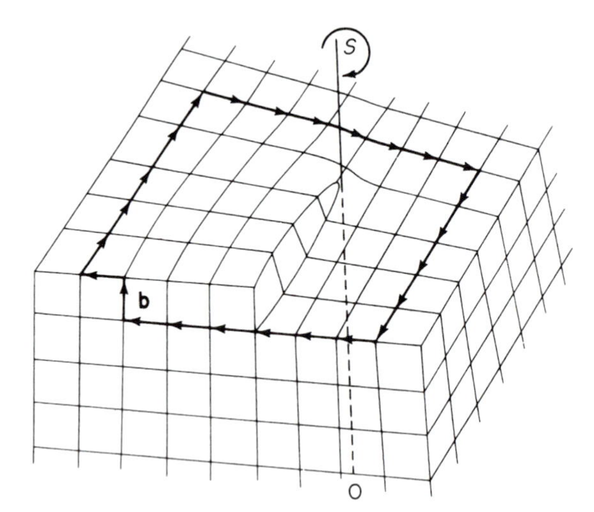

Figure 3-51 A screw dislocation. (Arthur L. Ruoff, *Materials Science* © 1973. Reprinted by permission of Prentice-Hall, Inc., Englewood Cliffs, N.J.)

near-perfect crystals or ***whiskers*** has indicated that such whiskers contain great strength. The reason why a relatively little force is able to deform a crystal structure with strongly bonded atoms or ions is that only a few atomic bonds need to be broken and re-formed when dislocations are present in the crystal structure. In view of our present inability to produce near-perfect crystalline solids, the problem resolves into one of determining how to control this movement of dislocations by hindering the movement (strengthening the solid) or facilitating it (temporarily weakening the solid for some purpose).

In summary, line defects have a great deal to do with the strength of a solid. An abundance of them will cause a mutual interference in their movement through a crystal, preventing the planes of atoms from slipping, thereby strengthening the material. The presence of a few dislocations increases the ductility of a crystalline solid.

Area defects (interfacial or two-dimensional) are the third type of imperfection and exist in the form of grain boundaries (Figure 3-52). As each crystal grows it establishes its own axis system, on which the atoms/ions orient themselves. Adjacent crystals with their differently oriented lattice structures close in on each other. The last atoms to take up position in a crystal find it more difficult to occupy normal lattice sites. Consequently, a transition zone is formed that is not aligned with any of the adjoining crystals. The atoms making up the grain boundary possess greater disorder, and hence greater energy, than their counterparts within the crystals themselves. Furthermore, the atoms are less efficiently packed together. These factors signify that the atoms in the grain boundaries are ready to act as sources of new crystal formation (nucleation sites) once the right conditions are met. Second, they assist in the diffusion of atoms through the solid. Third, they offer resistance to the movement of dislocations and therefore modify the strength and the ability of materials to plastically deform. Fourth, they act as sinks for vacancies. A solid with a large number of individual crystals (also called ***grains***) has more grain boundaries than a solid with a lesser number of grains. Fine-grained material will, at normal temperatures, be stronger than coarse-grained material. Figure 3-52 is a two-dimensional sketch showing grain boundaries with different orientations of grains and grain growth.

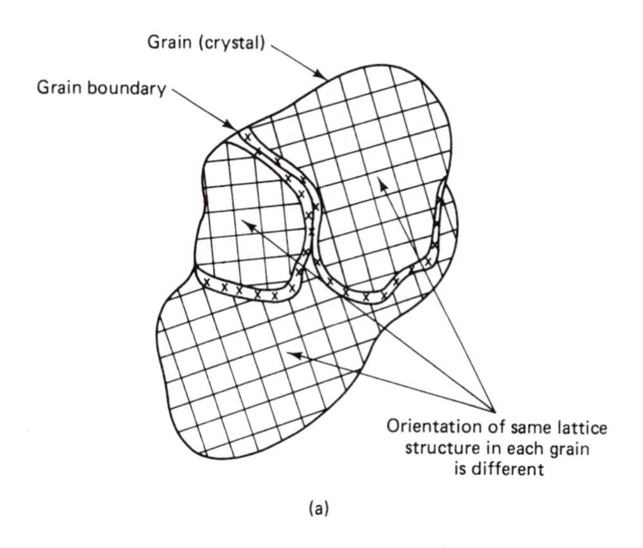

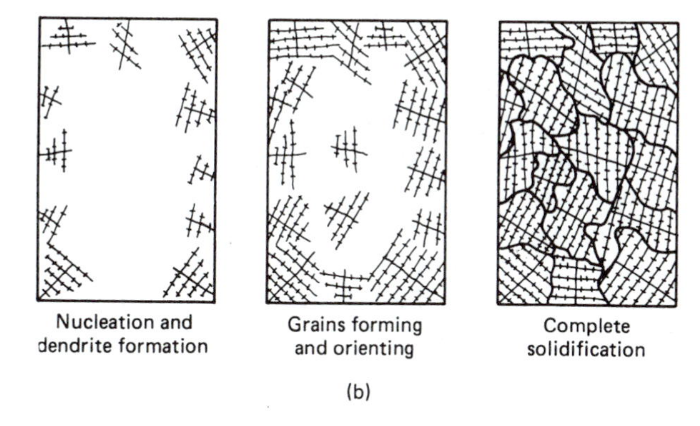

Figure 3-52 (a) Grain boundaries. (b) Grain growth.

It must be remembered that a solid contains crystals, each having the same lattice structure. What is different between the individual grains is the orientation of that structure within each grain. A technique for revealing the details of the grain structure of solids uses reflected light through a metallographic microscope (Figure 3-53).

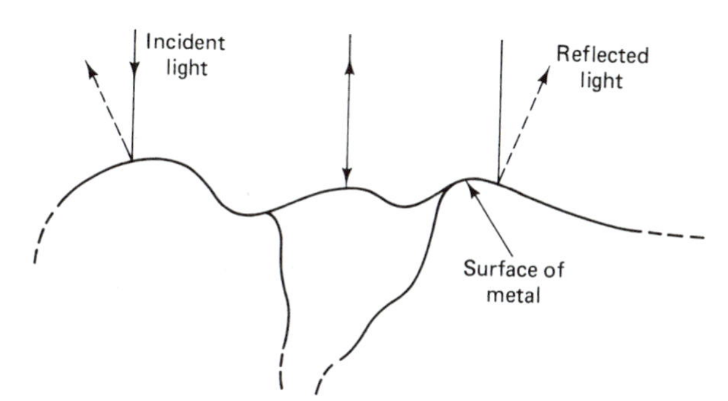

Figure 3-53 Identification of individual grains by reflected light.

3.5 PHASE DIAGRAMS

Phase or equilibrium diagrams serve as maps for finding one's way through the many solid-state reactions that occur in materials. Equilibrium conditions may be defined as slow heating and/or cooling of the many materials to permit any phase change to occur. ***Equilibrium,*** from a material's standpoint, has been defined as a condition of minimum energy. This condition might be compared to a metal spring that is neither stretched nor compressed. Once the spring is stretched or compressed, a force is set up inside the spring that tends to return the spring to its free-length position. The spring stores up the work done to stretch or compress it in the form of potential energy. Anywhere there is disorder in the crystalline structure of a solid (grain boundaries, imperfections, etc.) is a place of high energy similar to the example of the spring under action of some external force. The magnitude of the energy difference is proportional to the magnitude of the disorder.

To describe completely the conditions of equilibrium for a particular system, three externally controlled variables of temperature, pressure, and composition must be specified. Normally, phase diagrams record the data when the pressure is held constant under normal atmospheric conditions. Consequently, phase diagrams especially are graphical representations of a material system under varying conditions of temperature and composition. To determine the data necessary to plot these many phase diagrams, specialized equipment and techniques are used. Cooling curves for a pure metal, a pure iron, a metal alloy, and a nonmetal are sketched in the following pages to illustrate the phase changes that occur, if any. Note that for pure iron, three of the four phase changes occur while the iron is in the solid state. The α-iron stage existing at the higher temperatures is nonmagnetic.

A ***cooling curve*** is a graphical plot of the structure of a material (usually a pure metal or metal alloy) over the entire temperature range through which it cools under equilibrium conditions. Figure 3-54(b) and (c) shows the structures that the particular metal or alloy is in at any particular temperature and time. Three structures are indicated for a pure metal or alloy: (1) liquid phase (melt), (2) a combination structure consisting of both a liquid and solid phase, and (3) a final solid phase near or at room temperature.

For the temperature to remain constant, there must be a balance between the heat withdrawn from the metal and the heat supplied. In this instance, the metal is giving up heat (latent heat of fusion), which balances the heat removed. The result is a net change of zero, as measured by a thermometer. As a metal cools from a higher to a lower temperature, its electrons require less thermal energy as a result of the relatively more ordered, more dense, and more bonded positions occupied by the atoms. As the electrons surrender their excess thermal energy (in the form of heat), they move closer to their nuclei and slow down their motions to match the energy level of their new positions (*equilibrium* position). Cooling curves for nonmetals such as glass or plastics (Figure 3-55) show no clearly defined melting points.

Samples of an alloy are taken at various temperatures and quickly cooled to capture the structure at that particular temperature. Metallurgical microscopes using visible light study the grain structure, including the size, shape, and distribution of grains, with a maximum magnification of about 2000 ×. X-rays, gamma rays, and electron beams

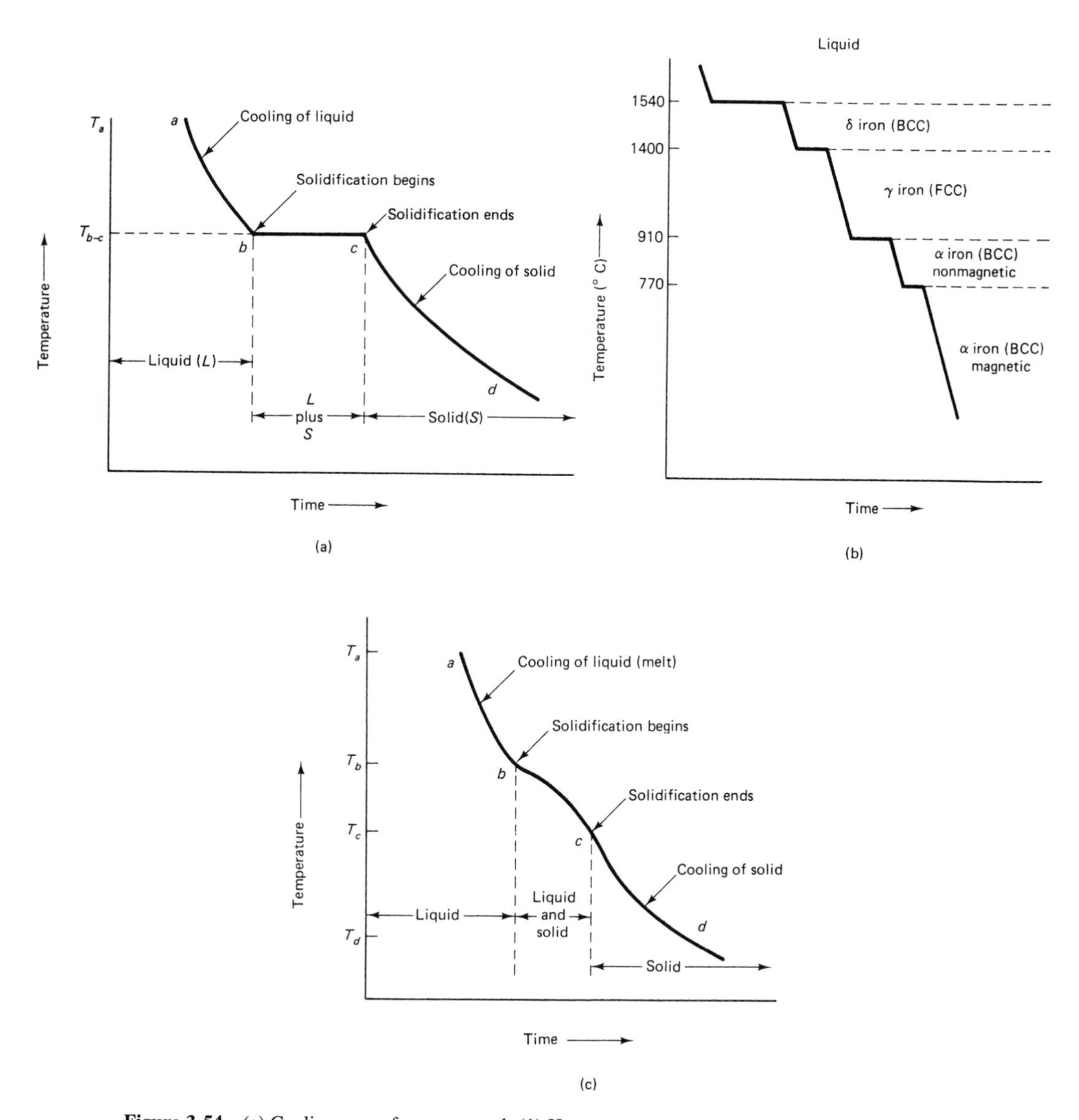

Figure 3-54 (a) Cooling curve for pure metal: (1) Heat pure metal to point T_a (liquid phase or melt); (2) cooling of liquid metal $a - b$; (3) at point b, pure metal starts to *precipitate* out of solution; (4) point c, *precipitation* (pure metal completely solid); curve from b to c straight horizontal line showing constant temperature $T_b - T_c$ because thermal energy absorbed in change from liquid to solid; (5) more cooling of solid pure metal from c to d and temperature begins to fall again; change in volume follows same pattern. (b) Cooling curve for pure iron. (c) Cooling curve for a metal alloy: (1) Two metals heated to point a (liquid phase or melt, with both metals soluble in each other); (2) cooling of alloy $(A + B)$ in liquid phase; (3) point b, solidification begins; (4) point c, solidification complete (both metals in solid form); sloped $b - c$ due to changing from liquid to solid over range of temperature T_b to T_c because metals A and B have different melting/cooling temperatures; (5) further cooling from c to d of solid-state metal alloy.

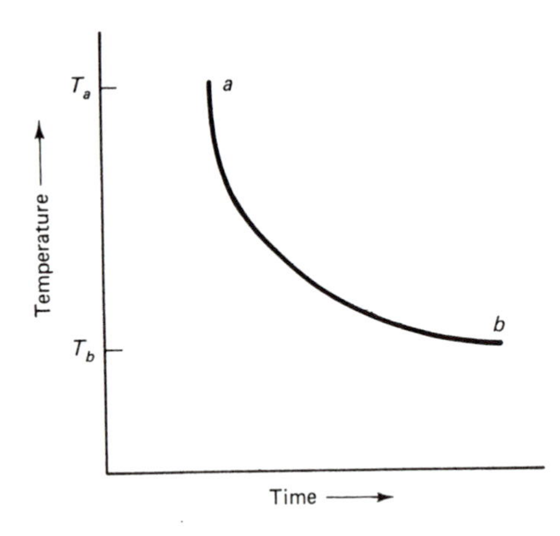

Figure 3-55 Cooling curve for a nonmetal. No phase transformation occurs. Nonmetallic heated to temperature T_a and allowed to cool to room temperature T_b. With no phase changes, continuous plot of temperature and time produces a smooth curve.

probe the crystal structure with great resolution and magnification. Electron microscopy permits the detailed chemical analysis of selected areas within the solid material, as well as the character of an individual dislocation. The data collected are used to construct various representations, particularly of the structure of the alloy under varying conditions of temperature, pressure, and composition. Most, if not all, of these graphical representations indicate the structure of a system under *equilibrium conditions.* Therefore, sufficient time must be allowed for atoms to diffuse and thus establish equilibrium conditions at any particular pressure, temperature, and composition. In practice, equilibrium conditions are normally departed from either because of the length of time involved (and attendant cost) or purposely to produce a solid phase that is in a ***nonequilibrium*** condition.

In our previous discussion we learned that atoms possess higher energies at elevated temperatures than at some lower temperature. This higher energy permits the atoms to diffuse more quickly. At lower temperatures, this kinetic energy is lessened. One can see that this reduced atomic mobility, coupled with the disequilibrium energy of a crystal structure at positions of high-energy content, presents a broad spectrum of possibilities for the diffusion (precipitation) of atoms producing a phase transformation in the solid state of a system.

The phase diagram for water makes a good departure point for an understanding of phase diagrams for systems comprising more than one component. The single component case of water is the compound H_2O. The system under study would then contain a definite amount of H_2O. The one-component diagram for a definite amount of water shown in Figure 3-56 is a graphical plot of pressure along the vertical axis versus temperature along the horizontal axis. These variables are expressed in SI units. This curve represents the physical changes that water (H_2O) undergoes. Note that, regardless of the physical changes (liquid to solid, i.e., water to ice), there is no alteration of the chemical composition of the water. Only its structure is changing. On the other hand, chemical change always alters a material's composition. Observe in Figure 3-57 how a change in temperature chemically changes the composition of the metal alloy plotted along the horizontal axis.

The pressure–temperature fixed composition diagram shown in Figure 3-56 actually contains three curves. The fusion curve is the line that separates the solid from the liquid

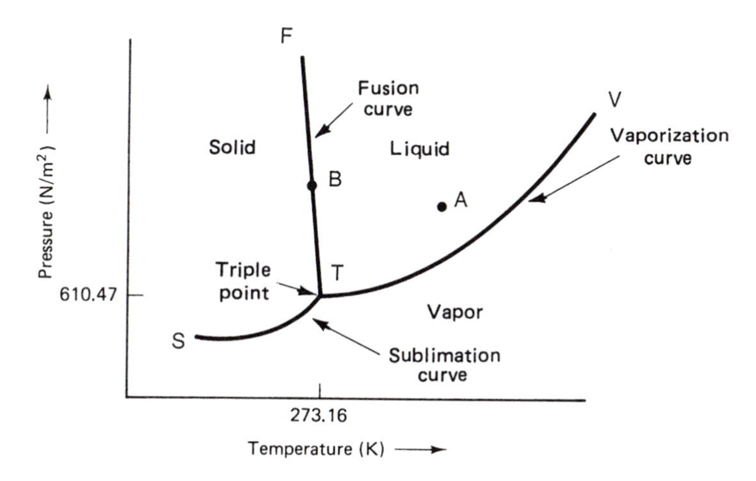

Figure 3-56 Unary phase diagram for water.

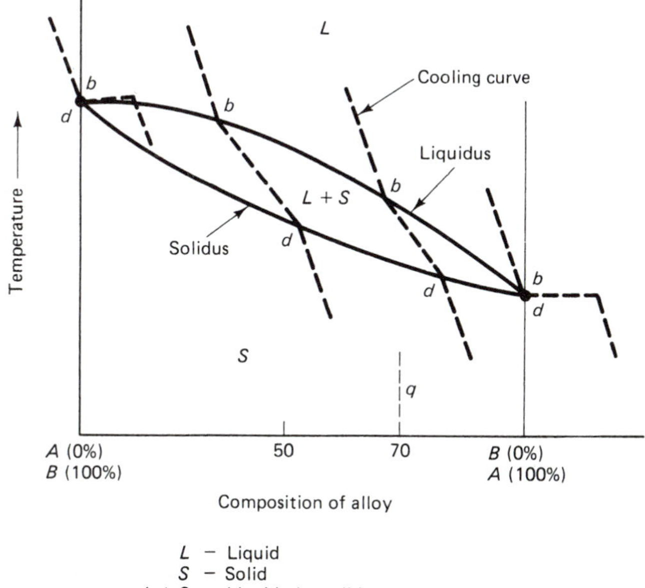

Figure 3-57 A set of hypothetical cooling curves for an alloy system phase diagram. If cooling curves for several combinations of two metals (metal alloy) are plotted as shown here, a phase or equilibrium diagram forms. Points *b*, in cooling curves where solidification begins, are connected to form a smooth line (*liquidus* line). Points *d*, where solidification is completed for each alloy, are also connected by a smooth line (*solidus* line). More discussion of phase diagrams for metals and ceramics follow in the text.

phase. The curved line separating the liquid from the vapor phases, called the ***vaporization curve,*** is a plot of the various pressures and temperature combinations at which the liquid and vapor phases coexist in equilibrium with each other. The same can be said of the sublimation curve separating the solid from the vapor phases. ***Sublimation*** is the process by which a solid changes directly to its vapor state without passing through its liquid phase. Each curve forms a boundary between two different phases. Special note should be made of the point where all three curves intersect. This point, known as the ***triple point,*** is where all three phases coexist. For water the triple point occurs at a temperature of 273.16 K (~0°C) and a pressure of 610.47 N/m^2.

In our study of metal alloys, not only do we have more than one component, but we want to study the solid-state reactions while varying the amounts of these components. To accomplish this, several approaches are used. Fortunately, in practice, most metallurgical processes are conducted at normal atmospheric conditions, which simplifies matters considerably.

If the cooling curves for many combinations of two metals (metal alloy) or ceramics were plotted, as in Figure 3-57, a phase, or equilibrium diagram, would be formed. Points *b* in the cooling curves, at which solidification began, can be connected to form a smooth line, or *liquidus*. Points *d*, at which solidification is complete for each alloy, are also connected by a smooth line, a *solidus*. There are various types of phase diagrams of metal alloys, depending on whether or not the two components are soluble, insoluble, or partially soluble in each other in the liquid or solid phase.

The ***liquidus*** is the name of the curve that represents the temperatures at which solidification begins. It marks the lower boundary of the liquid phase, and thus it is a solubility curve that shows the limit of solubility of the components A and B in the liquid. The ***solidus*** is the curve that passes through all points at which solidification is completed. It marks the upper boundary of the solid solution phase and therefore it is the solubility limit of both components A and B in the solid solution. Between these curves, where a two-phase slushy region of liquid plus solid exists, the process of solidification takes place. It is evident that the points where the liquidus and solidus curves meet represent compositions of 100% A or 100% B (i.e., pure metals). Between these two points are an indefinite number of varying compositions of the two pure metals A and B. Metal A could represent Cu and metal B, Ni.

3.6 SINGLE-CRYSTAL CASTINGS

Ordinary metal castings produce polycrystalline structures. These many crystals or grains of random orientation possess weak atomic bonds across grain boundaries since impurities gather at boundaries. Techniques to improve across-boundary strength have included improved mold material, alloying and producing refined ***equiaxed*** castings. Then casting technology worked to control the direction of grain boundaries and finally eliminate them for certain applications, as discussed in the Pause & Ponder of this module. ***Directional solidification*** (Figure 3-58) produces long grains growing continuously and parallel from one end of a casting to the other, with the longitudinal growth designed to be in the direction to withstand greatest stress. The next step involved selection of one grain and control of its solidification into a ***single-crystal*** casting (***monocrystal,*** Figure 3-58). The monocrystal eliminates the need for boundary strengtheners, which depress melting points and limit high-temperature use; allows high treatment temperatures; and improves high-temperature corrosion resistance. The monocrystal turbine blade seen in Figure 3-58 represents the growing acceptance for high-temperature use of single-crystal castings. Space-shuttle turbine blades for the main engines were originally cast as directional solidified and then as single crystal.

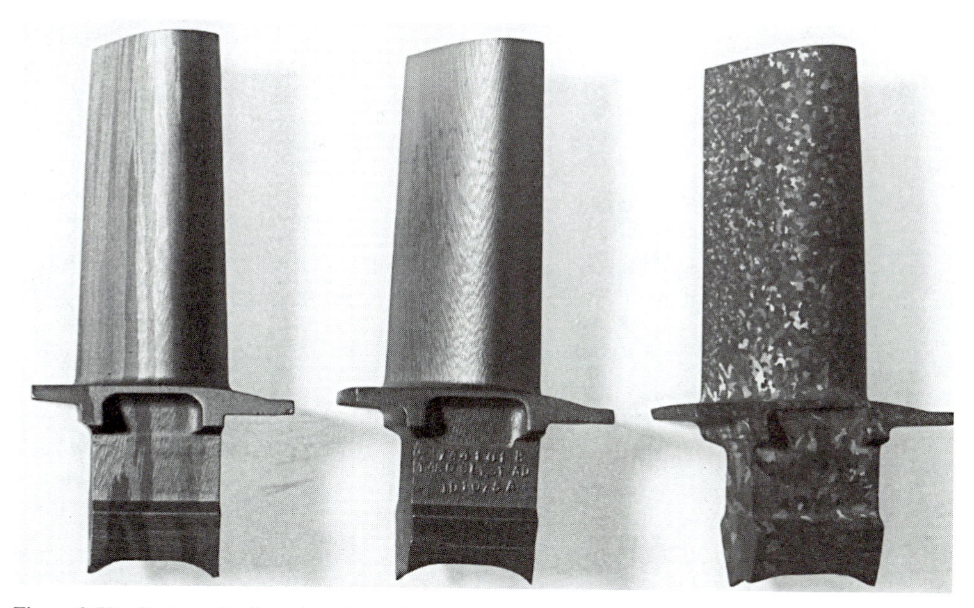

Figure 3-58 Photograph of metal castings of turbine engine airfoils showing the progress in grain control from equiaxed (right), to directionally solidified (left), to mono crystal (center). (Howmet Turbine Components Corporation.)

3.7 STRUCTURES RESULTING FROM NONEQUILIBRIUM CONDITIONS

A new breed of material, Metglas, developed in the late 1970s by Allied Corporation, is an amorphous metal alloy. It is produced by bringing a molten metal alloy at about 1000°F into contact with rapidly moving and relatively cool substrate (or chill block) in a continuous casting process. The drastic quenching operation is similar to that used in conventional solid-state thermal processing of metals such as steel to transform austenite to martensite. It differs because the cooling is so rapid that for most metallic glasses the minimum rate is 10^5 kelvin/second (K/s). To ensure a uniform cooling rate at all points within the metal, one dimension of the liquid layer must be kept to a minimum. Thus, foil, wire, or powder are the forms currently produced. The magnitude of the cooling rate (about 1 million degrees per second) allows but 1 millisecond (ms) for the metal to solidify. The end result is that atomic diffusion is prevented, which in turn precludes any nucleation and growth of crystals. A glassy state consisting of one chemically homogeneous phase, with atoms packed in a random arrangement similar to that of glass or liquid metal, is produced. Alloy compositions consist of transition elements such as Fe and Ni with small percentages of metalloids (B, C, Si, P). Also, mixtures of transition elements such as B and Ni are common. Initial applications of this material are in vacuum furnace brazing of engine parts for turbines in which the older, less desirable brazing transfer tape is replaced by Metglas brazing foils.

The ductility of Metglas foil is greater than that of transfer tape. Metglas foil is capable of being folded back on itself without fracture. This permits it to be converted to various preforms, as well as punched to exact shape to conform with various brazing-joint

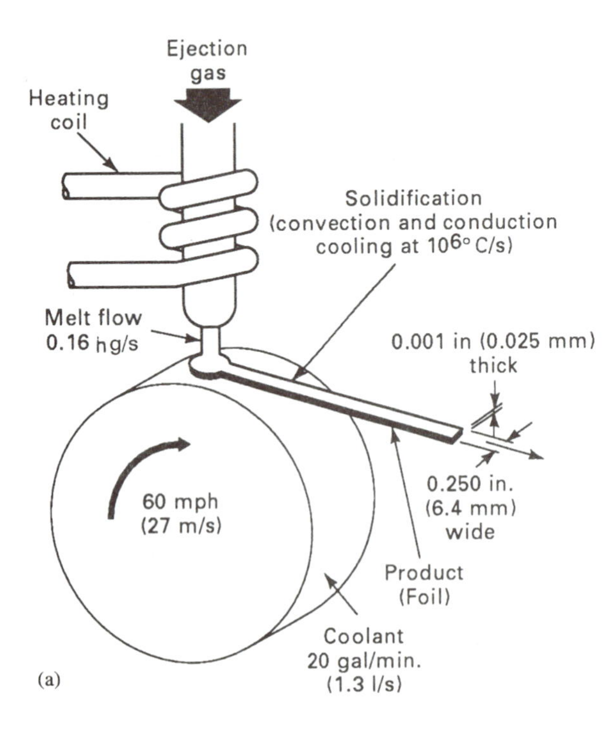

Figure 3-59 (a) A jet of molten metal is solidified rapidly when it strikes a liquid-cooled rotating drum. (b) This striking photo shows a melt-spinning device at the National Institute of Standards and Technology. Glowing metal, melted by coils at top right, hits a whirling wheel and flies off as rapidly solidified ribbon. Time photography produced multiple images. (Photo courtesy of NIST.)

(b)

designs. A great number of new techniques have been developed since the introduction of Metglas. Known by the name ***rapid-solidification processing,*** these techniques employ large departures from local equilibrium in the formation of homogeneous structures. Powders for rapid solidification are produced mainly by *inert gas atomization* and *centrifugal atomization.* Both processes produce spherical particles in the size range 20 to 100 μm in diameter by employing cooling rates up to 10^6 K/s. Other processes are *melt spinning* and *self-quenching,* which produce ribbons/filaments that are pulverized into gritlike powders (Figure 3-59). Powder consolidation can be accomplished by *hot isostatic processing (HIP), hot extrusion,* or *dynamic compaction.*

The materials produced by rapid solidification are finding applications in many areas, including as reinforcing filaments in ceramic-matrix composites and as surface-modification materials (surface alloying) for hard facing of tools, dies, valve seats, and turbine blade tips.

By adjusting the cooling rate, a variety of new materials are produced with highly desirable mixes of properties. Amorphous magnetic materials with atomic structures that impose no obstacles to the movement of magnetic domain boundaries, low hysteresis losses, and high permeability (see Section 9-3) have been developed for various electronic devices such as sensors and motors. New fine-grained homogeneous crystalline materials (Al and Mg alloys) for use as tough tool steels and nickel-based superalloys are now available. This new technology makes it possible to produce alloys that could never be produced with conventional ingot-casting techniques. The highly alloyed powders are sintered and pressed into bulk form. In some processes the powders are pressed into "green" compacts and sintered without ever coming into contact with the air (see Sections 5.12 and 7.3).

Ferrous amorphous metals are "soft" magnetic materials, meaning that they possess no preferred orientation of crystals and hence no easy direction of magnetization. In practical terms, these new materials have a vast potential for use as metal cores in electrical transformers in which billions of kilowatthours are lost in overcoming resistance to changes in the direction of magnetization of the presently used silicon metal alloy cores. Other uses for such a family of materials are being developed by a multitude of industrial concerns.

APPLICATIONS & ALTERNATIVES

In the early 1980s, using techniques such as rapid solidification, new structural arrangements of atoms in solids were discovered for the first time. One was thought to be impossible based on the existing laws of crystallography. Known as ***quasicrystals,*** whose atomic structure remains a mystery to this day, they have spurred the development of three theories, which will ultimately lead to a theory of quasicrystalline structure. The end result will be a theory that will give direction to materials scientists in the development of this new member of the subgroup of materials with a new ordered structure and far-reaching electrical properties. One method for producing these new materials is to melt together aluminum and manganese and subject them to rapid solidification. Such a shock treatment produces a variety of novel structures or phases (Figure 3-60). Quasicrystals do not form

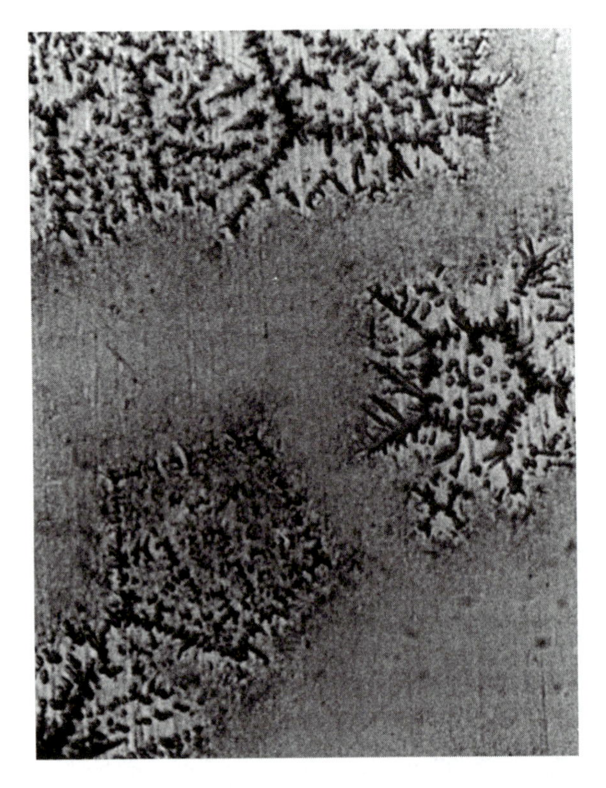

Figure 3-60 Quasicrystal (five-fold symmetry) of Al and Mg at 400 ×. (NIST)

unit cells as do regular crystalline materials. Nor are they amorphous, like glass. Instead, they form an ordered structure somewhere between the two recognized structures.

Where will quasicrystals lead? What new material structure is waiting around the bend? Your education must never stop. You must continue to prepare for new applications and alternatives of materials technology.

SELF-ASSESSMENT

3-1. From the section Pause & Ponder, what goals are sought in using metal casting such as thixotropic magnesium injection molding?

3-2. What are the synthesis/process capabilities of contemporary materials technology at the various scales (macro, micro, nano, and atomic)?

3-3. Explain the difference between a metallurgical microscope and a metallograph.

3-4. In an x-ray diffraction analysis, the x-ray source used produces an x-ray with a wavelength of 1.32×10^{-10} m. The angle of maximum diffraction (θ) is found to be 10°40′ with n of 1. What is the interplanar distance between the atomic planes that produced this diffraction?

3-5. Why are electron microscopes operated within a vacuum?

3-6. Glass lenses are used to focus the beam of light in an ordinary optical microscope. What types of lenses are used to focus electron beams? Explain why such lenses affect an electron beam.

3-7. Define the word *spectroscopy.* Is it possible to have a spectrum of something other than sunlight? If so, name some spectra other than light spectra.

3-8. What are some of the new instruments that allow visualization beyond the capabilities of optical and electron microscopes?

3-9. Name an example of qualitative characterization and one of quantitative characterization.

3-10. What distinguishes a cooling curve for a metal from a cooling curve for a nonmetal?

3-11. What is the volume of a piece of iron having a mass of 864 g?

3-12. Under what conditions can atoms or ions form an fcc crystal structure in which each atom is in contact with 12 other atoms or ions?

3-13. What is the coordination number (CN) for the fcc structure formed by ions of sodium and chlorine producing the chemical compound NaCl?

3-14. Table salt and water both form crystals. What, if any, is the difference in the manner in which these crystals are held together as a solid material?

3-15. Using a ball model (rubber or polystyrene balls), construct an fcc structure to show the closest-packed atom planes (potential slip planes). How many such planes are there?

3-16. A single crystal of zinc, like many other single-metal crystals, is fairly ductile. There is nothing to prevent slippage of planes of atoms over one another. Explain why some of these metals, like zinc, become brittle when they are in the polycrystalline state, although they are ductile in single-crystal form.

3-17. What pressure is normally used in constructing a phase diagram?

3-18. What line on a binary diagram marks the upper limit of the solid solution phase?

3-19. Can a ceramic material be represented by a phase diagram?

3-20. What are the Miller indices for the basal plane (base of the box) in Figure 3-30?

3-21. What is the crystal direction for a line from the origin passing through point *A* in Figure 3-38?

3-22. Consult a table of properties of materials and record the melting-point temperatures of several common materials. Explain why there are differences in such temperatures when these solids change from a solid to a liquid.

3-23. What holds the atoms (ions) together in a compound such as NaCl?

3-24. Compare the ionic radii of negative ions with the radii of positive ions for the same elements. Tin, sulfur, or lead would be likely candidates. Explain the differences, if any.

3-25. The atomic radius of copper is 1.278×10^{-10} m. Express this radius in terms of nanometers (nm) and in inches (in.).

3-26. Whenever the density of a material is mentioned or recorded in a table, the temperature is also noted. What is the explanation for this?

3-27. What is the CN for a cph unit cell?

3-28. What is the maximum number of spheres that can surround and be in contact with a given sphere, provided that all of them are the same size?

3-29. Is there some relationship between the linear coefficient of thermal expansion for a solid and its melting point? List both the coefficient and the melting-point temperature for a metal, a ceramic, and a polymer and note the differences.

3-30. The highest atomic-packing factor is obtained when atoms are of the same size. What is its magnitude?

3-31. The ideal shear strength of a perfect crystalline structure of iron has been calculated to be around 10^{10} N/m^2. This theoretical strength has never been attained. "Whiskers" of iron grown in a near-vacuum in a laboratory approach this value. With the new capabilities provided by the NASA space program, discuss the future ramifications of being able to grow large, flawless crystals that are undistorted by their own weight as they form in space laboratory conditions.

3-32. Observe a piece of galvanized steel. Describe and sketch the zinc coating as you observe it with the naked eye.

3-33. In the cubic axes system, the measures of the angles α, β, and γ are how many degrees?

3-34. Sketch an fcc unit cell.

3-35. Using Table 3-2, determine the crystal structure of lead.

3-36. The rapid quenching of austenite in the heat treatment of steel produces a supersaturated and distorted crystal structure known as body-centered tetragonal (bct). Sketch the unit cell for such a structure.

3-37. What are the most likely directions for slip to occur in cph crystals? Sketch the unit cell.

3-38. Describe the solid state in terms of kinetics and energy.

3-39. Why is lead more ductile than tin?

3-40. What steps can be taken to reduce slip in metal crystals?

3-41. Obtain a rubber band. Stretch it as far as it will go without breaking while holding it in contact with your moistened lips. Observe the temperature change. Hold the band in this stretched condition for about 30 seconds. Note any change in temperature. Release the band suddenly to its original unstretched length and touch it to your lips. Note the temperature change. Rubber is an elastic polymer when stressed. The stressed molecules align themselves and local crystallization occurs. When the stress is released the molecules return to their original arrangement. Determine if energy is absorbed or released when an amorphous material crystallizes. Record all your observations and comments.

3-42. Mark the following statements as true (T) or false (F).

(a) Single-phase materials have a single-crystal structure.

(b) Single-crystal structured materials are single-phase materials.

3-43. What is the packing factor of pure iron when it is in **(a)** the bcc structure and **(b)** the fcc structure?

3-44. What happens to the atomic radius of pure iron as it is heated from room temperature (RT) to 910°C?

3-45. If the APF of a material increases when heated, what affect does this have on its volume?

3-46. Using Figure 3-42, sketch an isosceles right triangle described in the accompanying text, label its sides, and derive the equation $(r = R)^2 = 2R^2$.

3-47. Explain the difference between a short-range ordered material and a long-range ordered material. Given an example of each.

3-48. Diffusion of atoms through a solid takes place by two main mechanisms. One is diffusion through vacancies in the atomic structure. Describe another method of diffusion.

3-49. Name two possible substances formed by alloying two or more metals.

3-50. Would chromium atoms be a likely substitute for aluminum atoms when forming a substitutional solid solution? Ruby rods are doped with chromium atoms to convert the well-known ceramic (CrO_3 in Al_2O_3) into a laser material.

3-51. Disorder in the arrangement of atoms in solids is limited to the atomic level. Cite an example that refutes this statement.

3-52. Grain boundaries in crystalline solids play an important role in the movement of dislocations through a solid. Note at least three actions that take place in the grain boundary transition zone in a crystalline material.

3-53. List some examples of allotropic solids.

3-54. Metals are classified as crystalline materials. Name one metal that is an amorphous solid and name at least one recent application in which the use of it is saving energy or providing greater strength and/or corrosion resistance.

OBJECTIVE QUESTIONS

3-1. What type dendrites form in thixotropic metals?

a. tree-like **b.** spheroidal
c. BCC **d.** FCC

3-2. Ability of a material to exist in different space lattices.

a. Allotropic **b.** Crystalline **c.** Solvent **d.** Amorphous

3-3. Contemporary synthesizing/processing capabilities are best at which level of the materials technology scale?

a. Atomic (0.1 nm) **b.** Macro (cm—m)

3-4. Amorphous metals develop their microstructure as a result of

a. Dendrites **b.** Directional solidification
c. Slip **d.** Extremely rapid cooling

3-5. In an alloy, the material that dissolves the alloying element.

a. Solute **b.** Solvent **c.** Matrix **d.** Allotrope

3-6. Methods used to determine unique properties by identifying and analyzing (both quantitatively and qualitatively) features such as voids, roughness, grain size, coating, hardness.

a. Characterization **b.** Slip **c.** Diffusion **d.** Synthesis

3-7. The Pause & Ponder focused on changing structure through processing. Which processes produced the improved equiaxed grains?

a. Investment casting **b.** Shell molding
c. Thixotropic molding **d.** Injection molding

3-8. What advantage comes from equiaxed grains?

a. Lower processing costs **b.** Uniform strength
c. Directional strength **d.** Reduced density

Select the *best* match from the following techniques and instruments for questions 3-9 to 3-14.

a. Spectroscopy
b. X-ray diffractometer
c. Optical microscope
d. Electron microscope
e. Scanning tunneling microscope
f. Nanotechnology

3-9. Uses magnetic lenses to focus electron beams that provide over 200,000× magnification.

3-10. Provides magnification of about 2000× and is a relatively low-cost instrument.

3-11. A general term for techniques that utilize the differences in wavelength spectra produced by chemical elements.

3-12. A method to manipulate structures from hundreds to hundredths of a micrometer.

3-13. Used to measure the angles of electrons glancing off material specimens.

3-14. Uses very fine probes that move over the surface of a material and provide images of individual atoms.

3-15. Characterization technique involving visual clues to crafter or inspector.

a. Nanoscale **b.** Quantitative **c.** Qualitative **d.** Artificial

3-16. Characterization technique involving instruments and methods such as AI.

a. Nanoscale **b.** Quantitative **c.** Qualitative **d.** Artificial

3-17. What distinguishes a cooling curve for a metal from a cooling curve for a nonmetal?

a. Phase transformations occur with nonmetals.

b. No phase transformations occur with nonmetals.

3-18. What is the volume of a piece of iron having a mass of 864 g?

a. 10.8 cm^3 **b.** 109.78 cm^3 **c.** 20.8 cm^3 **d.** 200.6 cm^3

3-19. What is the coordination number (CN) for the fcc structure formed by ions of sodium and chlorine producing the chemical compound NaCl?

a. 6 **b.** 8 **c.** 14 **d.** 16

3-20. What pressure is normally used in constructing a phase diagram?

a. 100 psi **b.** Depends on material
c. Ambient **d.** Normal atmospheric pressure

3-21. What line on a binary diagram makes the upper limit of the solid solution phase?

a. Liquidus **b.** Eutectic **c.** Eutectoid **d.** Solidus

3-22. A ceramic material cannot be represented by a phase diagram.

T (true) **F** (false)

3-23. What are the Miller indices for the basal plane (base of the box) in Figure 3-30?

a. (0001) **b.** (001) **c.** (111) **d.** (101)

3-24. What is the crystal direction for a line from the origin passing through point *A* in Figure 3-38?

a. [110] **b.** [222] **c.** [120] **d.** [001]

3-25. What holds the atoms (ions) together in a compound such as NaCl are electrostatic forces between ____________.

a. Atom and ion **b.** Covalent bonds
c. Electrons and nuclei **d.** Neutrons

3-26. What is the CN for a cph unit cell?

a. 5 **b.** 8 **c.** 12 **d.** 24

3-27. What is the magnitude of the highest atomic-packing factor when atoms are of the same size?

a. 0.74 **b.** 0.81 **c.** 21 **d.** 81

3-28. How many directions of slip are there in cph systems?

a. 3 **b.** 4 **c.** 5 **d.** 6

3-29. Single-phase materials have a single-crystal structure.

T (true) **F** (false)

3-30. Single-crystal structured materials are single-phase materials.

T (true) **F** (false)

3-31. Diffusion of atoms through a solid takes place by two main mechanisms. One is diffusion through vacancies in the atomic structure. Another method of diffusion is ____________.

a. Cold **b.** APF **c.** Substitutional **d.** Interstitial

3-32. Two possible substances formed by alloying two or more metals are solid solution and __________.

a. Substance **b.** Interstitial **c.** Mixture **d.** Compound

3-33. Traditional ceramics have poor tensile strength due to what?

a. Ionic bonding within grains

b. Metallic bonding within grains

c. Poor bonding across grain boundaries

d. Good bonding across grain boundaries

3-34. Grain boundaries __________ movement of dislocations through a solid.

a. Improve **b.** Inhibit **c.** Do not affect

3-35. Iron can be alloyed with carbon because it is _________.

a. Crystalline **b.** Amorphous **c.** A mixture **d.** Allotropic

3-36. Metals can be cooled only to crystalline solids.

T (true) **F** (false)

3-37. Solid crystals that do not form unit cells as do regular crystals.

a. Allotropes **b.** Twins **c.** Quasicrystals **d.** Alloys

REFERENCES & RELATED READINGS

ASM INTERNATIONAL. "Advances in Characterization and Testing of Engineering Materials," special issue, *Advanced Materials and Processes,* November 1989.

ASM INTERNATIONAL. "Characterizing Thermal Spray Coatings," special issue, *Advanced Materials and Processes,* April 1992.

DAVIS, J. R., ed. *ASM Materials Engineering Dictionary.* Materials Park, Ohio: ASM International, 1993.

DREXLER, K. ERIC. *Nanosystems: Molecular Machinery, Manufacturing, and Computation.* New York: John Wiley & Sons, Inc., 1992.

FREDERICK, P. S., N. L. BRADLEY, and S. C. ERICKSON. "Injection Molding Magnesium Alloys," *Advanced Materials and Processes,* November 1988, pp. 53–56.

FULRATH, R. M. and J. A. PASK. *Ceramic Microstructures.* Melbourne, Fla.: R. E. Krieger, 1976.

HALL, CECIL E. *Introduction to Electron Microscopy.* Melbourne, Fla.: R. E. Krieger, 1983.

INGERSOL RAND, *Compressed Air,* October/November 1994, p. 16.

LOVELAND, R. P. *Photomicrography: A Comprehensive Treatise,* Vols. 1 and 2. Melbourne, Fla: R. E. Krieger, 1981.

MCDONALD, ALAN F. "Scanning X Microscopy," *National Educators' Workshop, NEW: Update 89, Standard Experiments in Engineering Materials, Science, and Technology,* NASA Langley Research Center, Hampton, Va., October 19, 1989.

MCLEAN, D. *Mechanical Properties of Metals.* Melbourne, Fla: R. E. Krieger, 1977.

MALLARDI, JOSEPH L. *From Teeth to Jet Engines.* Howmet Corporation, 1992.

ROY, RUSTUM. "New Materials: Fountainhead for New Technologies and New Science," *NEW: Update 92,* Oak Ridge National Laboratory, Oak Ridge, Tenn., November 12, 1992.

SEMTC. *Navy Manufacturing Letter,* February 1995, p. 6.

STEPHENS, PETER W., and ALAN I. GOLDMAN. "The Structure of Quasicrystals," *Scientific American,* April 1991, pp. 44–53.

WICKRAMASINGHE, H. KUMAR. "Scanned-Probe Microscopes," *Scientific American,* October 1989, pp. 98–105.

Module

4

Properties, Degradation, Failure, and Selection of Materials

After studying this module, you should be able to do the following:

4.1. Use diagrams, explanations, and calculations to determine physical, mechanical, chemical, and thermal properties of materials.

4.2. Determine the names and types of standard mechanical tests for the properties they measure.

4.3. Explain the types and uses of industrial standards in design and manufacturing, and the role of technical societies and organizations involved in developing standards for engineering materials.

4.4. Analyze service conditions (stresses and environments) to determine properties critical to the ability of parts or products to survive the conditions.

4.5. Explain reasons for the degradation and failure of materials and the role of failure analysis, factors of safety, post-process thermal treatments, and corrosion protection.

4.6. Recall purposes and methods for nondestructive testing, inspection, quality control, and quality assurance.

4.7. Relate properties of materials and the effect of cost in materials selection criteria.

(a)

(b)

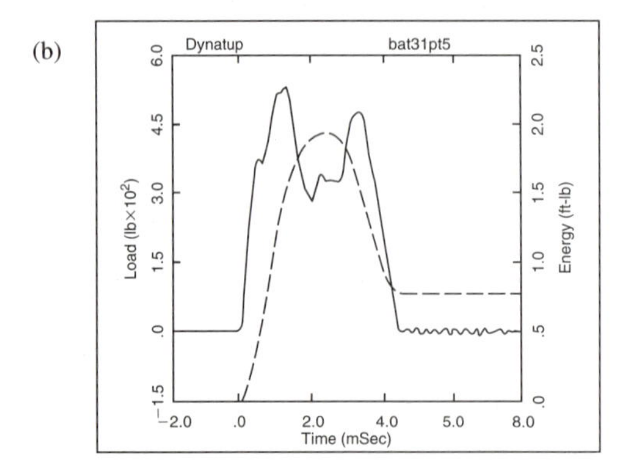

Figure 4-1 High-tech testing. (a) Testing aluminum bats to determine the sound produced, "ping," and to increase energy reflectivity or performance. The reflectivity of the bat depends on the properties of the selected aluminum alloy and the wall thickness of the hollow bat. The designer's goals are to increase the "sweet spot" area where the maximum force is transferred to the ball. (b) The curves generated by the impact tester provide a "sweet spot" response. This location (5″ from the end of the bat) is where maximum energy is transferred from the bat to the ball. (Courtesy GRC Instruments.)

PAUSE & PONDER

Do you ever wonder why most of the products that you purchase seem to withstand considerable punishment without ill effects? How can companies manufacture lightweight, aluminum baseball bats, like those shown in Figure 4-1, that do not break when the strongest hitter belts a home run? In fact, many products carry guarantees against breakage and malfunction. How can those manufacturers be so sure that their products will last 90 days, 50,000 miles, or for a lifetime?

The answer is, they spend considerable effort in "engineering" the products. *Materials engineering* involves designing and testing of materials, components, and products. *Materials testing* involves subjecting materials, components, and whole products to adverse conditions to determine the ***limits to failure,*** which provides an indication of the type of abuse they can be expected to withstand before failing.

Advances in technology have provided new tools to materials engineering and manufacturing to improve the confidence with which to predict the ***product reliability*** (the ability of the product to perform as specified by the designer). Companies may find it very difficult to compete in a global economy if they do not use the new technology and instead rely on only ***subjective*** (personal experience and feelings) rather than ***objective*** data (data that can be measured with instruments). Figure 4-1 depicts a high-tech, instrumented materials-testing setup that you might find in a typical manufacturing plant. Because of demand by consumers, quality must be built into products. The way to ensure built-in quality is to test every aspect of the product design and manufacture before releasing it to the public. Quality involves both performance and durability.

A baseball bat should have the proper weight and balance, be able to transfer energy from the batter to the ball efficiently, and be able to withstand the energy transfer from the strongest batter without breaking. To determine the batting quality of aluminum bats, manufacturers evaluated them with a drop-weight impact test, as seen in Figure 4-1(a). The falling tip inputs controlled energy to the bat, and a computerized data-acquisition system (wired into the testing machine) generates data curves showing the ratio of energy absorbed to energy reflected. The closer the impact to the "sweet spot," the less energy is absorbed by the bat. ***Sweet spots,*** as found on bats, tennis rackets, badminton rackets, and golf clubs, are the best zones for hitting a ball or birdy to achieve maximum control and distance. Less energy absorption by a bat means the ball should travel farther and with less strain on the batter. The curves shown on the monitor in Figure 4-1(b) indicate the falling tip has hit 12.5 centimeters from the fat end of the bat, which is the heart of the "sweet spot." Therefore, the load from the impact tip is reflected away with minimum absorption.

Information from computerized materials-testing equipment allows designers to make modifications in design and allows manufacturers to be sure they have obtained their quality goals. Prior to the development of new, computerized materials-testing technology systems, much of the design of sports equipment and many other products was subjective, based on trial and error with no quantitative analysis. In other words, the designer and user had to rely on the look and feel of parts, or maybe the endorsement of some star athlete, rather than actual objective test results. Knowledgeable consumers are beginning to ask for objective evidence like that found in statistics from materials and product testing.

Modern tools for engineering allow manufacturers of sporting goods, CD players, computers, mattresses, paint, eyeglasses, and many other products to control the manner in which they engineer materials, then apply the materials to products. This modern approach to design and manufacturing provides a wide selection to suit individual needs.

Before you move into the module, Pause & Ponder the last time you were in a sporting goods store. Did you look at the literature that may come with tennis rackets, fishing rods, bowling balls, baseball bats, or similar equipment? If so, you may have seen informative charts and diagrams for different models of tennis rackets. These charts and diagrams inform the consumer of the degrees of stiffness and levels of player control, which aid a player in choosing the racket most suited to his or her particular style of play (see Figure 4-5).

This module is designed not only to give you a background in properties of materials, but also to increase your knowledge of engineering and manufacturing by showing the relationships among design, materials selection, materials testing, and manufacturing. From this study you should gain insight into how engineering involves in-depth efforts to understand connections between the structure of materials and testing of their properties, and how this information provides manufacturers with the confidence to provide guarantees for a product's life expectancy. This information has value to you both as a consumer and in preparation for a technical career because materials selection and/or testing may be part of your job.

4.1 PHYSICAL PROPERTIES

The properties of a material are those characteristics that help modify and distinguish one material from another. Taken as a whole, these qualities define a material. All properties are observable and most can be measured quantitatively. Properties are classified into two main groups, physical and chemical properties. ***Physical properties*** involve no change in the composition of the material. Density, strength, and hardness are examples of such properties. ***Chemical properties*** are associated with the transformation of one material into another. Iron rusts when it combines to produce an iron oxide through a chemical reaction. Physical properties are, in turn, arbitrarily subdivided into many categories. These subdivisions bear names such as ***mechanical, metallurgical, fabrication, general, magnetic, electrical, thermal, optical, thermonuclear,*** and ***electro-optical.*** Regardless of the name of the subdivision, physical properties result from the response of the materials to some environmental variable, such as a mechanical force, a temperature change, or an electromagnetic field (which includes all the radiation in the electromagnetic radiation spectrum).

For purposes of this book, physical properties will be divided primarily into mechanical, thermal, electrical, magnetic, and optical.

4.2 MECHANICAL PROPERTIES

In selecting a material for a product such as a piston in an internal combustion engine, a designer is interested in properties such as strength, ductility, hardness, or fatigue strength. These are some mechanical properties of a material. ***Mechanical properties*** are defined as a measure of a material's ability to carry or resist mechanical forces or stresses. When any matter is at rest, the atomic or molecular structure is in equilibrium. The bonding forces in this structure resist any attempt to disrupt this equilibrium. One such attempt may be an external force or load.

Stress results from forces such as tension, compression, or shear that pull, push, twist, cut, or in some way deform or change the shape of a piece of material. Often, this

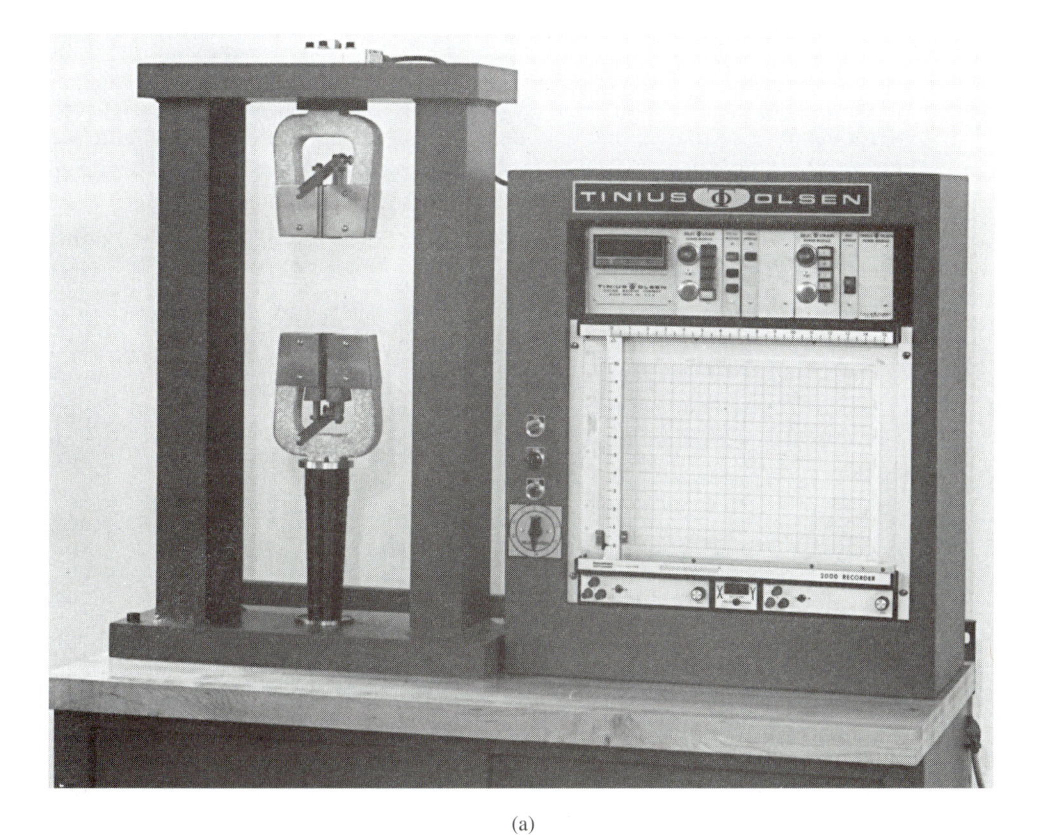

(a)

Figure 4-2 (a) Universal testing machine. The unloaded grippers would hold the specimen and strain indicators seen in (b) and (c).

(b)

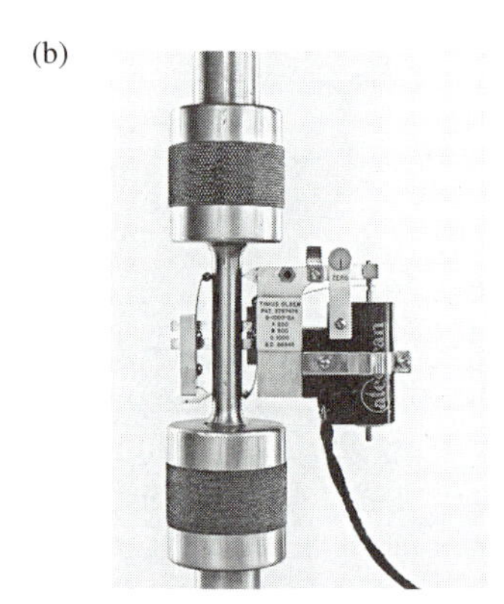

(c)

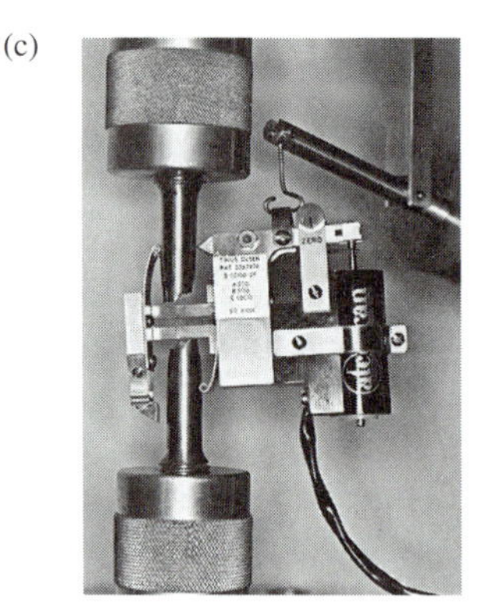

(d)

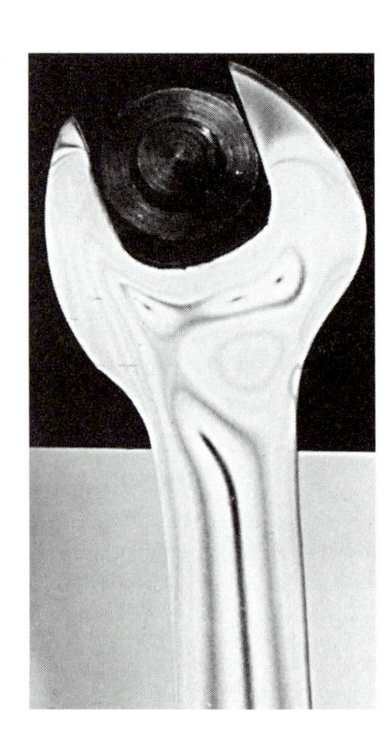

Figure 4-2 (b) Before stressed and (c) after breaking. (Tinius Olsen Testing Machine Co.) (d) Photoelastic model of wrench for analysis of stress. (Measurement Group Inc., Raleigh, N.C.)

deformation is so minute that only delicate instruments can detect it. Figure 4-2 shows a universal testing machine used to apply loads to material specimens and, with appropriate instrumentation, to detect minute deformation of materials under load.

4.2.1 Stress

Stress is defined as the resistance offered by a material to external forces or loads. It is measured in terms of the force exerted per area [pounds per square inch (psi)]. Normal stress is that applied perpendicular to the surface to which it is applied, i.e., tension or compression. The corresponding SI units are newtons per square meter (N/m^2) or pascal (Pa). One pascal (1 Pa) equals 1 N/m^2. Another way of defining it is to say that stress is the amount of force *(F)* divided by the area *(A)* over which it acts. Using σ (the Greek letter sigma) as the universal symbol for normal stress, we say mathematically that

$$\sigma = \frac{F}{A}$$

An assumption is made that the stress is the same on each particle of area making up the total area *(A)*. If this is so, the stress is uniformly distributed. When a load or force is applied to an object, we are unable to measure the stress produced by this force in the material. What we do is measure the force, identify the area over which the force acts, and measure it as well. These two quantities can then be used to calculate the stress produced in the material by the previous relationship. With the use of polarized light and models made of photoelastic plastic, it is possible to detect concentrations of stress, as seen in Figure 4-2(d).

Illustrative Problem

Calculate the stress on a 10-ft circular rod with a diameter of 1 in. that is in tension due to a pulling force of 50 lb. Sketch in two dimensions the rod showing the force and the area over which the stress is acting (to resist the external force). Express your answer in pounds per square inch (psi).

Solution

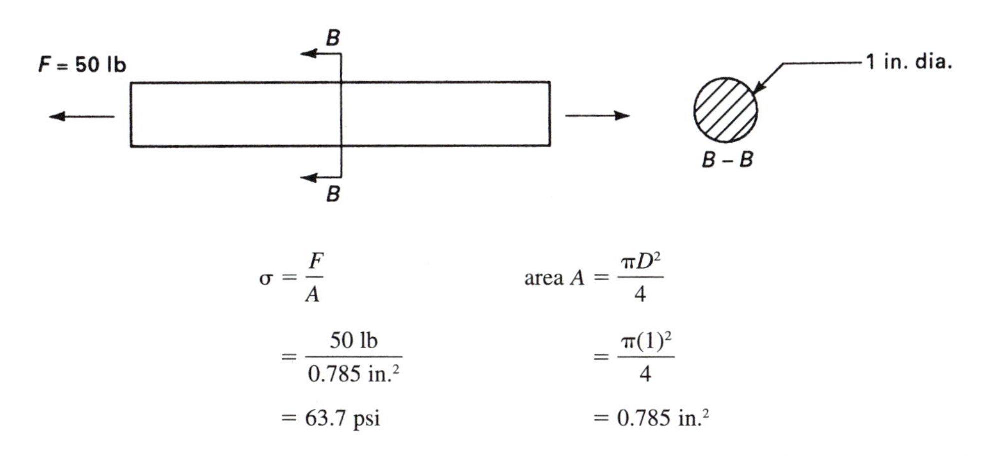

$$\sigma = \frac{F}{A} \qquad \text{area } A = \frac{\pi D^2}{4}$$

$$= \frac{50 \text{ lb}}{0.785 \text{ in.}^2} \qquad = \frac{\pi(1)^2}{4}$$

$$= 63.7 \text{ psi} \qquad = 0.785 \text{ in.}^2$$

4.2.2 Strain (Unit Deformation)

Strain, or ***unit deformation,*** is defined as the unit change in the size or shape of material as a result of force on the material. Many times we assume that a solid body is rigid; that is, when the body is loaded with some force, the body keeps its same size and shape. This is far from correct. Regardless of how small the force, a body will alter its shape when subjected to a force. In other words, the body will change its dimensions. The change in a physical dimension is called ***deformation*** (δ, Greek letter delta). Figure 4-3(a) illustrates a rod of original length (l_0) and original diameter (d_0) placed under a tensile load *(F)* and elongated (stretched). Its change in length (Δl) equals the difference in the two dimensions

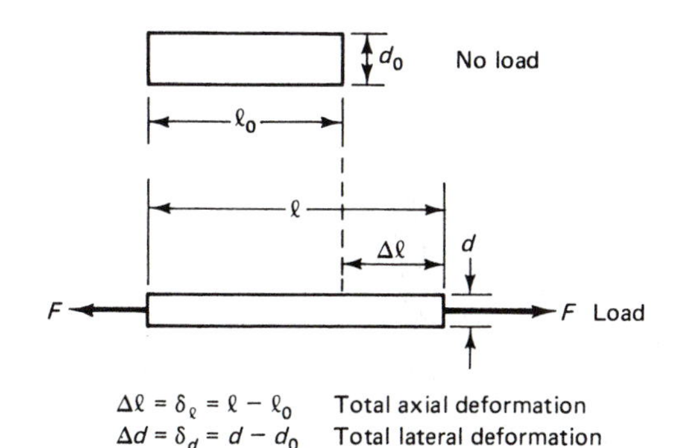

(a)

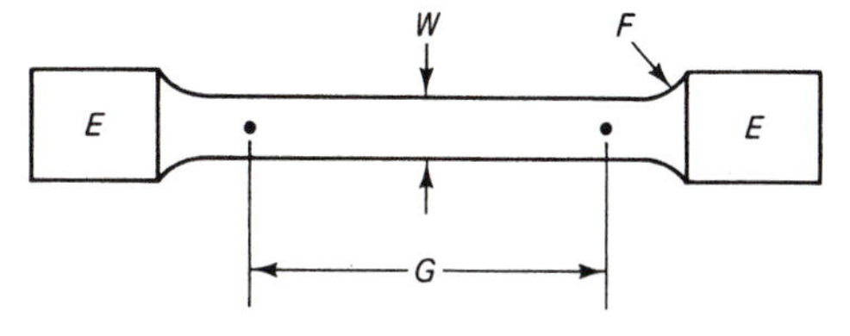

E – gripped ends, may be threaded, plain, or with hole for gripping by machine
W – reduced width to insure specimen breaks in middle – round on round specimens and flat on flat specimens
G – marked gage length to precisely measure the change in length before, during and after test
F – fillet to reduce stress concentrations

(b)

Figure 4-3 (a) Rod under a tensile load. (b) Standard tensile test specimen.

$(l - l_0)$ in the direction of the force. In this example, Δl is exaggerated for illustrative purposes. This is known as total deformation, δ (Δ, Greek capital letter delta).

The relationship between the total deformation and the unit deformation can be expressed mathematically. Using the symbols δ for total deformation (or Δl), l for length, and ϵ (Greek letter epsilon) for unit deformation, we can write the equation $\delta = \epsilon \times l_0$.

Note that the original diameter has been reduced in size and produces a corresponding change in the lateral direction (at right angles to the direction of the load). Note too that in one instance the dimension increased and the other decreased. The change in the length is called a ***total axial*** or ***longitudinal deformation.*** The change in the lateral dimension is known as a ***total lateral deformation.*** The ratio of the total axial deformation to the original length is known as the ***unit axial*** or ***longitudinal strain,*** ϵ. The linear units are not canceled and are kept as part of the term. Therefore, in mathematical terms,

$$\epsilon_{\text{long.}} = \frac{\Delta l \text{ mm}}{l_0 \text{ mm}} = \frac{\delta \text{ mm}}{l_0 \text{ mm}} \quad \text{(longitudinal unit deformation)}$$

$$\epsilon_{\text{lat.}} = \frac{\Delta d \text{ mm}}{d_0 \text{ mm}} = \frac{\delta \text{ mm}}{d_0 \text{ mm}} \quad \text{(lateral unit deformation)}$$

In summary, when a piece of material (a body) is subjected to a load, it will not only deform in the direction of the load (axial deformation), but it will also deform in a lateral direction (at right angles to the direction of the tensile or compression load). The ratio of the lateral unit deformation or strain ($\epsilon_{\text{lat.}}$) to the unit longitudinal deformation or strain ($\epsilon_{\text{long.}}$), given the symbol μ (Greek letter mu), is known as ***Poisson's ratio:***

$$\mu = \frac{\epsilon_{\text{lat.}}}{\epsilon_{\text{long.}}}$$

The unit longitudinal deformation is larger than the unit lateral deformation and therefore Poisson's ratio is less than 1. For steel, it is about 0.3.

In our discussion of deformation we demonstrated it by using a tensile load (Figure 4-3): If a body was loaded in compression, the length of the body would have decreased and its width increased. One major reason for conducting tension and compression tests using standardized equipment and specimens (see Figure 4-3) is to determine the data needed to plot stress–strain diagrams for the material under investigation so that other meaningful relationships can be determined.

4.2.3 Stress–Strain Diagrams and Hooke's Law

The stress–strain diagram is used to determine how a certain material will react under load. Figure 4-4 is a stress–strain diagram for a low-carbon (mild) steel. Strain values (mm/mm) are plotted along the horizontal axis (abscissa), and along the vertical axis (ordinate) are plotted the stress values (MPa, megapascal).

The straight-line portion of the diagram up to almost the yield point is known as the ***elastic region*** of the diagram. Within this range of stresses, the material will return to its original dimensions once the load, hence, the nominal stress, has been removed. (Nominal stress is the load divided by the original cross-sectional area. Actual stress can vary due to

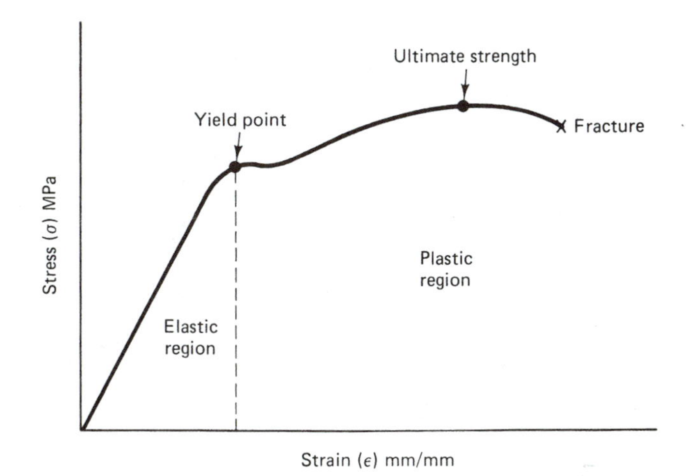

Figure 4-4 Stress–strain diagram.

changes in area caused by lateral deformation.) In the elastic region, each increase in stress will produce a proportionate increase in the strain. This statement is known as ***Hooke's law*** ($\sigma = E\epsilon$), with E the constant of proportionality (***elastic modulus***—the word *modulus* means a ratio or constant of proportionality).

Illustrative Problem

Using the same rod as in the Illustrative Problem in Section 4.2.1, determine how much the rod has elongated by the tensile force of 50 lb. Assume that the modulus of elasticity for steel with which the rod is fabricated is 30×10^6 psi.

Solution Using Hooke's law, $\sigma = E\epsilon$, and solving for ϵ, we find that

$$\epsilon = \frac{\sigma}{E} = \frac{63.5 \text{ psi}}{30 \times 10^6 \text{ psi}} = 2.12 \times 10^{-6} \text{ in./in.} \quad \text{and}$$

$$\delta = \epsilon \times l = 2.12 \times 10^{-6} \text{ in./in.} \times 10 \text{ ft}$$

$$= 2.12 \times 10^{-5} \text{ in./in.} \times 1 \text{ ft} = 212 \times 10^{-5} \text{ ft}$$

Beyond the yield point, the material will continue to deform, but with less stress than before, because the material has begun to yield. In this region, known as the ***plastic region,*** plastic deformation takes place and when the load is removed the material will not return to its original dimensions. It now has a ***permanent set.*** Note also that it takes less nominal stress to break the metal specimen than it does to reach the ultimate strength. Because the material has yielded, the original cross-sectional area of the specimen has been reduced in size so that less material is available to resist the load.

The *yield point* represents the dividing line or transition from the elastic to the plastic region of the curve. When the stress reaches the yield point, a large increase in strain occurs with no increase in stress. The ***modulus of elasticity, elastic modulus, tensile***

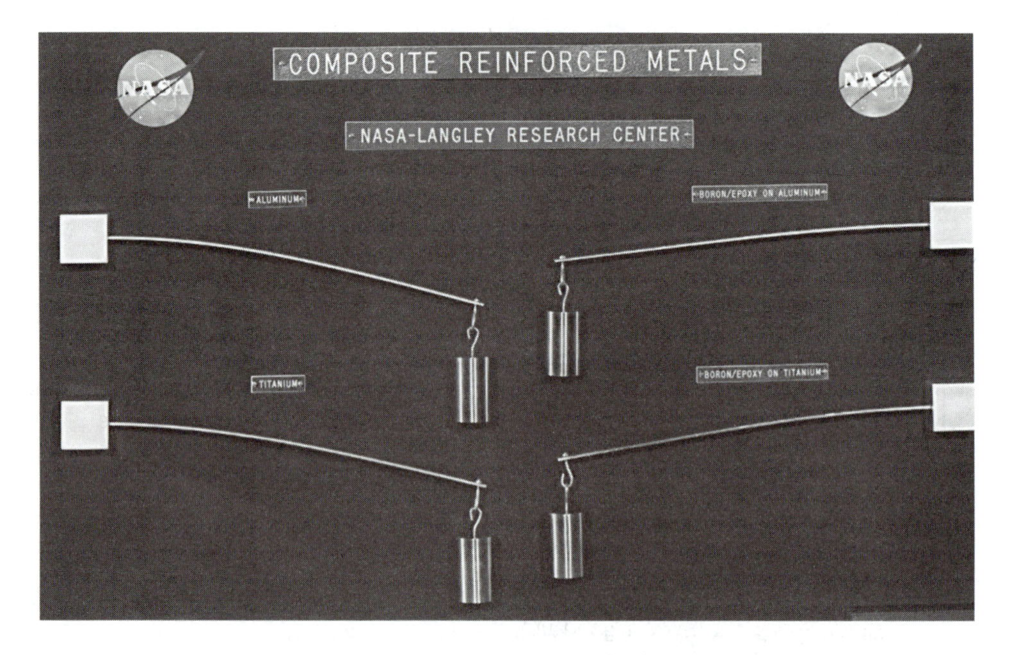

Figure 4-5 (a) The stiffness of pure metals versus metal composites. (NASA.)

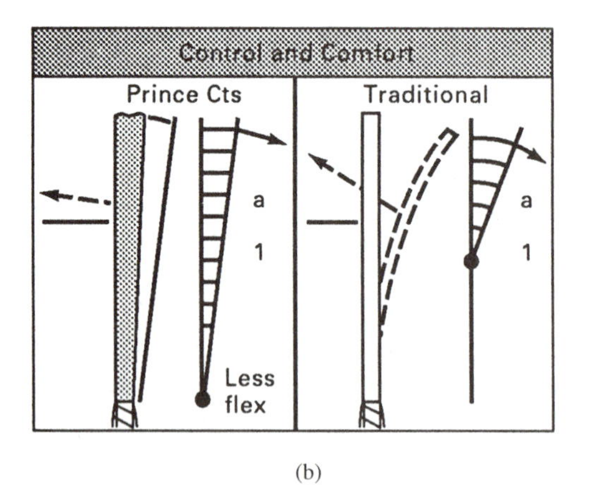

(b)

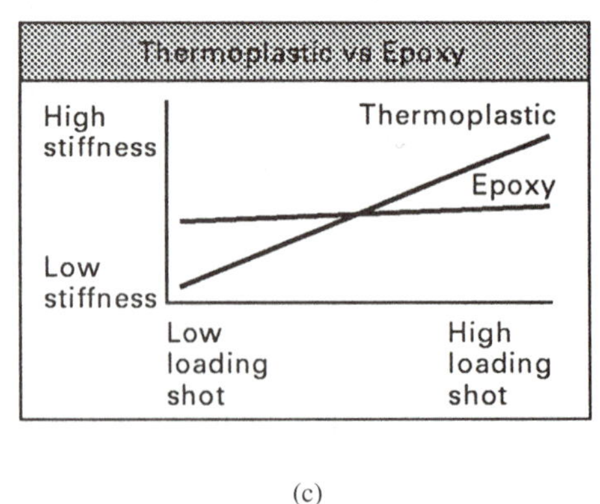

(c)

Figure 4-5 (b) The concept of stiffness has been incorporated into the design of tennis rackets. CTS design, shown on the left, uses a wider body frame to provide greater stiffness than the traditional design (right), thereby giving greater power. (c) Thermoplastic polymers in the form of liquid-crystal polymers (LCP) have the ability with low-loading shot (slower-moving ball and racket) to provide less stiffness, but with higher-loading shot (faster-moving ball and racket) to provide higher stiffness to give the player better control and power. (Prince Mfg. Company.)

modulus, Young's modulus, modulus of elasticity in tension, or ***coefficient of elasticity,*** given the symbol *E,* is the ratio of the stress to the strain in the elastic region of the stress–strain diagram.* The ***tensile modulus*** is approximately equal to the ***compressive modulus of elasticity*** within the proportional limit (elastic limit of the diagram). Note that this ratio expresses the slope of the straight-line portion of the curve. Regardless of the name, this modulus is an indication of the stiffness of the material when subjected to a tensile load. The stiffness of a material is defined as the ratio of the load to the deformation produced. The higher the value of Young's modulus, the stiffer the material, as demonstrated in Figure 4-5.

Not all materials produce stress–strain diagrams (Figure 4-6) on which there is a clear indication of the start of yielding as the load is increased. Cast iron is an example. This situation should not be interpreted to mean that cast iron does not exhibit elastic properties under moderate loads. In other words, such materials are elastic with strains returning to zero when moderate loads, hence stresses, are removed. The modulus of elasticity for these materials is sometimes taken as the slope of a tangent to the stress–strain curve at the origin.

4.2.4 Ultimate Strength or Tensile Strength

Ultimate strength or ***tensile strength*** is the maximum stress developed in a material during a tensile test. It is a good indicator of the presence of defects in the crystal structure of a metal material, but it is not used too much in design because considerable plastic

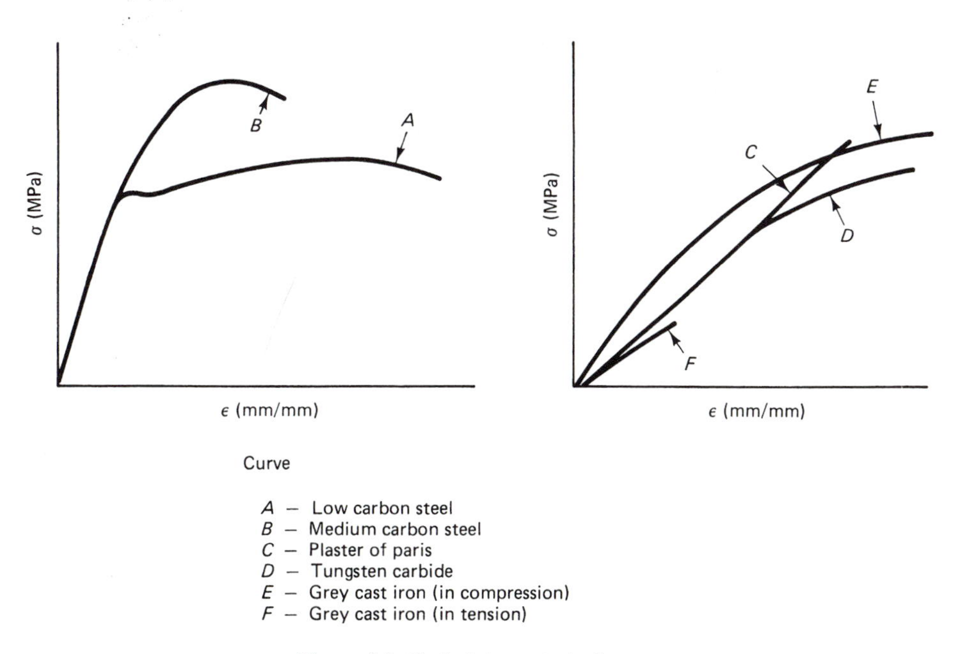

Figure 4-6 Typical stress–strain diagrams.

*The SI unit for *E* is the pascal (Pa).

deformation has occurred in reaching this stress. Plastic deformation is not all bad. However, in many applications the amount of plastic deformation must be limited to much smaller values than that accompanying the maximum stress. The ultimate shear strength (Section 4.2.7) is about 75% of the ultimate tensile strength.

4.2.5 Yield Strength

Many materials do not have a yield point. Low-carbon steel is one of just a few that exhibit a point where the strain increases without an accompanying increase in stress. This poses a problem in deciding when plastic deformation begins for such materials. By agreement, a practical approximation of the elastic limit, called the ***offset yield strength,*** is used. It is the stress at which a material exhibits a specified plastic strain. For most applications, a plastic strain of 0.002 in./in. can be tolerated, and the stress that produces this strain is the ***yield strength.*** This is sometimes expressed as 0.2% strain. The yield strength is determined by drawing a straight line, called the offset line, from the 0.2% strain value on the horizontal axis parallel to the straight-line portion of the stress–strain curve. The stress at which this offset line intersects the stress–strain curve is designated as the yield strength of the material at 0.2% offset. In some cases the offset can be specified as 0.1% or even 0.5%.

Figure 4-7 shows a typical stress–strain curve for an aluminum alloy with no pronounced yield point. The offset, offset line, and point of intersection of the offset line with the stress–strain curve are shown. When reporting yield strength, care must be taken to include the amount of offset as well as the value of the stress at the intersection with the data. In general, the yield strength of metals is much higher than that of other materials. For brittle materials, yield strength differs very little, if at all, from tensile strength. As an example, for class 40 gray iron, both strengths are 40,000 psi.

Ductile materials show a wide difference in these strengths. Figure 4-6 compares typical stress–strain curves for various materials. The yield strength in shear for a ductile material is determined to be about one-half (0.577) the yield strength in tension.

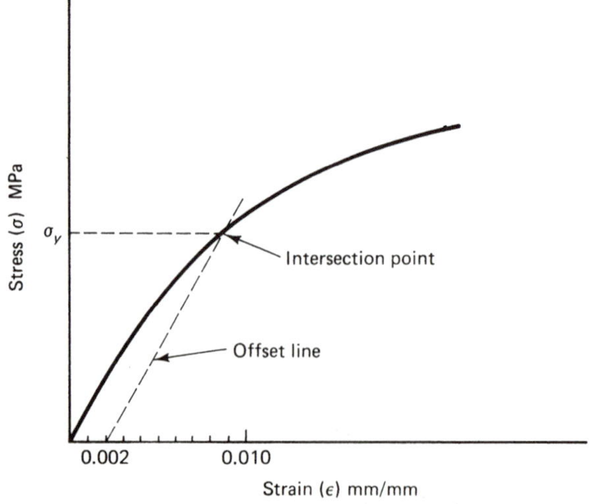

Figure 4-7 Determining offset yield strength using a stress–strain diagram.

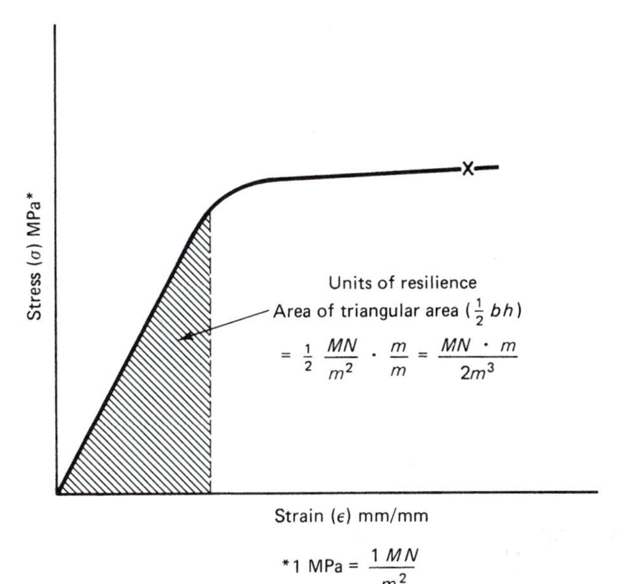

Figure 4-8 Resilience of a material as determined by a stress–strain diagram.

4.2.6 Resilience

The ***modulus of resilience** (R),* represented by the area under the straight-line portion (elastic region) of the stress–strain curve, is a measure of the energy per unit volume that the material can absorb without plastic deformation. If the unit of volume is one meter cubed, the SI units of resilience are $MN \cdot m/m^3$. Figure 4-8 shows that this area is a right triangle.

4.2.7 Shear Stress

A second family of stresses is known as ***shear stress*** or ***shearing stress.*** The symbol τ (Greek letter tau) represents a shear stress. A shearing force produces a shear stress in a material, which, in turn, results in a shearing deformation. Figure 4-9 shows a shearing load acting. In Figure 4-9(a), the shear force F produces an angular deformation of a block of material and not a lengthening or shortening. This deformation is exaggerated for illustrative purposes. In Figure 4-9(b), a block of material is subjected to a shear force that, if larger than the shear strength of the material, will shear a section out of the block. A hole in a metal plate can be produced by the action of a punch and hammer that delivers an impact blow, causing the metal in the plate to fracture by shear.

Shear strain, γ (Greek letter gamma), then is the deformation (δ_s—read as "delta sub s") produced by the shear force F [see Figure 4-9(a)] divided by the dimension h, or

$$\gamma = \frac{\sigma_s \text{ in.}}{h \text{ in.}}$$

Note that this ratio is also an expression for the tangent (ratio of the opposite side of an angle in a right triangle to the adjacent side of that angle) of the angle labeled γ in radians.

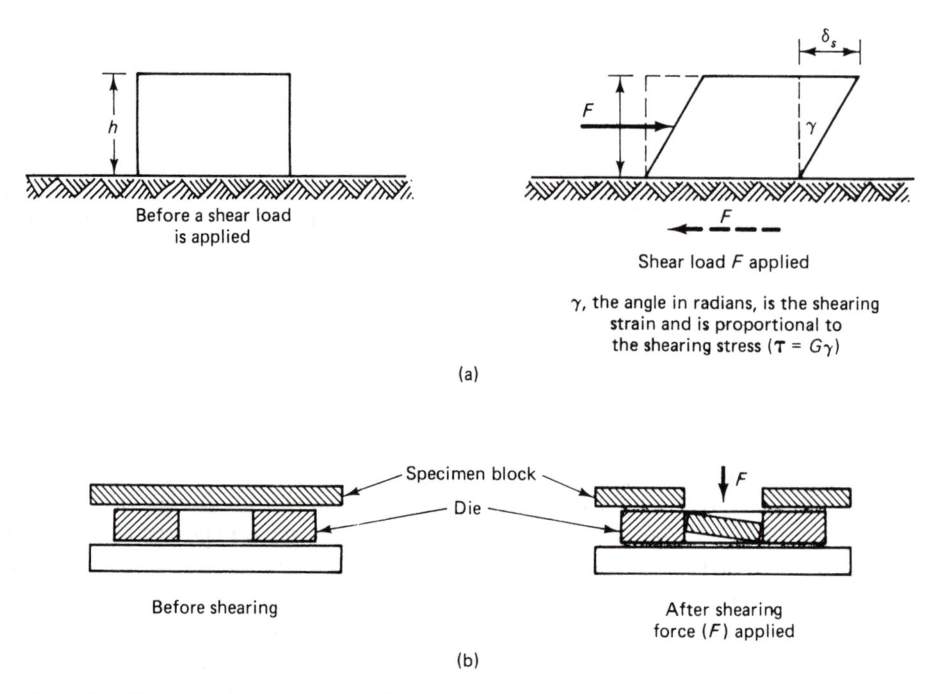

Figure 4-9 Shearing stress and strain. (a) Before shear load. (b) After shear load.

Illustrative Problem

A small pad subjected to a shearing force is deformed at the top of the pad 0.006 in. [see δ_s in Figure 4-9(a)]. The height of the pad *(h)* is 2 in. What is the shearing strain on the pad?

Solution Using $\gamma = \delta/h = 0.006$ in./2 in. $= 0.003$ radian.

Note: $\gamma =$ is an angular measurement in radians or degrees (π radians $= 180°$).

A shear force acts parallel to the area over which it acts, producing a shearing stress (τ) and a shear deformation (δ). A torque wrench used to tighten a nut on a bolt produces a shear force on the bolt that acts along a circular path. The atoms along this path resist a tendency to slide past one another. The shear stress, called ***torsion,*** is discussed more thoroughly later in this module. (Figure 4-2d shows this stress in a model of a plastic wrench, and Figure 4-27 shows a failure produced by torsion.)

A stress–strain diagram can be plotted using shear stress and shear strain. Such a diagram will show a definite straight-line portion (elastic region) in which the shearing stress is directly proportional to the shearing strain. Like the normal stress–strain ratio, the ratio of the two shear quantities, *G,* is known as the ***modulus of rigidity*** or ***shear modulus*** of elasticity. In mathematical terms,

$$G = \frac{\tau}{\gamma}$$

with units of psi or pascals. Finally, we can state, as we did with normal stress and normal strain, that the two quantities can be set equal to each other (Hooke's law), and the preceding equation can be written $\tau = G\gamma$, where G, the constant of proportionality, is the shear modulus. In addition to these simple loads, more complex forces are often imposed on materials, producing corresponding stresses, such as a bending force (see Figure 4-29) and a twisting force (see Figure 4-26).

Illustrative Problem

A piece of structural steel with a modulus of rigidity of 12×10^6 psi is subjected to a shearing stress of 15,000 psi. Find the strain on the steel.

Solution Using Hooke's law and solving for shearing strain yields

$$\gamma = \frac{\tau}{G} = \frac{15{,}000 \text{ psi}}{12 \times 10^6 \text{ psi}} = 0.00125 \text{ rad}$$

4.2.8 Ductility

A material that can undergo large plastic deformation without fracture is called a ***ductile material.*** A ***brittle material,*** on the other hand, shows an absence of ductility. Consequently, a brittle material shows little evidence of forthcoming fracture by yielding as a ductile material would do. A brittle fracture is a sudden fracture. A ductile material, by yielding slightly, can relieve excess stress that would ultimately cause fracture. This yielding can be accomplished without any degradation of other strength properties. Figure 4-10 shows a stress–strain curve for both a ductile and a brittle material. Note the difference in the amount of plastic deformation shown by each curve prior to fracture.

Figure 4-10(c) and (d) shows an application of the plastic deformation of a material undergoing a metal-forming operation called ***deep drawing.*** Flat sheets of metal (in this case, steel) are drawn and stretched by a punch and die set into the shape of a can. The metal is subjected to a variety of tensile and compressive forces that must be controlled to not

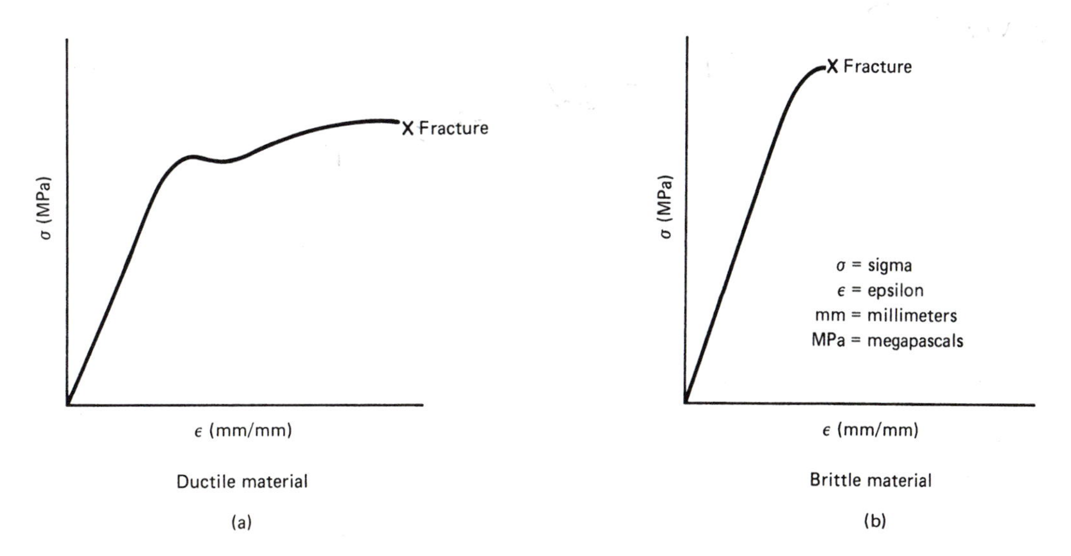

Figure 4-10 (a) Stress–strain curves for an (a) ductile and a (b) brittle material.

(c)

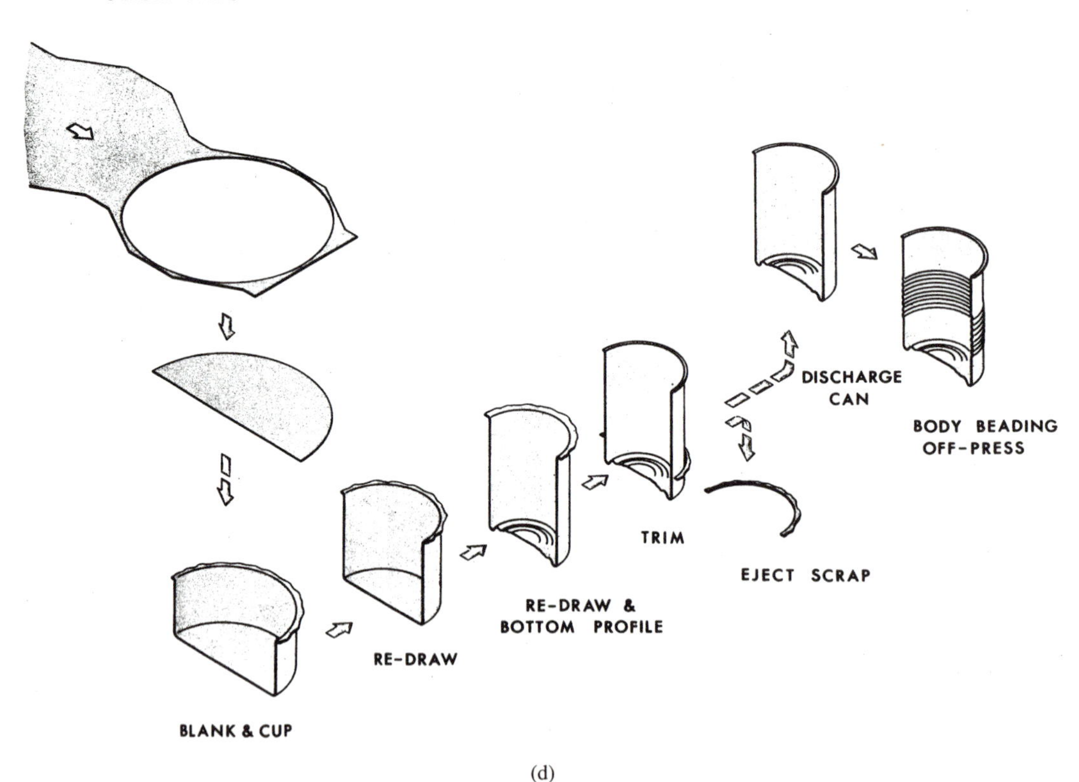

(d)

Figure 4-10 (c)Ductile and malleable metals allow complex forming. Photograph of the various stages in deep drawing a tin can. Beginning with a blank of 8.3 to 8.5 inches in diameter, a tin can is drawn with final dimensions of 3³⁄₁₆ in. diameter and 4³⁄₈ in. tall. (American Can Co.) (d) Diagram showing the stages required to produce a two-piece can. (American Can Co.)

exceed the strength of the metal. Other effects of this cold working is discussed in Module 5. If a failure of the metal occurs through tearing or excessive thinning of the can walls, the operation fails. Redrawing, resorted to when containers (shells) are too difficult to draw in one operation, produces longer sides in the container with reduced diameters. Greater can lengths can be produced by multiple drawing that involves drawing, annealing, and then redrawing to a greater depth. The term ***annealing,*** to be explained further in Module 5, is a heat-treating operation that, among other things, restores ductility to metals after they have been cold worked. Experience gained in deep drawing reveals that ***cold-rolled (cold-worked)*** sheets of metals with the desired grain orientation contain an ***anisotropy*** (material properties that differ in direction) conducive to successful drawing operations. In determining this anisotropy, tensile specimens are cut from cold-rolled sheets, subjected to an elongation of 15–20%, and strains are measured to determine the degree of anisotropic properties.

Ductility is measured in either of two ways. In the first method, the ***percent elongation*** is the ratio of the change in length of a specimen from zero stress to failure, compared to the original length; the quotient is then multiplied by 100%. In terms of a mathematical relationship, the factors above can be written as

$$\% \text{ elongation} = \frac{l_F - l_0 \times 100\%}{l_0}$$

Illustrative Problem

A test specimen with an extensometer clamped to it is undergoing a tensile test in a universal testing machine. [Figure 4-2, (a) to (c)]. Its original length is 2.0 in. and its final length is 2.8 in. What is its percent elongation?

$$\% \text{ elongation} = \frac{l_F - l_0}{l_0} \times 100\% = \frac{2.8 - 2.0}{2.0} \times 100\% = 40\%$$

The second method, ***percent reduction in area,*** measures the change in the cross-sectional area of a specimen, compares it to the original cross-sectional area, and multiplies the quotient obtained by 100%:

$$\% \text{ reduction in area} = \frac{A_0 - A_F \times 100\%}{A_0}$$

It is customary to consider a specimen that has 5% or less elongation as a brittle material. Brittle materials should not be considered as having inferior strength. Such materials lack the ability to plastically deform under load. Figure 4-11 is a pair of photographs showing the effects of increased ductility on the impact resistance of a specimen of boron–aluminum composite material. Pure metals can undergo elongations of 40–60% without rupture, whereas typical superalloys, because of their different size atoms, have ductilities of less than 20%.

4.2.9 Toughness

The ability or capacity of a material to absorb energy during plastic deformation is known as ***toughness.*** The ***modulus of toughness (T),*** equal to the total area under the stress–strain

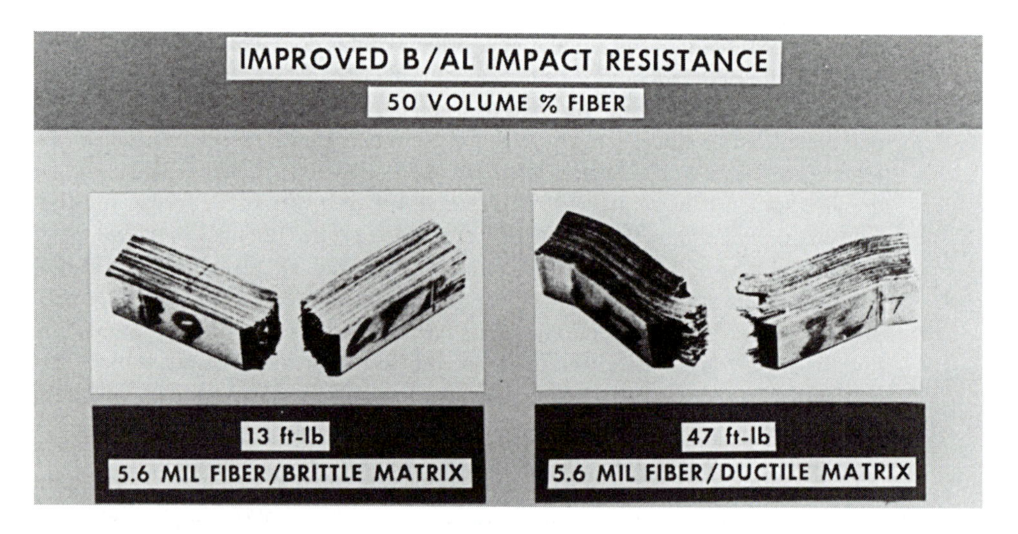

Figure 4-11 Photographs of impact specimens of fiber composites. (NASA.)

curve up to the point of rupture (see Figure 4-12), represents the amount of work per unit volume of a material required to produce fracture under static conditions.

Toughness can also be expressed in terms of the ease or difficulty in propagating a crack. It can be measured by the amount of energy absorbed by a material in creating a unit area of crack. A tough material would have no defects in its microstructure (not normally attained). Some crack-stopping mechanisms, such as inserting fibers in a metal matrix composite, are used to increase its toughness (Figure 4-11). Having a good amount of ductility makes many materials tough because the plastic flow permitted by the ductility prevents a concentration of stress at the tip of a crack.

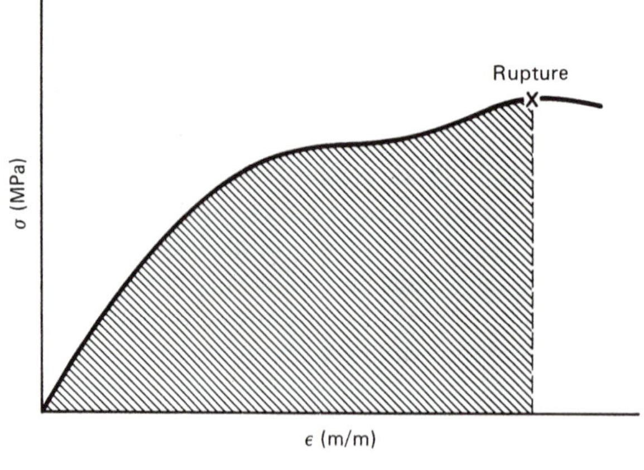

Figure 4-12 Modulus of toughness.

Impact is defined as a sudden application of a load confined to a localized area of a material. Exemplified by the striking of a material with a hammer, this relatively quick application of force, as opposed to a slow or static loading of a material, can cause considerable damage to a material that cannot adequately redistribute the stresses caused by the impact.

Ductile materials usually survive impact due to their microstructure, which allows slip to take place. Most metals have good toughness and thus have good impact resistance. Due to their inherent nature as compounds of metals and nonmetals, ceramics do not possess the ability to redistribute stresses and plastically deform. Consequently, they have poor toughness, poor impact resistance, and poor fracture toughness.

Impact tests are also used to give an indication of the relative toughness of a material. Notched-bar impact tests using either of the two standard notched specimens, the ***Charpy*** or the ***Izod,*** reveal the material's behavior in sustaining a shock load. These notched specimens of the material under test have either a keyhole or a V-notch cut to specifications. Figure 4-13 shows a universal impact tester for conducting Charpy, Izod, and tension impact tests on materials. Figures 4-12(b) and 11 are pictures of impact failures. Impact testing is considered ***dynamic testing,*** in contrast to the slower, ***static testing,*** such as tensile or compression testing. Impact tests serve the building construction industry, which must be concerned with how storms will affect building materials [see Figure 4-14(a)].

In contrast to the relatively small effect of temperature on the strength and ductility of metals, temperature has a marked effect on the impact resistance of notched bars. The Izod test is customarily used for nonmetallic materials. V-notched specimens are used to determine the resistance of the material to crack propagation. The impact energy absorbed by a specimen during failure is expressed in foot-pounds (ft-lbs) or joules (J), both of

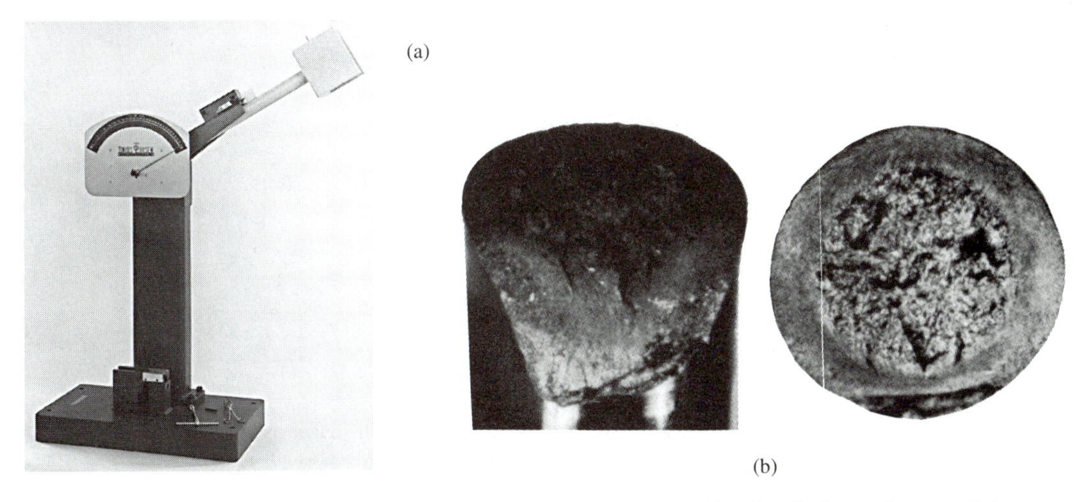

Figure 4-13 (a) Impact testing machine. (Tinius Olsen Testing Machine Co.) (b) Impact fractures. (John Deere Co.)

(a)

(b)

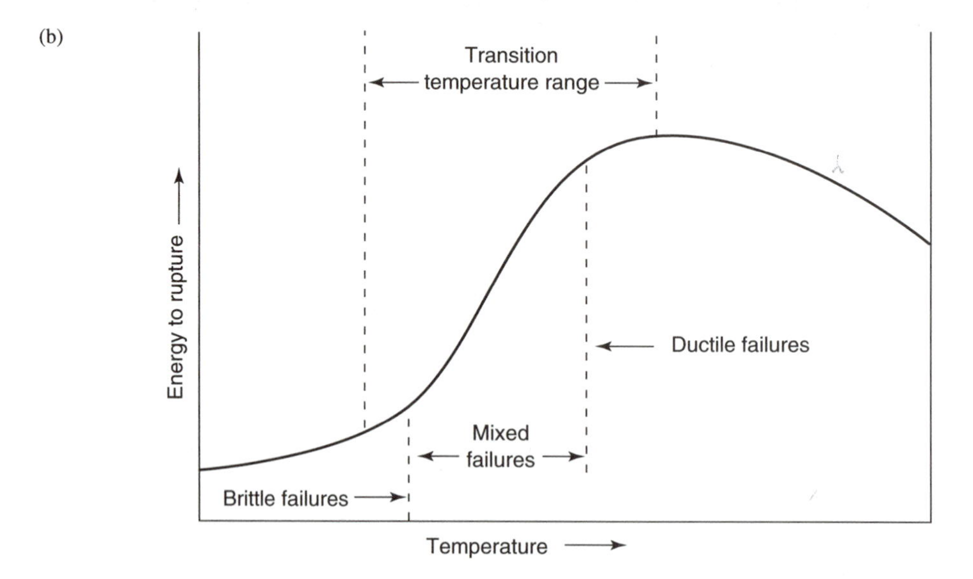

Figure 4-14 (a) Designers of the Hard Rock Cafe located at Universal Studios in Orlando, Florida, chose polycarbonate plastic sheet for the dramatic, covered arched walkway. To ensure that the plastic glazing would endure Florida's hurricanes, the complete window system underwent dynamic impact tests in which projectiles were fired from an air cannon. The plastic glazing passed local building codes for hurricane resistance and met Occupational Health and Safety Administration (OSHA) guidelines for fall-through protection for skylights and other overhead glazing applications. (G.E. Plastics.) (b) Variation of temperature versus energy to rupture in impact testing of metals.

which are used to describe the toughness of a material. Figure 4-11(b) shows, in general form, the effects of temperature over a wide range on the impact resistance of metals. Below some critical temperature, a particular metal will experience a brittle failure with little absorption of energy. Above some critical temperature, the failures are ductile with a much greater absorption of energy. Between these temperatures there is a zone of temperatures in which these failures are mixed [see Figure 4-14(b)]. In some steels the ***transition temperature range*** may be abrupt, while with other steels it may extend over a considerable range. Coarse grain size, strain hardening, or the addition of certain elements tend to raise the transition range of temperature, whereas fine grain size, heat treatments, and the addition of certain elements (nickel) tend to enhance notch toughness even at rather low temperatures. Not all materials have a distinct transition temperature. Metals with an fcc crystal structure do not, whereas bcc metals do have such a characteristic.

This knowledge was not known to materials engineers during World War II, when Liberty ships were built with steel that had a high ductility transition temperature; that is, this type of steel could not absorb low energy fractures at low temperatures, with the result that cracks continued to propagate until the ship broke apart. Many of these ships sailing on the Murmansk supply route bringing supplies from the United States to the Soviet Union have never been recovered after sinking somewhere in the North Atlantic.

4.2.10 Malleability

Malleability, workability, and ***formability*** are some terms related to ductility that describe, in a general way, the ability of materials to withstand plastic deformation without the occurrence of negative consequences (rupture, cracking, etc.) as a result of undergoing various mechanical processing techniques. Terms such as ***weldability, brazability,*** and ***machinability,*** although more properly classified under *processing properties,* are mentioned here as additional examples of terms used to generally describe the reaction of materials to various manufacturing and/or fabricating processes in industry.

4.2.11 Flexural or Bending Strength

A ***beam*** is a structural member that bends or flexes when subjected to forces perpendicular to its longitudinal axis. This axis in beam strength determination is called the beam's ***neutral axis*** (NA) because no normal stresses are assumed to exist along its length. The neutral axis is the edge view of a neutral plane. Figure 4-15(a) is a sketch of a ***simple supported beam*** (a beam supported at its ends with a pin and a roller). The transverse load or force P bends the beam (causes it to deform, i.e., deflect), thus producing both shear and normal stresses in the beam. The load in this example would cause the beam to deflect, as in Figure 4-15(b), resulting in normal stresses in compression near the top surface and normal stresses (tensile) at the bottom of the beam. These bending stresses are normal to any cross-sectional area through the beam, which makes their direction parallel to the neutral axis of the beam. The maximum bending stress in tension at failure is called the ***bend strength*** for ceramic materials and is often referred to as the ***modulus of rupture*** (MOR). The question of the presence of flaws, their distribution in a ceramic material, and their

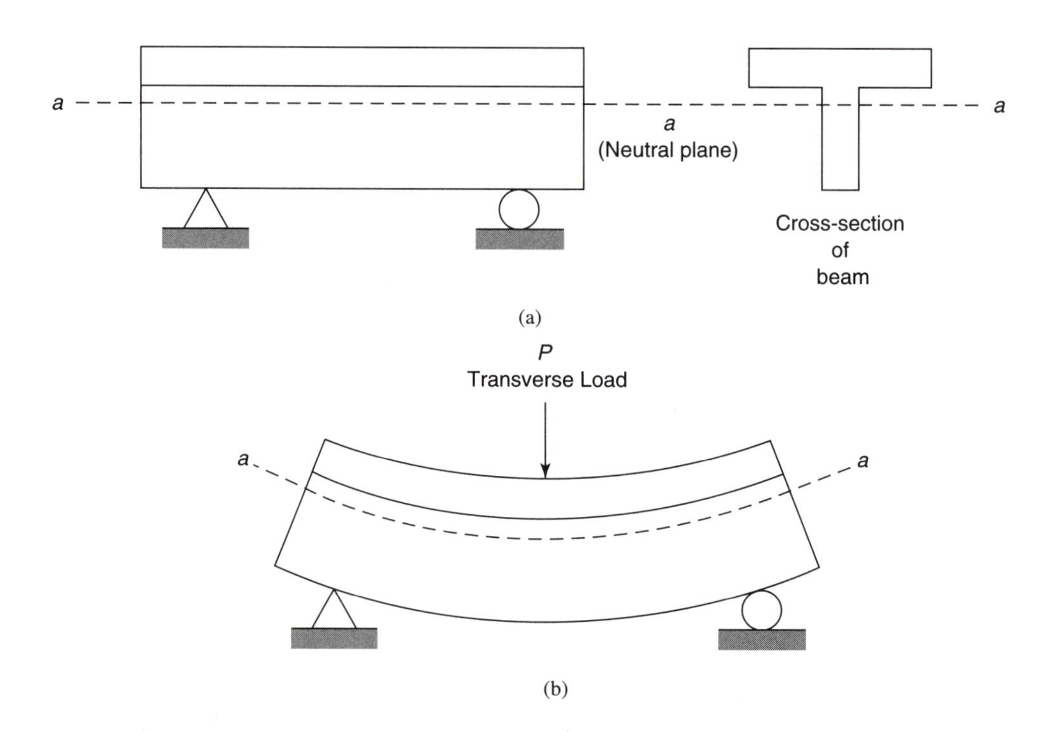

(c)

Figure 4-15 (a) Simple-supported beam with a transverse load. (b) Loaded beam showing deformation (deflection). (c) Wood specimen undergoing a static bending (flexural) test and showing evidence of failure due to tension. (Forest Products Laboratory, Forest Service, USDA.)

attendant adverse effects on the strength of the material are not taken into consideration when using this indicator of strength.

Assuming that the beam material is ***homogeneous*** (same material nature throughout) and ***isotropic*** (same properties in all directions throughout the beam material), the normal stresses will be at a maximum near the top and bottom surfaces of the beam. These normal stresses (both compressive and tensile) are known as ***flexural, fiber,*** or ***bending stresses.*** The ***flexure formula,***

$$\delta = \frac{Mc}{I}$$

relates these stresses (σ) to the ***bending moment*** *(M),* the maximum distance *(c)* from the beam's neutral axis (where bending stresses are zero) to the outer surfaces of the beam, and the rectangular ***moment of inertia*** of the cross-sectional area of the beam *(I).* A *flexure test* is performed using a simple-supported beam loaded as shown in Figure 4-15c. The maximum bending stress and deformation (deflection) are recorded for increments of load *P.* These data are plotted to obtain a stress–strain diagram. The maximum bending stress developed at failure is known as the ***flexural strength.*** For those materials that do not crack, the maximum bending or flexural stress is called the ***flexural yield strength.*** A bend test used to determine the ductility of certain materials should not be confused with this flexure test.

Illustrative Problem

Referring to the beam in Figure 4-15(b), the bending moment *(M)* produced by the force *(P)* at the center of the beam is 3530 ft · lb. The beam is a T-beam with a distance of 5 in. *(c)* from its centroid to the bottom surface of the beam. Calculate the stress at the bottom surface of the beam midway between the supports. The moment of inertia for the beam's cross-section is 136 in^4. Also indicate in your answer whether the stress is tensile or compressive.

Solution

$$\sigma = \frac{Mc}{I}$$

$$= \frac{3530\ \cancel{\text{ft}} \cdot \text{lb} \times 5\ \cancel{\text{in.}}}{136\ \text{in.}^2}\left(\frac{12\ \cancel{\text{in.}}}{1\ \cancel{\text{ft}}}\right)$$

$$= 1560\ \text{psi} \quad \text{(tensile)}$$

4.2.12 Fatigue (Endurance) Strength

Service conditions commonly involve many repetitions of applied stress or reversals of stress. We know that when a ***simple beam*** is subjected to a downward-acting, transverse load (see Figure 4-16), the material in the top half of the beam will be compressed (the stress s_c is compression) and the lower half of the beam will be subjected to a tensile stress (s_t).* This is demonstrated in Figures 4-17 and 4-18. Roller and pin supports are termed

Note: σ is the universal symbol for stress. However, lowercase s is sometimes substituted for σ, as seen in Figures 4-17 and 4-18.

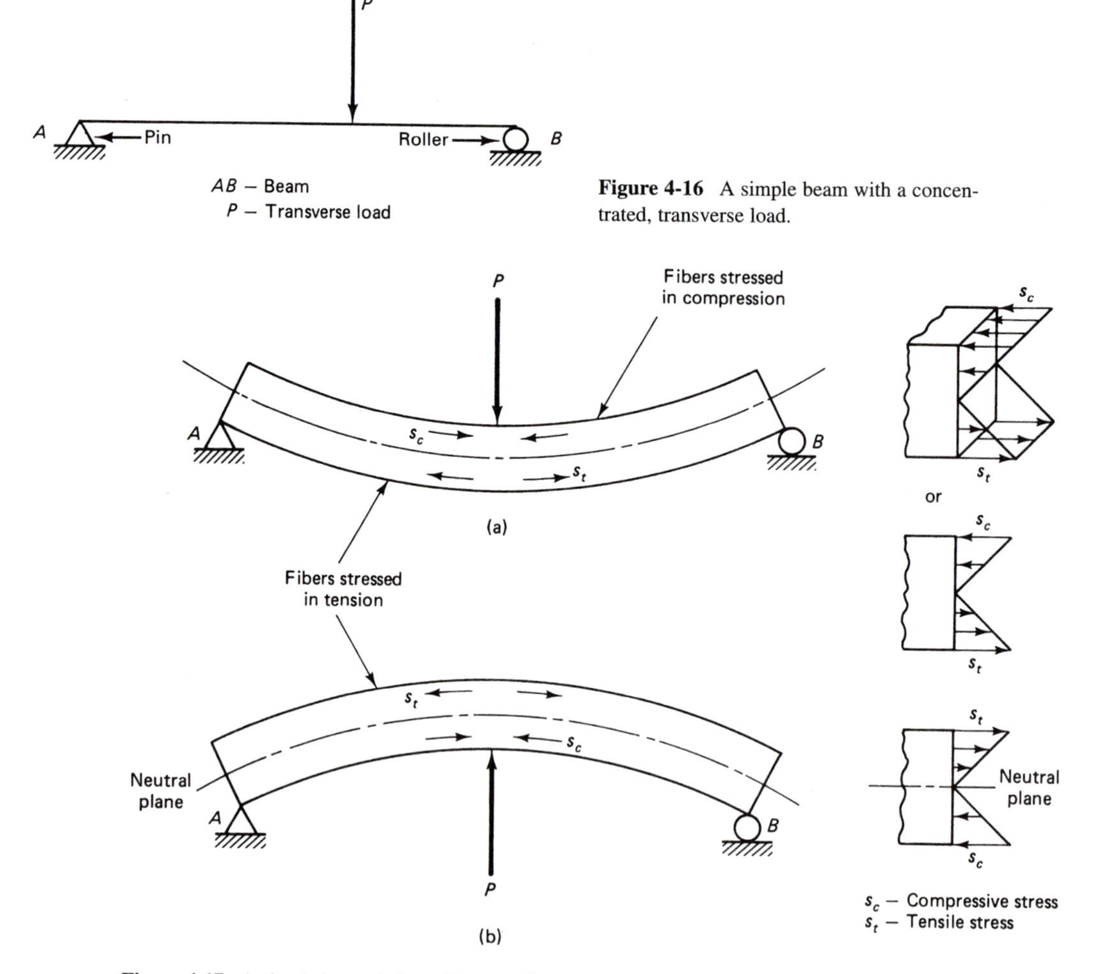

Figure 4-16 A simple beam with a concentrated, transverse load.

Figure 4-17 A simple beam deflected by a cyclic, transverse load.

simple supports to differentiate them from fixed supports, which can also resist rotational forces. A shaft with a pulley or gear can be compared to the horizontal beam referred to above. The load or force, in this instance, is a pulley or gear force that causes the shaft to deflect (bend), particularly if sufficient bearing supports are not provided. Now, if a transverse load is applied and removed in some cyclic fashion, the material in the beam or shaft would go from a condition of zero stress to a stressed condition and back to a condition of zero stress. This type of cyclic loading is called *repetitious.* If the stress in the material changes due to the loading from compression to tensile, or vice versa, this is known as ***stress reversal.*** The latter condition is easier to visualize with a point on the surface of a rotating shaft turning at 1000 revolutions per minute whose bending stress changes from compression to tension 1000 times a minute. In practice, many failures have occurred un-

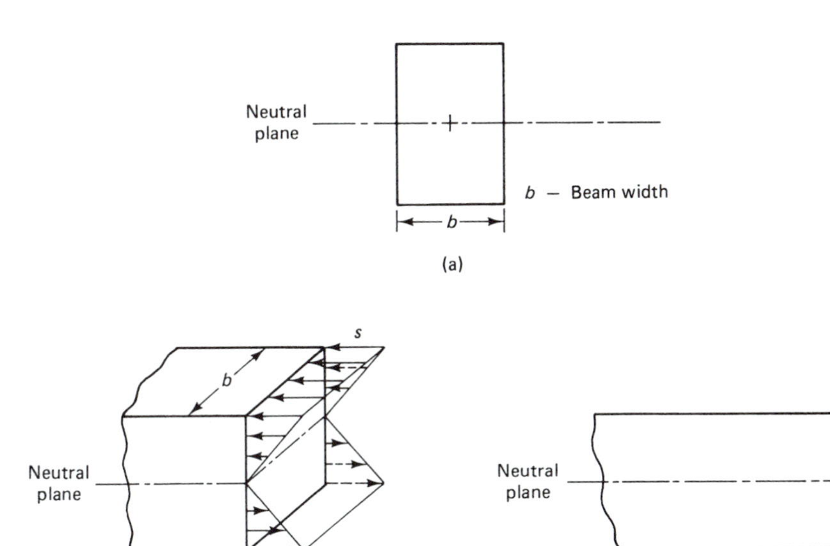

Figure 4-18 (a) End view of beam showing cross-section. (b) Front view (three-dimensional) of a portion of beam showing distribution of tensile and compressive bending stresses acting on a transverse plane. (c) Front view (two-dimensional) showing same stress distribution as in (b).

der such conditions when the stresses developed were well below the ultimate stress and frequently below the yield strength. These failures are called ***fatigue*** (or ***endurance***) ***failures*** (see Figure 4-19).

A fatigue failure starts as a tiny crack whose origin is often traced to an inspection stamp, tool mark, or other defect on the surface. The crack produces a stress concentration that assists in the growth of the crack until eventually the area of material remaining to withstand the stress is insufficient, which results in a sudden fracture (see Figure 4-19). Much empirical research is done using fatigue testing machines to determine the strength of materials under fatigue loading. Machines developed by the National Aeronautics and

Figure 4-19 Combined fatigue fracture due to combined bending and torsional loads on an axle. The outer surface was hardened. A crack began at the top surface. The rough surface in the center indicates the last area to fracture instantly when the axle broke. (John Deere Co.)

Space Administration use computers that simulate aircraft or spacecraft flights to test the material until it fails under varying conditions of load and ambient temperatures. The results of these many tests are recorded on semilog or log-log paper to produce *s* (stress)–*N* (cycles) diagrams. Typical ***s–N diagrams*** are shown in Figure 4-20 for a fiberglass composite, in Figure 4-21 for a typical low-carbon steel, and in Figure 4-22 for some nonferrous metals and some plastics. Figure 4-23 is a photograph of a typical fatigue fracture showing two very distinct surfaces.

The *s–N* curve for a low-carbon steel (see Figure 4-21) shows an abrupt break, or *knee,* at which point the curve tends to approach a horizontal line. In other words, with any stress below the fatigue or endurance limit, a low-carbon steel can be cycled continuously without fracturing. Aluminum and other nonferrous materials fracture at relatively low stresses after many cycles. For many ferrous alloys the endurance limit is about one-half the tensile strength of the metal. Furthermore, they exhibit no fatigue limit, which means that there is no stress below which they will not fracture. Therefore, when speaking of the fatigue of nonferrous materials, it is necessary to express both the stress and the number of cycles in describing the life of the material.

Figure 4-24 is a diagram of the DISMAP system (measurement of micro*dis*placements by *ma*chine vision *p*hotogrammetry) which is related to fatigue testing and an image processing system for stereoimaging measurement of displacements from photographs. Stereoimaging is a technique for measuring displacements and computing strains. True stereoimaging uses two photographs of the same object taken at different locations, whereas stereoimaging uses two images that are intentionally distorted in relation to each other and taken at the same location. This new system makes it possible to obtain seven values of strain at each of several hundred locations in a photographic field within a few minutes of conducting an experiment. Coupled with SEM technology, DISMAP has resulted in a detailed understanding of the micromechanics of deformation and fracture of high-temperature alloys and composites systems and has improved considerably the

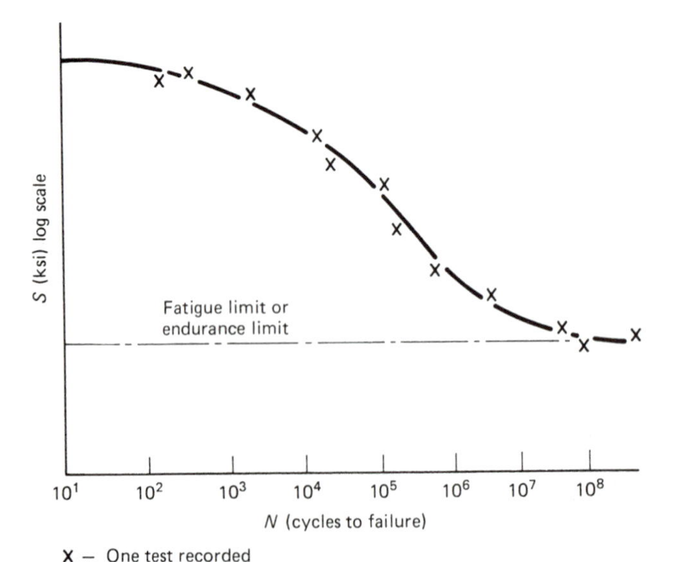

Figure 4-20 *s–N* diagram for a typical fiberglass composite material.

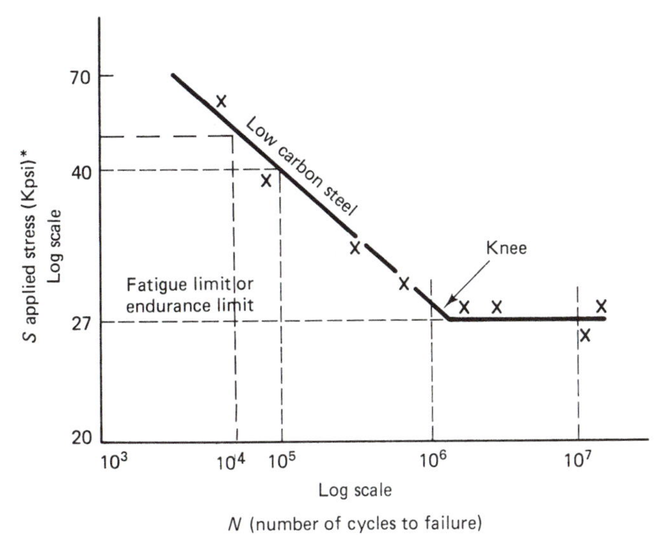

Figure 4-21 *s–N* diagram for a typical low-carbon steel.

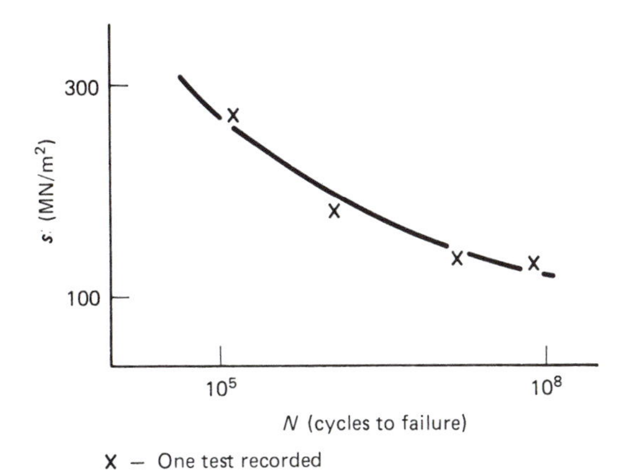

Figure 4-22 *s–N* diagram for nonferrous metal (e.g. aluminum and copper) and some plastics (e.g. nylon).

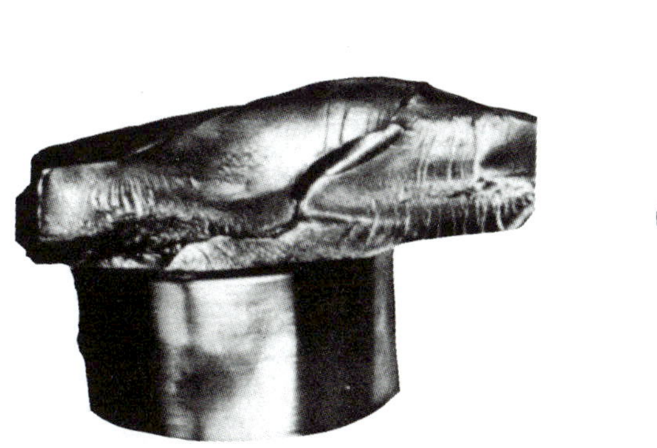
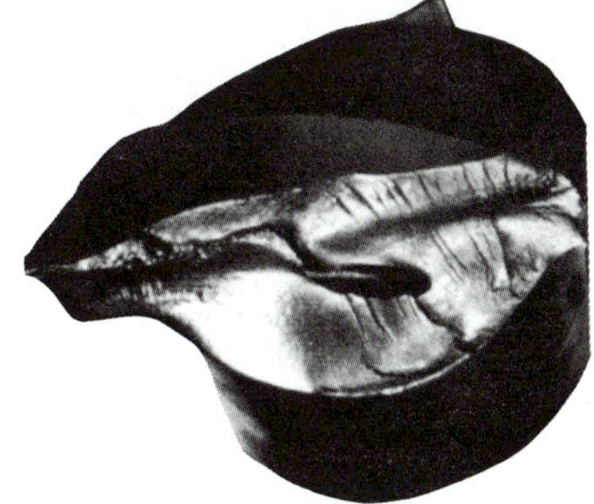

Figure 4-23 Fatigue fracture due to progressive fatigue cracking, with cracking progressing across most of the section before the final overload fractures the remaining metal. (John Deere Co.)

(a)

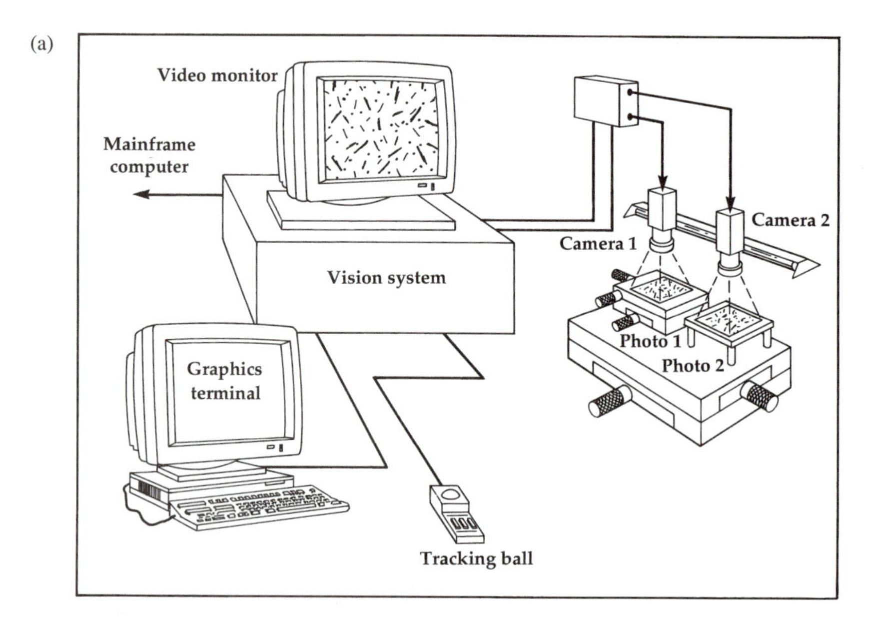

(b)

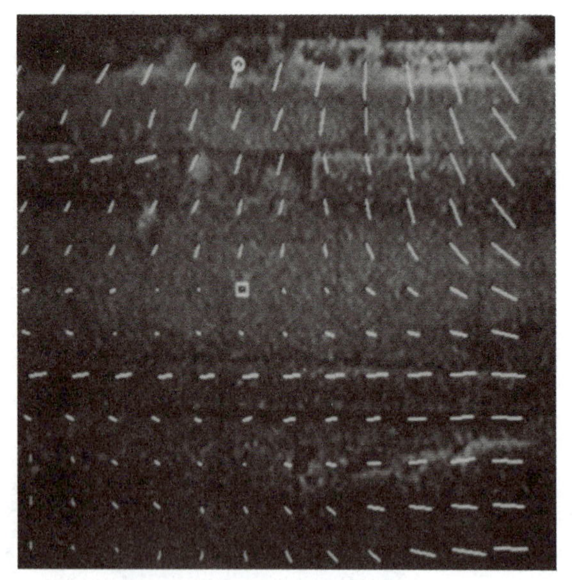

Figure 4-24 (a) DISMAP image processing system developed by Southwest Research Institute is used for stereoimaging measurement of displacement from photographs. (b) Displacements measured at 6-μm intervals are superimposed on the microstructure of a glass-ceramic composite specimen deformed at 800°C (1470°F). *(Advanced Materials & Processes.)*

understanding of high-temperature failure mechanisms. These combined testing and analytical methods are used for tensile deformation, fatigue-crack initiation and growth, creep, creep-fatigue interaction, and fracture toughness testing of numerous metallic alloy, composite, and ceramic specimens.

The ***fatigue*** or ***endurance limit*** is the maximum stress below which a material can presumably endure an infinite number of stress cycles (Figures 4-20 and 4-21). ***Fatigue***

strength is the maximum stress that can be sustained for a specified number of cycles without fracture. In other words, fatigue strength can be any value on the ordinate of the *s–N* diagram. To make it meaningful, the corresponding *N* value must be reported also. In Figure 4-21 the fatigue strength of about 40 kpsi corresponds with an *N* of 10^5 cycles. The fatigue limit, as determined empirically, is generally below the yield strength. Most design stresses are lower than the fatigue or yield strength of a material primarily because of the adverse effects of surface conditions on the strength of materials.

Another term used in describing failures is ***endurance ratio,*** which is the quotient of endurance limit to tensile strength. For many ferrous alloys the endurance limit is about one-half the tensile strength of the metal. In view of the expense involved in running fatigue tests, particularly when attempting to duplicate service conditions in a laboratory, a fatigue or endurance ratio is sometimes used instead. This ratio varies between 0.25 and 0.45, depending on the material.

4.2.13 Creep (Creep Strength)

Creep is a slow process of plastic deformation that takes place when a material is subjected to a constant condition of loading (stress) below its normal yield strength. Creep occurs at any temperature. However, at low temperatures, slip (movement of dislocations) is impeded by impurity atoms and grain boundaries. At high temperatures, the diffusion of atoms and vacancies permits the dislocations to move around impurity atoms and beyond grain boundaries, which results in much higher creep rates. The word *creep* implies, then, that a material plastically deforms or flows very slowly under load as a function of time. After a certain amount of time has elapsed under constant load, the ***creep strain*** (plastic deformation) will increase and some materials will rupture. This rupture, or fracture, is known as ***creep rupture.*** Aluminum alloys begin to creep at around 100°C. Aluminum aircraft wing panels are creep formed. During ***creep forming,*** a workpiece is loaded against a die or tool for a specified time. The load is normally applied using pressure or vacuum at temperatures around 160°C. The process results in predetermined shapes that require little or no machining. The amount of creep resulting from such operations must be predetermined through numerous creep tests. In addition, springback must be taken into consideration. When the forming or bending force is removed at the end of the deforming operation, elastic energy remains in the deformed part, causing it to recover partially to its original shape. This elastic recovery is called ***springback,*** which is defined as the increase in the included angle of the bent part relative to the included angle of the forming tool after the tool is removed.

Although some materials will creep at low or room temperatures, this type of plastic deformation is usually associated with high temperatures. Polymeric materials creep at room temperature, but this low-temperature creep is called ***cold flow.*** Both steam and gas turbines used for propulsion of ships or the generation of electricity operate, by necessity, at high temperatures over many years. The extremely close fits between turbine blades and their casings prevent the escape of steam or gas past the blades. If these turbine blades were allowed to change their dimensions at any time during their expected service life, during which time they are under large centrifugal loads (rotating at extremely high speeds) at extremely high temperatures for long periods, possibly weeks, the failure would be catastrophic. Consequently, the material from which these turbine blades are fabricated must possess, among other necessary properties, high creep resistance. Continuing research over the years has produced materials for medium- and high-temperature service such as the many titanium alloys and the "superalloys" composed of iron, nickel, and/or cobalt. Module 5 discusses investment casting of such alloys.

Tensile creep tests develop data at a constant level (stress) and at a constant temperature. The amount of creep strain when plotted against time, as in Figure 4-25, indicates that the creep curve formed can be divided into several stages, with the last stage ending with a rupture of the material. The *creep rate* is determined, at any point, by the slope of the curve (drawing a tangent to the curve at the point). The creep rate, the slope of the middle portion of the curve, is nearly constant and represents the minimum rate. Once the creep enters the last stage, rupture soon follows. Creep rate is usually expressed in percent creep strain per hour. A typical rate might be $10^{-4}\%$ per hour.

Engineering materials must perform satisfactorily throughout their ***service lives***. In many applications, service life can extend well beyond 20 years. A determination must be made as to the maximum allowable deformation that can be tolerated during the expected service life of a material before the material is chosen for the particular application.

Research in polycrystalline materials with varying amounts of grain-boundary areas has shown that, at high temperatures, ***coarse-grained materials*** possess more creep resistance than do fine-grained crystalline materials. At low temperatures, ***fine-grained materials*** offer more creep resistance. These statements, including the phenomenon of creep itself, can be explained in terms of the movement and stoppage of mobile dislocations (linear arrangements of atoms making up imperfections in the atomic structure of solid materials) throughout the crystal structure. Creep, then, like other properties of materials, is dependent on the structure of materials.

4.2.14 Torsional Strength

Torsion describes the process of twisting. A body such as a circular rod (e.g., a shaft for transmitting power), as shown in Figure 4-26, is under torsion as a result of a force acting

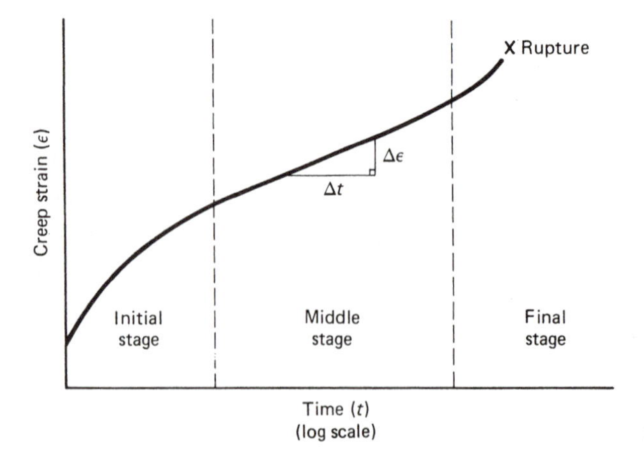

Figure 4-25 Typical creep curve.

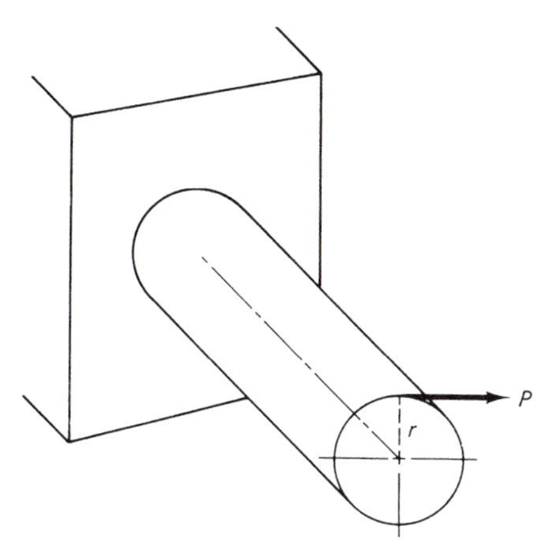

Figure 4-26 Circular rod under torsion.

to turn one end around the longitudinal axis of the rod while the other end is fixed or twisted in the opposite direction. The material resists this twisting action by generating a similar twist internally. The product of the force P and the radius r perpendicular to the line of action of the force is called a ***torque.*** The external torque produces both a stress and a deformation of the rod. The stress is classed as a shear stress, which causes the atoms to twist past each other. The ***shear deformation*** is measured in terms of the angle γ (Greek letter gamma) and the angle θ (Greek letter theta), as indicated in Figure 4-27. As a result of the torque T, the point C on the surface of the rod moves to point C'. This deformation can be measured by angle γ on the surface of the rod or by angle θ shown on the cross-sectional area of the rod. The angle θ is known as the angle of twist. Both angles are exaggerated for illustrative purposes. Forces can combine to cause failures, as seen in Figures 4-19 and 4-20(b).

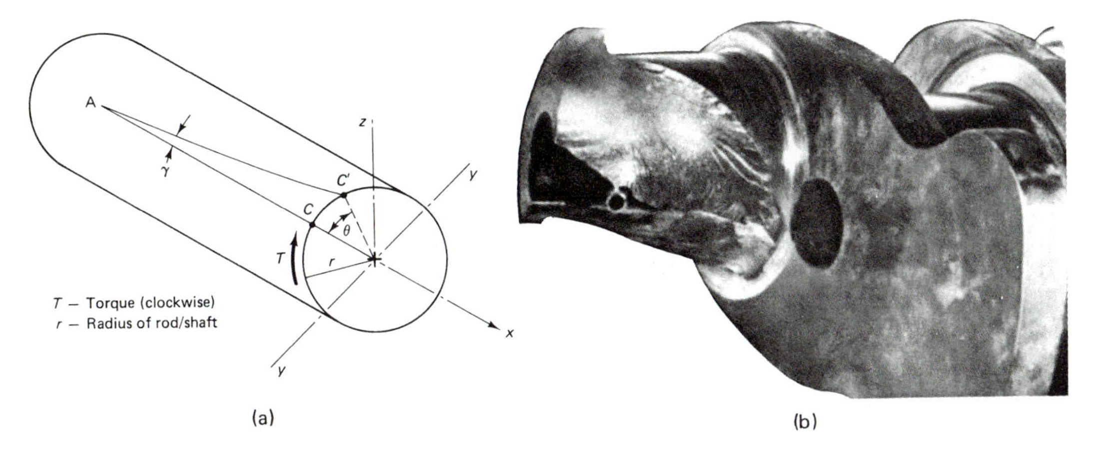

Figure 4-27 (a) Circular rod undergoing a torque, with resulting deformation. (b) Torsional fracture. Torsional (twisting) loads produce spiral types of failure. Note the curved line from (A) to (C') is part of a helix.

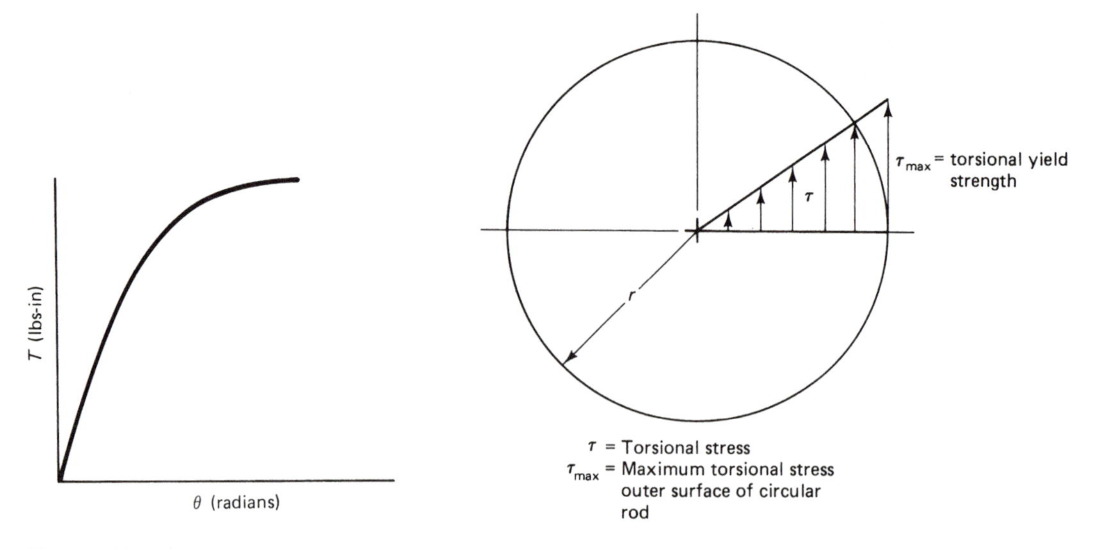

Figure 4-28 Typical torque-twist diagram.

Figure 4-29 Distribution of torsional stress in a circular rod.

A torsion-test machine measures the torque applied to a specimen of material along with the corresponding ***angle of twist* (θ)**. The results can be plotted as a *torque–twist diagram* (see Figure 4-28), which resembles an ordinary stress–strain diagram as obtained by the usual tensile-test procedure. The torsional stress is the shear stress (τ) produced in the material by the applied torque and is calculated using the torsion formula for circular shafts:

$$\tau_{max} = \frac{Tr}{J}$$

where r is the radius of the cross-sectional area and J is the centroidal ***polar moment of inertia.*** The maximum torsional stress occurs when r is a maximum, as indicated in the preceding equation and illustrated in Figure 4-29. The maximum torsional stress occurring at the outer surfaces of the circular rod is called the ***torsional yield strength***. It relates to that point on the torque–twist diagram where the curve begins to depart from a straight line. Torsional yield strength roughly corresponds to the yield strength in shear. The ***ultimate torsional strength*** or ***modulus of rupture*** expresses a measure of the ability of material to withstand a twisting load. It is roughly equivalent to the ultimate shear strength. The ***torsional modulus of elasticity*** as determined by the torque–twist diagram known as the ***modulus of rigidity*** is approximately equal to the shear modulus or the modulus of elasticity in shear *(G)*.

Illustrative Problem

How much torque *(T)* is needed to produce a shear stress of 79.6 MPa on the surface of a 40-mm-diameter shaft with a polar moment of inertia *(J)* equal to 25.12×10^{-8} m^4?

Solution

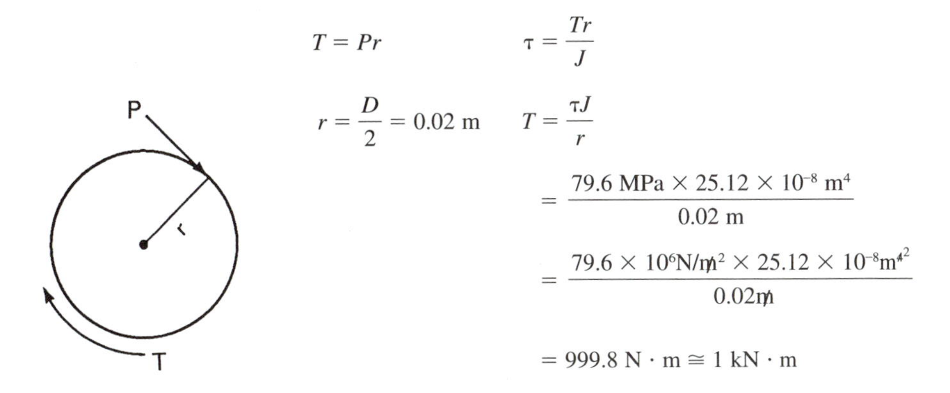

$$T = Pr \qquad \tau = \frac{Tr}{J}$$

$$r = \frac{D}{2} = 0.02 \text{ m} \qquad T = \frac{\tau J}{r}$$

$$= \frac{79.6 \text{ MPa} \times 25.12 \times 10^{-8} \text{ m}^4}{0.02 \text{ m}}$$

$$= \frac{79.6 \times 10^6 \text{N}/\cancel{\text{m}}^2 \times 25.12 \times 10^{-8} \text{m}^{\cancel{4}2}}{0.02\cancel{\text{m}}}$$

$$= 999.8 \text{ N} \cdot \text{m} \cong 1 \text{ kN} \cdot \text{m}$$

4.2.15 Hardness

Hardness is a measure of a material's resistance to penetration (local plastic deformation) or scratching. One of the oldest and most common hardness tests, based on measuring the degree of penetration of a material as an indication of hardness, is the Brinell. ***Brinell hardness numbers*** (HB) are a measure of the size of the penetration made by a 10-mm steel or tungsten carbide sphere with different loads, depending on the material under test. The indentation size is measured using a macroscope containing an ocular scale. ***Vickers hardness numbers*** (HV) employ a diamond pyramid indentor. Otherwise, the two tests are basically similar. ***Rockwell hardness testers,*** using a variety of indentors and loads with corresponding scales, are direct-reading instruments (i.e., the hardness is read directly from a dial). The hardness number, for example, 65 HRC, indicates the reading came from the C scale using a diamond cone indentor and a 150-kg load. It is therefore important in reporting Rockwell hardness readings to include the scale letter so that the person wishing to use the information knows the type of indentor as well as the size of load used in the test. (See Table 4-1 and Figure 4-30.)

Other tests used to report hardness include a file test (resistance to scratching), a ***Scleroscope*** HSc test (measures the rebound of a small weight bounced off the surface of the material), and a comparison test that also uses scratch resistance. The latter test compares a material's hardness to some 10 known minerals arranged in order of hardness. Mostly used by mineralogists, the ***Mohs scale*** classes hardness of all materials between 1 (the hardness of talc) and 10 (hardness of diamond). The scale is based on the ability of a hard material to scratch a softer material (Table 4-1).

Materials such as very thin materials (e.g., coatings, foils, plated surfaces), very brittle materials (e.g., glass or silicon), and very small parts (e.g., gears in a wristwatch) require special care in hardness testing due primarily to their thinness and/or size. Furthermore, laboratory research in materials necessitates hardness testing on a microscale in determining the differences in hardness over the minute area of a single grain of metal or between the middle of a crystal and the grain boundary area. For such purposes a microhardness tester [Figure 4-30(a)] finds application with loads and indentations that are so small that indentations require microscopic viewing with appropriate scales for accurate measurement.

TABLE 4-1 HARDNESS SCALES COMPARISONS

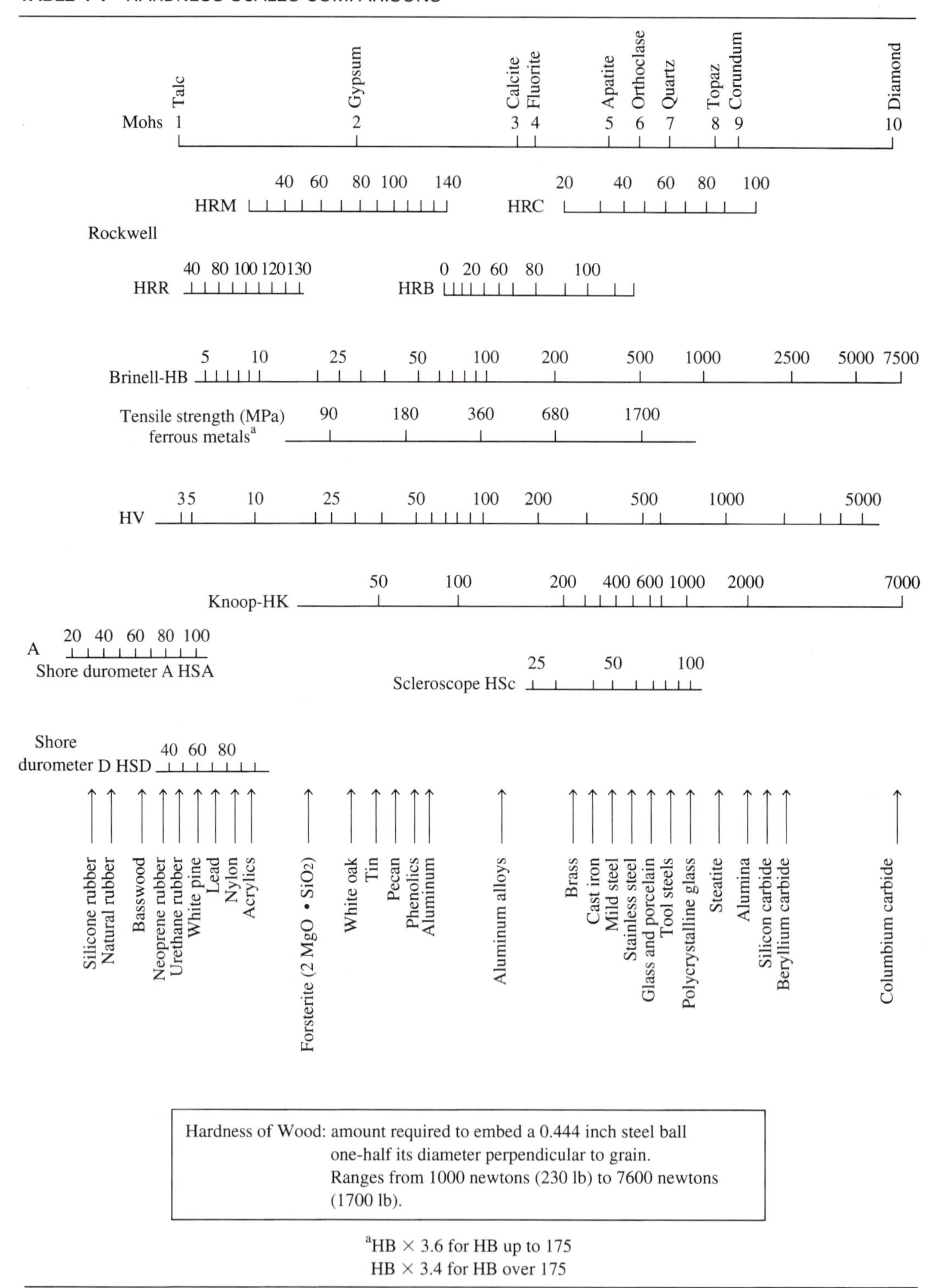

Hardness of Wood: amount required to embed a 0.444 inch steel ball one-half its diameter perpendicular to grain. Ranges from 1000 newtons (230 lb) to 7600 newtons (1700 lb).

[a]HB × 3.6 for HB up to 175
HB × 3.4 for HB over 175

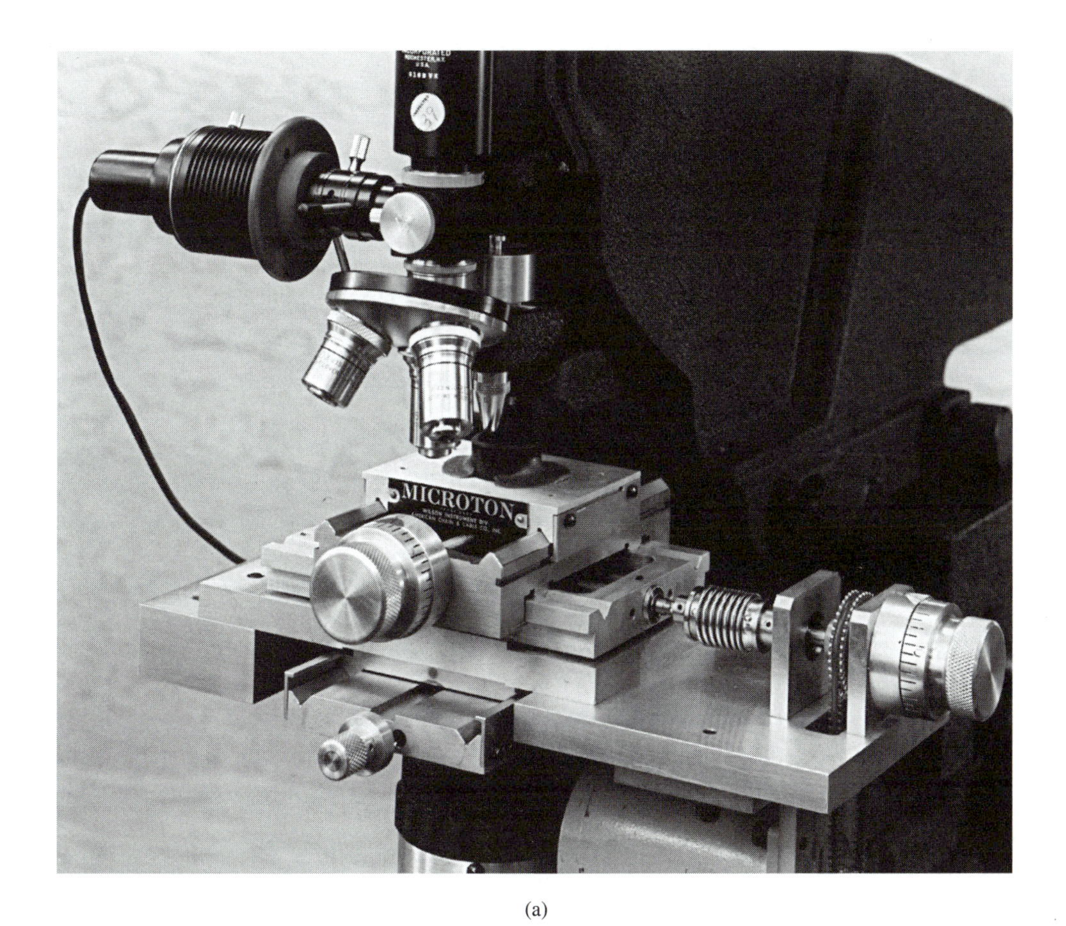

(a)

(b)

(c)

Figure 4-30 (a) Photograph of Tukon microhardness tester. (Measurement Systems Div., Page Wilson Corp.) (b) Photograph of Rockwell hardness tester. (Measurement Systems Div., Page Wilson Corp.) (c) Photograph of Air-O-Brinell metal hardness tester with digital readout of Brinell values. (Tinius Olsen Testing Machine Co.)

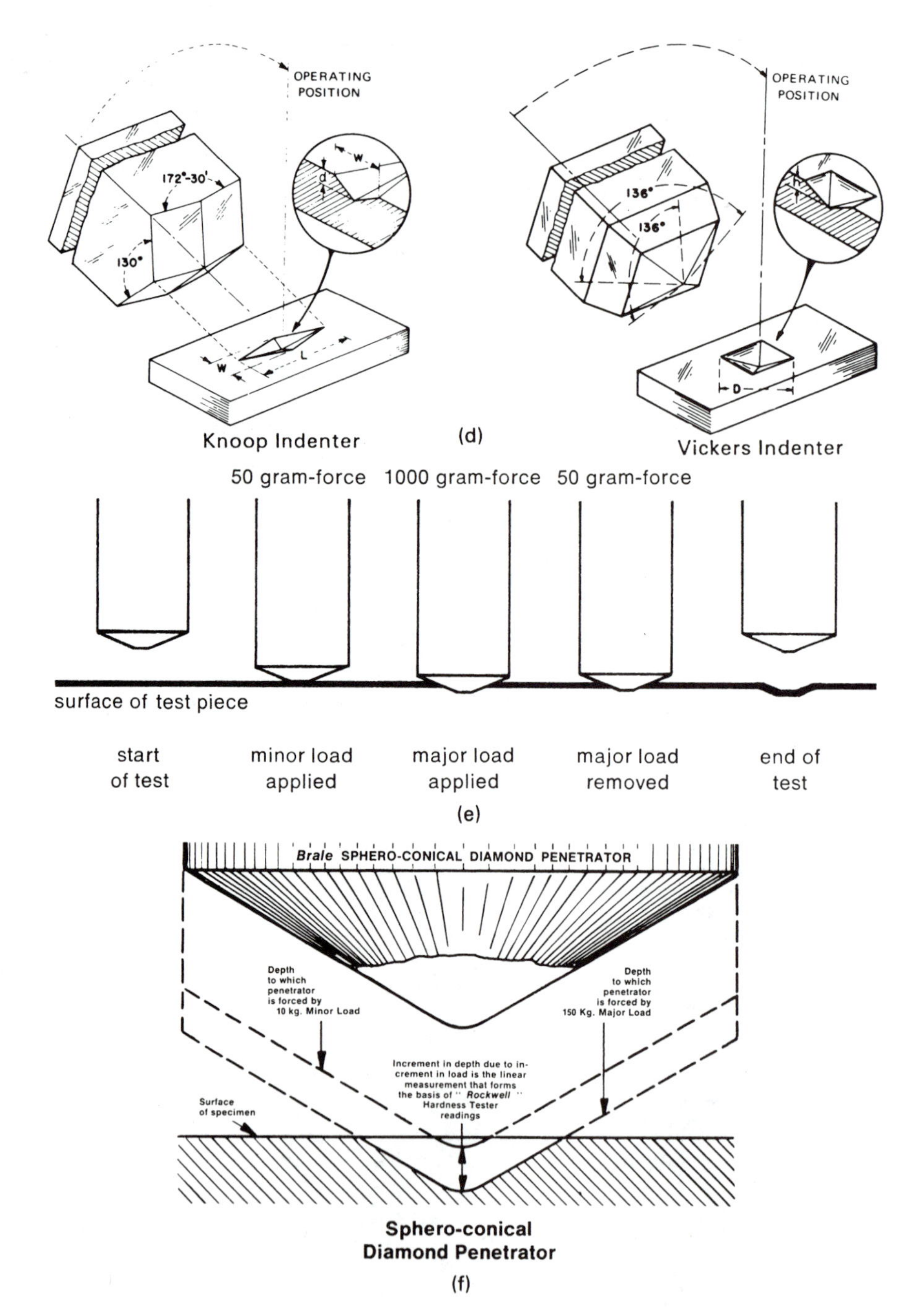

Figure 4-30 (d) Microhardness penetrator (Knoop and Vickers) indentations. (Wilson Instrument Division of ACCO.) (e) Various standard loads for the Rockwell hardness tester. (Wilson Instrument Division of ACCO.) (f) Brale sphero-conical diamond penetrator. (Wilson Instrument Division of ACCO.)

Table 4-1 provides a comparison of the approximate hardness of a variety of materials using 11 different hardness scales. These readings may be compared to the tensile strengths of steel given in Table 10-4. Some correlation exists between hardness and tensile strength, but it is only approximate. For example, the tensile strength of steel (but not other materials) is about 500 times the HB as listed in the table. As a general rule, the tensile strength of a given ductile metal can be estimated only from the hardness reading within an error of less than 10%.

Illustrative Problem

Given the formula for determining the Brinell hardness number (HB), where P is the applied load (kg), D is the diameter of the steel ball (mm), and d is the diameter of the indentation (mm),

$$HB = \frac{P}{(\pi D/2)(D - \sqrt{D^2 - d^2})}$$

what is the Brinell hardness number if a 10-mm ball with a load of 3000 kg produces an indentation with a diameter of 2.88 mm?

Solution

$$HB = \frac{3000}{\pi(10/2)\,[10 - \sqrt{100 - (2.88)^2}]}$$

$$= \frac{600}{\pi(10 - 9.576)} = \frac{600}{1.333} = 450$$

4.3 FAILURE ANALYSIS

Up to this point in the module, we have discussed many of the physical properties involved in mechanical failures of materials. In addition, other properties of materials produce failures in materials, such as thermal or chemical. Failures of materials not only cost vast sums of money annually but contribute to loss of human life. Many failures in equipment, structures, and systems are not all attributable to material failures but are the results of other factors not discussed in this book. When a failure occurs, failure analysis is undertaken to help determine whether the failure was due to a design or material deficiency. An examination of the fracture tries to reconstruct the sequence of events and the cause of the fracture. Much information can be determined by analysis of the path a crack makes as it propagates through the material, the origin of the crack, and the cause of the crack initiation (i.e., material flow, oxidation/corrosion, tensile stress overload, thermal shock, etc.). Often, most of these data can be determined from visual observation by a trained investigator. If not, low-power microscopy, scanning electron microscopy, electron probe, and other sophisticated instruments might be needed to analyze the chemical makeup of the fracture surfaces to aid in determining where responsibility for fracture lies. This highly technical area within the broad scope of ***failure analysis*** is known as ***fractography.***

A finding of responsibility is critical not only as an element in liability suits, but is equally important in future design, specification writing, modifications, selection of materials, and the like. A classic example of failure analysis is the investigation of the Hyatt Regency Hotel disaster, in which 113 people were killed and many injured when concrete walkways collapsed in 1981 (Figure 4-31).

Factors of Safety

In the design of machine parts and structural members, it is the responsibility of the designer to provide a margin of safety, i.e., to ensure that a machine part is safe for operation under normal operating conditions. ***Factor of safety*** (or design factor, N) is defined as the ratio of ultimate strength to allowable stress. ***Allowable stress,*** also called ***design stress, working stress,*** or ***safe stress,*** is the maximum level of stress that a part will be permitted to endure under operating conditions. In determining the allowable stress, considerations of the type of load on the part (static, impact, fluctuating, reversed, etc.) and the material from which a part is made are paramount. A material's ductility, as well as its strengths and stiffness, must be evaluated because the mode of failure of a ductile material varies dramatically from a brittle material such as gray cast iron, some plastics, or some heat-treated steels. For example, experience warrants that, for fatigue loading (reversal of stress from tension to compression on a repetitive basis), a ductile material be selected with a percent elongation of 10 percent or higher to ensure against a brittle-type failure. The need for a factor of safety is most evident when knowledge is sketchy about 1) possible unexpected high loads on a structure, 2) the degree of residual stresses within a material due to manufacturing processes, 3) homogeneity of the internal structure of a material throughout its entirety, 4) loading history of the material since its fabrication,

(a)

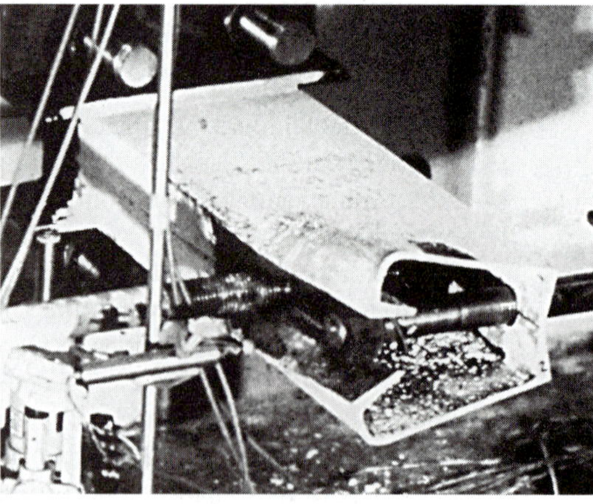

(b)

Figure 4-31 (a) The National Institute for Standards and Technology (NIST) investigated the Hyatt Regency Hotel disaster by using a mock-up and actual samples of box beams that supported the failed walkways. (b) Close-up shows joint pulling loose.

and 5) numerous but necessary assumptions used in the analysis and design procedures that can lead to appreciable errors. Computing the required factor of safety can be quite involved. However, in some cases the factor of safety is specified in codes promulgated by organizations such as the American Institute of Steel Construction (AISC) and the American Society of Mechanical Engineers (ASME), as well as in building codes and company policies as explained in Section 4.6.

4.4 DEGRADATION, OXIDATION, AND CORROSION

As used in materials science, engineering, and technology, ***degradation*** normally refers to swelling, dissolving, or breaking of internal polymer chains. Polymers include plastics, rubbers, paints, woods, and adhesives. ***Corrosion*** or ***oxidation*** are broader terms that include chemical degradation of polymers as well as electrochemical attack on metals by the environment. ***Rust*** is a term used for the deterioration of iron alloys, including steel.

4.4.1 Chemical Properties

Chemical properties are a measure of how a material interacts with gaseous, liquid, or solid environments. Common examples include the ability of iron to resist rusting when exposed to air and moisture, the resistance of wood to rotting, and the ability of rubber to withstand sunlight (ultraviolet rays) without drying and cracking. Many conditions or environments threaten materials. Polluted air is filled with elements emitted from gasoline engines, home furnaces, and industrial plants; these elements combine with materials to cause damage and shorten their service life. Some of our ancient architectural treasures, such as the Parthenon (an ancient Greek temple) and copper and stone statues, are rapidly deteriorating from corrosive polluted air. Salty water from ocean spray or road salt used on icy roads promotes corrosion in automobiles. The reaction of some materials to high temperatures and fire is hazardous. Many plastics emit poisonous gases when burned. Sometimes it is desirable for materials to deteriorate or be biodegradable so that the natural environment can break down the material, thus reducing solid waste.

4.4.2 Oxidation and Corrosion

The most familiar chemical property is corrosion resistance. ***Corrosion resistance*** is the ability to resist oxidation. Most often, corrosion resistance is defined in terms of aqueous environments. ***Oxidation*** is the interaction of oxygen with elements in a material to cause structural changes due to the movement of valence electrons in the atoms of the material. An oxidized material loses electrons from atoms or ions. The opposite of oxidation is ***reduction:*** the gaining of electrons. For example, iron (Fe) releases electrons (e) during oxidation and ends up with positive ions (Fe^{2+}). The chemical reaction is shown as $Fe \longrightarrow Fe^{2+} + 2e$ [see Figure 4-32(a)]. In reduction the process is reversed as electrons are consumed: $Fe^{2+} + 2e \longrightarrow Fe$. The reduction process is used to convert iron ore (iron oxide) into iron. The rusting of iron and steel results from iron's tendency to revert to the natural state and thus seek equilibrium [Figure 4-32(b)]. The same is true of most metals because in the refined form they are more prone to oxidation.

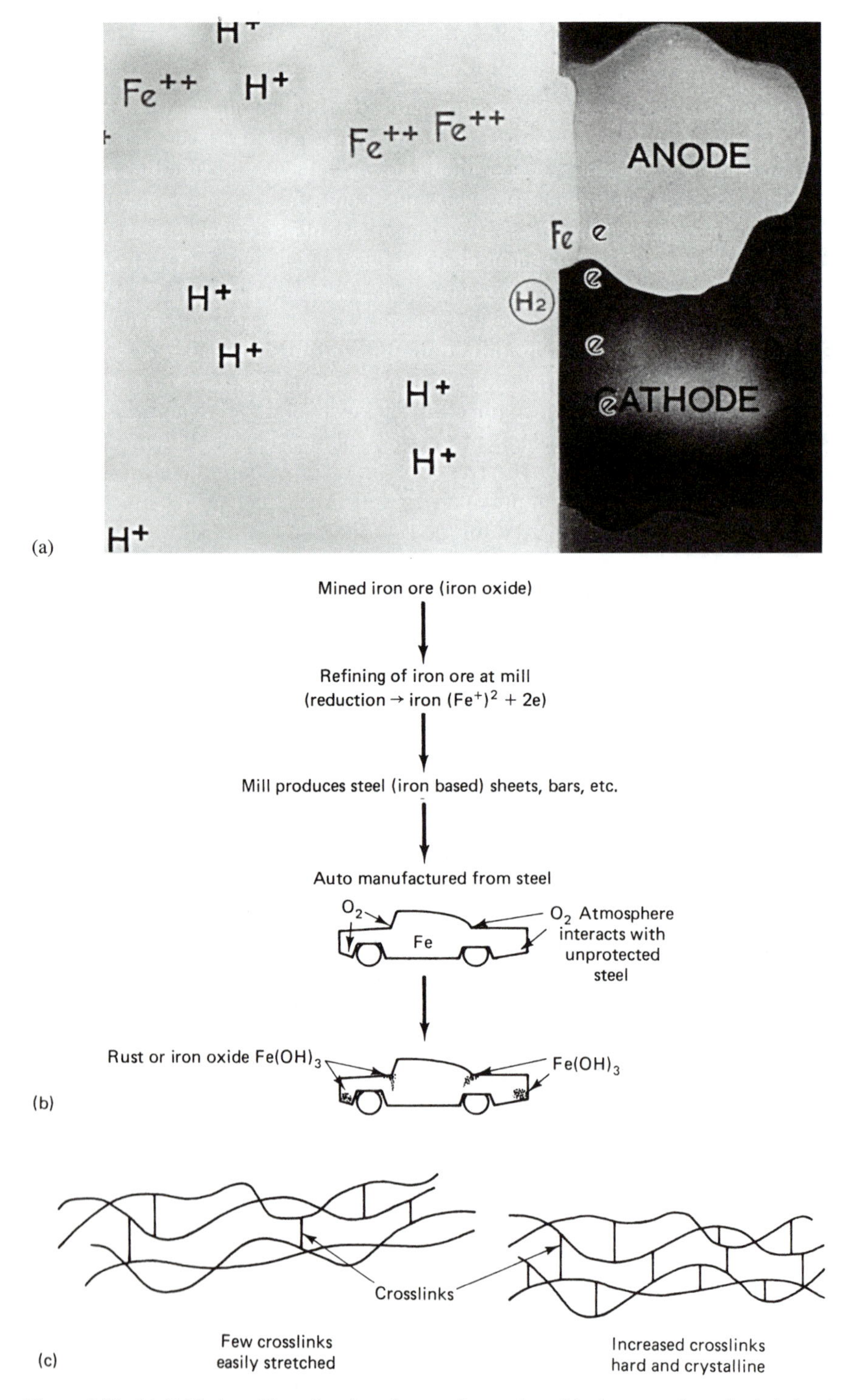

Figure 4-32 (a) Oxidation of iron. Iron ions form at the anode and hydrogen at the cathode in local cell action. (INCO.) (b) Cycle of iron and steel seeking natural equilibrium. (c) Typical thermosetting polymer. Increasing number of cross-links produces a network structure.

Old pencil erasers become hard and rubber bands become stiff and crack due to oxidation. The reaction of the rubber is common for many polymers exposed to sunlight, heat, and other conditions that speed up oxidation. Figure 4-32(c) shows a typical thermosetting polymer with ***cross-links*** that bond molecules together but also allow stretching; the cross-links increase due to oxidation, causing the rubber or plastic to lose its ability to stretch and to become hard and crystalline.

4.4.3 Outdoor Weatherability

Outdoor weatherability is a chemical property involving the ability of a material to withstand moisture, heat and ultraviolet rays from the sun, and pollutants in the air. Each of these factors affects oxidation. Rubber tires on automobiles and bicycles will develop cracks with age if they are not designed for good outdoor weatherability. A less than 1% increase in oxygen can cause severe damage to rubber.

Oxidation is normally enhanced by heat, and heat increases with oxygen. This is shown when oxygen is used to burn steel. The metal burning is done with an oxyacetylene torch, which causes a rapid melting away of the metal as it oxidizes. Even though oxygen is not a fuel, it is highly dangerous in the presence of fire or sparks and is capable of causing most materials to burn. In the fire triangle [Figure 4-33(a)], removal of any element (air, fuel, or ignition) can stop the fire. Occasionally, increased temperatures reduce corrosion in stainless steels and with copper alloys. The mechanism of the basic process of corrosion applies to all metallic materials in varying degrees. Corrosion can be prevented if only one of the basic design principles summarized in Figure 4-33(b) is removed or neutralized. Common ceramic materials such as silicon glass and aluminum oxide are

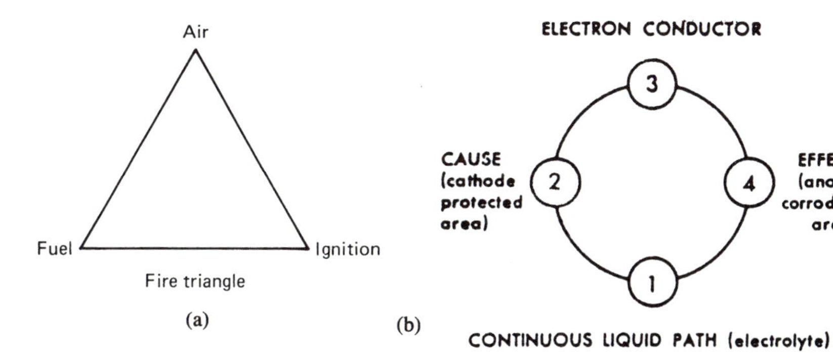

1. Electrolyte (continuous liquid path, usually water in the form of condensate, salt spray, etc.)
2. Cause (cathode—the cause of corrosion—the area through which electricity flows.)
3. Electron conductor (in a structure, usually a metal-to-metal contact, e.g. rivets, bolts, spot welds, etc.)
4. Effect (anode—the surface or object which corrodes.)

Figure 4-33 (a) Oxidation. (b) The corrosion circle.

oxidation resistant because they have already reacted to oxygen to form oxides. In the same way, an oxide film that forms on aluminum protects it.

4.4.4 Electrochemical Corrosion

Oxidation occurs in metals through an electrochemical process similar in operation to a car battery. In ***electrochemical corrosion,*** electrons flow through an electrolytic solution from one piece of the metal to another. The electrolytic solution (electrolyte) is a liquid such as water (H_2O) that contains ions. Recall that ions are charged atoms. Cations are positively charged ions and anions are negatively charged ions. When metals (especially dissimilar metals) are placed in the solution, one metal becomes the ***anode*** and the other the ***cathode,*** just as in a battery. Cations of one metal (anode) enter the solution to join anions. Left behind are electrons, which travel through the metal (conductor) to the cathode. Figure 4-34 illustrates electrochemical corrosion of iron that is bolted to copper. The positive ions carrying two positive charges (Fe^{2+}) enter the water to form bonds with the negative hydroxide ions (OH^-). Metallic bonds are broken to form ionic bonds. Electrons (e^-) left behind in the iron travel through the bolt to the copper, where they meet positive hydrogen ions (H^+) and become neutral hydrogen atoms. The anode (Fe) is pitted as it loses atoms. This type of corrosion is also known as ***galvanic action.***

Bosich (see references) lists eight types of corrosion: uniform attack, galvanic, concentration cell, pitting, dezincification, intergranular, stress, and erosion. Other authors use different labels and groupings. An explanation of each of these types of corrosion is beyond the scope of this book but can be gained from the books listed in the references at the end of the module.

Some authorities consider corrosion only in terms of metallics and their deterioration through chemical and electrochemical reaction to an environment. Rusting is applied

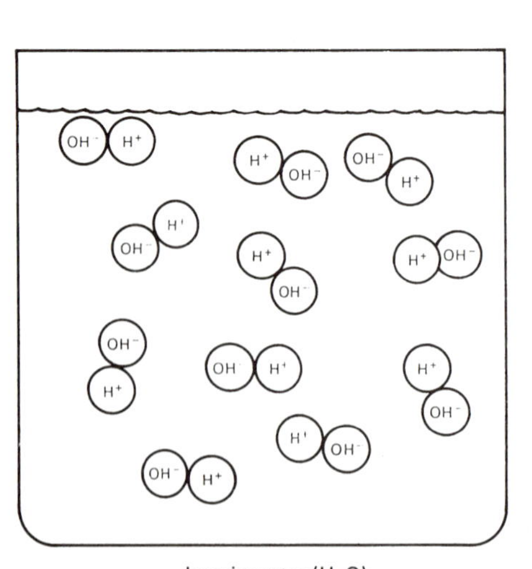

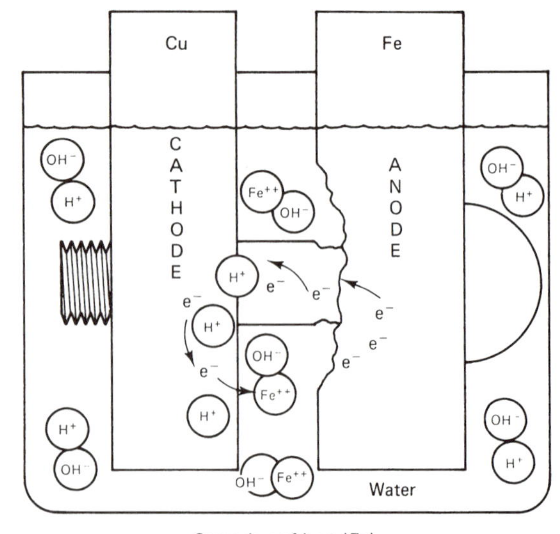

Figure 4-34 Electrochemical corrosion depicting ionized water and the corrosion of iron (Fe).

only to corrosion of iron and iron alloys. Others classify metallic corrosion as not only deterioration but any interaction of metallics with an environment, whether good or bad. Our use of ***corrosion*** refers to a deterioration of all materials and how the material reacts with its environment.

4.4.5 Electromotive-Force Series

It is possible to determine how active a metal will be by referring to the electromotive series. Table 4-2 shows the order in which dissimilar metals produce electromotive force (electron flow). The metals on top are stronger oxidizers and thus are active anodes. The bottom metals are less active and become reducing agents (cathodes). Greater galvanic action results from joining metals that have greater separation on the chart.

While aluminum oxidizes quickly, in doing so it forms a hard oxide layer on the surface that prevents oxygen from reaching the metal below. Iron and steel do not oxidize as rapidly as aluminum, but the scale on iron is soft and porous, thus allowing oxygen to penetrate farther and farther into the metal until it erodes away to nothing. A low-alloy steel known as Cor-Ten uses a small percentage (less than 0.5%) of copper to produce a compact oxide; therefore, the oxide coating on Cor-Ten also serves as a barrier and reduces corrosion to less than one-fourth times that of regular steel.

Oxidation can occur on a single piece of metal due to the varied energy states of the atoms of the metal. These states result from stresses due to machining, forming, and welding; from grain boundaries; from lack of homogeneity due to alloying and casting; and from cracks and other surface irregularities. In fact, there is no way to stop corrosion completely, but it can be slowed considerably. The high-energy areas of grain boundaries

TABLE 4-2 ORDER IN WHICH DISSIMILAR METALS PRODUCE ELECTROMOTIVE-FORCE (GALVANIC) SERIES

	Electromotive series (corrosion potential)		
Greater oxidizing agent ↑	Magnesium 2+[a]	Active or Anodic ↑	(greatest corrosion) more active
	Berlyllium		
	Aluminum 3+		
	Zinc 2+		
	Chromium 3+		
	Iron (steel) 2+, 3+		
	Cadmium 2+		
	Tin 2+		
	Nickel 2+		
	Cobalt		
	Lead 2+		
	Brass		
	Copper 1+		
	Silver 1+		
	Titanium 1+		
	Platinum 2+		
Greater reducing agent ↓	Gold 3+	Noble or Cathodic ↓	(least corrosion) more passive

[a]Oxidation numbers.

become anodes, giving up atoms to the lower-energy levels. Pitting and erosion of the metal occur at the anode, while rust buildup occurs at the cathode (Figure 4-35).

The atmosphere provides excellent electrochemical mechanisms to promote corrosion. Different corrosion rates result from varied atmospheres. Oxygen and ***ozone*****, water, and pollutants such as sulfur dioxide from burnt fuel oil and coal, ammonia, hydrogen sulfide, dusts, and salts exist in varying amounts depending on location (marine, industrial, or rural); these active chemicals produce differences in rates of corrosion for the materials exposed to them. While fluctuations in temperature do not make significant changes in cor-

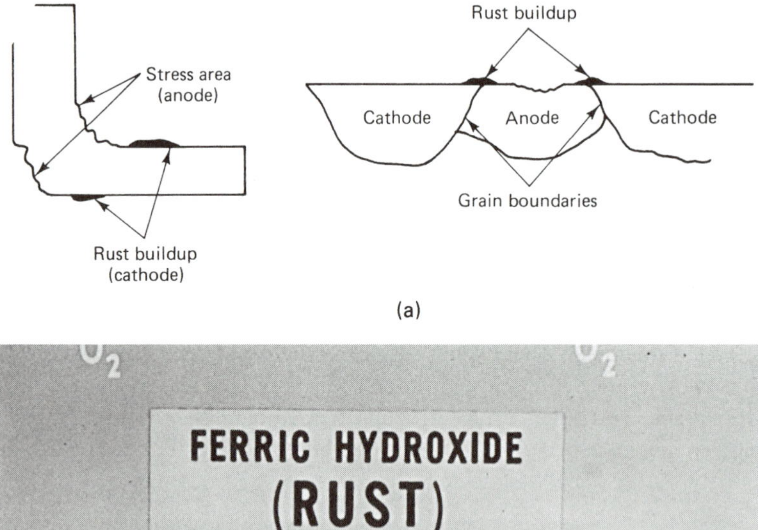

(b)

Figure 4-35 (a) Corrosion from high-energy levels. (b) Conversion of ferrous hydroxide into ferric hydroxide $Fe(OH)_3$ by the action of the oxygen. (The International Nickel Co.)

* *Ozone* (O_3)—a highly oxidizing allotrope of oxygen. It contributes to smog at the earth's surface, but provides a protective layer in the upper atmosphere.

rosion rates, wind can cause dramatic shifts in airborne chemicals that enhance corrosion. Abrasive sand carried in strong winds can erode protective coatings and promote corrosion. Table 4-3 reveals the possible rates of corrosion for certain environments on metals and alloys. Industrial environments normally produce greater corrosion rates due to the many corroding chemicals. Next, the high moisture and salt in a marine environment produce more corrosion than a rural area. Rural areas in relatively dry states such as Arizona and New Mexico would produce less corrosion than rural areas in coastal states such as Virginia and North Carolina.

4.4.6 Chemical Attack

Chemical attack involves the dissolving of a material ***(solute)*** by a chemical ***(solvent).*** A common example is salt (solute) dissolving in water (solvent). Polymers present a problem in use with chemicals because of the different properties of polymers of even the same name. Therefore, it is important to follow manufacturers' recommendations when making material selections. While ceramics such as glass and concrete are very stable, even they are subject to chemical attack. Concrete will expand and crack when subjected to sulfates

TABLE 4-3 POSSIBLE CORROSION RATES FOR METALS IN THREE GENERAL ATMOSPHERES

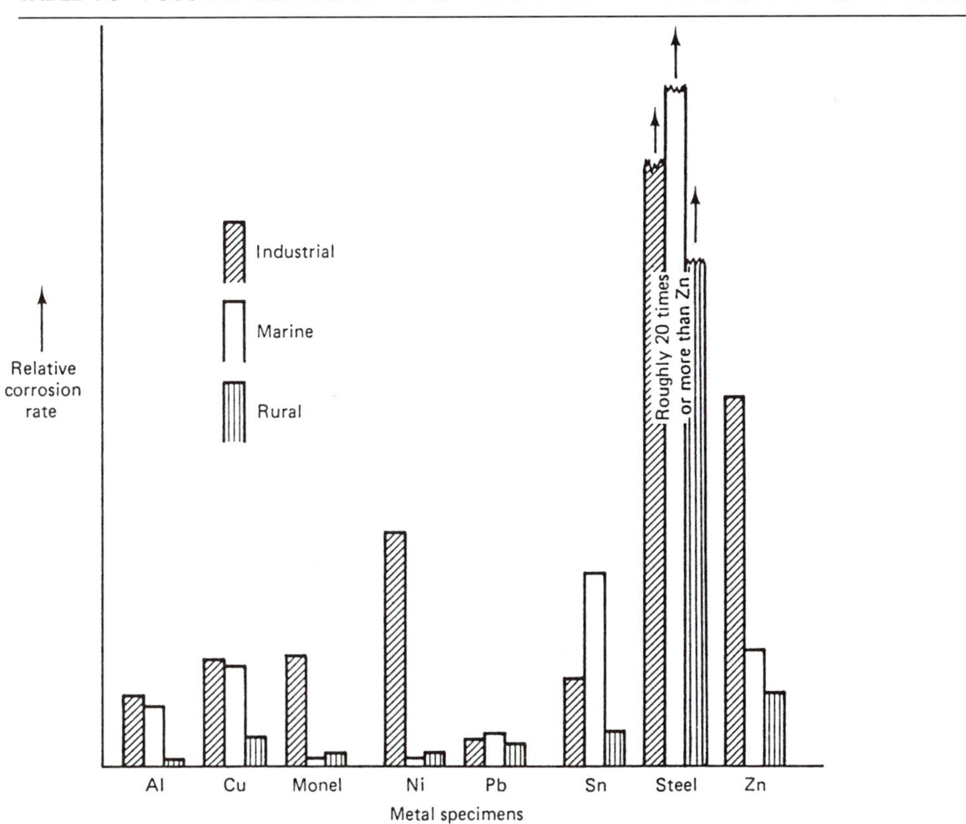

commonly found in the soil and from the rusting of rebar [Figure 4-36(a)]. Changes in temperatures and chemical concentrations also affect the reaction of a chemical to a material (see Section 7.1.5).

4.4.7 Corrosion Protection

The large variety of environments that a material can be subjected to makes corrosion protection quite complex. On the other hand, polymers have been developed that are so resistant to our natural environment that they become nearly indestructible and are solid waste pollutants. Some common methods of corrosion protection for metals include cathodic protection, protective coating, stress relieving, insulation, alloying, materials selection, and design. Figure 4-36 shows that materials are tested and then selected because of their corrosion resistance. Many rules of good design are intended to reduce the opportunity for corrosion to occur.

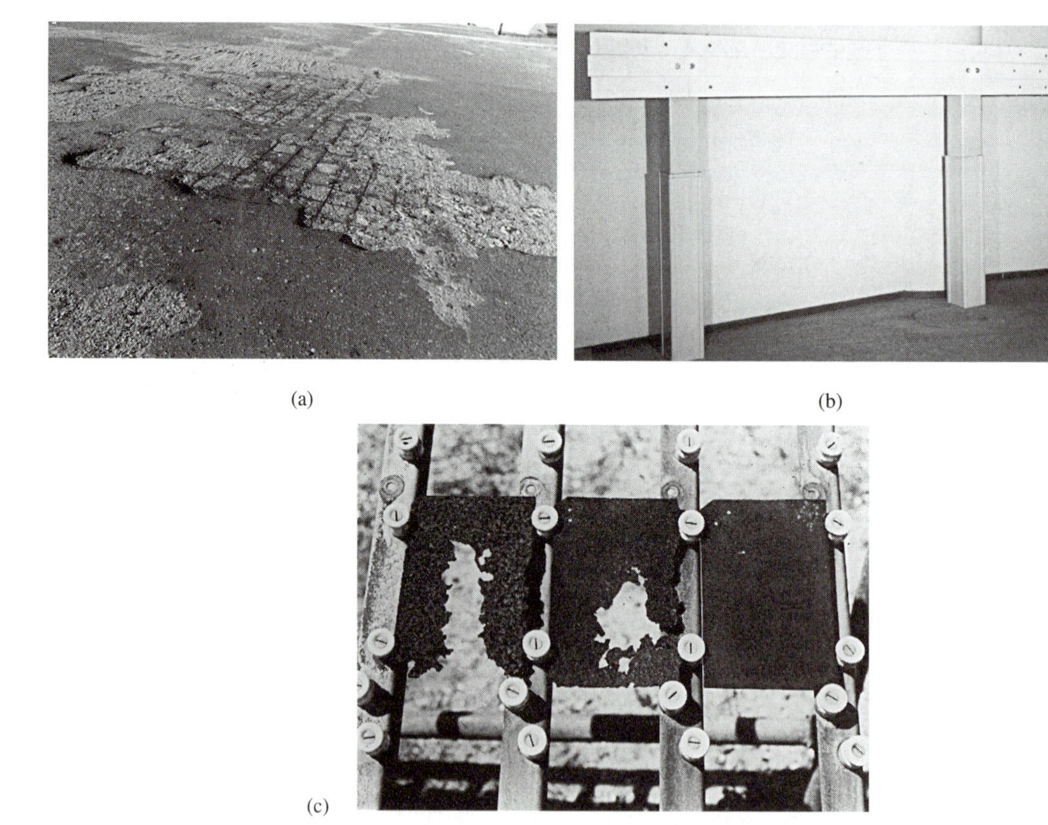

Figure 4-36 (a) Severe spalling and delamination of concrete roadway caused by corrosion of embedded rebar (steel reinforcing) initiated by deicing salts. (W.R. Grace & Co.). (b) Plastic-glass composite highway guardrail made of pultruded glass-fiber-reinforced polyester is noncorrosive, maintenance-free, and stronger than similar structures of steel and steel-reinforced concrete; they are also comparable in cost. *(Advanced Materials & Processes.)* (c) Corrosion of steels in a marine atmosphere. Left: Low-copper steel. Center: Ordinary steel. Right: Nickel-copper-chromium steel. (The International Nickel Co.)

It is conservatively estimated that well over 50% of the failures occurring under service conditions are fatigue failures. This explains in part the necessity for providing steel with some form of corrosion resistance such as paint or galvanizing (zinc) coating. If a metal part is to be subjected to a corrosive environment in service, special efforts must be made to provide corrosion resistance. Without it, the part is subjected to fatigue failure at levels of stress even lower than the fatigue limit. Corrosion is a source of tiny cracks in the surface, which grow with time and produce fractures. Some fractures are catastrophic, giving little, if any, warning. Shot peening and nitriding are other processes used to strengthen the outer layers of a metallic part. In view of the expense involved in running fatigue tests, particularly when attempting to duplicate service conditions in a laboratory, a fatigue or endurance ratio is sometimes used instead. This ratio, varying between 0.25 and 0.45 depending on the material, compares fatigue limit to tensile strength.

Cathodic protection involves the use of a sacrificial metal such as magnesium or zinc, which acts as an anode, thereby turning the metal being protected into a cathode with a reduced corrosion rate. Metals used as anodes are the more active metals, those with a greater tendency for reducing than those less active or lower on the electromotive series (Table 4-2). Common examples include the use of magnesium with steel water pipes or zinc on ships. Electric currents that pass electrons into a metal can achieve the same results.

Protective coating is the most familiar method to prevent corrosion. Paint, varnish, oil, and a variety of polymeric and ceramic coatings prevent oxygen and moisture from reaching the metal. Automobile bodies are dipped into protective coating solutions. Zinc-coated steel (galvanized steel) is an example of both protective coating and cathodic protection. Electrocoating and electrophoresis are two similar surface protective processes. ***Electrocoating*** is a painting concept in which water-borne paint is applied electrically, in a dipping process. ***Electrophoresis*** deals with the transport of charged colloidal or macromolecular materials in an electric field. The paint and resin particles are designed to carry an electrical charge. Paint thicknesses as thin as 10 μm are possible, and they provide good corrosion resistance without affecting negatively heat transfer in the substrate materials. Copper/brass automobile radiators with a 10-year lifespan are now fabricated. They are 30 to 40% lighter than previous radiators manufactured with the same materials.

Stress relieving through heat treatment, structural changes, and design considerations provides more homogeneous metals with nearly equal energy levels. ***Insulation*** to stop metal-to-metal contact of dissimilar metals through the use of polymers or ceramics reduces galvanic action. ***Alloying*** of elements such as aluminum or chromium with steel produces oxide layers that make the metal ***passive*** or resistant to further oxidation. Conversely, alloying of certain dissimilar metals produces *galvanic cells* within the metal, which enhances corrosion.

In addition to heat treating metals, alternative techniques are employed to produce stress relief. The following process is a nonthermal stress relief treatment. The **Meta-Lax** process with subresonance stress relieving is used by auto and marine engine builders. It is very effective in reducing the resonant frequency, hence the magnitude of the internal stress, built up as a consequence of the manufacturing processes that produce engine parts. Every body (structure) has a natural frequency. When acted on by periodic impulses, the body vibrates (forces vibration) at a frequency that may or may not be the natural frequency. When the period of the forced vibration is the same as that of the free vibration, the two effects reinforce each other and large amplitudes of the vibrating body may result. This effect is called ***resonance.*** When a body capable of vibrating is displaced from its equilibrium position and suddenly released, it executes a series of free vibrations and eventually comes to rest. The frequency of free vibrations is determined by the properties and condition of the body itself and is known as the natural frequency of the body. The first step in the Meta-Lax process is to determine the resonant frequency of the part. This step is followed by subjecting the part (bolted to a shaker table) to vibrations at a frequency below the resonant frequency. A point is reached where the vibrations no longer produce a lessening of the resonant frequency (hence the lessening of the residual internal stress) and the process ends. The results are an extended service life before failure and enhanced engine reliability.

Magnetic Processing Systems, Inc., has developed a ***pulsed magnetic treatment*** that relieves stress and improves the fatigue properties of metal parts having diameters up to 300 mm. This nonthermal technology applies high-energy pulses through an electromagnetic field. The technology reportedly modifies both surface and subsurface defects inherent to metals, thereby improving fatigue properties.

A design technique to reduce residual stresses in a new alloy joint has recently been publicized. A process called ***self-propagating high-temperature synthesis*** (SHS) is reported to lessen residual stresses in a new alloy joint recently developed to join a ceramic and a metal. This joint, called a functionally gradient material joint, is made up of materials that gradually change chemical composition and properties across a dissimilar materials joint, bridging mismatches and reducing thermally induced stress. The materials used are elemental powders of nickel, titanium, aluminum, and carbon, which are pressed in a uniaxial press. The dissimilar materials to be combined in the joint are silicon carbide and a nickel-base superalloy. They are heated in a graphite die to a temperature just above the ignition temperature of the lowest-temperature reaction and held at that temperature and slowly cooled. A new microstructure is thus formed. This process uses temperatures that reduce residual stresses in the joint and especially in the brittle ceramic.

Materials selection must consider the placing of materials into the proper environments to prevent oxidation, electrochemical reaction, or chemical attack. ***Design*** should incorporate the preceding methods, in addition to avoiding surfaces that will collect liquids and debris.

Protective coatings and surface engineering receives considerable attention throughout this book. Often, the preparation of materials for coatings and application of coatings, plus disposal of gases, liquids, and solids used in coatings, cause harmful effects on the environment. Heavy metals, chlorofluorocarbons (CFCs), and volatile organic compounds (VOCs) have been targeted among the polluters. A new coating known as ***PremAir***™, developed by Engelhard Corporation and Ford Motor Company, offers the potential for a new system of "smog-eating radiators." The catalyst system may turn car radiators into smog-eating, air-filtration systems to complement the catalytic converters now used in exhaust systems. The system may have the effect of cleaning all sources of smog from the atmosphere as air flows over vehicle radiators. Not only will this help animal life but it will also reduce corrosion. The developers' testing indicates the system will destroy 97% of VOCs, 90% of nitrogen oxides (NO_x), and 96% of carbon monoxide (CO) from the air it contacts (see Figure 4-37).*

* In Chapter 5 of Karl Sabbagh's *21st Century Jet: The Making and Marketing of the Boeing 777* (Scribner), you find an excellent case study involving many of the concerns in this module such as corrosion, fatigue, alloying, and their related materials selection issues.

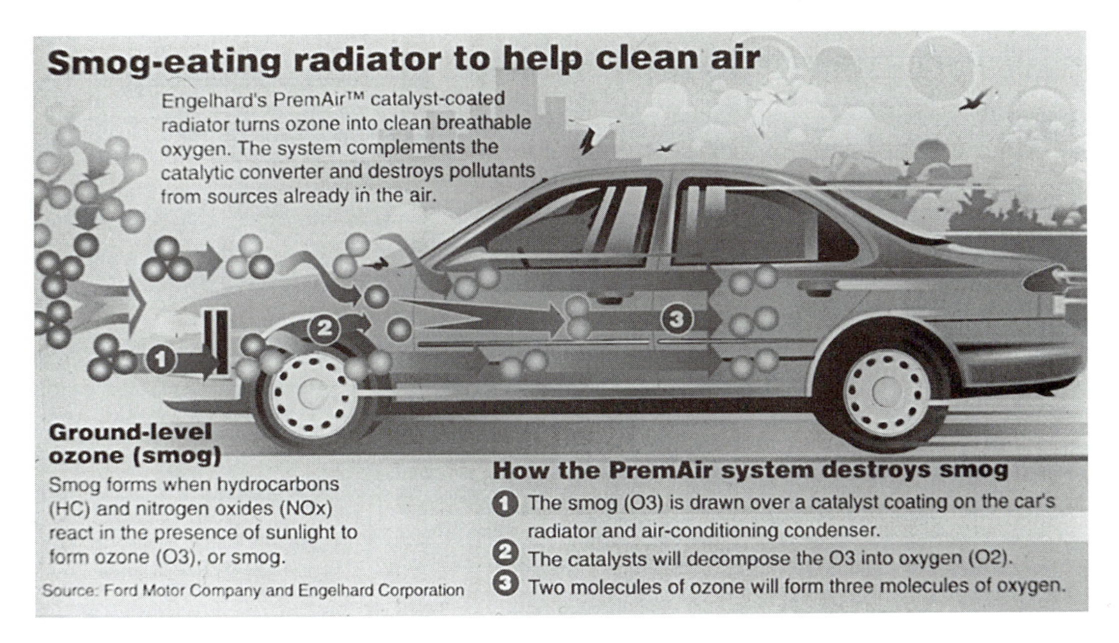

Figure 4-37 Polluted air not only harms animal life but also accelerates corrosion. Smog-eating radiators use PremAir™ coatings to convert ozone (O_3)—the main component of smog—into breathable oxygen (O_2). (Ford Motor Company.)

4.4.8 Water Absorption and Biological Resistance

Natural and synthetic polymers are subject to biological attacks and the absorption of water. When left untreated, dried wood will absorb moisture and then serve as a good environment for the growth of fungi (small plants) and insects; they feed on cellulose and lignin and cause deterioration. Unprotected dry wood is also a host for termites and other insects. Some synthetic polymers swell through water absorption, which causes deterioration and provides a good environment for damaging microorganisms. The corrosion of metals through oxidation can result from organisms such as barnacles living on the material.

4.4.9 Wear, Erosion, and Abrasion Resistance

The worldwide degradation of engineering materials caused by the insipid attack by chemicals and/or wear amounts to a staggering financial loss, not to mention the large drain on the world's limited resources. Previous paragraphs in this module describe corrosion that is the chemical removal of atoms. ***Wear,*** including some of its various types such as abrasive, adhesive, and erosive, also takes its toll on materials by removing part of the materials by mechanical action using solids, liquids, or gases. Rates of wear are difficult to measure beforehand, even though there are many different types of tests for materials under various testing conditions. ***Tribology,*** the science and technology dealing with interfacing surfaces, focuses on design, friction, lubrication, and wear behavior to avoid problems with mating parts.

Figure 4-38 shows four common tests to try to determine the rate of wear for some common materials under attack by oxidation, acid corrosion, and erosion by mechanical

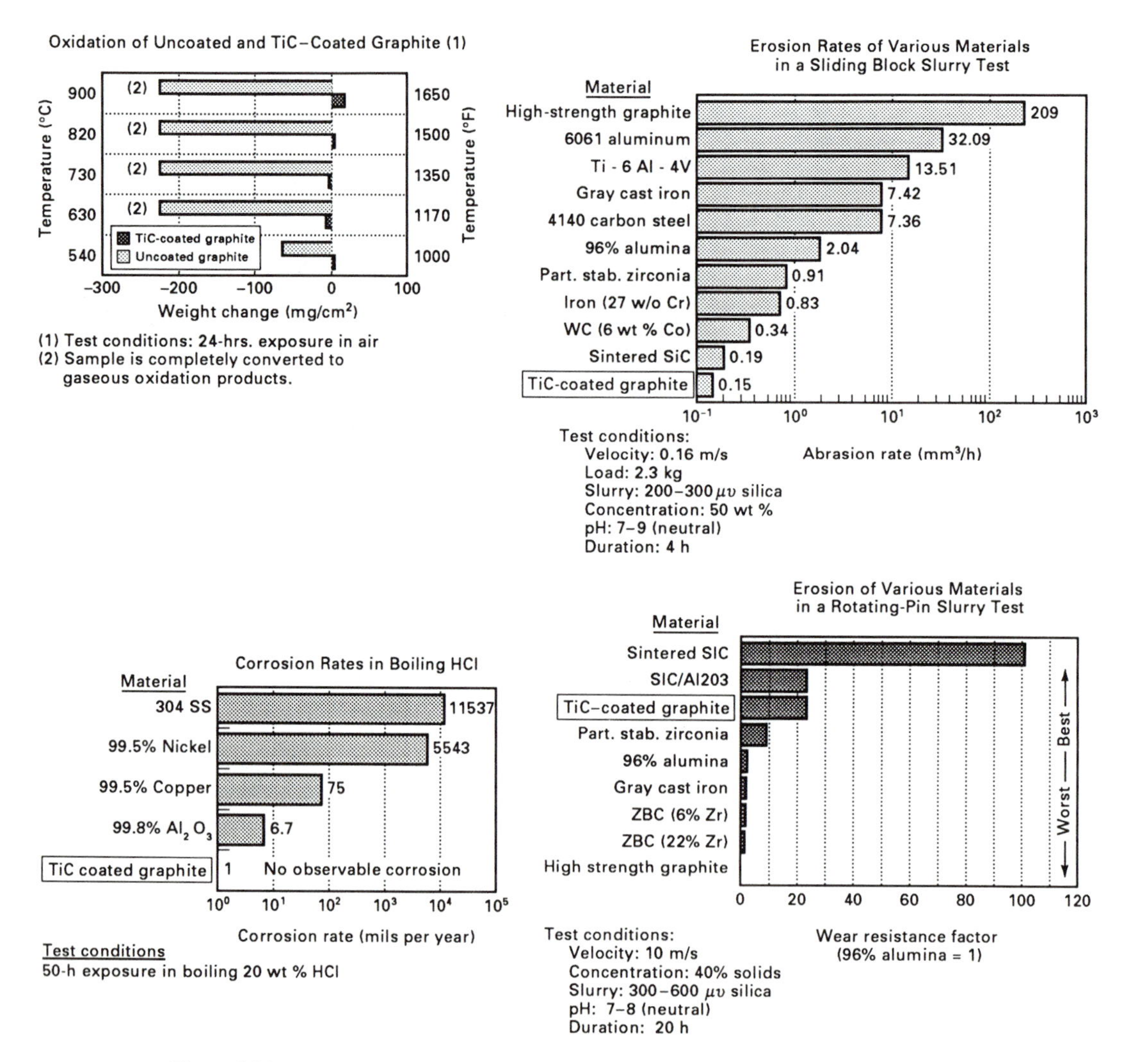

Figure 4-38 Corrosion wear rates for various materials under different source test conditions. (Alanx.)

testing. This figure points out that one factor in the struggle to control wear in materials is to protect their surfaces from the above type of attacks. In this case, ceramic (TiC) is used to coat the material (graphite) with a hard surface impervious to chemical attack, wear, or their combination. Further discussion of some of these techniques to provide wear resistance are found in each of the succeeding modules (see Section 7.1.3).

4.5 THERMAL PROPERTIES

A knowledge of thermal properties is vital to those who work with materials to make parts that will serve at temperatures other than as fabricated, or that perform some heat-transfer function. For example, when heated deliberately or by hot working, a metal rod exhibits three thermal effects: (1) the rod absorbs heat, (2) it expands, and (3) it transmits heat. Each of these situations has a property that measures and describes the degree to which each effect is present. ***Absorption*** of heat is characterized by the property's specific heat capacity (c_p). *Expansion* is usually described by the *coefficient of thermal expansion* (α). *Heat transmission* is identified by *thermal conductivity (k).*

4.5.1 Heat Capacity

The efficiency of a material to absorb thermal energy is known as ***heat capacity.*** Heat capacity (c_p) is the number of joules needed to raise 1 unit mass (1 gram mole) of a substance 1 K at constant pressure. Most solids have about the same c_p value at room temperature. If the heat capacity of a substance is compared to the heat capacity of water, the ***specific heat of a substance*** *(s)* is obtained. The c_p value for water is 1 calorie/gram · °C or 4.184 joule/gram · °C. Water is a substance that requires a large expenditure of energy to produce a change in temperature. Substances with high specific heat values do not change their temperature appreciably. Using units of cal/g · K at room temperature, specific heats *(s)* are: for silver, 0.06; iron, 0.1; glass, 0.14; and polyethylene, 0.54.

Another thermal property, ***thermal mass,*** is expressed in terms of specific heat capacity (c_p) in the following expression: thermal mass = apparent density (ρ) × specific heat *(s)*. This property gives an indication of the heat required for a unit volume of materials to achieve a prescribed temperature. Units of thermal mass are cal/cm³ · °C.

4.5.2 Thermal Expansion

Nearly all solid materials expand when heated and contract when cooled. Isotropic materials expand or contract equally in all directions. This temperature deformation results from the changes in the distances between adjacent atoms due to changes in thermal energy. The ***coefficient of linear thermal expansion*** (α, Greek letter alpha) describes this effect in the following equation for unit deformation (ϵ):

$$\epsilon = \frac{\Delta l}{l_0} = \alpha \Delta t$$

where Δl is the change in length of some dimension, l_0 is the original length of that dimension, and Δt is the change in temperature. The expression $\Delta l/l_0$ is another way of indicating unit deformation (ϵ, strain) brought about by temperature changes. Typical units of α are 10^{-6} (mm/mm · °C^{-1}). For 1020 steel, $\alpha = 11.7 \times 10^{-6}$ °C^{-1}; for graphite, $\alpha = 5 \times 10^{-6}$ °C^{-1}; and for nylon, $\alpha = 100 \times 10^{-6}$ °C^{-1}. A family of low-expansion-coefficient alloys has been developed based on the introduction in 1912 of Invar, an alloy of 64% iron, 36% nickel. Its coefficient of thermal expansion is 0.9×10^{-6} K^{-1}, which is invariant with respect to temperature.

In metals there is a close correlation between the coefficient (α) and the melting point (T_m). The lower the coefficient, the higher the T_m value. Knowing the T_m value gives an indication of the strength of the bonding forces. A bimetallic strip exemplifies the use of the coefficient of thermal expansion in temperature-control relays. With a temperature change, two metals bonded together and having different coefficients will bend, making electrical contact with a switch.

In the design of materials systems, the deformation of materials due to temperature must be taken into account. Changes in temperature are indications of heat transfer. An example of such a case is the use in a composite structure of two different materials that might have excessive differences in their coefficients of expansion. What is a possible outcome? What force or condition is recognized by ship and road builders who install expansion joints in their structures? What is the cause of the repetitious clinking sound produced by metal wheels running on steel railway tracks? In this discussion so far, all examples indicate that there is no resistance to the expansion of materials due to temperature changes. What happens to the materials if the expansion or contraction is resisted by an external force?

Illustrative Problem

A steel rod 2 m long is fastened at one end to a fixed support and is free to expand or contract. If the ambient temperature rises by 25°C, how much will the rod expand or contract?

Solution

$$\alpha_{\text{steel}} = 11.7 \times 10^{-6}\ {}^\circ\text{C}^{-1} \quad \text{(See page 187.)}$$

$$l_0 = 2\ \text{m} \quad \Delta t = 25^\circ\text{C}$$

$$\Delta l = l_0 \alpha\, \Delta t$$

$$= 2\ \text{m}(11.7 \times 10^{-6}\ {}^\circ\text{C}^{-1})(25^\circ\text{C})$$

$$= 0.585\ \text{mm} \quad \text{or} \quad 5.85 \times 10^{-4}\ \text{m}$$

Polymers have small bonding forces (van der Waals bonds) and consequently larger expansion coefficients than metals. In summary, the covalent and ionic bonded materials have the lowest coefficients (strongest bonding), polymers the highest coefficients (weakest bonding), with metals somewhere between these limits.

4.5.3 Thermal Conductivity

The ability of a nonmetallic solid to transmit heat, its ***thermal conductivity,*** depends on ***phonons*** (quanta of energy). The phonons or "particles" behave somewhat like gas atoms and transfer the solid's lattice vibrations from a region of higher energy (high temperature) to a region of lower energy. The phonons encounter more collisions as they move with the speed of sound through a material as the temperature increases. This results in a lowering of the thermal conductivity with the increase in temperature.

In metals, the free, or conduction, electrons also serve as carriers of thermal energy received from the phonons. Elements added to metals reduce the thermal conductivity, much as porosity and other crystal imperfections do in all materials. Polymers lacking free electrons and having a less crystalline structure (orderly atom arrangement) have poor thermal conductivities. Such a characteristic makes polymers good thermal insulators.

Thermal conductivity *(k)* is a measure of heat flow through a material. It relates heat flow (the flow of heat energy per unit area, per unit time) to the temperature gradient, causing the heat to flow. The term ***temperature gradient*** (in symbols, $\Delta t/\Delta d$) describes a temperature difference per unit distance. Using symbols, the heat flow can be expressed as q/A, where q represents the heat energy in units such as watts, Btu/hr, or calories per second (cal/s); A is the area through which the heat passes, in square units such as square meters or square feet. The change in temperature (Δt) can be expressed in °C, °F, or K. Using the symbols described above, thermal conductivity can be written

$$\frac{q}{A} = k\frac{\Delta t}{\Delta d}$$

This relationship shows that k is a constant of proportionality. Solving for k gives

$$k = \frac{q}{A} \cdot \frac{\Delta d}{\Delta t}$$

Substituting SI units into this equation, the units of k are

$$k = \mathrm{W/(m \cdot K)} \text{ or}$$

$$k = \mathrm{W \cdot m^{-1} \cdot K^{-1}}$$

Expressing k in these units, silver has a value of 4.1; steel (average), 0.5; glass (average), 0.01; and polyethylene, 0.004. Various other units have also been substituted into this equation. Some examples of k found in the literature are

Soda-lime glass	$k = 0.0023$ cal/cm · s · K
Copper	$k = 390$ W/cm · K
Stainless steel	$k = 0.015$ W/m²/°C/min
Brass	$k = 18{,}000$ Btu/day · °F × in./ft²

Illustrative Problem

An insulating material placed inside the walls of a home is 6 in. thick. The heat flow is calculated to be 4.2 Btu/hr when the inside temperature is 70°F and the outside temperature is 0°F. Determine the thermal conductivity of the insulating material as the heat flow passes through 1 ft² of the wall. In addition to expressing k in units of Btu/(hr · ft · °F), convert your units to SI units using watts, meters, and Kelvin temperature.

Solution

$$k = \frac{q}{A} \cdot \frac{\Delta d}{\Delta t}$$

$$= \frac{4.2\ \text{Btu}}{1\ \text{ft}^{2}\cdot\text{hr}} \cdot \frac{6\ \text{in.}}{70°\text{F}}\left(\frac{1\ \text{ft}}{12\ \text{in.}}\right)$$

$$= 0.03\ \text{Btu/(hr} \cdot \text{ft} \cdot °\text{F)}$$

Using conversion Table 10-7, we have

$$1 \text{ Btu/ft} \cdot \text{hr} \cdot {}^\circ\text{F} = 1.729 \text{ W/m} \cdot \text{K}$$

$$k = \frac{0.03 \text{ Btu}}{\text{hr} \cdot \text{ft} \cdot {}^\circ\text{F}} \left(\frac{\text{ft} \cdot \text{hr} \cdot {}^\circ\text{F}}{1 \text{ Btu}} \cdot \frac{1.729 \text{ W}}{\text{m} \cdot \text{K}} \right)$$

$$= 0.052 \text{ W/m} \cdot \text{K}$$

4.5.4 Thermal Resistance

Insulating materials have low thermal conductivities and can retard the transfer of heat. The thermal conductivity of most materials is temperature dependent. In other words, the value changes with a change in temperature. Refractory materials, on the other hand, have thermal conductivities that have a minimum dependence on temperature. Porous materials such as textiles, rock, wood, cork, foamed plastics, and human-made insulating tiles are good insulators partly because of their ability to trap air, which itself has a low thermal conductivity. Furthermore, the entrapped air is free from circulating currents, which aid the transfer of heat. Numerous applications for insulating materials exist in today's society, particularly in the area of heat (energy) conservation. Figure 4-39 shows two vital applications that exemplify several thermal properties mentioned in this brief discussion.

4.6 TESTING, STANDARDS, AND INSPECTION

4.6.1 Testing

Materials are tested to determine their basic properties. In particular, the testing of materials has as its objective the performance of tests to determine numerical values for properties. As stated previously, most properties of engineering materials can be placed into the following major classes: physical, mechanical, chemical, thermal, electromagnetic, and optical. An example of innovative testing techniques are those developed to investigate the high-temperature properties of advanced materials approaching 2200°C (4000°F). Current metallic alloys are capable of withstanding temperatures to only about 700°C (1300°F). These high-temperature techniques fall into two categories: macromechanical testing such as static/cyclic biaxial loading of a hollow specimen, and micromechanical such as the DISMAP image processing system mentioned previously in this module. ***Macromechanical*** systems provide data on the strength, fatigue, and tribological properties of materials mainly for engineering design of components, whereas ***micromechanical systems*** permit detailed experiments for determining and describing specific mechanisms of failure. The terms *testing* and *inspection* are often used interchangeably, but they should not be.

Table 4-4 provides a summary of the general types of tests used to determine the properties discussed in this module. Materials testing is widespread throughout industry and government. Our modern society relies on materials testing of (1) raw materials,

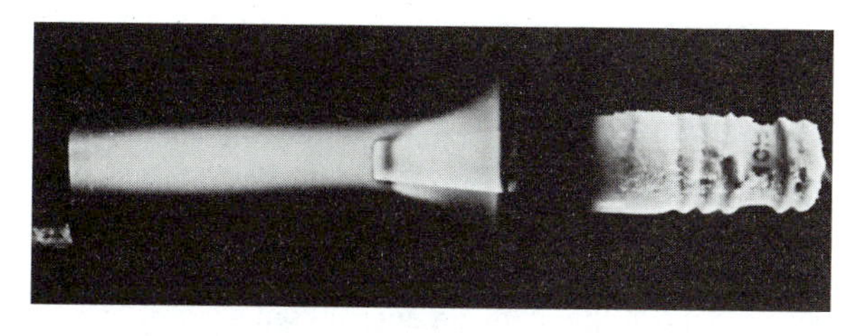

(a)

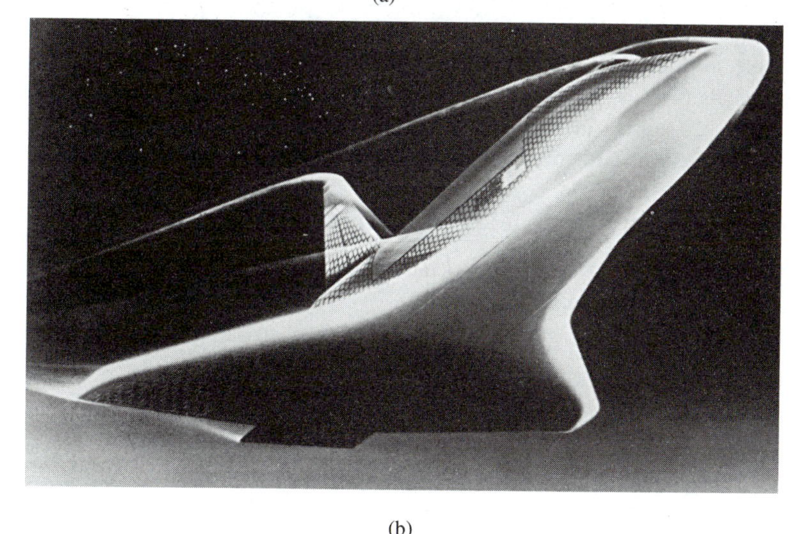

(b)

Figure 4-39 (a) Thermal resistance. High alumina (Al_2O_3) content spark-plug insulators can withstand thermal shocks from below zero (− 73°C) to white hot heat (over 1000°C). Note the frost still on the top of the insulator. (American Ceramics Society.) (b) Various thermal insulating materials systems protect space vehicles from the high heat of reentry into the earth's atmosphere. (Corning Glassworks.)

TABLE 4-4 GENERAL TYPES OF MATERIALS TESTING

Mechanical Tests

Used to evaluate mechanical properties of materials where there is a measure of the material's ability to carry or resist mechanical forces such as tension, compression, shear, torsion, and impact. Tests can be ***static,*** in which the stress is applied slowly, or ***dynamic,*** in which the stress is applied suddenly. Testing usually results in permanent damage to the specimen.

Nondestructive Tests

Used to examine materials and components in ways that do not harm the future usefulness and service. They detect, locate, measure, and evaluate discontinuities, defects, and other imperfections to assess integrity, properties, and composition, and to measure geometric characters.

Physical Tests

Used to determine physical properties of materials when there is no change in composition of the material. They include tests to determine density and electrical, optical, thermal, and mechanical properties.

Chemical Tests

Used to determine chemical properties of materials where there is transformation of one material into another. They include tests of the abilities of materials to resist oxidation, corrosion, and rust.

(2) product subassemblies or components, (3) products, and (4) systems. Who does the testing? Everyone! Read the caption in Figure 4-40. Think about the liability of Arrow Dynamics, the company who designed and built the Drachen Fire, and Busch Gardens, the company who operates this thrill ride that moves through spiral rolls and travels 60 miles per hour! Every aspect of this amusement must meet rigid quality standards to ensure the safety of riders while protecting the designers, builders, and operators of the ride from product liability. This includes the steel structural pipes and other structural members that make up the track. Also involved are the wheels, bearings, and suspension system that connects the cars to the track. Quality standards extend to the restraint system that holds the "brave souls" in the cars. It includes the composite body that shrouds the cars and makes the seats. But who gets involved in the many tests to ensure this quality?

4.6.2 Standards and Testing Organizations

Table 4-5 provides categories of agencies, societies, and organizations who (1) develop standards for materials testing and/or (2) conduct materials testing. ***Materials testing stan-***

Figure 4-40 Taming the beast. Brave souls challenge the mighty Drachen Fire, the fearsome roller coaster at Busch Gardens in Williamsburg, Virginia. As one of the world's largest roller coasters, it stands 150 feet tall, reaches speeds of more than 60 mph, and features many first-of-a-kind thrill ride elements. Hurtling along 3550 feet of electric-blue colored track, riders are turned upside down six times. (Courtesy Busch Gardens and Arrow Dynamics, Inc.)

TABLE 4-5 WHO SETS STANDARDS FOR MATERIALS TESTING AND CONDUCTS MATERIALS TESTING?

1. Technical societies
2. Governmental agencies and research centers
3. Private laboratories
4. Manufacturing companies
5. Wholesalers and retailers
6. Consumer groups
7. Individual consumers

dards are rules and procedures established for the testing of materials to ensure objectivity and common practices for making judgments about materials. Just as the standard measurements for the layout of lines on a tennis court ensures that everyone is playing within standard boundaries, it's vital that everyone have clear, standard rules on how to test materials so they have clear meaning. This sounds simple and straightforward, but development of materials testing standards is a complex topic that we will only touch on here. Many people make full careers of developing these standards that the entire civilized world relies on for commerce and safety.

4.6.2.1 Technical societies. These groups of people, made up of technical specialists from private industry, government, and universities plus individuals with common interests, come together and agree on what standards are best for their particular industry. Some examples of technical societies are the Society of Plastic Engineers (SPE), Society of Automotive Engineers (SAE), Society of Manufacturing Engineers (SME), ASM International (ASMI), American Society for Testing and Materials (ASTM), American National Standards Institute (ANSI), and International Standards Organization (ISO). These societies often cooperate in standard development. For example, ASTM publishes the *Annual Book of ASTM Standards,* which involves coordinating 140 main and 2034 subtechnical committees and can include representatives from ASMI, SAE, and ANSI. The *Annual Book* consists of approximately 60 volumes and thousands of standards that serve as guides to materials testing and other related concerns. These ASTM standards are used around the world to set legal criteria and objective means for analysis of materials. For example, ASTM devotes about forty standards to impact testing. One of these is *ASTM E23, Izod Impact Test of Metallic Materials* (see Figure 4-11).

4.6.2.2 Governmental agencies and research centers. Federal, state, and local government agencies become involved in many forms of materials testing. Other testing and development of materials are conducted at regional research centers that are under contract with federal agencies, such as the Department of Energy's Oak Ridge National Laboratory (ORNL) and Pacific Northwest Laboratory (PNL). The National Institute of Standards and Technology (NIST), formerly the National Bureau of Standards (NBS), is a federal agency that falls under the jurisdiction of the U.S. Department of Commerce. NIST is deeply involved in developing materials testing standards and conducts a broad range of tests of materials. For example, they set up a test of the structural connection of a concrete walkway that collapsed in a hotel, killing 113 people (see Figure 4-31). NIST was involved in testing the walkway connection because their experts in materials testing could help explain the fault in the walkway system and thus avoid similar disasters and perhaps develop new standards. The National Aeronautics and Space Administration (NASA), another federal research agency, has research centers across the United States that perform materials testing on all sorts of materials relating to both space and aeronautics and also to many associated technologies, such as aircraft runways or satellites.

4.6.2.3 Private laboratories. Many private laboratories test materials for both private businesses and government agencies. For example, Underwriter Laboratories (UL)® tests and approves appliances, tools, motors, and similar products. In addition to large private labs like UL, you can find small, private materials companies operating in many communities that test a wide range of materials and products.

Figure 4-41 illustrates the use of standards for building construction materials. The CoreGuard security wall system was designed for high-security areas such as hospital corridors, schools, computer centers, libraries, or anywhere high-impact materials are required. Note the UL® label; the panel system was designed and tested according to UL 263 and ASTM E-119 standards for one-hour and two-hour fire-rated wall assemblies.

4.6.2.4 Manufacturing companies. Large manufacturing companies rely on their materials testing labs to test all sorts of materials, ranging from raw materials and material stock from outside vendors (to ensure that they meet specifications), to the compo-

Figure 4-41 Meeting building standards. CoreGuard's one-piece panel incorporates 0.081-in. opaque recycled polycarbonate sheet and ⅝-in. or ¾-in. fire-rated gypsum wallboard. The panel provides at least 600 ft/lb of impact strength. (G.E. Plastics.)

nents and products they make (to be sure they will meet quality standards). For example, they may test nuts and bolts to ensure that they have sufficient strength, and they may crash test a car to see how the materials and components respond to improved design. Smaller manufacturers may not be able to afford their own testing lab, so they will rely on private labs. Even large manufacturers use private labs such as UL to certify their products.

4.6.2.5 Wholesalers and retailers. To ensure product quality, products wholesale distributors and retail stores may have their own lab or use private labs to test samples of products that they will sell. This quality testing is important to ensure reliable products that will keep the confidence of customers. Can you think of examples of retailers and the products they might test?

4.6.2.6 Consumers groups. Many nonprofit consumer organizations also conduct tests of products. A few months before the Christmas season, you may see consumer groups on TV talking about toys considered dangerous as a result of their testing. Consumers Union tests a wide variety of consumer products and publishes the results in various publications, such as *Consumer Reports.* This magazine is a useful resource to learn more about materials and product testing.

4.6.2.7 Individual consumers. The ultimate testers of materials and products are consumers. Manufacturers try to stay in touch with their customers to ensure satisfaction. As newer materials, such as composites and advanced ceramics, appear on the market in new products, consumer acceptance is of great concern to producers. Even though extensive testing goes into these materials and products, the final and most important testing is by you, the consumer, whose feedback to the manufacturers helps them to improve products as well as to validate their testing procedures.

4.6.3 Inspection

Inspection differs from testing in its objectives. The ***objectives of inspection*** are to examine parts of materials for the presence of *discontinuities* and *defects.* In other words, inspection estimates the degree to which a product conforms with design specifications. Materials characterization, as discussed in Section 3.3, may be part of the inspection procedure. The American Society for Nondestructive Testing (ASNT) defines a ***discontinuity*** as an interruption in the normal physical structure or configuration of a part, such as a crack or porosity. A discontinuity may or may not be detrimental to the usefulness of a part. A ***defect*** is a discontinuity whose size, shape, location, or properties adversely affect the usefulness of the part or exceed the design criteria for the part. ASNT has five major inspection systems for detecting such discontinuities. These systems are (1) radiographic testing (RT), (2) ultrasonic testing (UT) (see Figure 4-42), (3) eddy-current testing (ET), (4) magnetic-particle testing (MT), and (5) liquid-penetrant testing (PT). In addition to these five major systems, inspection techniques using holography, microwaves, liquid crystals, infrared radiation, and leak testing are available where circumstances dictate. Use of *computer simulation* of materials processing is developing into a reliable means for improving quality processes. Sophisticated simulation allows process engineers to look at

Figure 4-42 Ultrasonic-testing setup. (NASA.)

most variables prior to building molds for casting and other process components. As is evident from the preceding information, inspection techniques involve tests, most of which are classed as ***nondestructive testing (NDT)*** and ***nondestructive evaluation (NDE).***

PAUSE & LOOK

Look around your room, lab, or office and locate the familiar UL label on an electrical cord or piece of equipment. Discuss the type of material and product testing that may have been done on it. Do some research into UL testing to determine how it helps you, the consumer.

Two terms used in conjunction with a discussion of testing and inspection are ***quality control*** (QC) and ***quality assurance*** (QA). The objective of QC is to determine *statistically* how much testing and inspection are required to ensure that products will meet design specifications and service life expectations. For example, in the production of an automobile engine, each part cannot, because of high cost, be given a complete inspection to see if it meets exact dimensional tolerances, if the metal used is free of all external and internal defects, and if each part has the necessary strength. Rather, a ***statistical sample*** is taken of a batch of parts as they are produced. Through statistical analysis, the sample is determined to be representative (probably the same) as all parts made in the process. Whole engines are also chosen statistically to be samples of the engine-making process. They are tested to failure (destructive test) as a means of providing ***quality assurance*** (QA) for the engine itself. QA, the goal of any QC program, is used to refer to the total set of operations and procedures used in manufacturing a product whose goals are conformance of a product to design specifications.

4.7 MATERIALS SELECTION

In the section Pause & Ponder and throughout this book, new materials and processes are described. Often, 10 to 20 years elapse before new materials come into common use. It requires that much time for development of a reliable ***database*** to ensure that the potential (properties, processing techniques, cost, etc.), as well as any risk assessment, is well known. The normally high cost of committing to production with new materials dictates an *evolutionary* rather than a *revolutionary* procedure for introducing new materials. Computer simulation coupled with rapid prototyping (see Module 6) is helping to shorten the time from concept to reliable materials technology.

Advanced composites serve to illustrate the point. In Module 8, we develop the many benefits available from advanced composites, including the favorable properties studied in this module. However, these materials have been slowly accepted, even in the aerospace market, where their benefits (especially strength-to-weight ratios) are so desirable, as seen in the potential applications shown in Figure 4-43. Among the reasons why aerospace designers have moved cautiously is that material failure in a commercial

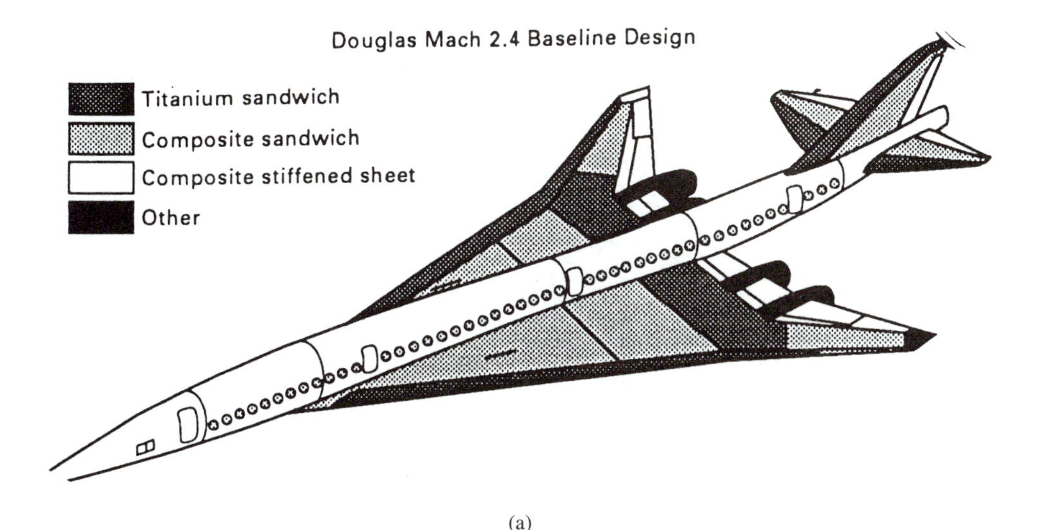

(a)

Figure 4-43 (a) Application of composite materials to aircraft.

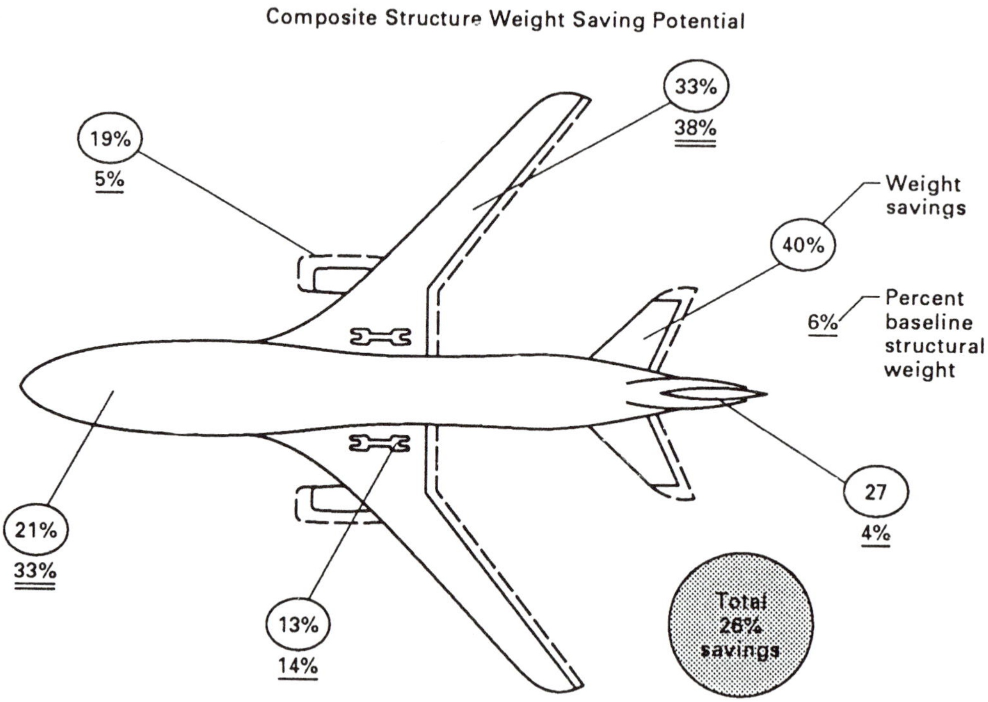

Composite Structure Weight Saving Potential

(b)

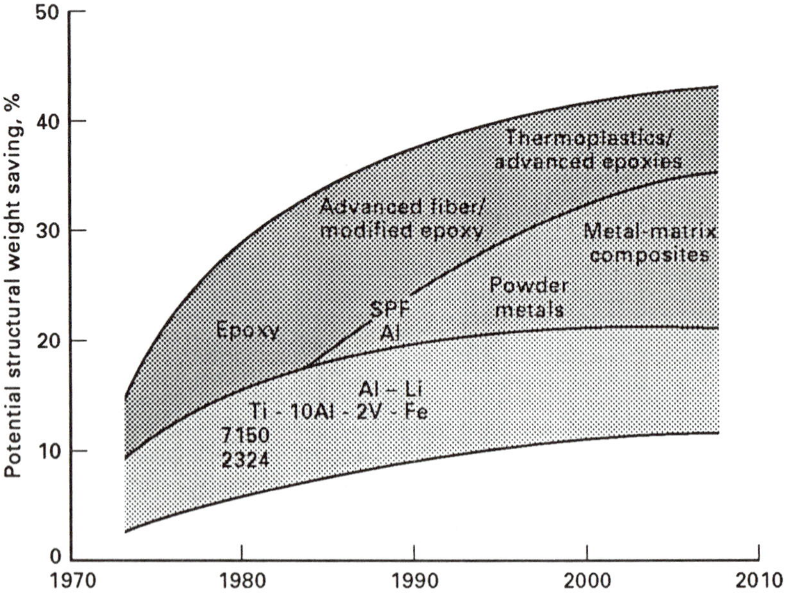

(c)

Figure 4-43 (b) Structural weight savings. (c) Potential weight savings trend for future structural materials. (NSAS.)

passenger aircraft carries enormous consequences. Another reason is the refinement of new design technology and automation of the fabrication of composites. Costs for advanced composites have declined. Graphite fibers were around $35 a pound in the early 1980s and dropped to around $20 per pound in the early 1990s. However, fabrication costs will remain high until automated fabrication of composites, with the help of robotics, is developed further. The section Application & Alternatives that follows demonstrates why the properties available in advanced composites make them the material of choice when cost is not an overriding consideration. Still, in applications where cost is a major concern, expect to see the trend toward composites grow [Figure 4-43(c)] because their properties make them so attractive that databases, design parameters, and processing/manufacturing technology will quickly advance. In rebuilding the world's transportation infrastructure, low-cost polymer matrices that are environmentally durable and safe will gain acceptance. As you will see in later modules, automobiles, trucks, trains, and oceancraft are also moving toward use of these composites because they offer the right properties to meet new demands related to energy efficiency and environmental impact.

As discussed in the Pause & Ponder for Module 3, low-energy materials and processes must also be considered. This is true due to cost and environmental issues. Also, in developing countries, which may lack natural resources and sophisticated technology, wood and other polymers and cement and other ceramics will be favored.

Later you will learn how properties of polymers, metals, ceramics, and composites qualify these materials to be selected for critical parts in a variety of products and systems. With a knowledge of the properties of materials and practice in using this knowledge, wise material selection should result.

APPLICATIONS & ALTERNATIVES

Sports and recreation are big businesses throughout the world. Both professional and amateur athletes continue the search for the winning edge. College courses offered in materials engineering for sports equipment focus on improving performance through engineered materials and new designs. As engineering materials evolved for the aerospace industry, designers began to apply the new materials to a wide variety of equipment—bicycles, snow skis, race cars, baseball bats, hang gliders, and tennis rackets, just a few of the products that have profited from such innovations.

The Prince Vortex racket shown in Figure 4-44 provides an example of how designers apply principles of mechanics, aerodynamics, and strength of materials, combined with properties of materials and manufacturing methods, to seek the optimum racket. Figure 4-5 depicted concepts of stiffness in racket frames. Contemporary frames no longer use wood and steel but use instead space-age composites of epoxy matrices reinforced with fibers such as boron, graphite, silicon carbide, liquid-crystal polymers, aramid, and fiberglass. Frame designs include string areas of 90 (midsized) to 110 (oversize) square inches in widebodies (Figure 4-5). The polymer-matrix composite (PMC) in the Vortex involves mixing viscoelastic polymers with graphite and epoxy to adjust to the force of the ball

Figure 4-44 Vortex Tennis Racket exemplifies the search for the ideal sports equipment through application of properties of materials. (Prince Mgf. Co.)

striking the strings. The frame design stiffens the top of the racket for greater power and allows more flexibility in the shaft near the handle. Other features include a different stringing pattern and cushion handle. The composites were "engineered," not selected, for the Vortex racket. The ideal tennis racket should provide power and control to the player, allow for a variety of shots, and reduce vibration.

Visit a sporting goods shop and notice the wide variety of tennis rackets now available. Read the literature and compare the prices. While in the shop, observe the other equipment to see how materials technology is meeting the desires of sports and recreational enthusiasts.

SELF-ASSESSMENT

4-1. A large percentage of the world's infrastructure—its buildings, roads, bridges, and tunnels—is badly deteriorated, posing a threat to the safety of people using these facilities. What is the main cause of this degradation? Cite one example of how this deterioration of materials is overcome.

4-2. What qualities define a material?

4-3. Define a chemical property.

4-4. Define *stress* and express it in appropriate SI units.

4-5. Which quantity, stress or strain, can be measured directly in a stress–strain tensile test?

4-6. How can you distinguish a ductile material from a brittle material with the aid of a stress–strain diagram?

4-7. Why should the test conditions under which some material property was determined be included along with the property?

4-8. Do all materials have a yield point?

4-9. Factors of safety are defined either in terms of the ultimate strength of a material or its yield strength. In other words, by the use of a suitable factor, the ultimate or yield strength is reduced in size to what is known as the design stress or safe working stress. Which factor of safety would be more appropriate for a material that will be subjected to repetitious, suddenly applied loads?

4-10. Stiffness of a material is readily apparent in studying the material's stress–strain curve. Explain.

4-11. Product liability court cases have risen sharply in recent years because of poor procedures for selecting materials for particular applications. Assuming that a knowledge of a material's properties is a valid step in the selection process, cite two examples where such lack of knowledge could or did lead to failure or unsatisfactory performance.

4-12. A material may be tough at room temperature. In technical terms, describe what this means. What happens to this toughness in most materials as the ambient temperature is decreased?

4-13. Make a sketch and fully dimension an Izod impact test specimen.

4-14. At what level of yield stress do most fatigue failures occur?

4-15. From your experience, what typical automobile part requires complete fatigue testing prior to development?

4-16. Identify the origin of most fatigue failures and explain the significance of your answer in terms of handling a metal workpiece in the machining process.

4-17. Torquing a nut on a bolt to specified values using a torque wrench is standard practice in many industries. Explain why.

4-18. A structural steel beam may deflect under its transverse loading. Is this considered bad practice? Is it allowed? Cite a case where deflection of beams (wood or steel) must be limited to a minimum.

4-19. When selecting a material that will be subjected to abrasion, what property would you expect the material to possess to a great degree?

4-20. How does the chemical industry protect its tanks and piping from corrosion due to the action of highly corrosive liquids?

4-21. What would your explanation be for the development of surface cracks on the surface of four-year-old automobile tires?

4-22. Name one method that large automobile manufacturers use to inhibit corrosion of the steel chassis parts of trucks and passenger vehicles.

4-23. A 1-in.-diameter steel rod 3 ft long is used between two immovable (rigid) walls. The rod is welded to the two walls at each of its two ends. The ambient temperature increases by 100°F. The modulus of elasticity for the steel is 30×10^6 in./in./°F.

(a) Make a sketch of the above problem using appropriate welding symbols. Note on the sketch that the walls do not move or deform. This sketch will show the rod welded at both ends to the walls.

(b) Describe what is happening to the rod. What is acting on the rod from the outside? What is happening to the atoms in the steel rod?

(c) If the rod were free to move, describe its movement and calculate how much it would move and in what direction.

(d) Make another sketch with the walls replaced by force vectors representing the walls acting on the rod.

(e) Calculate the stress acting inside the rod.

(f) Calculate the force on the rod that would be necessary to produce the same stress as calculated in part (e).

4-24. A 3000-kg load using a 10-mm indentor produces a 3.2-mm impression on a ferrous metal specimen that yields a Brinell hardness of 364 (HB = 364). Using the relationship of HB to the tensile strength (TS), determine the approximate TS value for the material. Also, knowing the relationship between TS and the endurance limit (EL), determine the EL value for the material.

4-25. Refer to Table 4-1 and the accompanying text. What prefix would be added to a Rockwell hardness reading taken from a major load of 150 kg with a diamond penetrator?

4-26. Compare the modulus of elasticity with the modulus of rigidity for the same steel and write a statement about the steel's capability to withstand a direct tensile load versus a shearing type of load.

4-27. Thermal resistivity is as important as thermal conductivity. Illustrate both properties by citing an example of each from your own experience or knowledge.

4-28. Using magnetic chucks in a machining operation, how does a knowledge of magnetic properties of metals pay big dividends?

4-29. How does quality control differ from quality assurance?

4-30. Why does a deep drawing operation, as illustrated in Figure 4-14(b), require so many phases?

4-31. Solid objects are three-dimensional. A bar with a rectangular cross-section under load will deform in how many directions? If it deforms in more than one direction, are the amounts of deformation the same? Which direction would have the greatest deformation?

4-32. Knowing that a common steel has a HB of 207, determine its approximate tensile strength.

4-33. Explain why water is chosen to cool automobile engines. What is the effect of using more than the recommended percentage of antifreeze in the automobile's cooling system?

4-34. Even though heat and work are different forms of the same quality (energy), custom has continued to express them using different units. Name two customary units used to measure heat energy.

4-35. The coefficient of thermal linear expansion may be expressed in units such as m/m/°C. What must you do to convert these units to °F? Would you do the same thing if you were given units of m/m/K?

4-36. Is the coefficient of linear expansion the same for iron and iron alloys (steels)? If not, how do you explain the difference, and what is the implication in designing steel structures?

4-37. Both the coefficient of thermal expansion and the thermal conductivity of most materials are temperature sensitive. Explain this sentence in terms of any changes in their respective values. Can you find a material with a coefficient of thermal expansion that is nearly zero?

4-38. For the major heading Properties, develop a list of at least two properties. Next to each, write a service condition from your own experience in which the property is important.

4-39. Express one cubic inch in terms of cubic centimeters.

4-40. Using Tables 10-7, 10-7A, and 10-11, calculate the thermal mass for type 405 stainless steel in cal/cm^3 · °C (refer to Section 4.5.1).

4-41. When determining if a heat treat oven has sufficient heating capacity, one can calculate the heat required per some stated period of time to raise the temperature of a material a certain number of degrees. Does this calculation represent the complete answer or are there other factors involved?

4-42. The specific heat for iron as stated in Section 4.5.1 is 0.1 cal/g · K. Express the specific heat in units of Btu/lb · °F.

4-43. Calculate the heat required to heat 1000 pounds of steel to 400°F in one hour. Assume the specific heat of steel is 0.125 Btu/lb · °F and the ambient temperature is 70°F.

OBJECTIVE QUESTIONS

4-1. Cite one property of a material that cannot be observed.

a. Conductivity **b.** Hardness
c. None can be observed **d.** All can be observed

4-2. Give two examples of a mechanical property.

a. Thermal resistance **b.** Wear resistance **c.** Hardness **d.** Strength

4-3. Scissors used in the home cut material by concentrating forces that ultimately produce a certain type of stress within the material. Identify this stress.

a. Bearing stress **b.** Shearing stress **c.** Compressive stress

4-4. Given μ for a material as 0.25 and the strain in the lateral direction as 1.50×10^{-6} mm/mm, what is the axial deformation expressed in inches?

a. 6×10^{-6} in./in. **b.** 0.6×10^{-5} in./in. **c.** Both (a) and (b)

4-5. An aluminum rod 1 in. in diameter ($E = 10.4 \times 10^6$ psi) experiences an elastic tensile strain of 0.0048 in./in. Calculate the stress in the rod.

a. 49,920 ksi **b.** 49,920 psi **c.** 49,920

4-6. Express the stress in question 4-5 in SI units with an approved prefix.

a. 0.3442 GPa **b.** 3.442×10^8 Pa **c.** 34.42 MPa

4-7. A 6-ft steel bar is deformed 0.01 in. in an axial direction. What is the unit deformation?

a. 1.389×10^{-4} **b.** 1.389×10^{-4} in./in. **c.** 1389×10^{-7} in./in.

4-8. A shaft supported by thrust bearings placed 4 ft apart will fail if the total deformation exceeds 0.00025 in. What is the maximum allowed strain?

a. 3.7×10^{-6} in./in. **b.** 5.2×10^{-6} in./in.
c. 15.6×10^{-5} in./in. **d.** None of the above

4-9. The minimum yield stress for a material is 48,000 psi. What is the factor of safety if the allowed working stress is 24,600 psi?

a. 0.51 **b.** 1.95 **c.** 3.90

4-10. Express the yield stress units of question 4-9 in megapascals, the recommended units in the SI system. Use appropriate abbreviations.

a. 331 MPa **b.** 331 GPa **c.** 331 MPA **d.** All of the above

4-11. A 1-in.-diameter steel circular rod is subject to a tensile load that reduces its cross-sectional area to 0.64 in.^{2}. Express the rod's ductility using a standard unit of measure.

a. 18.5% **b.** 1.85% **c.** 18.5 **d.** (a) and (c)

4-12. If a force (push or pull) of 40 lb is used to turn a wrench such that the 40-lb force is applied at right angles to the wrench at a distance of 16 in. from the center of its jaw, what torque is produced?

a. 640 inch pounds **b.** 640 #-in **c.** 640 pound-inches **d.** (b) and (c)

4-13. An angle of twist of 0.48° can be expressed as how many radians?

a. 8.4×10^{-3} **b.** 84 radians **c.** 0.0084 radians **d.** (a) and (c)

4-14. A standard-weight pipe with a nominal diameter of 1 in. has an outside diameter (OD) of 1.315 in. and a thickness of 0.133 in. Calculate its polar moment of inertia.

a. 0.247 $in.^3$ **b.** 0.247 $in.^4$ **c.** 2.47×10^{-3} $in.^4$ **d.** All of the above

4-15. Determine the bending stress on the top surface of a rectangular beam whose height is 11½ in. if its moment of inertia is 951 $in.^4$ and its bending moment is 82,856 pound-inches.

a. 499.34 psi **b.** 499.34 pounds/sq. in.
c. 499.34 $\#/in.^2$ **d.** All of the above

4-16. The specific heat (specific heat capacity) of water is 1.0 $kcal \cdot kg^{-1} \times K^{-1}$. Write this value using a different set of SI units.

a. 4184 J/kgK **b.** 4184 J kg^{-1} K^{-1}
c. 4.184×10^3 J/kg K **d.** All of the above

REFERENCES & RELATED READINGS

AMERICAN SOCIETY FOR TESTING AND MATERIALS. *1996 Annual Book of ASTM Standards.* Philadelphia: ASTM, 1996.

AMERICAN SOCIETY FOR TESTING AND MATERIALS. *UNSearch, Version 2.0,* software, Philadelphia: ASTM, 1992.

ANDERSON, DAVID B. "Corrosion Data," *ASTM Standardization News,* July 1992, pp. 42–48.

ASM INTERNATIONAL. *ASM Handbook,* Vol. 1, *Properties and Selection: Irons, Steels and High Performance Alloys*; Vol. 2, *Properties and Selection: Nonferrous Alloys and Special Purpose Materials*; Vol. 8, *Mechanical Testing*; Vol. 11, *Failure Analysis and Prevention*; Vol. 13, *Corrosion*; Vol. 17, *Nondestructive Evaluation and Quality Control.* Materials Park, OH: ASM International, 1990, 1991, 1985, 1985, 1987, 1989.

ASM INTERNATIONAL. "Materials Selection," *Advanced Materials and Processes,* special issue, June 1993.

BERKE, NEAL S., and LAWRENCE R. ROBERTS. "Reinforced Concrete Durability and ASTM," *ASTM Standardization News,* January 1992, pp. 46–51.

BLAU, PETER, ed. *ASM Handbook,* Vol. 18, *Friction, Lubrication, and Wear Technology.* Materials Park, OH: ASM International, 1992.

BOSICH, JOSEPH F. *Corrosion Prevention for Practicing Engineers.* New York: Barnes & Noble, Inc., 1970.

CUBBERLY, W. H., and RAMON BAKERJIAN, eds. *Tool and Manufacturing Engineers' Handbook.* Dearborn, Mich.: Society of Manufacturing Engineers, 1989.

DEXTER, S. C., ed. *Biologically Induced Corrosion.* Houston, NACE, 1986.

DIXON, J. I. *Failure Analysis: Techniques and Applications.* Materials Park, Ohio: ASM International, 1992.

FERRARA, MICHAEL. "Stress Relief Meta-Lax Extends Life of Engine Components." *Turbo & Hi-Tech Performance,* May 1995, pp. 62–65.

JOHNSTON, NORMAN J. "Introduction to High Performance Composites," *National Educators' Workshop, NEW:Update 92,* Oak Ridge National Laboratory, Oak Ridge, Tenn., November 13, 1992.

NATIONAL ASSOCIATION OF CORROSION ENGINEERS. *Corrosion of Metals in Concrete.* Houston: NACE, 1988.

OBERG, E., et al. *Machinery's Handbook,* 24th ed. New York: Industrial Press, 1992.

PRINCE MANUFACTURING COMPANY. *Facts About the Physics of Tennis.* Princeton, NJ: Prince, 1992.

PROVAN, JAMES W. "An Introduction to Fatigue," *Journal of Materials Education,* Vol. 11, No. 1/2, 1989, pp. 1–105.

RUMBLE, JOHN. "Computerizing ASTM Test Methods," *ASTM Standardization News,* July 1992, pp. 34–38.

WULPI, DONALD J. *Understanding How Components Fail.* Materials Park, OH: ASM International, 1985.

Periodicals

Advanced Materials and Processes
ASTM Standardization News
Corrosion
Geotechnical Testing Journal
Journal of Composites Technology and Research
Journal of Materials Education
Journal of Materials Engineering and Performance
Journal of Testing and Evaluation
Materials Performance
Surface Engineering

EXPERIMENTS & DEMONSTRATIONS IN TESTING & EVALUATION

NEW: Update 88 NASA Conference Publication 3060

SASTRI, SANKAR. "Fluorescent Penetrant Inspection."

SASTRI, SANKAR. "Magnetic Particle Inspection."

SASTRI, SANKAR. "Radiographic Inspection."

NEW: Update 89 NASA Conference Publication 3074

CHOWDHURY, MOSTAFIZ R., and FARIDA CHOWDHURY. "Experimental Determination of Material Damping Using Vibration Analyzer."

CHUNG, WENCHIANG R. "The Assessment of Metal Fiber Reinforced Polymeric Composites."

STIBOLT, KENNETH A. "Tensile and Shear Strength of Adhesives."

NEW: Update 90 NIST Special Publication 822

AZZARA, DREW C. "ASTM: The Development and Application of Standards."

BATES, SETH P. "Charpy V-Notch Impact Testing of Hot Rolled 1020 Steel to Explore Temperature Impact Strength Relationships."

CHOWDHURY, MOSTAFIZ R. "A Nondestructive Testing Method to Detect Defects in Structural Members."

CORNWELL, L. R., R. B. GRIFFIN, and W. A. MASSARWEH. "Effect of Strain Rate on Tensile Properties of Plastics."

GRAY, STEPHANIE L., KRISTEN T. KERN, WYNFORD L. HARRIES, and SHEILA ANN T. LONG. "Improved Technique for Measuring Coefficients of Thermal Extension for Polymer Films."

HALPERIN, KOPL. "Design Project for the Materials Course: To Pick the Best Material for a Cooking Pot."

KUNDU, NIKHILL. "Environmental Stress Cracking of Recycled Thermoplastics."

PANCHULA, LARRY, and JOHN W. PATTERSON. "Demonstration of a Simple Screening Strategy for Multifactor Experiments in Engineering."

TAYLOR, JENNIFER A. T. "How Does Change in Temperature Affect Resistance?"

WICKMAN, JERRY L., and SCOTT M. CORBIN. "Determining the Impact of Adjusting Temperature Profiles on Photodegradability of LDPE/Starch Blown Film."

WIDENER, EDWARD L. "It's Hard to Test Hardness."

WIDENER, EDWARD L. "Unconventional Impact–Toughness Experiments."

NEW: Update 91 NASA Conference Publication 3151

ALLEN, DAVID J. "Stress–Strain Characteristics of Rubber-Like Materials: Experiment and Analysis."

BUNNELL, L. ROY. "Tempered Glass and Thermal Shock of Ceramic Materials."

CORNWELL, L. R. "Mechanical Properties of Brittle Material."

DAHL, CHARLES C. "Computer Integrated Lab Testing."

DENTON, NANCY L., and VERNON S. HILLSMAN. "Isotropic Thin-Walled Pressure Vessel Experiment."

GORMAN, THOMAS M. "Designing, Engineering, and Testing Wood Structures."

KARPLUS, ALAN K. "Determining Significant Material Properties, a Discovery Approach."

LUNDEEN, CALVIN D. "Impact Testing of Welded Samples."

SPIEGEL, F. XAVIER, and BERNARD J. WEIGMAN. "An Automated System for Creep Testing."

STREHLOW, RICHARD R. "ASTM: Terminology for Experiments and Testing."

NEW: Update 92 NASA Conference Publication 3201

BUNNELL, L. ROY. "Temperature-Dependent Electrical Conductivity of Soda-Lime Glass and Construction and Testing of Simple Airfoils to Demonstrate Structural Design, Materials Choice, and Composite Concepts."

MARPET, MARK I. "Walkway Friction: Analysis and Testing."

MARTIN, DONALD H. "Application of Hardness Testing in Foundry Processing Operations: A University and Industry Partnership."

MASI, JAMES V. "Experiments in Corrosion for Younger Students by and for Older Students."

NEEDHAM, DAVID. "Micropipet Manipulation of Lipid Membranes: Direct Measurement of the Material Properties of a Cohesive Structure That Is Only Two Molecules Thick."

PERKINS, STEVEN W. "Direct Tension Experiments on Compacted Granular Materials."

SHIH, HUI-RU. "Development of an Experimental Method to Determine the Axial Rigidity of a Strut-Node Joint."

SPIEGEL, F. XAVIER. "An Automated Data Collection System for a Charpy Impact Tester."

TIPTON, STEVEN M. "A Miniature Fatigue Test Machine."

WIDENER, EDWARD L. "Tool Grinding and Spark Testing."

NEW: Update 93 NASA Conference Publication 3259

BORST, MARK A. "Design and Construction of a Tensile Tester for the Testing of Simple Composites."

CLUM, JAMES A. "Developing Modules on Experimental Design and Process Characterization for Manufacturing/Materials Processes Laboratories."

DENTON, NANCY, and VERNON S. HILLSMAN. "An Introduction to Strength of Materials for Middle School and Beyond."

DILLER, T. E., and A. L. WICKS. "Measurement of Surface Heat Flux and Temperature."

FISHER, JONATHAN H. "Bridgman Solidification and Experiment to Assess Boundaries and Interface Shape."

GRAY, JENNIFER. "Symmetry and Structure Through Optical Diffraction."

KARPLUS, ALAN K. "Knotty Knots."

KOHNE, GLENN S. "An Automated Digital Data Collection and Analysis System for the Charpy Impact Tester."

OLESAK, PATRICIA J. "Scleroscope Hardness Testing."

SPIEGEL, F. XAVIER. "Inexpensive Materials Science Demonstrations."

WICKMAN, J. L. "Plastic Part Design Analysis Using Polarized Filters and Birefringence."

WIDENER, EDWARD L. "Testing Rigidity by Torque Wrench."

NEW: Update 94 NASA Conference Publication CP-3304

BRUZAN, RAYMOND, and DOUGLAS BAKER. "Density by Titration."

DAHIYA, JAI N. "Precision Measurements of the Microwave Dielectric Constants of Polyvinyl Stearate and Polyvinylidene Fluoride As a Function of Frequency and Temperature."

DAUFENBACH, JODEE, and ALAIR GRIFFIN. "Impact of Flaws."

FINE, LEONARD W. "Concrete Repair Applications and Polymerization of Butadiene by an 'Alfin' Catalyst."

HILLSMAN, VERNON S. "Stress Concentration: Computer Finite Element Analysis versus Photoelasticity."

HUTCHINSON, BEN, KIM GIGLIO, JOHN BOWLING, and DAVID GREEN. "Photocatalytic Destruction of an Organic Dyd Using TiO_2."

JENKINS, THOMAS J., JOHN H. COMTOIS, and VICTOR M. BRIGHT. "Micromachining of Suspended Structures in Silicon and Bulk Etching of Silicon for Micromachining."

KARPLUS, ALAN K. "Paper Clip Fatigue Bend Test."

Kilduff, THOMAS F. , and JACOBS, JAMES A. "Mathematics for Engineering Materials Technology Experiments and Problem Solving."

KOHNE, GLENN S. "Fluids with Magnetic Personalities."

LIU, PING, and TOMMY L. WASKOM. "Ultrasonic Welding of Recycled High Density Polyethylene (HDPE)."

MARTIN, DONALD H., HERMANN SCHWAN, and MICHAEL DIEHM. "Testing Sand Quality in the Foundry (A Basic University-Industry Partnership)."

SHULL, ROBERT D. "Nanostructured Materials."

WERSTLER, DAVID E. "Introduction to Nondestructive Testing."

WHITE, CHARLES V. "Glass Fracture Experiment for Failure Analysis."

WICKMAN, JERRY L., and NIKHIL K. KUNDU. "Failure Analysis of Injection Molded Plastic Engineered Parts."

WIDENER, EDWARD L. "Dimensionless Fun with Foam."

NEW: Update 95 NASA Conference Publication 3330

BROWN, SCOTT. "Crystalline Hors d'Oeuvres."

KARPLUS, ALAN K. "Craft Stick Beams."

KERN, KRISTEN. "ION Beam Analysis of Materials."

KOZMA, MICHAEL. "A Revisit to the Helicopter Factorial Design Experiment."

POND, ROBERT B., SR. "Recrystallization Art Sketching."

RUSTUM, ROY. "CVD Diamond Synthesis and Characterization: A Video Walk-Through."

SAHA, HRISHIKESH. "Virtual Reality Lab Assistant."

SPIEGEL, F. XAVIER. "A Novel Approach to Hardness Testing."

SPIEGEL, F. XAVIER. "There Are Good Vibrations and Not So Good Vibrations."

TOGNARELLI, DAVID. "Computerized Materials Testing."

WICKMAN, JERRY L. "Cost Effective Prototyping."

Module

5

Metallic Materials

After studying this module, you should be able to do the following:

5-1. Describe basic alloying, forming, casting, and thermal processing, plus surface engineering methods used by industry to change metals' microstructures and thus vary properties.

5-2. Using cooling curves as an aid, recall the effects of solidification in pure metals, alloys, and nonmetals.

5-3. Analyze phase diagrams and photomicrographs to determine thermal processes, basic structures, and properties of ferrous and nonferrous metals.

5-4. Explain the effect on grain size, metal properties (tensile strength, hardness, and toughness) as a result of strain hardening and various thermal processing. Describe how degrees of both hardness and toughness are achieved through thermal treatments.

5-5. Differentiate among ferrous metal alloys, structures, processing, and properties. Cite examples of properties and applications of major classes of ferrous metals and smart metals.

5-6. Differentiate among nonferrous metal alloys, structures, processing, and properties. Cite examples of properties and applications of major classes of nonferrous metals.

5-7. Explain the corrosion resistance mechanism in metals and methods for preventing corrosion in metals and alloys.

5-8. Recall the fabrication techniques for powdered metal products. Explain some of the properties available and applications for powdered metal (P/M).

5-9. Explain changes involving metal structures, properties, and processing used by the metals industry to compete with other materials groups.

5-10. Use tables in the text, other handbooks, and periodicals to compare alloy content, cost, properties, processability, and applications for selecting metallic materials.

5-11. Describe some of the thrusts and barriers to the use of advanced metals and metals processing such as metal matrix composites (MMCs), superplastic forming (SPF), and light aluminum alloys.

Figure 5-1 Power stroke concept truck. Luxury, comfort, and most of the conveniences of a top luxury car can be found in this burly, tough Ford concept truck. Designed to provide the riding comfort of a car with leather-covered, captain's chair, bucket seats, this dual-rear-wheel truck can perform off-road duties. Concealed in the front air dam below the bumper, an 8000-pound-capacity single-line pull winch increases the utility of this work/play vehicle. Increased use of plastics and composites on trucks aims to reduce overall weight; but metals, especially steel, continue to provide necessary properties in key applications, including the following: *steel* for the winch and its stranded cable, bumpers, bolts, chassis, and bed; *aluminum* for polished truck-type, eight-lug wheels; and *chrome* plating on side-view mirrors, bumpers, and other trim. (Ford Motor Company.)

PAUSE & PONDER

Pause for a moment and think about recent reports dealing with changes in pickup trucks and automobiles on issues such as improved engines, electric vehicles (EVs), or improved fuel efficiencies. Figure 5-1 offers a glimpse of the future of pickup trucks, which have become popular because of their multipurpose abilities. The auto and light-truck industry acts as a major economic player in developed countries. Because these vehicles affect society in so many ways, ranging from ease and comfort of travel to environmental pollution, they continually undergo research and development for improvements.

Have you heard anything new on "Super Car"? The concept vehicle is a goal of a Partnership for a New Generation of Vehicles (PNGV). This industry/government consortium, composed of the U.S. auto industry and the federal government, aims to improve fuel efficiency to 80 mpg (miles per gallon). Researchers for PNGV seek to simultaneously reduce vehicle weight by as much as 40%, increase engine efficiency by 40 to 55%, implement regenerative braking, and increase energy storage by 90%.

These goals came from the United States Council for Automotive Research (USCAR). The group was founded in 1992 to facilitate, monitor, and promote a growing number of precompetitive, cooperative research and development programs among Chrysler, Ford, and General Motors. Many of the research programs organized under the USCAR umbrella are aimed at environmental improvements, fuel economy, and safety for occupants. Super Car focuses on fuel efficiency and will incorporate other aims of USCAR.

Figure 5-2(a) shows how current vehicles lose energy from systems problems such as engine, drivetrain, idling, accessories, aerodynamics, braking, and rolling losses. Addressing these losses to achieve the target of 80 mpg will require considerable redesign of current cars. Figure 5-2(b) depicts estimates of percentages of reductions in auto components (comparing current to Super Car weights) to accomplish the 40% weight reduction. Both a change in component design and materials substitutions will be necessary. Figure 5-2(b) shows how use of materials (in pounds) in cars has changed from 1975 to 1990, and shows projections for materials options beyond the year 2000 to meet Super Car goals. Note an anticipated increase in lighter metals: aluminum, magnesium, and perhaps titanium, along with polymer-based composites and high-strength steel.

Materials substitutions are complex issues that embrace concerns far beyond simple mass reduction. Aerospace technology presently used for many of the candidate materials has proven to be feasible at low production rates for airplanes and spacecraft. This technology must be modified for the much higher production rates required of cars and light trucks. This also means modifying aerospace manufacturing techniques to permit complex molding or machining, joining of dissimilar materials, and improvements in assembly methods: all at competitive costs in line with consumer demands.

This module will provide you with valuable insights into metals, their structures, properties, methods of processing, and applications so that you can contribute to solving the type of problems faced by USCAR. Enjoy the ride!

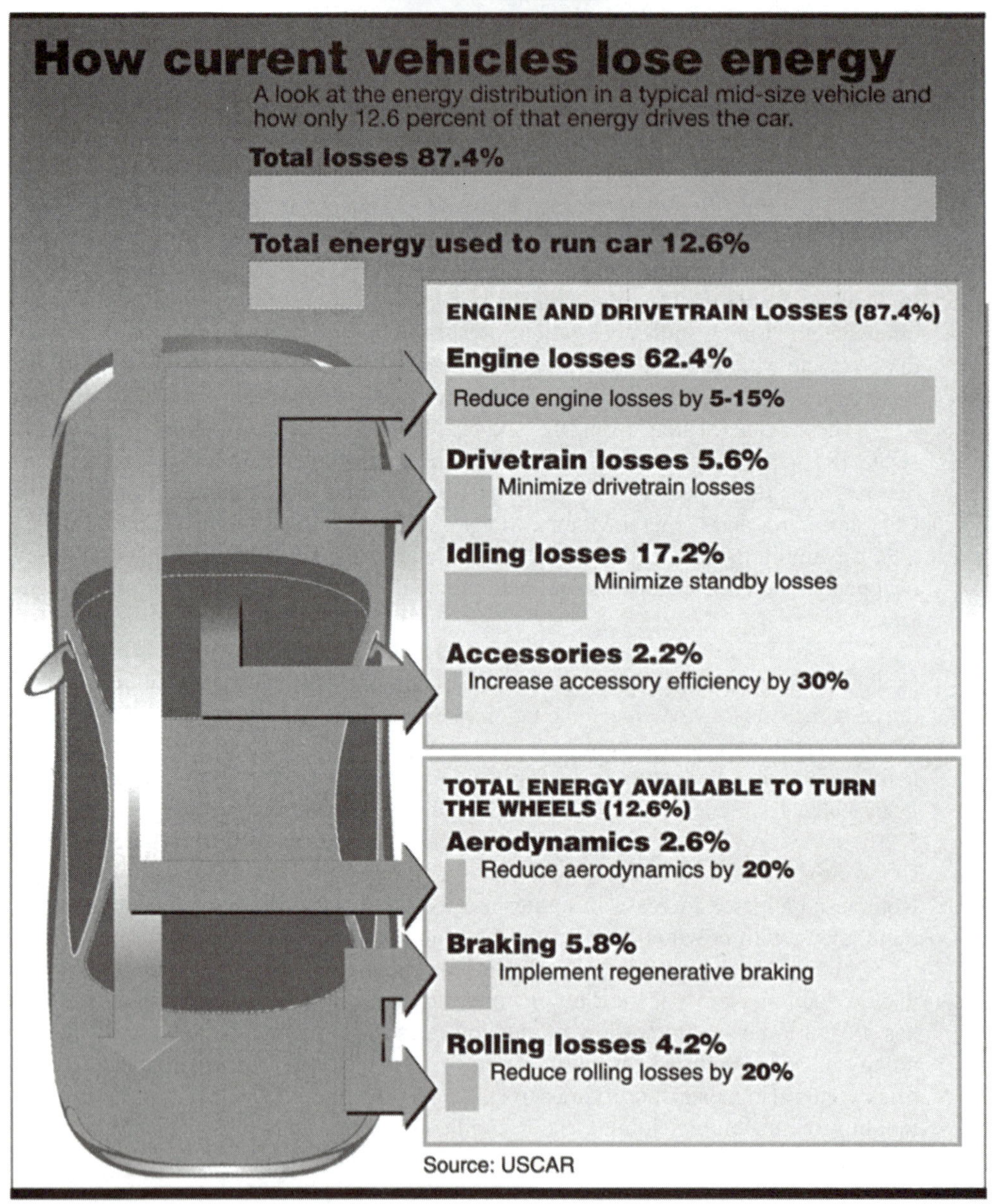

Figure 5-2 (a) How current vehicles lose energy.

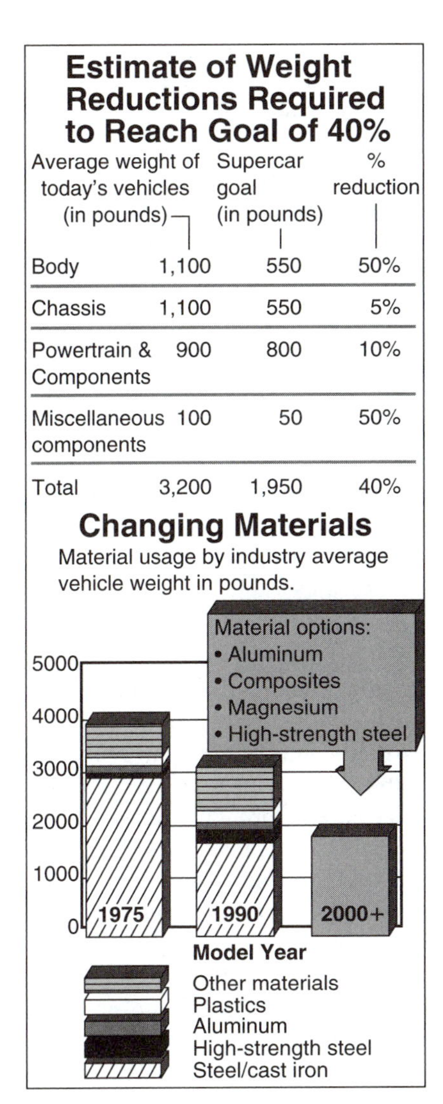

Figure 5-2 (b) Estimate of spot weight reduction required to reach goal of 40% total weight reduction. (Bob Graham, United States Council for Automotive Research.)

5.1 NATURE OF METALS

Metals have been useful in humanity through the ages because they are "strong" when subjected to the external forces encountered under service conditions, yet they become "soft" enough to yield to a machine cutting tool or to a compressive shaping force. Above a certain temperature, they melt and become liquids capable of being shaped by casting. Only in recent times have we realized that the properties of all types of solid materials, including metals, arise from their atomic architecture, that is, from the manner in which their atoms arrange themselves into a crystalline order, from the number and types of imperfections found in this structure, and from the bonding forces that keep the collection or structure of atoms bound or joined together.

The "softness" quality of metals can be explained by an understanding of the atomic structure and metallic bond of the metal atoms to form a crystalline structure. We recall that the electrons in the metallic bond are free to move about their positive ions in an electron cloud or gas, which acts to glue or bond the ions together. This free movement, within limits, also allows for the movement of the atoms under the influence of an external load. This slight movement, visible only under the most powerful microscopes, is called ***elastic deformation*** or ***elastic strain.*** Once the external force, such as a bending force, is removed, the internal electrical forces that cause the atoms to move will decrease, allowing the atoms to return to their normal position; they leave no sign of ever being moved. If you bend a piece of spring steel such as a machinist's rule or vegetable knife, it will return to its original shape, thus experiencing elastic deformation.

If you were not careful and applied too much external force by excessively bending the rule or knife, the atoms might move too far from their original positions to be able to move back again when you released the external force. Consequently, the rule or knife would be permanently bent and no longer fit for use. This permanent deformation is known as ***plastic flow, plastic slip, plastic deformation,*** or ***permanent set.*** When automakers stamp out a metal car body from low-carbon steel in a huge die press, they use this softness quality of metals. The term *cold working* (defined later) is applied to this stamping operation and many other metalworking processes that produce plastic deformation in a metal. Cold-working operations include rolling, heading, spinning, peening, bending, pressing, extruding, drawing, and others. The steel framing of the skyscraper seen in Figure 5-3 consists of structural steel produced by hot working; steel and other metals, shaped by cold working, serve in the construction and furnishing of the typical office building.

The microstructure of metals can be modified in a number of ways. By now we know that this last statement can be interpreted to mean that, through advances in metals technology, we can affect the atomic structure of metals in a precise, controlled manner in the design of metal alloys with the desired properties. In this module we classify, for learning purposes, the basic methods of changing a metal's properties into the following categories: (1) work or strain hardening, (2) thermal processing under equilibrium conditions (solid solution hardening), (3) thermal processing under near-equilibrium conditions (annealing and grain refinement), and (4) thermal processing under nonequilibrium conditions. Of the four categories, only the first does not involve primarily thermal processing. Prior to discussing thermal processing techniques, we need to understand the concepts involved in the internal structure of solids, as presented in Module 3, and have a working knowledge of phase and phase diagrams, covered in the present module.

Work hardening is a way to change a metal alloy's structure to alter its properties by performing work ***(cold working)*** on the metal itself. Work is a form of energy. If we can find a way to deliver energy to these metal atoms, we can give them the energy necessary for the atoms to increase their movement and thus their ability to diffuse through the metal structure. Figure 5-4 shows four industrial cold-working methods, and the photomicrographs demonstrate grain structure before and after work hardening.

In our opening remarks about the usefulness of metals, we mentioned that metals have the ability to yield to a mechanical force that can form them into desired shapes. An external, mechanical shaping force is a force that causes a metal to exceed its elastic deformation limit and ***deform plastically without fracturing.*** This force and the deformation

Figure 5-3 The Steel Triangle, the United States Steel Corporate Center in Pittsburgh. High-strength, low-alloy steel (HSLA), Cor-Ten, forms a tight adherent, dark russet, ferrous-oxide surface coating and requires no other protective coating. The partially completed building exposes steels such as USS "T-1," which is nearly three times the strength of carbon steel and bears great loads as interior columns, and USS Ex-Ten, an HSLA steel that serves as floor beams and some core columns. Much carbon steel finds use in areas not requiring maintenance and high strength. (United States Steel Corporation.)

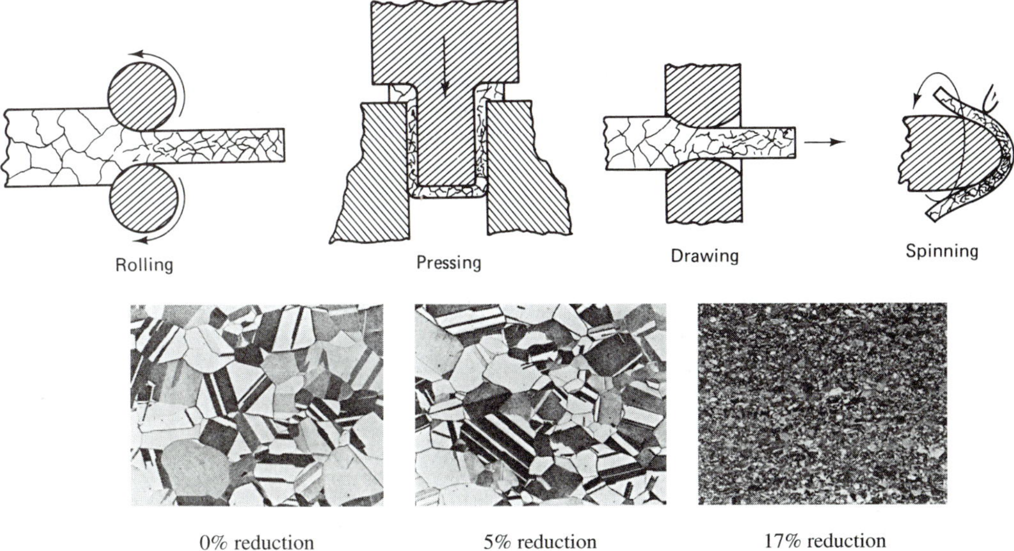

Figure 5-4 Some work-hardening (cold-working) processes.

produced are the means for transferring sufficient energy to the atoms to allow them to flow or move plastically. This movement of atoms, one row at a time along planes of close-packed atoms, shifts the positions of the atoms in relation to each other. From our study of crystal structures, we know that the spacing between atoms is not only critical, but varies with the particular crystal structure. Fcc structures have the closest-packed atoms and the greatest number of closest-packed planes of atoms, which take less energy to allow for the slip of the atoms. As more and more slip takes place, more dislocations are produced. The greater the slip, the greater the dislocations and the more distortion of the lattice structure. The end result is that the deformed metal is *stronger* than the original undeformed metal and offers greater resistance to further deformation.

Figure 5-4 also shows that the processing of materials, particularly the operations of rolling and drawing, can affect the orientation of metal grains (crystals). This is another example of the fact that the structure determines (in the main) the properties of a material. In this example, due to the ***preferred orientation*** of the grains, the metals will have different properties in the direction of the orientation than in other directions. This is known as ***anisotropy*** (see Section 5.5.4 for further discussion). Metal grains that are random in their orientation produce ***isotropy,*** the same properties in all directions.

In summary, through cold working we have (1) reduced the metal's ductility, (2) reduced the effectiveness of the metal atoms to slip, (3) reduced movement of dislocations in the structure, (4) created distortion of the lattice structure, and (5) ended up with a stronger metal that requires a greater force or greater amount of energy to deform it further. At the same time, changes in other properties, such as electrical conductivity, have also occurred. This condition of noticeable increase in energy for further deformation or increase in the metal's yield strength is known as ***work*** or ***strain hardening***.

5.2 PHASES

A ***phase*** is a homogeneous part or aggregation of material that differs from another part due to difference in structure, composition, or both. The difference in structures forms an interface between adjacent or surrounding phases. Some solid materials have the capability of changing their crystal structure with varying conditions of pressure and temperature. These varying conditions cause these materials to change phase. Water can exist as a gas, liquid, or solid. Note that each of these general phases contains the same *components* (i.e., the same basic chemical elements or chemical compounds). The single component present in water is the compound H_2O, made up of hydrogen and oxygen.

Water in the solid phase can have different phases because it forms different crystal structures under different conditions of temperature and pressure. At the triple point of water (see Figure 5-9) all three distinct phases can coexist with identical compositions, each having unique atomic structures and properties.

Being allotropic, like many metallic elements, iron can have different crystal structures and hence different phases (see Figure 5-5 and also Figure 3-54). At room temperature iron has a bcc structure. Heated to above 910°C (1670°F), iron's structure transforms to fcc. These allotropic or polymorphic forms are also referred to by Greek letters: α (al-

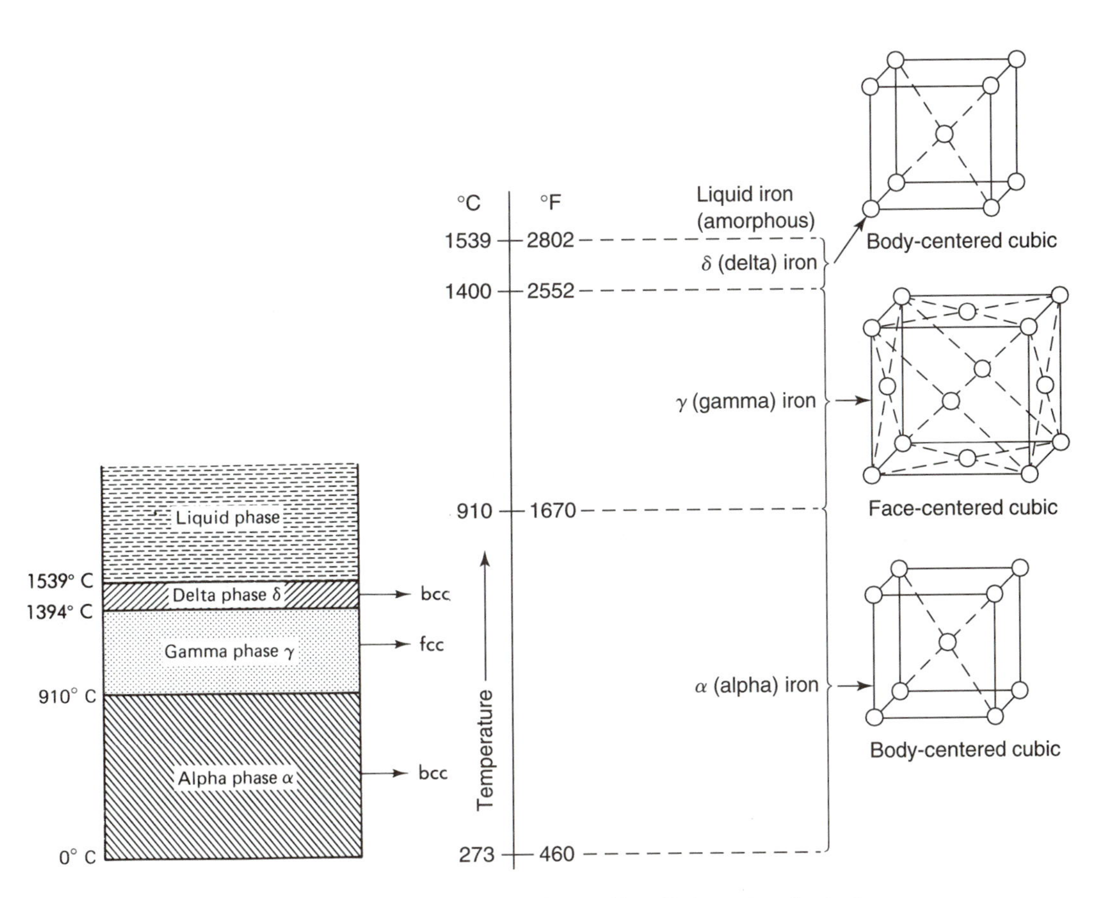

Figure 5-5 Allotropic forms of iron (3 phases: bcc, fcc, bcc).

pha), γ (gamma), and δ (delta). Then, on heating above 1394°C (2541°F), iron's structure changes back to bcc.

It is worthwhile to pause briefly to recall that *phases,* being physically distinct with their own characteristic crystal structure, must of necessity possess different properties. No elements found in nature are 100% pure. During the processing of metals, additional impurities are introduced unintentionally, due partly to the high costs involved in trying to eliminate the impurities.

What are the effects of adding other substances to a pure element? The results produced can be many and varied. One result of this addition of impurities might be the formation of three different solid phases. Second, solid solutions may result. (A ***solid solution*** is simply a solution in the solid state that consists of two kinds of atoms combined in one type of space lattice.) A third possibility is that one or more compounds could form. Finally, two or more of the preceding results could coexist, depending on the prevailing conditions of pressure, temperature, and degree of concentration of the components of the system. Some examples that illustrate the diversity of the results follow. An fcc structure of solvent Cu atoms will substitute Ni atoms in a solid solution of copper and nickel, known as a ***substitutional solid solution***. The substitution of Zn atoms for Cu atoms in an fcc structure forms brass, a disordered or random substitutional solid solution [see Figure 5-6(a)]. If the ordering were 100%, a compound would be formed. The alloy Cu–Au [Figure 5-6(b)] has Cu atoms occupying the face-centered sites and Au atoms occupying the corner sites of the fcc unit cell.

Ordered solid solutions form generally at lower temperatures, or they come into existence when a disordered solid solution becomes unstable at a lower temperature. The fundamental lattice structure may or may not change during this particular transformation. It is important to point out again that this modification in the arrangement of the atoms brings on an alteration in physical properties. In some instances the ordered arrangement is harder and has greater electrical conductivity than the disordered arrangement.

Another type of crystal structure is formed when some atoms take up positions between the regular solvent atoms, that is, in the vacant spaces between lattice points. These solute atoms and the solid solutions formed are ***interstitial.*** The carbon atoms in ferrite occupy the interstices (spaces) in this bcc allotropic form of iron. For solute atoms to form interstitial solid solutions [Figure 5-6(c)], they must be small enough to fit into the interstices between the normal lattice points.

It is time to ask, Is it possible to determine ahead of time whether an element will form a separate phase or a solid solution when added to another element? In other words, can we predict the degree of solubility occurring in certain systems? Certainly, this lies at the heart of metals technology in developing alloys (solid solutions) that possess ductility for subsequent forming operations or two-phase materials with greater strength and hardness than those of the pure substances alone. A set of general rules known as the

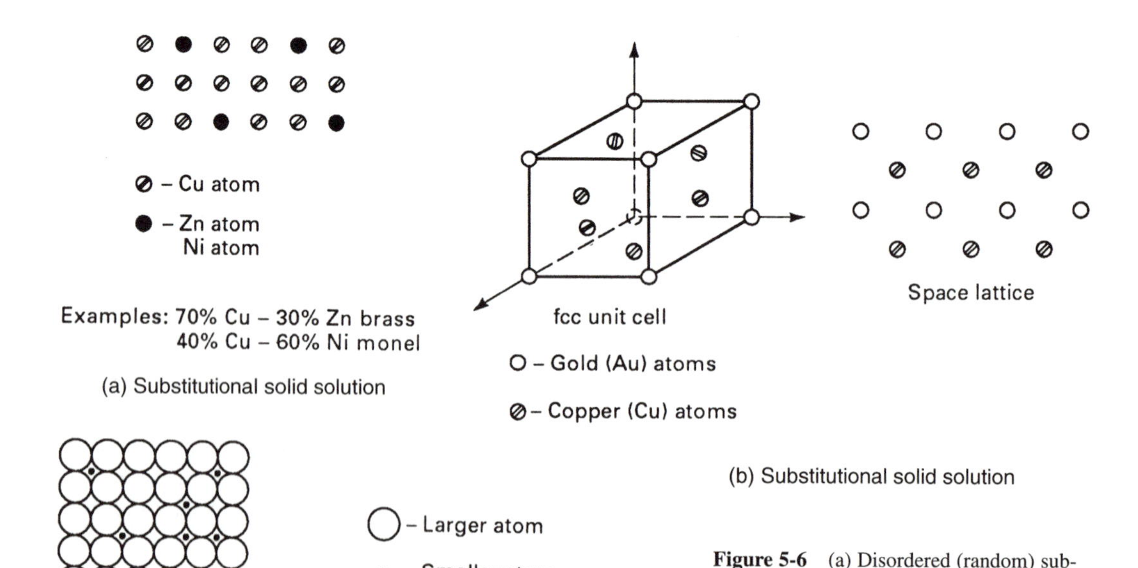

Figure 5-6 (a) Disordered (random) substitutional solid solution. (b) Ordered crystalline compound (Cu_3Au). (c) Interstitial solid solution.

Hume–Rothery solubility rules provides this guidance, and like all such statements is not to be considered foolproof. In brief, these rules state that *to form a solid solution* (refer to Table 3-2):

1. The difference in atomic diameter between the solvent atom and the solute atom should be less than 15%.
2. Elements that do not readily form compounds (near neighbors in the periodic table, such as Fe and Co) have a tendency to form solutions in one another.
3. Elements that have the same lattice structure tend to form a complete range of solid solutions.

A pause to discuss these rules with regard to other previous definitions of solid solutions is worthwhile at this time. Assume two metals are soluble in each other to some extent. If the size difference in their atoms is less than 8% and the other conditions are satisfactory, there is almost complete solid solubility. In other words, the two metals are soluble in each other in all possible proportions. Examples of this are Monel (Ni–Cu) and brass (Cu–Zn) [Figures 5-6(a), 5-7(a) and (b)]. Referring to Table 10-1B, copper's atomic radius is 1.57Å (1Å = 10^{-10}m) and nickel's is about 1.62Å. Expressed in nanometers, these radii are 0.157 nm and 0.162 nm, respectively. Both have an fcc structure. Note their position in the periodic table. They are elements 29 and 28, respectively. The fcc Cu–Ni alloys can range from near 0% Ni to almost 100% Ni.

There are some 25 alloys used in industry that are based on this mutual solubility of copper and nickel in all proportions. These nickel alloys, known as ***Monel alloys,*** contain about 30% copper and have combinations of high strength and good corrosion resistance. Some high-strength Monels, in addition to being nonmagnetic, are equivalent to heat-treated steels having tensile strengths approaching 200,000 psi. In contrast to pure Cu, which has relatively low strength, the alloys of Cu and Ni permit much higher strength

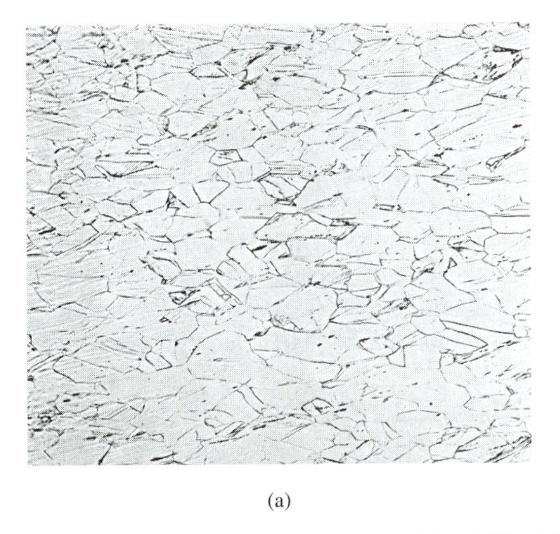

(a)

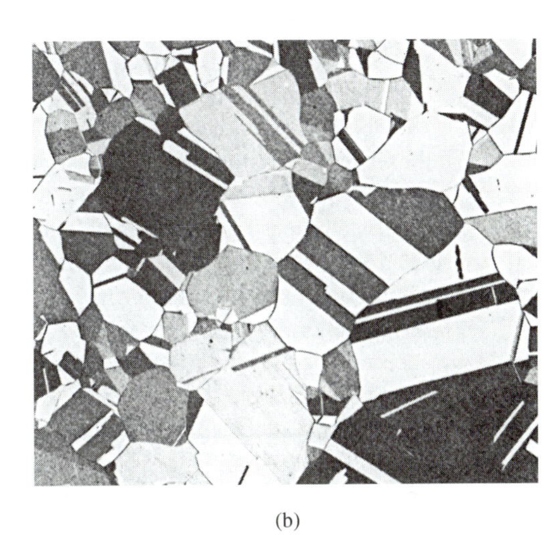

(b)

Figure 5-7 Solid solutions. (a) Monel (Ni, 67.5%; Cu, 30.18%). (b) Brass (Zn, 70%; Cu, 30%). (Buehler Ltd.)

because of the interaction between the different atoms in the solid solution producing increased resistance to slip.

Another common alloy, ***brass,*** is a solid solution of (Cu) and (Zn) [Figure 5-7(b)]. Zinc's atomic radius is about 0.153 nm. It is element 30 in the periodic table. Cu and Zn atoms differ in size by less than 15%. In this system, Zn atoms can replace Cu atoms up to a maximum of 50% and produce disordered substitutional solid solutions. We can add tin (Sn) to Cu; Sn, the fiftieth element of the periodic table, has a radius of about 0.146 nm. This indicates that we are approaching the 15% limit. In fact, only about a maximum of 10% of the Cu atoms can be substituted by Sn atoms to produce the single-phase alloys known as ***bronzes.***

For an interstitial solid solution to form, the added atoms must be sufficiently small to fit into the interstices between the solvent atoms. From our study of atoms and the periodic table, we know that the atom's radius is measured from the center of the atom to the outermost electron. We also know that this diameter decreases as we move from left to right in a period of the table (Table 3-2). Taking these facts into consideration, the five elements with radii less than 0.1 nm are hydrogen (H), carbon (C), boron (B), nitrogen (N), and helium (He). A classic example is the interstitial solid solution formed by C in Fe in the production of steel. Pure Fe has a bcc structure at room temperature. This phase is called α (alpha) iron. In this form the Fe is relatively weak and incapable of being shaped for commercial use. Above 912°C, iron, being allotropic, changes to an fcc structure known as γ (gamma) iron. From our study of unit cells of the various lattice structures, we know that the interstices between atoms in a bcc structure are quite small. In the fcc unit cell, a relatively large interstice exists in the center. Carbon added to α-Fe allows only 0.025% of the C atoms to take up positions in the iron bcc structure. If, instead, the α-Fe is heated to above 912°C and then the C atoms are added, the γ-Fe will accommodate up to 2% of the C atoms. This interstitial solid solution is given the name ***austenite.*** The formation of steel and cast iron will be treated extensively later when phase equilibrium diagrams and heat treating are discussed.

In studying the formation of compounds in liquid or gaseous states, it may be fair to conclude that most chemical compounds are nonmetallic. Second, they involve the exchange or sharing of electrons. Third, one of the elements involved must have a positive valence and the other a negative valence (oxidation states). Fourth, the algebraic sum of these oxidation states must be zero in the compound formed. Experience indicates that two metals show no inclination to join together in a chemical way to produce compounds. The latter statement is true for metals in the liquid and gaseous states. We are now discussing solid-state reactions. In solids, two metals can combine to form ***intermediate alloy phases.*** Their compositions are intermediate between the two pure metals, with crystal lattices differing generally from those of the pure metals. The intermediate phase can occur alone or accompanied by the pure metal or their solid-solution phases. These later phases are then called terminal phases.

Many intermediate phases have great technological importance in the strengthening of solid materials. They will be eventually called on, as an example, in a heat-treating process to precipitate out at some lower temperature to impede the slippage of planes of atoms over one another, thus producing an increase in strength and hardness of steel. (Intermediate phases are discussed in more detail in Section 5.3.)

The liquid (or molten) state of a metal may consist of a single pure metal. It may comprise a solution of two or more metals that are completely soluble in each other. A third possibility is that the liquid state represents two *immiscible* metals. Finally, it could contain two insoluble liquid solutions. Note that the word ***insoluble*** is synonymous with ***immiscible.*** Substances that are not soluble (that are insoluble) are immiscible. Water and oil have no solubility; that is, they are insoluble, or immiscible.

Very few metals are insoluble in each other in the liquid state. Aluminum and lead come close to being immiscible in both the liquid and solid states. In the solid state they solidify into two separate layers with a clear line of contact showing no appreciable degree of diffusion. Most metals are soluble in each other to some degree in their liquid phases. When these metals are cooled under equilibrium conditions from their liquid phases, they solidify and produce various solid phases. Assuming that they are completely soluble in the liquid phase, on cooling they may be:

1. Completely miscible in the solid state
2. Insoluble in the solid state
3. Partly soluble in the solid state

Other products may be formed, such as intermediate phases. Also, if the liquid phases are not completely soluble, other results may be produced in the solid state. To understand why and how these various results occur requires a good knowledge of the construction and use of *phase diagrams* (see Section 3.5).

The process of solidification begins with the formation of seed crystals or embryos from the liquid. Seed crystals, embryos, or nuclei (do not confuse these with the nucleus of an atom) subsequently grow into full crystals. For these nuclei to form, ***supercooling*** or undercooling of the liquid phase is necessary; that is, a temperature decrease below the equilibrium temperature is needed to make the phase transformation proceed at a measurable rate. This supercooling is an indication of the removal of the latent heat of solidification. Again, we can call on our example of the spring that, once deformed, stores up energy within itself. Supercooling is comparable to the amount the spring is deformed. For a nucleus to form, a process called ***nucleation,*** energy is needed to create the nucleus by overcoming surface tension effects. Nucleation can be aided by the presence of impurity atoms, which are almost always present. As each of these nuclei grows larger, it forms dendritic, or treelike, crystals with a random distribution of crystallographic planes relative to one another (see Figure 5-8). These crystals, called ***grains,*** are surrounded by regions of

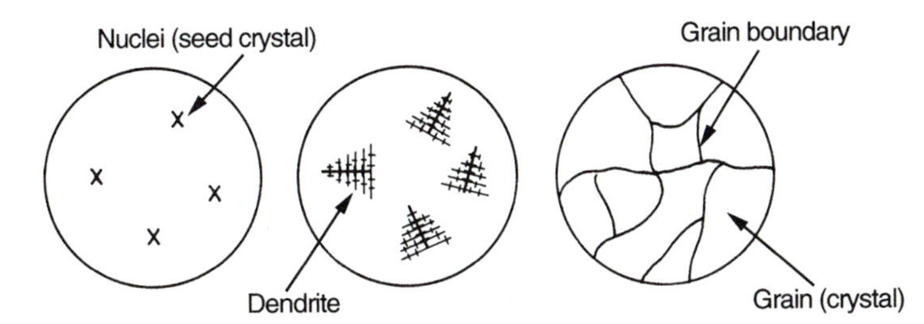

Figure 5-8 (a) Grain formation.

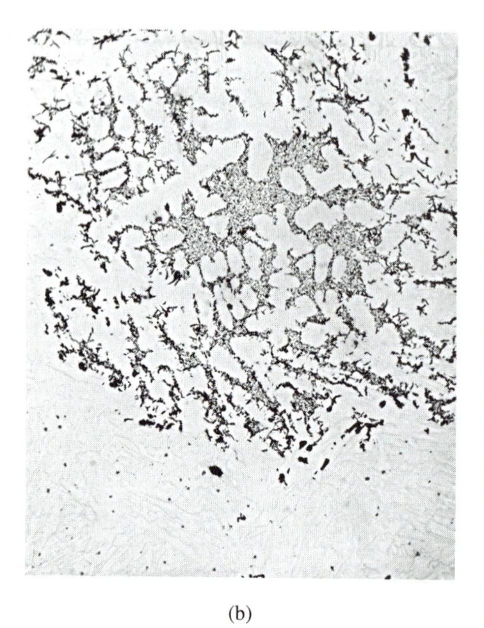

(b)

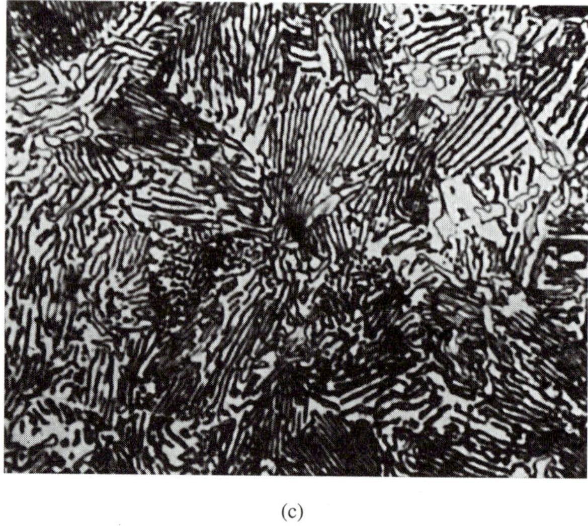

(c)

Figure 5-8 (b) Dendrites of "tough pitch" copper, dendrites outlined by copper oxides (200×). (Buehler Ltd.) (c) Grains of "high-carbon" steel. (American Iron and Steel Institute.)

high energy known as ***grain boundaries.*** If many nucleation sites are formed, a fine grain size in the solid state is the result. To obtain this fine grain, a very minute amount of some element is purposely added. The particles of the element added act as impurities, on which the nuclei can form. In other words, the average grain size is proportional to the degree of supercooling.

5.3 PHASE DIAGRAMS (EQUILIBRIUM DIAGRAMS) FOR METALLIC SYSTEMS

Building on our knowledge of phase diagrams presented in Section 3.5, we will describe "road maps" that guide and direct those involved in the development, fabrication, heat treatment, and design of alloys using some simple diagrams to illustrate their main features. Before beginning this task it must be noted that various rules must be observed to ensure correctness of the diagrams. These rules are derived from Gibbs' formulations developed over 100 years ago, the most important of which is the ***Gibbs phase rule.***

The construction of phase diagrams is based on the Gibbs phase rule, which is expressed in equation form as

$$P + F = C + 2$$

where P is the number of phases in equilibrium; F is the variance or number of degrees of freedom, or the number of variables such as pressure, temperature, or composition that can be varied without affecting the number of phases in equilibrium; and C is the number of components (elements, compounds, or solutions) in a particular system. The number "2" in the equation stands for temperature and pressure, the two variables that can be allowed to change.

The phase diagram for water is plotted in Figure 5-9, with the pressure as ordinate and the temperature as abscissa. This is a ***unary phase diagram,*** meaning that it consists of one component—in this instance, water (H_2O). The series of curves in the diagram show the division between the three phases in which water can exist depending on temperature and pressure conditions.

Point A in the liquid region is defined by a certain combination of pressure and temperature. The number of components (C) at this point is 1, a unary diagram. The number of phases (P) is 1 because only liquid water exists in this region. The phase rule tells us that the number of degrees of freedom is

$$\begin{aligned} F &= C - P + 2 \\ &= 1 - 1 + 2 \\ &= 2 \end{aligned}$$

The "2" means that, within the limits of the liquid phase, the pressure, temperature, or both can be changed and the phase would still be liquid. No phase changes occur and equilibrium is maintained.

Point B is on the fusion curve (T–F), the boundary between the liquid and solid phases of water. The number of components has not changed and therefore $C = 1$. The number of phases is $P = 2$ because anywhere along this curve, liquid and solid phases of water coexist in equilibrium. Using the phase rule to find the number of variables that can be changed without changing any phases in equilibrium, we have

$$\begin{aligned} F &= C - P + 2 \\ &= 1 - 2 + 2 \\ &= 1 \end{aligned}$$

This single degree of freedom means that if the pressure is changed, the temperature must also be changed to stay anywhere on that boundary line. We can no longer change either the pressure or the temperature independently and stay on the boundary line where liquid and solid phases of water coexist in equilibrium.

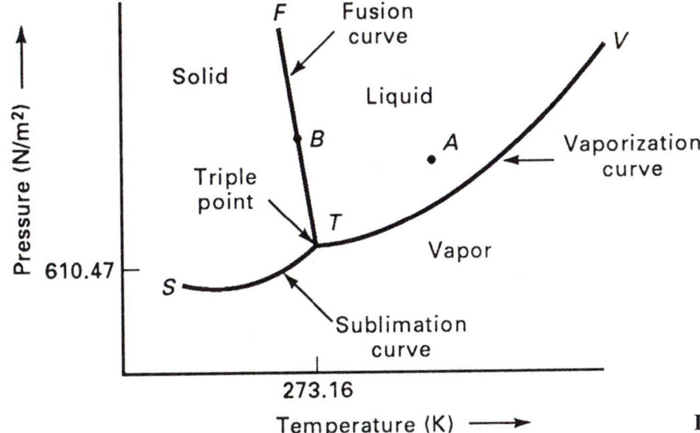

Figure 5-9 Unary phase diagram for water.

Point T, at the intersection of the three curves, is known as the triple point of water. Here the three phases of water coexist. The phase rule tells us that the number of degrees of freedom for this position on the phase diagram is

$$\begin{aligned} F &= C - P + 2 \\ &= 1 - 3 + 2 \\ &= 0 \end{aligned}$$

Zero degrees of freedom means that we cannot change pressure or temperature. If we did, the three phases would no longer coexist in equilibrium. It also means that the pressure and temperature are fixed for this condition to exist. The word *invariant* is also used to describe this particular point.

Point V, which is known as a ***critical point,*** means that no matter how high the temperature and pressure rise beyond this point, the vapor phase will never change to a liquid.

Materials science deals primarily with the solid and liquid phases of materials. Second, the pressure variable has only a small effect on materials. Third, metallic phase diagrams are usually limited to showing the interactions between only two components and thus are known as ***binary diagrams.*** For example, the iron–iron carbide diagram or the phase diagram for brass has only two components, copper and zinc. Consequently, phase diagrams deal with temperature changes versus changes in the composition of materials. This means that the phase rule can be simplified as follows:

$$P + F = C + 1$$

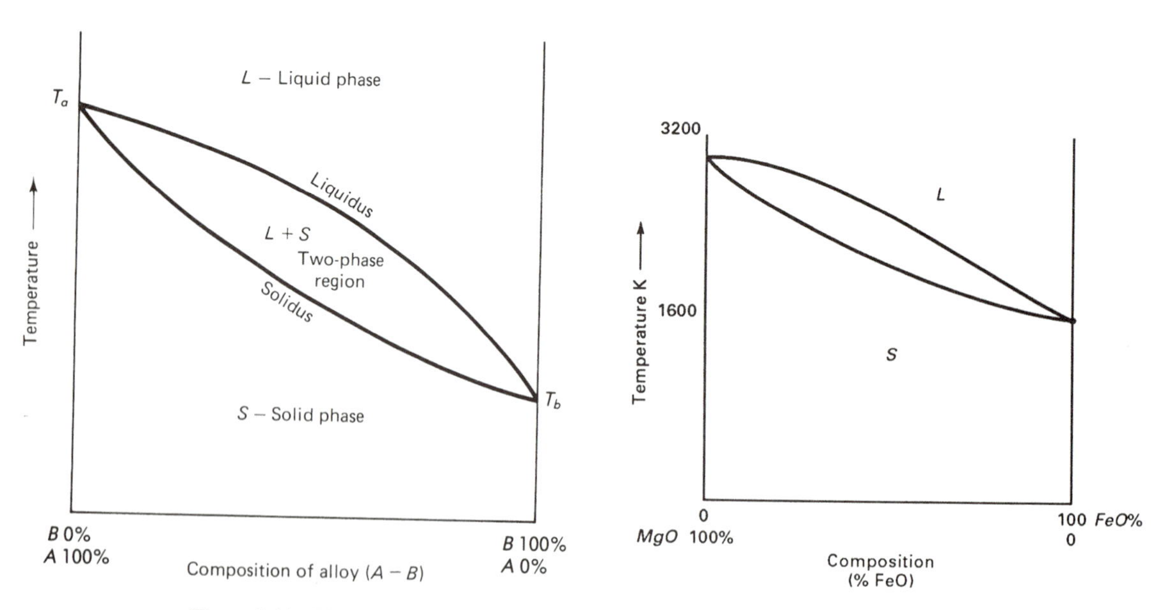

Figure 5-10 Phase diagrams of metals and ceramics. (a) Metals A and B completely soluble in each other in both liquid and solid phases. (b) Phase diagram representing two ceramic materials completely soluble in each other in the liquid and solid states.

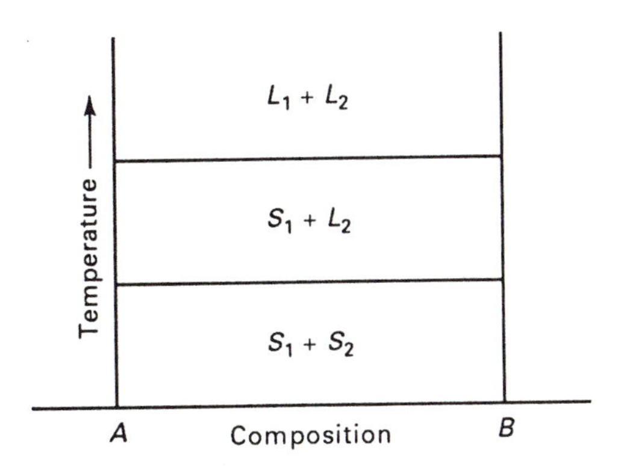

Figure 5-11 Phase diagram for metals A and B completely insoluble in each other in both liquid and solid phases (hypothetical).

We may now wish to know what the *actual composition of the two-phase region* (Figure 5-10) is for a particular alloy at a specific temperature. Figure 5-10(a) shows a typical phase diagram for two metallic elements, and Figure 5-10(b) shows ceramic materials that are completely soluble in each other in both liquid and solid phases. The vertical axis at the left represents a pure metal A with a melting point of T_a. The vertical axis at the right represents metal B and has a melting temperature less than metal A. The upper curved line of the cigar-shaped $L + S$ region is the liquidus and the lower line is the solidus. The pure metals and any composition of the two are in the liquid (melt) phase upon heating above the liquidus. Upon cooling to the solidus, they solidify. When an alloy is at a temperature above the solidus but below the liquidus, it exists as part liquid and part solid in a two-phase region labeled $L + S$. Figure 5-11, for comparison purposes, is a hypothetical phase diagram for two metals that are completely insoluble in each other, both in the liquid and solid phases.

Using an alloy with a composition of 60% Ni–40% Cu, we will plot this composition on Figure 5-12 to determine first its chemical composition at any given temperature.

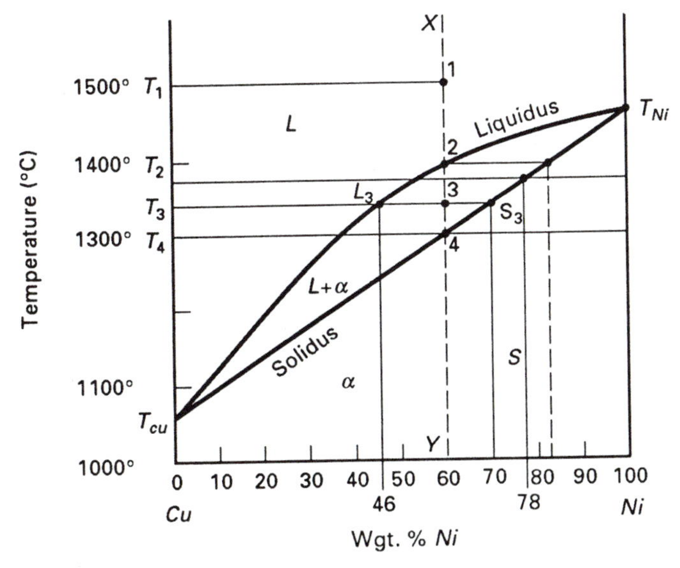

Figure 5-12 Cu–Ni phase diagram. Alloy X–Y is 60% Ni–40% Cu.

Vertical line X–Y represents our alloy. At point 1 the alloy is at temperature T_1 in the liquid solution phase. Upon slow cooling, point 2 is reached at a temperature of T_2. Our alloy is now entering the two-phase region. Part of our alloy is liquid and part has formed a solid solution. To determine the composition of these two phases at any point, such as at 3, a horizontal temperature line called a ***tie-line*** is drawn through point 3. This tie-line will cross both the liquidus and solidus curves at points labeled L_3 and S_3, respectively. From both these points, drop vertical lines to the abscissa (horizontal axis) to determine the composition of these two phases at the temperature T_3. The liquid phase formed so far has a composition of 54% Cu and 46% Ni, determined by proceeding from point 3 along the tie-line to the liquidus L_3, then vertically down to the horizontal axis at the point labeled 46% Ni. Similarly, the composition of the solid phase is found to be 70% Ni–30% Cu. This time we proceeded along the tie-line to the solidus at S_3 and vertically downward to the horizontal axis to read 70%. Note that the tie-line drawn across the two-phase region ties in or connects the adjacent phases. In this instance, the solid phase is on the right and the liquid phase is on the left of the two-phase region. This is an excellent way of identifying any two-phase region. As the alloy continues to cool to temperature T_4, observe, using the same procedure, that both phases continue to increase in the percentage of copper. Upon reaching the solidus at temperature T_4, the last liquid, very rich in Cu, solidifies at the grain boundaries. Through diffusion of the atoms, all the solid solution will be at an overall composition of 60% Ni–40% Cu. The solid solution is designated α solid solution.

The tie-line procedure plays an important role in identifying two-phase regions in a phase diagram. In Figure 5-13, a phase diagram that roughly resembles the Al–Si phase diagram shows the two-phase regions labeled. Greek letter symbols are commonly used to identify those solid phases first to form. By drawing a tie-line across these regions and knowing what phases are present at the extremities of the tie-line, the two-phase region can readily be labeled. Note the complete solidus line, showing that any phase below it is a solid phase. Lines BE and DF are solvus lines, or ***lines of maximum solubility*** of the ter-

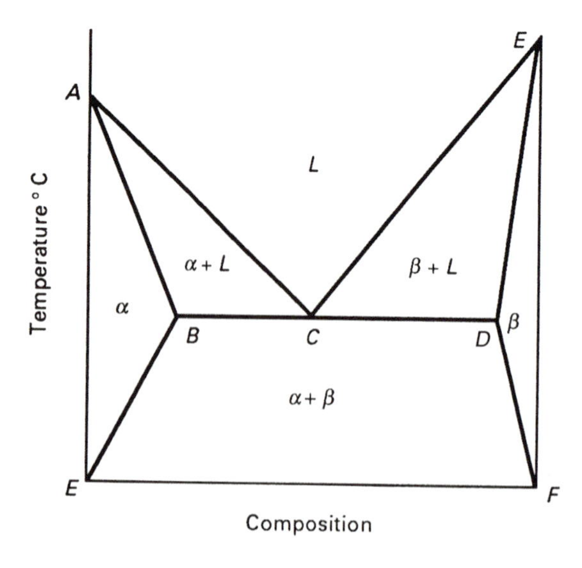

Figure 5-13 Two-phase regions of a phase diagram showing partial solid solubility.
ACE—liquidus
ABCDE—solidus
BE and *DF*—solvus lines
C–eutectic point

minal or first-formed solid solutions. Note the reduction in the limits of solubility of these solutions to absorb other phases as the temperature decreases.

The next problem to be tackled is to determine the *relative amounts of liquid and solid existing at a specific temperature in a two-phase region.* So far we know that at temperature T_3 (see Figure 5-12) two phases exist: a solid phase with a composition of 70% Ni–30% Cu and a liquid phase with a composition of 54% Cu–46% Ni. How much of the liquid has solidified upon being slowly cooled to the temperature T_3? To determine this percentage, we use what is known as the ***lever rule*** (also known as the ***inverse-ratio law,*** as we shall soon demonstrate). The law gets its name from treating the tie-line as though it were a lever supported (fulcrum) at the point where the vertical line representing our alloy X–Y crosses the tie-line. Figure 5-14 is an enlarged sketch of the tie-line in Figure 5-12.

The complete tie line from L_3 to S_3 represents the total weight (100%) of the two phases present at temperature T_3. The vertical line (X–Y) representing our alloy cuts the tie-line into two parts or lever arms; one part forms L_3 to 3, and the second forms 3 to S_3. The length of the first lever arm, L_3 to 3, represents the amount of solid phase present; the right lever arm, 3 to S_3, represents the amount of liquid phase present. Note that the left arm or part contains L_3 on the liquidus line; yet it represents the amount of solid phase. The right lever arm contains a point S_3 on the solidus, but this arm represents the amount of liquid phase present. Thus, the lever rule states that these two lengths are inversely proportional to the amount of the phase present in the two-phase region. Observe that point 3 is closer to the solid-phase region, and therefore the quantity of the liquid phase is greater than the quantity of the solid phase at that temperature. The lengths of these two parts or lever arms of the tie-line are measured in units of composition.

Now let's apply this lever rule to our alloy 60% Ni–40% Cu. Figure 5-15 is another sketch of the L_3-3-S_3 tie-line in Figure 5-12 (a repeat of Figure 5-14). The tie-line's length

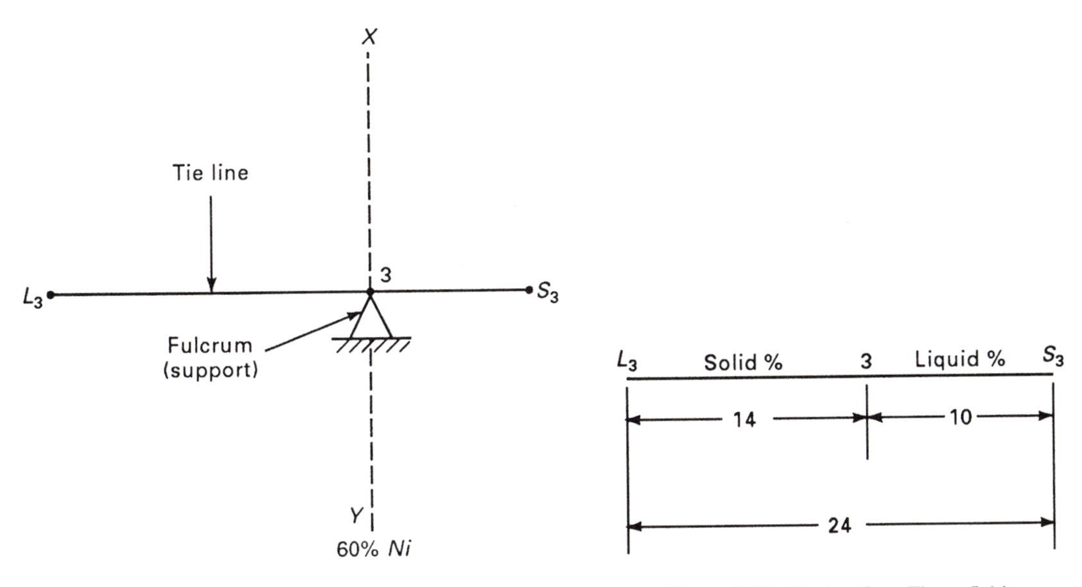

Figure 5-14 Tie-line from Figure 5-12.

Figure 5-15 Tie-line from Figure 5-14.

in terms of composition is 24, or $70 - 46 = 24$, leaving out the percent sign. The length of L_3 to 3 (left lever arm) is 14, leaving 10 for the length of the right arm. Next ask yourself: What percentage of the entire line is the right lever arm? To find the answer, you would take the length of 3 to S_3 and divide it by the total length of the tie-line, L_3 to S_3, and multiply the result by 100%.

$$\frac{10}{24} \times 100\% = 42\% \qquad \text{(liquid formed)}$$

Using the same procedure, we find that the left part (left arm) represents the percentage of solid formed:

$$\frac{14}{24} \times 100\% = 58\% \qquad \text{(solid formed)}$$

Let's take another look at Figure 5-12 to note another metallurgical process that is taking place: ***diffusion.*** For our purposes, note that the solid formed at T_2 consists of a composition of about 82% Ni–18% Cu. Assuming slow equilibrium cooling, our alloy will ultimately solidify completely, with a composition of about 60% Ni–40% Cu. What this tells us is that sufficient time must elapse for the atoms to migrate through the solid phase to satisfy the principles of equilibrium; that is, given sufficient time, atoms will diffuse away from places of high concentration such that a uniform distribution of atoms in the crystal structure is once again achieved. The reheating of a solid metal to give atoms greater energy to diffuse or move toward equilibrium (hence more uniform composition) is termed ***homogenization.*** Homogenizing eliminates the segregation of phases, such as coring, that form during solidification. Such segregation leads to nonuniform properties. Annealing, specifically, normalizing (see Section 5.5), is an example of a homogenization process involved in the heat treating of metals.

Research into the diffusion process discloses that the rate of diffusion slows as the difference in concentration of different atoms is reduced. Second, diffusion in a solid slows with reduced temperature. Therefore, in practice it is extremely difficult to achieve equilibrium cooling. Also, slow cooling produces large grain size, which is usually undesirable. Faster cooling rates are the rule, not the exception, and they prevent complete diffusion. The end result is that the initial crystals formed are of one composition. As the crystal grows in a dendrite fashion by atoms attaching themselves to the original crystal, the composition of the total crystal changes. The original higher-melting-point central portion is surrounded by lower-melting-point solid solutions. This condition is known as ***coring*** or ***dendritic segregation.*** Segregation is always present to some extent when metals are melted and subsequently solidified in the making of steel (steel ingots) and in the casting of metal parts. A metallurgical process that attacks the problem of cored structures will be discussed later.

Illustrative Problem

Determine the relative amounts of liquid and solid existing at temperature T_3. (See Figure 5-12.)

Solution Using the lever rule and the previously determined compositions of the two phases existing at that temperature, draw a horizontal tie-line at $T_3 = 1336°F$. The complete line (Figure 5-14) from L_3 to S_3 represents the total weight (100%) of the two phases present at this temperature. The vertical line (X–Y) representing the alloy cuts the tie-line into two parts or lever arms. Lever arm L_3 to 3 represents the amount of solid phase present. Lever arm 3 to S_3 represents the amount of liquid phase present. Note that these two lengths are inversely proportional to the amounts of phase present. The length of arm 3 to S_3 is a bit smaller than arm L_3 to 3, hence, one would expect the amount of solid to be a bit larger than the amount of liquid. Referring to Figure 5-15, the estimated lengths of the arms are determined to be L_3 to 3 = 60 − 46 = 14, and 3 to S_3 = 70 − 60 = 10.

Solid phase: $$\frac{14}{24} \times 100\% = 58\%$$

Liquid phase: $$\frac{10}{24} \times 100\% = 42\%$$

Illustrative Problem

Determine the composition of alloy X–Y (40% Cu–60% Ni) at point 2 in Figure 5-12.

Solution The tie-line through this isotherm (at T_2) intersects the solidus at about 60% Ni–40% Cu giving the composition of the liquid phase. The solid phase at T_2 is 82% Ni–18% Cu.

Illustrative Problem

Determine the composition of the last remaining liquid phase of alloy X–Y in Figure 5-12 just prior to the alloy completely forming a solid solution.

Solution The tie-line constructed through point 4 (T_4 − 4) intersects the liquidus. Reading vertically downward, the percentage composition of the alloy X–Y at this temperature immediately prior to forming a solid solution is about 67% Cu–38% Ni.

To continue our study of phase diagrams, the next step is an understanding of the *eutectic reaction.* The word ***eutectic*** is taken from the Greek and means "to melt well." Figure 5-16 labels the intersection of the liquidus and solidus as the eutectic point *(E).* The temperature that corresponds to it is labeled T_E, which is the lowest temperature at which a liquid solution will remain completely liquid. The alloy 57% *A*–43% *B* has the lowest melting point of any alloy in the system *AB*. Not only does it have the lowest melting temperature, but this eutectic, a mechanical mixture, solidifies completely at this temperature rather than over a range of temperatures. Alloy 80% *A*–20% *B* begins to solidify at temperature T_1 and completes its solidification at temperature T_E. The two-phase region to the left of the eutectic point is composed of metal *A* and liquid. The two-phase region to the right of the eutectic point is made up of metal *B* and liquid. Once they reach the solidus, all these alloys will solidify as separate phases. Thus, the region below the solidus consists of a mixture (not a solution) of two solid phases. This eutectic reaction for the eutectic alloy, in our example 57% *A*–43% *B,* can be written in equation form as follows:

$$L \rightleftharpoons S_1 + S_2$$

The double arrow indicates reversibility. Cooling the liquid phase produces two solid phases (S_1, S_2). Heating the two solid phases produces a liquid (melt) *(L).* For the lowest-melting composition of the system *AB,* the eutectic composition 57% *A*–43% *B,* its two solid phases

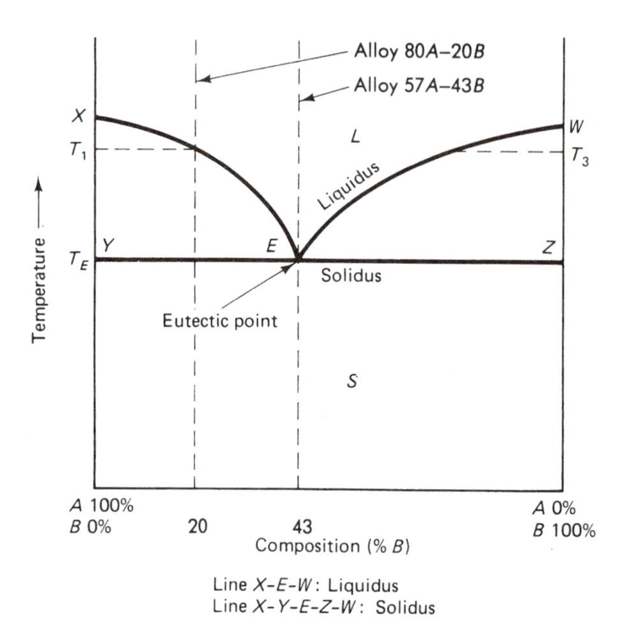

Figure 5-16 Metals *A* and *B* completely soluble in the liquid phase and completely insoluble in the solid phase (hypothetical).

will be completely distinguishable from each other when viewed under a metallurgical microscope. The parallel wavy-line pattern formed, as viewed by the microscope, is known as a lamellar type (Figure 5-17). It can be compared to the stripes of a zebra.

Referring to Figure 5-16, alloys to the left of the eutectic mixture are known as ***hypoeutectic*** mixtures. Those mixtures to the right of the eutectic are known as ***hypereutectic.*** Those alloys closer to the eutectic composition will contain more eutectic mixture in the solidified alloy. All phase diagrams drawn depicting the complete insolubility of two metals in the solid state have been labeled as hypothetical. (See Figure 5-11.) Most metals are soluble in each other to some degree. An alloy that comes close to being completely insoluble is aluminum silicon (AlSi); see Figure 5-18. Silicon is only slightly soluble in

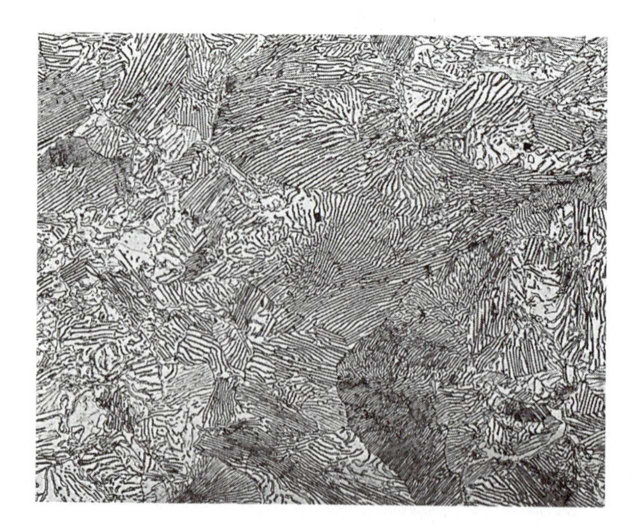

Figure 5-17 Two-phase lamellae microstructure. (Buehler Ltd.)

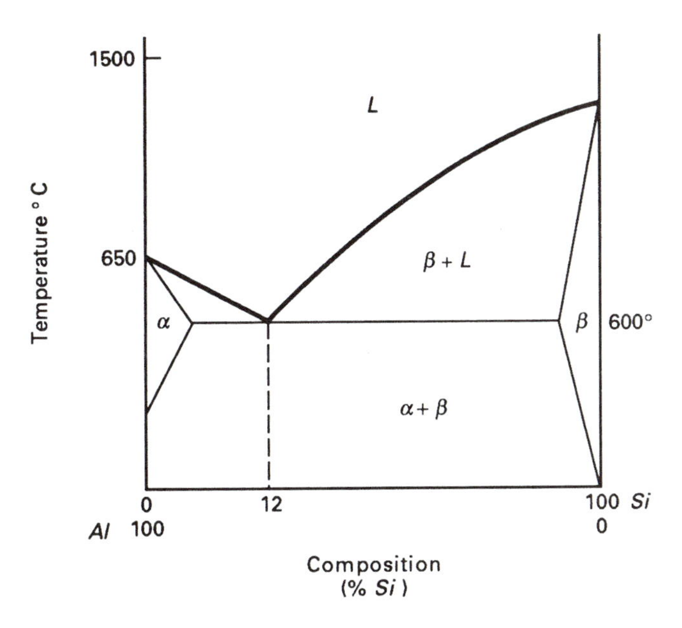

Figure 5-18 Aluminum–silicon phase diagram.

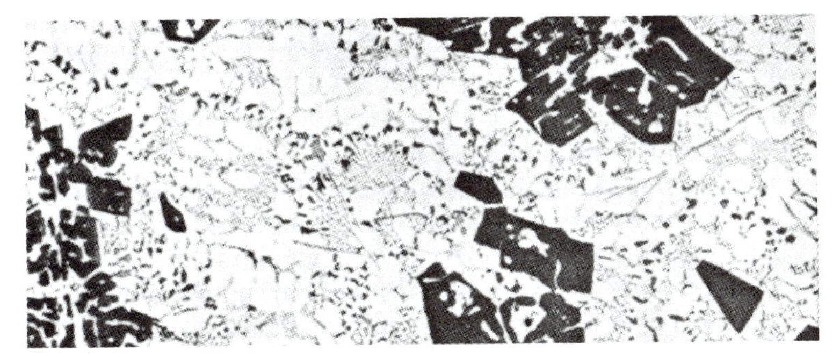

Figure 5-19 Microstructure of Al–Si alloy as cast ×250, unetched. (Buehler Ltd.)

aluminum. Figure 5-18 is a sketch of the Al–Si phase diagram. This family of alloys forms the basis for a number of commercial alloys. Figure 5-19 is a photomicrograph of an Al–Si alloy.

Our interest now turns to the more realistic situation in which the metals are to some degree soluble in each other in the solid state. This situation is depicted in Figure 5-20. Notice that this diagram is basically the same as the hypothetical phase diagram, Figure 5-16, with the addition of points *Y* and *Z* and the lines extending from these points. Also note the similarities of Figures 5-18 and 5-20. Lines *Y–S* and *Z–T* are *solubility curves,* also known as ***solvus lines.*** Before learning more about these lines, note that there are three areas of single-phase solutions, three regions of two-phase solid solutions, and one eutectic mixture.

The solid alpha (α) phase is composed mostly of metal *A* with some metal *B*. The solid beta (β) phase is rich in metal *B* along with some metal *A*. The α phase, which is called a terminal phase, differs from pure metal *A* by the presence of some metal *B*. Using a tie-line, the two-phase region between α phase and the liquid (L) phase is labeled α + *L*. The main

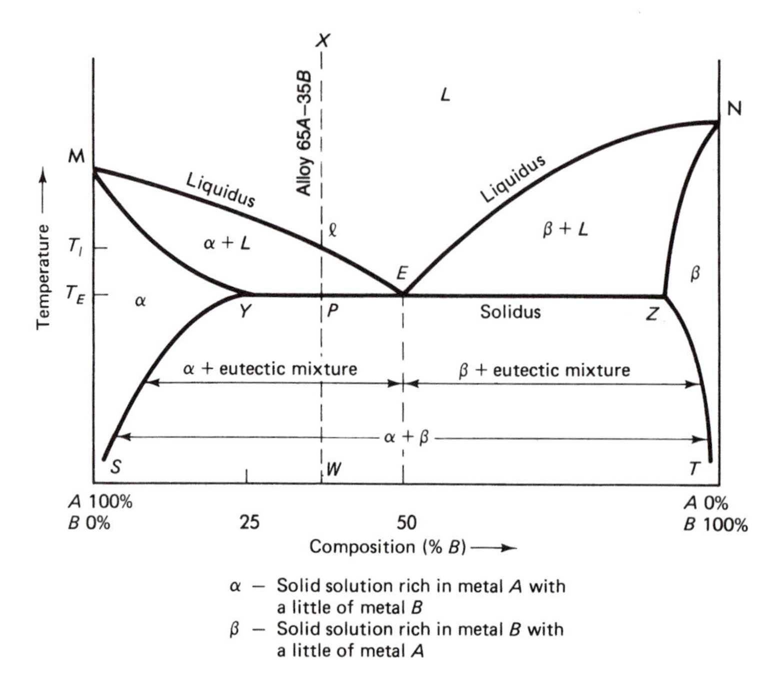

Figure 5-20 Typical eutectic-type phase diagram: metals *A* and *B* completely miscible in the liquid phase and partially miscible in the solid phase.

solid phase, α + β, is determined in the same manner. Figure 5-20 shows the typical shape of a phase diagram where *A* and *B* are partially soluble in each other. As the temperature increases, the more solid (composition *B*) can be dissolved in the liquid. Note that the solvus lines end at points *Y* and *Z* in Figure 5-20. Point *Y* is the limit of solubility of metal *B* and solid solution. At the lowest temperature the degree of solubility is a minimum (98% *A*–2% *B*); that is, only 2% *B* is soluble in *A*, but as the temperature increases this degree of solubility of *B* in *A* increases to a maximum limit (75% *A*–25% *B*).

Now it is time to follow a particular alloy (65% *A*–35% *B*) and explain briefly the phase transformations that take place as it is cooled from somewhere in the liquid phase. At point *l* (Figure 5-20) on vertical line *X–W* and temperature T_l, the liquidus line is reached. Solid solution α is beginning to form. As the alloy cools, this solid solution becomes richer in metal *B* until the solidus is reached at point *P* and temperature T_E. The last liquid to solidify has a composition of 50% *A*–50% *B*. Assuming that diffusion has kept pace with nucleation and growth of crystals, the solid phase has a composition of 75% *A*–25% *B*. Applying the lever rule, we could determine the relative amounts of these two phases present. Note that the last liquid to solidify has the eutectic composition and therefore begins to solidify and form a eutectic mixture consisting of alternate layers of crystal of α phase and crystals of β phase. At this point we are just below temperature T_E. Here is where the solvus lines come into the action. If you were to draw in the tie-line for this temperature, it would be pretty close to the line *Y–Z*. As the temperature continues to lower, the tie-lines that you might draw get longer and longer. Add to this observation the fact that the solvus line *S–Y* shows the maximum solubility of metal *B* in metal *A* at various temperatures, and that this solubility decreases as the temperature is reduced to room temperature. Metal *B* is in solution in the α solid phase. Therefore, the excess β phase must precipitate out of solution. At room temperature our alloy will consist of α phase with some excess β phase precipitated within it, plus the eutectic mixture. This eutectic mixture is also made up of both α and β phases.

Intermediate Phases

The phase diagrams discussed up to this point are binary alloy phase diagrams, such as copper-nickel, or binary eutectic phase diagrams similar to Figure 5-20, which could easily represent a lead–tin alloy system. In both these types of phase diagrams, there are only two solid phases α and β, called terminal solid solutions, that exist over a range of compositions near the extremities of the diagram. Other alloy systems contain ***intermediate solid solutions*** or ***intermediate phases.*** These phases are formed at other than the two extremes in composition. The copper–zinc system (not shown) has, in addition to its two terminal phases, four intermediate phases or intermediate solid solutions. None of these four phases extends to the extremities of the phase diagram.

In some solid systems discrete intermediate compounds that have distinct chemical formulas are formed. Intermediate compounds formed in metallic phase diagrams are known as ***intermetallic compounds*** or ***intermetallics.*** They have a narrow range of composition (see mullite in Figure 7-11), while some have a fixed composition that is represented on a phase diagram as a single vertical line (see alloy *AB* in Figures 5-21 and 5-22). Their atomic bonding can be either mixed ionic/metallic or mixed covalent/metallic. Their properties include brittleness coupled with good high-temperature resistance and a high strength-to-weight ratio. Consequently, intermetallics Ni_3Al and Ti_3Al are finding applications in the aerospace industry.

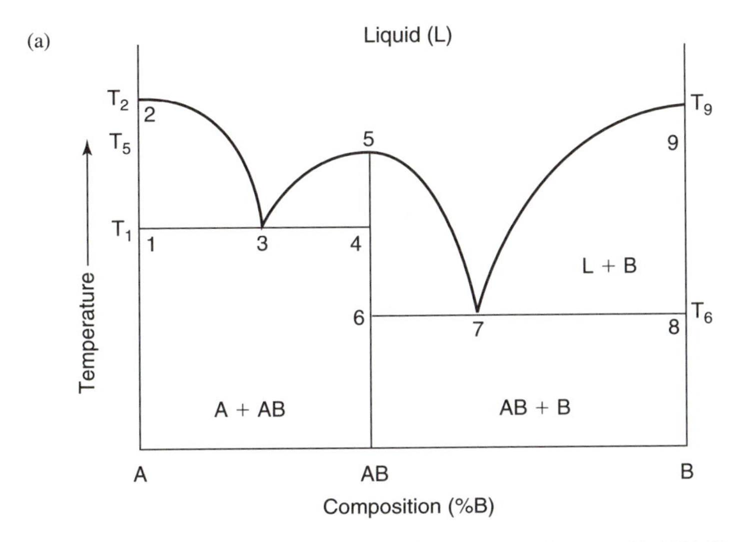

Figure 5-21 (a) Binary phase diagram with an intermediate, congruent intermetallic *(AB)*. *Note:* Points 3 and 7 are eutectic points.

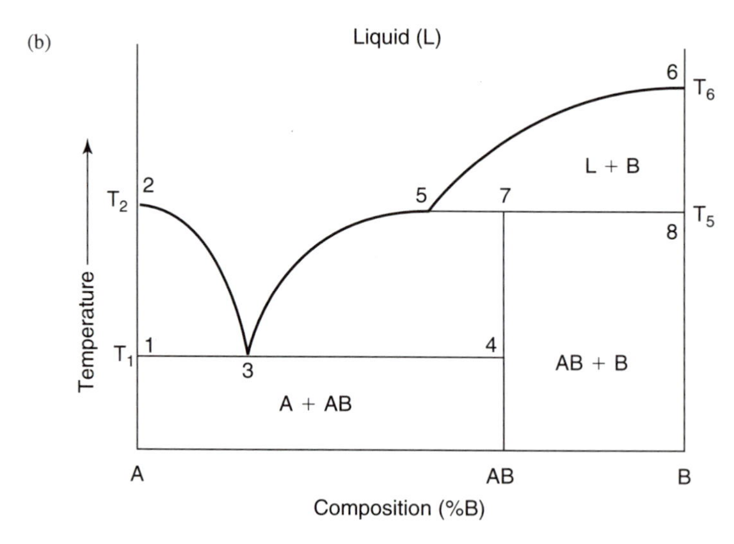

Figure 5-21 (b) Binary phase diagram with an intermediate, incongruent intermetallic *(AB). Note: AB* never reaches the liquidus. Point 5 is a peritectic point.

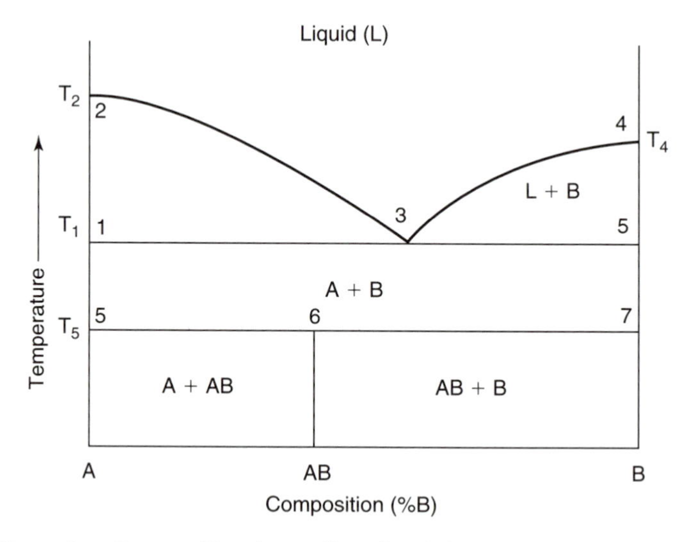

Figure 5-22 Binary phase diagram with an intermediate, dissociating intermetallic stable only over a limited temperature range. *Note:* T_5 represents the upper limit of stability upon heating.

These compounds follow the rules set forth in ***stoichiometry.*** Intermetallics are stoichiometric compounds with characteristic metal atom ratios such as CuZn, AlLi, Ni_3Al, Al_3V, AlSb, Mg_2S, and Ti_3Al. Stoichiometry (pronounced "stoy-key-om-eh-tree") is the calculation of the quantities of chemical elements or compounds involved in chemical reactions. Chemical equations balance the starting materials (reactants) needed to produce a certain amount of product. In addition, the cost of the starting materials determines the cost of the final product. Stoichiometry allows one to calculate the amount of starting materials necessary to obtain the correct amount of product. In ceramic materials, stoichiometry is used to define the ionic state for ionic compounds, wherein there is an exact ratio of cations to anions as predicated by the chemical formula. A ceramic is nonstoichiometric if there is any deviation from this exact ratio. Using calcium chloride as an example, each calcium ion (cation) has a +2 charge (Ca^{2+}) and associated with each chlorine ion (anion) is a single negative charge Cl^{-1}. Thus, there must be twice as many Cl^{-1} anions as there are Ca^{2+} cations in the compound with the formula for calcium chloride, namely, $CaCl_2$. The positive–negative attraction is what holds the ions together. In this compound a Ca atom gives up its two outermost electrons, transferring one electron to each chlorine atom. The calcium atoms become the positively charge Ca^{2+} ions (cations). Each chlorine atom has an extra electron, producing a negatively charged chlorine, Cl^{-1}, ion (anion). In other words, a stoichiometric ceramic is one in which all crystallographic lattice positions are filled according to the normal chemical formula. The chemical formula of a compound indicates the ratio of cations to anions, or the composition that results in the balancing of the charges.

In general, intermetallics possess strong ionic/covalent bonding, which gives them nonmetallic properties such as poor ductility and electrical conductivity. An intermetallic might contain but a single component. Therefore it is represented, on a phase diagram, as a vertical line and labeled by a chemical formula such as NiAl or MgSn.

Congruent phase changes occur when one phase changes into another phase at constant temperature (isothermal) and without any change in chemical composition. Pure metals and some intermediate compounds change phase congruently. In Figure 5-21(a), the congruently melting compound is a solid from room temperature up to its melting point (T_5). In other words, a congruent intermetallic melts immediately once it reaches the liquidus line. An ***incongruent*** melting compound [Figure 5-21(b)] would not melt directly but would decompose into a liquid plus one of the pure metals. In ceramics some compounds are not stable through the complete temperature range. In other words, the compound may form only over a limited temperature range (Figure 5-22). The word ***dissociating*** is used to describe these compounds. Both congruent and incongruent compounds are formed in ceramics.

One last solid-state reaction will be discussed. The ***eutectoid*** reaction takes place within the solid state. The suffix ***oid*** means "resembling" or "like." Therefore, the eutectoid reaction is like the eutectic reaction. As you recall, the eutectic reaction involves the transformation of a liquid phase into two solid phases ($L \rightleftharpoons S_1 + S_2$). The eutectoid reaction then is the transformation of a solid phase (S_l) into two new solid phases (S_2 and S_3). An appropriate equation describing this reaction is

$$S_1 \rightleftharpoons S_2 + S_3$$

Figure 5-23 shows the many similarities of this type of reaction to the eutectic reaction (see Figure 5-20). In industry the eutectoid reaction is most important because it forms the basis for heat treatment of many metallic materials. Point E is the eutectic point. Point H is the eutectoid point located well below in the solid region of the diagram. As shown, the gamma solid solution at point H transforms into the eutectoid mixture, made up of two solid phases. The prefixes ***hypo-*** and ***hyper-*** are attached to particular compositions of

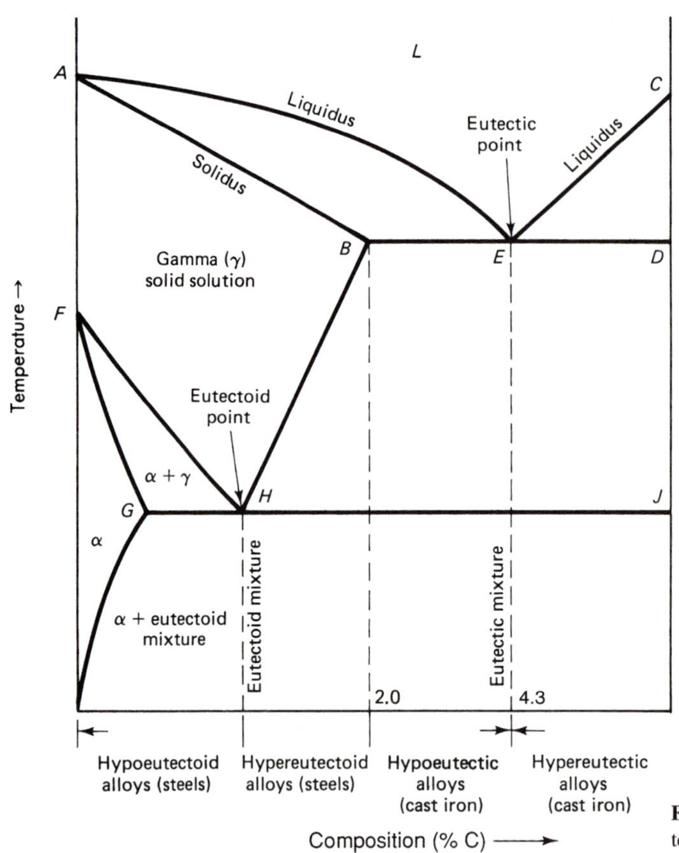

Figure 5-23 Phase diagram showing a eutectoid reaction.

metals on either side of the eutectoid mixture in a fashion similar to those compositions related to the eutectic composition. This very crude diagram will be refined and studied in more detail in Section 5.5. At that time the diagram will be called the iron–iron carbide phase diagram—the heat treater's main road map to the processing of steel and cast iron.

Multiphase metal alloys. So far our discussion has involved solute and solvent atoms that have the same type of crystal structure and thus form relatively pure ***homogeneous*** mixtures or solutions. Now we advance to alloys that are classified as mixtures of more than one phase that differ from each other in their composition or structure. Under the microscope, the boundaries between phases can be seen in some metal alloys (Figure 5-31). Solder is an example of this type of multiphase metal alloy, with the metals tin and lead present in two separate phases. Steel is a more complex multiphase mixture of different phases, some of which are solid solutions. There are many more ***multiphase metal alloys*** than single-phase solid solution alloys, due primarily to the greater flexibility of their properties, with more variables to control to produce differing properties. In other words, the properties of multiphase metal alloys such as steel depend not only on the phases present, but on the structure, amount, shape, distribution, and orientation of these

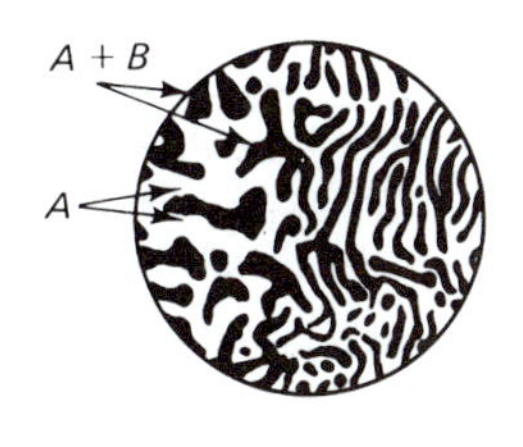

Figure 5-24 Microstructure of a multiphase metal alloy. Two different compositions (*A* and *A* + *B*) in the solid state. 65 Al–Cu alloy as cast ×250, $Fe(NO_3)_3 \cdot 9H_2O$. (Buehler Ltd.)

phases (Figure 5-24). The ability to control these phases allows us to produce desired properties in metals. For example, low-carbon ductile steel is used on auto bodies, and harder and more brittle steel is used in metal files.

Phase or structural transformations. Just how can we bring about these transformations to produce a metal alloy with desirable properties? First, let's consider the atomic world once more. We know that a few metal atoms can change their structural arrangement as a result of a change in temperature. Remember the name of this phenomenon? *Allotropy.*

We know that there is a definite limit of solubility of one solid material in another and that this limit also depends on temperature. If this limit is reached, we see atoms coming out of solution to form a multiphase metal alloy. Remember the name of this phenomenon? *Precipitation.*

We know that atoms can move about in a solid just as cigarette smoke moves in a room, but certainly not as fast. Remember this phenomenon? *Diffusion.* And the rate at which the atoms move varies with the temperature. The atoms move at faster rates because they have more energy. What energy? *Thermal energy.* Where does this energy come from? From an external source of heat. The more heat, the more the energy and the greater the movement. By now you should have a pretty good idea about the answer to the next question. In looking over the various ways that we can change atomic structure in a metal alloy, what is the one common quality or characteristic in all of them? Yes, it is thermal energy.

5.4 THERMAL PROCESSING: EQUILIBRIUM CONDITIONS

By controlling the thermal energy, that is, by controlling the supply or removal of heat, we can make metal atoms move. This art and science of controlling thermal energy for the purpose of altering the properties of metals and metal alloys is known as ***thermal processing*** or ***heat treating.***

As with most subjects in materials science, thermal processing concerns a metal's atomic and crystal structure. So far in our study of materials technology, we have accomplished many learning objectives dealing with the "inner workings and hidden mechanisms" in the world of a metal's atoms. Crystal structures, bonding, diffusion, crystal imperfections, dislocations, allotropy, solid solutions, phases, multiphase solid mixtures, solute, solvent, and saturated solid solutions are but some of the many terms that now have meaning for us.

When we studied phases and solid solutions, we used cooling curves for pure metals and metal alloys. These curves are graphic pictures showing the relationships between the variables of time and temperature. Many of these single cooling curves are plotted to give a graphical picture of the possible phases a metal alloy could be in at any particular temperature and at any particular composition. Such phase diagrams are also known as ***equilibrium diagrams. Equilibrium*** means that the time variable is not controlled but is allowed to run its course. In other words, *equilibrium* implies that phase changes shown on equilibrium diagrams are produced under conditions of slow cooling with no restraints on time. This allows the atoms that are diffusing through a solid material the time to seek and find the equilibrium position of lowest energy level. Phase diagrams show us the following:

1. The phases that are present for a particular alloy composition and temperature
2. The extent of solubility between two metals
3. The maximum solubilities of each metal in the other
4. The alloy composition with the lowest melting point (eutectic)
5. The melting points of the pure metals making up the alloy

But phase diagrams do not show what happens to a solid system of materials in equilibrium when a change in temperature, pressure, composition, or any combination of these variables occurs. We know that the system will seek another state of equilibrium with its lowered level of energy (free energy). We also know that this process takes time. In many cases this new state of equilibrium is extremely slow, particularly at low or room temperatures. During this time the system is said to be in a nonequilibrium condition or *metastable* state. This metastable state may persist indefinitely. Some steel and nonferrous alloys are in metastable states brought on by heat treatments whose goal is to increase their strength or other desired property. The heat treatments change their microstructures not only in a metastable state during the process, but also in the final or complete solid material.

To learn how thermal processing can produce phase changes, we use iron and its alloy steel as examples of ferrous metals. We will not discuss other industrial materials that are heat treated, such as glass, but leave this important area of study for independent research.

Being an alloy of iron and carbon, steel is also allotropic, existing in several forms. Referring to the steel equilibrium diagram (Figure 5-25), which is a graphical record of the various phase transformations of steel in the solid state, and using the eutectoid composition, we see that above 727°C steel is in a solid state with a structure called *austenite,* or gamma (γ) iron, a single-phase fcc solid solution. Figure 5-25(a) is a simpler version of the complete diagram shown in Figure 5-25(b). Austenite is characterized by its ability not only to be deformed but also to absorb carbon. The fcc unit cell of iron atoms contains interstices that are large enough for the small carbon atoms to occupy, producing an interstitial solid solution. Many ***hot-working*** operations in industry take place with steel heated to a temperature that produces this austenite phase. Furthermore, the austenite region is the starting point for many of the thermal processes about to be discussed.

At temperatures below 910°C, pure iron changes to a stable phase called alpha ferrite, alpha (α) iron, or ***ferrite.*** This ferrite phase can accommodate up to a maximum of 0.02% carbon (by weight), which produces a solid solution with a bcc structure. The bcc structure contains interstices in its unit cell that are too small to accommodate the carbon atom. Another component in this diagram is ***cementite,*** an interstitial compound with the chemical formula Fe_3C. Cementite is orthorhombic, with twelve iron atoms and four carbon atoms in the interstices among the larger iron atoms in the unit cell. The ratio of iron atoms to carbon atoms is 3 to 1 and it does not change. (Refer to Sections 3.4.1 and 3.4.2, Figure 3-32, and Table 3-3.) *Note:* Cementite is also discussed briefly on pages 113 and 276. Also called iron carbide, this brittle substance contains 6.67% carbon and has an HB of 700. When steel with the eutectoid composition forms at 727°C, it produces a lamellar two-phase mixture of ferrite and cementite called ***pearlite*** (see Figure 5-27). Using tie-lines across the two-phase regions of Figure 5-27, a hypoeutectoid steel can be described as a mixture of ferrite and pearlite (Figure 5-28), eutectoid steel as having a pearlitic structure, and hypereutectoid steel as a mixture of pearlite and cementite. Such structures can be discerned under a microscope. At this point it should be evident that there are countless alloys of carbon and iron (carbon steels) with varying amounts of carbon, producing a corresponding variety of steels with different properties to serve the demands of industry. Later in this book other elements will be discussed that also play a part in producing numerous specialty steels (alloy steels) that find ever-increasing use.

It is most important to understand that steel is possible only because iron is allotropic and carbon atoms are small enough to fit between the iron atoms in an fcc austenite structure (interstitial solid solution). The many different properties of carbon steel are produced as a result of changes in the amount of carbon and the difference in the abilities of ferrite and austenite to dissolve carbon. Austenite can dissolve almost 100 times more carbon than can ferrite. But austenite does not exist at room temperature under equilibrium conditions, whereas ferrite does. It is worthwhile to note on the iron–iron carbide phase diagram (Figure 5-25) the locations of wrought iron (almost pure iron), steels, and cast irons (greater than 2% carbon). Most important of all is the realization, first, that everything we have discussed about the allotropic forms of iron, the diffusion of carbon atoms, phase transformations, austenite, ferrite, cementite, and pearlite takes place in the *solid state.* Not once did we get even close to the melting point of iron, 1538°C. Second, all phase transformations took place under *equilibrium conditions.*

5.5 THERMAL PROCESSING: NEAR-EQUILIBRIUM CONDITIONS

Several thermal processing techniques used with ferrous metals approach equilibrium conditions, as represented by the iron–iron carbide equilibrium, or phase diagram. By definition, such processes bring about a change in a material's properties. Therefore, they must

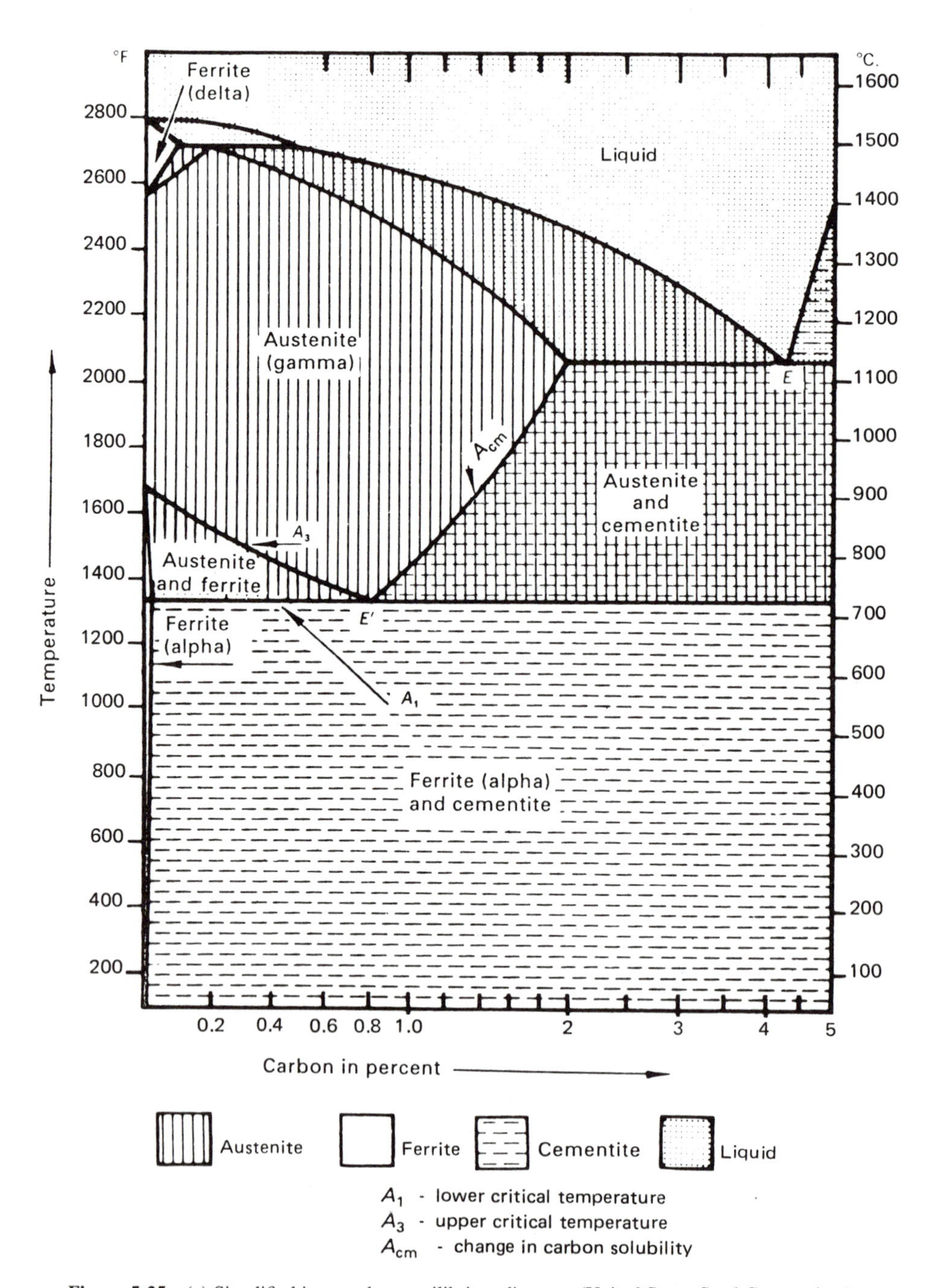

Figure 5-25 (a) Simplified iron–carbon equilibrium diagram. (United States Steel Corporation.)

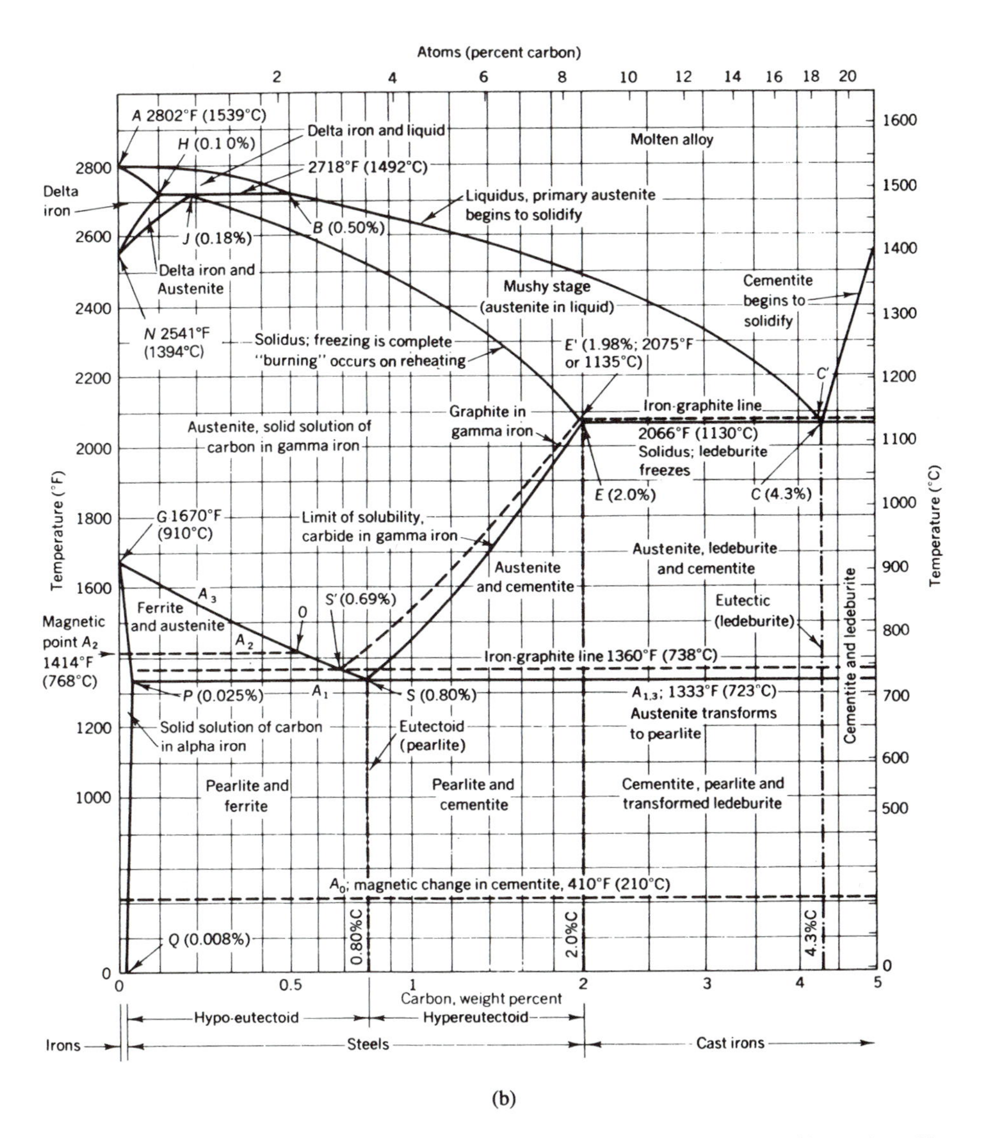

(b)

Figure 5-25 (b) Iron–carbon equilibrium diagram. Dashed lines show true equilibrium of iron and graphite. Solid lines show a metastable phase diagram of iron and iron carbide (Fe_3C), also called cementite. The metastable diagram is used in the same manner as a true equilibrium diagram. The distinction between the two is negligible above 2100°C. (ASM International.)

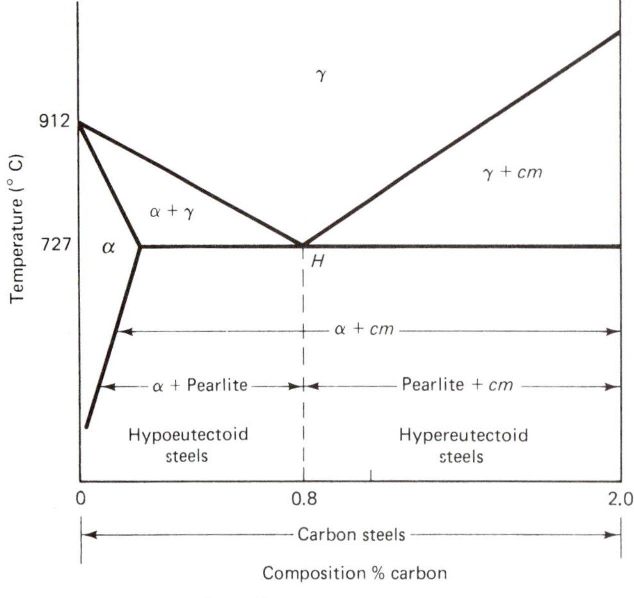

Figure 5-26 Simplified steel equilibrium diagram.

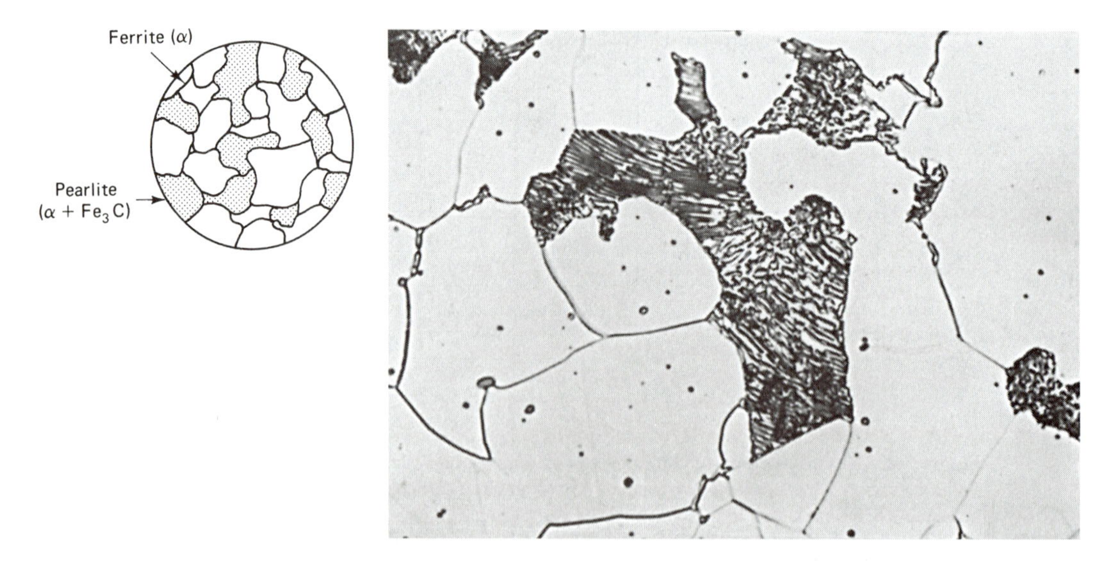

Figure 5-27 Pearlite. Lamellar two-phase mixture of ferrite and cementite. The white areas in this hypoeutectoid steel are grains of proeutectoid alpha ferrite. The dark areas are pearlite colonies consisting of alternate lamellae of alpha ferrite and Fe_3C (cementite). The resolution of the two phases in the pearlite region depends on magnification and orientation and varies from area to area. (Buehler Ltd.)

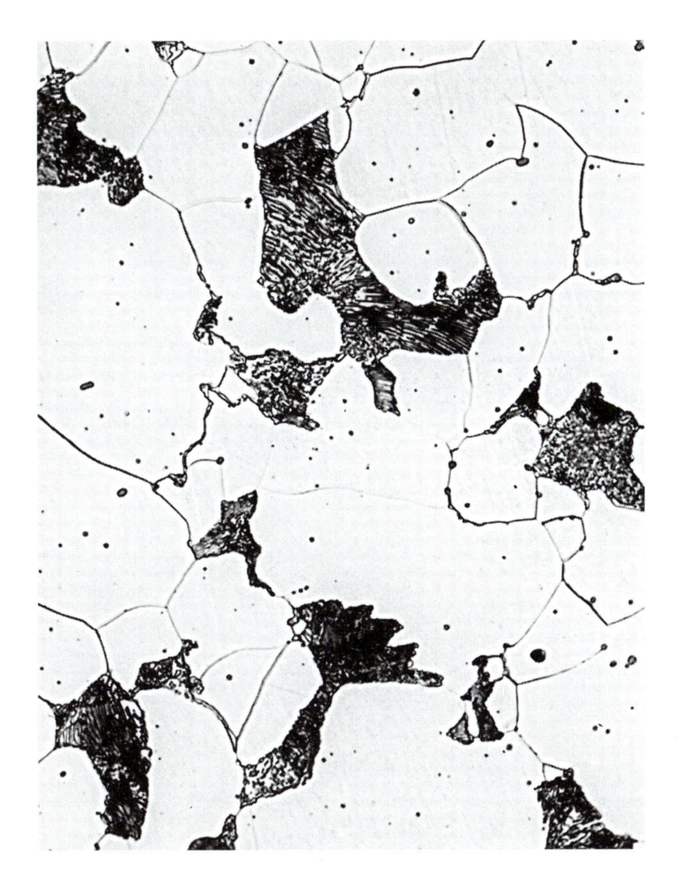

Figure 5-28 Slowly cooled, hypereutectoid steel 1000×. White areas are cementite and dark areas are pearlite. (United States Steel Corporation.)

affect the material's atomic structure. The first thermal process to be discussed is generally known as ***annealing.*** Annealing processes bring about changes mostly by producing phase transformations that result in a rearranged, stable atomic structure with less distorted grains. Therefore, the initial step in most annealing techniques is to heat the steel above its critical temperature to form austenite.

Annealing is broken down into a number of related operations whose overall purpose is to reduce hardness, refine grain structure, restore ductility, remove internal stresses left over from some industrial forming process, or to improve machinability or some other property. These operations—full anneal, normalizing, stress relief, process anneal, and spheroidizing—are shown on a simplified equilibrium diagram for steel (see Figure 5-29). A ***full anneal*** consists of heating steel to about 30°C above the line marked A_1 (the eutectoid temperature line), depending on the carbon composition; holding it at that temperature to obtain a homogeneous structure of austenite; and slowly cooling it in a furnace. The result is, for a hypoeutectoid steel, a transformation of austenite to a coarse lamellar pearlite that is soft, stress free, and a fine ferrite. The word ***annealing*** used alone when referring to ferrous alloys implies full annealing. When referring to nonferrous alloys, the word is used to describe a heat treatment designed to soften a cold-worked structure by recrystallization and/or subsequent grain growth. Annealing of age-hardened alloys implies

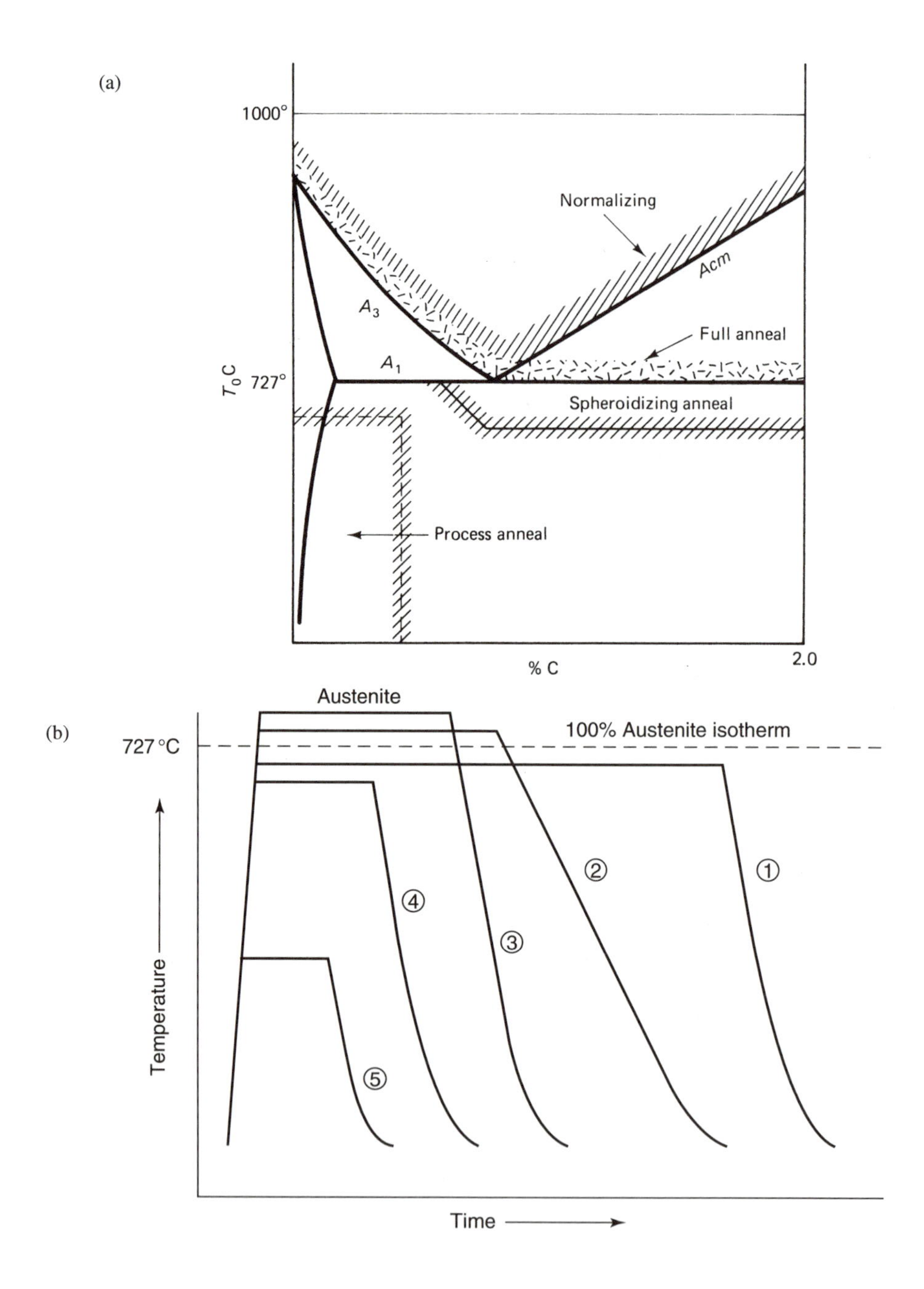

Figure 5-29 (a) Simplified steel equilibrium diagram showing various types of anneals. (b) Schematic of heat-treat annealing processes for steel: 1) spheroidize, 2) full anneal, 3) normalize, 4) process anneal, and 5) stress relief.

(c)

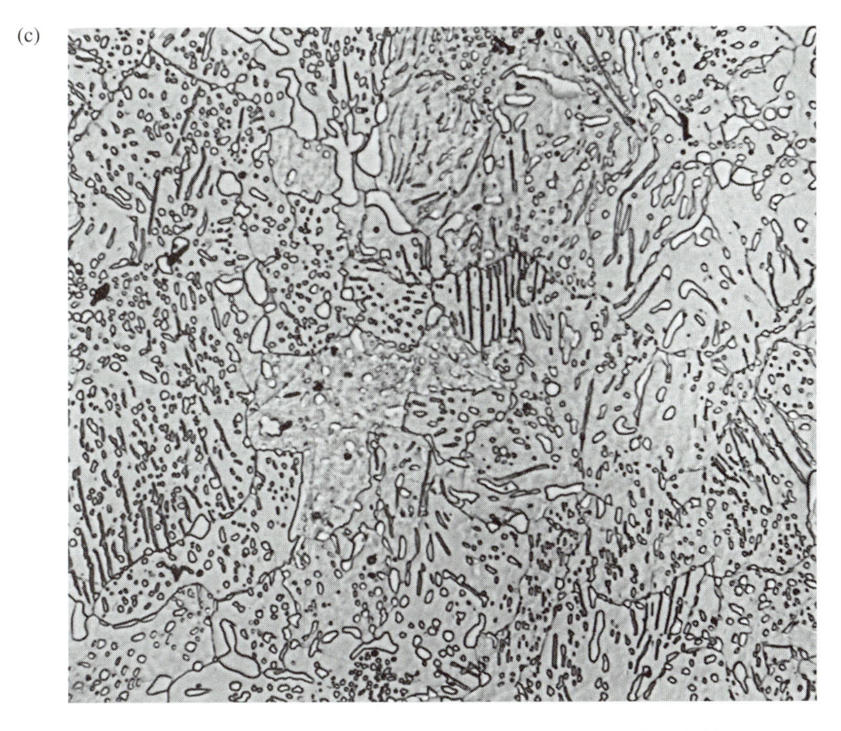

Figure 5-29 (c) Spheroidized 1045 steel, 400×. (Buehler Ltd.)

softening of the alloy by a nearly complete precipitation of the second phase. If the annealing is intended for the sole purpose of reducing stresses, it should be designated as stress relieving. Forged parts are annealed to refine the microstructure produced by forging, which is very hard, so that machining operations such as trimming can be performed. Annealing of gray iron, especially small castings that require a large amount of machining, is noted with an example of ASTM class 35 gray iron, which ends up as a class 20 iron due to the reduction in strength as a consequence of the treatment. An example of the annealing treatment taken from the *ASM Handbook* of a heat-resistant casting made of Alloy Castings Institute (ACI) alloy HA, or ASTM A217 iron–chromium heat-treatable alloy used in the oil industry, is cited. These alloys are heated to an annealing temperature of 1625°F, followed by slow cooling in the furnace at about 50°F per hour to below 1300°F. For improved strength these castings are normalized by heating to 1825°F and air cooled to below 1250°F.

Because it is time consuming and expensive (due primarily to the use of furnace cooling), full annealing in many cases is replaced by ***normalizing.*** Normalizing of steels through using higher temperatures does not require furnace cooling. Instead, all cooling is in still air, and a fine pearlite structure is obtained. Steel is normalized to obtain greater hardness than that of a full anneal. Normalizing does not approach equilibrium cooling conditions as closely as a full anneal. The tensile strength of an annealed alloy is 95,000 psi versus 107,000 psi for the same alloy if normalized. Nodular cast iron castings are usually heat treated to improve their properties. A typical heat treatment to obtain a pearlitic structure is to heat to 1650°F, hold at temperature, cool to 1450°F in the furnace, and then air quench. The quench may be followed by tempering to obtain greater hardness and strength. This additional treatment consists of heating to between 1600° and 1650°F, oil quenching, and tempering. The yield strength for normalized and tempered nodular cast iron castings is 76,000 psi, compared to 124,000 psi for normalized only, and the HB increases from 255 to 321. Normalizing of steels through using higher temperatures (about 60°C above the A_3 line; see Figure 5-29) does not require furnace cooling. Instead, all cooling is in still air, and a finer and more abundant pearlite structure is obtained than from a full anneal. However, its main purpose is to homogenize the microstructure, particularly if further hardening is required. Heating to high temperatures is critical. If too high a temperature is used or if the steel is held for too long at a high temperature, the most likely result is undesirable grain growth. In general, slow heating is more desirable, particularly when working with highly stressed materials (i.e., cold worked), to avoid distortion. Parts with varying cross-sections (thick and thin sections) call for slow heating.

A ***stress relief anneal*** requires temperatures around 600°C, which are below the critical temperature at which austenite begins to form upon heating. The primary objective of this technique is to relieve residual stresses in all metals and steels as a result of a welding, cold-working, or casting operation. These stresses are eliminated or reduced, even though there is no change in the metal's microstructure.

Process anneal is allied with cold working, in which the metal, after heating, is cooled slowly in a furnace down to room temperature. The cooling phase distinguishes this process from normalizing. The end product has higher ductility and lower strength than if normalized. Process annealing is used to "soften" metals, particularly steel sheet and wire products, for further cold working. An electrical transmission line made of copper would become brittle after several drawing operations and, without a process anneal, would fracture. This technique, as well as stress relief, does not involve a phase transformation (no austenizing). The structure of the metal involved, however, is affected. A process called recrystallization is involved, which will be explained later in this module.

Spheroidizing anneal is used to improve the machinability of high-carbon steel. Hypereutectoid steels provide good wear resistance. A spheroidizing anneal helps to toughen them by providing more ductility. The hard and brittle cementite network present in hypereutectoid steel also makes machining to close tolerances difficult. Heating the metal for a longer duration near the critical temperature, followed by slow cooling, is one technique used to produce a spheroidal or globular form of cementite in the ferrite matrix. This entire structure is called spheroidite or spheroidized pearlite; it has a lower hardness, higher ductility, and higher toughness than does the original metal (Figure 5-29).

Most of the above processes may be used to improve machinability but the carbon content of the steel usually is the determining factor. Hypoeutectoid steels with less than 0.3% carbon are normalized, those with up to 0.6% carbon are annealed, and spheroidizing is reserved for steels with 0.6 to 1.0% carbon.

5.5.1 Grain Refinement

Grains that have become plastically deformed as a result of cold working can be given enough energy through thermal processing to permit an orderly rearrangement of the atoms to take place with less deformed grains. The thermal process involved is annealing, and a certain stage of annealing in which new crystals form, known as recrystallization, is worthy of separate comment.

Recrystallization is the formation of a new, strain-free grain structure from that existing in a cold-worked metal upon heating the metal up through a critical temperature for a suitable period of time. When referring to cold-worked metals (mostly all nonferrous metals), annealing can be described as consisting of three stages: recovery, recrystallization, and grain growth.

The recovery stage (see Figure 5-30), the lowest-temperature region, provides the deformed atoms with sufficient energy to rearrange themselves, thus reducing distortion and its attendant stress. The properties recover toward their original values with no visible change in their microstructure. The recovery is due to the cancellation or annihilation of the point defects and dislocations in the deforming metal. A stress-relief anneal, as described previously, requires temperatures in the lowest-temperature region of the recovery stage. As temperatures increase, a point is reached where the nucleation of new crystals appears in the microstructure. These new crystals have the same structure and composition as those of the original undeformed crystals before the metal was cold worked. The grain boundaries and

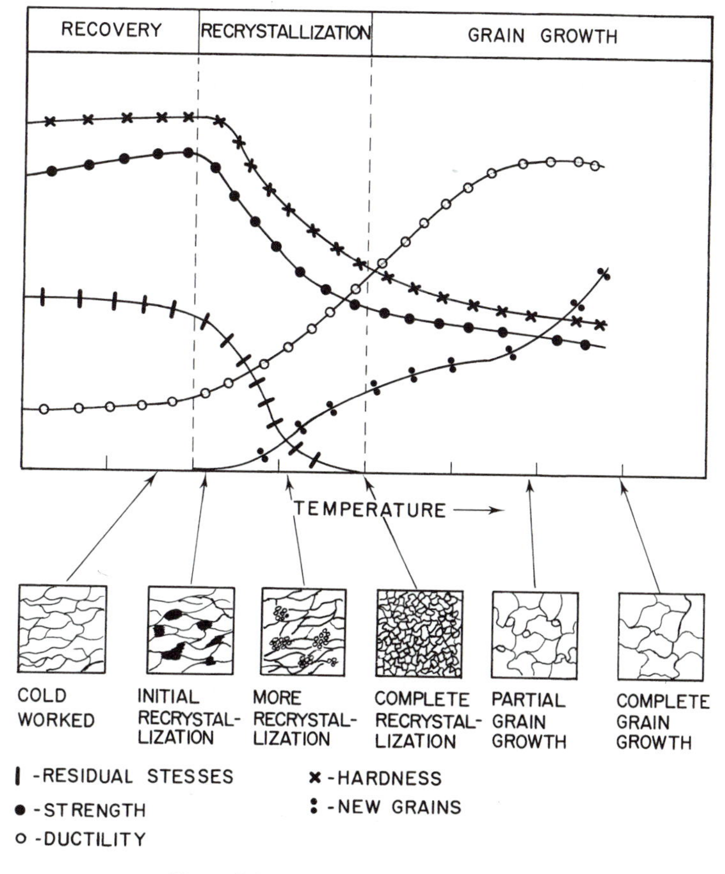

Figure 5-30 Cold-work annealing treatment.

slip planes with the maximum distortions caused by the cold working contain the majority of these new crystals, and eventually they grow and spread throughout the metal structure. A short time after the appearance of the new crystals, the properties of the metal change rather drastically. Hardness and strength drop off, while ductility increases. As the temperatures increase, the region of grain growth is reached, about 0.3 to 0.6 of the melting point of the metal in Kelvin, which points out that control of grain size can be exercised to some extent by a wise choice of annealing temperatures. Recrystallization temperatures vary with metals and with the degree of cold working. The number and size of crystals that develop in the recrystallization region result from the amount of cold working done to the metal. If a metal is not cold worked prior to heating, recrystallization will not occur and the grains will grow in size. To transfer sufficient energy to the deformed crystals, a minimum plastic deformation of about 7% is required before any change in grain size takes place.

From our knowledge of cold working, we know that a highly cold-worked metal contains very little capacity for further plastic deformation. Consequently, the metal is heated to a temperature that permits the atoms to realign and diffuse to a more stable position. The temperatures over which a marked softening occurs (a drop in tensile strength) are in the recrystallization zone. This range or zone of temperatures falls between one-third and one-half of a metal's melting temperature expressed in absolute units (Kelvin). The average recrystallization temperature is not the same as the critical or transformation temperature. It depends on the amount of cold working, as well as the duration of the heating. The more a metal is cold worked, the more energy the atoms possess. Therefore, they need less energy transfer from an external source than does a less-cold-worked metal. The recrystallization temperature is used to distinguish between cold working and hot working of metals. The cold working of copper at 95°C is at a higher relative temperature than the hot working of zinc at 20°C. This is so because the recrystallization temperatures of pure copper and zinc are about 120° and 10°C, respectively. A point to remember is that these changes do not come about as a result of any phase change. When a metal is not allotropic, several processes are needed to change its properties. First, it must be cold worked to bring about a minimum deformation of the crystal structure. Second, it must be heated sufficiently to produce recrystallization. Third, after slow cooling the emergence of new, stress-free, small, uniformly dimensioned grains will result in the properties desired. Finally, it must be realized that if, while in service, a metal is subjected to a temperature that is higher than its recrystallization temperature, it will recrystallize, lose its strength and hardness, and possibly cause a failure of some magnitude.

5.5.2 Grain Size

Heat treating can alter the size of grains, which produces corresponding changes in properties. A grain, or individual crystal, varies in size from one metal to another. Zinc crystals in galvanized steel sheet and crystals in some brasses can be seen with the naked eye, but these are exceptions to the general rule. Most crystals are visible only with the use of a microscope.

Even within one metal, the grain size can vary from one region to another. Grain size has a direct influence on the properties of the metal. For example, as grain size *decreases*, the yield strength of the metal increases. Another fact is that fine-grained metals are stronger and tougher than coarse-grained metals under low or room temperatures. At high temperatures, the reverse is true; coarse-grained metels are stronger and tougher than fine-grained metals. Grain-size ratings, as contained in the American Society for Testing and Materials (ASTM) specifications, are made by comparison with a standard chart of sizes numbered from 1 to 10. When a match is arrived at, the grain size is then designated by the number corresponding to the index number *(n)* of the matching chart. These matching standard-grain-size charts represent a microstructure projected at a magnification of 100×. The comparison method described above (ASTM E112) is an estimation at best.

Steel is ordered on the basis of either fine grain or coarse grain. Grain size must be specified for metals that are to be cold worked. If the grain size is too coarse, the worked surface will be uneven, and if too fine, the metal will be too hard to work satisfactorily. Several different methods, in addition to the ASTM comparison method, are used in industry to estimate grain size.

Temperature alone causes grain growth and not grain size reduction or refinement. If you wish to reduce the size of grains to refine them, you must destroy the original grains and produce new ones with the desired size. One way of doing this is by heating an allotropic metal sufficiently to produce a phase change with its initially small grains. In steel, the fcc grains of austenite are large, but on transforming to bcc crystals, the grains are small. Another way to refine grains is to deform the metal plastically by cold working. This refinement in grain size is not the primary purpose of cold working. However, due to the smaller size of the grains, the cold-worked metal is found to possess properties different from those before it was cold worked.

5.5.3 Grain Boundaries

We know that the atoms that make up the grain boundaries are the last to solidify. As a result, they are not arranged in the same orderly way as are the same atoms in the interior of the crystal. We also know that any time we observe a disordered arrangement of atoms, we look for distortion, increased stress, and resistance to further plastic deformation or slip—all of which result in stronger material in the grain boundaries (Figure 5-31). Arrows point to some of the many grain boundaries shown in Figures 5-31 and 5-32.

The more small grains present in a metal, the more grain-boundary material per unit of volume of metal. The conclusion drawn is that fine-grained metals are stronger than coarse-grained metal at room temperatures. At high temperatures, the heat supplied to the fine-grained metal atoms that make up the grain boundaries increases their energy. This reduces their strained condition, which allows them to spread out and become less densely packed, which in turn results in a decrease in strength. Based on this reasoning and coupled with knowledge gained from experience, coarse-grained metals are required for high-temperature applications where resistance to creep is critical. Can you recall the definition of *creep*? A good example where knowledge is needed is in the selection of the correct metal for the manufacture of turbine blades that revolve at very high speeds and temperatures, with extremely small clearances between the moving tips of the blades and the stationary parts of the turbine.

Cold working results in the formation of more grain boundaries due to the coming together of the large number of dislocations with their disorderly arrangement of atoms. With more grain boundaries come smaller grains; reduced ductility; and increased hardness, tensile strength, and electrical resistance. These conditions describe the terms ***work***

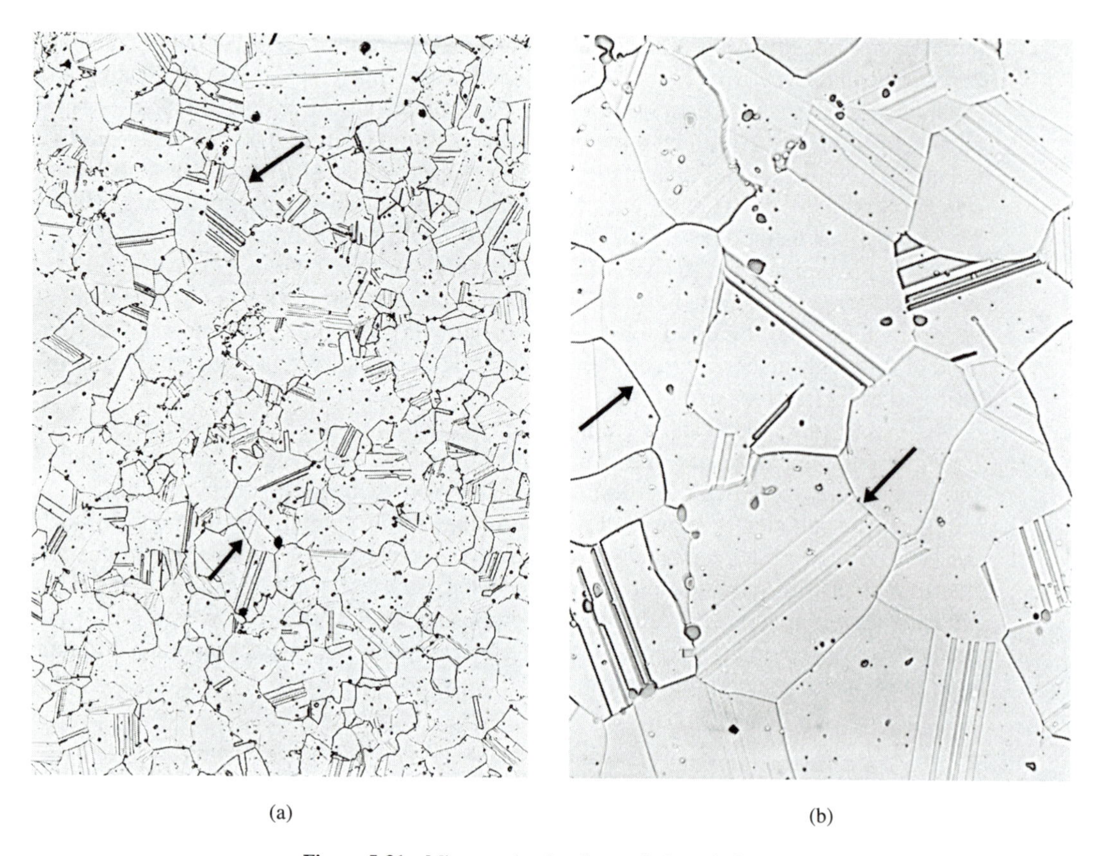

(a) (b)

Figure 5-31 Micrographs showing grain boundaries. (Buehler Ltd.)

or ***strain hardening*** produced by cold working. A good example of cold working is the drawing of copper wire. ***Drawing*** has two meanings in materials engineering. In heat treatment, it means tempering. In cold working, the process of drawing refers to the forming of a recess in sheet metal with dies (see Figure 4-10) and the pulling of metal through dies, as in the manufacture of wire. A copper wire transmission line is cold worked to increase its strength so that it can support itself without excessive sagging over a moderate span.

5.5.4 Grain Shape

Cold and hot working produce a distorted grain structure; that is, the shape of the grain is deformed or becomes elongated in the direction of the metal flow. Figure 5-32(a) and (b) compares hot- and cold-worked grains of copper. Such deformed metals no longer possess uniform properties in all directions (isotropic). Instead, the metal generally shows higher-strength properties in certain directions, with a corresponding decrease in those same properties in other directions (anisotropic).

To summarize the preceding discussion, the set of photomicrographs in Figure 5-32(c) illustrates the concepts of cold working, recrystallization, and grain growth. The

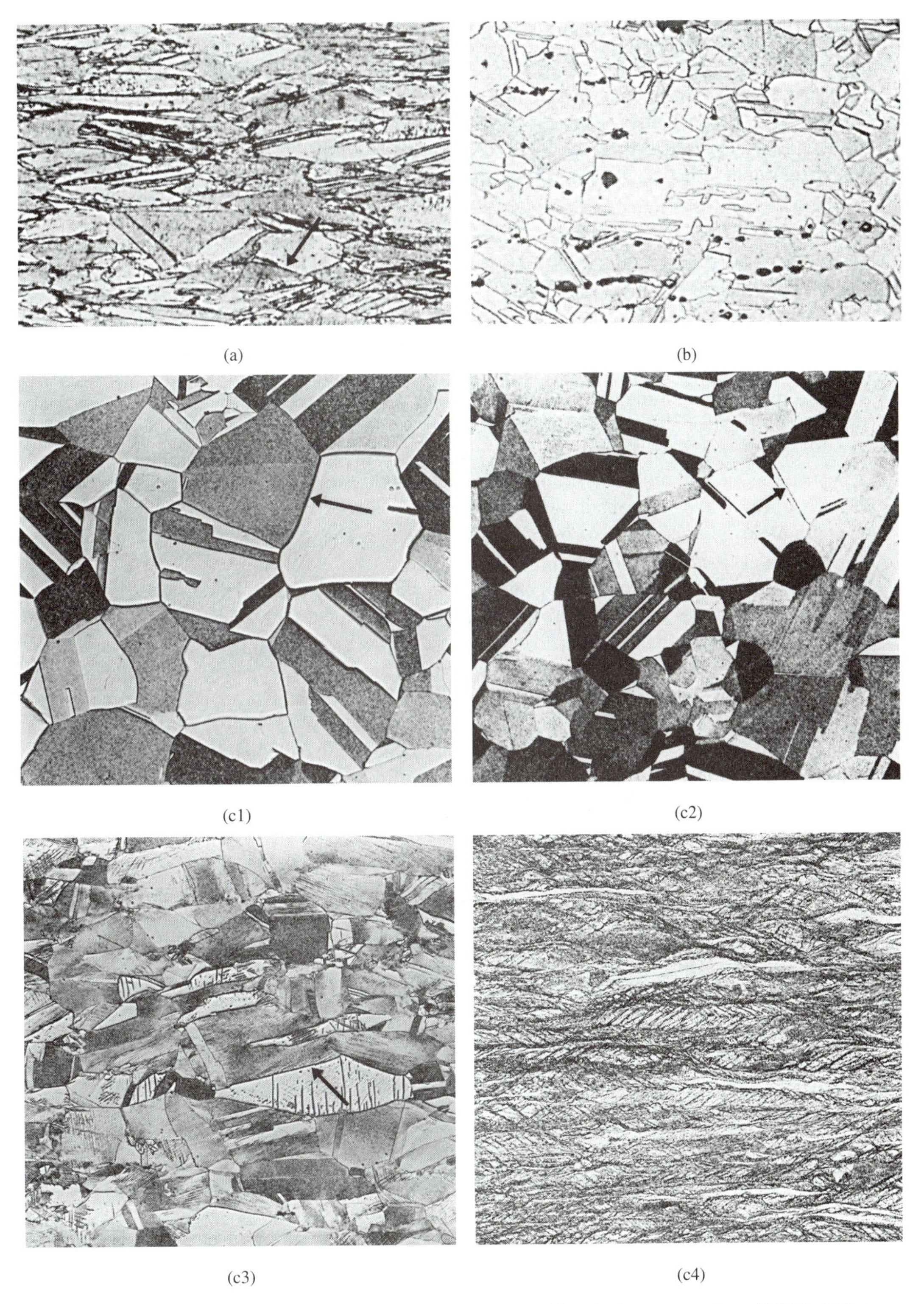

Figure 5-32 Micrographs showing: (a) hot-worked, then cold-worked, copper specimen (100×); (b) hot-worked copper specimen; (c) 70/30 brass 300×. Reduction: 1) 0%, 2) 5%, 3) 25%, 4) 75%. (Buehler Ltd.)

specimens are from annealed, commercially available 70% Cu–30% Zn nonleaded brass, initially 0.128 in. thick. Reductions were accomplished by small reduction rolling passes while maintaining the strip at room temperature at all times. All photomicrographs of the strip samples are shown at 300× for the best possible comparison of microstructures.

5.6 THERMAL PROCESSING: NONEQUILIBRIUM CONDITIONS (MARTENSITE STRENGTHENING)

By varying the amount of carbon dissolved in the single-phase solid solution of iron and carbon (austenite) and cooling the austenite under equilibrium conditions, the austenite transformed into a multiphase mixture known as carbon steel. Depending on the amount of carbon, steel alloys of ferrite and pearlite or cementite and pearlite are formed, which have different hardness and strength properties. But we now reach a limit. Low-carbon steel (0.25% carbon) has a strength of about 44,000 psi; eutectoid steel (0.8% carbon) has about 112,000 psi. To obtain steel with greater strengths, we must depart from equilibrium conditions primarily in the cooling of the austenite. This departure, in the form of a rapid cooling, is known as ***quenching.***

The degree of cooling depends mostly on the quenching medium used. For drastic quenching, water or brine is used. Less severe cooling rates are achieved with such media as oil, molten salt baths, still air, or molten sand. Most liquid media require agitation of the liquid to help reduce the gaseous layer formed adjacent to the metal as a result of the vaporization of some of the liquid in contact with the hot metal, whose temperature is above the boiling point of the liquid quenching medium.

Rapid cooling of steel produces a metastable phase called ***martensite*** (Figure 5-33), which is defined as an interstitial supersaturated solid solution of carbon in iron having a bct lattice structure. The transformation of martensite results from a three-dimensional me-

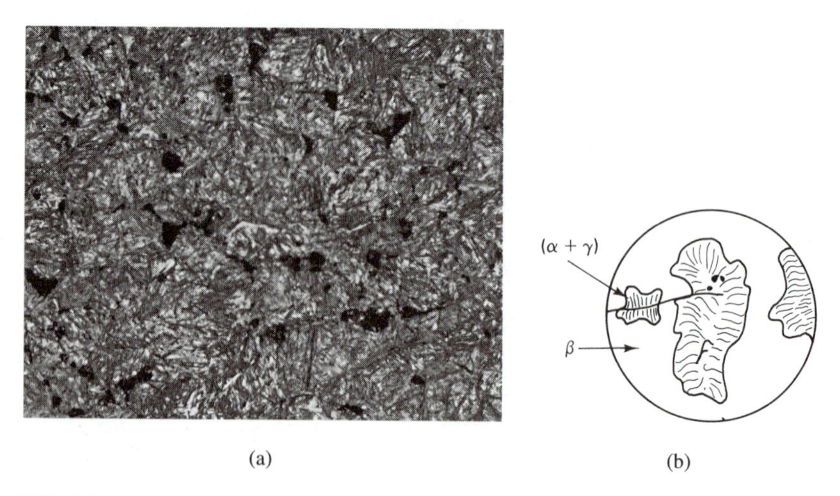

Figure 5-33 Micrograph of 1095 steel martensite. (a) 1095 steel. (b) Microstructure of martensite (100×). (Buehler Ltd.)

chanical shearing action that distorts the structure, producing hardness and brittleness. Martensite transformations occur in other metals and alloys, such as titanium, lithium, iron, nickel, and copper-aluminum.

To summarize what we have said about equilibrium and nonequilibrium conditions of cooling, our initial step was to heat the basic components of steel (into the *austenite-forming* region) to produce austenite, a single-phase solid solution of carbon in gamma (γ) iron. Once we formed austenite, we could cool it under equilibrium conditions to produce a two-phase mixture of ferrite and cementite called ***pearlite.*** The carbon in the cementite distorts this structure, producing qualities of strength and hardness. The ferrite contributes ductility. If we depart from equilibrium conditions and cool the austenite more quickly, we can form pearlite, with greater hardness. If we cool it even more rapidly, we produce a transformation product called ***martensite,*** which is extremely brittle, strong, and hard. It far surpasses the hardness and strength of pearlite. Steel with 100% martensite has limited use due to its sensitivity to fracture by impact. Therefore, we call on another heat-treating process known as ***tempering*** to heat the martensite to selected levels and to transform some or all of the martensite to other microstructures with differing properties. Tempering is sometimes known by another name, ***drawing,*** due to the fact that, through controlled heating, hardness can be drawn from the material. Actually, tempering, like any process that supplies energy to atoms through heating, depends on the diffusion of these atoms. The rate of diffusion depends not only on the temperature, but also on the time allowed for diffusion to take place. In allowing the carbon atoms to proceed once more on their paths through the atomic structure, varying microstructures and grain sizes are formed with just the correct hardness and strength desired for some industrial application. One application might require a good amount of toughness with very little brittleness; another might demand greater hardness. As in many situations, we cannot have the best of two worlds; to gain greater strength and toughness, we have to trade a little hardness.

In discussing nonequilibrium conditions in the thermal processing of steel, we make use of an ***isothermal–transformation (IT)*** or ***transformation–temperature–time (TTT)*** diagram for a particular steel composition. This diagram is derived by plotting hundreds of isothermal cooling curves for samples of a particular steel composition. The isothermal cooling curve sketched in Figure 5-34 shows the percentage of austenite remaining and the percentage of the austenite transformed phase plotted against time. The isothermal temperatures are all below the critical A_e (or eutectoid) temperature at which the austenite becomes unstable and begins to transform (see Figure 5-35). The data thus collected from these isothermal curves are then used to construct the isothermal–transformation diagram for that particular steel. The diagram shows the beginning and end of the austenite transformation as well as the transformed product at various isothermal temperatures.

This procedure is illustrated in Figure 5-35 using just one set of data. A characteristic S-shaped curve is formed that represents the transformation zone or region at various temperatures. Figure 5-36 is a copy of such a diagram for eutectoid steel, indicating the Brinell and Rockwell hardness readings alongside photomicrographs of the various products formed from austenite at the various temperatures. Figure 5-37, another sketch of a TTT diagram, shows the S-shaped austenite-to-pearlite transformation zone, the austenite-to-martensite transformation zone, and a subdivision of the pearlite into coarse and fine pearlite and bainite, a transformed product of austenite that has characteristics of both

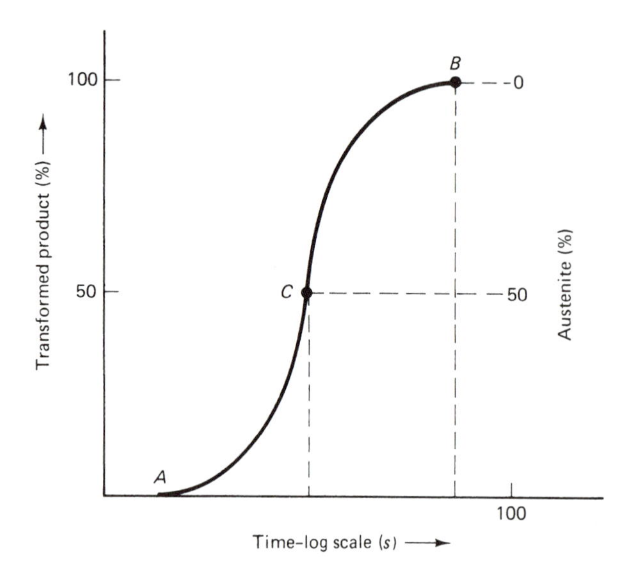

Figure 5-34 Isothermal–transformation curve at temperature T°C.

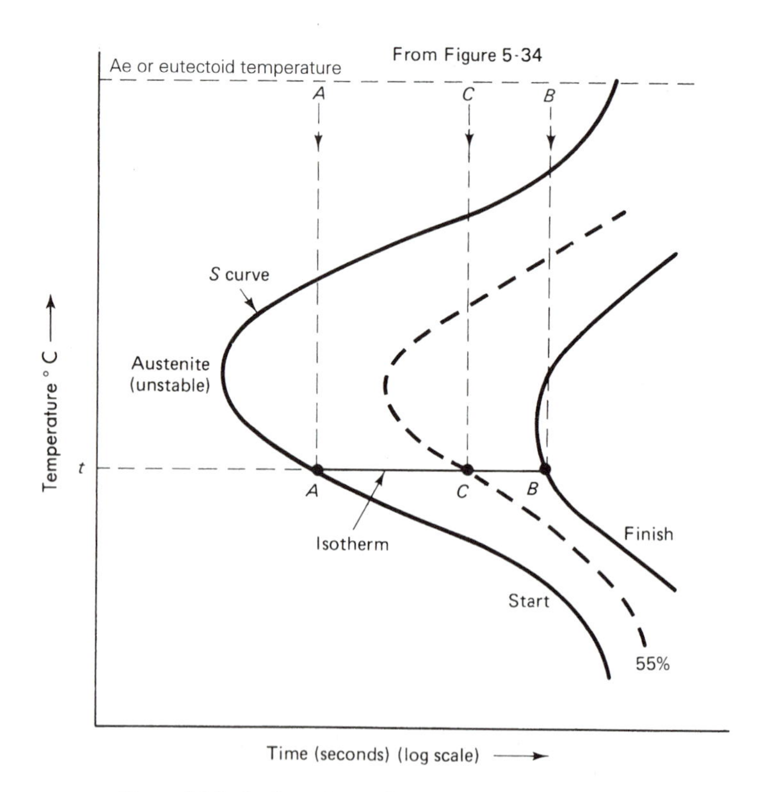

Figure 5-35 Isothermal–transformation diagram (IT or TTT).

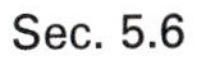

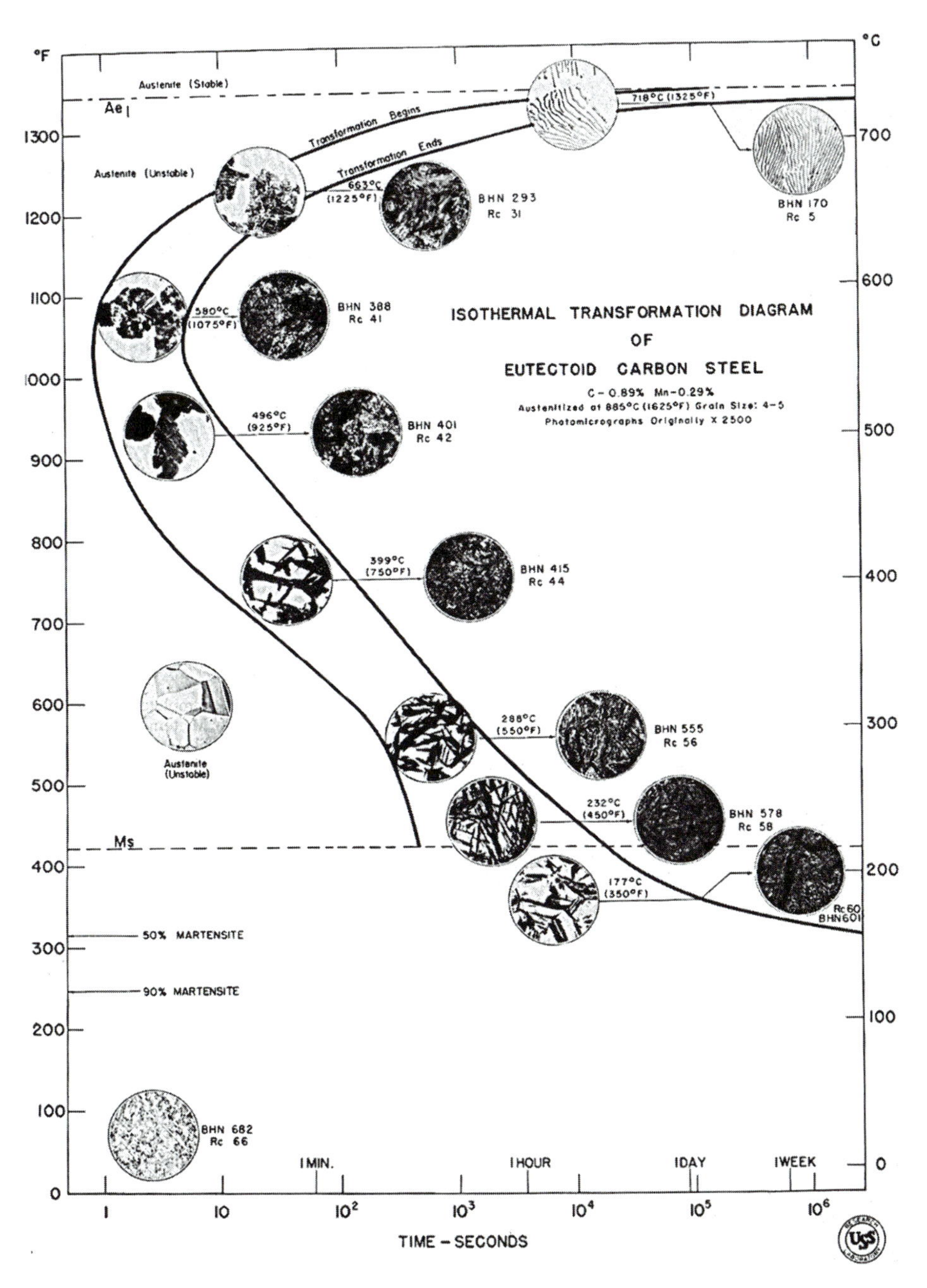

Figure 5-36 Isothermal–transformation diagram of a eutectoid steel. (United States Steel Corporation.)

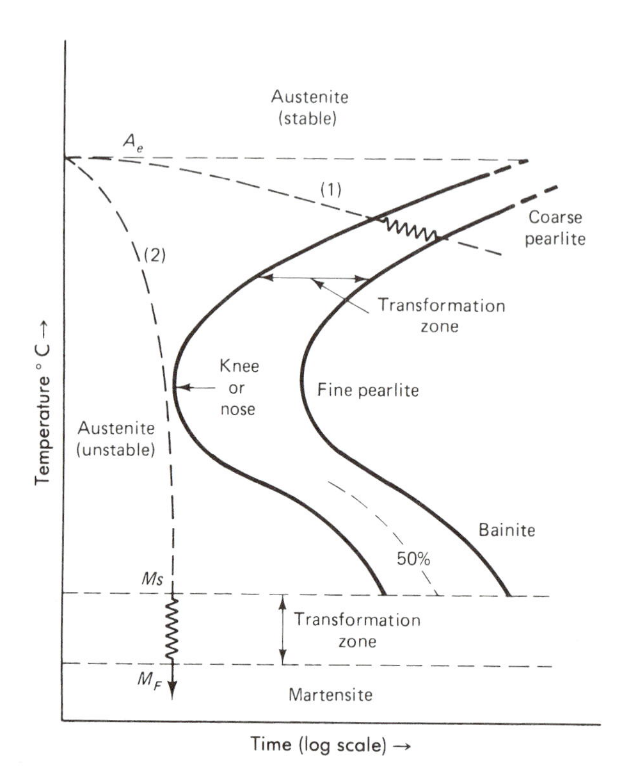

Figure 5-37 IT diagram showing transformation zones, two cooling curves, and products formed.

pearlite and martensite. In addition, two cooling curves are plotted. The first curve, labeled 1, represents the transformation of austenite to coarse pearlite. Such a curve has a relatively small slope, which indicates a slow cooling rate. The second curve has a much steeper slope and consequently a much faster cooling rate. The latter curve is shown passing to the left of the leftmost portion of the pearlite transformation zone, known as the knee or nose. As a consequence, all the austenite transforms into martensite.

It should be evident at this point that the goal of thermal processing under nonequilibrium conditions is to produce steels with greater strengths than those produced by solution hardening under equilibrium conditions (very slow cooling). The mechanism used to accomplish this goal is a phase transformation. Such a mechanism is possible because steel is allotropic. In the next few pages we will describe briefly some major thermal processing techniques used by industry to strengthen steel under nonequilibrium conditions.

5.6.1 Conventional (Customary) Heat, Quench, and Temper Process

The previous statements on the formation of austenite, quenching to form martensite, and the final heating to temper the martensite describe the conventional process of making steel parts with desired properties. Study Figure 5-38 for the customary quench/temper cooling rates. The log scale along the bottom indicates time. One serious drawback to this process is the possibility of distorting and cracking the metal as a result of the severe quenching required to form the martensite without transforming any of the austenite to pearlite.

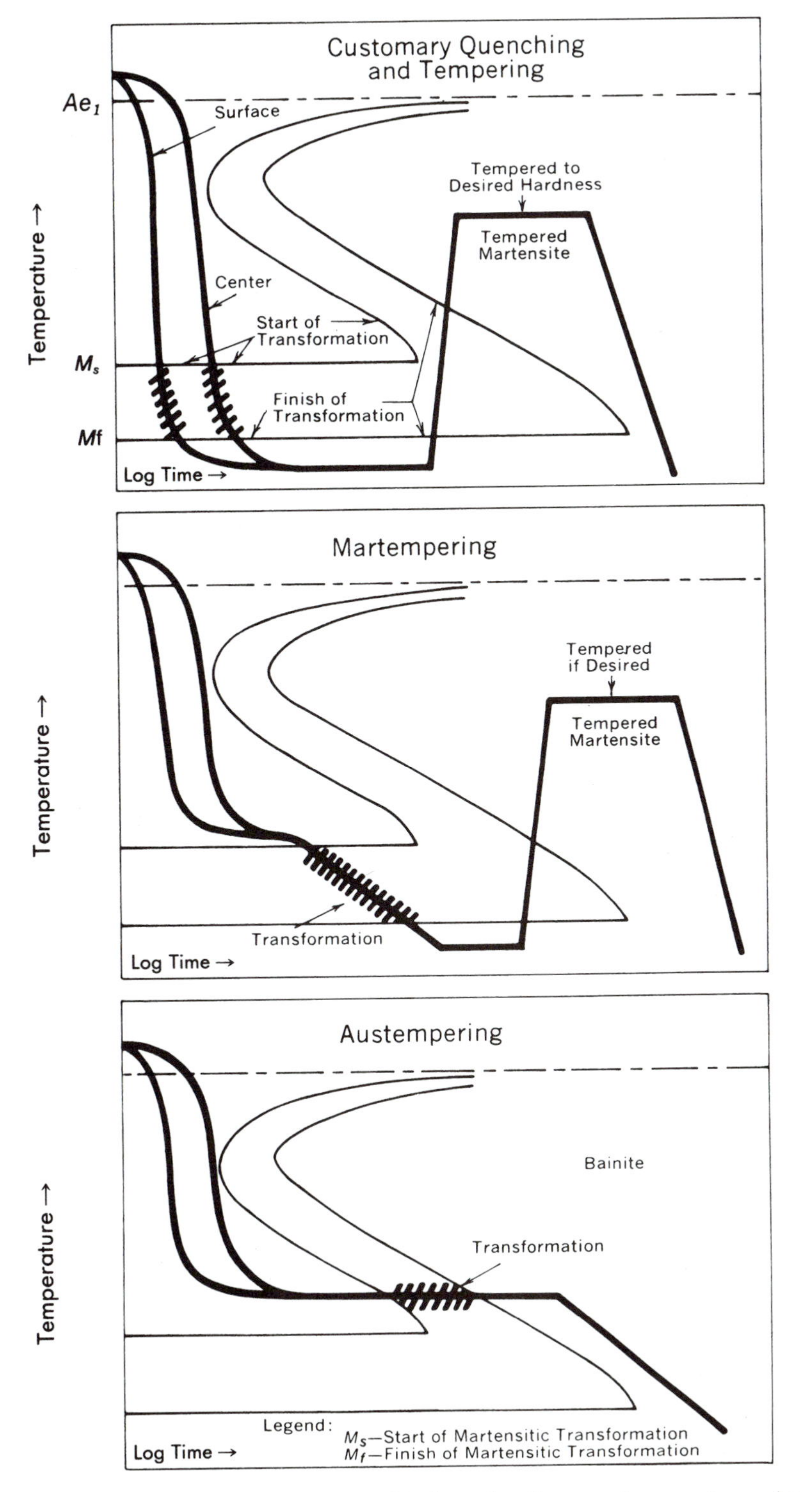

Figure 5-38 Comparison of heat-treating processes. Heating and cooling operations superimposed on a typical isothermal–transformation diagram. The temperature range at which the transformation to the hard products bainite or martensite occurs has been indicated. (United States Steel Corporation.)

As a metal object is quenched, the outer area is cooled more quickly than the center. Thinner parts are cooled faster than are parts with greater cross-sectional areas. What this means is that transformations of the austenite are proceeding at different rates in a single metal object. As we cool a metal object, it also contracts and its microstructure occupies less volume. Put these statements together and the conclusion is that extreme care is necessary to prevent undue distortion and/or fracture. Extreme variations in the size of metal objects complicate the work of the heat treater and should be avoided in the design of metal parts. This means that there is a limit to the overall size of parts that can be subjected to such thermal processing.

5.6.2 Martempering

To overcome these restrictions, two other thermal processes are used. The first, ***martempering*** or ***marquenching,*** permits the transformation of austenite to martensite to take place at the same time throughout the structure of the metal part. This is shown graphically in Figure 5-38. By using an interrupted quench, cooling is stopped at a point above the martensite transformation region to allow sufficient time for the center to cool to the same temperature as the surface. Then cooling is continued through the martensite region, followed by the usual tempering.

5.6.3 Austempering

A second method of interrupted quenching is called ***austempering.*** Figure 5-38 shows this process graphically. The quench is interrupted at a higher temperature (200 to 375°C) than for marquenching, to allow the metal at the center of the part to reach the same temperature as the surface. By maintaining that temperature, both the center and the surface are allowed to transform to bainite and are then cooled to room temperature. The advantages of austempering are (1) even less distortion and cracking than marquenching, due primarily to the higher transformation temperatures, and (2) no need for final tempering. However, austempering has the disadvantage of requiring more time, even though it requires no tempering treatment. Also, parts with large sections of thickness cannot be processed. Sections with a maximum thickness of ½ in. are cooled sufficiently fast to permit the transformation of austenite to bainite without the formation of pearlite.

5.7 PRECIPITATION STRENGTHENING/HARDENING (AGE HARDENING)

Nonferrous metals do not undergo the significant phase transformations possible with steel. Consequently, most thermal processing of nonferrous metals is used to relieve stresses in a single-phase microstructure and/or to produce recrystallization. These techniques do not result in significant strengthening of a metal's structure. Instead, the most effective thermal processing technique for increasing the strength of such metals is *precipitation hardening,* or *age hardening.* Precipitation hardening involves two steps. [See Figure 5-40(b).] The first is *solution treatment,* the heating (annealing) of an alloy that exists at room temperatures as a two-phase solid solution to produce a single-phase solid solution at the elevated temperature. The second step is the subsequent rapid cooling ***(quenching)*** of this single-phase solid solution to form an unstable supersaturated solid solution. Some alloys require a further heating process to expedite the strengthening of the metal.

To explain how this strengthening comes about, it is first necessary to review earlier material in this chapter, specifically that concerning diffusion (pages 227–228, 264) and the effects of solid solubility curves (solvus lines) using Figure 5-20. Such a review will assist the reader in understanding our explanation of precipitation hardening. As stated previously, the partial solubility of one solid metal in another is determined by the slope of the solid solubility curve (solvus line). The solvus must slope so that it indicates there is greater solubility of one metal in another at a higher temperature than at a lower temperature.

Figure 5-20, similar to Figure 5-39, is a schematic of the equilibrium diagram for silver (Ag)–copper (Cu) alloys. At each end (the silver-rich and the copper-rich), the solvus indicates the type of partial solubility desired for precipitation hardening. Using the silver-rich end of the diagram (at the left of Figure 5-40a), an alloy consisting of about 92.5% Ag and about 7.5% Cu (sterling silver) is shown as a vertical dashed line. At room temperature this alloy exists as a metal consisting of two distinct phases (α and β). Phase α represents a silver-rich solid solution with some dissolved Cu, and β is a solid solution of almost pure Cu with very little dissolved Ag. As the alloy is heated, it crosses the solvus at about 760°C. The β phase dissolves at this temperature and diffuses uniformly to form a solid solution of α. If, then, this same alloy were slowly cooled, the β phase would precipitate out of the α solid solution because the solubility of β (copper) in the α (silver) decreases from about 8.5% at 780°C to less than 1% at room temperature, as shown by the solvus line.

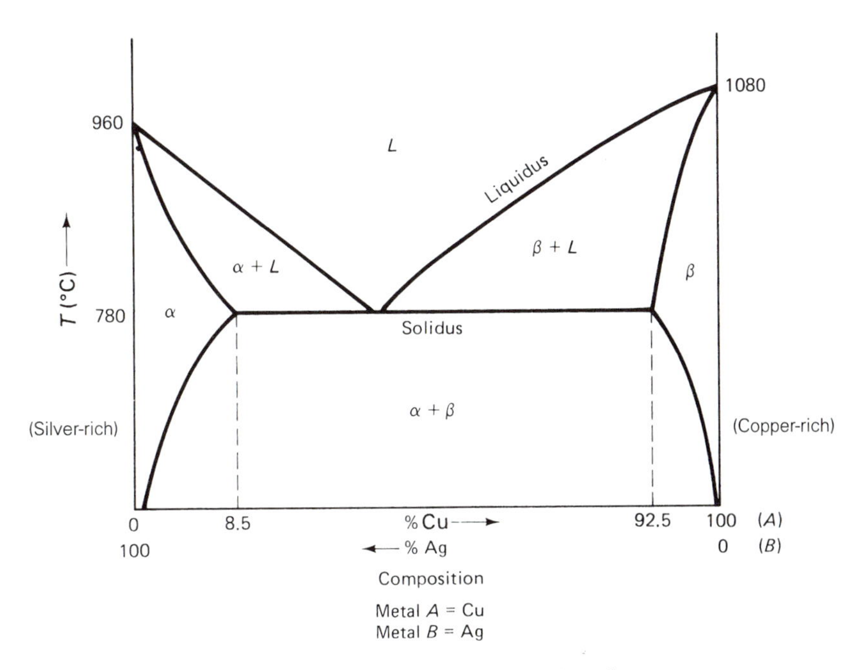

Figure 5-39 Ag–Cu alloys equilibrium phase diagram.

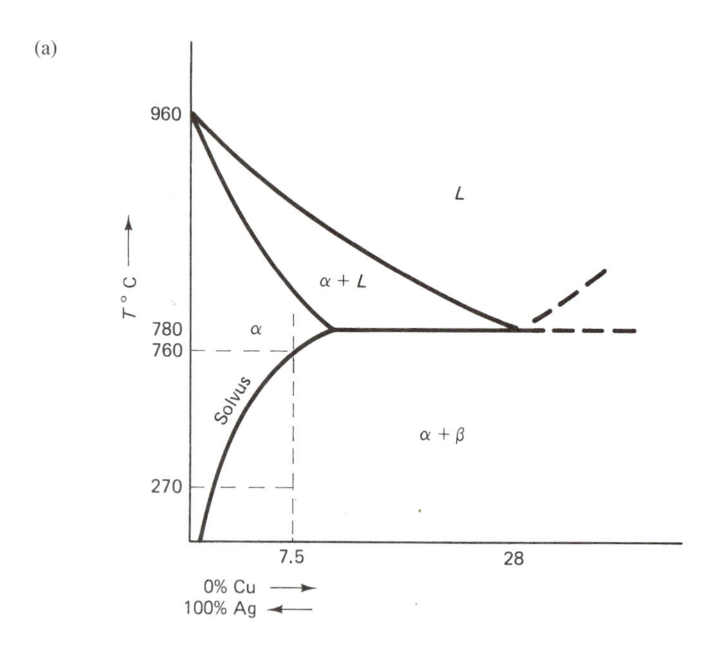

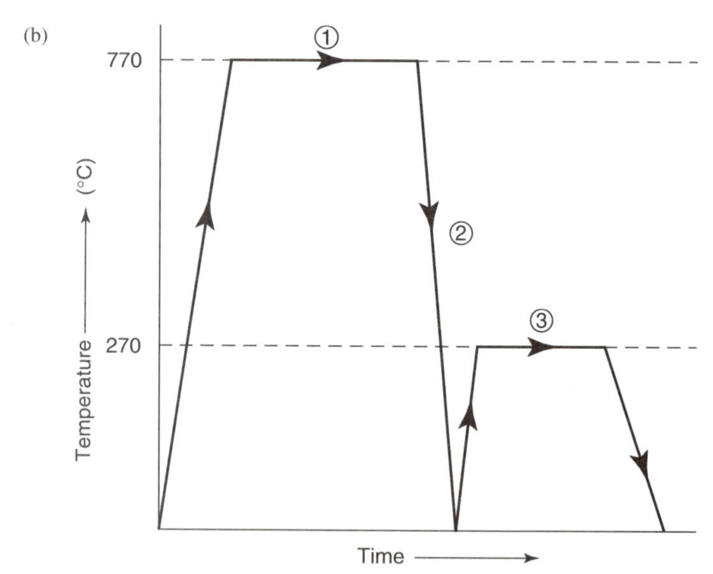

Figure 5-40 (a) Schematic of the Ag-rich end of Ag–Cu equilibrium diagram. β phase is pure copper (Cu) plus a little silver (Ag). (b) Temperature/time plot showing heat-treat cycles for precipitation strengthening.

Now if after the solution is heated to ① around 770°C [see Figure 5-40(b)], this alloy were cooled quite rapidly (quenched ②), there would be insufficient time to permit the Cu atoms to diffuse, which would produce equilibrium-type precipitation of β atoms. The result is a retained α phase that is unstable, with its excess Cu atoms trying to precipitate out to form the β-phase crystal structure. The unstable, supersaturated solid solution formed is quite ductile.

With the metal in such a crystal structure, it can be worked, straightened if distorted, or machined much more easily than when the metal was in the stable, original form at room temperature. Once the metal is worked, the strengthening process can continue. Other alloys require heating to a temperature below the solvus line to cause this precipitation. Sterling silver can be heated to around 300°C for about 30 minutes and the hardness doubles in value. The increase in strength and hardness is explained by precipitation of the trapped β metal out of the α-metal structure at the grain boundaries and at the sites of impurities in the lattice structure of α metal, which produces resistance to the movement of dislocations. If the precipitation occurs at room temperatures within a few days' duration, it is known as ***natural aging.*** If additional heating is required, as it is in sterling silver, to cause precipitation and full strengthening, the process is called ***artificial aging.*** This additional heating/cooling cycle is also referred to as the precipitation heat-treating cycle [Figure 5-40(b) ③]. At this stage the supersaturated α solid solution is heated to some intermediate temperature [see the 270°C isotherm in Figure 5-40(b)] to increase the rate of diffusion of β participate phase in the form of finely dispersed particles of metal *A* or, in this case, copper atoms.

It is possible to stop the precipitation entirely by lowering the temperature sufficiently through refrigeration. Once the precipitation has produced a saturated normal dual phase at or near room temperature, the metal reaches its greatest hardness and strength. As fine precipitates form, they have a tendency to grow with time, and with natural-aged alloys this growth of the precipitate will result in a decrease in the strength. Such a phenomenon is known as ***overaging.*** Artificially aged alloys are less subject to overaging because the growing process can be stopped effectively by a simple quench. It must be pointed out that the strength property was used more in this discussion than was any other property. We know that other properties are also affected by distortions of a lattice structure brought about by the precipitation of atoms of another element. The main difference between precipitation strengthening and the thermal processing under nonequilibrium conditions that produces martensite (see Section 5.6) is the formation of a new crystal structure in the transformation of martensite.

To round out our treatment of precipitation hardening, the popular aluminum–copper alloy known as Duralumin (2017) and a part of its equilibrium diagram are used to illustrate the steps taken to bring out the strength of the alloy in industry. Figure 5-41 is a schematic of the aluminum-rich end of the Al–Cu equilibrium diagram. Alloy 2017 contains about 96% Al and 4% Cu. Naturally or artificially aged, this alloy finds many applications, one of which is for aircraft rivets. The vertical dashed line (which intersects the solvus line) in Figure 5-41 represents this particular alloy. Upon solution treatment to a temperature of about 550°C, the single-phase solid solution designated *k* is formed in which all the Cu atoms are diffused into the crystalline structure of the Al atoms. Next it is quenched to form the unstable, supersaturated *k* phase, and the alloy becomes very

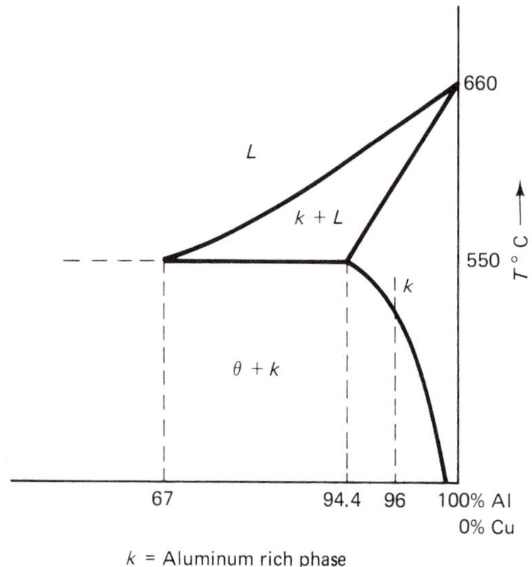

Figure 5-41 Schematic of the Al-rich end of Al–Cu equilibrium diagram.

ductile and can be maintained if it is refrigerated to retard the eventual precipitation of Cu atoms from the *k* phase. Once sufficient energy is received through heating to room temperature or beyond, the aging process, and hence an increase in strength, take place. Rivets made from this alloy taken from refrigerated storage can then be driven easily. Once driven, the aging process begins as room temperature is reached, which brings on the attendant growth in strength and hardness.

5.8 HARDENABILITY

Not to be confused with hardness, ***hardenability*** is the measure of the depth to which hardness can be attained in a metal part using one of the standard thermal processing techniques designed primarily to increase hardness. We know that carbon is vital to steel making. Steels are even classified by their carbon content. Hardness, as an indicator of strength, depends on carbon content. The higher the carbon content, the higher the maximum hardness that can be attained at greater depths within a part, assuming that other factors remain the same. Alloys also play a significant role in producing steels with higher degrees of hardenability. We know that one way to change a metal alloy's microstructure is to purposely add impurities. The impurities are elements that will alloy or join with the main elements or ingredients to form solid solutions—substitutional and/or interstitial. These alloying elements, such as nickel, boron, copper, or silicon, produce effects similar to the effects of impurities found in the original metal when mined and/or those added, out of necessity, during the processing of the metal. The presence of atoms of the alloying element (1) distorts the lattice structure; (2) increases the metal's resistance to plastic slip; (3) hinders the diffusion of atoms past these areas of distortion; (4) effectively increases the brittleness, hardness, and strength of the metal alloy; and (5) increases significantly the metal's hardenability. Adding alloys or increasing carbon content shifts the location of the S-curve (transformation zone) of austenite to pearlite or bainite to the right on the TTT diagram. The result is that a steel need not be quenched as rapidly to achieve the desired hardness. One disadvantage is that alloys are costly, but benefits usually outweigh such costs. One benefit in using a less drastic quench is the reduction in the danger of cracking and/or warping of steel parts, particularly the parts that are, of necessity, complex in shape or size or that require great dimensional stability.

It is essential to point out here the necessity of following proper design standards, which help to reduce such things as residual stresses, warping, and cracking. Such techniques as maintaining, if possible, uniform cross-sections or thicknesses and using fillets and rounds make the job of the heat treater less difficult.

The Jominy end-quench test performed on steels using standardized specimens and procedures determines the necessary nonequilibrium cooling rates to attain a certain hardenability in a particular steel part (see Figure 5-42). The 25-mm-diameter specimen with a length of 100 mm represents a metal alloy austenized at a prescribed temperature and for a prescribed time. After removal from a furnace it is clamped into a fixture and quenched at a specified flow rate and temperature. After cooling to room temperature, shallow flats are machined along its length and Rockwell hardness readings are taken at prescribed intervals over the first 50 mm. Each hardness reading is plotted to produce a hardenability curve similar to Figure 5-42(b). Each steel alloy has its own unique hardenability curve. A steel alloy that has a high hardenability is one that hardens (forms martensite), not only at the surface, but generally throughout the interior. Hence, ***hardenability*** is defined as the ability of an alloy to be hardened by the formation of martensite as a result of a given heat treatment.

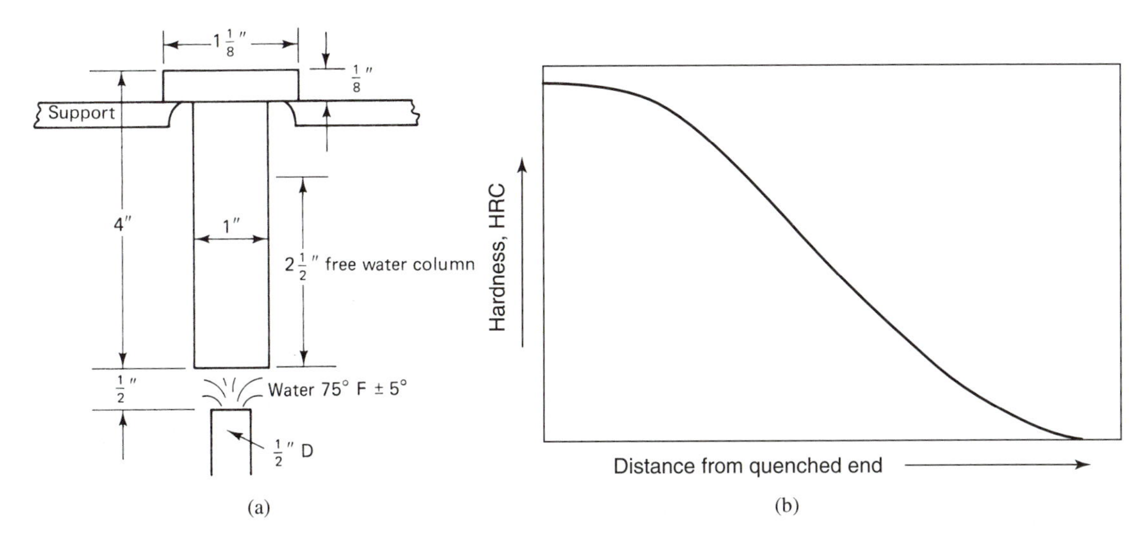

Figure 5-42 (a) Jominy end-quench hardenability test. (b) Hardenability plot of hardness (Rockwell C-scale) versus distance from quenched end of specimen.

5.9 SURFACE HARDENING AND SURFACE MODIFICATION

In all our discussions about hardening and strengthening metal alloys, our objective was to obtain the same hardness and strength throughout the part that we were heat treating. It would be impractical, if not impossible in many instances, to harden the entire part, that is, the interior and the exterior, to that hardness demanded by wear-resistant surfaces. Even if we could attain this degree of hardness throughout, the part would have little *toughness* (the ability to withstand impact loads). Little application is found in industry for such a part.

Most steel parts require machining, and this operation is best performed when the part is not hard. Once machining is completed and the part is within the required tolerances, it may be hardened on the surface to provide a strong, hard, wear-resistant surface. Figure 5-43(a) is a case-hardened bar. Figure 5-43(b) shows a gear with case-hardened teeth.

5.9.1 Case Hardening

Hardening of the surface only is known as ***case hardening,*** and the surface layers that are actually hardened make up the *case*. The inner surfaces that are not hardened are known as the *core* (Figure 5-43). The first three mentioned case-hardening methods are referred to as chemically modifying methods or diffusion-coating methods. They provide the needed elements that are lacking in the surface layers to permit the hardening reactions to take place. We have learned that carbon is an essential element in steel if we are to harden it (about 0.3% carbon in steel is the minimum for hardening steel). Many hot-working operations result in a removal of the carbon from the surface. This is known as ***decarburization*** [Figure 5-44(a)].

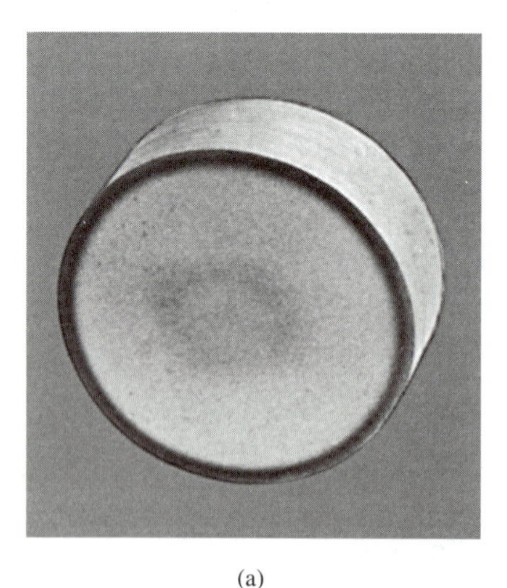

(a)

(b)

Figure 5-43 Polished section of 8620 steel bar carburized to a depth of 0.060 in., measured to a 0.40 percent carbon content. (Republic Steel Corporation.)

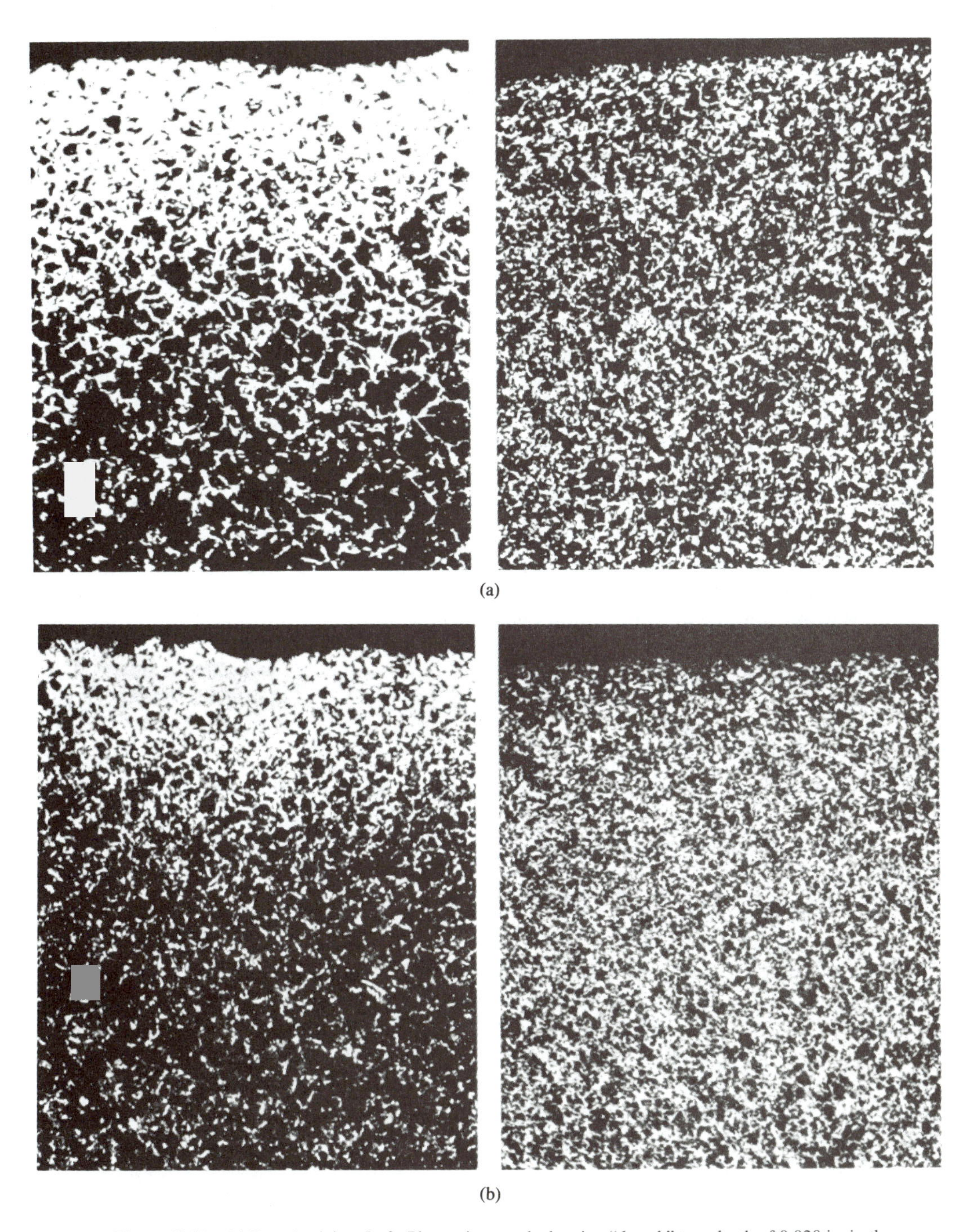

Figure 5-44 (a) Decarburizing. Left: Photomicrograph showing "decarb" to a depth of 0.020 in. in the surface of a hot-rolled bar of 1050—magnified 100 times. Right: Same bar after carbon correction annealing. Note the restoration of carbon on the outer surface and also the refinement of grain structure. (b) Carbon correction. Left: Photomicrograph of AISI 5046 before carbon correction—magnified 100 times. Right: Same bar after carbon correction annealing with surface carbon restored. (Republic Steel Corporation.)

5.9.2 Carburizing

To harden the steel, the carbon content must be restored. As the carbon content in steel increases, the heat treater's ability to create greater hardness also increases. The purposes of carburizing are to restore lost carbon and/or to increase the amount of carbon in the outer layers of the metal part. Basically, the metal part is surrounded by a high-carbon-content solid material or gas; by supplying the necessary energy through high furnace temperatures, the carbon atoms diffuse from this high-carbon-content medium into the crystal structure of the metal part to be case hardened.

After carburization, the steel part is ready for nonequilibrium thermal processing, which transforms the high-carbon-content austenite into martensite. This processing is followed by various quenches and tempers, depending on the degree and depth of hardness as well as the degree of grain refinement desired in both the case and the core [Figure 5-44(b)].

5.9.3 Nitriding

In our study of the various phase changes produced by heat treating steel, we learned that some iron reacts with carbon to produce a very hard, brittle compound, iron carbide (FeC), that precipitates out during a phase change, giving the steel hardness and strength. Aluminum alloyed with copper forms another compound ($CuAl_2$), with similar results. In addition to Ti, Al, and Fe, there are several other elements, such as chromium (Cr) and molybdenum (Mo), that react with nitrogen (N) in steel to produce compounds called ***nitrides*** that are also hard and brittle like carbides. The purpose of nitriding is to provide the necessary nitrogen in the steel so that these hard nitrides may form, producing the necessary hardness. The rich nitrogen medium may be ammonia gas. As with carburizing, heat is called on to provide the energy for the nitrogen atoms to diffuse into the steel case. The heat also helps produce the necessary chemical changes, and therefore additional heat treating is not required.

5.9.4 Cyaniding and Carbonitriding

The methods discussed above also provide cases that are rich in carbon and nitrogen. ***Cyaniding,*** an old process, might be considered as providing both carbon and nitrogen. Cyanide (CN) decomposes in a liquid bath of molten salts containing sodium cyanide, and the nitrogen reacts with iron to form hard iron nitride. The carbon diffuses to assist in further hardening operations, as discussed previously. ***Carbonitriding*** is very similar to gas carburizing because it uses a medium consisting of a mixture of gases rich in carbon and also nitrogen. The presence of nitrogen in the austenite also permits slower transformation to martensite, and therefore lower temperatures are required, followed by less drastic quenching. These conditions result in less distortion and less danger of cracking.

5.9.5 Flame Hardening and Induction Hardening

Flame hardening and induction hardening do not change the chemical makeup of the steel case. Through the application of heat, the outer layers of the metal part are first transformed into austenite, followed by quenching to transform the austenite into martensite. The steel to be case hardened must contain all the hardening ingredients. Any decarburized surface would have to be removed by machining or grinding prior to induction or flame hardening. ***Flame hardening*** uses oxyacetylene torches as sources of heat, as shown in Figure 5-45. ***Induction hardening*** uses heat generated by the resistance of the metal to the passage of high-frequency-induced electrical currents.

Surface-hardened steel products are truly ***composite materials*** containing two or more integrated components. Reinforced concrete is a composite material whose integrated components are steel rods strong in tension and a concrete matrix strong in compression. They work together in an integrated system to provide strength properties not obtainable by either one acting alone. With surface-hardened steel parts, the integrated components are the hard case and the tough core, together providing superior performance not obtainable by their use alone.

Even steels that are through-hardened, such as fine-pearlite or tempered-martensite steels, are strong and relatively hard as a result of the distribution of fine particles of a second harder phase throughout a soft ferrite matrix (see dispersion hardening). The relatively soft matrix material, the solvent component of an aluminum–copper alloy, though deformable, cannot deform independently of its harder phase. The point is that there is a striking similarity among all human-made materials—steel, aluminum alloys, prestressed or reinforced concrete, fiber-reinforced plastics, or surface-hardened metal products. They all contain integrated components working together to provide the right mix of material properties needed for the almost endless uses in today's highly technological society. We provide further details on composite materials in Module 8.

5.9.6 Other Methods of Surface Modification (Surface Engineering)

Throughout this book we describe means of modifying or engineering surfaces both for tribological reasons, where interfacing surfaces may cause friction, wear, or erosion, and cor-

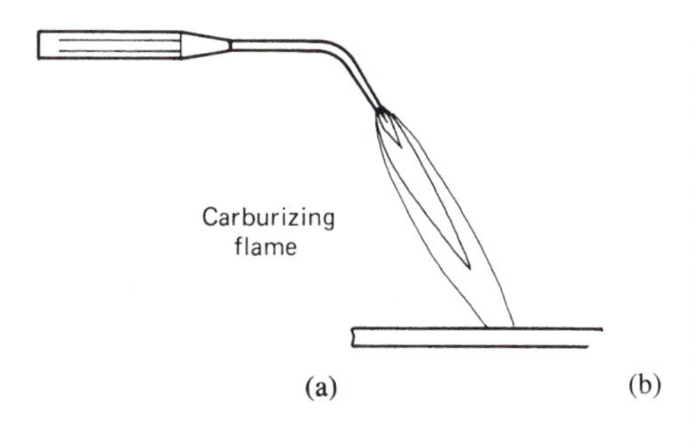

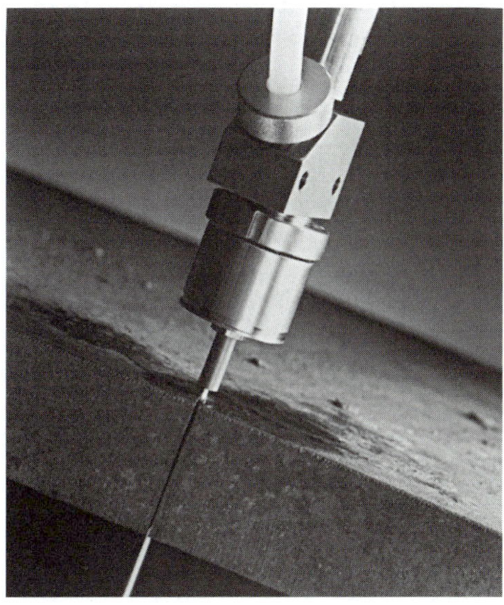

Figure 5-45 (a) Flame hardening. (b) Water jet cutting allows thick metals to be cut without the thermal effects produced by flame cutting. This Waternife is cutting three-inch steel plate. (Flow Systems.)

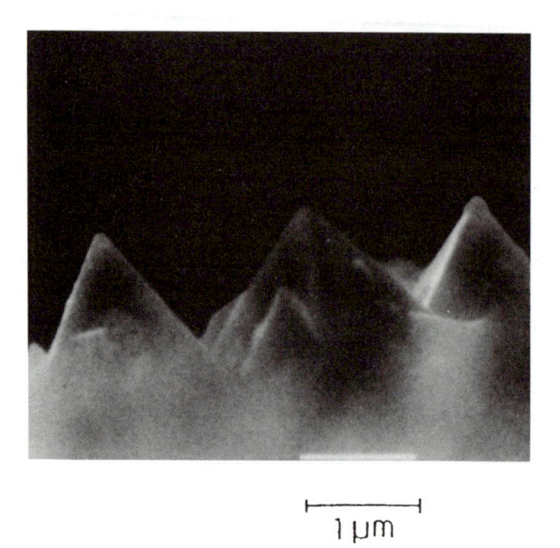

Figure 5-46 SEM photomicrograph of crystal-oriented Pb-8Sn-2Cu. "Side view" of deposit shows the pyramidal surface morphology that enhances wettability (defined in Section 6.6.1) by motor oil, which results in improved bearing performance. *(Advanced Materials & Processes.)*

rosion prevention. In Section 3.3 you learned about material characterization and may now wish to look back at Figure 3-17 to reexamine surface characteristic features. In the ceramics module you will read about additional surface engineering techniques such as diamond film and titanium coatings for cutting tools and other high friction/abrasion interfaces situations. Similarly, discussions in the polymer and composites modules deal with surface modifications to improve tribological characteristics and aspects of components.

The automotive industries' never-ending quest to improve engine efficiencies continues to modify materials and materials surfaces. One such example involves the improved tribological aspects of sliding, plain bearings for automotive crankshafts. The modification involves a lead-alloy electroplated overlay. The goal of the improvement was to produce an overlay surface that would best retain lubrication. Thus, an optimum surface is produced by controlling the crystallography of the electroplated alloy deposit. The solution was a designed surface morphology produced by controlling the orientation of the deposits with highly oriented crystal texture in primarily the [200] and [400] planes (refer back to Section 3.4.3). As seen in Figure 5-46, the scanning electron microscope (SEM) photomicrograph reveals a surface of the overlay made up of relatively uniform, pyramid-shaped crystals approximately 1.5 μm across, having bases approximately 2 μm across.

SELF-ASSESSMENT

5-1. In the section Pause & Ponder, we discussed applications of metals on trucks. List three metals that are used in trucks and name their uses.

5-2. Why are metals used so often as engineering materials?

5-3. Name four basic methods for changing a metal's properties.

5-4. What other name does the term *work hardening* go by?

5-5. Using the variable time, describe what the term *equilibrium conditions* means for phase changes in an allotropic metal.

5-6. Describe the differences between the words *eutectic* and *eutectoid.*

5-7. Name four metals, with an application for each, that are out of normal view.

5-8. Name the four methods used to change metal microstructure to achieve different properties.

5-9. Name at least four cold-working operations.

5-10. Name two other terms used to describe permanent deformation of a material.

5-11. Which crystal structure for metals has the closest packed plane of atoms?

5-12. How many types of space lattice does a solid solution contain?

5-13. Referring to Figure 5-9, with two degrees of freedom, explain what happens if the temperature is changed very slightly for liquid water existing at point *A*.

5-14. What is the other name for the inverse-ratio law used in phase diagrams?

5-15. Identify the process that takes place when a solid metal is reheated to allow the atoms to diffuse or move toward equilibrium.

5-16. Copper–zinc phase diagrams contain how many intermediate solid solutions?

5-17. What is the name of the chemical calculation that allows one to determine the quantities of chemical elements and/or compounds involved in a chemical reaction?

5-18. Cite one example of a multiphase metal alloy in which the elements exist in two separate phases.

5-19. When referring to nonferrous alloys, what does the term *annealing* signify?

5-20. What is the usual consequence of heating steel to too high a temperature?

5-21. What heat-treating process uses cooling in still air?

5-22. What is the main difference between precipitation strengthening and thermal processing under nonequilibrium conditions that produces martensite?

5-23. Describe what *high hardenability* means.

5-24. Using Figure 5-20 and an alloy of about 60% *A* in a liquid phase, what solid is formed on cooling when this alloy reaches the liquidus line?

5-25. A carbon–steel alloy containing 0.20% carbon when cooled from the austenite region crosses the A_3 line in Figure 5-25. What phase begins to form at the austenite grain boundaries at this point?

5-26. Referring to Figure 5-29, determine approximately the temperature to which you would heat a hypereutectoid steel to anneal it fully.

5-27. Name two annealing processes that do not involve phase changes.

5-28. Describe how to reduce the grain size in a piece of metal.

5-29. When grinding metal for sharpening purposes, operators are cautioned not to grind the metal too long; otherwise, the temper in the metal will be drawn. What is meant by this warning, and how do you prevent losing the metal's temper in this situation?

5-30. Figure 5-36 is a diagram of the transformation of austenite in eutectoid steel. How would you redraw this diagram if the steel were a hypereutectoid steel?

5-31. Hardness and hardenability are related properties. Discuss their differences.

5-32. If a metal part needs to be hard and at the same time possess a high degree of toughness, how are these two properties acquired through thermal treatment?

OBJECTIVE QUESTIONS

5-1. Adding elements to replace base metal atoms in a solid solution.

a. Interstitial **b.** Elutriation **c.** Addition **d.** Substitutional

5-2. The ability of a metal to exist in more than one space lattice.

a. Isotopic **b.** Amorphous **c.** Allotropic **d.** Equilibrium

5-3. Heat treatment that draws out some (but not all) hardness to improve toughness.

a. Tempering **b.** Spheroidizing **c.** Case hardening **d.** Annealing

5-4. Heat treatment that relieves stress and produces softer and more ductile metal.

a. Tempering **b.** Spheroidizing **c.** Case hardening **d.** Annealing

5-5. Heat treatment that produces a very hard shell but a tough inner core.

a. Tempering **b.** Spheroidizing **c.** Case hardening **d.** Annealing

5-6. The measure of depth to which hardness can be obtained in a piece of steel.

a. Hardenability **b.** Hardness **c.** Toughness **d.** Ductility

5-7. Steel that is work hardened.

a. Hot rolled **b.** Forged **c.** Cold rolled **d.** Plated

5-8. The type of heat treatment process in which the metal is rapidly cooled.

a. Annealing **b.** Equilibrium **c.** Nonequilibrium **d.** Allotropism

5-9. Grain structure that provides good creep strength and toughness at high temperatures.

a. Fine **b.** Large **c.** Polymorphic **d.** Serrated

5-10. Regions within a metal that are high energy and form as a result of metal cooling into a polycrystalline solid.

a. Boundaries **b.** Lattice **c.** Void **d.** Interstices

5-11. Homogeneous phase with all solute atoms dissolved into the solvent.

a. Multiphase **b.** Uniform **c.** Single phase **d.** Substitutional

5-12. What is Fe_3C?

a. Cementite **b.** Pearlite **c.** Ferrite **d.** Austenite

5-13. Permanent change in a metal as a result of force.

a. Elastic flow **b.** Elutriation **c.** Tempering **d.** Plastic flow

5-14. Metals or alloying elements that are insoluble in each other.

a. Amorphous **b.** Immiscible **c.** Morphologic **d.** Diffused

5-15. A diagram used to determine quenching points for nonequilibrium thermal processing.

a. TTT **b.** Alpha **c.** Gamma **d.** Jominy

5-16. Metal designation system used to indicate the alloy and percentage of alloy in steel.

a. AA **b.** ASTM D2000 **c.** AMS **d.** AISI-SAE

5-17. A scale that expresses temperature in absolute units.

a. Celsius **b.** Kelvin **c.** Fahrenheit **d.** Centigrade

5-18. An element with a radius of less than 0.1 nm.

a. Mn **b.** Mg **c.** C **d.** Fe

5-19. A line of maximum solubility in a phase diagram.

a. Tie-line **b.** Solvus line **c.** Isotherm

5-20. A complex multiphase mixture of different phases, some of which are solid solutions.

a. MMC **b.** Steel **c.** Wrought iron

5-21. A two-phase mixture of ferrite and cementite.

a. Austenite **b.** Ledeburite **c.** Pearlite

5-22. Annealing that is intended for the sole purpose of relieving stresses.

a. Normalizing **b.** Full anneal **c.** Stress relief **d.** Tempering

5-23. Spheroidizing of steels with a carbon content.

a. Less than 0.1% **b.** 0.3% to 0.6% **c.** 0.6% to 1.0%

5-24. A eutectoid steel transformed from austenite to fine pearlite at about 750°F.

a. HRC 31 **b.** HRC 44 **c.** HRC 58

Top: New MacGold steel bars are precision hot-rolled by a reducing mill installed within the continuous casting and mill production line. Bottom: The mill line includes a three-strand continuous caster. (MacSteel—Fort Smith, Arkansas.)

5.10 FERROUS METALS

Iron and its many alloys, including cast irons and a nearly limitless variety of steels, comprise the ferrous metals group. Even with the wide acceptance of aluminum and polymeric materials, the iron-based alloys dominate all other materials in the weight consumed annually for manufactured products. Ten times more iron (mainly in the form of steel) is used than all other metals combined. Figure 5-47 and 5-48 show the processes that iron ore, coal, and limestone undergo in the production of iron. Coke is made from coal; many other products also come from this source.

5.10.1 Cast Iron

As shown in the iron–iron carbon equilibrium diagram (Figure 5-25), cast iron has between 2% and 4% carbon, compared with less than 2% for steel. Other elements in cast iron are silicon and manganese, plus special alloying elements for special cast irons. Many cast-iron products are used as they are cast, but others require changes in properties, which are achieved through heat treatment of the cast parts.

5.10.1.1 Gray cast iron. Gray cast iron is a supersaturated solution of carbon existing in a ***pearlite*** (two-phase structure) matrix. This carbon is mostly in the form of ***graphite*** flakes (soft form of carbon known as ***elemented carbon***). It is the familiar metal formerly used as the engine block of automobiles and for other internal combustion engines. Figure 5-49 shows photomicrographs of two "as-cast" gray cast-iron specimens, one of low strength and one of medium strength. The amount of carbon, 3.2%, exceeds the solubility limits of iron, and the carbon precipitates out of solution with ferrite (carbon precipitates out in graphite form). Inoculation with small amounts of iron–silicon alloys helps produce finer graphite flakes, thus improving strength. The graphite promotes machinability and lubricity of this metal. The damping ability of this alloy provides excellent absorption of vibrations and noise, which leads to its selection as piano sound-board frames, engine blocks, and machine tool bases. These combined properties have not only made gray cast iron a popular gearing material, it is also the least expensive of all metallic materials.

The ASTM system of designation for gray iron places it into classes 20 to 60 based on the minimum tensile strength for each class. For example, class 30 would have a minimum tensile strength of 30,000 psi (207 MPa), while a class 60 gray iron would be 60,000 psi (414 MPa). This classification is often preceded by "48" (ASTM A48 class 40), which designates the specification used to determine the mechanical properties of representative samples. Brinell hardness numbers (HB) range from 160 to 200 for ASTM 48 class 20, to 212 to 248 for ASTM 48 class 60.

Machinability is a term used to attempt to describe the ease of machining steel to the size, shape, and surface finish required commercially. From this term comes ***machinability rating,*** with cold-drawn B1112 grade steel rated at 100%. This standard test for B1112 machining requires turning with a suitable cutting fluid at 180 ft per min under normal conditions. All other steels are rated above or below this 100% level. Such ratings have proved to be undependable primarily due to unintended variations in the chemical content

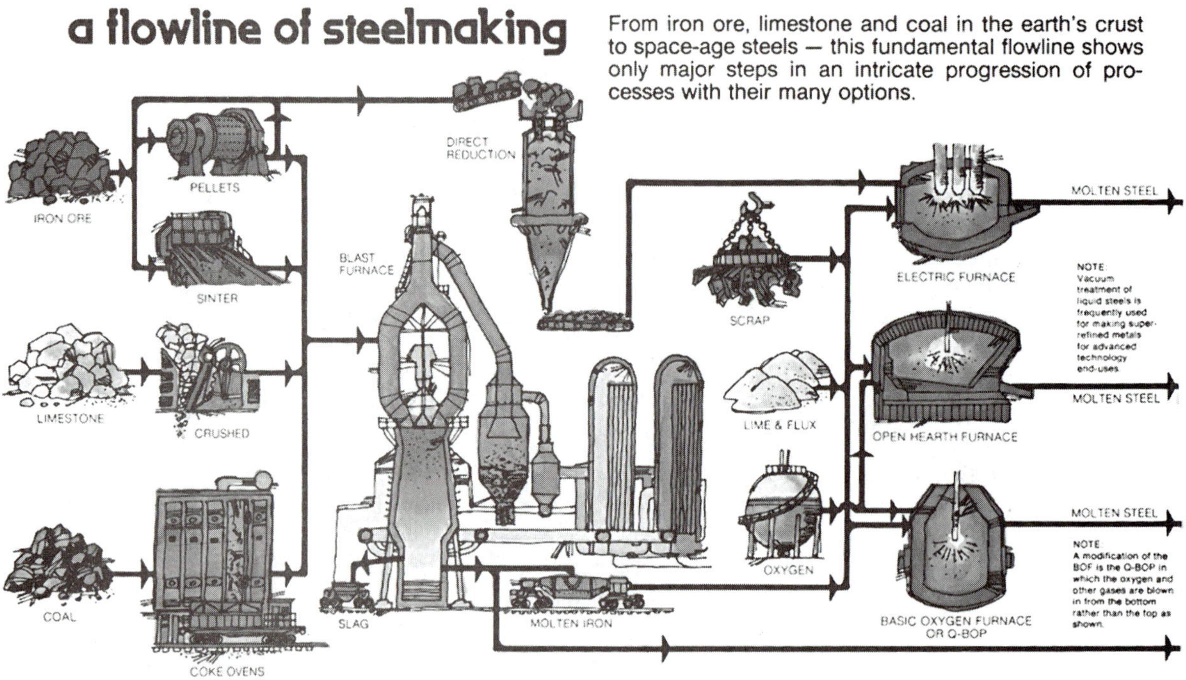

Figure 5-47 A flowline of steel making. (American Iron and Steel Institute.)

Figure 5-48 Tapping a heat of steel from a state-of-the-art electric furnace. (MacSteel.)

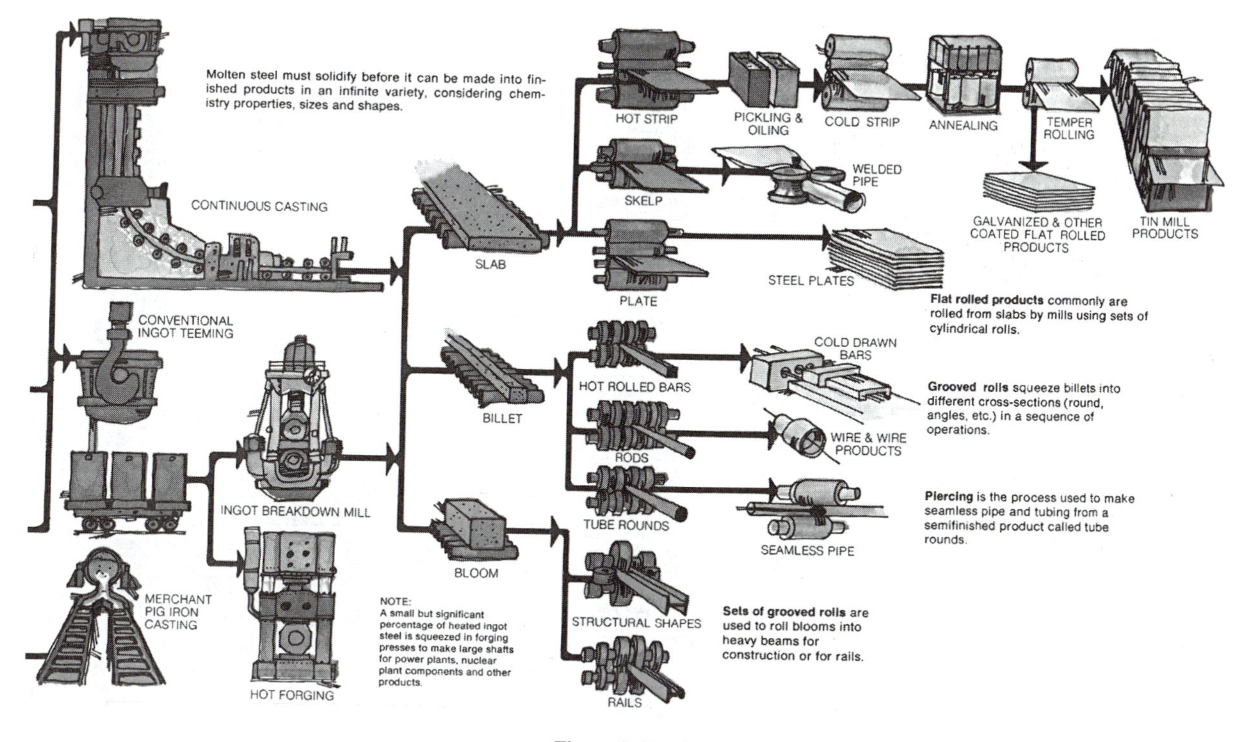

Figure 5-47 (continued)

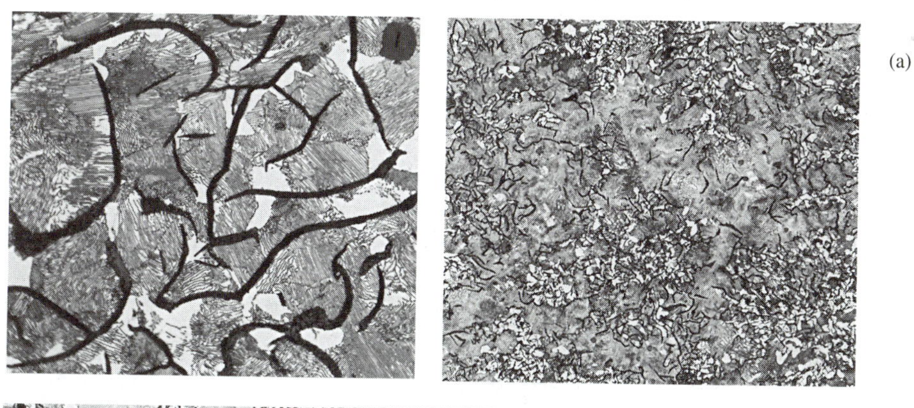

(a)

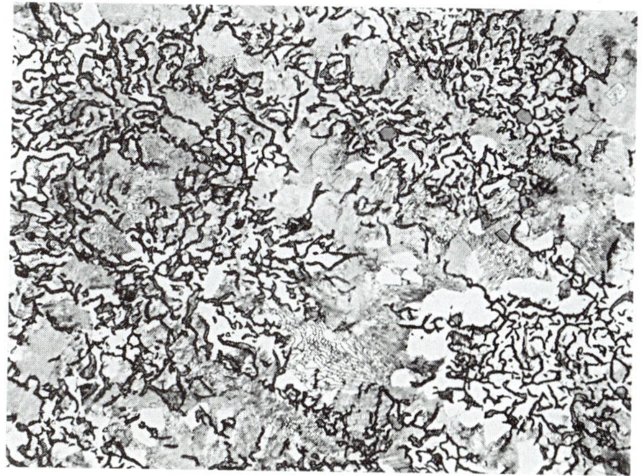

(b)

Figure 5-49 Gray cast iron. (a) The low-magnification photomicrograph at left illustrates the graphite distribution, type, and size. At 1000× (at right), the pearlite colonies and small ferrite grains adjacent to the graphite flakes are clearly distinguishable. The large, round gray particles are manganese–sulfide inclusions. The surface of the specimen contains products of transformation of a faster cooling rate. (b) The distribution of fine graphite flakes in this sample results in an increase in strength of the iron. (Buehler Ltd.)

of steel. Further, it must be pointed out that there is a poor correlation between the hardness of steels and their machinability ratings. Gray cast iron is also available alloyed with nickel, chromium, and molybdenum to improve resistance to wear, corrosion, and heat while improving strength. Flame and induction hardening allow for increased surface hardness with a slightly tougher core.

5.10.1.2 White cast iron. Through slow cooling in sand molds, chilling of specific portions of a casting, and alloying, graphitic carbon is stopped from precipitating out of solution with the ferrite to produce a white cast iron. Most of the carbon exists as cementite instead of graphite, as is the case with gray cast iron. This is due to the low silicon content (less than 1%). The name ***white iron*** comes from the white color produced in the fracture surface of the alloys. Figure 5-50 shows a photomicrograph of white cast iron to contrast with the photomicrograph of gray cast iron in Figure 5-49, which reveals graphite flakes. The carbon composition of 3.5% for unalloyed white iron has 0.5% silicon. The structure is an interstitial compound of carbon and iron known as ***cementite,*** plus a layered two-phase solution of ferrite and cementite known as ***pearlite.*** The castings are very brittle, with Brinell hardness values from over 444 to 712. White cast iron has very good compressive strength, above 200,000 psi (1380 MPa), with tensile strength around 20,000 psi, and good wear resistance; it finds applications as rolls for steel making, stone and ore crushing mills, and brickmaking equipment. Heat treatment can reduce brittleness, and as with gray iron, other properties are possible with alloying, but both of these add cost to the castings.

5.10.1.3 Nodular or ductile cast iron. With 3.5% C and 2.5% Si, the addition of small amounts of magnesium (Mg), sodium (Na), cerium (Ce), calcium (Ca), lithium (Li), or other elements to molten iron will cause tiny balls, or ***spherulites,*** of graphite to precipitate out. As the name implies, the spherulitic structure improves the elongation or ductility while yielding superior tensile strength (150,000 psi, or 1034 MPa) and machinability (similar to gray iron).

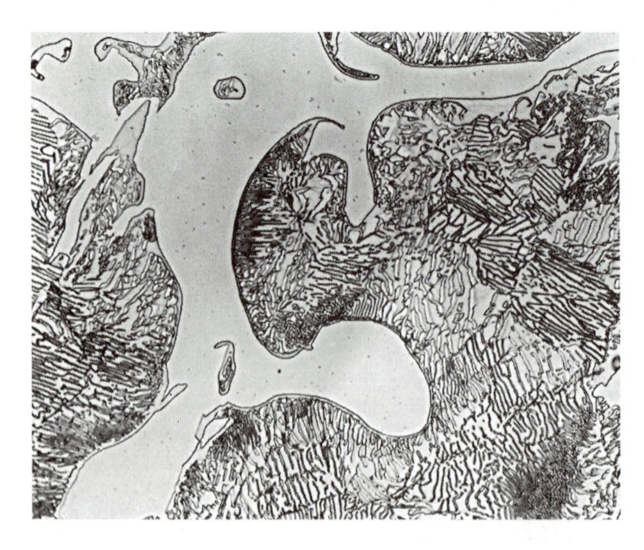

Figure 5-50 White cast iron. This specimen shows a hypereutectoid structure of pearlite and massive cementite. The dark areas are pearlite colonies surrounded by a network of cementite. At higher magnification, the alternate lamellae of alpha ferrite and Fe_3C are clearly resolved. (Buehler Ltd.)

ASTM specifications for nodular iron indicate minimum tensile strength, minimum yield strength, and minimum percentage of elongation in 2 inches. For example, ASTM 120-90-02 would have a minimum tensile strength of 120,000 psi (828 MPa), minimum yield strength of 90,000 psi (621 MPa), and 2% minimum elongation in 2 inches (50.8 mm). In internal combustion engines, nodular iron finds application as crankshafts, rocker arms, and pistons. It is also used for cast gears, pumps, and ship propellers, and with caustic-handling equipment. It is more expensive than gray iron in terms of weight, but its specific strength makes ductile iron more economical. However, it lacks the damping ability and thermal conductivity of gray iron.

5.10.1.4 Malleable iron.

The annealing of white iron castings causes nodules (large flakes) of soft graphitic carbon to form through the breakdown of hard and brittle cementite (Fe_3C). Two basic types of malleable iron are possible by varying the heat-treatment cycle. ***Pearlitic malleable*** iron is strong and hard, whereas ***ferritic malleable*** iron is softer, more ductile, and easier to machine. Malleable iron has 2.2% carbon and 1% silicon. In pearlitic malleable iron, 0.3% to 0.9% of the carbon is combined as cementite and allows for selective hardening of portions of a casting.

According to ASTM specifications A47-52 and A197-47, three grades are available: 35018, 32510, and cupola malleable iron. The 35018 and 32510 grades are ferritic, with the latter lower in silicon and consequently more ductile. Cupola malleable iron has a higher carbon and lower silicon content than the other grades, which yields lower strength and ductility. Basic properties for the three grades are shown in Table 5-1.

Applications of the ferritic grades include machined parts (120% machinability rating), automotive power trains, and hand tools such as pipe wrenches that take hard beatings. Applications for the stronger and harder pearlitic malleable iron include parts that require high surface hardness (up to HRC 60 or HB 163-269), such as bearing surfaces on automobiles, trucks, and heavy machinery.

Table 10-11 provides a comparison of properties of cast irons and wrought iron. As shown in the photomicrograph in Figure 5-51, ***wrought iron*** is an iron of high purity (less than 0.001 part carbon) with the slag (iron silicate) rolled or wrought into it. The ferrite matrix encloses iron silicate fibers shaped in the direction of rolling, which makes it an easy material to form. It is not a common metal today, but before the development of cast iron and steel making, a cruder form of wrought iron served as weapons, tools, and architectural shapes. The Eiffel Tower in Paris was constructed of wrought iron in 1872.

TABLE 5-1 PROPERTIES OF THREE CAST IRON GRADES

Grade	Minimum tensile strength [psi (MPa)]	Minimum yield strength [psi (MPa)]	Minimum elongation [% in 2 in. (50.8 mm)]
35018	53,000 (365)	35,000 (241)	18
32510	50,000 (345)	32,500 (224)	10
Cupola	40,000 (276)	30,000 (206)	5

Figure 5-51 Wrought iron. The matrix of this specimen consists of ferrite grains similar to that of ingot iron. The elongated stringers are inclusions of slag composed largely of FeO and SiO_2. At higher magnification, small, dark particles within the ferrite grains are visible. These are finely dispersed impurities, apparent only after etching. (Buehler Ltd.)

5.10.2 Steel

As the most widely used engineering material, steel is available in an almost limitless variety. Several groups can be used, such as cast steel and wrought steel. Wrought steel covers the largest group and is the steel most common to consumers. Steel is cast into ingots when it comes from steel-making processes such as the open-hearth furnace or basic oxygen furnace. These ingots are processed further while in the hot "plastic" state to produce a variety of ***wrought*** or ***hot-rolled steel*** (HRS) products, such as bars, angles, sheet, or plate. Further working of HRS sheet or bar stock at below the recrystallization temperature of the steel is known as ***cold working*** or ***cold finishing. Cold-rolled steel*** (CRS) is a harder steel because its grains have work hardened. Further classifications of steel are the carbon steels and alloyed steels.

The classification of steels takes a variety of forms. A very common system developed by both the Society of Automotive Engineers (SAE) and the American Iron and Steel Institute (AISI) uses four or five digits and certain prefix and suffix letters to cover many steels and steel alloys. As shown in Figure 5-52, the numbers reveal the major alloying element, its approximate percentage, and the approximate amount of carbon in *hundreds of 1 percent,* commonly called ***points of carbon.*** Table 5-2 shows the major groupings of carbon and alloy steels under the SAE–AISI classification. The alloy and carbon contents given in this four- or five-digit system are approximations. Complete specifications are available from SAE, AISI, handbooks such as *Machinery's Handbook,* and Table 10-11. For example,

TABLE 5-2 SAE–AISI SYSTEMS OF STEEL CLASSIFICATION

Digit designation[a]	Types of steel
10xx	Plain carbon
11xx	Sulfurized (free-cutting)
12xx	Phosphorized
13xx	High manganese
2xxx	Nickel alloys
30xx	Nickel (0.70%), chromium (0.70%)
31xx	Nickel (1.25%), chromium (0.60%)
32xx	Nickel (1.75%), chromium (1.00%)
33xx	Nickel (3.5%), chromium (1.50%)
34xx	Nickel (3.00%), chromium (0.80%)
30xxx	Corrosion and heat resistant
4xxx	Molybdenum
41xx	Chromium–molybdenum
43xx	Nickel–chromium–molybdenum
46xx	Nickel (1.65%), molybdenum (1.65%)
48xx	Nickel (3.25%), molybdenum (0.25%)
5xxx	Chromium alloys
6xxx	Chromium–vanadium alloys
81xx	Nickel (0.30%), chromium (0.30%), molybdenum (0.12%)
86xx	Nickel (0.30%), chromium (0.50%), molybdenum (0.20%)
87xx	Nickel (0.55%), chromium (0.50%), molybdenum (0.25%)
88xx	Nickel (0.55%), chromium (0.50%), molybdenum (0.35%)
93xx	Nickel (3.25%), chromium (1.20%), molybdenum (0.11%)
98xx	Nickel (1.10%), chromium (0.80%), molybdenum (0.25%)
9xxx	Silicon–manganese alloys

[a]x's indicate that numerals vary with the percentage of carbon in the alloy.

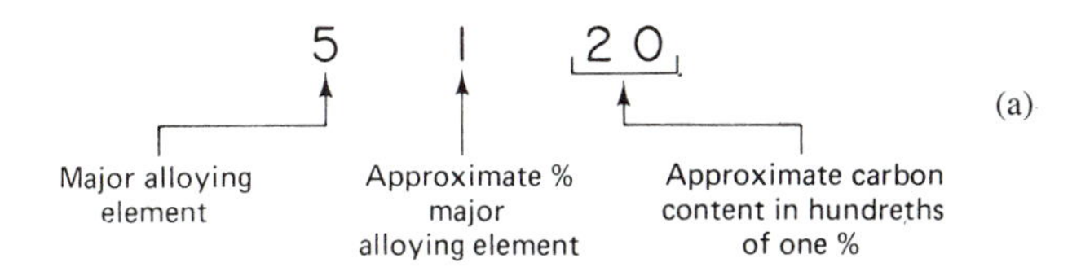

(a)

Examples:

Shown above – chromium steel alloy with about 1% chromium and 0.20% (0.002 parts or 20 points) carbon

1015 – plain carbon steel with 0.15% (0.0015 parts or 15 points) carbon

E52100 – chromium steel alloy with about 20% chromium and 1 point carbon produced in an electric arc furnace

Figure 5-52 (a) SAE–AISI steel designation.

(b)

Figure 5-52 (b) The many alloys of steel. The Boeing Commercial Airplane assembly plant has the capacity to assembly seven of the widebody 777 planes per month. This plant is the world's largest building by volume, covering 98.36 acres (39.84 hectares). Many ferrous metal alloys (mostly steels) are seen in this picture, including the skeletal structure of a variety of structural steel shapes that frame the building. Various steels are also used in the elaborate fixtures used for the assembly. These fixtures are designed by tool engineers whose responsibility is to provide tooling to assemble the many parts and components that make up this large commercial aircraft. Various steel alloys are also chosen for the many tools used in assembling and maintaining the aircraft. (The Boeing Company.)

the chromium–steel alloy 5120 has the following composition: C = 0.17 to 0.22%, Mn = 0.70 to 0.90%, Cr = 0.80 to 1.10%, P = 0.040%, and Si = 0.20 to 0.35%. Prefixes can indicate the process in making the steel, such as E for electric-arc furnace; and suffixes further clarify, for instance, H for hardenability guaranteed. Other societies, such as ASTM and ASME (American Society of Mechanical Engineers), have specifications for specialty steels, such as tool steels for dies, cutting tools, and punches or structural bolts and plates.

5.10.2.1 Carbon steel. This group of steels, also known as ***plain carbon*** and ***mild steel,*** dominates all other steels produced and is essentially iron and carbon with other elements that occur naturally in iron ore or result from processing. These elements are held to certain maximum levels: manganese (Mn), 1.65%; silicon (Si), 0.60%; and copper (Cu), 0.60%. Carbon steel may be cast or wrought. Typically, cast steels have more uniform properties since wrought steel develops *directional properties* as a result of rolling it into shape. See Table 10-11.

5.10.2.2 High-strength low-alloy (HSLA) steel. HSLA steel is a product of recent technology aimed at producing strong, lightweight steel at a price competitive with that of carbon steels. Although the price per pound of HSLA steel is greater than carbon steel, thickness is reduced due to a higher strength of 414 kPa (60,000 psi) versus 276 kPa (40,000 psi) for carbon steel; consequently, overall cost may be better for the HSLA, and significant weight savings are realized. The transportation industry, especially the automotive section, has employed HSLA steel in numerous structural applications. Not as malleable as carbon steel, sheet HSLA steel could not be used in auto bodies, but a modification resulted in a dual-phase steel acceptable for the small bending radii required on auto bodies.

5.10.2.3 Alloy steel. The classification of alloy steel is applied when one or more of the following maximum limits are exceeded: Mg, 1.65%; Si, 0.60%; Cu, 0.60%; or through the addition of specified amounts of aluminum (Al), boron (B), chromium (Cr, up to 3.99%), cobalt (Co), niobium (Nb), molybdenum (Mo), nickel (Ni), titanium (Ti), tungsten (W), vanadium (V), zirconium (Zr), or others. Alloy steels are grouped as low-, medium-, or high-alloy steel, with high-alloy steels encompassing the stainless steel group. Table 5-2 shows the SAE–AISI classification systems used for certain alloy steels. Elements added to steel can dissolve in iron to strengthen ferrites or α-iron (bcc) and form with carbon in the austenite or γ-iron phase (fcc) to produce carbides to improve hardness. High-temperature (heat resistance) alloys must maintain corrosion resistance, mechanical strength, creep resistance, stress–rupture strength, and toughness at temperatures greater than 425°C. Chromium is most important for oxidation resistance; cobalt, aluminum, silicon, and the rare earth elements [the 14 elements in the sixth row of the periodic table called the lanthanides, such as cerium (Ce), atomic number 58] also contribute to producing a stable oxide surface layer. Nickel provides strength, stability, and toughness; and tungsten and molybdenum increase high-temperature strength. The new high-temperature alloys generally contain high levels of chromium and cobalt, which are found outside the United States. Consequently, newer alloys are being developed that eliminate completely or reduce the amounts of these two elements.

Chromium is effective in increasing strength, hardness, and corrosion resistance. Copper forms in austenite to reduce rusting. Manganese is an austenite former that, much like carbon, increases hardness and strength. Vanadium forms with ferrite to improve hardness, toughness, and strength. Molybdenum combines in carbide to improve high-temperature tensile strength and high hardness. Silicon dissolves in ferrite to improve electromagnetic properties, plus toughness and ductility. Nickel is an austenite former that both improves high-temperature toughness and ductility and provides rust resistance. Aluminum is effective as a ferrite former in reducing grain size, thus giving improved mechanical properties. Cobalt dissolves in austenite to improve magnetic properties and high-temperature hardness. Tungsten dissolves in ferrite to increase both hardness and toughness at elevated temperatures. Table 10-11 provides a comparison of selected alloys.

A recent example of the advances in the technology of alloy steels is AerMet 100, a nickel-cobalt alloy steel strengthened by carbon, chromium, and molybdenum from Carpenter Technology Corp. The patented alloy, with a designation AMS (Aerospace Materials Specification) 6532, has an nominal analysis of 13.4 Co, 11.1 Ni, 3.1 Cr, 1.2 Mo, 0.23 C, with the balance Fe. This alloy steel has the highest fracture toughness of any commercially available steel. Through heat treating it can obtain 1930–2700 MPa (280–300 ksi) tensile strength and exceed a fracture toughness of 110 MPa · $\sqrt{m}$ at 1930 (100 ksi · in. at 280 ksi). Stress-corrosion cracking and fatigue resistance are two other attributes. The superb combination of high strength and hardness coupled with high fracture toughness and ductility make it a superior alloy steel for applications outside the aerospace industry, for which it was developed. Figure 5-53(a) shows one aircraft application of this steel's unique combination of properties that make it lighter and tougher and reduce the size without sacrificing strength.

5.10.2.4 Stainless steel. This group of high-alloy steels contains at least 10.5% chromium; it is more correctly called ***corrosion-resistant steel (CRES)***. The 10.5% chromium does not ensure that the steel will not rust; a sufficiently high content of carbon or other alloys may negate the passivity of the chromium (see Figure 5-53). As with other steels, stainless may be wrought or cast. Wrought stainless is grouped by its structure as ferritic, martensitic, austenitic, or precipitation hardening (PH). Cast stainless may be classified as heat resistant or corrosion resistant.

The ASTM and SAE, along with other groups, have developed the Unified Numbering System (UNS), a five-digit designation with an S prefix to replace the AISI designations for stainless steels. Stainless steels in the S30000 (AISI 300) series are nickel–chromium steels, while the S40000 (AISI 400) series have chromium as the major alloy. Series S20000 (AISI 200) are austenitic alloys, with manganese and nitrogen replacing some of the nickel—far less expensive alternatives when high formability and good machinability are not required. ***Austenitic stainless steel*** is a single-phase solid solution that has good corrosion resistance.

Martensitic stainless steels in the S40000 (AISI 400) series have high carbon content, up to 1.2%, with 12% to 18% chromium. The higher carbon content allows formation of more γ iron, which quenches to a hard martensitic (up to 100%) steel, but the high carbon content reduces some of the corrosion resistance. If an austenitic stainless steel is heated sufficiently so that carbon precipitates out of solid solution as chromium carbide (which leaves less than 12% chromium in some segments of the alloy), it promotes intergranular corrosion.

The ***ferritic stainless steels*** have low carbon (0.12% or less) content and high chromium content (14% to 27%) in a solid solution and do not harden by heat treatment. They are in the S40000 (AISI 400) series and, unlike martensitic and austenitic, are magnetic. The ferritic grades have good formability, machinability, and corrosion resistance above martensitics. Specific properties of each type of stainless steel are found in standard references on steels. Table 10-11 shows selected stainless steels and their properties.

(a)

(b)

Figure 5-53 (a) A superior alloy steel is the AMS 6532, seen used for the landing gear of the carrier-based McDonnell Douglas F/A-18 because of the alloy's high strength and stress–corrosion cracking resistance. (b) Stainless steel. Cookware of nickel–chrome stainless. (International Nickel Company.)

5.10.2.5 Other steel alloys. Beyond and including high-alloy steels and stainless steels previously discussed, there are many specialty alloys. The ASM publishes several volumes of a *Metals Handbook; Volume I, Properties and Selection of Metals,* gives in-depth coverage of most metals. Included in the steel alloys is a range of tool steels, high-yield-strength (HY) steels, magnetic and electrical steels, ultrahigh-strength steel (see Table 10-12 for properties), maraging steels, low-expansion alloys, and ferrous powdered metals (PMs).

SELF-ASSESSMENT

5-33. Carbon steel is an alloy of carbon and iron. Is stainless steel an alloy of carbon steel? Is it an alloy steel?

5-34. How does the composition of ferrite differ from pearlite?

5-35. Write the chemical symbol for cementite and attach another name to this ingredient of steel.

5-36. Wrought iron is mainly pure iron mixed with a slag composed mostly of iron silicate. Specify an excellent application for wrought iron. Could wrought iron be alloyed with another metal? If so, which metal would you choose, and why?

5-37. What are some barriers to the introduction of advanced materials?

5-38. Plain-carbon steels (i.e., low-, medium-, and high-carbon steels) contain up to a maximum of about 1% carbon. These steels provide much of the steel that goes into welded parts, bolted and riveted structures, springs, and free machining parts. Why is it necessary to develop more sophisticated specialty steels containing alloying elements other than carbon?

5-39. What crystalline phase of steel is referred to as a very fine platelike (lamellar) mixture of ferrite and cementite?

5-40. What is the maximum solubility of carbon (in percent) in the interstitial solid solution of carbon dissolved in gamma iron?

5-41. Modern steels are sometimes referred to as examples of human-made composite materials. Explain the significance of this statement.

5-42. The production of metals and the casting of metal parts take place under the influences of the earth's atmospheric pressure and gravitational force. Speculate about how the crystalline structures of metals, and hence their properties, would change if these processes could be accomplished with the absence of gravity.

5-43. What is the base metal in all ferrous metals, and what raw materials are used in extracting this metal from ore?

5-44. **(a)** What is the carbon content of cast iron? **(b)** Explain the designations ASTM 48 class 35 and 32510.

5-45. List one application, the key structural difference, and the best quality for each of the following cast irons: **(a)** ductile cast iron, **(b)** gray cast iron, **(c)** white iron, **(d)** malleable iron, **(e)** wrought iron.

5-46. **(a)** How is steel different from cast iron? **(b)** How is CRS superior to HRS?

5-47. Explain the following designations: **(a)** SAE 1020, **(b)** ASTM S 40000, **(c)** AISI 400, **(d)** SAE E 52100, **(e)** 6016 T 2, **(f)** CRES.

5-48. **(a)** What drawback does HSLA present to auto-body makers? **(b)** What steel is normally used for auto bodies?

5-49. What is the major function of the following in alloy steel: **(a)** copper, **(b)** vanadium, **(c)** chromium, **(d)** aluminum, **(e)** lead, **(f)** nickel, and **(g)** tungsten?

5-50. Describe the major advantage of each of the following stainless steels: **(a)** martensitic, **(b)** austenitic, and **(c)** ferritic.

5-51. What is the name of the 14 elements in period 6 of the periodic table that contribute to a stable oxide layer on alloy steels?

5-52. Name two physical properties needed in a naval fighter aircraft's landing gear.

5-53. From Tables 10-11 and 10-12, name the two highest hardness steel alloys.

OBJECTIVE QUESTIONS

5-25. The carbon content in a 1020 steel.

a. 20% **b.** 0.20%

c. 2% **d.** 0.20

5-26. Steel with alloying elements to resist corrosion.

a. CRS **b.** HRS

c. CRES **d.** TTT

5-27. Metal with no iron.

a. Alloy **b.** Ferrous

c. Nonferrous **d.** Magnetic

5-28. As the steel-making flowline shows, ____________ is a major source of raw material.

a. Scrap **b.** Lead

c. Aluminum **d.** Silica

5-29. Carbon content of cast iron.

a. 0.6% to 1.0% **b.** 2% to 4% **c.** 4% to 6%

5-30. Graphite in a metal promotes ____________.

a. Lubricity **b.** Malleability

c. Toughness **d.** Conductivity

5-31. Silicon added to gray cast iron helps produce ____________.

a. Strength **b.** Malleability

c. Ductility **d.** Magnetism

5-32. The least expensive of all metallic materials.

a. Carbon steel **b.** Gray cast iron

c. Wrought iron **d.** CRES

5-33. Structural steel shapes are produced from ____________.

a. Billets **b.** Slabs

c. Ingots **c.** Blooms

5-34. Another name for nodular cast iron.

a. Malleable **b.** Gray
c. Ductile **d.** Brittle

5-35. What ferrous material was used to construct large buildings prior to the development of steel in the late 1890s?

a. Cast iron **b.** Wrought iron
c. Ductile iron **d.** Stainless

5-36. Slag is found in ____________.

a. Cast iron **b.** Wrought iron
c. Low carbon steel **d.** CRES

5-37. Another name for plain carbon steel.

a. Low carbon steel **b.** Mild steel
c. Wrought steel **d.** HSLA

5-38. To compete with lighter metal and plastics the steel industry offers ____________.

a. Low carbon steel **b.** Mild steel
c. Wrought steel **d.** HSLA

5-39. Copper forms in austenite to reduce ____________.

a. Rust **b.** Spheres
c. Graphite **d.** Strength

5-40. In addition to corrosion resistance, chromium also gives ____________ to steel.

a. Magnetism **b.** Formability
c. Castability **d.** Strength

5-41. A steel alloy with nickel-cobalt and rated with the highest fracture toughness.

a. AISI-SAE 2430 **b.** AerMet 100
c. HSLA **d.** AISI 400

5-42. Metal designation system used to indicate the alloy and percentage of alloy in steel.

a. AA **b.** ASTM D2000
c. AMS **d.** AISI-SAE

5-43. SAE 4000 steel contains the major alloying element ____________.

a. Manganese **b.** Molybdenum
c. Nickel **d.** Lead

5-44. Aluminum, as an alloying element, improves the steel's ____________.

a. Magnetic properties **b.** High-temperature properties
c. Mechanical properties **d.** Processing properties

5-45. A steel with alloying elements to resist corrosion.

a. CRS **b.** HRS **c.** CRES **d.** TTT

5-46. Two major alloying elements needed for high-temperature steel that are not mined in the United States.

a. Lead and tin **b.** Nickel and silicon
c. Chromium and cobalt **d.** Iron and aluminum

5-47. Martensitic stainless steels have at least ____________ of Cr.

a. 0.8% **b.** 4%

c. 12% **d.** 21%

5-48. From Table 10-11, what is the hardness capability of type 302 stainless?

a. 85 HRB to 40 HRC **b.** 75 HRB

c. 325 HB **d.** 180 HB

5-49. From Table 10-12, what ultrahigh-strength steel would provide these properties: tensile strength, 2140 MPa; yield strength, 1650 MPa; hardness, 54.5 HRC?

a. UNS G41300 **b.** UNS K44220

c. UNS G61500 **d.** UNS T20811 Mod

777 Uses Advanced Materials for Reliable and Economic Operation

Advanced alloy applications

- Advanced alloys improve durability
- Advanced alloys save 3,200 lb of structural weight

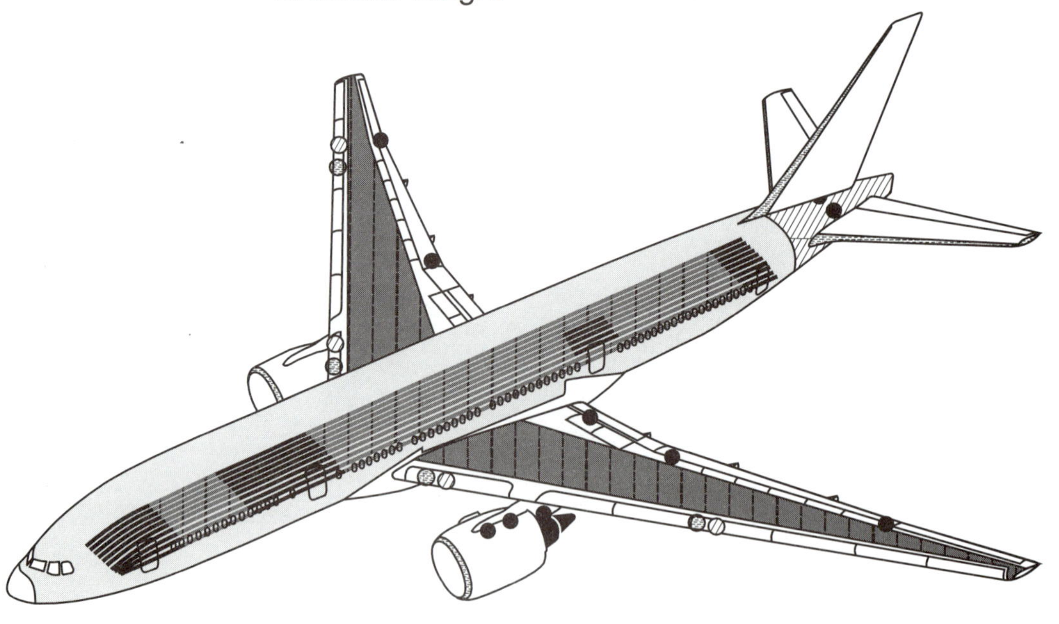

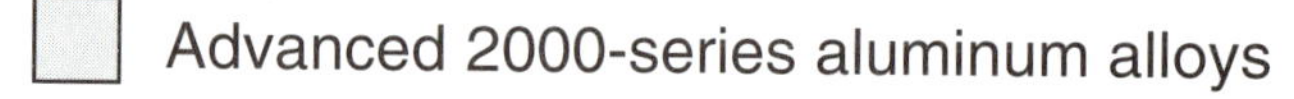

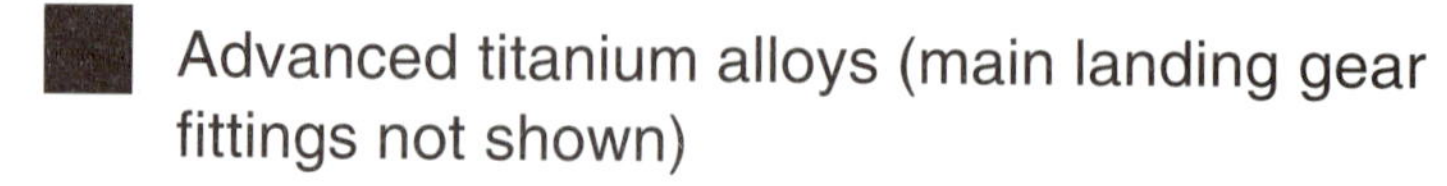

Advanced nonferrous alloys for the Boeing 777. Various aluminum and titanium alloys (shown with symbols) used on the 777 improve durability while saving on structural weight. (Courtesy of the Boeing Company.)*

*The Public Broadcasting System (PBS) series "Twenty-first Century Jet" provides an excellent in-depth case study of the product development process, from design through testing, manufacturing, and delivery. The series includes many examples of engineering materials technology. Call PBS at 1-800-344-3337 to obtain further information on the video series.

5.11 NONFERROUS METALS

Although the ferrous alloys are used much more than other metals, nonferrous metals comprise three-fourths of the known elements. However, the problem of extracting certain metals from the earth or ocean and the practical commercial value limits the number of metals used in significant quantity. The most commonly used nonferrous metals include aluminum, copper, zinc, nickel, chromium, tin, magnesium, beryllium, tungsten, lead, molybdenum, titanium, tantalum, and the noble (or precious) metals such as gold, platinum, silver, and rhodium. With demands for superior properties through advances in technology, each metal receives careful consideration for its potential value in unique circumstances. Catalytic converters used in automobile exhaust systems contain platinum to aid in the removal of pollutants in our environment. Ruthenium is an example of a metal that has little practical application alone, but as an alloy with platinum, it is valued in thin-film circuitry (ceramic substrates with printed circuits) for solid-state electronics. In addition to the properties of a material, selection for design bears heavily on the processibility of the material. Table 5-3 provides a simplified comparison of metals to common processes.

TABLE 5-3 PROCESSIBILITY OF COMMON METALS

	Steel	Iron	Aluminum	Copper	Nickel	Magnesium	Zinc	Titanium	Tin
Castability									
Centrifugal	■	■	■	■	■				
Continuous	■	■	■	■					
Investment	■		■	■	■				
Permanent mold	■	■	■	■	■	■	■	■	■
Die casting			■	■		■	■		■
Formability									
Cold	■		■	■	■	■	■	■	
Hot	■		■	■	■	■		■	
Machinability	■	■	■	■	■	■	■	■	
Powder metal Compacting	■	■	■	■	■				
Weldability									
Gas	■		■		◢			■	
Inert arc	■		■		◢	■		■	
Electrical Resistance	■		■	■	◢	■		■	

■ – Common on most alloys

◢ – Used on some alloys

Source: Adapted from *Machine Design,* March 15, 1976.

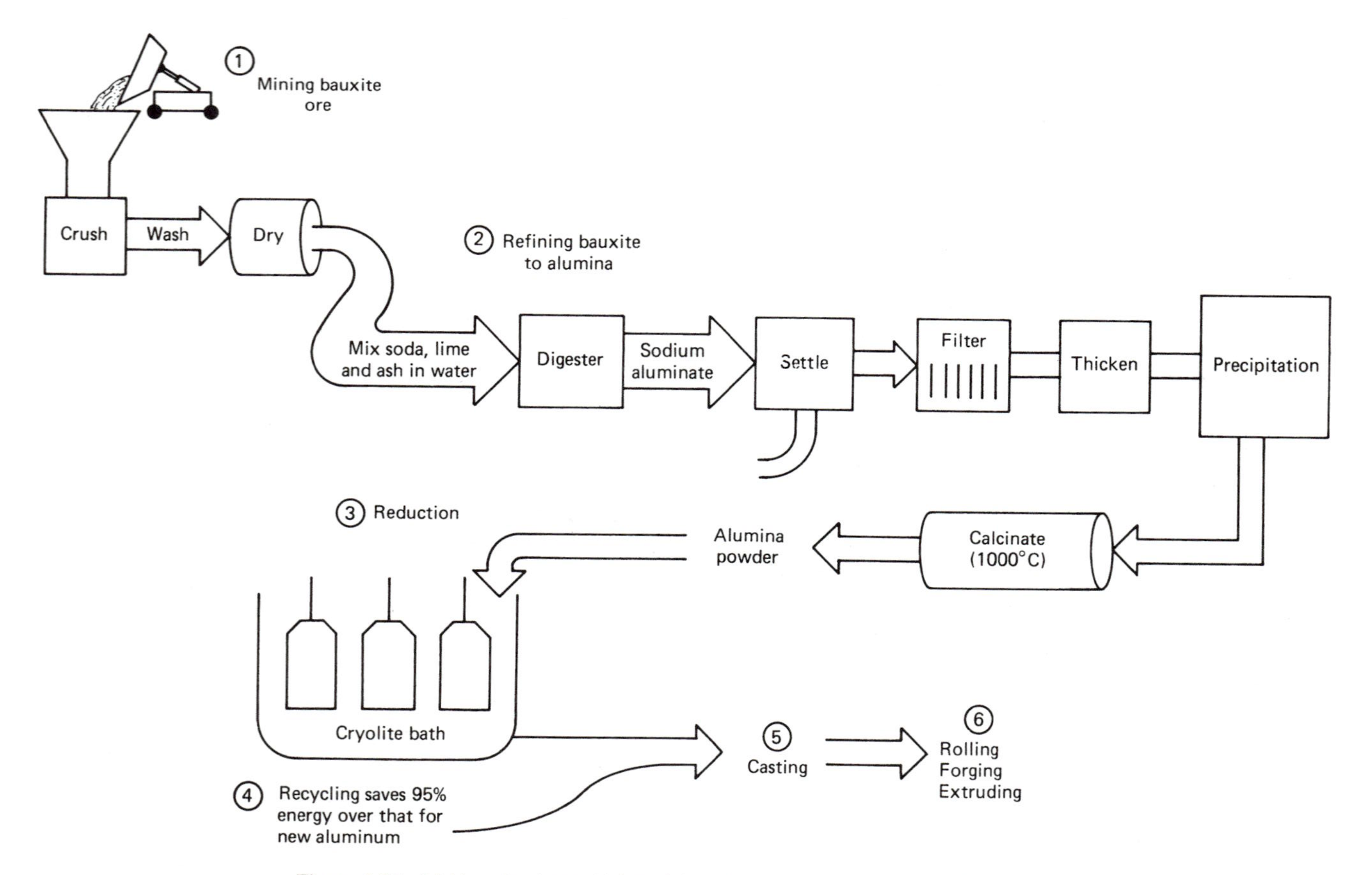

Figure 5-54 Making aluminum. (Adapted from the Aluminum Association flowcharts.)

5.11.1 Aluminum

The most abundant of metals, aluminum is locked up with other elements in the form of bauxite ore. In 1852, this metal was only rarely available in usable form and was prized by Emperor Napoleon as dinnerware; it sold for $545 a pound. The development by Charles Hall in Ohio and by Paul Heroult in Paris, France, of an electrolytic process for economical extraction of bauxite occurred simultaneously in 1886. Even though the price had dropped to 24 cents per pound by 1964, the cost of electrical energy keeps up the price of virgin aluminum. About 4% of our metal-processing energy is consumed in the production of aluminum. However, recycling of aluminum drastically reduces the cost of the metal.

Figure 5-54 shows the steps in the production of aluminum from bauxite. Aluminum has many favorable properties, including excellent thermal conductivity, good strength, corrosion resistance, and light weight (one-third that of steel). Figures 5-55 and 5-56 show products that stem from aluminum properties. Table 5-3 shows that it is processible by the major methods. Wrought aluminum and cast aluminum are designated by two different numbering systems. The Aluminum Association uses four digits to specify wrought alloy groups. The meaning of each digit is shown in Figure 5-57. Wrought and cast designations are similar; the last digit in cast aluminum indicates cast (0) or ingot (1 or 2). Temper designations follow the four digits; they indicate thermal treatment (T through T10) and solution heat-treatment (W), as-fabricated (F), annealed (0), or strain-hardened (H).

Figure 5-55 Energy-saving and low-maintenance requirements are two important reasons why aluminum building products, like the siding being applied here, are used today. (Reynolds Metal Company.)

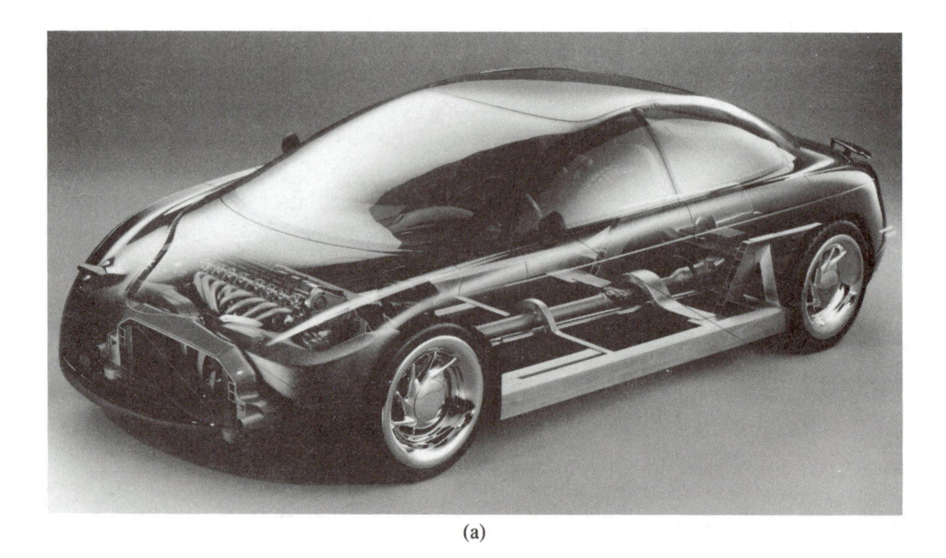

(a)

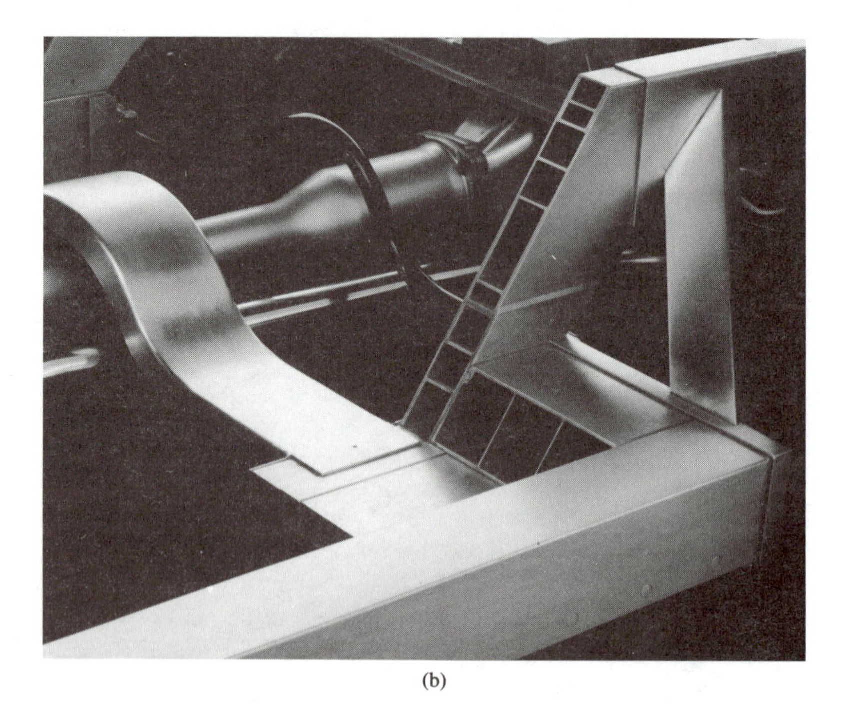

(b)

Figure 5-56 (a) The Ford Contour concept car uses an extruded-aluminum spaceframe that reduces weight by 50% over a steel frame, while providing savings in time and materials over steel stamped frames. A pound of aluminum is twice as expensive as a pound of steel, but CAFE (corporate average fuel economy) mileage standards and virtually no scrap make Al competitive as frame material. (b) Additional cost and weight savings will come from adhesive bonding of the Al extrusions, onto which composite plastics body panels will be bolted. No welding will be involved. (Ford Motor Company.)

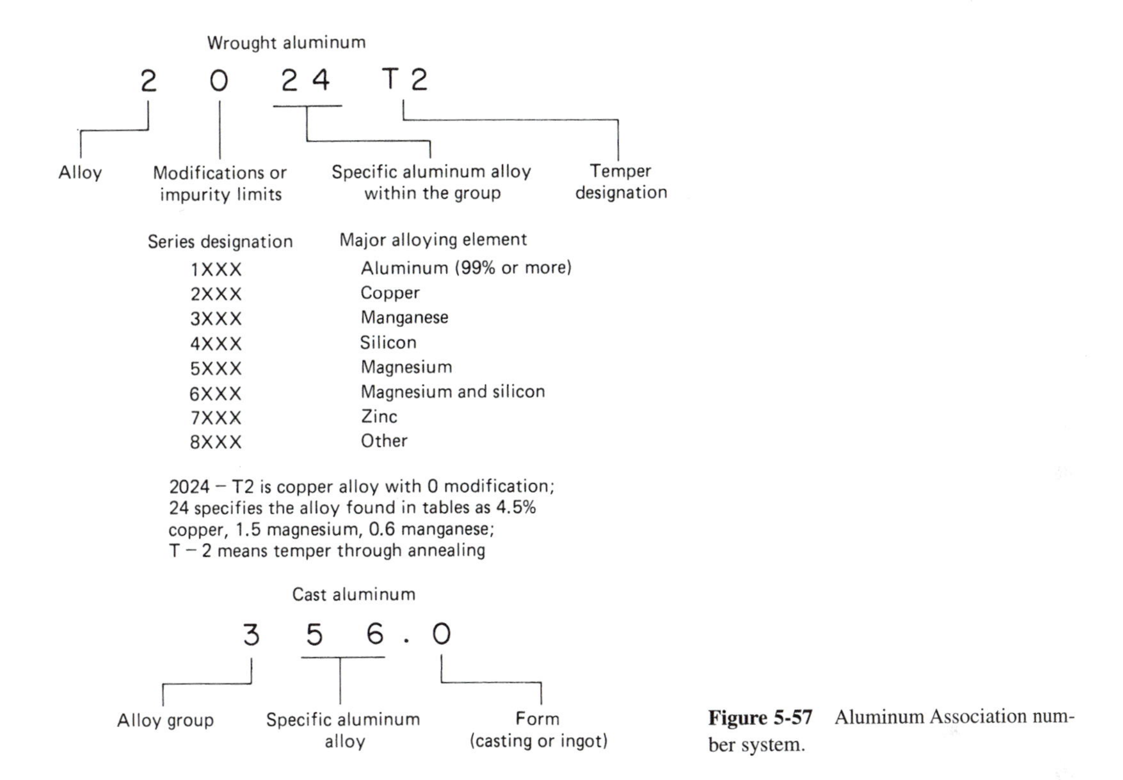

Figure 5-57 Aluminum Association number system.

Aluminum of high purity (1000 series), or at least 99% purity, finds applications as thin foils for electronics applications; as plate and bar stock for chemical apparatus, fuel tanks, and cooking utensils; and as paint pigments and in gasoline production. Aluminum alloyed with copper (2000 series) extends the tensile strength (greater than mild steel), improves machinability, and proves good specific strength but has reduced corrosion resistance and weldability. Aluminum–manganese alloys (3000 series) provide good strength, workability, formability, and weldability, and the high corrosion resistance makes this a popular alloy for applications such as highway signs, furniture, cooking utensils, architectural shapes, and truck panels.

The aluminum and silicon alloys (4000 series) have high wear resistance and relatively low coefficients of thermal expansion; they permit good castings and forgings and find applications as welding and brazing rods, anodized architectural shapes, marine equipment, and automotive pistons. Aluminum alloyed with magnesium (5000 series) yields moderate to high strength, good corrosion resistance in marine environments, and good weldability and formability for uses as extrusions, ship and boat parts, and automotive structural parts. The 6000 series alloys of aluminum, magnesium, and silicon have excellent corrosion resistance, plus good weldability, machinability, and formability; they find uses as bridge rails, automotive sheets, piping, and extrusions. The aluminum–zinc series 7000 alloy is heat treatable with tensile strength up to 606 MPa (88,000 psi) and has good corrosion resistance in rural areas but is poor in marine environments. Uses for the

7000 alloys are chiefly as aircraft structures and other equipment requiring high specific strength. The 8000 series includes alloys of combinations not included above, for example, titanium or zirconium. Table 10-11 provides comparisons of various aluminum alloys.

Corrosion resistance in aluminum is a result of the natural occurrence of an aluminum oxide film that prevents the metal from further corrosion. Through an ***anodizing*** process, it is possible to build up the oxide layer and even add color to the aluminum surface. Concerns about the toxicity of materials and their effect on the environment have led to an improved process for anodizing aluminum. The older process used chromic acid and sodium dichromate, which contained hexavalent chromium, a carcinogen. The newer process uses sulfuric acid and nickel acetate, which produces thinner coats (2 to 5 μm thick) that offer better corrosion and fatigue resistance than the older method. Aluminum oxide formed in the anodizing process is quite brittle and rough, creating a poor wear surface. Adding zirconium by ion implantation after anodizing greatly improves its wear resistance (see Section 7.10). Alloying, cold working, and heat treatment can reduce the corrosion resistance of aluminum. For example, intergranular corrosion can occur in a precipitation-hardened aluminum because of the lack of homogeneity (i.e., some areas have a high copper content and become anodic to copper-depleted areas).

5.11.1.1 Superplastic forming (SPF) and diffusion bonding (DB). When deformed in tension at elevated temperatures, some advanced aluminum alloys are also capable of achieving large, uniform elongation (>500%) at relatively low stresses prior to localized thinning and fracture. The property of superplasticity joined with diffusion bonding provides a space-age composite. A metal alloy that is not considered very ductile at or near room temperatures but at certain higher temperatures possesses extremely high ductility is said to exhibit ***superplasticity.*** Titanium is such a metal with normal plastic deformation around 20%, but at certain high temperatures it can be deformed as much as 2000%. Diffusion bonding or hot pressing is a widely used technique for metal–fiber composite fabrication. Examples of such composites are aluminum and titanium alloys reinforced with boron or Borsic fibers. The matrix is usually in sheet form, with the reinforcing filaments or wires mechanically spaced and oriented between the sheets to form alternating layers of filaments and matrix. Surface coatings such as carburizing, nitriding, and bimetallic castings are also examples of diffusion bonding, but they do not use metal filament reinforcement.

The first step in diffusion bonding is to wind the fiber over a metal-foil-covered drum. The resulting mats are cut and made into tapes by diffusion bonding under high temperature and pressure. The tapes, in turn, are cut into plies, stacked in a die, and consolidated by further diffusion bonding.

Complex structural shapes can be formed in a single operation by the large forming strains produced by SPF. Superplastic forming/diffusion bonding (SPF/DB) uses several alloys to fabricate major structural components of aircraft. Figure 5-58 depicts the SPF process. The older fabrication method [Figure 5-58(a)], using machining and chemical milling, resulted in 80% of the starting plate machined away as scrap. The newer built-up structure approach, using an SPF beaded-web stiffener that was joined to the skin material by resistance spot-welding techniques, produced only a 15% loss in scrap material while reducing the material cost for fabrication [Figure 5-58(a)]. SPF increases the efficiency of

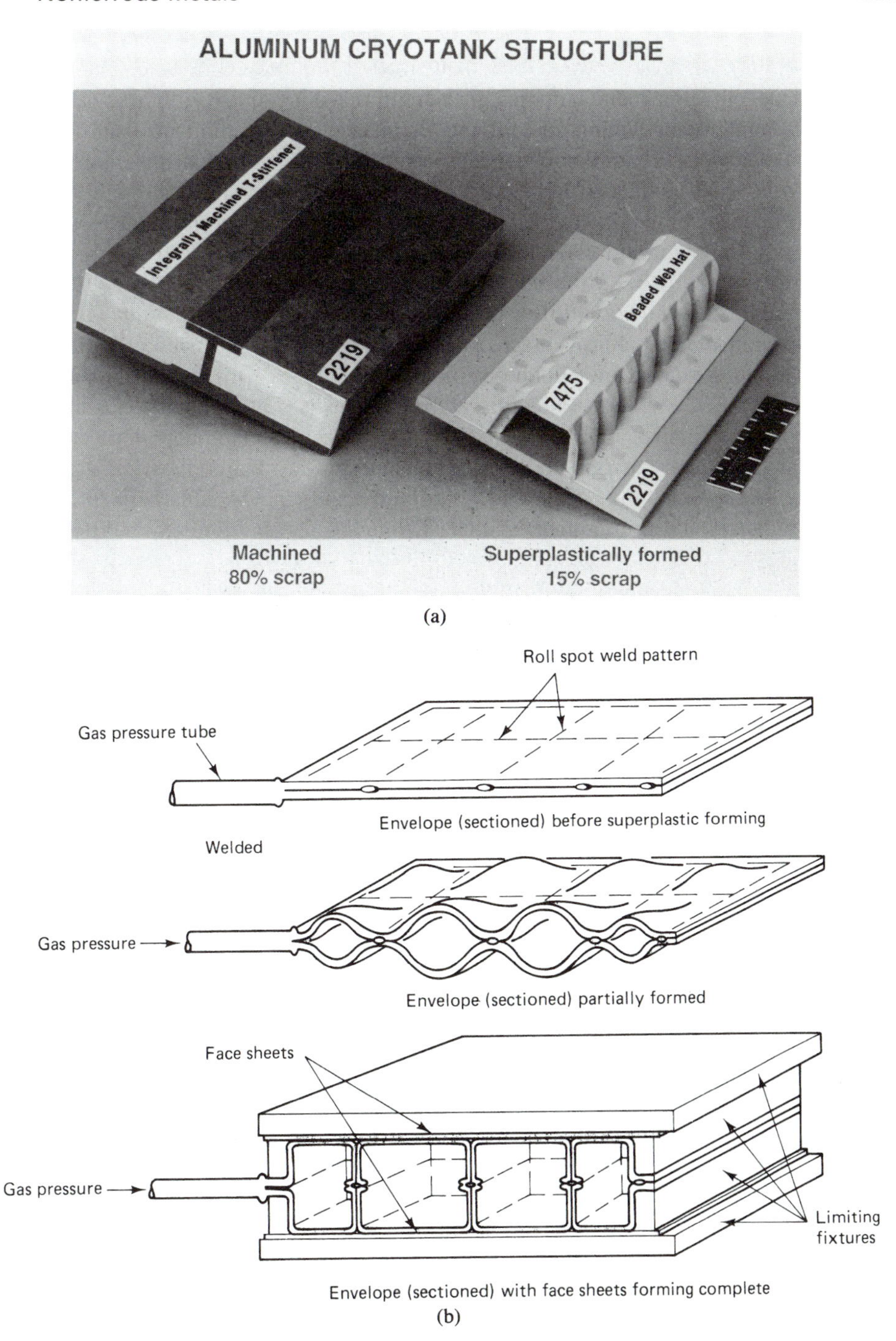

Figure 5-58 (a) An aluminum cryotank structure for the space shuttle produced by old machining methods (left) and newer SPF/DB (right). (b) Steps for superplastic forming with diffusion bonding. (NASA.)

hat stiffeners by as much as 60% by using more complex configurations such as the beaded web, which increases stiffness. With these reductions in cost and increase in efficiencies, the use of more costly aluminum–lithium alloys will permit further reductions in weight and cost to be realized.

Compared to steel or aluminum, titanium alloys have a high strength-to-weight ratio, particularly at elevated temperatures. One main drawback to their use is difficulty in cutting, milling, and forming. This new process, with many of the details still secret, uses one or more titanium sheets laid on top of each other and held together by a jig while being heated in a furnace [Figure 5-58(b)]. Argon gas is blown through tiny holes drilled in the sheets at desired locations. When the titanium reaches a soft and supple superplastic state, the gas assists the titanium to expand into cells, with the other sheets remaining flat. The internal structure takes on whatever shape is intended. Through this process, the outer titanium sheets flow together under pressure to form a complex expanded sandwich unit that has the strength of a single piece.

5.11.1.2 Aluminum in MMCs. As explained in Module 8, metal–matrix composites (MMCs) can involve fiber reinforcing with boron, carbon, and silicon. Aluminum alloys with ceramic-particle reinforcing are prime candidates for castable MMCs. Ingots of aluminum MMCs can be used for many casting processes (Figure 5-59).

5.11.2 Copper

Both as an alloying element in alloys such as steels and aluminum and as a base for copper alloys, copper is a very valuable metal. Included in the copper-based alloys are brass (copper and zinc), bronze (copper and tin), cupronickel (copper and nickel), leaded copper, leaded brass, aluminum bronzes, nickel silver (copper, nickel, and zinc), and high cop-

Figure 5-59 Aluminum alloys with ceramic particle reinforcement can yield near-net shapes. Representative aluminum-alloy parts cast from MMC foundry ingot include, clockwise from top left, an investment-cast pressure vessel; a green-sand cast automobile-brake rotor; permanent-mold cast pistons; shell-investment cast airbus components; a dry-sand cast automobile control arm; and high-pressure, die-cast bicycle sprockets. At center is an investment-cast, aircraft-camera gimbal. (*Advanced Materials & Processes,* ASM International.)

per. Table 5-3 shows that copper and its alloys are processed easily by most common methods, except certain types of welding. Several systems of designation exist for copper and its alloys but are beyond the scope of this book. Besides its value as an alloy, copper has its greatest value as an electrical conductor. High purity of over 99.9% is sought to attain the best conductivity and formability. Due to its scarcity and weight, the trend is, when possible, to move away from copper as electrical conductors. Copper and copper alloys develop a protective film that will not corrode in water or nonoxidizing acids. It will corrode if liquids, solids, or gases break down the protective film or if other conditions cause electrolysis to develop. In brass the presence of oxidizing chemicals can result in dezincification, in which the zinc dissolves, leaving a porous metal sponge.

Copper provides an excellent example of new applications of traditional engineering materials to satisfy modern needs as a consequence of the never-ending changes in the technology of materials brought about by the emphasis on industrial research and development efforts. Recent breakthroughs in brazing, external corrosion protection and laser welding are making new copper radiators for automobiles and commercial vehicles competitive with their aluminum counterparts. These new processing techniques, combined with copper's superiority in thermal conductivity, corrosion resistance, and mechanical strength, will result in radiators that are significantly lighter, more compact, and more durable than those presently in wide use throughout the world. Table 10-11 lists several copper alloys with corresponding designations, compositions, and properties.

5.11.3 Nickel

This metal gains widest use as an alloying element. It provides resistance to both atmospheric and chemical corrosion while imparting other properties to alloying systems such as formability and strength. Table 5-3 shows that several of the common methods of processing are not recommended for nickel. While alloyed in steel, copper, and other materials, nickel is also electroplated onto objects for decorative and corrosion-resistance purposes. After development at the end of the nineteenth century, nickel–cadmium batteries have undergone tremendous refinement, to the point that these rechargeable batteries are found in hand electronic calculators, aircraft-engine starters, emergency power equipment, and portable medical equipment. Figure 5-60 shows the construction of nickel-metal hydride batteries and how a typical battery works. One type of nickel-metal hydride battery is being developed, with the involvement of the U.S. Battery Consortium (USBAC), for possible use in the next generation of electric vehicles.

Common alloys of nickel include Monel (nickel–copper), Inconel (nickel–chromium), Hastelloy (nickel–molybdenum–iron), Duranickel (nickel–aluminum), and illium (nickel–chromium–molybdenum–copper). Nickel is magnetic up to 360°C and is alloyed with iron, aluminum, cobalt, and copper in the powerful permanent alinco magnet. Table 10-1 shows the composition and properties of certain nickel alloys.

5.11.4 Magnesium

A chief source of magnesium is seawater. Magnesium is abundant in supply and 100% recyclable. About 6 million tons of magnesium are available for processing in each cubic mile of seawater, which makes magnesium an abundant metal. But the need for electrolysis to

Inside the Power Cell

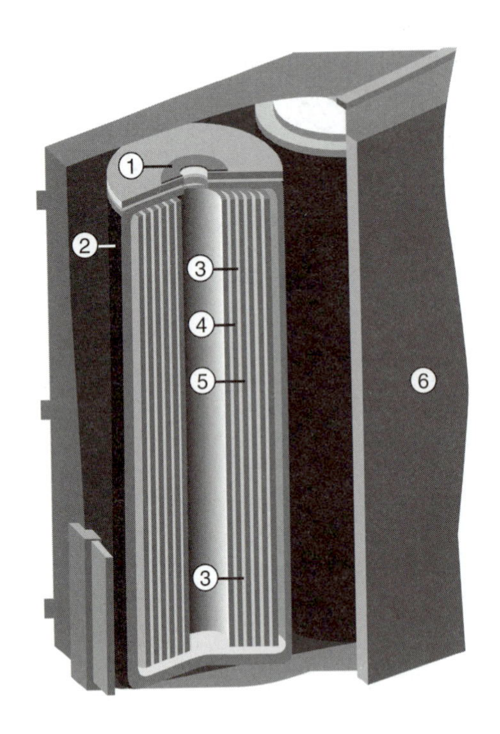

Nickel-Metal Hydride Battery

1. Positive Cap
Connected to the nickel electrode; serves as the positive terminal

2. Can
Connected to the metal-hydride electrode serves as the negative terminal

3. Separator
Prevents direct internal short circuits

4. Cathode
Composed of nickel oxide; serves as the positive electrode of the cell

5. Anode
Composed of metallic alloy; serves as negative electrode of the cell

6. Battery Case
Plastic container which holds interconnected cells with contacts and internal electronics

Figure 5-60 (a) Nickel-metal hydride battery.

extract the metal from mineral deposits or seawater increases its cost. The lightest of all structural metals, with a specific gravity of 1.75, magnesium weighs 1.5 times less than an equal volume of aluminum and 4 times less than zinc. It has good strength, stiffness, and dimensional stability, and in pure form and alloyed with other elements, it provides a high strength-to-weight ratio. Magnesium alloys have relatively high thermal and electrical conductivities; good energy absorption characteristics, i.e., its ability to damp vibration compares with cast iron; and nonmagnetic properties. Often used as an alloying element in engine parts, magnesium is used to produce wheels for racing cars.

A recent lightweight magnesium seat frame made from tubular extrusions and stampings [see Figure 5-60(c)] was developed by Findlay Industries for use in automotive and/or aircraft industries. The purpose of this development was to design a frame that reduced the overall weight of an existing tubular steel frame by 40% to 50%, retained its integrity to meet federal safety standards, and met customer requirements. Other material candidates considered were HSLA steel, aluminum, carbon fiber, and Asdel (compression-molded plastic). Carbon fiber and compression-molded plastic were eliminated from consideration because of the high investment cost for prototype samples. The HSLA steel provided a 25% weight reduction, and a 20% weight reduction if the extruded tube shape was

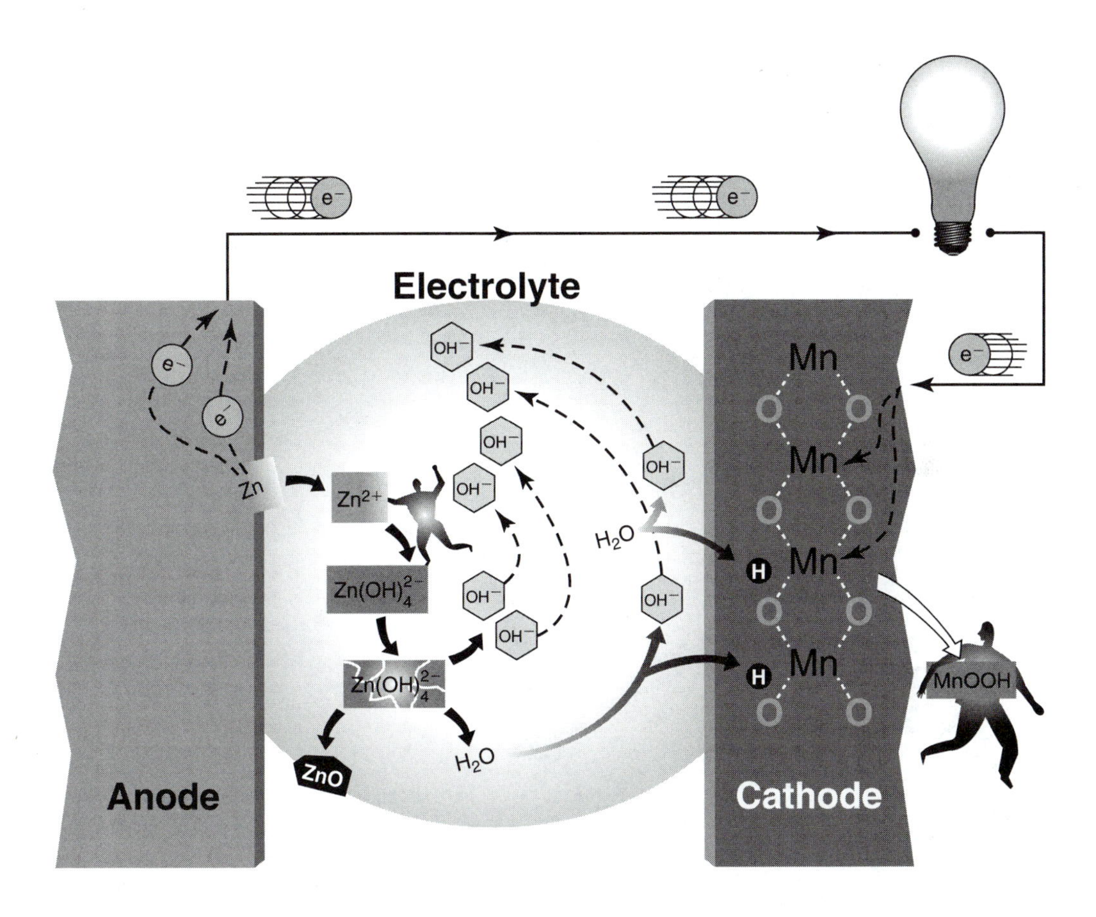

Batteries may seem simple, but delivery of packaged power is a very complicated electrochemical process. Electric current in the form of electrons begins to flow in the external circuit when the device, in this case a light bulb, is turned on. At that time, the anode material, zinc (Zn), gives up two electrons (e^-) per atom in a process called ***oxidation,*** leaving unstable zinc ions (Zn^{2+}) behind. After the electrons do their work powering the light bulb, they re-enter the cell at the cathode, where they combine with the active material, manganese dioxide (MnO_2), in a process called ***reduction.*** The combined processes of oxidation and reduction could not occur in a power cell without an internal way to carry electrons back to the anode, balancing the external flow of current. This process is accomplished by the movement of negatively charged hydroxide ions (OH^-) present in the water solution called the electrolyte. Every electron entering the cathode reacts with the manganese dioxide to form $MnOO^-$ ion. Then, $MnOO^-$ reacts with water from the electrolyte. In that reaction, the water splits, releasing hydroxide ions into the electrolyte and hydrogen ions (H^+) that combine with $MnOO^-$ ion to form MnOOH. The internal circuit is completed when the hydroxide ions produced in this reaction at the cathode flow to the anode in the form of ionic current. There, they combine with unstable zinc ions, which were formed at the anode when the electrons were originally given up to the external circuit. This produces zinc oxide (ZnO) and water (H_2O).

Figure 5-60 (b) How a typical power cell (battery) works. (Duracell International Inc., *The Story of Packaged Power.*)

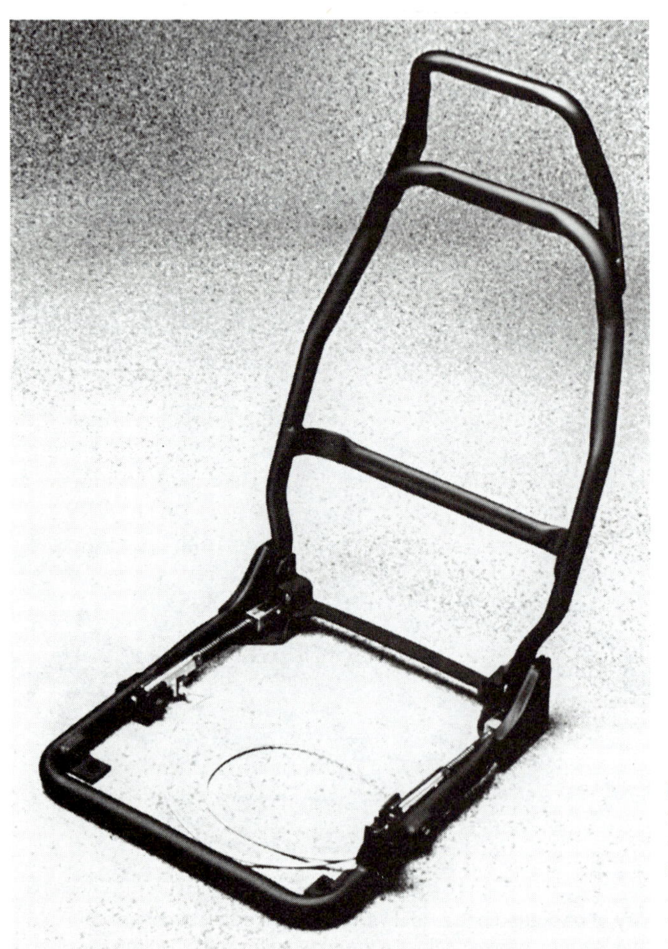

Figure 5-60 (c) Extruded magnesium seat frame represents an advanced, lightweight (3.6 kg) seat system. It is made from magnesium tubular extrusions and stampings, and it reduces seat frame weight by 40%, while retaining its strength integrity. (Findlay Industries.)

altered. Aluminum also showed substantial weight savings, but certain manufacturing problems had to be solved. For example, the aluminum tubing of the same size specified would provide a 35% weight reduction, but the tube was too brittle to bend to the required shape. Heat treating the entire frame, along with increasing the thickness of the tubing 100%, was too impractical, especially when the overall weight reduction would be only 15% to 20%. Welding problems were also encountered. Magnesium AZ61 and AZ31B alloys offered the best weight savings. These alloys are 4.51 times lighter than steel, but AZ31B won out because of its greater ductility. The seat frame weighs only 3.6 kg (7.9 lbs) in bucket seat form, which is about half the weight of a steel frame. Other advantages of using magnesium for this application are that it is noncorrosive and weldable, it has greater design flexibility to accommodate customer preferences and styling concepts, it has excellent design and performance/packaging characteristics (i.e., improved comfort, better vehicle packaging, less foam required), and it is applicable to front and rear seats (bucket or bench type). Finally, the design meets all federal standards (FMVSS 201, 202, 207, and 208).

5.11.5 Titanium

As with aluminum and magnesium, titanium (Ti) is a very abundant metal in its mineral form. As a raw material element, it ranks fourth in availability of the structural metal elements. Unfortunately, the fabrication processes are costly. With a density between aluminum and stainless steel, titanium, "the world's most glamorous metal," was discovered in 1795. It was not until 1930 that a practical method of extracting the metal from titanium oxide (TiO_2), its ore, was commercially feasible. This process produces titanium sponge, which is then further refined in an electric furnace to form ingots, which, in turn, are further remelted to form a final pure titanium. Boeing's 777 (see page 288) uses the largest amount of titanium ever in a Boeing aircraft. In fact, commercial and military aircraft now use 80% of American-made titanium. A major effort is under way to reduce the weight of such aircraft by substituting lighter-weight, advanced materials such as continuous-fiber ceramic composites (CFCCs) for titanium alloys, thus permitting larger payloads and greater hauling distances. ***Investment casting*** of the versatile titanium alloys results in very complex shapes with excellent surface finishes and unique properties often not possible with other processes [Figure 5-61(a)].

Titanium's alloys offer superior specific strength (up to 26.5×10^3 in.) in high temperatures (over 590°C) and low temperatures (−253°C), which makes it a popular structural metal in ultrahigh-speed aircraft and accounts for its use on the space shuttle. The *superplastic* nature of titanium allows it to deform over 2000% without nicking or cracking when it is heated to around 925°C in a process known as superplastic forming. Advances in new technologies such as *superplasticity forming* and *diffusion bonding* (illustrated in Figure 5-58) have furthered the potential of titanium. Titanium is nonmagnetic and has a lower linear coefficient of expansion and lower thermal conductivity than steel alloys or aluminum. Like zirconium and beryllium, titanium is allotropic and exists in a cph structure (alpha) at below 885°C and a bcc structure (beta) above that temperature (see Table 3-2). Various alloying elements alter the effect of the structure and stabilize the alpha or beta phases. For example, aluminum stabilizes the alpha structure, raising the temperature of the alpha–beta transformation. Table 10-11 shows some titanium alloys, compositions, and properties.

Biomaterials are materials that are compatible with human and animal systems, allowing the material to be implanted or manipulated in people and animals. Titanium is an example of a biomaterial that serves as implants for joint replacements and dental reconstruction [Figure 5-61(b)].

5.11.6 White Metals

The general term ***white metal*** includes zinc, tin, lead, antimony, cadmium, and bismuth, which all have relatively low melting temperatures (Table 5-4). ***Zinc*** is a readily available and inexpensive metal that can easily be applied to steel to serve as the sacrificial protection coating, as discussed previously. Zinc is also die cast for housings and decorative trim; used as additives in rubber, plastics, and paint; and used in wrought forms that can achieve

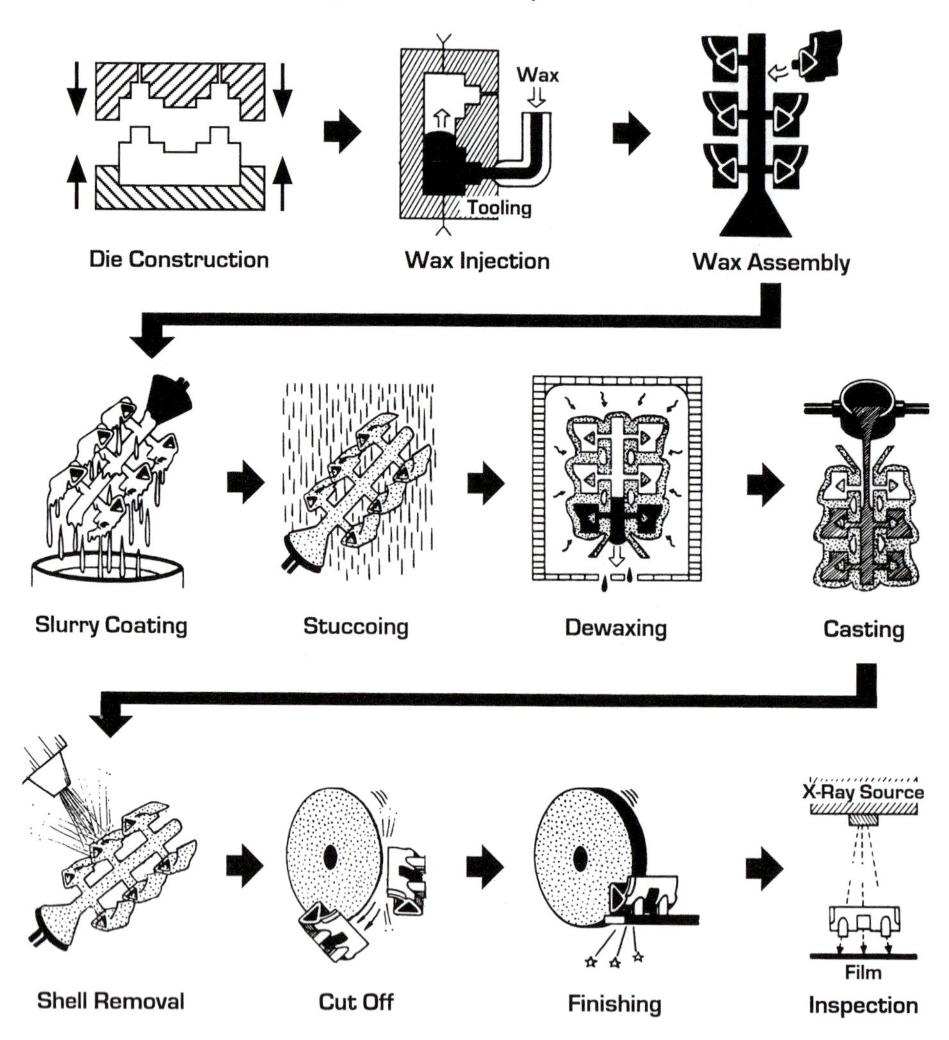

Figure 5-61 (a) Investment casting. From teeth to jet engines come products made from a process that evolved from the ancient art of "lost wax casting." Investment casting serves many fields ranging from medicine for dental and bone implants, to aircraft with intricate gas turbine engine components (Figure 3-58). Constant innovations in the process came with improved nonferrous metal alloys such as titanium and Vitallium (cobalt, chromium, molybdenum, and nickel), refinements in ceramic mold materials, control of metal grain structure, and the ability to produce larger castings. (Howmet Corporation.)

THE DENTAL IMPLANT SYSTEM

Sunken titanium implants that become one with the jawbones and serve as miniature pilings for the dentures are now available in Hampton Roads.

Sinus cavity
Gums
Jawbones
Area shown

3 Prosthetic tooth* is permanently attached to the abutment with a gold screw, and the hole above the screw is filled.

2 Titanium abutments are attached to jaw fixtures with the abutment screw and left to heal for about two weeks.

1 Hollow titanium metal fixtures are inserted into jaw bone and allowed to heal for about four months.

3 Gold screw
Prosthetic tooth*
Metal prosthesis base
2 Abutment screw
Titanium abutment
Gum tissue
Jaw bone
Titanium insert fixture
1

* Or bridge with several teeth

Figure 5-61 (b) Titanium is a biomaterial compatible with the human system for dental and joint implants. The large-scale sketch shows the interface of bone tissue with the titanium and TiO_2 boundary layer. *(Ledger-Star.)*

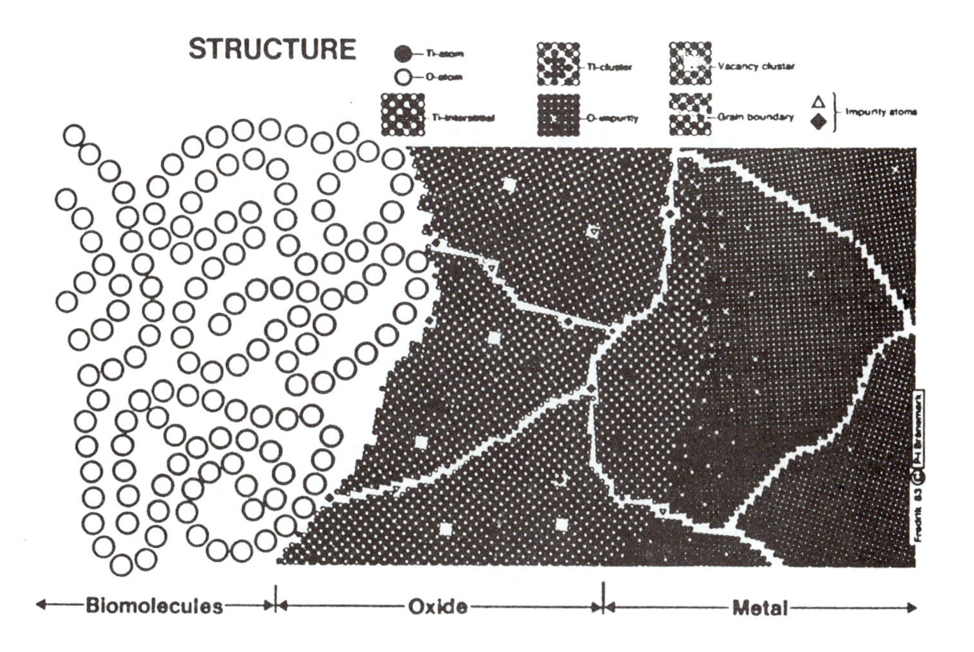

osseointegration

TABLE 5-4 MELTING TEMPERATURE OF WHITE METALS

Metal	°C	Metal	°C
Antimony (Sb)	631	Lead (Pb)	327
Bismuth (Bi)	271	Tin (Sn)	232
Cadmium (Cd)	321	Zinc (Zn)	420

high surface hardness of over 70 HRC. *Tin* has long been used as a coating for "tin cans" made of tin plate and also finds application as a valuable alloying element with lead as soft solder, in tin babbit, and as antimonial tin solder. Solder balls as small as 2 mils (0.002) hold semiconductors onto metal-pinned ceramics that are used for computers. ***Pewter*** (alloyed with lead) was a popular alloy for cooking and eating utensils, plus ornamental casting, in colonial America. The modern tin-based pewter is lead-free with 91% tin, 7% antimony, and 2% copper and is often used to replicate colonial dishes. ***Lead*** is not only an important alloy in solder, but when evenly dispersed in leaded steels at 0.15% to 0.35%, it provides built-in lubrication to ease machining and produce tightly curled chips. Lead is also important in electrical batteries as terminals and, alloyed with antimony, for grid plates (Figure 5-62). Its resistance to many corrosive chemicals, x-rays, and gamma rays and its sound-damping capacities coupled with its high density find unique applications in the medical, chemical, and nuclear industries. It is compounded with ceramic glazes. Small portions of lead built up in the human body cause lead poisoning (plumbism). Consequently, it is desirable to eliminate lead from paint on children's furniture, in gasoline, and in soldered tin cans and to reduce exposure by industrial workers. ***Cadmium*** is harder than tin and serves as a corrosion-resistant coating electroplated on steels, especially fasteners. A coating of 0.0008 mm has the equal protection of a 0.025-mm zinc coat. It also serves as an alloy in copper to improve hardness.

Battery technology, spurred on by the desire for greater mobility and long life for heart pacemakers, telephones, portable radios/CD players, space vehicles, and notebook computers with multiple functions (e.g., fax/modem, CD-ROM, and sound), continues to make strides. State legislatures gave battery technology a boost by mandating a percentage of electric vehicles (see Applications & Alternatives) for their respective state to offset the air pollution from gasoline vehicles. A variety of metals in chemical systems serve as batteries for a wide range of applications in both primary batteries and rechargeable batteries. They include the following: lithium/alkaline-manganese, manganese dioxide, zinc air, silver oxide, lithium ion (see Figure 5-62), nickel-metal hydride, and nickel–cadmium.

5.11.7 Other Metals

Beryllium is a high-cost metal that serves as an alloying element in copper for age-hardening to enhance oxidation resistance and refined grains. ***Refractory metals*** are those with melting points above 1980°C and include tungsten (W) (3370°C), tantalum (Ta) (2850°C), columbium (Cb) or niobium (Nb) (2415°C), and molybdenum (Mo) (2620°C), which are commonly alloyed with many of the metals discussed above to impart strength and hardness at high temperatures, such as type 6–6–2 molybdenum–tungsten high-speed steel. Because of the high temperatures and fabrication difficulties, refractories are often formed as powdered metal parts. Numerous nonferrous alloys known as ***superalloys*** have been developed to meet the requirements of high-technology innovations, such as jet and rocket engines, for high strength and for corrosion and creep resistance at extremely high temperatures (exceeding 1100°C). These superalloys use iron and nickel, nickel, or cobalt as the base metal. Examples of such alloys are Inconel (nickel base), Incoloy (iron–nickel base), and S-816 (cobalt base). Possessing such high strength, most of the alloys are cast or use powder metallurgy techniques in their manufacture. Where machining is required, chipless-type machining methods are mandated.

Smart metals, the ***shape-memory alloys (SMAs)*** such as ***Nitinol*** (nickel–titanium alloy), have been around for many years and are beginning to find wide application as designers learn more about them. One application is in eyeglass frames. The high-quality, shape-memory frames made of this "smart material" return to their original shape when bent. SMAs are also implanted in "smart structures," and are used as controls for robot fingers and dental braces.

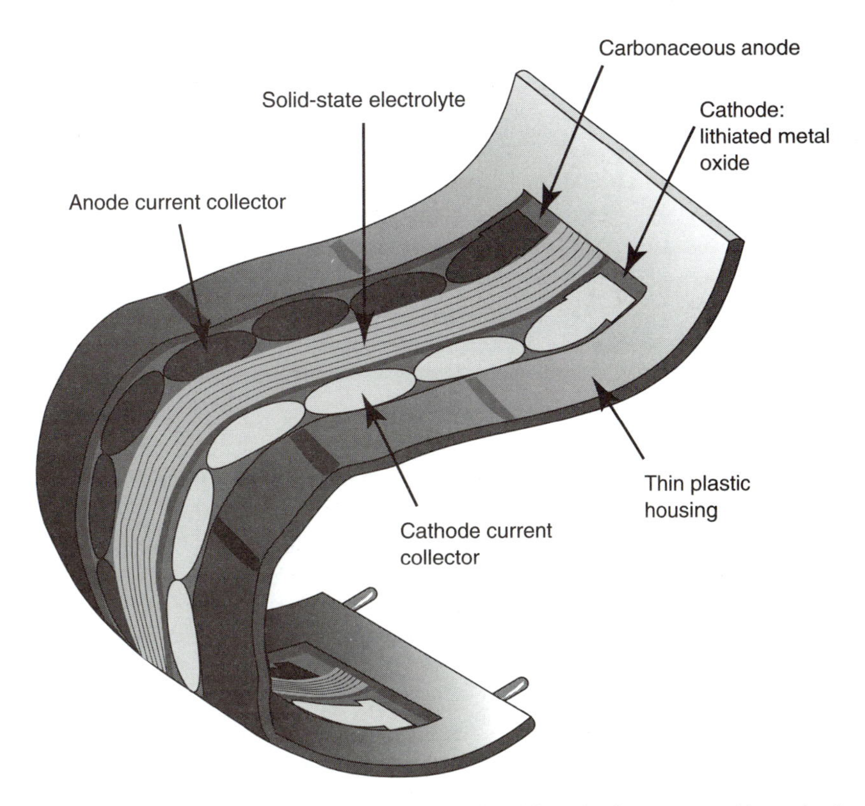

Figure 5-62 Lithium-ion batteries have replaced the nickel-based batteries for many portable notebook computers. (a) These new batteries combine lithium technology with solid polymer electrolytes for a thin, flexible, lightweight, and high-energy-density battery that can be shaped to fix most computer and portable electronics appliances.

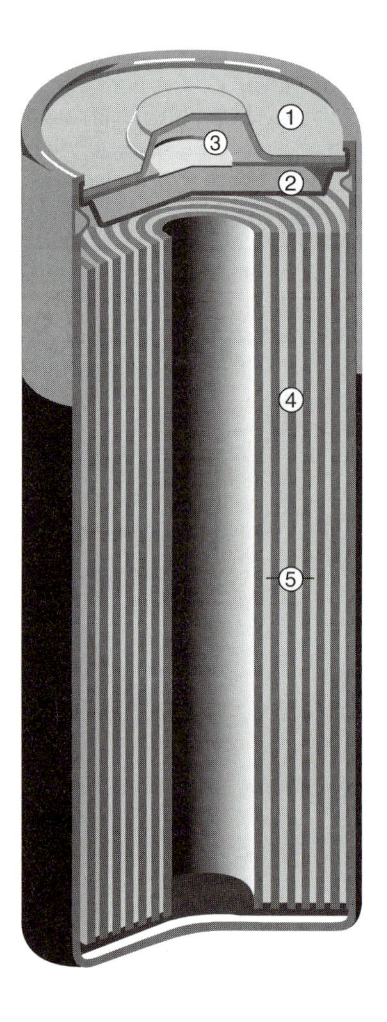

Lithium Cell

1. Positive Cap
Formed nickel-plated steel disk which serves as positive terminal

2. Seal
Molded plastic part which contains cell materials and provides a safety vent

3. PTC Device
Polymeric device which shuts off current in a cell that has been accidentally short-circuited

4. Electrode Jelly Roll
Assembly of positive and negative electrodes which are separated by microporous polymer separator

5. Electrolyte
Solution of complex salts which are dissolved in organic solvents

Figure 5-62 (b) Construction of a standard (AA, C, D, etc.) lithium cell. (Duracell International Inc., *The Story of Packaged Power.*)

SELF-ASSESSMENT

5-54. Cobalt, chromium, titanium, tungsten, manganese, and platinum are some of the strategic metals that the United States must import to produce the vital specialty alloy steels needed in our advanced technological society. Most of these metals are found in the developing nations. Stockpiling aside, what does this situation signify to the many materials scientists, engineers, and technicians working in the metals industries?

5-55. Name three noble metals and give an application of each.

5-56. Explain why aluminum, magnesium, and titanium are expensive when they are so readily available as raw materials. Give a reason for recycling aluminum.

5-57. Explain the corrosion-resistance mechanism of copper and aluminum. Which metal offers the widest range of processibility?

5-58. Give three applications of nickel and explain why nickel is used in these applications.

5-59. What are **(a)** brass, **(b)** bronze, **(c)** Monel, **(d)** Inconel, and **(e)** refractory metals?

5-60. Name one barrier and two advantages for advanced metals technologies such as MMCs, near-net-shape forming with SPF and SSF, and light aluminum alloys.

5-61. **(a)** What common characteristic do white metals possess? **(b)** Provide a typical application for each of these metals. **(c)** Why is lead eliminated from paint, plumbing, gasoline, ceramics cookware, and similar uses?

5-62. Name one refractory and one white metal.

5-63. Describe the reasons for four material substitutions in engines like that used in Ford's Contour concept car. Emphasize specific environmental and economic factors.

5-64. What are the advantages of the new anodizing process for aluminum using sulfuric acid?

5-65. List four properties of magnesium alloys. Cite one example of an application where a magnesium alloy was selected because it possessed one of these properties contained in your list.

5-66. Explain why pure titanium is expensive to produce.

5-67. List the sequences of operation for the investment casting process. Explain how a typical casting is moved from one sequence to another in a modern casting operation.

5-68. The recrystallization temperature of about 600°C for Monel metal is the approximate temperature at which a highly cold-worked Monel part completely recrystallizes in an hour. Other recrystallization temperatures are low-carbon steel, 540°C; zinc, 10°C; tin and lead, −4°C. Working of tin, lead, or zinc at room temperature (about 20°C) is considered hot working. Explain.

5-69. Describe the steps in diffusion bonding of metal-fiber composites.

5-70. Refractory metals have melting points above 1980°C. With the present demand for materials with high strength-to-weight ratios, and high mechanical strength and hardness at service temperatures above 1650°C, what is the main reason why refractory metals are not used to any great extent for these types of applications. Explain how you would overcome this problem, assuming the technology was available to you.

5-71. Use Table 10-11 to select a metal alloy for the axle of a racing bike. Cite properties in explaining your selection.

OBJECTIVE QUESTIONS

5-50. Advanced alloys saved ____________ pounds of structural weight in the Boeing 777.

a. 1000 **b.** 2000 **c.** 3200 **d.** 5400

5-51. Nonferrous metals comprise ____________ of the known elements.

a. ½ **b.** ⅔ **c.** ¾ **d.** ⅘

5-52. The temper designation (W) that follows the four digits specifying wrought alloy groups for the Aluminum Association stands for

a. Solution heat treatment **b.** As fabricated **c.** Strain hardened

5-53. Recycling saves ____________ energy over that for new aluminum.

a. 60% **b.** 20% **c.** 80% **d.** 95%

5-54. Titanium, at certain temperatures, can be deformed as much as ____________.

a. 100% **b.** 200% **c.** 500% **d.** 2000%

5-55. What metal is added to the surface of anodized aluminum to improve its wear resistance?

a. Beryllium **b.** Silicon **c.** Zirconium

5-56. A pound of aluminum is ____________ as expensive as a pound of steel.

a. Two times **b.** Three times **c.** Four times

5-57. ____________ is a metal and an example of a biomaterial that serves as implants for joint replacements in humans.

a. Titanium **b.** Steel **c.** Copper **d.** Lead

5-58. The lightest of all structural metals, with a specific gravity of 1.75, is ____________.

a. Lithium **b.** Manganese **c.** Molybdenum **d.** Magnesium

5-59. ____________ is added to copper to improve its oxidation resistance.

a. Aluminum **b.** Brass **c.** Niobium **d.** Beryllium

5-60. Brass is a metal alloy made from two elements: ____________ and ____________.

a. Tin, lead **b.** Nickel, copper **c.** Zinc, copper

5-61. Monel is a metal alloy consisting of two elements: ____________ and ____________.

a. Tin, lead **b.** Nickel, copper **c.** Zinc, copper

5-62. Inconel is a metal alloy consisting of two elements: ____________ and ____________.

a. Nickel, copper **b.** Nickel, chromium **c.** Nickel, aluminum

5-63. Which metal listed below is capable of being superplastically deformed 2,000%?

a. Nickel **b.** Aluminum **c.** Titanium **d.** Magnesium

5-64. Which metal listed below is a white metal?

a. Zinc **b.** Nickel **c.** Aluminum **d.** Magnesium

5-65. Superalloys are made from a particular base metal, such as one of the following:

a. Cobalt **b.** Iron **c.** Aluminum **d.** Beryllium

(a)

(b)

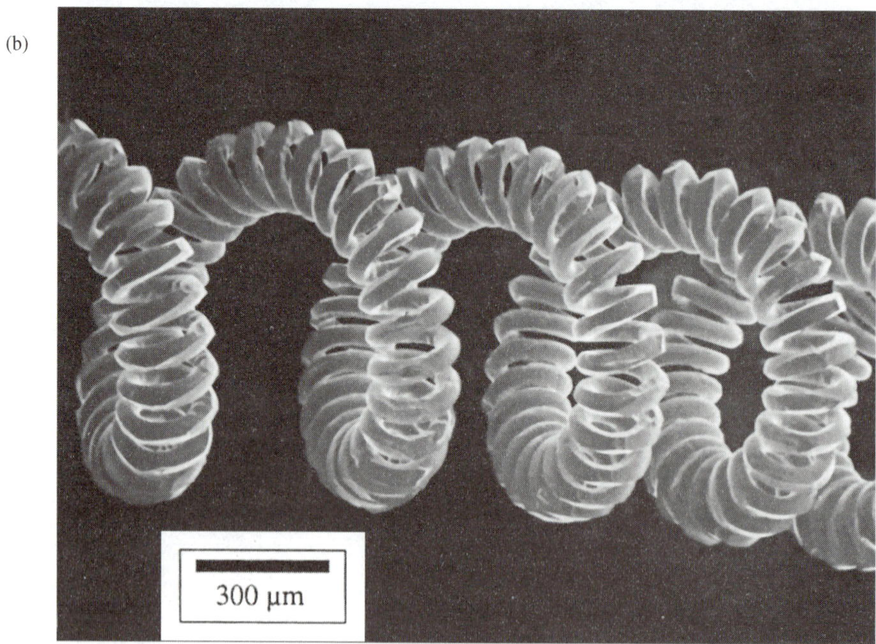

Scanning electron photomicrograph of a nonsagging AKS tungsten (W) filament after several hundred hours of operation at 2500°C in a common 60-watt light bulb. (a) The "coiled coil" geometry of the filament is possible because of the availability of ductile, Coolidge-process P/M, tungsten wire. (b) A closeup reveals some of the tungsten evaporated, thus exposing atomic facets of the individual W grains and characteristic interlocking grain boundaries (chevron features). This grain boundary structure prohibits creep deformation (sagging) through grain boundary sliding, which results in longer lasting filaments. (Jerry P. Wittenauer, Lockheed Missiles and Space Co., and *Advanced Materials and Processes*.)

5.12 METAL POWDERS

Through a variety of techniques, powders of selected materials are produced for ***powder metallurgy (P/M)*** applications. While there is evidence that P/M technology existed at least 5000 years ago, only recently has the technology made significant strides. Developments from the eighteenth century through World War I resulted primarily from the inability of metallurgists to fuse metals such as platinum and tungsten, so they turned to powder compacted and heated to allow diffusion bonding. This last step is known as ***sintering,*** and P/M parts are often referred to as *sintered metals.* Current powder metallurgy deals with most metals and a host of processes. Figure 5-63 shows the flow of the raw powder through compacting ***(briquetting)*** to finished P/M parts. Some parts undergo secondary operations to improve tolerances or finishes, to apply lubricants or coatings, or for heat treatment. A major advantage of the P/M part is the reduction or elimination of scrap and machining, so if too many secondary operations are required, this advantage can be reduced. Once the part is compacted and sintered, it may undergo secondary operations such as *coining* to squeeze the part to closer tolerances; ***infiltration,*** in which a lower-melting-

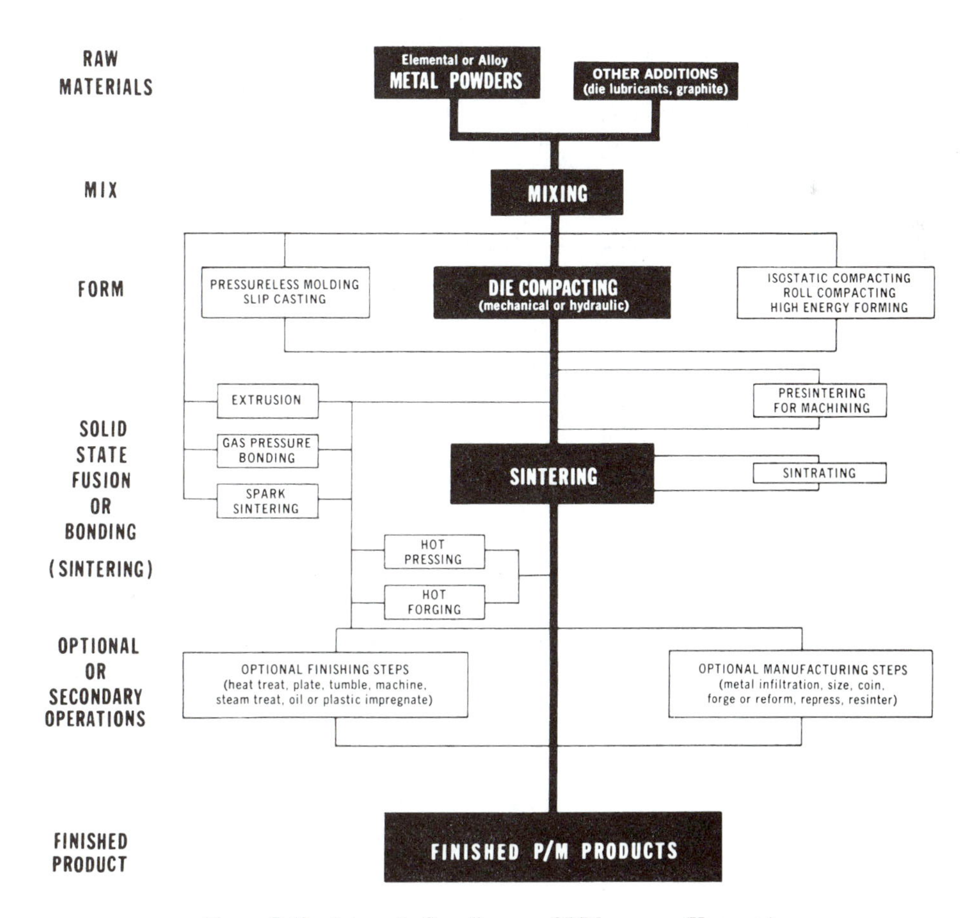

Figure 5-63 Schematic flow diagram of P/M process. (Hoeanaer).

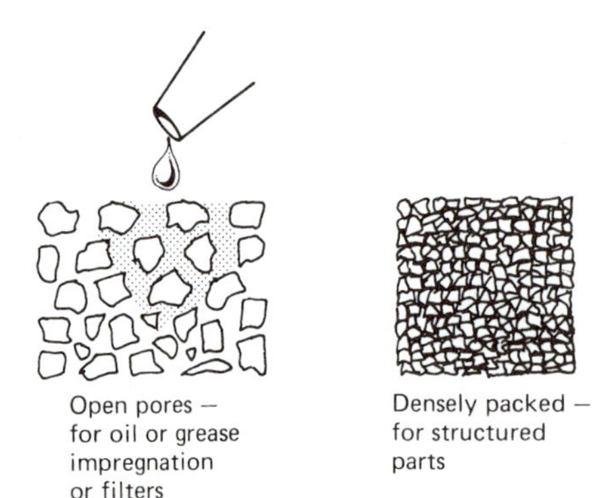

Figure 5-64 Impregnation.

point metal is added to close pores and increase density; ***impregnation*** (Figure 5-64), which adds a lubricant for self-lubricating bearings; or *machining* to achieve a profile not practical through the pressing operation. Not all metal powder is compacted and sintered. Iron powder is added to human and livestock food as a dietary supplement. Iron powders are also used in nondestructive testing (NDT) and magnetic-particle testing, for welding electrodes, and in oxyacetylene cutting and scarfing (exothermic reaction increases the heat of flame to melt oxides formed on high alloys or refractories). Platinum powder is a catalyst used in making gasoline, copper powder goes into marine paints, carbon powder is used in automotive ignition wires and in making xerographic paper copies, and aluminum powder aids in paper manufacture.

In addition to the advantages given, P/M allows the manipulation of a part's density through control of particle size and degree of compaction. Particles range from 0.2 to 2000 μm in size. Smaller particles can pack tightly for increased density; or if open pores are required, as in filters or oil-impregnated, self-lubricating bearings, larger particles would be compacted together (Figure 5-65). P/M structural parts of steel powder have tensile strengths from 310 MPa (45,000 psi) to over 1200 MPa (175,000 psi), which makes them competitive or superior in terms of strength with cast iron, steel, and nonferrous alloys in certain applications. Although conventional P/M processing can produce parts up to 35 lb,

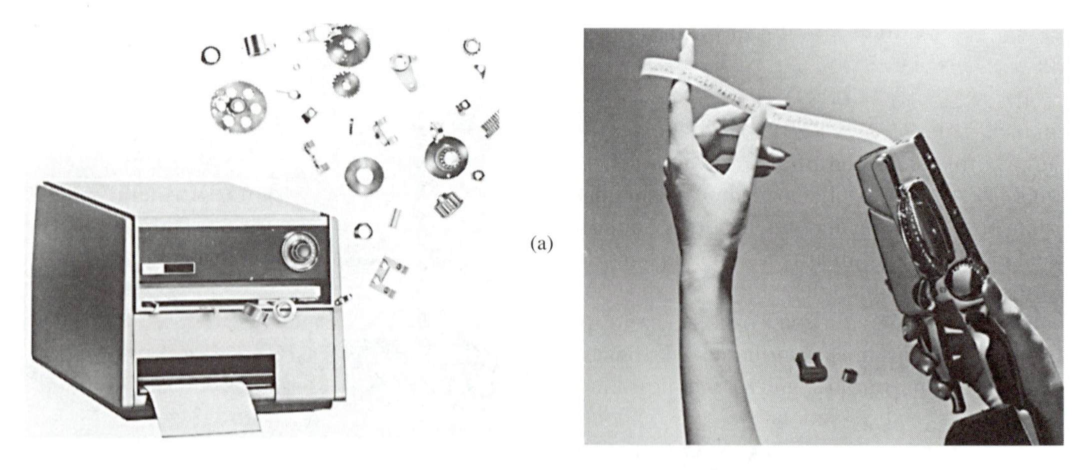

Figure 5-65 (a) Typical P/M parts (over 60 in an office copier).

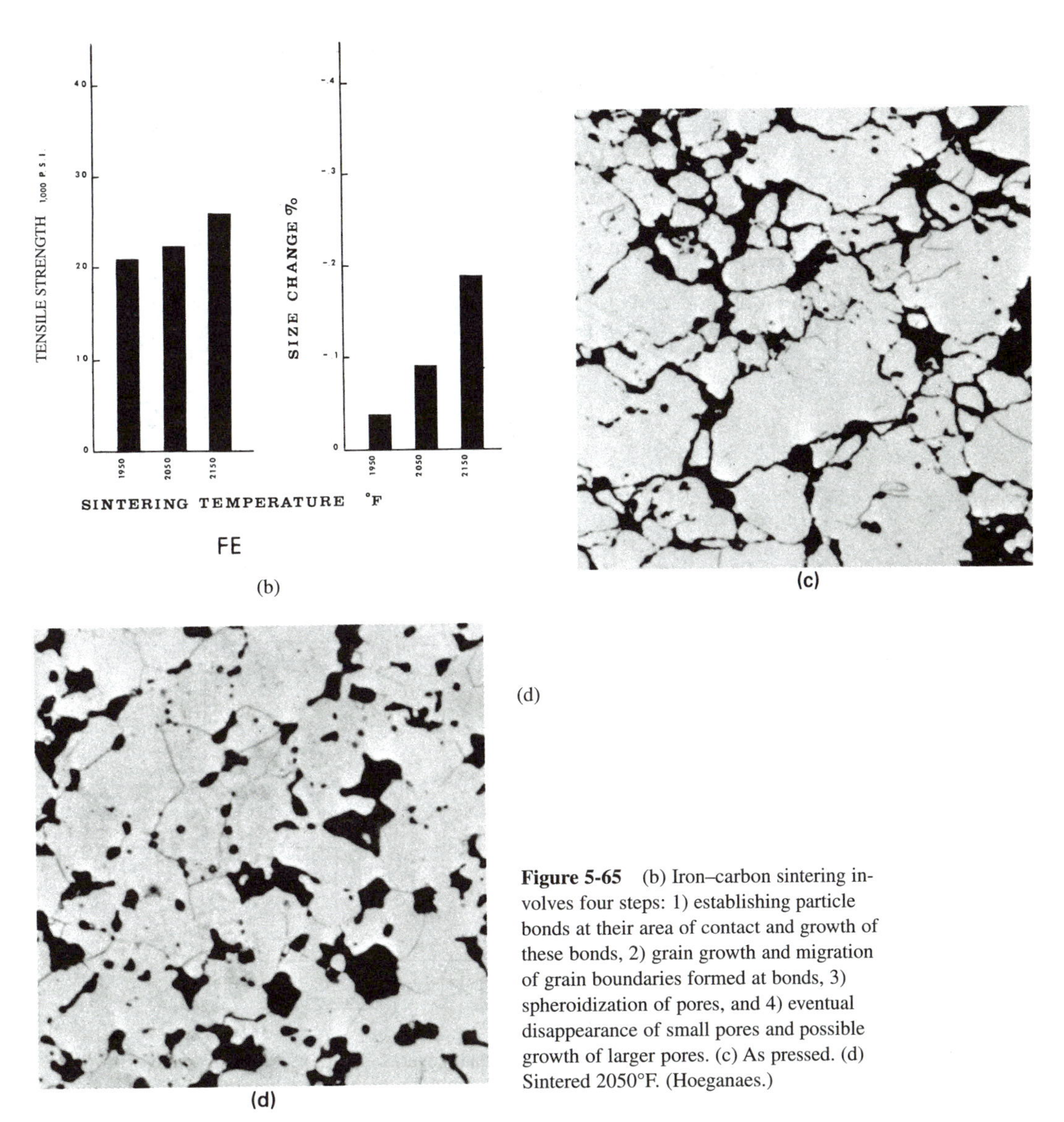

Figure 5-65 (b) Iron–carbon sintering involves four steps: 1) establishing particle bonds at their area of contact and growth of these bonds, 2) grain growth and migration of grain boundaries formed at bonds, 3) spheroidization of pores, and 4) eventual disappearance of small pores and possible growth of larger pores. (c) As pressed. (d) Sintered 2050°F. (Hoeganaes.)

most are smaller and weigh less than 5 lb. The most common P/M materials are iron, steel, copper and copper alloys, stainless and other alloy steels, nickel alloys, aluminum alloys, and refractories such as tungsten carbides, but most other metals find certain unique P/M applications. Tungsten light bulb filaments are made with a P/M process (see page 310).

5.13 MATERIALS SELECTION

Throughout this book and in many technical publications, advanced materials receive attention because of favorable properties and processing possibilities. Figure 5-66 shows the forecast for the market for advanced materials in the United States and Japan for the year

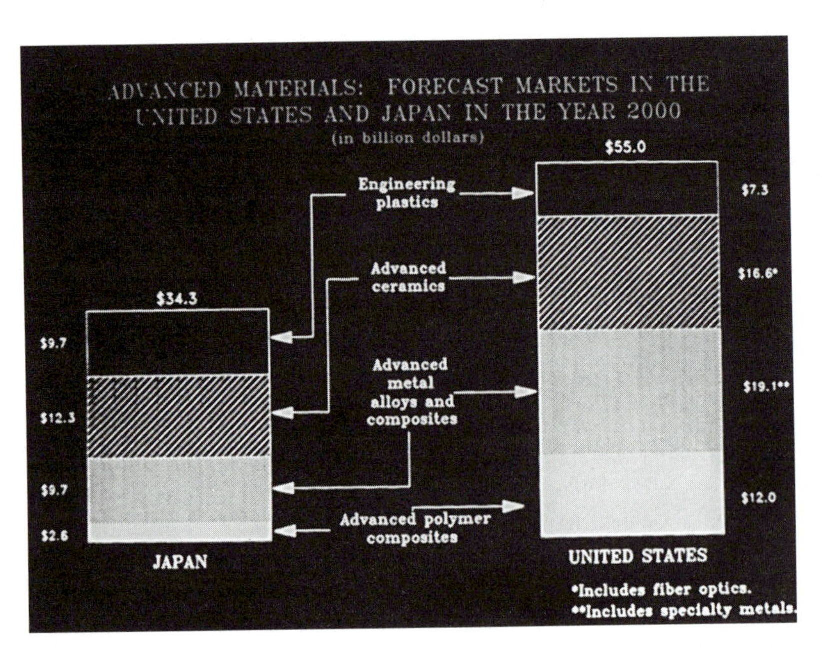

Figure 5-66 Advanced materials: Forecast markets in the United States and Japan in the year 2000.

2000. The cross-sectional view in Figure 5-67 of a concept gasoline engine that uses lightweight components brings together several of the advanced-materials components illustrated in this book and provides support for the market forecast for advanced materials that encompass advanced metals alloys and metal-matrix composites (MMCs), engineering plastics (especially thermoplastics that are easier to recycle), advanced ceramics, and advanced composites. Requests for advanced-materials technology come from many areas, including demand for increases in fuel efficiency with lighter, hotter-running engines; lighter, faster, safer transportation vehicles; more powerful, smaller, friendlier computers; cleaner, cheaper, more abundant energy; and on and on.

A number of examples of materials substitutions for engine parts are in other parts of the book and show how materials selection and engineered materials can make more efficient engines. Various cast iron alloys have been widely used for a wide variety of engines since the beginning of automobiles' manufacture. Figure 5-68 compares two versions of a timing-belt tensioner-pulley pivot bracket; the earlier one made of nodular cast iron and the redesign of ***semisolid forged (SSF)*** aluminum. SSF uses the principles of thixotropic metal processing (refer to the discussion for Figures 3-3 through 3-5) to make alloys for forging. The SSF aluminum bracket was redesigned using CAD with finite element analysis. Weight was reduced from 0.31 kg (0.69 lb) for nodular iron to 0.16 kg (0.36 lb) for the SSF aluminum. The SSF process provides near-net shape since it enables the wear pad and the pulley mounting bolt to be integrally formed into the pivot brackets. This eliminated the machining operation required for the wear pad and pulley mounting bolt areas on the nodular iron version.

High cost offers the main barrier to advanced materials. Usually, they cost several times to tens of times more per pound than traditional engineering materials. A forecast for the twenty-first century (Figures 5-66 and 5-69) results from the expectation that materials

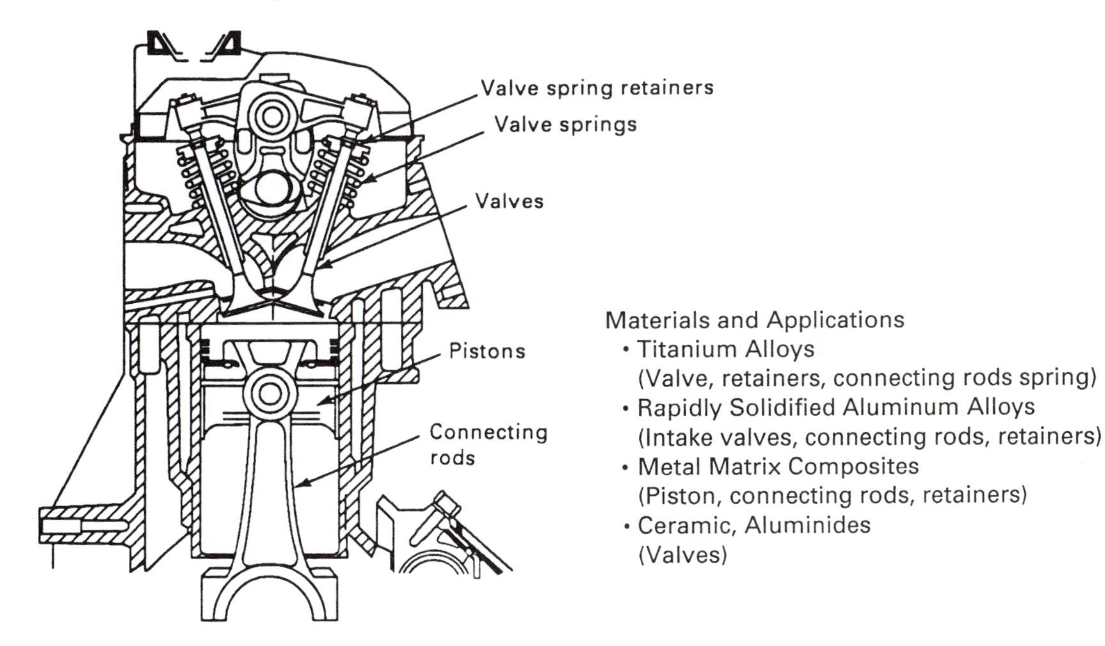

Figure 5-67 Gasoline engine: Lightweight components and advanced materials applications. (Ford Motor Company.)

Figure 5-68 Part at left is the original, designed in cast nodular iron. Part at right shows SSF aluminum design. Note the forged-in-place wear pad and pulley mounting stud. The only machining area on the SSF part is the large hole where the pivot bolt mounts. *(Advanced Materials & Processes.)*

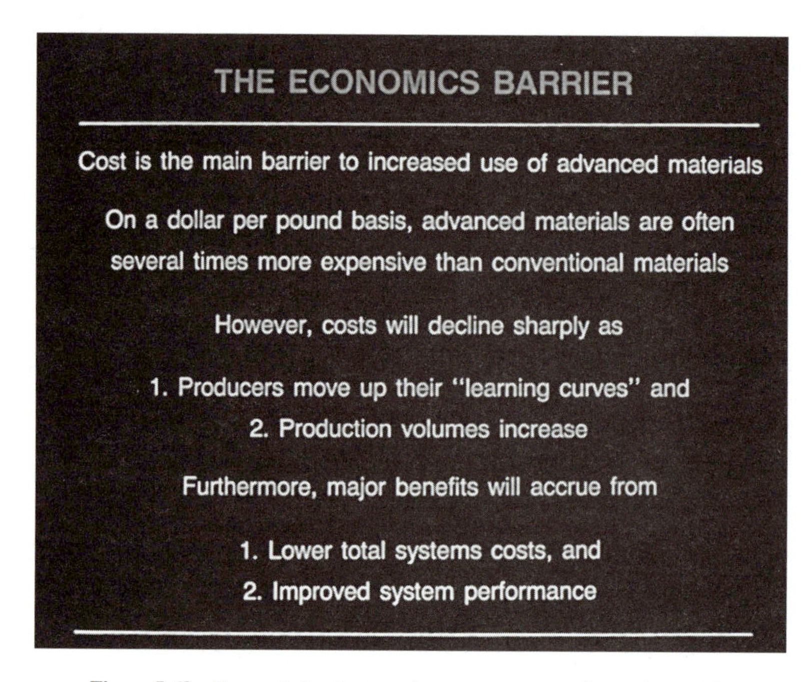

Figure 5-69 Economic barrier to and reason to accept advanced materials.

costs will decline as materials producers and product fabricators gain more experience with the new materials, thereby increasing the demand for greater production rates of the materials. Because advanced technology brings with it lower ***total-systems cost*** and improved systems performance, these forecasts can be made with confidence. However, the changeover to advanced-materials technology will evolve with traditional engineering materials technology coexisting. For example, the gasoline engine that uses advanced-materials components (Figure 5-67) is still expected to have an engine block of cast iron or aluminum for some time. As a materials user, designer, manufacturer, tester, or in whatever role you interact with materials, expect to continue to use traditional metals for many years. Table 10-11 offers only a tiny sampling of the wide variety of traditional metals available. Many handbooks and volumes of references, some listed throughout this book, will enable you to make wise metals selections.

Failure to consult reliable references can meet with tragic consequences. Twelve people lost their lives when a Swiss swimming-pool-building roof collapsed. Failure analysis revealed that 207 stainless-steel hooks had corroded and could no longer support the concrete roof, which weighed 166 tons. The hooks corroded over 13 years from the pool's aggressive chloride atmosphere, which caused intergranular stress-corrosion cracking.

APPLICATIONS & ALTERNATIVES

Environmentally friendly vehicles, long sought to ease the polluting affects of gasoline and diesel cars and trucks, seem to be still a long way from reality. In this module, you learned about the full range of ferrous and nonferrous metals. Every day we see applications of metals in clear view, but many remain hidden and perform their duties in various systems that are out of sight, such as plumbing and heating, electrical, and structural systems. As mentioned in the discussion on batteries, metals also work as part of a chemical system to generate power.

With several states enacting laws aimed at reducing the number of gasoline-powered vehicles, automakers turned to alternative power systems. This short case study looks at the ***sodium–sulfur (Na–S)*** battery-powered vehicles, designed by Ford Motor Company as ***electric vehicles (EVs)*** intended to provide a superior power source to the common lead-acid battery. The two Ford EV concept vehicles, Ecostar and Connecta, serve as platforms to evaluate Na–S battery and related EV technology. The Ghia Connecta, seen in Figure 8-45(b), and its motor, shown here in Figure 5-70(a), was designed as a "family taxi" concept EV. The Ecostar [Figure 5-70(b)] is built on the chassis of the European Escort van. These two vehicles demonstrate the challenges of producing EVs to compete with gasoline vehicles.

The Ecostar uses a 3-phase, AC induction motor that produces 75 horsepower with up to 13,500 rpm and 140 lb-ft torque. The 770-pound Na–S battery, a Ford invention, provide three to four times as much energy as the familiar lead-acid storage battery used on current vehicles. However, it suffers from major limitations, including cost, need for temperature control, and long recharging time.

As an experimental battery the Na–S costs $46,000, with a projected cost of $15,000 for normal production volumes. It would require replacement every few years. The battery operates at 290°C to 350°C (555° to 660°F) and requires a cooling system consisting of a radiator, pump, and expansion tank to transfer away the heat. With the motor turned off, the battery needs a heating system to keep the Na–S in a molten state. If left unheated for a couple of days, it requires two days to reheat and recharge the battery. Charging the battery with normal household current (120v) requires 18 to 24 hours to charge up to 100% from an 80% drain. Obviously, these limitations make the Na–S system undesirable for many applications. With laws requiring EVs, however, they may be practical to serve some commercial applications such as delivery and utility trucks within a city. Still, much remains to be researched with this battery technology.

The ***United States Advanced Battery Consortium (USABC)*** was formed in 1991 to pursue research and development of an advanced energy system capable of providing future generations of electric vehicles with practical solutions. The stated goals of a long-term battery are the following: specific power of 400 watts per kilogram, specific energy of 200 watts per kilogram, and 10 years of useful life at a cost of less than $100 per kilowatt-hour. The nickel-hydride battery is a strong candidate to power electric cars. It can double the driving capacity of current nickel–cadmium batteries and it requires only an hour to fully recharge. However, cost is still a problem.

This brief case study of EVs serves as an example of the complexities of solving involved problems. Many underlying issues can also serve as points for thought and class discussion. For example, the lead in current lead-acid batteries is a hazardous material. What about nickel-hydride and nickel–cadmium; what hazards to they present? While using EVs may reduce the air pollution problems in areas where they are driven, what about the air quality of areas with the electric generating plants? Does the use of lightweight metals (aluminum, magnesium, and newer steels) plus other technology in USCAR's Supercar negate the need for EVs? Are those states with laws mandating EVs expecting too

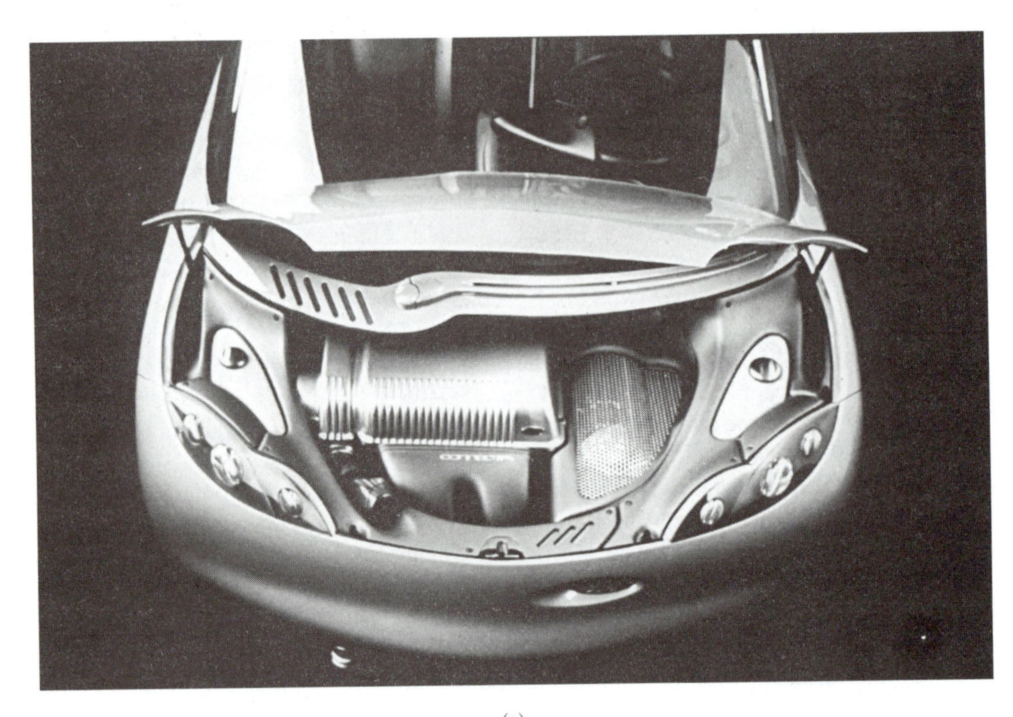

(a)

(b)

Figure 5-70 (a) The Connecta motor, powered by experimental Na–S batteries, serves as an electric vehicle–concept "family taxi." (b) The Ecostar also uses Na–S battery technology, but it is built on the popular European Escort van platform and offers a 100-mile range on a full charge. (Ford Motor Company.)

much too soon from a still-evolving battery technology? What other solutions might you offer to this social/technological issue involving transportation? You might wish to write a position paper, placing emphasis on the related materials technology, to support your choice of several vehicle power technologies.

As citizens we have responsibilities to be informed of the ever-increasing dilemmas facing civilization living on the fragile spaceship Earth. Your education should help to prepare you not only for a career, but also aid with your responsibilities as a world citizen. The issues become more and more technical. Your technical education should help contribute to solutions!

SELF-ASSESSMENT

5-72. **(a)** Provide three reasons for the use of powder metal. **(b)** List three applications of P/M.

5-73. Discuss the possibilities, if any, of using P/M with high-melting-point materials.

5-74. Explain why it is more economical to use P/M for parts that require very close dimensional tolerances.

5-75. Discuss the relationship between compacting and sintering. Are they the same operation? State the objectives of each, and discuss where diffusion bonding actually takes place with the ultimate result of the final elimination of remaining voids between particles in a structure.

5-76. Discuss the reasons behind the continual drive to produce advanced materials.

5-77. List reasons why the production of near-net-shape parts by P/M is economical.

5-78. Cite an example in which the poor selection of a material resulted in tragic circumstances, and explain how better procedures in selection can reduce or eliminate such failures in the future.

OBJECTIVE QUESTIONS

5-66. Another name for sintered metals is ______________.

a. Fused **b.** Compacted **c.** P/M

5-67. In P/M the powder is compacted and heated to allow for ______________.

a. Diffusion bonding **b.** Reduced grain growth **c.** Lower density

5-68. Name another use for iron powders (other than P/M).

a. Making steel **b.** Destructive testing **c.** Incendiaries **d.** Welding electrodes

5-69. The primary purpose of briquetting is to ______________.

a. Facilitate handling **b.** Diffusion bond **c.** Sinter

5-70. Which of the following is the main barrier to the adoption of advanced materials?

a. Lack of database **b.** High cost **c.** Lack of experience with material

5-71. Generally speaking, lightweight materials are selected for what particular area of application?

a. Building construction **b.** Machine tools **c.** Transportation

5-72. Name one source of information about materials that serves as a valuable aid in their selection.

a. Handbooks **b.** CD ROM **c.** Internet **d.** Textbooks

5-73. The following operation is not an example of a secondary operation in the P/M process.

a. Mixing **b.** Coining **c.** Impregnation

5-74. A P/M part designed as a self-lubricating bearing requires ____________.

a. Extremely close-packed particles **b.** Loose-packed particles

c. High diffusion bonded particles

5-75. Sintering implies ____________.

a. High pressure **b.** High temperature **c.** Both high pressure and high temperature

REFERENCES & RELATED READINGS

ALUMINUM ASSOCIATION. *The Story of Aluminum.* New York: AA, 1987.

AMERICAN SOCIETY FOR TESTING AND MATERIALS. *1993 Annual Book of ASTM Standards.* Philadelphia: ASTM, 1993.

ASM INTERNATIONAL. *ASM Handbook,* Vol. 1, *Properties and Selection: Irons, Steels and High Performance Alloys;* Vol. 2, *Properties and Selection: Nonferrous Alloys and Special Purpose Materials;* Vol. 3, *Alloy Phase Diagrams;* Vol. 4, *Heat Treating.* 1990, 1991, 1993, 1994.

ASM INTERNATIONAL. *SteCal,* software for heat treating. Materials Park, OH: ASM International, 1992.

COHEN, ARTHUR. "Copper and Copper Alloys," *ASTM Standardization News,* October 1992, pp. 62–71.

CUBBERLY, WILLIAM H., and RAMON BAKERIJIAN. *Tool and Manufacturing Engineers' Handbook,* desk edition. Dearborn, Mich.: Society of Manufacturing Engineers, 1989.

DURACELL INTERNATIONAL INC. *The Story of Packaged Power.* Bethel, CT: Duracell International Inc. 1995. To request free copy, call 1-800-551-2355.

EMMONS, JOHN R., and K. J. KALLENBORN. *Anodizing and Selecting Aluminum in Nonchromated Solutions.* Marshall Space Flight Center. NASA Tech Briefs, May 1995, pp. 87–88.

"The Featherweight Titan." *Compressed Air Magazine,* Vol. 99, No. 1, Jan./Feb. 1994, p. 40.

FELLERS, WILLIAM O. *Materials Science, Testing and Properties for Technicians.* Englewood Cliffs, N.J.: Prentice-Hall, 1990.

INCO. *A Quick Refresher on Stainless Steel.* New York: Inco.

KALPAKJIAN, SEROPE. *Manufacturing Engineering and Technology,* 2nd ed. Reading, MA.: Addison-Wesley, 1992.

KUHN, HOWARD A. *Metalworking Technology Update.* National Center for Excellence in Metalworking Technology, 3rd quarter, 1992.

LEFFLER THOMAS J. Manager of Advertising and Projects, Findlay Industries, Crooks Road, Troy, MI 48084. Letter dated June 7, 1995.

MALLARDI, JOSEPH L. *From Teeth to Jet Engines.* Howmet Corporation, 1992.

METAL POWDER INDUSTRIES. *Powder Metallurgy.*

MIELNIK, EDWARD M. *Metalworking Science and Engineering.* New York: McGraw-Hill, 1991.

NOEVER, DAVID A. *Improved Tennis Racquets Have Tapered Strings.* Marshall Space Flight Center. NASA Tech Briefs, March 1995, pp. 88–89.

OSTWALD, PHILLIP F. *Engineering Cost Estimating,* 3rd ed. Englewood Cliffs, N.J.: Prentice-Hall, 1992.

SAMUELS, L. E. *Metals Engineering: A Technical Guide.* Materials Park, Ohio: ASM International, 1988.

SIMANAITIS, D., and KIM REYNOLDS. "Ecostar Electric." *Road & Track,* August 1994 reprint.

SIMON, JOHN G. "Advanced Materials for the 21st Century," talk to Hampton Roads Chapter of ASM International, Hampton, VA., April 21, 1992.

Welding Design and Fabrication, May 1988, p. 18.

WEST, C. W. *Stahschlussel [Key to Steel],* 16th ed. Materials Park, Ohio: ASM International, 1992.

WITTENAUER, JERRY P., and JEFFERY WADSWORTH. "Refractory Metals Forum: Tungsten and Its Alloys," *Advanced Materials and Processes,* September 1992, pp. 28–37.

Periodicals

Advanced Materials and Processes
Compressed Air Magazine
Journal of Materials Education
Machine Design

EXPERIMENTS & DEMONSTRATIONS IN METALS FROM NATIONAL EDUCATORS' WORKSHOP

NEW: Update 88 NASA Conference Publication 3060

NAGY, JAMES P. "Sensitization of Stainless Steel."

NEVILLE, J. P. "Crystal Growing."

POND, ROBERT B. "A Demonstration of Chill Block Melt Spinning of Metal."

SHULL, ROBERT D. "Low Carbon Steel: Metallurgical Structure vs. Mechanical Properties."

NEW: Update 89 NASA Conference Publication 3074

BALSAMEL, RICHARD. "The Magnetization Process: Hysteresis."

BEARDMORE, PETER. "Future Automotive Materials: Evolution or Revolution."

BUNNELL, L. ROY. "Hands-On Thermal Conductivity and Work-Hardening and Annealing in Metals."

KAZEM, SAYYED M. "Thermal Conductivity of Metals."

NAGY, JAMES P. "Austempering."

NEW: Update 90 NIST Special Publication 822

BATES, SETH P. "Charpy V-Notch Impact Testing of Hot Rolled 1020 Steel to Explore Temperature Impact Strength Relationships."

CHUNG, WENCHIANG R., and MARGERY L. MORSE. "Effect of Heat Treatment on a Metal Alloy."

RASTANI, MANSUR. "Post Heat Treatment in Liquid Phase Sintered Tungsten–Nickel–Iron Alloys."

SPIEGEL, F. XAVIER. "Crystal Models for the Beginning Student."

YANG, Y. Y., and STANG, R. G. "Measurement of Strain Rate Sensitivity in Metals."

NEW: Update 91 NASA Conference Publication 3151

COWAN, RICHARD L. "Be–Cu Precipitation Hardening Experiment."

KAZEM, SAYYED M. "Elementary Metallography."

KREPSKI, RICHARD P. "Experiments with the Low Melting Indium–Bismuth Alloy System."

LUNDEEN, CALVIN D. "Impact Testing of Welded Samples."

MCCOY, ROBERT A. "Cu–Zn Binary Phase Diagram and Diffusion Couples."

PATTERSON, JOHN W. "Demonstration of Magnetic Domain Boundary Movement Using an Easily Assembled Videocam–Microscope System."

WIDENER, EDWARD L. "Heat-Treating of Materials."

NEW: Update 92 NASA Conference Publication 3201

DAHIYA, JAI N. "Phase Transition Studies in Barium and Strontium Titanates at Microwave Frequencies."

RASTANI, MANSUR. "A Thermal Treatment for High-Strength Steels to Give Higher Fracture and Notch Toughness with Resulting Identical and Even Slightly Higher Strength."

WALSH, DANIEL W. "Visualizing Weld Metal Solidification Using Organic Analogs."

NEW: Update 93 NASA Conference Publication 3259

GUICHELAAR, PHILIP J. "The Anisotrophy of Toughness in Hot-Rolled Mild Steel."

MARTIN, DONALD H. "From Sand Casting to Finished Product (A Basic University-Industry Partnership)."

PETIT, JOCELYN I. "New Developments in Aluminum for Aircraft and Automobiles."

SMITH, R. CARLISLE. "Crater Cracking in Aluminum Welds."

NEW: Update 94 NASA Conference Publication 3304

GABRYKEWICZ, TED. "Water Drop Test for Silver Migration."

KAVIKONDALA, KISHEN, and S. C. GAMBRELL, JR. "Studying Macroscopic Yielding in Welded Aluminum Joints Using Photostress."

KREPSKI, RICHARD P. "Exploring the Crystal Structure of Metals."

MCCLELLAND, H. THOMAS. "Effect of Risers on Cast Aluminum Plates."

WEIGMAN, BERNARD J., and STAMOS COURPAS. "Measuring Energy Loss Between Colliding Metal Objects."

NEW: Update 95 NASA Conference Publication 3330

ELBAN, WAYNE L. "Metallographic Preparation and Examination of Polymer-Matrix Composites."

SHIH, HUI-RU. "Some Experimental Results in the Rolling of Ni_3Al Alloy."

Module

6

Polymeric Materials

After studying this module, you should be able to do the following:

6-1. With the aid of diagrams, explain synthesis of polymers in terms of raw materials, bonding, and by-products, and explain how specific types develop their properties.

6-2. Analyze applications of polymers to determine the effects on structure as a result of branching, degree of crystallinity, molecular weight, and networking on the required properties needed for the finished product.

6-3. Calculate molecular weight for a given compound and explain its significance in polymers.

6-4. Explain the techniques to improve or change polymer properties such as optical, mechanical, chemical, electrical, color, thermal, and density.

6-5. Analyze the properties of various plastics to determine how specific plastics would compete with other plastics, wood, glass and other ceramics, and metals for specific design applications and cost factors. Use tables and handbooks to select plastics, elastomers, and wood to meet specified environments and stresses. Calculate the specific strength of materials to make comparisons.

6-6. Recall the characteristics and applications for general-purpose, engineering, and cellular plastics, and determine selection criteria for plastics.

6-7. Recall the types of natural and manufactured fibers; describe their properties and typical uses.

6-8. Diagram thermoplastic and cross-linked rubber; describe the effects of rubber bonds (lengthening and straightening); recall types, properties, and comparative cost of elastomers; and describe rubber additives and hardness.

6-9. Use Marra's adhesive joint model to explain the principles of bonding, name the conditions and events required for an adhesively bonded joint, and describe main methods by which adhesives work. Explain why interfacial technology is gaining wider study and application.

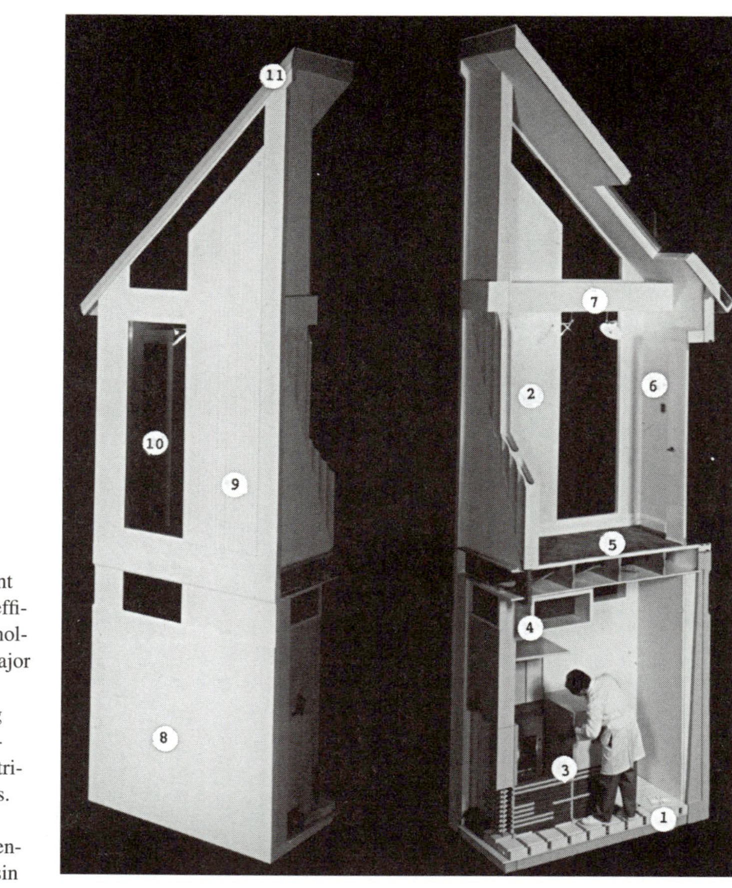

Figure 6-1 (a) Smart house. This prototype house uses smart plastics, intelligent structures, and smart electronic systems to provide the homeowner with energy-efficient, high-tech comfort and convenience. Teams of specialists in computer technology, materials engineering, industrial design, and architecture want to become major players in the home construction industry by utilizing advances in technologies. (General Electric Plastics.) (b) Cutaway model of "living environments" showing unique features. On the right is interior Technopolymer Structures (TPS) of high-strength composites (glass fiber mat and Azdel resin). TPS is used for (1) the distribution floor made of Azdel skin and low-density foam and (2) radiant wall panels. Azdel is a glass-fiber-reinforced thermoplastic resin. The hollow floor encases plumbing and electrical components, as well as storage bins. (3) Total Environmental Control (TEC) are modules made of Noryl (modified polyphenylene oxide resin that has excellent mechanical and electrical properties) and Alloy (composite of amorphous polymers or semicrystalline alloys and blends, reinforced with glass fiber) for heating, cooling, hot water, air purification, and energy recovery; the number of modules depends on house size and climate. (4) Heating and air-conditioning ducts made of TEC Altum (polyetherimide resin, structural foam, sheets, and film). Altum is exceptionally strong and flame and heat resistant. If burned, it produces a low amount of smoke; it comes in clear or opaque. (5) Floor panels of Azdel and Alloy floor joist. (6) Doors of Noryl low-density foam. (7) Plastic extrusions form lighting tracks that fit into wall panels. Left cutaway of exterior; (8) pre-cast concrete foundation wall panels with integral low-density plastic foam wall inside; plastic forms for casting concrete can be left in place. (9) Vinyl siding. (10—Windows) Lexan (polycarbonate resin) with liquid crystals (LCD) are energy-efficient and photochromic (change from transparent to translucent). (11—Roof) Roof of low-density plastic foam can look like cedar shakes or ceramic tile, and can also make use of photovoltaic solar panels to generate electricity. Silicone rubber sealing creates an air and moisture barrier. Modular components such as roof systems could be fabricated in a factory to assemble solar panels, skylights, gutters, and roofing, which would then be transported to the building site and placed over roof trusses. (General Electric Plastics.)

6-10. Recognize the value and uses for protective coatings, sealants, smart polymers, and other important polymers.

6-11. Analyze wood structure and explain how its physical and mechanical properties develop. Cite typical uses for hardwood, softwood, composite wood, and wood by-products. Give reasons why wood is being rediscovered as a valuable engineering material.

6-12. State some popular myths and rules regarding municipal solid waste (MSW). Describe recycling of scrap automobiles and compare that to the discussion of the use of plastics in vehicles in terms of "greening" the Materials Cycle. Explain the savings gained in recycling metals, plastics, paper, and glass. Describe trends in materials technology with regard to improved design, materials selection, and the environment.

PAUSE & PONDER

Do you think much about your shelter? The home you live in, the classroom building, office buildings, and other constructions of shelter play a major role in our comfort, safety, and privacy. Many people in the world struggle for adequate shelter. The availability of materials and the ability to afford them affects the type of shelter people use.

Pause for a minute, look around you, and try to identify the materials in the shelter that you know. Jot down the names of some of them.

Shelter is one of our basic needs, and for most people, their home will account for the largest outlay of money over their lives, whether they buy or rent. Affordable shelter, including homes, is a major goal of all societies. So it is wise to give careful consideration to your home, especially if you plan to purchase a house. The materials selected for a house are very important in terms of their initial cost; the materials will also affect the ease and long-term cost to live in and maintain a house.

House-construction technology has evolved over the history of civilization. Currently, the average house in the United States is a custom home, framed on its site (stick construction) with wood, using plastic or masonry siding, glass windows, and wood or metal doors. House designers and materials manufacturers feel that it is possible to make tremendous improvements in house building to achieve more cost-efficient and enjoyable homes. The *smart house* shown in Figure 6-1(a) reflects the desire of teams of specialists in computer technology, materials engineering, industrial design, and architecture to utilize advances in technologies for improvements in the home construction industry.

Prototype homes [shown in Figure 6-1(a)] are designed to provide the homeowner with energy-efficient, high-tech comfort and convenience. This "living environments" concept house was built to convince the house-construction industry that plastics can have a beneficial impact on houses in a manner similar to that for automobiles. The concept house, built by General Electric (GE) and other business partners, serves as a research tool and a showplace for design, materials engineering, manufacturing and construction processes, and home systems (e.g., heating, air conditioning, security, entertainment, and appliances).

Approximately 30% of the "living environments" concept house is plastic, including the roof, windows, siding, plumbing system, foundation, and electrical and mechanical systems. Plastics are used in harmony with the traditional materials (wood, metal, and ceramics). Many of the plastics are engineering thermoplastics. Because plastics generally do not biodegrade and their processing may lead to pollution, they may present environmental concerns. The GE project also focuses on demonstrating that plastics can be environmentally friendly. Recycled engineered plastics should keep many plastics out of municipal landfills. For instance, it is possible to reuse discarded plastic packaging (bottles or boxes) to produce automotive parts. Once the automotive parts are discarded, they can be reclaimed for materials in house construction.

Another potential of the concept house comes from building factory-manufactured houses rather than traditional site-built houses. Factory-built houses can be more affordable, of higher quality, and less expensive to run and maintain, and they should have features that make for more enjoyable living. Among the unique features in the house shown in the cutaway [Figure 6-1(b)] are the modular wall and flooring systems. Walls include foam-polymer sections that radiate heat into the room and serve in a system for heating, air conditioning, humidity control, and air filtration, with the capacity to recover wasted heat. Hollow composite plastic sandwich floors provide easy access, and they encase plumbing and electrical components as well as providing storage bins. Roofing is done with shingles of lightweight fiber-reinforced plastic that eliminate underlayment.

Plastics provide low emissivity and therefore conserve energy that might otherwise escape through roofs and walls. Instead of ceramic tiles, a modified polystyrene foam serves as floor tiling that is held in place with Velcro. Other components, such as doors, drawers, and cabinets, snap together, which allows for easy assembly and disassembly. Acrylic windows can be changed from clear to white because of embedded liquid crystals. This affords privacy and blocks the sun to avoid bleaching of colors. Replacing traditional baseboards are hollow raceways that carry all wiring: telephone, coaxial cable, and low-voltage electrical.

6.1 POLYMERICS

Section 2.2.3 provided an introduction to polymeric materials (often classified simply as polymers). The emphasis in this module is on the main polymeric materials used in engineering, manufacturing, and construction, and excludes natural polymers such as skin; hide (leather); bone; or natural fibers such as wood, cotton, silk, and jute. Plastics, woods, elastomers, and adhesives are all polymers. While plastics, adhesives, and synthetic rubber are human-made, wood and natural rubber are naturally occurring materials that are used in industry with various alterations to their natural state. Polymers can include many types of industrial materials, as seen in Figure 6-2.

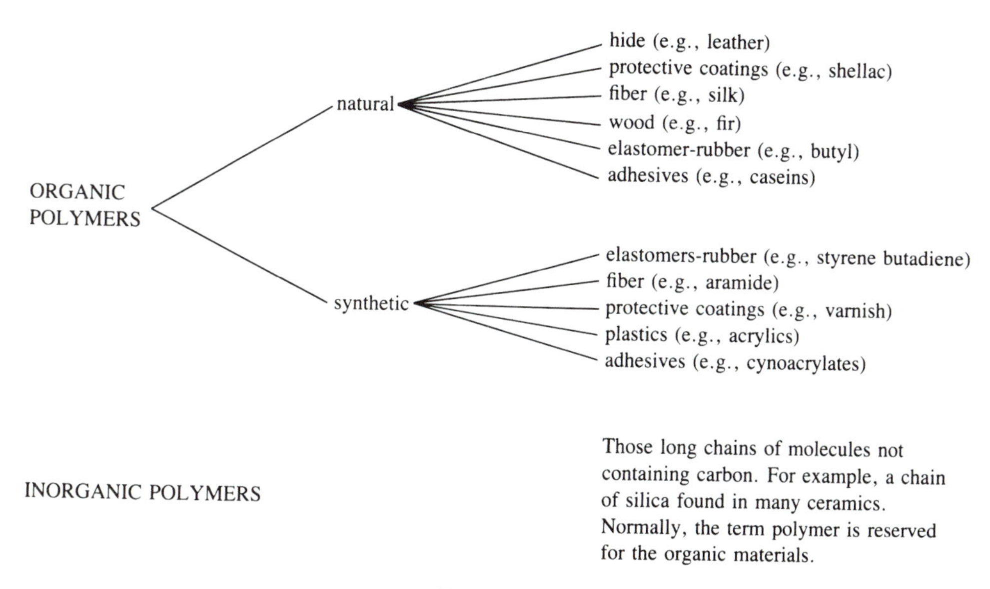

Figure 6-2 Industrial polymers.

6.2 NATURE OF POLYMERS

The basic makeup of polymers consists of smaller units ***(mers)*** joined together either naturally or synthetically to produce polymers or macromolecules. This section on structure deals with the structures common to most polymers, and most of the examples cited are plastics, due to their abundant use in industry. The discussion will concentrate on the synthetic methods of producing polymers known as *polymerization* and will not deal with the biological process by which the natural or biological polymers are grown.

The initial step in polymer production is achieved by petrochemists or petrochemical engineers, engineering technicians and engineering technologists, or others with chemistry in their education and training. This initial step involves breaking down raw materials such as crude oil, coal, limestone, salt, natural gas, and air. With cellulose plastics, the raw material is cotton or wood; however, they are not broken down, as are most other polymer ingredients. Whether it be the cracking of natural gas, distillation of coal or crude oil, or the esterification of cotton or wood, processing the basic organic (carbon base) raw materials yields the element carbon together with hydrogen. Other basic elements are also obtained in the initial stages, including oxygen, nitrogen, silicon, fluorine, chlorine, and sulfur. Basic ingredients of polymers such as ethylene and naphtha are but two of the many products obtained in the cracking and distilling processes. An ***ester*** is an organic compound that is a colorless liquid, insoluble in water, with a very pleasant aroma. Esters find use in perfumes and in the manufacture of flavoring agents for the soft drink industry. When an acid and an alcohol chemically react ***(esterification),*** the product is an ester, such as the polyester resin. From 100 gallons of crude oil, at a market value of around $25, it is possible through distillation and further processing to produce products such as gasoline, motor oil, propylene, and a large variety of polymers—styrene–butadiene rubber, polyethylene plastics, polypropylene fibers for fabrics, and others—that combine to make products with a market value of over $3500. While oil is the major raw material for plastics, plastics take only a very small percentage (about 1.3%) of the world's petroleum production. Yet shortages in oil cause problems in the plastics and polymers industry, so alternative raw materials continue to receive study.

6.2.1 Polymerization

Scientists observed nature's methods of joining elements into chains and duplicated that natural process to produce macromolecules or polymers. Module 2 covered the principles of bonding, which provide a basic understanding of polymerization. ***Polymerization*** is the linking together of smaller units (monomers) into long chains. The repeating units (mers) of some polymer chains are identical, as in polyethylene [Figure 6-3(a)], polystyrene, and poly(vinyl chloride); these are labeled ***homo***polymers [Figure 6-3(b)]. *Co*polymers contain two different types of monomers, such as poly(vinyl chloride) mixed with vinyl acetate to produce poly(vinyl acetate) [Figure 6-3(c)], and *ter*polymers such as ABS (acrylonitrile–butadiene–styrene) contain three types of monomers. ***Isomers*** [Figure 6-3(d)] are variations in the molecular structure of the same composition. Isomers are found not only in hydrocarbons (HCs) but also in polymeric molecules. One class of isomers is known as ***stereoisomers,*** which, in turn, is divided into three categories. The first category is called ***isotactic*** stereoisomerism. Using a single carbon (C) chain of atoms with some hydrogen (H) atoms being replaced by another atom or group of atoms, denoted by the symbol **R,** one can demonstrate that the **R** group atoms are situated on the same side of the carbon chain. In a ***syndiotactic*** stereoisometric arrangement, the **R** group of atoms are located on opposite sides of the carbon chain in an alternating fashion. The third category is called ***atactic*** stereoisomerism, in which the **R** groups are in a random position. Refer to Figure 6-3(e), (f), and (g). It is important to note that copolymers and terpolymers consist of units from each contributing mer and are not an alloy of mers. If polymerization permitted only the production of homopolymers, the properties of polymers would be severely limited.

Most polymers are produced by ***unsaturated hydrocarbons,*** which means that they have one or more multiple covalent bonds, such as ethylene:

$$\begin{array}{cc} H & H \\ | & | \\ C & = C \\ | & | \\ H & H \end{array}$$

or adipic acid, which is used in nylon synthesis and has two hydroxyl monomers:

$$\begin{array}{c} \quad O \qquad\qquad\quad O \\ \quad \| \qquad\qquad\quad \| \\ HO—C—(CH_2)_4—C—OH \end{array}$$

Saturated hydrocarbons have all single bonds. To achieve the polymerization process, monomers must be capable of reacting with at least two neighboring monomers or be ***bifunctional.*** Because of copolymerization, terpolymerization, or other multicomponent

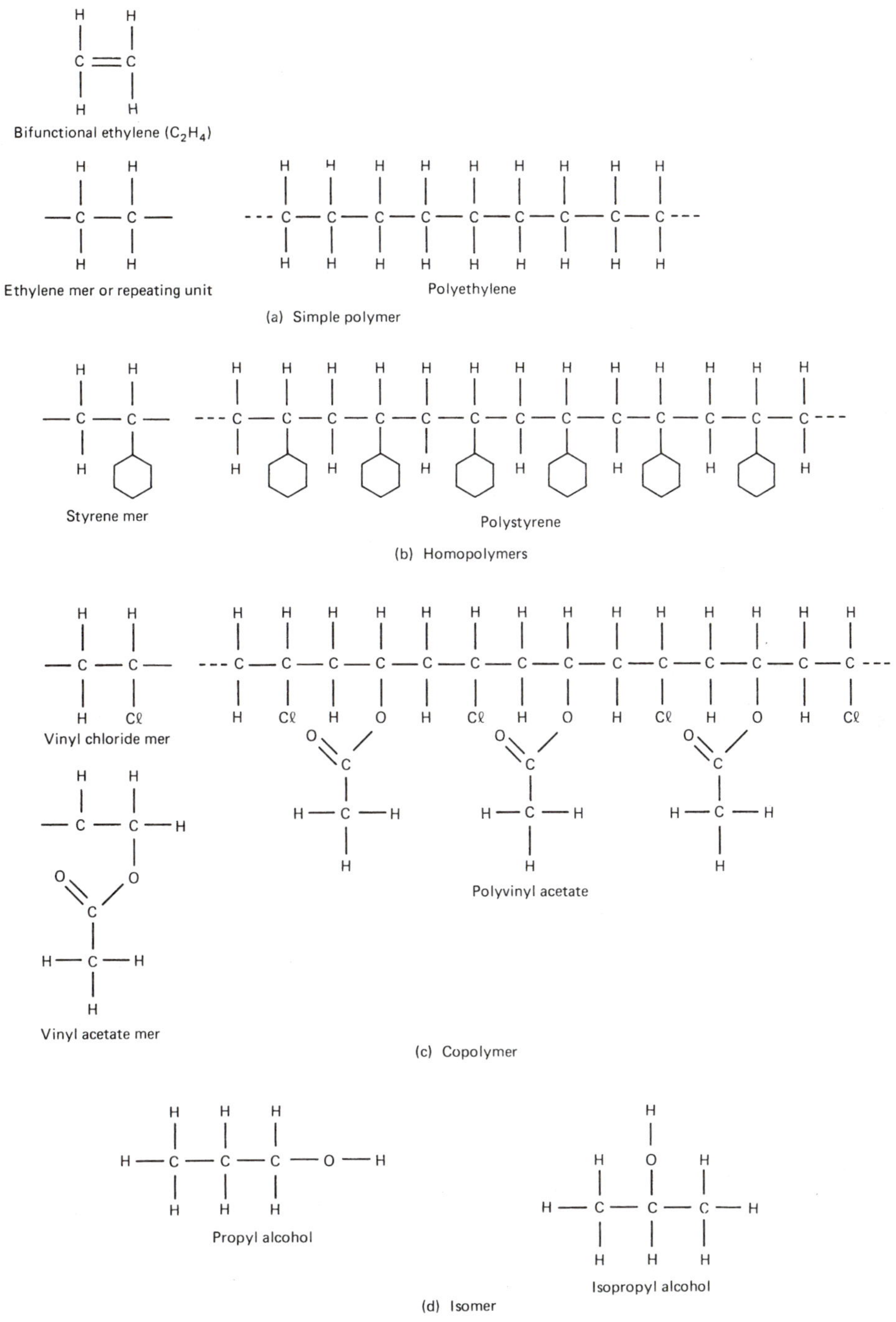

Bifunctional ethylene (C_2H_4)

Ethylene mer or repeating unit

Polyethylene

(a) Simple polymer

Styrene mer

Polystyrene

(b) Homopolymers

Vinyl chloride mer

Vinyl acetate mer

Polyvinyl acetate

(c) Copolymer

Propyl alcohol

Isopropyl alcohol

(d) Isomer

Figure 6-3 Types of polymers: (a) Simple polymer. (b) Homopolymers. (c) Copolymers. (d) Isomers.

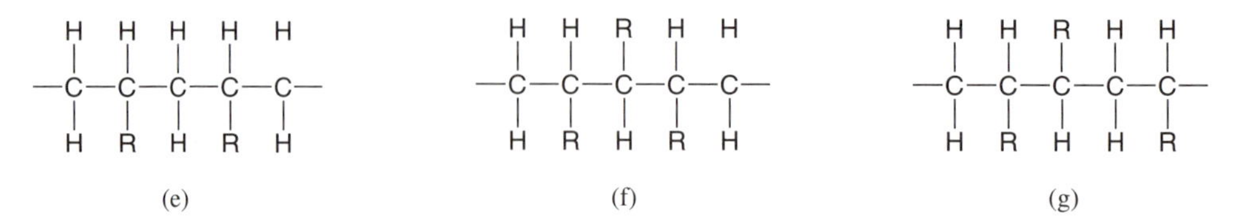

Figure 6-3 (e) Isotactic stereoisomer. (f) Syndiotactic stereoisomer. (g) Atactic stereoisomer.

polymerizations, a large variety of polymers is available. One example is styrene, a brittle polymer that has limited toughness and poor chemical resistance. But through copolymerization of acrylonitrile with styrene, it is possible to obtain a chemically resistant, more rigid, and stronger plastic. Copolymerization of butadiene with styrene yields an elastomer; and terpolymerization of acrylonitrile, butadiene, and styrene produces ABS plastics that are tough and elastomeric, and have good chemical resistance (see Table 10-9).

Catalysts begin the polymerization process. In addition, these chemicals serve a variety of purposes in chemical processing, ranging from serving as molecular sieves to agents for extracting heating oil, gasoline, and ethane from crude oil, to serving as a molecular matchmaker that holds chemicals together for a reaction to occur such as forming long-chain polyethylene polymers from molecules of ethane. ***Zeolites*** are catalysts composed of inorganic grains of alumina (Al_2O_3) and silica (SiO_2) through which molecules of crude oil are strained to gain oil by-products. Activated carbon, platinum, and nickel are also used to catalyze synthetics. ***Aerogels*** are open-cell foams that have ultrafine cell or pore size less than 50 nm, high surface areas (500 to 1000 m^2/g), and density approaching that of air. They are the lightest, most transparent human-made solids. Known as the world's best insulators, they do not conduct heat. With a density of 15 mg/cm^3 and a melting point of 1552°F, the properties of these materials approach those of glass. Aerogels of organic and inorganic systems of fibrous chains provide catalyst support. The development of scanning tunneling microscopes (STMs) has allowed nanoscale examination of catalysts. Thereby, new designer catalysts can catalyze products more efficiently without creating polluting by-products.

The main polymerization processes are ***addition (chain reaction) polymerization*** and ***condensation (step reaction) polymerization.*** The addition process is the simpler of the two. By use of heat and pressure in an ***autoclave*** or reactor, double bonds of unsaturated monomers (Figure 6-4) break loose and then link up into a chain. These addition reactions (in addition polymerization for unsaturated HCs) are atoms or groups of atoms that attach themselves to the carbon atoms at the sites of multiple bonds. No products other than the polymer are formed. Saturated hydrocarbons undergo substitutional reactions in which hydrogen atoms are replaced by other atoms or groups of atoms. The products of addition polymerization, also referred to as chain reaction polymerization, include polypropylene (PP), polyethylene (PE), poly(vinyl acetate) (PVA), poly(vinyl chloride) (PVC), acrylonitrile–butadiene–styrene (ABS), and polytetrafluoroethylene (PTFE).* *Chain polymers* generally fit into the ***thermoplastic*** (soften when heated) group. Most

*ASTM abbreviations.

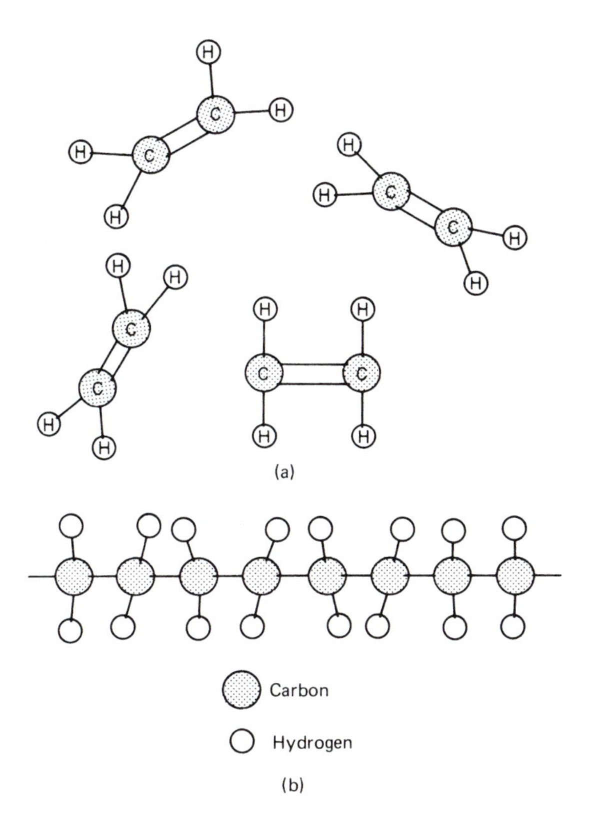

Figure 6-4 (a) Monomer of ethylene (C_2H_4). Carbon atoms with four valence electrons have four arms (shared electrons) for covalent bonding. Two shared electrons hold another carbon atom in a double bond. Two other shared electrons, each covalently bonded, hold one shared electron of hydrogen, which has only one valence electron for single bonds. Double bonds are not stable and form ***reactive sites.*** (b) Polyethylene, chain of ethylene monomers. During polymerization, heat and pressure break the hold of one shared electron on each of the double-bonded (reactive sites) carbon atoms, thereby leaving each carbon atom of the monomers free to grasp another carbon atom (covalent bond) from other ethylene monomers, which form into a chain of thousands of ethylene monomers (polyethylene). The single bonds satisfy the carbon bond arrangement and produce the most stable saturated polymers because they have no reactive sites.

plastics that cannot be resoftened when heated (***thermosetting polymers***) come from condensation polymerization, also known as ***step reaction polymerization.*** This group gets its name from the by-product (***condensate***) of the polymerization, which is often water, but it may be a gas. Phenolic (PF), polyester (PET), silicon (SI), and urethanes (PUR) are typical thermosets from the step reaction synthesis, while nylon (PA) and polycarbonates (PC)* are thermoplastic resins synthesized through step polymerization. Figure 6-5 diagrams the addition process for the production of some typical polymers, and Figure 6-6 shows the flow for production of nylon through condensation polymerization.

The term ***resin*** covers both solid and semisolid organic polymers. Resins are often considered as the uncompounded ingredients or monomers that are mixed but not yet polymerized. For example, thermosetting resins or pellets are molded into thermosetting plastic or elastomeric parts. Sometimes the term *resin* is used synonymously with plastics (e.g., acetal resins of acetal plastics, or thermoplastic resin instead of thermoplastic plastics).

6.2.1.1 Polymer chain lengths. Considerable variation in polymer chains results during polymerization. The ***chain lengths*** play an important role in processing

*ASTM abbreviations.

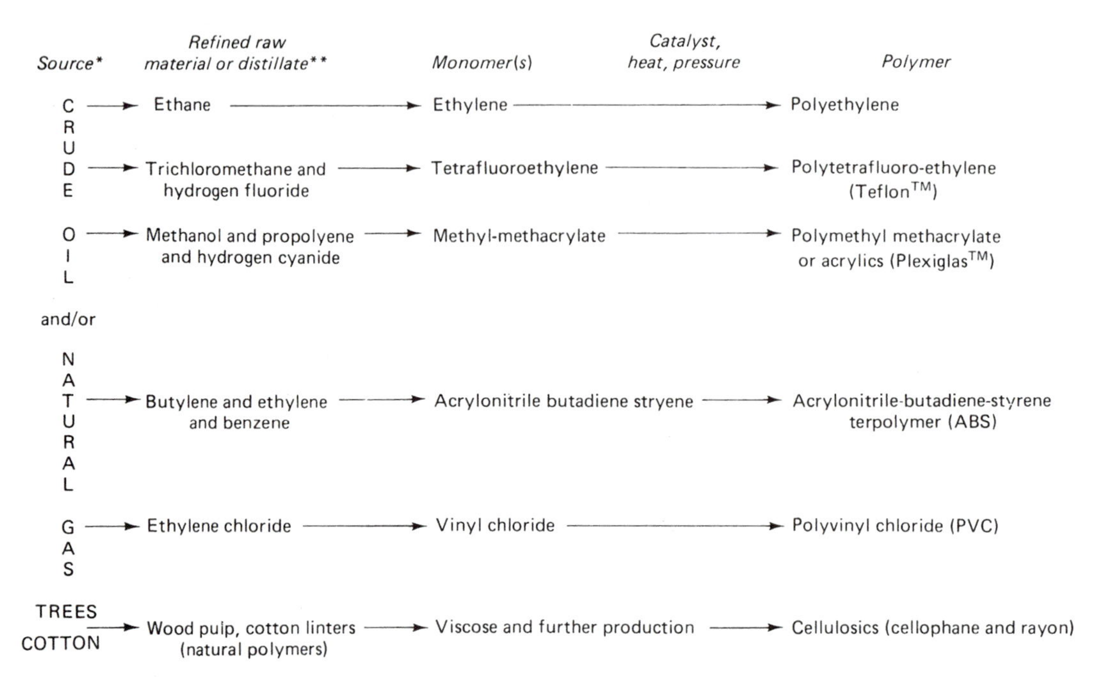

*LRG (liquefied refinery gas) is obtained from refining crude oil and can be further refined to yield the butadiene, ethylene, methane, and propylene. Coal and coke also serve as raw materials for production of gases used to produce polymers.

**In addition to petroleum raw materials which provide the carbon and hydrogen base, other elements such as oxygen, nitrogen, sulfur, chlorine, hydrogen, and fluorine are mixed to obtain the monomer.

Figure 6-5 Typical polymers produced through addition (chain) polymerization.

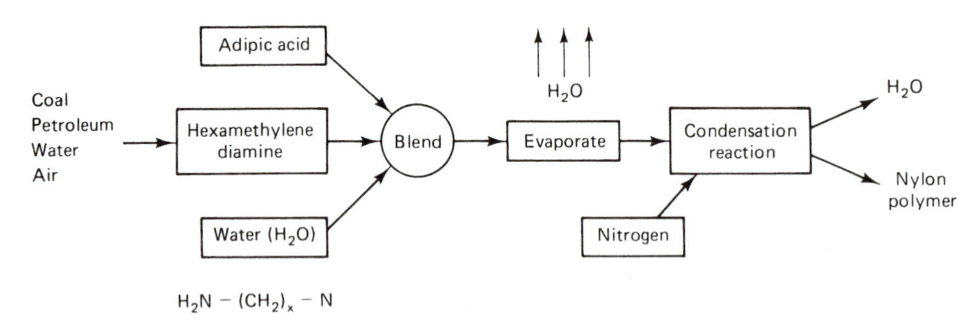

Figure 6-6 Graphical illustration of condensation (step) polymerization of a polyamide (nylon).

polymers and their final properties. While we cannot see the length of the polymers, the molecular weight provides a good indication. Refer to Module 2 for an introduction to molecular structure and bonding.

Molecular weight provides a weight for a single unit or molecule in a compound. To determine the ***molecular weight of the compound,*** it is necessary to multiply the weight of a molecule by the number of molecules or monomers. We said polyethylene typically has about 2000 monomers, so we multiply 2000 times the molecular weight of one monomer to obtain 56,000:

Illustrative Problem

Molecule or monomer	Molecular weight		Number of monomers		Molecular weight of compound
C_2H_4	28	×	2000	=	56,000

Involved chemical analysis is required to determine polymer chain lengths, and such analysis reveals that chains are not uniform in length. Consequently, the value given for a particular resin would be either its ***average molecular weight*** or ***degree of polymerization*** (DP), which indicates the number of repeating mers in the material. The degree of polymerization is the ratio of the molecular weight of the molecule to its mer's weight. Long-chained polymers have a greater molecular weight, which causes more entanglement of the chains and thereby increases ***viscosity*** (resistance to flow). The degree of viscosity affects the processing of the polymer, so manufacturers must take it into account.

6.2.2 Crystalline Structures

In addition, polymerized neighboring chains are held together by weak secondary (intermolecular) bonds known as van der Waals forces, as described in Module 2. As the chains grow during polymerization, they intermingle in a random pattern that will lead to an ***amorphous*** (disorderly) ***structure.*** Bear in mind that the growth of chains is three-dimensional and not simply flat, as normally diagrammed. Visualize a bowl of cooked spaghetti with its uneven strands entangled; that gives a good idea of the arrangement of an amorphous polymer structure.

Actually, polymers are semicrystalline in varying degrees, with the amorphous structured polymers having only slight regularity, while other polymers may have a high degree of crystallinity (Figure 6-7). Metals achieve crystallinity due to the uniform nature of the unit cells of their space lattices. Crystallinity in polymers can alter strength and toughness. The twisting, coiling, and branching of polymer chains causes disorder, and amorphous regions develop in polymers as they solidify. Chain configuration and rate of cooling affect the orderliness. Slow cooling allows chains to move into alignment. While metals are normally fully crystalline, the degree of crystallinity in polymers can range from mostly amorphous to mostly crystalline. Mostly amorphous polymers will be less dense than mostly crystalline polymers because the chains pack closely together.

Molecular motion of short polymer chain segments and polymer branches cause amorphous polymers to be stiff, hard, and brittle at room temperatures. Increases in temperature cause greater molecular motion through thermal mixing of the atoms and molecules. This causes an increase in the volume of the material. The thermal mixing from higher temperatures also increases the spacing between molecular segments and permits increased flow of the materials. A discussion of glass transition later in this module deals further with the concept of molecular motion and flow in polymers.

The degree of crystallinity is determined by structural regularity, compactness, and amount of flexibility, which allows packing of chains. Reduction of random chain lengths

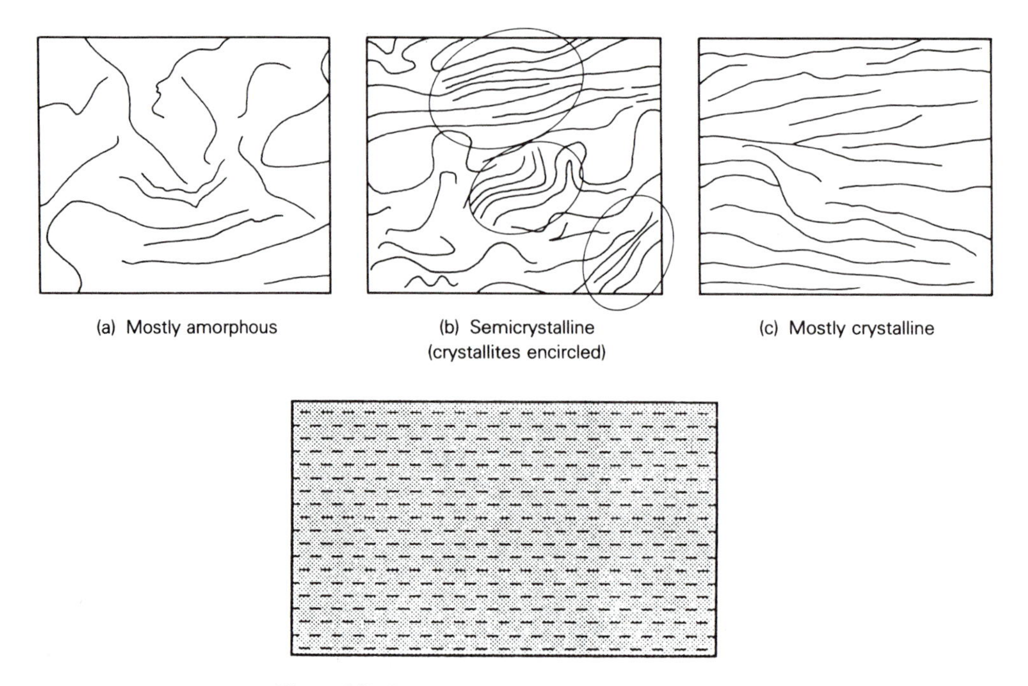

Figure 6-7 Degrees of crystalline structure of polymers.

provides regularity. Stronger secondary forces allow greater compacting. ***Liquid-crystal polymers*** (LCPs) develop highly oriented (rod-like) molecules for increased directional strength [Figure 6-7(d)]. Other chemical factors affect crystallinity, including configuration and tacticity [see Figure 6-3(e)–(g)]. See the references at the end of the module for more information.

Polyethylene serves as a good example of a polymer capable of a high degree of crystallinity because (1) the linear structure of the chains is conducive to packing, and (2) its molecular pattern is flexible, which provides easy packing even though it has weaker secondary bonding. Secondary bonds (van der Waals forces) normally promote crystallinity.

Polyethylene (PE) has the potential for a wide range of properties because of technology's ability to control its molecular weight and crystallinity. Low-density polyethylene (LDPE) has a molecular weight below 10,000, while ultrahigh-density polyethylene (UHMWPE) has molecular weights much above 1.0×10^6 [see Figure 6-8(a)].

Crystallinity of polymers is also achieved through processes such as extrusion and drawing. ***Oriented polymer fibers*** are obtained by drawing an amorphous polymer through a die, which improves their strengths or ***orientation*** in the direction of drawing [see Figure 6-8(b)]. This concept can be illustrated by stretching a polyethylene sandwich bag. The method of manufacturing orients the polymers in one direction. By pulling on the bag in perpendicular directions it is possible to notice the greater resistance from the oriented direction; also, as the bag is stretched, it becomes stronger due to further orientation.

The cooling rate and processing during cooling affect the crystal patterns. ***Spherulites*** form, in which nuclei generate a spherical pattern that grows until several

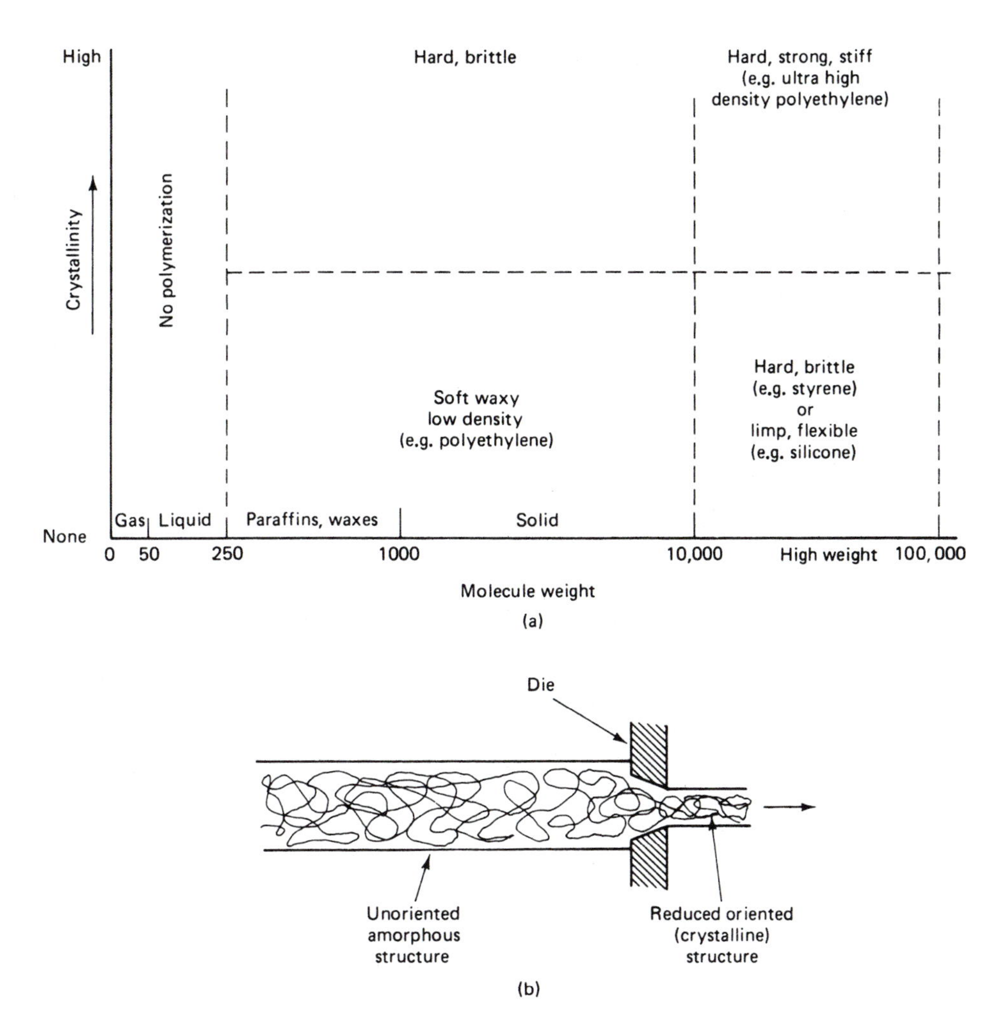

Figure 6-8 Polymers and crystallinity. (a) Properties: relation to molecular weight and crystallinity. (b) Effect of pulling an amorphous polymer through a die.

spherulites melt at boundaries, as a result of supercooling. When large spherulites are allowed to grow, weakness results. Heat treatment of polymers can change their crystal structure; for example, the annealing processes reheat the polymer for a specified time to permit crystal thickening. In Figure 6-9, electron micrographs show polyethylene in the amorphous state, then as a single linear crystal, and then as an annealed crystal.

6.2.3 Branched and Network Structures

As seen in Figure 6-10(c), some polymers do not simply grow linear chains [Figure 6-10(a) and (b)]; rather, the chains develop branches much like those of a tree. As a result, the branched polymers are not as symmetrical and consequently will not achieve the degree of crystallinity obtained in the more regular linear polymers. Numerous techniques

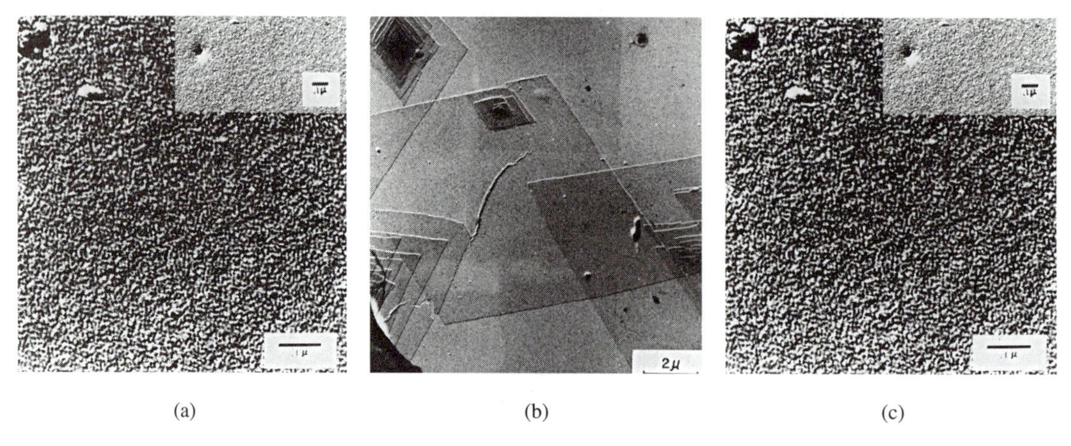

Figure 6-9 Crystallinity in polymers. (a) Electron micrograph of amorphous polyethylene. (b) Single crystal of polyethylene. (c) Single crystal of linear polyethylene annealed at 120°C for 30 minutes. (Dr. Phillip H. Geil, University of Illinois.)

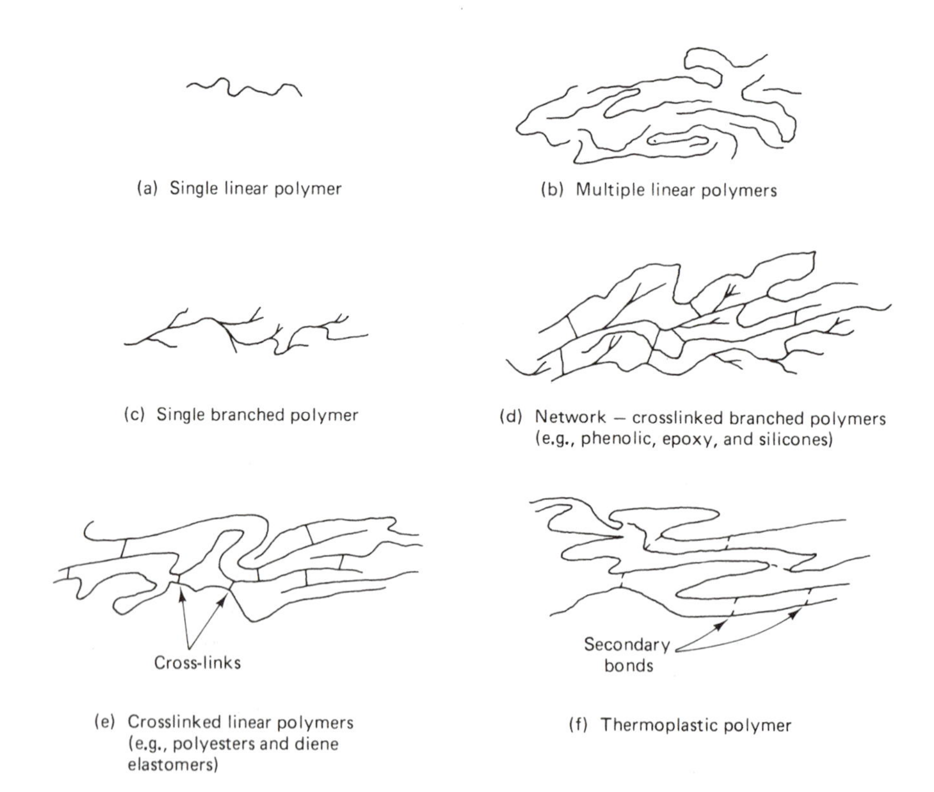

Figure 6-10 Various polymer structures. (a) Single linear polymer. (b) Multiple linear polymers. (c) Single branched polymer. (d) Network: cross-linked branched polymers, e.g., phenolic, epoxy, and silicones. (e) Cross-linked linear polymers, e.g., polyesters and diene elastomers. (f) Thermoplastic polymer.

are employed by polymer chemists and materials technologists to vary both the length of linear chains and the degree and uniformity of branching.

Networks of polymers develop, and primary bonds form between chains. As contrasted to the weaker secondary forces in linear polymers, ***cross-linking*** is the development of covalent bonds between chains, as seen in Figure 6-10(d) and (e). The cross-linked network polymers do not soften when heated, as do the ***thermoplastic*** polymers that are held together by van der Waals forces [Figure 6-10(f)]. Heating of thermoplastics causes these weaker bonds to lose their hold, and thus movement of the mers and softening of the polymer. Thermosetting polymers are like hard-boiled eggs. Once cooked, the egg cannot be softened to flow into another shape. But a thermoplastic material such as a candle can continually be reheated and molded into new shapes. Normally, polyethylene is a thermoplastic material, but through electron irradiation cross-links are developed. Epoxy can exist as a thermoplastic polymer, or through the connections of end groups, it will become a thermosetting polymer. Sulfur is used in rubber to form numerous cross-links in the ***vulcanization*** process. The more cross-links, the harder the rubber is. A reference to cross-linking is the ***netting index,*** which designates the number of cross-links per 100 linear bonds.

The number of double bonds in a polymer provides more sites for reaction than those with fewer double bonds. Natural rubber has hundreds more double bonds than butyl rubber, which has more tightly placed, single-bonded molecules. A hard rubber has a netting index of 10 to 20, while a hard thermosetting plastic such as phenolic has an index of about 50.

SELF-ASSESSMENT

6-1. Refer to the Pause & Ponder at the beginning of this module. What was the goal of the living environments concept house? Name some of the special features.

6-2. What is the name for a chemical that serves as a molecular matchmaker, holding chemicals together for a reaction? Provide examples of these chemicals and some of the specific reactions in which they are used.

6-3. Define the term *polymer.* List the major groups of polymers, and cite examples of materials in each group.

6-4. Discuss the sources of raw materials used in the production of synthetic polymers. Explain how the depletion of fossil fuels can affect synthetic polymers. What alternatives can meet the needs served by present-day synthetic polymers?

6-5. Explain the necessity of unsaturated hydrocarbons and bifunctional monomers for polymerization. Describe the two main polymerization processes and list the typical products of each.

6-6. What type of polymer develops highly oriented (rodlike) molecules for increased directional strength?

6-7. Discuss the effect of crystallinity on polymers and explain methods of achieving crystallinity.

6-8. Calculate the molecular weight for poly(vinyl chloride), polystyrene, and nylon 6.

6-9. Calculate the average molecular weight of a polyethylene compound with 10,000 monomers. Would this be classified as a low-, medium-, or high-molecular-weight polymer?

6-10. What mechanism produces network polymers? Do thermoplastic polymers form networks?

6-11. Does condensation polymerization usually produce thermosetting or thermoplastic polymers? What are the exceptions? What terms describe cross-linking in rubber? What chemical produces this cross-linking?

6-12. What provides for greater cross-linking in polymers? What is the nature of a polymer with a netting index of 50?

OBJECTIVE QUESTIONS

6-1. Which is *not* a goal of the living environments concept house?

a. More affordable housing

b. Increased use of stick construction

c. Demonstration of the benefits of plastics in construction

d. Exhibition of factory-manufactured houses

6-2. Chemical that serves as "molecular matchmaker" necessary to begin polymerization reaction.

a. Monomer **b.** Isotope **c.** Catalyst **d.** Filler

6-3. Polymer with one or more multiple covalent bonds, such as ethylene.

a. Unsaturated hydrocarbons **b.** Saturated hydrocarbons

c. Aerogels **d.** Zeolites

6-4. Adding sulfur to rubber to strengthen it.

a. Condensation **b.** Vulcanizing **c.** Wetting **d.** Addition

6-5. Multiple covalent bonds that provide active sites for cross-linking.

a. Isotopes **b.** Single bonds **c.** Double bonds **d.** Ablation

6-6. Highly oriented (rodlike) molecules that increase directional strength in polymers.

a. Liquid-crystal polymers **b.** Linear polymers

c. Amorphous polymers **d.** Branched polymers

6-7. What does the *netting index* indicate about a polymer?

a. Degree of hardness **b.** Ability to weave fiber

c. Degree of branching **d.** Level of adhesion

6-8. Allowing large crystals to grow in polymers will usually do what?

a. Strengthen **b.** Weaken

6-9. What effect will drawing polymers through dies yield?

a. Increases amorphous structure **b.** Increases crystallinity

c. Develops cross-link **d.** Develops branches

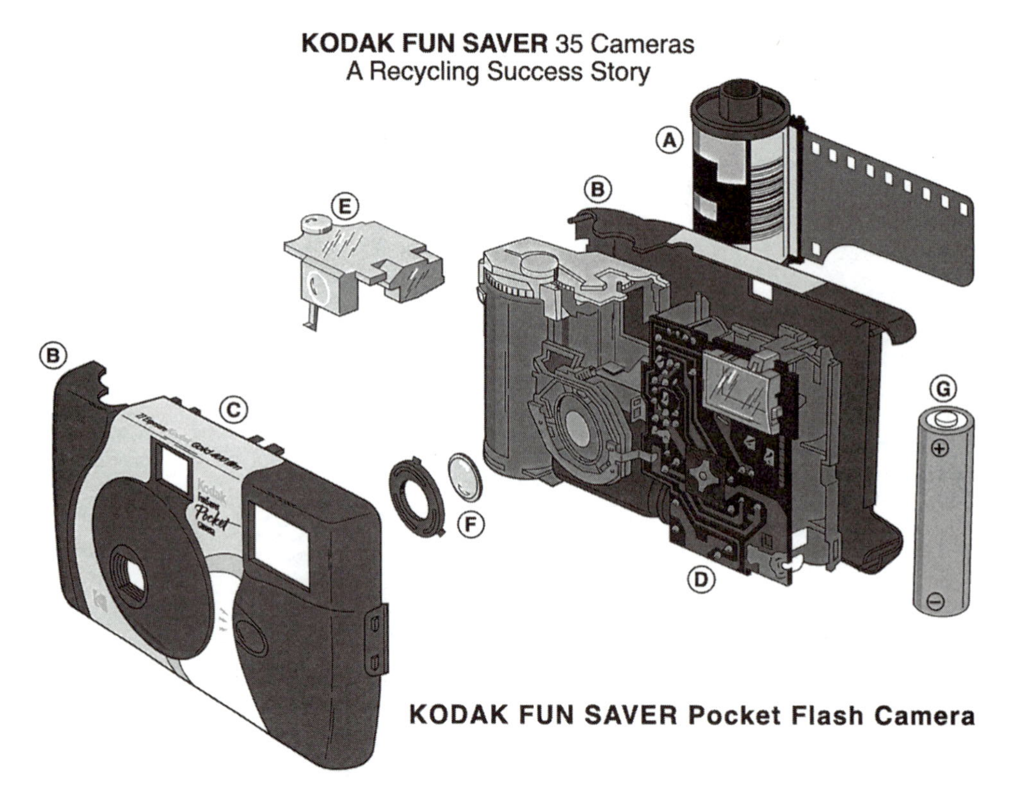

Plastics in the materials cycle. Kodak's Fun Saver 35 "single-use" cameras seem to present disposal problems. However, Kodak has set up a "closed-loop" recycling system. The cameras are returned to the photofinisher, who has the option of sending them back to Kodak. Kodak reimburses the photofinisher for each camera, plus shipping costs. The components of the camera shown here are recycled in the following manner: A: Film—processed by photofinisher. B: Covers—polystyrene ground up and processed for new covers; paperboard outershell made of recycled paper; polycarbonate plastic shells sold to recyclers for nonphotographic materials. C: Label—ground up for recycling with covers. D: Camera mechanism and electronic flash system are tested, inspected, and reused. E: Viewfinder—polycarbonate plastic is reground and recycled into new internal camera parts. F: Lens—to ensure optical purity, new lenses are installed on each camera; acrylic lenses are ground up and sold to an outside company for raw materials on other products. G: Battery—donated to charity, or photorefinisher may use it. (Courtesy © Eastman Kodak Company.)

6.3 PLASTICS

The Society of Plastics Industry defines ***plastics*** as

> any one of a large and varied group of materials consisting wholly or in part of combinations of carbon with oxygen, hydrogen, nitrogen, and other organic or inorganic elements which while solid in its finished state, at some stage in manufacture is made liquid, and thus capable of being formed into various shapes, most usually through the application, either singly or together, of heat and pressure.

This definition is very broad and rather awkward, but it points out that plastics come from such wide ranges of raw materials and processes, have so many varied properties, and take such diverse forms that they almost defy definition. Even natural and synthetic rubber fit this description.

6.3.1 Classification of Plastics

Several methods are used in the classification of plastics. The older and perhaps obsolete system of thermosetting plastics and thermoplastic plastics groups is still commonly used; however, many plastics, and someday possibly all plastics, can fit into either grouping. Using a grouping of thermosetting or thermoplastic is useful as a ***processing classification*** because it indicates what types of processes a certain plastic can undergo. To illustrate, a thermoplastic sheet can be used as a finished product or heated to be reshaped. Once a thermoset part has been molded, it can no longer be reheated for further shaping. Classifying plastics as either ***linear*** or ***cross-linked network*** provides a more descriptive system that is useful to the designer in determining general properties. For example, linear plastics can continually be remolded and usually have low heat resistance, whereas the cross-linked network plastics have greater heat and chemical resistance. Another grouping system involves the *nature of the material:* rigid, flexible, or elastic. Closely connected with the nature of the material classification are the *uses:* general purpose or engineering. ***General purpose*** would include the bulk of plastics that we encounter daily, such as cellulosic, acrylics, and vinyls. ***Engineering plastics*** are those that substitute for traditional engineering materials such as steel and wood. Some engineering plastics are considered ***structural plastics*** capable of bearing supporting loads. Much overlap exists for plastics in these classification systems.

Cellular (foam) plastics are listed by the Society of Plastics Industry (SPI) as a plastics group. Cellular plastics are those that have been foamed or had gas and/or air entrapped in the polymer resins to reduce the density of the finished product [Figure 6-11(a)]. Foaming agents include gases (nitrogen, carbon dioxide, or air) and liquids (chlorinated aliphatic hydrocarbons, alcohols, or ethers). Hollow spheres of glass or resin are also used to create voids. Whipping action introduces air, carbon dioxide is generated from chemical reactions in the resin, and volatile liquids vaporize into gas through ***endothermic*** reactions (absorbed heat when resin bonds break). Water is produced through ***exothermic*** chemical reactions (heat given off) and it volatilizes to produce cells. Aerogels, which include both polymer and ceramic open-celled structures, have nanoscale matrices composed of interconnecting colloidal-like fibrous chains with diameters of about 10 nm. They serve as refrigerator insulators and solar windows, with the potential for many similar uses

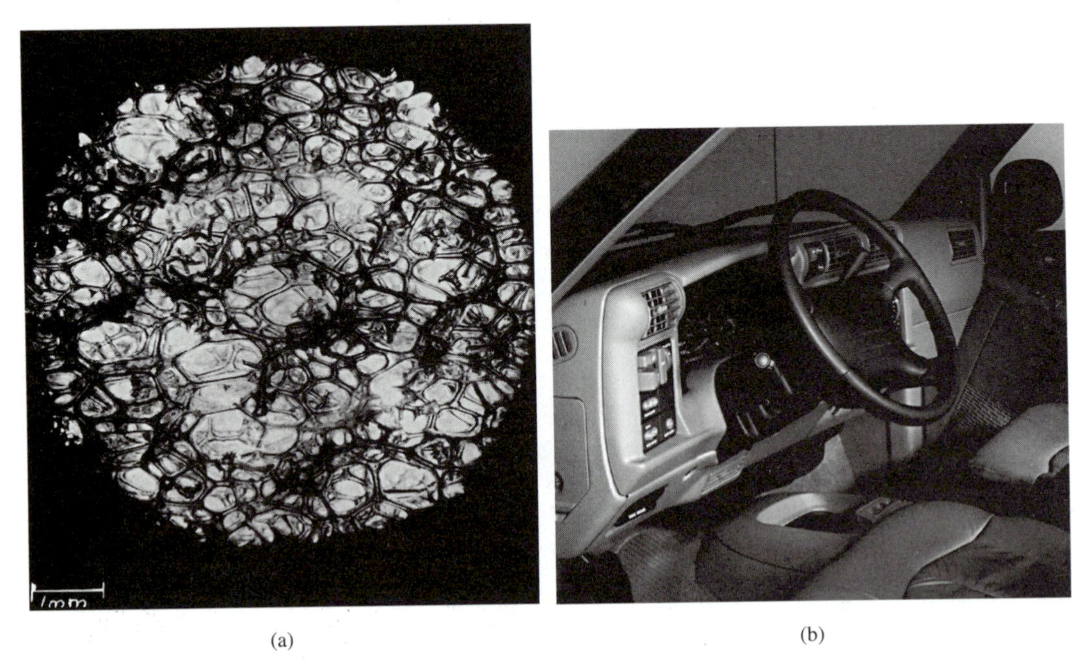

(a) (b)

Figure 6-11 High-density, low-density, and energy absorbing cellular foam polymer. (a) Photomicrograph enlargement of foamed-polymer reveals cells created by gas. (b) Cellular plastics and elastomers are found on many vehicles, as seen here for the seats and steering wheel. The heat-resistant, high-impact, dimensionally stable thermoplastic polycarbonate plastics are also seen on the instrument panels for GM pickup trucks and sports utility vehicles. The ML6143H grade of Lexan™ provides ease of processing plus the use of NVOCs (nonvolatile organic compounds) and water-based painting. (GE Plastics.)

due to their density (a few times that of air) and low thermal conductivity (1% of full-density glass). Practically all thermosetting and thermoplastic resins can be foamed.

Advantages offered by cellular plastics include weight reduction, increased bulk, improved thermal insulation values, greater shock absorption, increased strength-to-weight ratio, and ability to duplicate wood in terms of texture, density, and feel. Cost savings come with foaming. The addition of glass microspheres or bubbles to ***sheet molding compounds*** (SMC)* can reduce weight while improving impact strength because the hollow spheres dissipate energy through shock-absorbing action (dampening). The tiny bubbles, which are from 10 to 200 μm in diameter, also improve mechanical and chemical properties while facilitating processing. Cellular plastics are available in many densities (1.6 to 96 kg/m^3) as rigid, semirigid, or flexible compounds, and with or without color. The auto industry has accepted many cellular plastics for their numerous favorable properties, especially their weight-saving value. The polyurethane (PUR)* foams produced through reaction injection molding (RIM)* composited with reinforcing fibers (FR)* serve as hoods, seat frames, or doors.

*ASTM abbreviation.

Cellular plastics are widespread and include polystyrene foams for building construction, packaging, and appliance insulation; cellular polyethylene for electrical wire insulation; polyimide foams with high heat resistance for aircraft; epoxy foams for flotation devices; cellular silicone for electrical and electronic encapsulation, chemical-resistant fillers, and structural insulation; flexible and rigid poly(vinyl chloride) (PVC)* for cushioning, upholstery, carpet backing, and toys; and foamed acrylics for decorative materials. Syntactic glass-filled foams of polyester are used in sports equipment for composite tennis rackets, bowling balls, floating golf balls, helmets, skateboards, and skis.

Manufactured fibers are polymer strands classified as either regenerated cellulosics or noncellulosics (i.e., synthetics). ***Natural fibers*** such as jute, hemp, wool, and cotton have been used since early times and still are important commercial materials. Fibers are woven into fabrics and used for line, string, and rope (e.g., nylon-monofilament fishing line and tennis string or polypropylene rope). In Module 8, we will further define fibers and related terms and develop their importance in composites.

Because cellulose does not melt or dissolve, it must be processed (regenerated) into a syrup; and from there it is formed into fibers such as rayon and acetate. Synthetic fibers created by chemical synthesis methods (condensation polymerization and addition polymerization) include polyamide, polyester, and polyurethane fibers. Blending of natural fibers such as cotton and wool will provide desired properties, as may be sought in woven fabrics for clothing and upholstery. Engineering of fibers allows desired properties to develop and is accomplished by varying molecular weight, varying cross-section, and orienting molecules through drawing and stretching. The ability of fibers to ***wet*** (absorb or ***wick*** moisture) is important in coloring them or in using them in clothing. Olefins wick well and make good sports clothing; acrylic fibers do not absorb moisture well. Fibers treated with poly(ethylene glycol) (PEG), a polytherm coating, are ***enthalpic*** (absorb, store, and release large amounts of heat) and ***hygroscopic*** (release moisture as vapor, not droplets), so as woven fabric they make excellent sporting and cold-weather clothing.

Newer formulations of synthetic fibers have produced the superstrong, lightweight aramid fibers that are woven into bulletproof vests and serve as reinforcers for tires, fire-fighting and racing helmets, and clothing. DuPont's aramid fiber, Kevlar, is a liquid-crystal polymer (LCP) fiber. Poly-*p*-benzamide can be produced only as fibers; it cannot be molded into solid bulk shapes. Even superior to aramid, the polyolefin fiber Spectra consists of ultrahigh-molecular-weight polyethylene polymers processed to yield highly oriented fibers. This polyolefin fiber is woven into sails and space-suit/boot/glove structures because of its superior abrasion resistance, high flexural endurance, chemical resistance, low moisture absorption, and superior specific strength.

Typical of advanced materials, the engineered Gore-Tex fiber possesses a string of properties that makes it suitable for a wide range of applications. The expanded polytetrafluoroethylene (PTFE) [Figure 6-12(a)] is chemically inert (not affected by any common chemical), has a low friction coefficient, can function in a wide temperature range, does not age, and is weather durable. Such properties have resulted in its selection for wide-ranging applications such as clothing, joint sealants, nonshedding filters to provide micro-contamination control for computer disk drives and manufacturing clean rooms, and

*ASTM abbreviation.

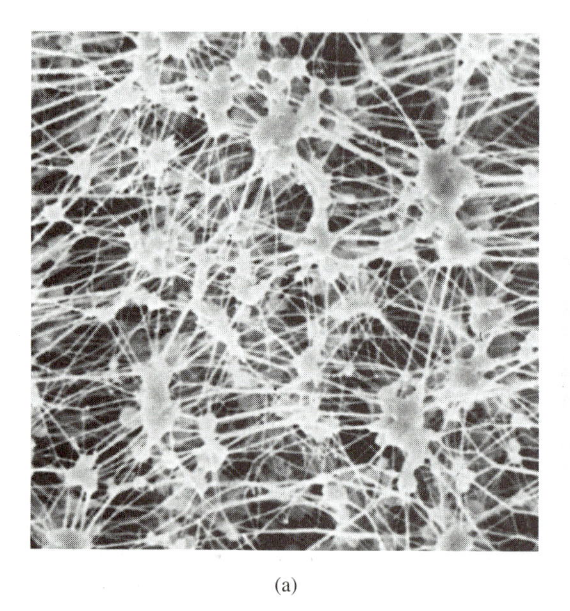

(a)

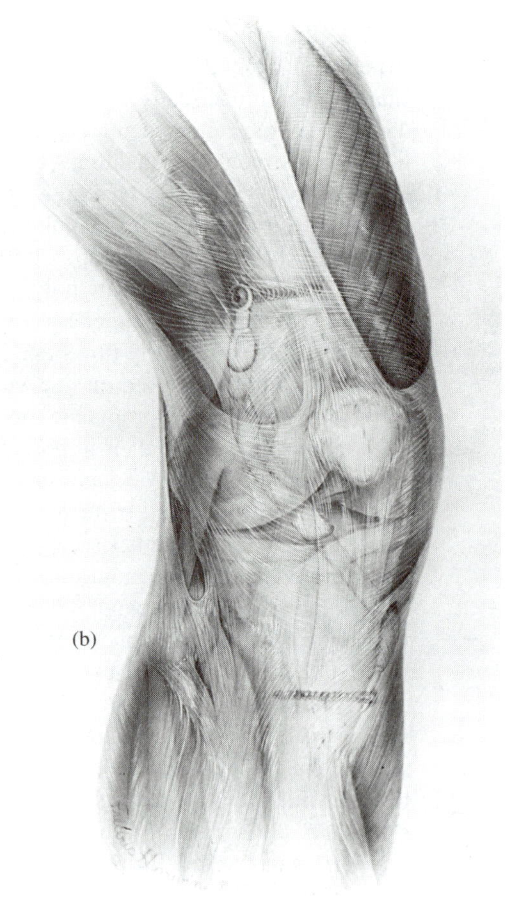

(b)

Figure 6-12 (a) Expanded PTFE Gore-Tex® fiber. Photomicrograph reveals porous microstructure. (b) PTFE fibers for synthetic knee ligament. (W.L. Gore & Associates, Inc.)

membranes that offer microfiltration of bacteria and other particulates from air or liquids. Because Gore-Tex materials have superior strength and are biocompatible, hydrophobic, porous, and air permeable, they have been selected for clothing for sports, outdoors, and walking in outer space. They are used in surgery for threadlike sutures to sew together tissue, as heart patches, for tubes to replace diseased arteries, and as synthetic knee ligaments that provide stability for replacing the natural anterior cruciate ligaments [Figure 6-12(b)].

PBI (polybenzimidazole) is woven into fabrics that will not burn in air. Sulfar resists melting until 285°C and acts as an excellent filter that, through reverse osmosis, can separate toxins and heavy metals from water. ***Osmosis*** is the passage of a liquid from a dilute to a more concentrated solution through a semipermeable membrane. This type of membrane allows passage of the liquid but not the dissolved solids in the concentrated solution. ***Reversed osmosis*** is the forced reversal of this natural phenomenon by the application of enough pressure to the concentrated solution to overcome the pressure (osmotic) of the less concentrated solution.

Fiber-reinforced plastics (FRP), along with many laminated and filled plastics, comprise a large group of plastics that fit diverse engineering and general-purpose applications. For both woven fabrics and as reinforcing fibers in FRP, the ***aramid fiber*** is a plastic of superior qualities. Under tensile loads, this polymer equals steel when compared to similar sizes, but has a superior strength-to-weight ratio. The superior strength develops as a result of the stringing out of the carbon atoms, which line up rather than coiling like a spring in most polymers. The built-in stiffness of the C—C bond greatly resists stretching under tensile loads. See Module 8 for additional discussion of this remarkable fiber with the trade name Kevlar.

Table 10-9 provides comparisons of various plastic classification systems. When using the table it is important to realize that these are general data, and any given plastic may have considerable grade variations due to fillers, additives, and structure. The table also provides the basic molecular structure, chemical formula, American Society for Testing and Materials (ASTM) abbreviations, some selected trade names, and typical uses of the plastics listed. Note that carbon (C) is the backbone for all those plastics shown except silicone, which has an inorganic silicon (Si) and oxygen (O) backbone. In the polyolefins group, ionomers come from ethylene gas combined with inorganic compounds consisting of sodium (Na), zinc (Zn), magnesium (Mg), or potassium (K) salts that bond both covalently and ionically. Ionomers are resilient, oil resistant, and tough, while having very high transparency.

6.3.2 Properties of Plastics

In earlier discussion, the internal structure of polymers was shown to dictate properties. The nearly infinite variations of structure in plastics provide a wide range of properties that continue to grow annually as the demands of technology prompt new discoveries in the manipulation of polymeric structures. These properties fall under the categories of mechanical, chemical, thermal, electrical, and general. Additives and fillers so significantly affect the properties of plastics that one must turn to a handbook or manufacturers' specifications for a given plastic to ascertain its properties. (See Figure 6-13.)

6.3.2.1 Additives (modifiers). Most plastics are truly composites because of the many additives compounded with the resin to enhance its properties. Beyond the catalysts (also known as accelerators or initiators) that are added in minute quantities to start polymerization, additives serve to increase processibility, reduce oxidation (corrosion), add color, reduce molecular weight, increase flexibility, reinforce, retard flammability, or increase electrical conductivity. Most polymers will absorb foreign elements that can cause degradation. ***Stabilizers*** are added to prevent this. Ultraviolet radiation and oxygen (ozone) can alter the chemical bonds of plastics so that ***free radicals*** develop within chains. A ***free radical*** is an atom, ion, or molecule with an unpaired electron. The Pauli exclusion principle states that no two electrons in an atom can have the same four quantum numbers. For that reason, only electrons with opposite spins can pair, thus creating a covalent bond. The free radicals are segmentations within the polymer that can easily combine with other elements to cause a breakdown of the plastics structure and severely affect properties. Cross-links may form to produce a more brittle plastic or cause the disruption of the chain structure, which will limit tensile strength. ***Carbon black,*** produced through the burning of ***carbonaceous*** materials such as gas or oil, will block out ultraviolet light while

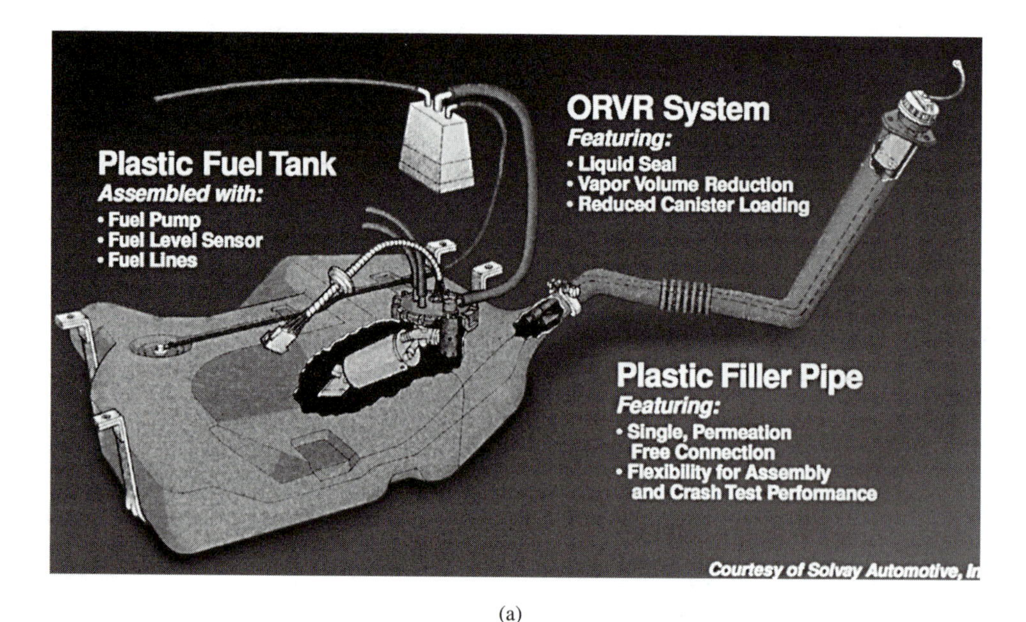

(a)

Figure 6-13 Engineering thermoplastics. (a) The complex, integrated, plastic molded fuel system: the tank houses components for a fuel system consisting of six layers of coextruded high-density polyethylene (HDPE) plus two layers of ethylene vinyl alcohol (EVOH). The EVOH acts to provide on-board fueling vapor recovery (ORVR) that emits less than 0.1 gram of hydrocarbons from the gasoline per day. A projected 60% of the 15 million vehicles produced in North America will utilize plastic fuel systems by 1998. Advantages offered by these systems include reduced hydrocarbon emissions, design flexibility, light weight, and lower cost. (Solvay Automotive, Inc.)

strengthening the plastic. Metals such as barium, cadmium, and lead in compounds are used as stabilizers. Amine and phenol chemicals serve as sunscreens as they interact with photons.

Colorants. Both organic ***dyes*** and inorganic (metal-based) pigments add color to plastics, and some serve dual roles such as stabilizing. Generally, the inorganic pigments will withstand high temperatures without charring or fading. ***Pigments*** disperse rather than dissolve in plastics and reduce the transparency of the material. The pigments also hide flaws such as air bubbles, making it difficult to judge the quality. While nearly an infinite range of color, transparency, translucency, and opacity is possible, the Food and Drug Administration (FDA) allows only certain types of colorants for plastics that make contact with food and drugs.

Plasticizers. Plasticizers are additives that increase flexibility, while cross-linking agents such as organic peroxides cause hardening to produce free radicals. Plasticizers reduce the attraction (secondary valence bonds) between polymer chains and are normally solvents such as alcohol. Poly(vinyl chloride) (PVC) can be very hard and rigid for water and sewage piping. The addition of a plasticizer such as carbon alphates alcohol produces a resin of low volatility that is used in the slush or dip molding of very flexible products such as tool handle coatings, rain boots, or doll parts. *Flame retardants* have become more

(b)

Figure 6-13 (b) Polyphenylene oxide (PPO), Noryl®, offers many advantages over traditional roofing materials, including light weight, easy installation, very good impact resistance, weatherability, walkability, and fire performance. Because of the easy moldability, it can be molded to resemble cedar shake, slate, and barrel tile for the construction industry. (GE Plastics.)

important in the United States as our search for greater product safety broadens. Elements such as boron, nitrogen, chlorine, antimony, and phosphorus reduce the flammability of plastics by preventing oxygen reactions and improving charring. ***Charring*** is seen when wood burns and leaves a protective residue or ash that slows burning; this is known as ***ablation.*** Flame retardants can cause problems by reducing flexibility, tear resistance, tensile strength, and heat deflection. Fluorocarbons, polyvinyls, and polyimide plastics can offer low- or nonburning properties without the use of flame retardants.

Fillers. Fillers improve plastics by increasing bulk, tensile strength, hardness, abrasion resistance, and rigidity; they improve electrical and thermal properties, appearance, and chemical resistance while either increasing or decreasing specific gravity. Common fillers are woodflour, quartz, glass spheres, talc, wollastonite ($CaSiO_3$), calcium carbonate ($CaCO_3$), carbon black, clay, and alumina trihydrate (ATH). For example, glass spheres, clay, or calcium carbonate are added to sheet-molded compounds (SMC) or bulk-molded compounds (BMC) to decrease cost while increasing the rigidity of polyester resins.

Other additives include ***antistatic agents*** and ***coupling agents*** to aid bonding between plastics and inorganic materials in composites, ***foaming agents*** to produce cellular

TABLE 6-1 RANGES[a] OF TENSILE STRENGTH AT ROOM TEMPERATURE FOR SELECTED PLASTICS COMPARED TO OTHER MATERIALS

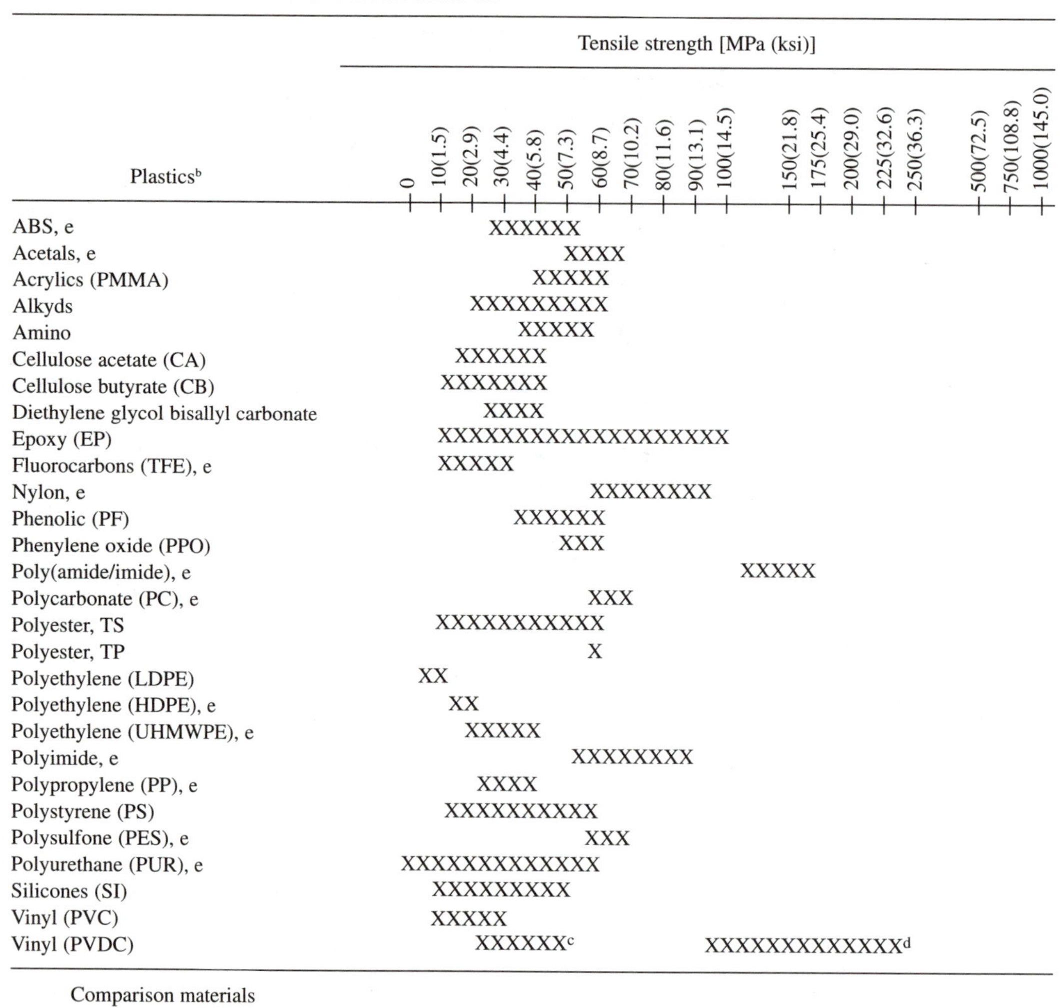

Tensile strength [MPa (ksi)] scale: 0, 10(1.5), 20(2.9), 30(4.4), 40(5.8), 50(7.3), 60(8.7), 70(10.2), 80(11.6), 90(13.1), 100(14.5), 150(21.8), 175(25.4), 200(29.0), 225(32.6), 250(36.3), 500(72.5), 750(108.8), 1000(145.0)

Plastics[b]	Tensile strength [MPa (ksi)]
ABS, e	XXXXXX
Acetals, e	XXXX
Acrylics (PMMA)	XXXXX
Alkyds	XXXXXXXXX
Amino	XXXXX
Cellulose acetate (CA)	XXXXXX
Cellulose butyrate (CB)	XXXXXXX
Diethylene glycol bisallyl carbonate	XXXX
Epoxy (EP)	XXXXXXXXXXXXXXXXXXX
Fluorocarbons (TFE), e	XXXXX
Nylon, e	XXXXXXXX
Phenolic (PF)	XXXXXX
Phenylene oxide (PPO)	XXX
Poly(amide/imide), e	XXXXX
Polycarbonate (PC), e	XXX
Polyester, TS	XXXXXXXXXXX
Polyester, TP	X
Polyethylene (LDPE)	XX
Polyethylene (HDPE), e	XX
Polyethylene (UHMWPE), e	XXXXX
Polyimide, e	XXXXXXXX
Polypropylene (PP), e	XXXX
Polystyrene (PS)	XXXXXXXXXX
Polysulfone (PES), e	XXX
Polyurethane (PUR), e	XXXXXXXXXXXXX
Silicones (SI)	XXXXXXXXX
Vinyl (PVC)	XXXXX
Vinyl (PVDC)	XXXXXX[c] XXXXXXXXXXXXX[d]
Comparison materials	
Alumina	XXXXX
Aluminum alloys	XXXXXXXXXXXXXXXXXXXXXX
Brass	XXXX
Cast iron	XXXXXXXXXXXXXXXX→
Fiber-reinforced plastic	XXXXXXXXXXXXXXXXXXXXXXXXXXXX→
Glass	XXXXXXXXXXXXXXXXXXXX E & S fibers →
Stainless steel	XXXX→
Steel—plain carbon	XXXXX→
Wood	X

[a]Ranges reflect the varieties of the named plastic. Does include certain filled varieties but not fiber-reinforced plastics (FRP). Ranges for other materials are general.

[b]e, Engineering plastic; TS, thermoset; TP, thermoplastic.

[c]Unoriented.

[d]Oriented.

plastics, ***heat stabilizers*** for processing and end product durability, ***lubricants*** to decrease friction and resin melt during processing, ***mold release agents, preservatives*** to retain physical and chemical properties and in some cases prevent growth of bacteria and algae, and ***viscosity depressants*** to reduce viscosity during processing.

6.3.2.2 Mechanical properties. As discussed previously [see Figure 6-8(a)], variations in properties are achieved through variations of molecular weight and crystallinity. With mechanical properties, an increase in crystallinity or density, plus an increase in molecular weight, produces corresponding increases in tensile strength, hardness, creep resistance, and flexural strength, but decreases the impact resistance and percentage of elongation. Additives, fillers, and reinforcers also change mechanical properties to a great extent.

Table 6-1 lists a few selected plastics that reveal the range of *tensile strength* for plastics and also includes other engineering materials for comparison. Because many different varieties and grades are available within a certain plastics group, there is a wide range of tensile strength for plastics of the same name. For example, some flexible cast epoxies have a tensile strength lower than 10 MPa, while a cycloaliphatic epoxy casting resin exceeds 100 MPa. The values in Table 6-1 are for specimens tested at standard temperature and pressure (STP). Many plastics lose strength rapidly at relatively low temperatures (20°C), while others, such as polyimides, can maintain full strength near 500°C over short periods while also withstanding cryogenic temperatures around −300°C. Figure 6-13 shows applications of engineering thermoplastics for automotive use and building construction. Note the integration of various plastics that form a materials system for the fuel tank in Figure 6-13(a). Figure 6-14 provides another example of an engineering thermoplastic, polyethersulfone (PES), seen in this case used for fiber optic (FO) connectors. Radel®, a PES blend from Amoco, was selected for use as FO connectors because of the following properties: allows precise injection moldings; facilitates concurrent engineering

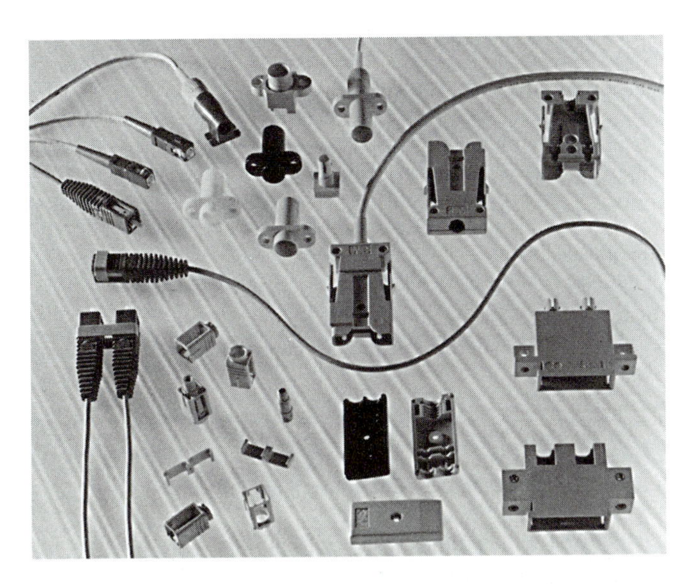

Figure 6-14 Polyethersulfone engineering thermoplastic allows for a wide range of injection molded fiber optic connectors and components. Intricate parts can be molded to extremely close tolerances to meet the critical alignment requirements for FO cables. *(Advanced Materials & Processes.)*

due to the ease of design that does not require metal inserts or glass reinforcers; makes tough, chemical resistant, and dimensionally stable components in temperature ranges from −40° to +85°C (−40° to +185°F); and permits use of organic colorant systems to meet industry standard color codes for FO connectors. Table 6-1 shows that other engineering materials have far greater tensile strengths than the engineering plastics, but the low density of plastics often yields a competitive or higher strength-to-weight ratio. Using ***specific strength*** as the comparison factor, we have

$$\text{Specific strength (meters)} = \frac{\text{tensile strength}}{\text{density (weight)}}$$

Polyester pultrusion, fiber-reinforced plastics (FRP) have specific strengths around 35,000 m, while stainless steel has a specific strength around 25,000 m. Other FRPs, dealt with in Module 8, have even higher specific strength ratios (e.g., up to 90,000 m for epoxy reinforced with carbon fibers). Table 6-6 can be used with the tensile strengths given in Table 6-1 to calculate other specific strengths for comparisons.

Illustrative Problem

Use Table 6-1 and Table 6-6 to determine the mid-value in the range of tensile strengths and densities of plain-carbon steel and polyimide. Substitute the values into the formula above and determine which has the higher specific strength. Use Table 10-7 for conversion of units.

Solution

$$\text{Specific tensile strength} = \frac{\text{tensile strength}}{\text{density (weight)}}$$

From Table 6-1,

Plain-carbon steel	680 MPa
Polyimide	76 MPa

From Table 6-6,

Plain-carbon steel	8200 kgf/m^3
Polyimide	1260 kgf/m^3

Substituting in the formula above, we find that

$$\text{Specific tensile strength} = \frac{680\ \text{MPa}}{8200\ \text{kgf/m}^3}$$

Reduce units to meters (m) by using the following unit conversions from Table 10-7; 1 kgf = 9.807 N, 1 Pa = 1 N/m^2, and M = 10^6. The specific tensile strength for plain-carbon steel is:

$$\text{Specific tensile strength} = \frac{680\ \text{MPa}}{8200\ \text{kgf/m}^3}\left(\frac{1\ \text{N/m}^2}{1\ \text{Pa}}\right)\left(\frac{10^6}{\text{M}}\right)\left(\frac{1\ \text{kgf}}{9.807\ \text{N}}\right) = 8456\ \text{m}$$

The specific tensile strength for polyimide using the same conversion units is

$$\text{Specific tensile strength} = \frac{76\ \text{MPa}}{1260\ \text{kgf/m}^3}\left(\frac{1\ \text{N/m}^2}{1\ \text{Pa}}\right)\left(\frac{10^6}{\text{M}}\right)\left(\frac{1\ \text{kgf}}{9.807\ \text{N}}\right) = 6150\ \text{m}$$

Plastics compared to steel and most other engineering materials are much softer. Table 4-1 showed various hardness scales in an approximate comparison for matching hardness of selected materials. Hardness comparisons of most common plastics are made with the Shore Durometer A and D method and Rockwell M (HRM) tests. The problem with hardness tests of plastics is that they do not closely correlate to wear or abrasion resistance as do most other materials: polystyrene has a Rockwell M value of 72 but scratches easily; diethylene glycol bisallyl carbonate, marketed as CR39® (allyl diglycol carbonate) by PPG Industries, has a Rockwell M value of 95 to 100 but has abrasion resistance approximating that of glass. This makes it a competitor with glass because it has optical properties approaching those of glass. Methyl methacrylate, under trade names such as Plexiglas and Lucite, has an average hardness of 93 to 98 HRM but abrasion resistance 30 to 40 times less than CR39 using a modified Taber test. ***Abrasion resistance*** and ***wear resistance*** are measured by Taber and other tests that determine the ***percentage of haze*** (loss of clarity) due to marking, or ***percentage of material lost*** through rubbing with abrasives such as aluminum oxide. Table 6-2 shows the results of Taber tests on several plastics and their hardnesses.

Toughness or ***impact resistance*** is generally better in thermoplastics than thermosets. As seen in Table 6-3, polycarbonate is the toughest of the transparent rigid plastics at 12 to 14 pounds per inch and far exceeds polystyrene (10.4 ft·lb/in.) and acrylics (0.8 to 1.6 ft·lb/in.). Polyurethane, with an impact strength of 5 to 8 ft·lb/in., has gained acceptance as automobile bumpers and hoods. Both polyethylene and poly(vinyl chloride) (PVC) have very broad ranges of impact strength, as you may know from the thin-walled tough polyethylene milk jugs and PVC piping that have begun to replace much of the copper and cast-iron pipe used for water and sewage. The values given in Table 6-3 are notched specimens. Notching of most plastics severely limits their toughness, so designs should avoid sharp corners or conditions in which parts subjected to impact might develop scratches or cuts. Instead of the notched Izod test, which may not give the most reliable results, some designers prefer data obtained from impacts made with a falling weight on sheet plastic (ASTM D3029), film (ASTM D1709), and pipes and fittings (ASTM D2444). Figure 6-15 reveals that certain plastics compete well with metals when toughness is required.

6.3.2.3 Viscoelasticity.

Viscoelasticity is a property unique to polymers (plastics, elastomers, adhesives, and wood) and not found in metals or ceramics. Because of the viscoelastic property of plastics, engineering technologists must become involved in rhe-

TABLE 6-2 TABER-TEST RESULTS

Plastic	Hardness	Abrasion resistance (mg loss per 1000 cycles)
Nylon 6/6	114 HRR	5
Acetals	95 HRM	16
ABS	100 HRR	84
Polysulfone	120 HRR	20
Polyimides	98 HRM	0.08

TABLE 6-3 COMPARATIVE IMPACT (IZOD) STRENGTH OF SELECTED PLASTICS

Plastic[a]	Notched impact strength (ft·lb/in.)										
	0	2	4	6	8	10	12	14	16	18	20
ABS, e		XXXXXX									
Acetals, e		X									
Acrylics (PMMA)	XX										
Alkyds		XXX									
Aminos—melamines	X										
Cellulose acetate (CA)	XXXXXXXX										
Epoxy (EP)	X										
Fluorocarbon (TFE), e		XXX									
Nylon 6/6, e	XX										
Phenolic (PF)	X										
Phenylene oxide (PPO), e			X								
Poly(amide/imide), e	XXX										
Polycarbonate (PC)							XXXXXX				
Polyester (PBT)	X										
Polyester (PET)	X										
Polyethylene (LDPE)										XXX→	
Polyethylene (HDPE)	XXXXXXXXXXXXXXXXXXXXXXXXXXXXXXXX										
Polyimide (GRP), e									XXX		
Polypropylene (PP), e	XXX										
Polystyrene (PS)	X										
Polysulfone (PES), e	XX										
Polyurethane (PUR), e			XXXXXXXX								
Silicones (SI)	X										
Vinyl (PVC)	XXXXXXXXXXXXXXXXXXXXXXXXXXXXXXX										

[a]e, Engineering plastic.

(a) (b)

Figure 6-15 Tough plastics. (a) Nylon resin, among the toughest of *engineering plastics,* serves as motorcycle sprockets with wear characteristics that surpass conventional sprockets made of steel and aluminum. (b) This new swim mask made with tough polycarbonate provides optical clarity, antifogging due to the coating, and durability. The mask's neoprene rubber skirt conforms to the face for swimmers ages 8 and up, and Velcro tabs provide precise adjustability. (G.E. Plastics.)

ology (study of flow and deformation of matter) to determine why plastics will react to mechanical stresses in a different manner than metals. The property ***viscoelasticity*** incorporates two properties: viscosity and elasticity. ***Viscosity*** refers to the nature of a liquid's resistance to flow. Motor oil used in an automobile engine is rated by its viscosity; heavy oil, such as 40 SAE (Society of Automotive Engineers), is thick syrup and is used in hot weather, while lighter oil (10 SAE) that flows much like water is used in temperatures below zero. As with oil, plastics become less viscous or flow more easily with increased temperatures. This makes them easier to process but also means that they lose strength with heat.

Elasticity (discussed in Module 4) refers to the ability of a material to return to its original size and shape once a load is removed. With plastics, the two factors come into play so that a sudden impact on the plastic will not result in immediate strain or immediate and full recovery when the stress is removed. Rather, there is a viscous flow internally as the force is absorbed through shearing of molecular bonds, much the way that a shock absorber works in an automobile. The shock absorber is a combination of ***dashpot*** (force gradually released through a slow bleeding of fluid) and spring. Even after all load is removed, the deformation is not fully recovered as with a metal that experiences elastic loading. The effect of heat on viscoelastic properties is discussed under thermal properties. Figure 6-16 illustrates the mechanical models (Maxwell and Voigt) that simulate viscoelasticity.

Viscoelastic properties are determined by measuring the stress relaxation (Section 6.3.2.4) of a material subjected to a constant strain at different temperatures as specified by ASTM D2991. The reduction in stress is measured until it reaches equilibrium. Materials that offer the best strength retention and have good viscoelastic properties are needed for many applications where, for example, a force is required to be maintained for contact

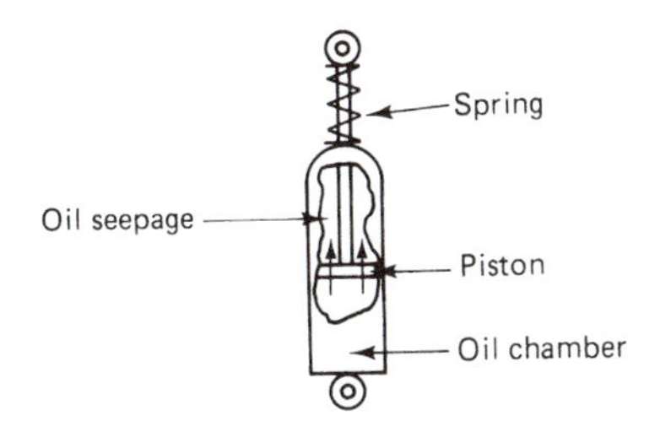

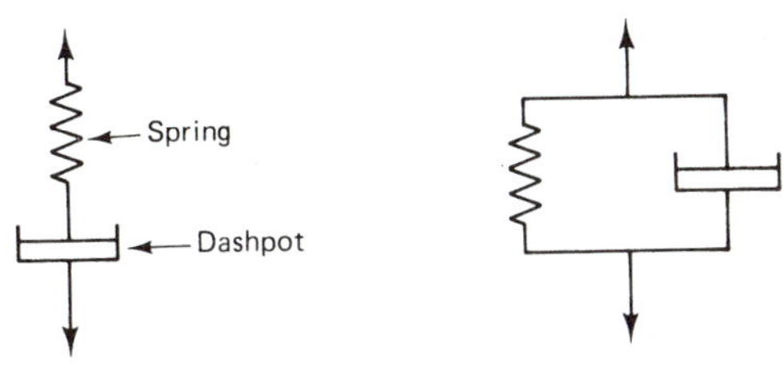

Figure 6-16 Viscoelasticity. (a) Shock absorber. (b) Maxwell series model. (c) Voigt parallel model.

purposes or for clip strength. Figure 8-43 (in Module 8) exemplifies such an application for the electronics industry.

6.3.2.4 Thermal properties. Many plastics lose their strength at relatively low temperatures. Continuous-service-temperature comparisons of plastics reveal most common plastics can endure temperatures no more than 150°C when under low or no stress. Slight increases in stress would reduce the continuous-service-temperature even more. Newer plastics, such as the polyimide thermosetting resins, resist intermittent heat up to 500°C with low stress and up to 250°C for thousands of hours. Graphite-reinforced polyimides can withstand flexural stress to nearly 69,000 Pa at 250°C. Plastic composites have replaced aluminum pistons in some race cars because the aluminum with high thermal conductivity acts as a heat sink and begins to soften at around 150°C. Plastic composites of polyimide and epoxy with glass and graphite reinforcement withstand temperatures approaching 290°C and tremendous mechanical stress, yet they offer weight savings over metals.

Glass-transition temperatures or ***glass point*** *(T_g)* is the point at which polymers act as glass or become viscous liquids (see Figure 6-17). The glass-transition is a reversible change that occurs when a resin polymer is heated to a certain temperature *(T_g)*, resulting in a sudden change or transition from a rigid polymer to a flexible, rubbery material or a viscous liquid. Such a material suffers a major loss in strength when subjected to a temperature near its T_g value, leading to catastrophic failure even if the service (use) temperature exceeded the T_g value for only a short time.

In cooling, the thermal mixing of the atoms and molecular chains slows, and the volume of the plastic decreases because they pack closer together. In packing, many polymers do not form crystals as do metals; instead, they freeze into an amorphous glassy structure. Of course, highly crystalline polymers such as high-density polyethylene (HDPE) would

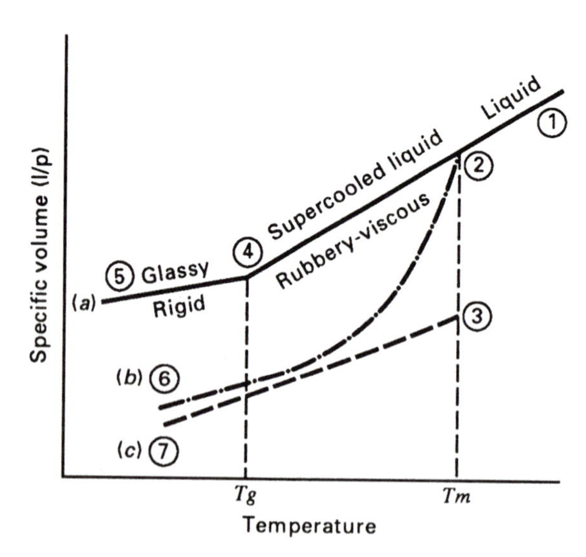

Figure 6-17 Specific volume versus temperature for polymers. A) Linear amorphous polymer [path (a) 1,2,4,5]. B) Partially crystalline polymer [path (b) 1,2,6]. C) Crystalline polymer [path (c) 1,2,3,7]. D) Path through points 1,2,3,7 is a theoretical path approaching a crystalline solid such as a metal. E) T_g (glass temperature) and T_m (melting temperature) for a crystalline polymer.

not follow this pattern, and the degree of crystallinity of plastics would control T_g. The result of cooling to the glass point is that noncrystalline polymers become brittle. This occurs at or above room temperature for certain plastics, for example, for polystyrene and acrylic, at about 100°C. In crystalline plastics with low T_g, such as polypropylene at −10°C, flexibility and impact strength are maintained. In reheating an amorphous thermoplastic, it passes from the brittle glassy structure through its T_g point, and then turns into a tough rubbery form, progressing to a softer and more pliable stage into a viscous liquid upon reaching the melting point (T_M).

Exact melting points are determined optically in crystalline and semicrystalline plastics when the ***birefringence*** (double refraction) is lost due to the change of structure from the crystalline to the glassy state. The earlier discussion on mechanical properties covered viscoelasticity. One can see how T_g will affect this property. A linear amorphous plastic will be quite brittle and not exhibit viscous flow below its T_g, but it will be quite rigid and elastic. As the degree of crystallinity increases in plastics, they become more viscous and less capable of recovering from deformation under load as they possess less elasticity.

Time becomes an important factor in the loading of plastics. A stressed linear amorphous polymer at temperatures above T_g over sufficient time will begin to deform without additional loading or heat as the molecular chains become untangled through viscous flow; this is ***stress relaxation.*** A network thermoset at relatively low temperatures that is under constant stress over a long time will begin to deform and continue deforming with progressively less stress; this is ***creep.*** The softening temperature of polymers can also be increased through an increase in average molecular weight or polymer chain lengths. Thermoplastics are far more prone to creep than thermosets.

Thermal conductivity is low in most plastics, which makes them valuable as thermal insulators. In the cellular or foamed state, air cavities, which do not carry heat or cold, improve even further the insulating properties. However, the rate of *thermal expansion* is quite high for most plastics, generally 10 times as much as metals. Table 6-4 provides a comparison of the rate of thermal conductivity and coefficient of thermal expansion for selected plastics and compares them with other materials.

Plastic films made of select plastics for packaging can withstand heat from conventional cooking ovens; most are not affected by microwaves but some melt from the heat of the food as it is cooked. A polyvinylidene chloride (PVDC) product such as ***Saran Wrap*** has oxygen protection like glass and will not melt with ***microwave cooking;*** crystallized polyester (CPET) will withstand conventional oven temperatures, to 425°F, and microwaves. Polyethylene (PE) products such as Handiwrap and Glad Wrap do not provide the oxygen protection of PVDC or CPET. They can be used in microwave cooking if the fat and sugar content of the food is low enough so that they do not melt the plastic. Polystyrene (PS) foam, often used to package food, is safe for heating liquids such as water, tea, and coffee but will melt if used to heat food such as hamburgers or french fries.

The range of *flammability* for plastics is quite wide. For plastics that sustain combustion, a burning rate test (ASTM D635), which measures the propagation of flame (in./min) on a given size specimen, permits a measure of comparability. For instance, phenolics, polyimides, and fluorocarbons without fire-retardant additives are considered nonburning, whereas the cellulosics are highly flammable. Certain plastics such as

TABLE 6-4 COMPARISONS OF THERMAL PROPERTIES FOR SELECTED PLASTICS AND OTHER MATERIALS[a]

Coefficient of thermal expansion	10^{-6} m/m/K (60 ... 120 ... 180 ... 240 ... 300)	
Plastics		
ABS, e	X	
Acetals (POM), e	X	
Acrylics (PMMA)	XXXXXXXXX	
Cellulose (CA)	XXXXXXXXXXXX	
Epoxies (EP)	XXXXXX	
Fluorocarbons (PTFE), e	X	
Nylon	X	
Phenolic (PF)	XXX	
Polycarbonate (PC)	XXXXXXXXX	
Polyester (PBT)	XX	
Polyethylene (LPDE)	XXXXXX	
Polyethylene (HDPE), e	XXXXXXXXXXXXXXXXXXXXX	
Polyimide	X	
Polypropylene (PP)	XXXX	
Polystyrene (SAN)	X	
Polysulfone, e	X	
Polyurethane (PUR)	X	
Silicone (SI), e	XXXXXXX	
Vinyls (PVC)	XX	
Alumina	X	. . .
Aluminum	XXX	. . . .
Cast iron	XX	. .
Copper	XX	. . . .
Glass	XX	
Rubber	X	
Silver	X	
Steel	X	
Stainless steel	X	
Wood	I	. . . .
Zinc	X	. . . .
Thermal conductivity	W/(m · K) 0.............10...........50..........100.........150.........200..........300.......400	

[a]e, Engineering plastic.

polyurethane and vinyls give off highly toxic fumes when burned. The vinyl siding used on housing ignites above 370°C, whereas wood siding ignites at about 270°C. Generally, thermoplastics are more flammable than thermosets. Thermoplastics can be ground into chips and melted for recycling. While thermosets have the potential for recycling through the use of chemical solvents, the cost may not be justified. However, thermosets can be ground up and used as a composite filler. The high thermal energy possible from thermosets, nearly 3 MJ/kg (megajoules per kilogram), has value as an energy source in which scraps are burned to generate heat for steam-turbine generators in the production of electrical energy. As with wood, plastics develop a char layer that serves as an ***ablative*** or protective shield, which insulates the unexposed area. Nylon and phenolic resins were used on NASA spacecraft as ablative heat shields for reentry into the earth's atmosphere. Although the polymers did burn when exposed to temperatures over 6500°C for several seconds, the charred ablative shield protected the spacecraft because of its thermal insulating properties.

6.3.2.5 Chemical properties. Plastics do not corrode like metals do, but they are subject to deterioration and chemical attack. Whereas the corrosion of metals is determined by weight loss, a thermoplastic's deterioration is measured by weight gain as the attacking chemical combines with the plastic. The result is usually discoloration, swelling, or crazing (fine cracks) with corresponding loss of tensile, impact, and flexural strengths. Polymer-matrix composites (PMCs) have undergone considerable development to arrive at the point where they can be used with confidence in the aerospace industry. Wide use of PMCs on the Boeing 777 (Figure 6-18) attests to the corrosion resistance and many other favorable properties of these polymer-based materials.

As with mechanical properties, ***chemical resistance*** will decrease in plastics as temperatures increase. Table 6-5 provides data for plastics tested at room temperature for their degree of resistance to weak acid, organic solvents, and water absorption. Some common strong acids include hydrochloric and nitric, and weaker acids are lactic, boric, and citric. Common organic solvents include gasoline, alcohol, and acetone. The ***weatherability*** comparison involves several chemical stresses, including heat, moisture, ultraviolet light, and chemicals in the air such as ozone (allotropic form of oxygen) and hydrochloric acid (see Figure 6-19). Notice that the fluorocarbons, with their inertness, have great resistance to weak acids, organic solvents, water absorption, and weathering. Acetal homopolymer resins have great resistance to acetone at room temperature, but at elevated temperatures (65°C) they are unacceptable for service and do not perform well when in environments with strong acids. The copolymerization of acetal resins greatly improves their resistance to inorganic liquids at elevated temperatures, so they are utilized as plumbing products such as valves and pumps. High-density polyethylene (HDPE) has great resistance to acids, water absorption, and weathering, which, coupled with its low cost, makes it a good candidate for numerous packing applications such as blow-molded bottles used for the home and industry; it also serves as electrical wire insulation, oil storage tanks, and recreational body vehicles. As seen in Table 6-5, acrylics (Lucite and Plexiglas) have great resistance to ***weathering,*** which provides good service as exterior windows because their impact resistance is better than that of soda-lime glass and they transmit light (92%) as well as fine optical glass. Certain plastics with little ***water-absorption resistance,*** such as nylon 6/6, are dried before processing so that the absorbed moisture will not cause corrosion of machinery. Moisture absorption causes swelling in plastics, which creates problems in holding dimensional accuracy. When high accuracy must be maintained, plastic of low moisture absorption is required. Polyester used as the resin in fiberglass has great water-absorption resistance and serves well in marine applications such as boat hulls and surfboards.

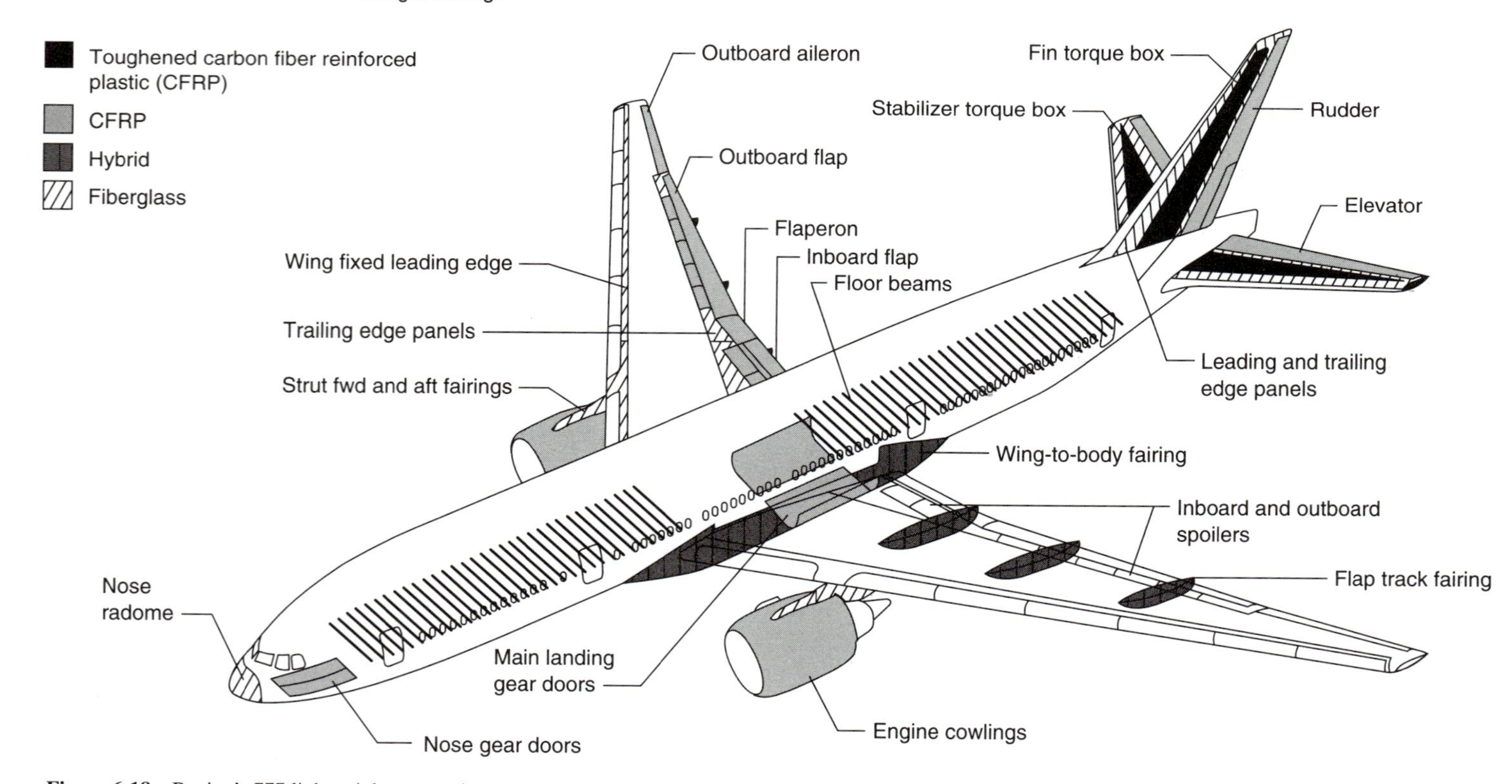

Figure 6-18 Boeing's 777 lightweight composite structure makes extensive use of polymer matrix composites (PMCs). After thorough testing of PMCs in military and research aircraft, polymers proved reliable for commercial aircraft. The "paperless" project utilized concurrent engineering (CE), in which design engineers and manufacturing engineers worked in design-build teams. In addition to advanced metals, the 777 structure used graphite, toughened graphite, fiberglass, and hybrid PMC composites, which were selected for corrosion and fatigue resistance among the other properties listed above. (Courtesy of the Boeing Company.)*

*The Public Broadcasting System (PBS) series "Twenty-first Century Jet" provides an excellent in-depth case study of the product development process, from design through testing, manufacturing, and delivery. The series includes many examples of engineering materials technology. Call PBS at 1-800-344-3337 to obtain further information on the video series.

TABLE 6-5 COMPARATIVE CHEMICAL PROPERTIES OF SELECTED PLASTICS AND OTHER MATERIALS[a]

	Weak acid resistance	Organic solvent resistance	Water absorption resistance	weatherability
Plastics[b]	Little Fair Great	Little Fair Great	Little Fair Great	Little Fair Great
ABS, e	XXXXXXX	XXXXXX	XXXX	XXXXX
Acetals (POM), e	XXXX	XXXXXXXXX	XXXX	XXXXXXXX
Acrylics (PMMA)	XXXXXXXXX	X	XXXXXX	XXXXXXXXX
Alkyds	X	X	XXXXXXXX	XXXXXXXX
Aminos—melamides	XXXX	XXXXXXXXX	XXXXXXXXX	
Cellulose acetate (CA)	XXXXXX	XXXXXX	X	XXXXXX
Cellulose butyrate (CAB)	XXXX	XXXXX	XX	XX
Epoxy (EP)	XXXXXXXXX	XXXXX	XXXXXXXX	XXXXXXXXX
Fluorocarbons (PTFE), e	XXXXXXXXX	XXXXXXXXX	XXXXXXXXX	XXXXXXXXX
Nylon 6/6, e	XXXX	XXXX	X	XXXX
Phenolic (PF)	XXX	XXX	XXXX	XXXX
Phenylene oxide (PPO), e	XXXXXXXXX	XXXXX	XXXXXXXXX	XXXXXXXXX
Poly(amide/imide), e	XXXXXXX	XXXXXXXX	XXXX	XXXXX
Polycarbonate (PC)	X	X	XXXXXXXXX	XXXXX
Polyester, TP	XXXXXXXX	XXXX	XXXXXXXXX	XXXXXXXXX
Polyester, TS	XXXX	XXXX	XXXXXX	XXXXXXXXX
Polyethylene (LDPE)	XXXXXXXXX	XXXXX	XXXXXXXX	XXXXXXXXX
Polyethylene (HDPE), e	XXXXXXXXX	XXXXX	XXXXXXXXX	XXXXXXXXX
Polyethylene (UHMWPE), e	XXXXXXXXX	XXXXX	XXXXXXXXX	XXXXXXXXX
Polyimide, e	XXXXX	XXXXX	XXXXXXXXX	XXXXXXXXX
Polypropylene (PP), e	XXXXXXXXX	XXXXXXXX	XXXXXXXXX	XXXXX
Polystyrene (PS)	XXXXXXX		XXXXXXXX	
Polysulfone, e	XXXXXXXX	XXXXXXXXX	XXXXXXX	XXXXX
Polyurethane (PUR), e	XXXX	XXXXXXXXX	XXXXXXX	XXXXXX
Silicones (SI)	XXXX	X	XXXXXX	XXXXXXXX
Vinyl (PVC)	XXXXXXXXX	XXXXXX	XXXXXXXXX	XXXXXXXXX
Vinyl (PVDC)	XXXXXXXXX	XXXXXX	XXXXXXXX	XXXXXXXX
Comparison materials				
Alumina	XXXXXXXX	XXXXXXXXX	XXXXXXXXX	XXXXXXXXX
Aluminum	XXXXXXXX	XXXXXX	XXXXXXXX	XXXXXXXX
Brass	XXX	XXXXXXXXX	XXXXXXXXX	XXXXXXXXX
Cast iron	X	X	XXXXXXXX	XXXXXXX
Fiber-reinforced plastics vary with plastic matrix from little to great.				
Glass	XXXXXXXXX	XXXXXXXXX	XXXXXXXX	XXXXXXX
Stainless steel	XXXXXXXXX	XXXXX	XXXXXXX	XXXXXXXXX
Steel—plain carbon	X	X	X	
Wood	XXXXXXXXX	XXXXXXXXX	X	XXXX

[a]Exposures at room temperatures except weatherability.

[b]e, Engineering plastic; TS, thermoset; TP, thermoplastic.

Figure 6-19 Chem suit used when entering chemical tanks uses face shields of optical-grade plastic made of CR39® monomer, which resists solvents, acids, scratching, and impact. (PPG Industries.)

6.3.2.6 Density. The density (kilograms per cubic meter) of plastic is generally lower than that of other engineering and general-purpose materials. Table 6-6 shows that only magnesium and some wood have as low mass per unit volume as plastics. Even glass-reinforced plastics (GRP) are lighter than aluminum and steel, while their strength approaches or, in some instances, equals that of these popular engineering metals. CR39® (allyl diglycol carbonate) (Figure 6-19), developed by Pittsburgh Plate Glass for competition with optical glass, has a density about one-half that of glass. This plastic is a practical substitute for glass in larger stylish eyeglasses and for industrial safety glasses because it combines light weight with impact resistance and is superior to acrylics in wear resistance.

6.3.2.7 Specific gravity. Specific gravity is the ratio of the mass of a measured volume of any material to the mass of an equal volume of water at a standard temperature. This ratio is often used for weight comparisons. Specific gravity ratios for plastics range from below 0.06 for foams to over 2.0 for fluorocarbons, compared to 0.5 for softwoods, 0.7 for hardwoods, 2.2 to nearly 4 for glasses and ceramics, nearly 3 for aluminum, and around 8 for steels.

6.3.2.8 Optical properties. A number of plastics have optical properties comparable to glass and offer impact strength superior to most glasses. In addition, they process more easily and, when broken, do not produce sharp splinters. Refer to Section 9.5 for the definition of optical properties used here. Opacity develops in plastics with an increase in crystallinity, whereas an amorphous structure produces transparency. Low-density polyethylene film, such as sandwich bags, is clear, while HDPE for detergent bottles is opaque. Acrylics are available in an unlimited range of colors and are as transparent as the finest optical glass (light transmission equals 92%), can be made opaque, or can be produced through a full range of translucencies. Acrylics have a ***refractive index*** of 1.49 compared to 1.52 for soda-lime glass, 1.47 for borosilicate glass, and 1.46 for 96% silica

TABLE 6-6 MASS WEIGHT DENSITY COMPARISONS OF SELECTED PLASTICS AND OTHER MATERIALS

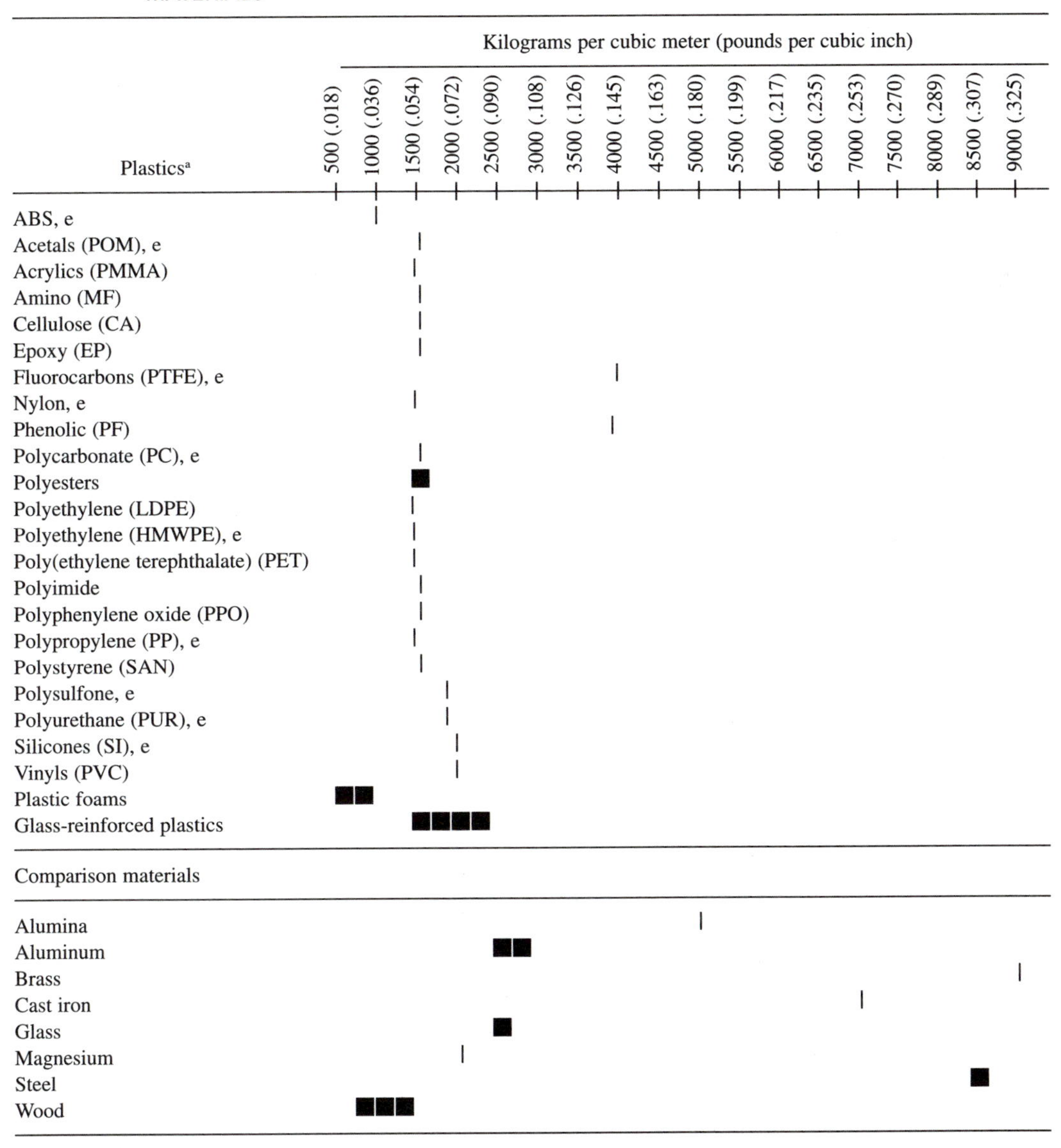

[a]e, Engineering plastic.

glass. (Refractive index indicates the ratio of the speed of light through the materials compared to light traveling through a vacuum. Higher values indicate greater bending of the light.) The limits of transparency of some other plastics follow:

Acrylics	92%	Polyethylene	80%
Cellulose	88%	Polypropylene	90%
Ionomer	92%	Polystyrene	92%
Polycarbonates	90%		
Amino	29%[a]	ABS	33%[a]

[a]Translucent.

Translucent plastics transmit light but objects cannot be clearly seen through them.

Certain plastics, such as acrylics, polyesters, cellulosics, and polystyrene, show colorful stress concentrations when viewed with a polarized-light filter. This photoelastic effect is used as a design tool. ***Photoelastic plastic*** is shaped to a specific design and stressed under polarized light to determine what areas will receive the greatest stress, thus allowing for added material or redesign to handle expected loads (Figure 6-20). As with glass, many plastics can effectively bend light and serve as ***light pipes*** for optical fibers used in medical applications, signs, telecommunications, and plastic art.

6.3.2.9 Electrical properties. Advances in materials have fostered many technological developments. The electrical and electronic developments over the past 75 years owe much to the continuing breakthroughs in plastic materials. Superior insulating properties coupled with good heat resistance of silicones and fluorocarbons lead to large reductions in the weight of electrical motors. Epoxies serve to encapsulate electronic components subjected to temperatures as high as 150°C, corrosive chemical environments, and high vibrations and shocks. Phenolics have the oldest history as electrical insulators and

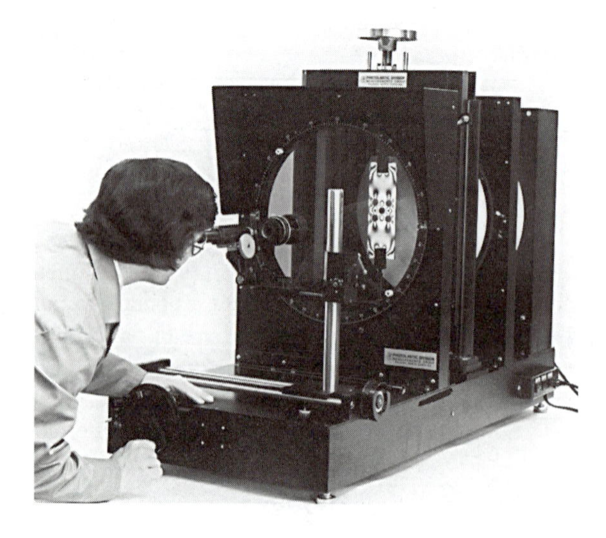

Figure 6-20 Photoelastic plastic. A demonstration model to show stress concentrations. (Measurement Group, Inc., Raleigh, N.C.)

find wide use as housings for automotive-ignition parts (coils and distributors), switches, receptacles, terminal blocks, and bulb bases.

The ***dielectric strength*** (maximum voltage that a dielectric can withstand without rupture) of most plastics makes them the logical choice as insulators. They range from 79×10^6 volts per meter (V/m) for fluorocarbons to 12×10^6 V/m for certain phenolics. These values compare to 19×10^6 V/m for porcelain and 12×10^6 V/m for alumina. The ***dielectric constant*** (ability to store electrical energy) of certain plastics puts them into the condenser category. Condensers in electrical circuits maintain voltage with less fluctuation. At 60 hertz, comparative dielectric constants *(K)* compare as follows: poly(vinyl fluoride), 8.5; cellulose acetate, 4.0 to 5.0; polyethylene (LDPE), 2.25 to 2.35; and polystyrene, 2.45 to 4.75; as compared to alumina, 8 to 10; mica, 5.4 to 8.7; and glass, 3.8 to 4.6. Sections 9.2 and 9.3 describe electrical properties discussed in this section.

Arc resistance is the ability of a material to withstand the arcing effect of an electric current. Insulators, terminals, and switches are subjected to arcing effects, which can burn through or damage the material's surface. The arc test (ASTM D495) measures the total elapsed time in seconds that an electrical current must arc to cause failure. The plastic may ***carbonize*** and become a conductor; burst into flames; produce thin, wiry lines or ***tracks*** (surface imperfections in the form of wiry lines) between electrodes; or become an incandescent (glowing hot) conductor. Arc resistance for unfilled polyimide is 125 s; melamines, 110 to 150 s; FEP fluorocarbon, over 300 s; macerated fabric and cord-filled phenolic show tracking but do not conduct; thermosetting cast polyester, 100 to 125 s; and polyethylene (LDPE), 135 to 160 s.

6.3.3 Plastic Selection

In the selection of a material, cost is of paramount concern. Assuming that properties are equivalent and other features such as appearance are similar when comparing materials, then cost is usually the overriding factor in the selection of one material over the other. Often, a trade-off is made when costs are higher for one material than another. For example, epoxy reinforced with graphite fibers offers tensile strength nearly three times that of stainless steel; however, the cost for stainless steel may be 10 or 20 times less than that of the epoxy composites. If weight is important, as in aircraft, the fact that the stainless steel is over five times as dense as the graphite composite can cause the selection of the reinforced plastic. These comparisons are for exotic materials. More common materials such as plain-carbon steel, which is used more frequently, find strong competition from low-cost plastics such as polyethylene, polypropylene, and vinyls.

High demand for a particular material may result in price drops if suppliers increase their production capabilities and competition results in improved processing methods. In the mid-1970s, graphite fiber composites cost about \$24 to \$29 a pound. To purchase a fishing rod or tennis racket of this new composite, you would have to pay \$150 to \$250. Now it's possible to buy even better designed rods and rackets of graphite composites for as low as \$35 to \$60. The processing technology, driven by demand and competition, yields benefits to the consumer. In turn, that demand should cause even better composite processing technology, which will drive prices down while improving composite quality. Looking at materials from more mature plastics processing technology reflects how low prices may go. Plastics in high demand, such as polyethylene and polyvinyls, sell for prices around $1 to $3 a pound.

Much of our materials innovation results from market pressures and legislation. Once a material is available, the next problem is informing designers of its availability. Some companies specialize in developing databases for a variety of materials by many producers. Producing companies also provide databases to customers. The PC–based *Plastics Materials DATA Digest* gives information on over 13,000 TP and TS resins, including properties, ASTM test descriptions, and manufacturers and suppliers.

Corporate average fuel economy (CAFE) is a federal legislative concept that mandates that automakers achieve set average fuel efficiency for the entire line of cars they produce. As the miles per gallon (mpg) values continually rise to meet CAFE, plastics come into greater use due to good strength-to-weight ratios. To encourage auto and truck designers to use its plastics, General Electric Plastics created the EDD-PC engineering-design database. It is coupled with an Alpha I multimaterial, multiprocessing machine for development of large, integrated thermoplastic parts such as bumper systems and automotive hoods. Customers can perform material evaluation 24 hours a day using the latest test and data-acquisition hardware and software. Figure 6-21 shows an example of a large, inte-

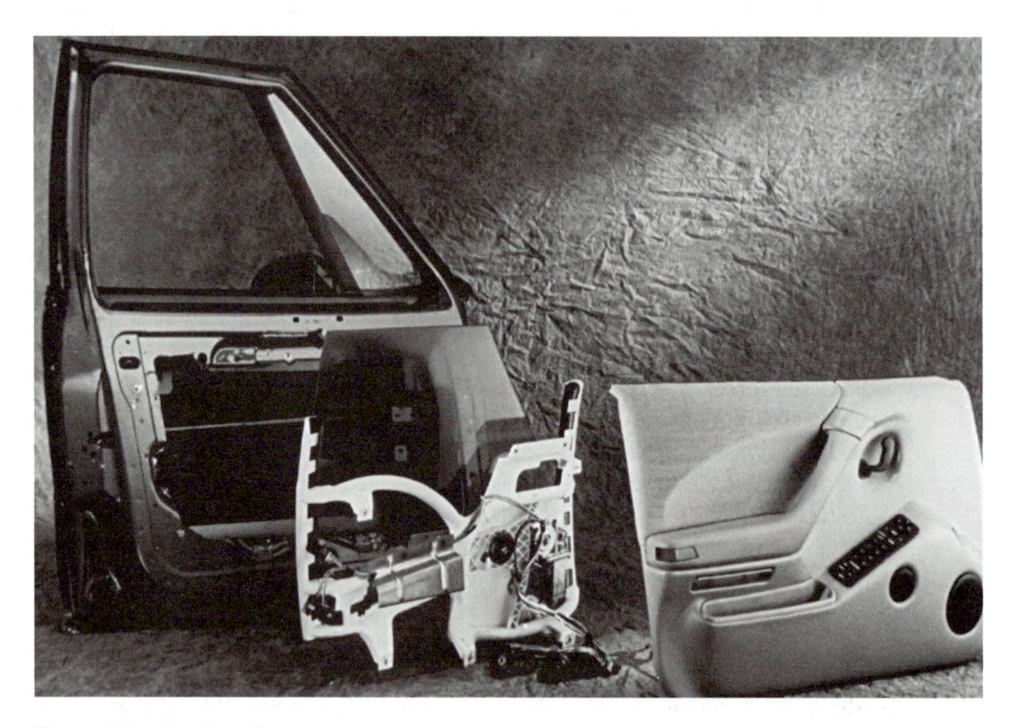

Figure 6-21 The Super Plug is a patented, "next-generation" door hardware module developed as a result of a special, five-year program between GE Plastics and Delphi Interior & Lighting Systems. The module, which can consolidate up to 61 separate door parts into one system, provides original equipment manufacturers (OEMs) with a reduction in system costs and weight and vehicle assembly time. GE Plastics custom compounded a new XENOY® PC/Polyester, 30% glass-fiber blend material for proprietary use in the Super Plug. (GE Plastics.) Also see Figure 1-8 for more detailed illustrations.

grated component that reflects the results of CAFE and industrial design/materials applications of new technology.

Figure 6-22 shows the use of plastics and polymer-matrix composites in vehicles based on past and projected use. Note that the use of PP, PVC, and PE is expected to remain constant, while the increased use of PU, ABS, nylon, SMC, PC, and thermoplastic polyester that began in the mid-1980s is expected to continue in this decade. This last group reflects the replacement of plastic for iron and steel. The selection of plastics in vehicles, for packaging, and in other products will be influenced by environmental concerns, as discussed in the section Applications & Alternatives at the end of the module.

6.3.4 Processing Plastics

The plastics industry has been very successful in developing a large variety of resins and numerous processes, and plastics have become so competitive with other materials that plastics are now found in most markets. Designers continue to seek lightweight, strong materials that allow improved forming capabilities, and plastic-matrix composites, reinforced with fibers and particles, and as laminates, are often favored as the engineering material of choice. Reduced costs are often realized through near-net and net-shape processes inherent in composite processes. Module 8 provides thorough coverage of plastic matrix composites.

Intelligent processing of materials (IPM) involves placing nondestructive evaluation (NDE) sensors in the processing equipment to allow computer monitoring and ***adaptive control*** (mixtures, temperatures, pressures, etc., adjustment in "real time" during process) to control the complex variables matched against the complex process model or expert system. IPM lends itself to the trend for ***total quality management (TQM)*** and ***statistical process control (SPC),*** techniques associated with improved quality in manufacturing products and reduced waste and overall costs. World competitiveness is driving the move for these processing improvements in ceramics and metals as well.

Manufacturing engineers working with designers, materials engineers, and technicians engage in ***concurrent*** or ***simultaneous engineering,*** which brings together team members from materials and manufacturing and design during design stages. The designers then provide ongoing engineering throughout production stages. The demand for shorter cycles from design to production, often coupled with low-volume products, favors plastics. Materials scientists and engineers are continually developing new resins to meet performance demands.

Thermosetting (TS) plastics are typically processed by casting, spraying, compression molding, transfer molding, and reaction injection molding (RIM) (Figure 6-23). Because thermosets cannot be resoftened, they are not suitable for recycling except as regrinds used for fillers. They are burned for energy production. Other processes are discussed in Module 8.

Thermoplastics (TP) plastics are much easier to recycle and are therefore favored as the plastics of choice in more and more applications. ***Liquid-crystal polymers*** (LCPs) are a newer class of engineering thermoplastics that include aromatic polyesters and phenylene polyamide. LCPs lend themselves to easier melt processing that develops highly oriented (rodlike) molecules [Figure 6-7(d)]. The resulting molded LCP parts possess

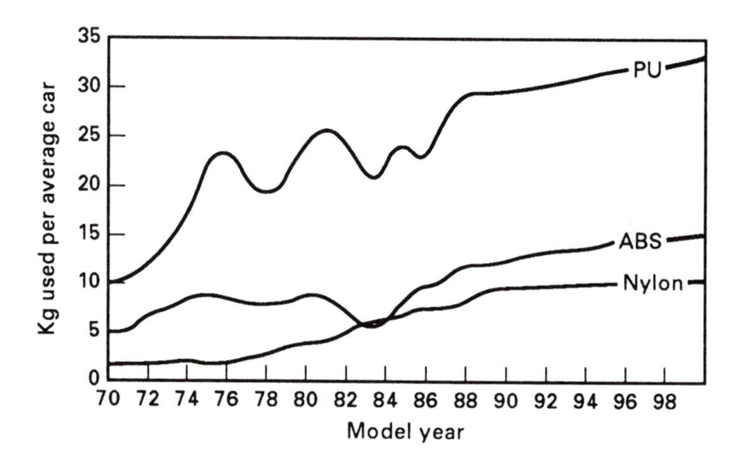

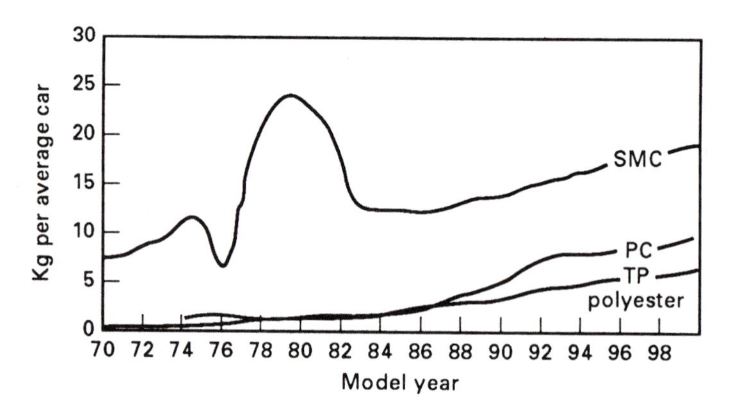

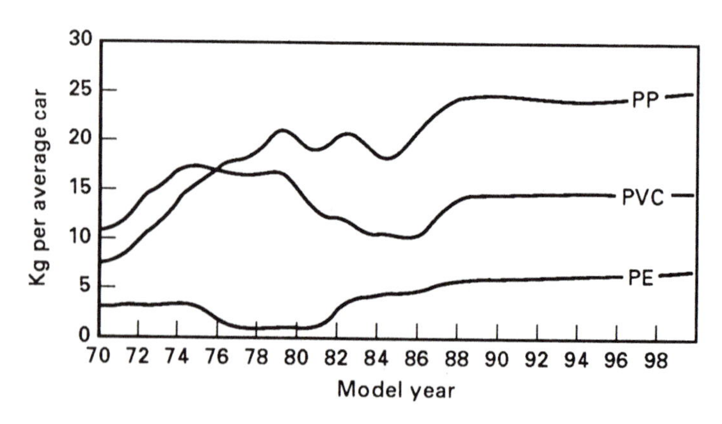

Figure 6-22 Use of plastics and polymer matrix composites (PMCs) in vehicles (Ford).

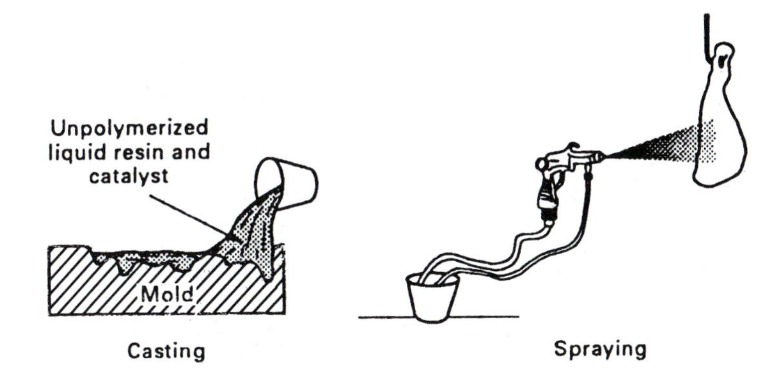

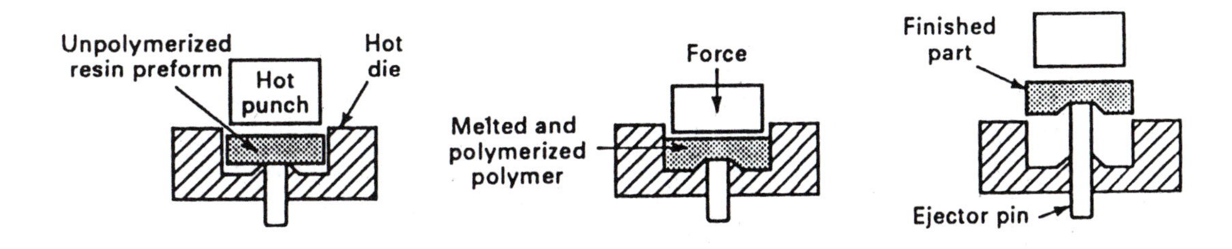

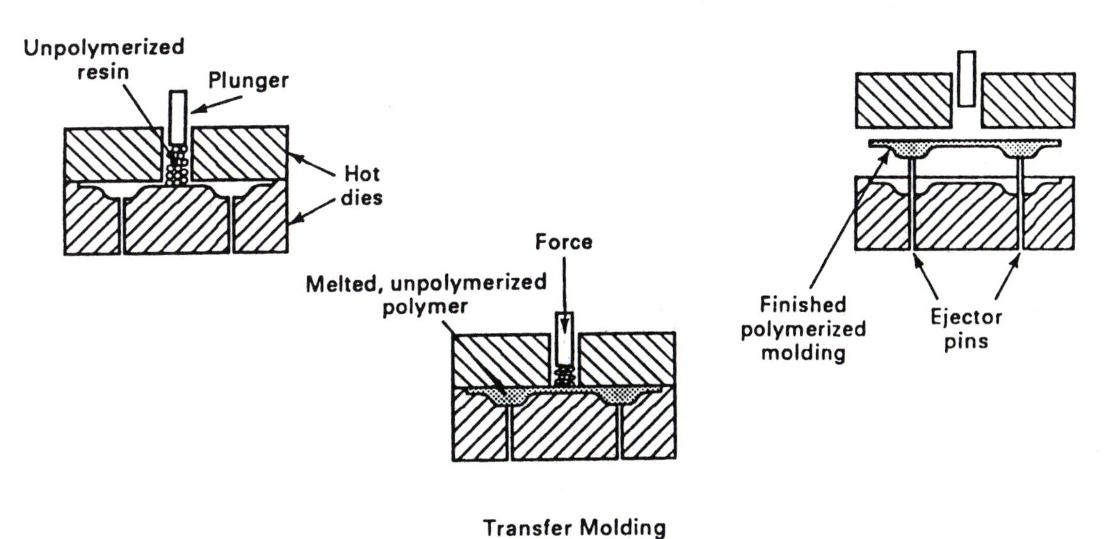

Figure 6-23 Typical thermosetting processes.

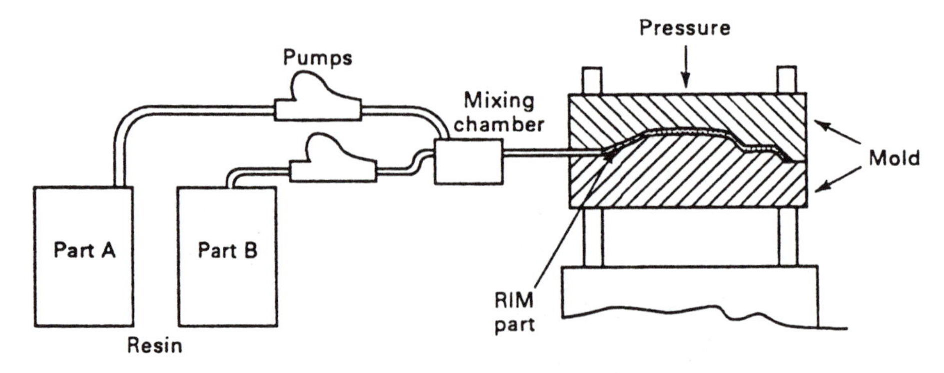

Figure 6-23 Typical thermosetting processes (continued).

anisotropic properties with chemical and thermal stability at service temperatures up to 190°C. In a manner similar to wood, which is composed of cellulose fibers that generate multidirectional strength, the rod-shaped LCP molecules provide self-reinforcement. Amoco's Xydar, an LCP copolyester resin, represents the newer engineering thermoplastic, which can be molded into complex shapes in injection-molding processes. Xydar has a heat-deflecting temperature of 350°C and a tensile strength of 115 MPa, making it suitable for automotive-engine parts. The LCPs can replace ceramics for chemical-processing equipment and find application as engineering plastics, including tennis rackets, electric-motor insulators, and printed circuit boards onto which components are surface mounted. The LCP can withstand temperatures about 40°C hotter than the molten solder.

Thermoplastics are typically processed by casting, extrusion, injection molding, blow molding, thermoforming, drawing, rolling, calendaring, rotational molding, resin transfer molding (RTM), foaming, and spinning (Figure 6-24). For FRP, other processes are discussed in Module 8. Because of the resoftening abilities, scrap thermoplastics can often be put back into the process. After service life, recycling works well.

Cyclic thermoplastic polymers, improved in the late 1980s by General Electric (GE), provided a breakthrough for the automobile industry, which was searching for a low-viscosity engineering thermoplastic that could easily wet fibers for high-production structural composites. The resulting FRP provided strength comparable to steel with about a third of the weight to provide significant gains in automobile fuel efficiency. Demonstration of the feasibility of this material, for the *Advanced Technology Program,* fell to a joint GE, Ford Motor, NIST venture as part of the big-three automakers' consortium, the *Automobile Composite Consortium* (ACC), which sought to improve U.S. automakers' world competitiveness. For the first time, this advance in cyclic polymer technology made it possible to use engineering thermoplastics in high-speed, low-cost, resin-transfer molding (RTM) composite-manufacturing processes.

To form a structural thermoplastic matrix of bisphenol-A polycarbonate, which is used to impregnate fibers, two stages of polymerization occur. First, an ***oligomer*** (*olig* is a Greek word for "few") of 2 to 20 mers forms into rings (cyclic) of low-viscosity resin that easily flows into the fiber within the mold. Next, a special "initiator" is added to cause the

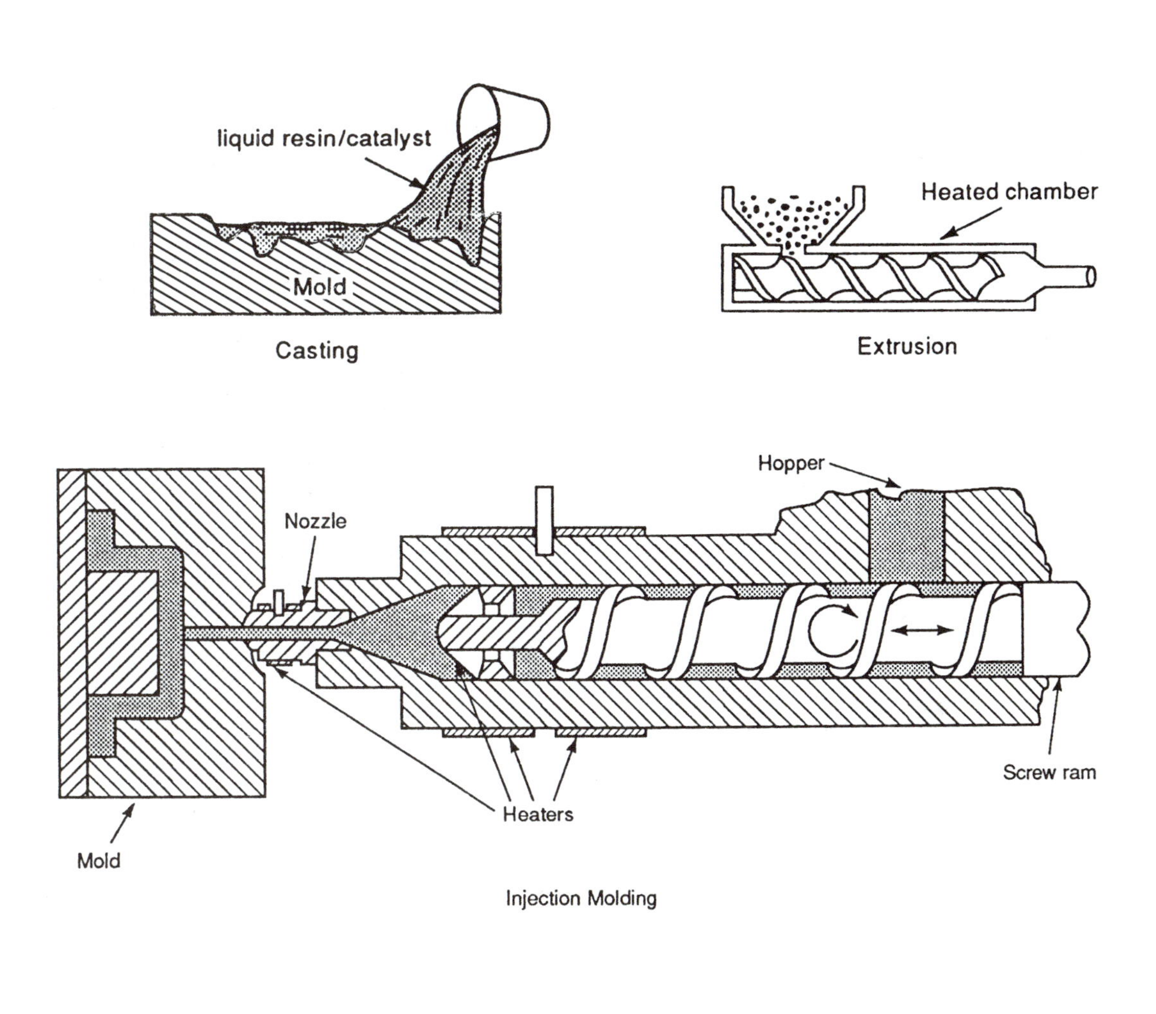

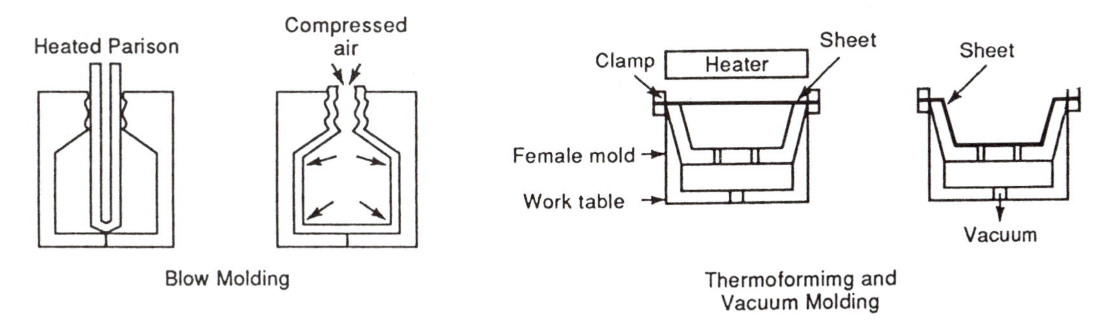

Figure 6-24 Typical thermoplastic plastic processing.

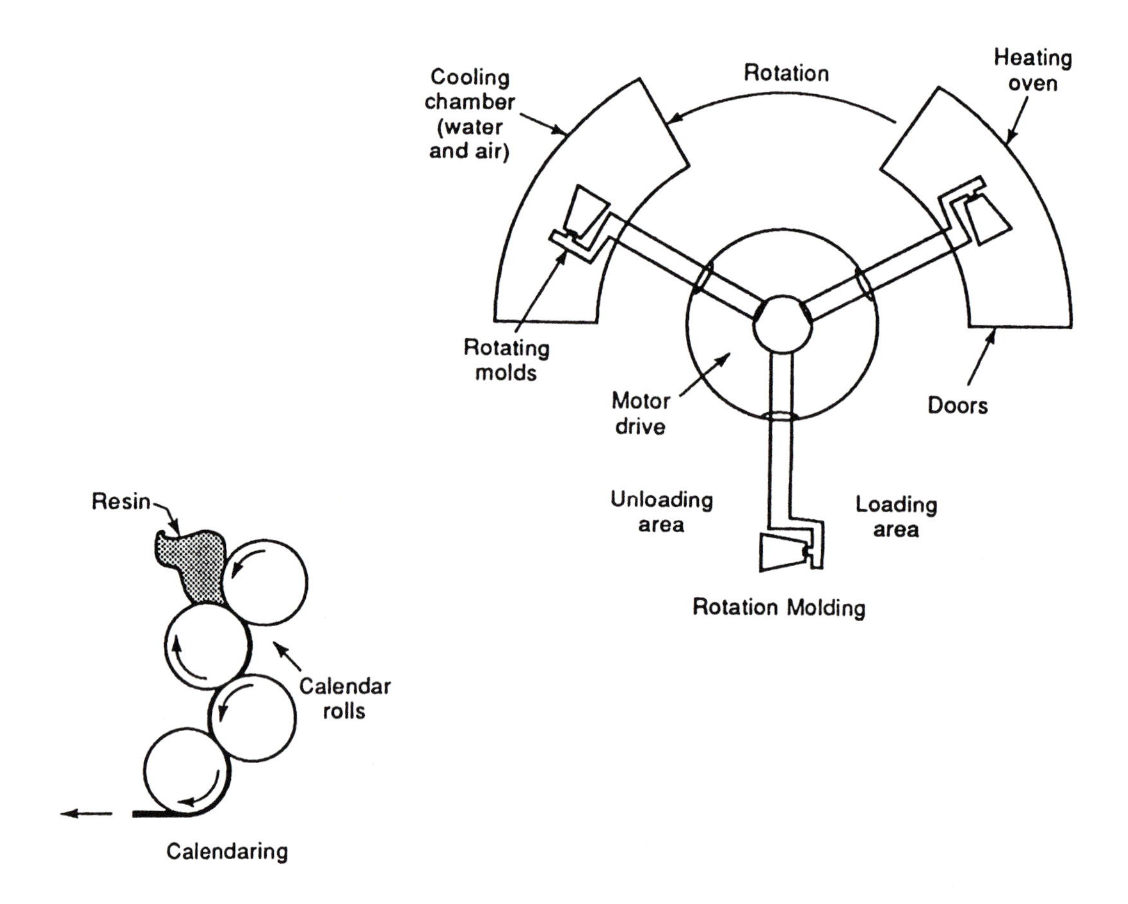

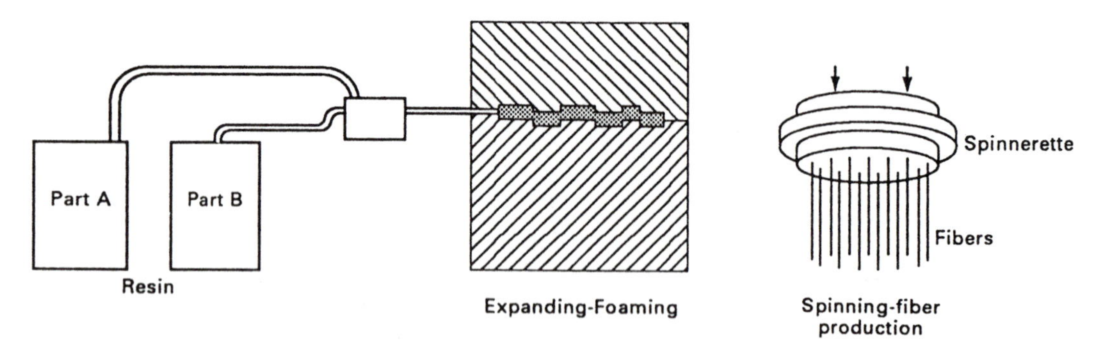

Figure 6-24 Typical thermoplastic plastic processing (continued).

linear chains to form. The processes for making cyclic bisphenol-A polycarbonates (BPAPC) involve production of an oligomer [low-molecular-weight (2 to 20) polymer], which has a consistency similar to water rather than the viscosity of cold molasses, as is normal in polycarbonates. The polymer ring (***cyclic polymer***) shown in Figure 6-25(a), with its low melt viscosity below temperatures of 300°C, flows like water around the reinforcing polymer preform for complete impregnation. Once that is accomplished, a special "initiator" triggers a polymerization reaction that rapidly transforms the rings into long, linear [Figure 6-25(b)], high-molecular-weight molecules (M_W 50,000 to 300,000) that give thermoplastics their characteristic strength.

In addition to excellent damage resistance and favorable processes characteristics, the cyclic ester material offers excellent resistance to common automobile solvents such as gasoline and brake fluids. This BPA polycarbonate is also environmentally friendly because it does not give off volatile or nonvolatile by-products in the condensation polymerization; it offers recycling opportunities because it is thermoplastic.

As an example of applying RTM in the auto industry [see Figure 6-26(a) and (b)], Chrysler chose resin-transfer molding for its Viper sports car because it offered a 40% weight savings over sheet steel, with reduced tooling time and tooling costs. Compared to sheet-molding-compound (SMC) composites, RTM gives lighter, stronger panels at lower molding pressure with less labor. With only minor hand finishing, the RTM provides a class A finish. Some problems of holding tolerances on the large complex hood slowed the production rate.

Applying new engineering materials technology that couples new analysis technology such as ***magnetic resonance imaging (MRI)*** with the ability to engineer materials can result in remarkable design and production innovations. For example, Toyota engineers produced a Super Olefin Polymer as a result of a desire to change the conventional bumper fascia material ***elastomer modified polypropylene (EMPP).*** Figure 6-26(c) provides a property comparison of the two modified polymers. The inner octagon is the EMPP and the outer polygon depicts the improved properties of the Super Olefin Polymer. EMPP is a blend of polypropylene (PP) and ***ethylene propylene rubber (EPR).*** Toyota's newly engineered polymer met a material goal for engineering a Super Olefin Polymer with 50%

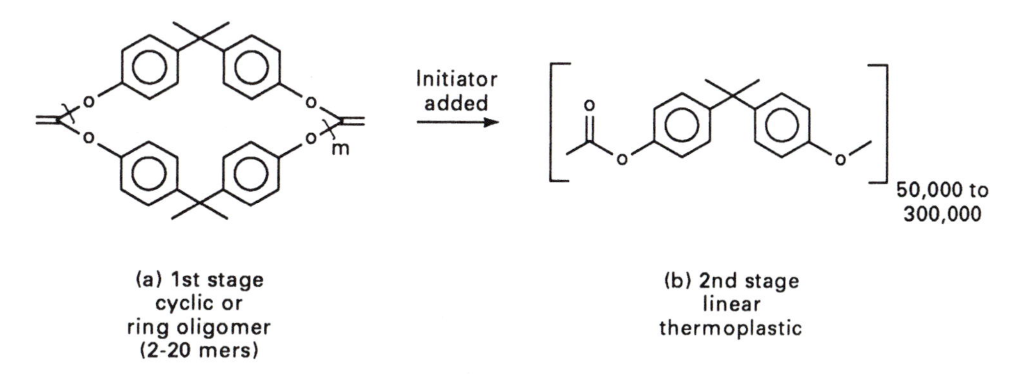

Figure 6-25 Steps in making bisphenol-A polycarbonates.

(a)

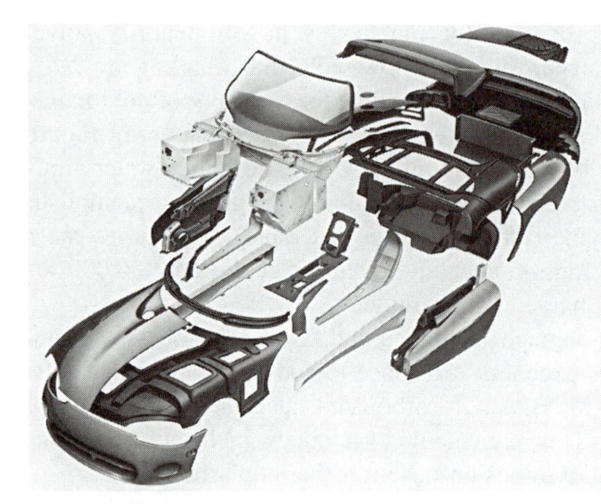

Figure 6-26 (a) The Dodge Viper was an experiment in new technologies, including concurrent engineering, low-volume and low-investment production, unique design, and numerous materials and processes innovations. (b) An exploded view of the Viper showing RTM plastic panels as the majority of outer and inner body panels; only the lower front body enclosure used SMC. (Chrysler.)

(b)

(c)

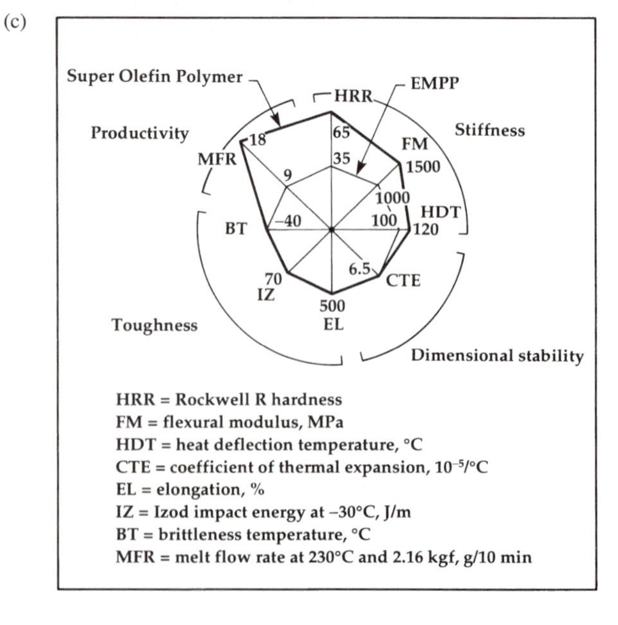

Figure 6-26 (c) Major advantages of Super Olefin Polymer over conventional EMPP are dramatized in this property comparison. Note the toughness values of brittleness, temperature, impact strength, and elongation were maintained. Productivity values increased with improved flow rate, as did stiffness values (flexural modulus and heat deflection temperature). There was a slight improvement with dimensional stability, which is a combination of coefficient of thermal expansion (COT or CTE) and stiffness. *(Advanced Materials & Processes.)*

minimum improvement in flexural modulus, melt flow rate, and hardness over the conventional EMPP, while retaining a low coefficient of thermal expansion (CTE) and high impact resistance at −30°C (−20°F) and below. The reengineered polymer composite used an EPR elastomeric matrix to provide the following basic performance characteristics required of a bumper fascia:

a) Maintained low-temperature impact resistance and a low CTE.

b) Introduced hard segments into the EPR phase to increase hardness, toughness, and stiffness. The segments form a microscale interpenetrating network (IPN) structure discussed below.

c) Reduced molecular weight and maximized the crystallization of the PP component to increase melt flow rate and hardness.

d) Enhanced the formation of the IPN-like structure by controlling its crystallization rate as well as the crystalline component of the PP.

e) Achieved a recyclable polymer component.

The comparison below reveals the dramatic improvements.

	Super Olefin Polymer	EMPP
Wall thickness, mm	3.8	4.8
Fascia weight, kg/car	6.8	8.7
Molding cycle time, s	85	115
Relative production cost	89	100
Molding temperature, °C	210	230

Because plastics offer so much to the designer, new processes coupled with improved resins will continue to emerge. Another recent development in plastics processing, ***solid imaging,*** is the use of lasers to polymerize plastic for net-shape processing (Figure 6-27). ***Photopolymer*** plastics are based on acrylic monomers and oligomers and are generally inferior to typical engineering plastics. Photoactive polymers have been used for decades in applications such as printing plates and inks, ultraviolet (UV) and electron-beam (EB) curable coatings, and photoresists, and for production of holograms.

As a ***rapid phototyping*** technique, UV laser beams follow paths generated by computer-aided design (CAD) data to create solid parts layer by layer as the lasers polymerize resins. As each layer is cured, a very accurate, solid plastic model emerges on an elevator from the vat of resin. Various techniques of rapid prototyping have been developed, allowing designers to shorten the time required to go from drawing to a solid model without the need for numerous molding, sculpting, cutting, joining, and similar processes traditionally involved in model making. Although photopolymers are suitable for model making, fully functional parts await new polymer development. Rapid prototyping is now being integrated with solidification modeling and computer tomography to produce investment-cast rapid tooling for the die-casting, permanent-mold, and investment-casting industries. Existing parts can be scanned with computer tomography to produce a CAD file. The CAD data can be used with solidification modeling software to determine the best

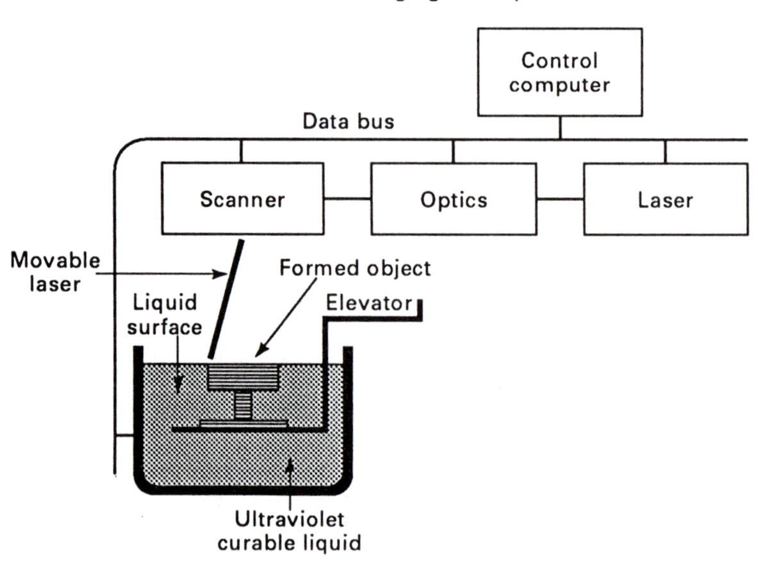

Figure 6-27 Photopolymers permit a solids imaging technique, known as rapid prototyping, that uses laser beams to process resin into solid models and prototypes.

method for casting a tool. Rapid prototyping equipment can be used to produce a pattern for investment casting, while computed tomography can be used again to characterize both the tool and the part it produces. Still another promising technique for processing plastics is the free electron laser, discussed in Module 9.

Dendrimers, another new class of oligomers (Figure 6-28), are spherical or rod-shaped nanostructures with hollow interiors that lead to easier engineering of desired properties. They offer the potential for superior protective coatings, nonlinear fiber optics, adhesives, conductive polymers, lubricants, and reinforcers for composite matrices.

Interpenetrating polyimide-network (IPN) plastics represent research aimed at having the best of both worlds: in this instance, thermosets (TS) and thermoplastics (TP). NASA research has developed a tough, high-temperature polyimide that resists microcracking. The IPN ***neat resin*** (no additives or reinforcers) makes a valuable composite matrix for graphite reinforcement suitable for aircraft engines and aerospace structural applications. By combining cross-linked (TS) polyimide with linear (TP) polyimide (Figure 6-29), the ease of processing TS is retained while the toughness of TP is achieved.

Block copolymers (identical mers clustered in blocks along the backbone chain polymer) are also valuable as composite matrices because of their ability to combine the stiffness of a plastic with the toughness of an elastomer. One example of this synergy is with plastic-resin polystyrene, which provides stiffness in a block copolymer with the elastomer polyisoprene that possesses toughness. This is illustrated in the TP rubber (refer to Figure 6-35).

Figure 6-28 Computer model of carbon-based dendrimers, which resembles the structure of atoms, but at the molecular level. The symmetry provides special physical and chemical properties. (George R. Newkome, University of South Florida.)

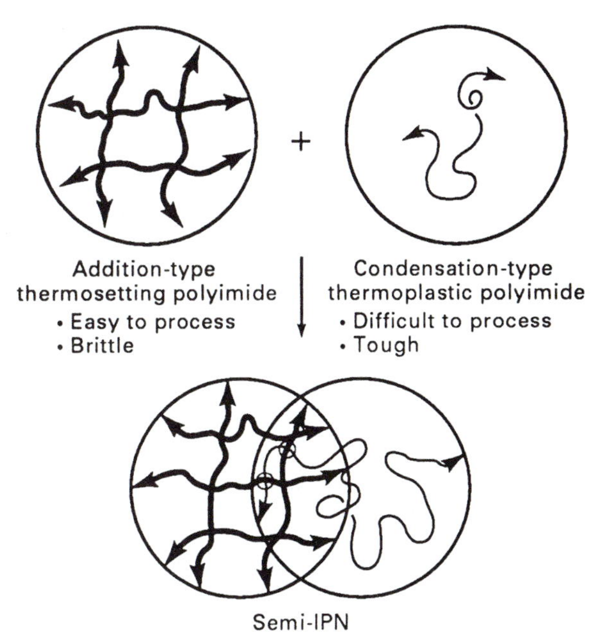

Figure 6-29 Formation of an IPN plastic matrix.

6.3.5 Smart Polymeric Materials

Because polymer chemists can synthesize a wide range of organic polymers, many opportunities for smart polymers wait to be exploited. Current examples of smart polymers include multigrade and synthetic lubrication oils. A 10W40 SAE motor oil will become more viscous with increased heat and less viscous as the temperature drops because of polymers that expand and contract with heat variations.

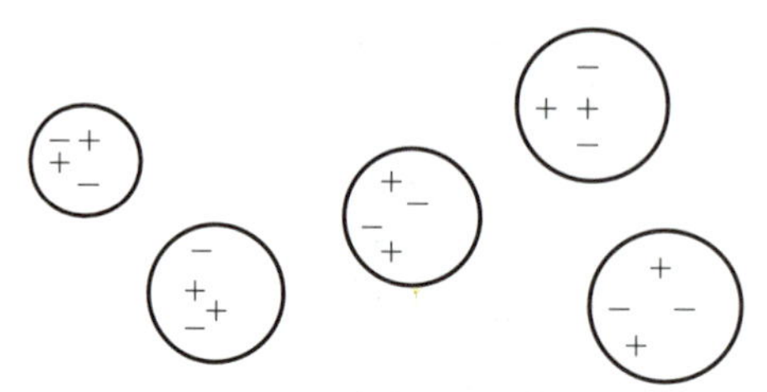

No electrical field present, charges are random, no particle attraction.

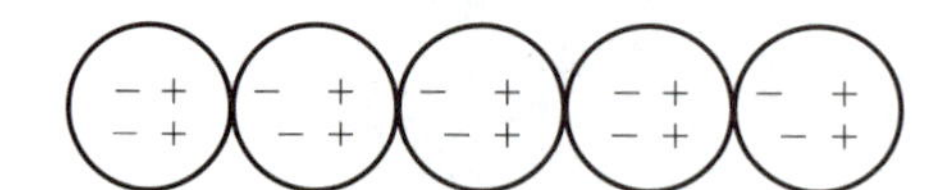

Probe energizing the fluid, particles linked together.

Figure 6-30 Smart polymers—electrorheological (ER) fluids. (From John A. Marshall, "Liquids That Take Only Milliseconds to Turn into Solids." National Educators' Workshop: Update 93—Standard Experiments in Engineering, Materials Science, and Technology. NASA Conference Publication 3259, April 1994, pp. 315–322.)

Electrorheological (ER) and ***magnetorheological (MR) fluids*** also change viscosity, plasticity, or elasticity with variations in the application of either electrical current or magnetic fields. ERs and MRs go from liquids to gels in milliseconds with the application of external forces. These ***smart fluids*** consist of polymers (oils and additives) and ceramic particles suspended in liquids. The liquid carrier must be a good insulator and the particles either good conductors or ferromagnetic. When the field is applied, the particles link together to form a gel. The particles assume their random pattern with removal of the external field (Figure 6-30). The smart fluids offer potential in applications such as robotics, hydraulic valves, power transmissions, and other automated equipment that can react to computerized controls.

Shape-memory polymers, much like the shape-memory alloy Nitinol, offer control after polymerization. Polynorbornene, a shape-memory polymer, is molded to a desired shape to which it will return when subjected to loads. Other smart polymers, including liquid crystals, change colors, opacity (see the living environments house in the Pause & Ponder section), and other characteristics with variations in external forces. These polymers find applications in diverse fields, such as medicine (color thermometers), sports (smart tennis rackets), and building construction (LCD heating display and smart windows).

SELF-ASSESSMENT

6-13. What does "organic" refer to in the SPI definition of plastics? What are the other major ingredients in plastics?

6-14. What advantages do most plastics offer compared to linear plastics?

6-15. Describe the following plastic classifications: **(a)** general purpose, **(b)** engineering, **(c)** cellular.

6-16. What term means that a material is capable of bearing supporting loads?

6-17. Name some natural fibers. Name some manufactured (synthetic) fibers and give typical applications.

6-18. Why are cellular plastics increasing in use? List typical applications of foam plastics.

6-19. What is FRP?

6-20. What element serves as the backbone for most plastics? Name one plastic that has an inorganic backbone.

6-21. What does ASTM stand for?

6-22. Describe the benefits of smart polymers. Explain the advantage of shape-memory polymers.

6-23. What additive would be valuable in a plastic that is exposed frequently to sunlight? How does it alter the plastic chemically?

6-24. List additives for plastics that moderate the following service conditions: **(a)** aircraft storage compartments subject to fire; **(b)** coating for a wrench handle; **(c)** vegetable-storage drawers in refrigerated bins; **(d)** flexible gloves; **(e)** stiff, lightweight automobile interior door panels.

6-25. Determine whether thermosetting or thermoplastic plastics are *generally* best for the following type of service conditions: **(a)** high impact, **(b)** creep, **(c)** chemical exposure.

6-26. Although a certain plastic might have lower specific gravity, better toughness, and equal optical properties, why might glass still be preferred for a service condition such as storm doors, eyeglasses, or lenses on instrument panels? What plastic competes strongly with optical glasses? List some of the competing factors.

6-27. What electrical property is of primary consideration when selecting an electrical insulator? How do plastics compare as electrical insulators?

6-28. When would it be undesirable to have a plastic with low water-absorption resistance? Name three plastics with good water-absorption resistance.

6-29. What are the combined stresses that act on a material subjected to weathering? Select the three plastics most capable of service conditions that involve weathering. How would the three compare to **(a)** plain-carbon steel, **(b)** alumina, and **(c)** glass? Give the coefficient of thermal expansion and the rate of thermal conductivity for each of the three materials.

6-30. List five properties essential for the rope used to pull water skiers. Select two plastics suitable for this service and explain the selections. Use the ASTM abbreviations.

6-31. Give the choice among PS, PMMA, ABS, and PF as possibilities for motorcycle helmets, select one and explain your choice.

6-32. Use Tables 6-1 and 6-6 to calculate the specific strength of aluminum, and compare it to the specific strength of an acetal plastic.

6-33. Select two plastics that would withstand the service conditions under the hood of an automobile. Use the ASTM abbreviation. Explain your selection.

6-34. Give some reasons why viscoelasticity offers a designer certain advantages and disadvantages when using polymers in product design.

6-35. What technique involves placing NDE sensors in processing equipment to allow computer monitoring and adaptive control (adjustment in mixtures, temperatures, pressures, etc., in real time during the process) to control the complex variables that are matched against the complex process model? What factors cause a demand for simultaneous engineering?

6-36. What two stages do cyclic thermoplastic polymers go through as they are processed into matrices for engineering composites?

6-37. From the tables in this module and current prices listed in the reference issues of periodicals such as *Machine Design* and *Modern Plastics,* calculate the specific strength and cost per unit volume for the following materials: **(a)** acrylic, **(b)** polycarbonate, **(c)** plain carbon, **(d)** steel, **(e)** wood, **(f)** epoxy, **(g)** aluminum.

OBJECTIVE QUESTIONS

6-10. What main advantage do cellular plastics offer?

a. Reduced density **b.** Improved crystallinity
c. Greater density **d.** Improved strength

6-11. The objective of adding microspheres to plastics is which of the following?

a. Increase weight **b.** Improve impact strength
c. Improve processibility **d.** Reduce crystallinity

6-12. Which group of polymer fibers provides superior strength and resistance to biological attack?

a. Manufactured **b.** Cotton **c.** Jute **d.** Hemp

6-13. In transportation vehicles, structural plastics are often not used because they lack good specific strength.

T (true)
F (false)

6-14. A property unique to polymers that gives them the ability to return to their original size and shape once a load is removed, much like a shock absorber handles bumps.

a. Viscoelasticity **b.** Voigt **c.** Specific strength **d.** Branching

6-15. A problem with many plastics is that they often creep at room temperature over time with fairly small loads.

T (true)
F (false)

6-16. Paper is a better packaging material because, in the modern landfill, paper will biodegrade.

T (true)
F (false)

6-17. Thermosetting plastics are not good for recycling because they cannot be softened for reprocessing.

T (true)
F (false)

6-18. Common processes for thermoplastic plastics are casting, extrusion, blow molding, and thermoforming.

T (true)
F (false)

6-19. The first stage of a cyclic thermoplastic plastic that makes processing easier.

a. Oligomer **b.** Linear **c.** Hevea **d.** RIM

6-20. Property found in wood that results from grain patterns.

a. Anisotropy **b.** Isotropy **c.** Homogenous **d.** Interface

6-21. A ratio of strength to weight.

a. Isotropic **b.** Tensile **c.** Specific strength **d.** Density

6-22. Which advantage does smart polymeric materials offer designers?

a. Eliminate selection problems **b.** Reduce raw materials costs

c. Adapt to environmental changes **d.** Reduce processing

6-23. Which application would benefit most from the properties of electrorheological fluids?

a. Multigrade engine oil **b.** Robotic power transmission

c. Superconductors **d.** Telephone switching

Polymers in cars. How many types of polymers can you think of in the typical car? This diagram depicts applications of rubber. The rubber reduces vibration and isolates against noise, as seen in (1) engine mounts, (2) strut mounts, (3) bushings, (4) couplings, (5) air sleeves, (6) insulation against impact, (7) insulating foams, (8) gaskets, (9) seat padding, (10) trim, (11) cushions for headrests, (12) door cushions, and (13) roof trim. It's easy to notice the rubber tires, which involve blending of various natural and synthetic rubber with reinforcement wire and fibers. Rubber is also used to make cooling and brakes hoses, floor mats, and fan belts. Each of the applications has special requirements that result in many rubber compounds. Other important polymers include sealants, lubricants, oils, natural leather and numerous adhesives. Of course, many plastics are also used in the modern automobile. (Bridgestone Firestone.)

6.4 ELASTOMERS: NATURAL AND SYNTHETIC RUBBERS

Elastomers refer to the group of polymers that exhibit rubbery or elastic behavior, including *natural rubber* and *synthetic rubber.* The American Society for Testing and Materials (ASTM) defines ***elastomers*** in ASTM D1566-66T as "macromolecular material that returns rapidly to approximately the initial dimension and shape after substantial deformation by a weak stress and release of the stress." Elastomers will stretch from 200% to over 900% of their original length at room temperature and return to their original state when the tensile stress is released. Natural rubber, known as ***hevea rubber,*** comes from the rubber tree *Hevea brasiliensis.* Columbus brought this bouncy material back to Europe with other curiosities from the New World, but nearly 300 years passed before it was given serious attention. The name ***rubber*** resulted from the material's ability to erase black pencil marks from paper by rubbing. Synthetic rubber, as with plastic, is a hydrocarbon with petroleum as its major source. Development of the processing of synthetic rubber into a method suitable for high-volume production did not fully materialize until the beginning of World War II, when the United States was cut off from the natural rubber supply in India and was forced to use large quantities of synthetic rubber.

6.4.1 Natural Rubber and Vulcanization

Natural rubber (NR), in the form of liquid resin secreted from the inner bark of the ***Hevea brasiliensis*** tree, is known as latex. Latex is not tree sap. It consists of isoprene molecules (Figure 6-31a). Polyisoprene, or rubber, is formed through a natural polymerization in the tree (Figure 6-31a). The liquid latex dries into a thermoplastic material. The structure is an amorphous mass of coiled and kinked chains with constant, thermally induced motion of the atoms and chains. Rather than stretch when heated, rubber shrinks because the thermal

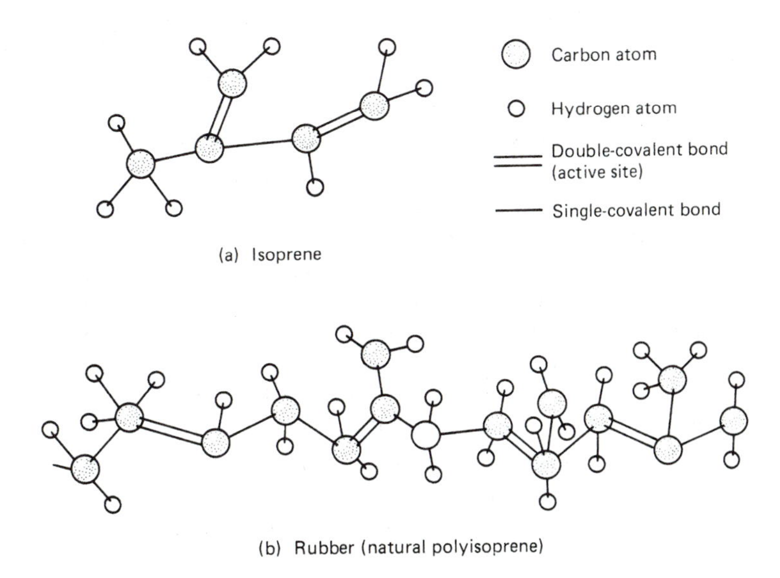

Figure 6-31 (a) Natural rubber.

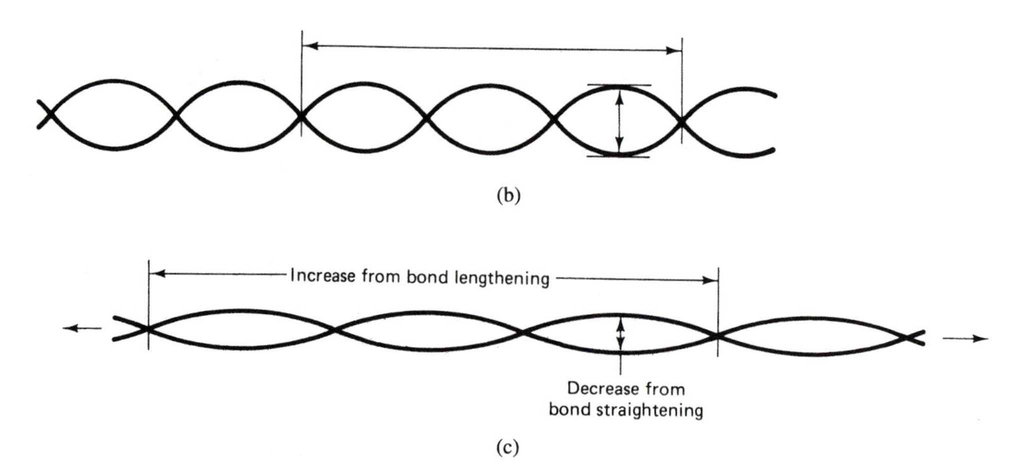

Figure 6-31 Effect of tension on bonding: (b) no load, (c) tensile load.

agitation causes the chains to entangle and draw up. The structure provides the ***resilience*** for rubber to spring back into shape when compressed or stretched. When a tensile load is applied, the structure changes as the chains straighten (bond straightening) and stretch (bond lengthening) [Figure 6-31(b) and (c)], and crystallinity is achieved in varying degrees depending on the amount of stress. With increased crystallinity comes a greater strength, increased rigidity, and increased hardness. The structural changes in rubber make it good for tires because the portion of the tire under high stress is crystalline and provides support for the vehicle it is carrying, while at the same time the portions of the tire not under high stress are still resilient and absorb shock from bumps in the road.

The problems with natural rubber are that it is too soft and has too many reactive sites (double bonds), which cause rapid oxidation and ***dry rot.*** It is also somewhat plastic and will not recover from high stress. In 1839, Charles Goodyear discovered, through the addition of sulfur to natural rubber compounds, that it was possible to increase hardness and reduce susceptibility to oxidation and reaction with other chemicals. The sulfur ***vulcanizes*** the rubber, or changes it into a thermosetting polymer, by linking together the molecular chains at their double bonds (Figure 6-32). Two sulfur atoms per pair of isoprene mers are needed, and with 5% of the mer pairs cross-linked, a flexible, resilient rubber exists. Further cross-linking increases hardness. Vulcanization of only sulfur to rubber requires several hours and temperatures around 145°C. However, accelerators and activators added to the compound will result in vulcanization within minutes.

In addition to natural rubber, sulfur cross-links polybutadiene (BR), polyisoprene (IR), and acrylate (ACM) elastomers. Along with sulfur, other cross-linking agents vulcanize certain synthetic elastomers. Butyl rubber (IIR) and ethylene–propylene copolymers (EPM) rely on sulfur and phenolic agents; zinc oxides form cross-links in polysulfide (PTR) and polychlorapene (CSM). Vulcanized natural rubber has excellent flexural strength or deformability, tensile strength, and abrasion resistance, but it is attacked by petroleum oil, greases, and gasoline. Its value in automobile tires comes from low heat buildup, discussed later. Natural rubber has superior overall engineering capabilities com-

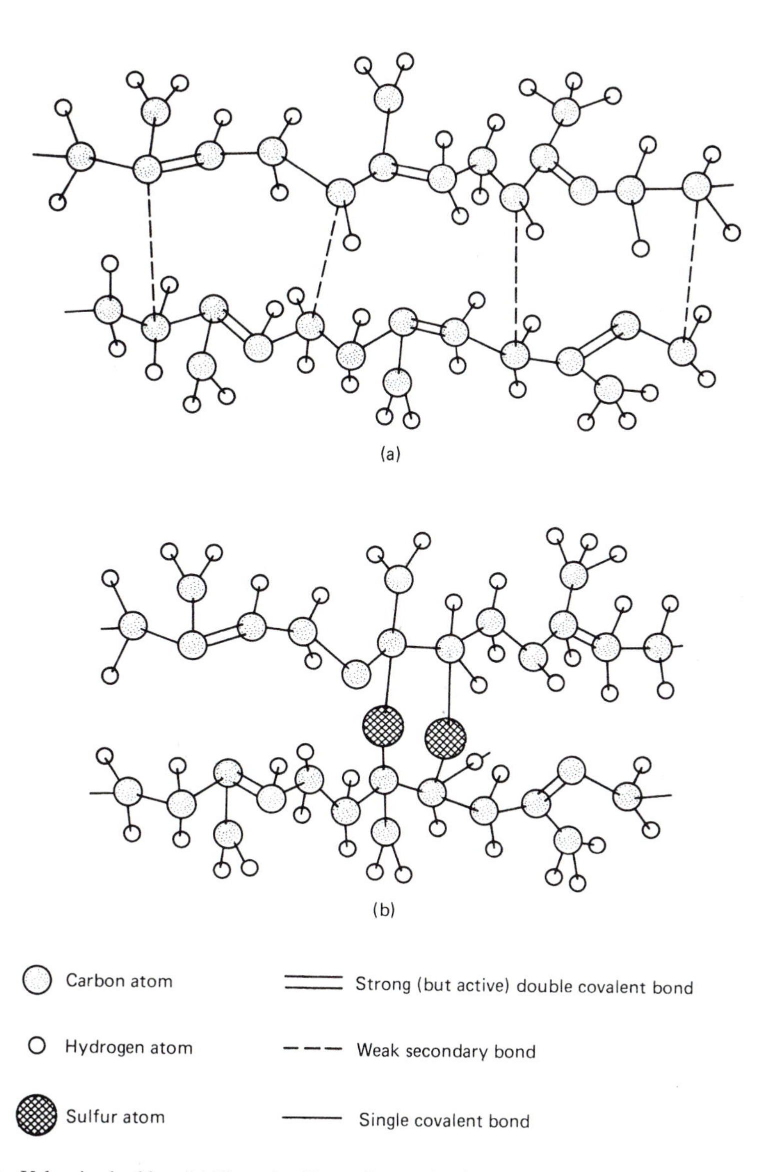

Figure 6-32 Vulcanized rubber. (a) Natural rubber—thermoplastic. (b) Vulcanized (cross-linked) rubber—thermoset.

pared to synthetics. The choice of a synthetic rubber over natural rubber or of one synthetic over another boils down to specific properties required, price, and processibility.

Because natural rubber comes from a renewable source and because it has excellent properties, natural rubber science and technology continue to make strides. Deproteinized natural rubber (DPNR) is an example of developments in improving this natural elastomer. By removal of ingredients with an affinity for water, such as protein, polyols, and inorganic salts, fatigue resistance and other mechanical properties are enhanced. Research has been renewed in the United States into the use of guayule shrubs to produce rubber; they were grown in California during World War II. Only 5% to 10% yield was possible then; a 20% yield is required to justify the use of this raw material.

6.4.2 Synthetic Rubber

Even with the improved properties obtained through vulcanization, natural rubber has poor resistance to aging. It is attacked by ultraviolet light, oxygen, and heat because it still has many reactive sites. Due to these shortcomings and because the rubber tree grows only in special environments that were vulnerable to political sanctions, the search for a substitute produced several synthetic rubbers in the early to middle part of the twentieth century. The number of synthetic rubbers has grown to include many special-purpose elastomers. Improvements in synthetic elastomers have also brought improvement in the additives and fillers for natural rubber (NR). Almost identical to NR is synthetic polyisoprene (IR), except that it has greater stretching ability. Synthetic rubbers have the same raw materials used in plastics. Starting with crude oil and natural gas, many formulations are possible. Styrene, the monomer for styrene–butadiene rubber (SBR), is derived from coal as vinyl benzene, $CH{=}CH_2$, a product of benzene and ethylene. Butadiene is derived from petroleum. The hydrocarbon is obtained in fractionation of cracking petroleum used for olefins, polymers, or gasoline. The 1-butene is separated and catalytically dehydrogenated in the vapor phase to produce butadiene.

Figure 6-33(a) shows the steps for modern automobile tire production. During the final processing step, heat and pressure are used to form the final tire shape, then vulcanize the rubber that was in a thermoplastic condition during forming. Note, at the beginning stage, the mixing of natural and synthetic rubbers with carbon black, sulfur, and other additives. A runflat tire and wheel system is shown in Figure 6-33(b).

6.4.2.1 Additives (modifiers). Additives (modifiers) to synthetic rubber are introduced at the initial processing stage and also during compounding. Soap and water are added to styrene and butadiene in the making of raw styrene–butadiene rubber (SBR) to produce an emulsion that keeps droplets of the monomers from separating out. ***Accelerators*** and ***activators*** speed up sulfur vulcanizing, while ***retarders*** prevent vulcanizing before it is required. For those saturated synthetic elastomers that have no double bonds, ***peroxides*** promote sulfur vulcanizing. ***Peptizers*** soften raw rubber, ***pigmenters*** add color, ***abrasives*** provide abrasing action for products such as erasers, ***hardeners*** increase rigidity, and ***antioxidants*** and ***antiozonants*** prevent aging or dry rot from the sunlight and ozone.

Fillers help to increase bulk to the compound while holding down cost. Inert fillers include talc, chalk, and clay. Reinforced rubbers use many of the same ***fillers*** as plastics, including metals, glass, and polymer fibers. ***Carbon black*** also serves as a strengthening filler and imparts hardness.

6.4.2.2 Thermoplastic rubbers or elastoplastics. Thermoplastic rubbers (TR) or elastoplastics do not cross-link as do most elastomers, even though their properties are similar to many rubbers. A copolymer of styrene and butadiene was introduced in 1965 as a thermoplastic rubber. The elastoplastics have the ability to soften when heated for

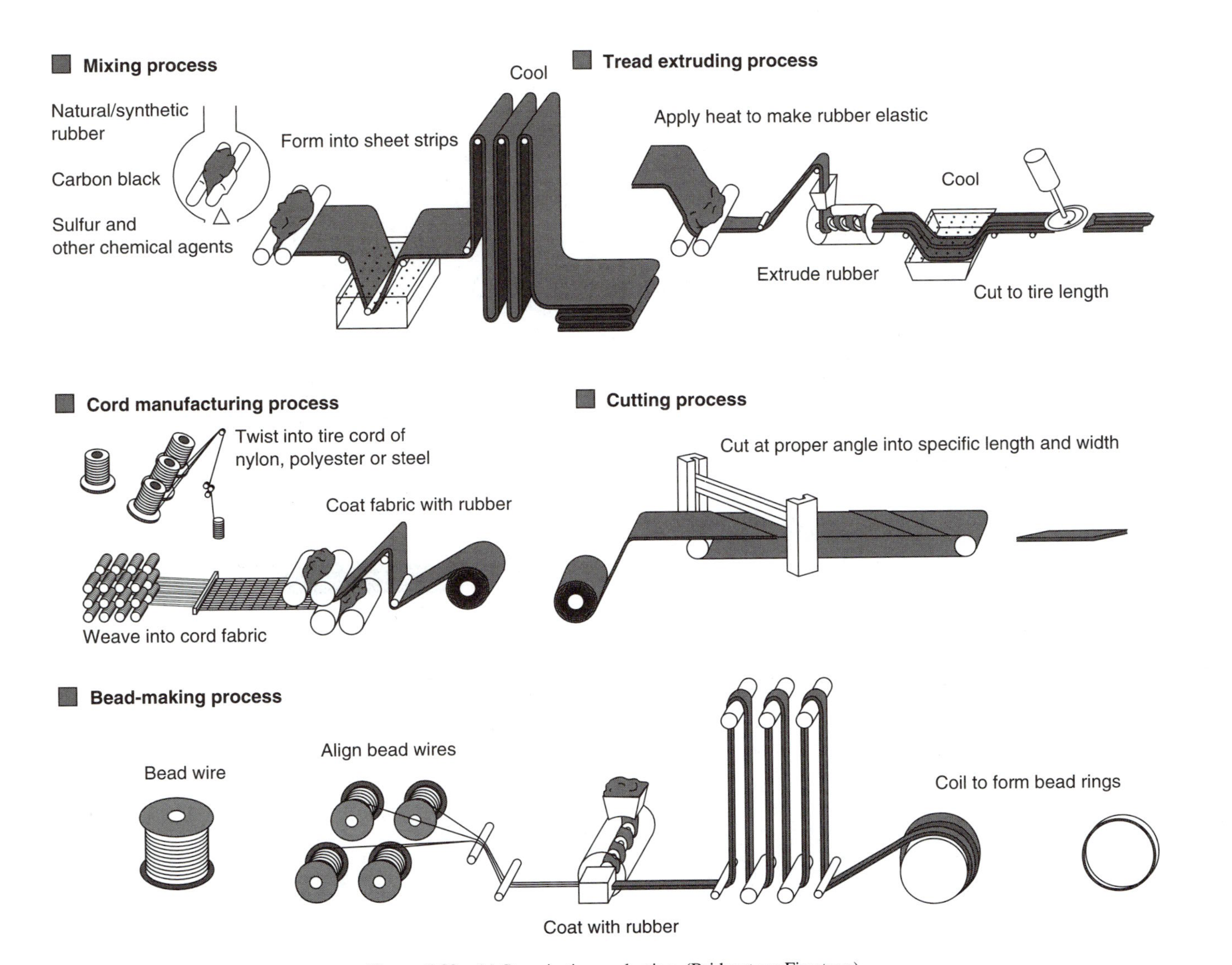

Figure 6-33 (a) Steps in tire production. (Bridgestone Firestone)

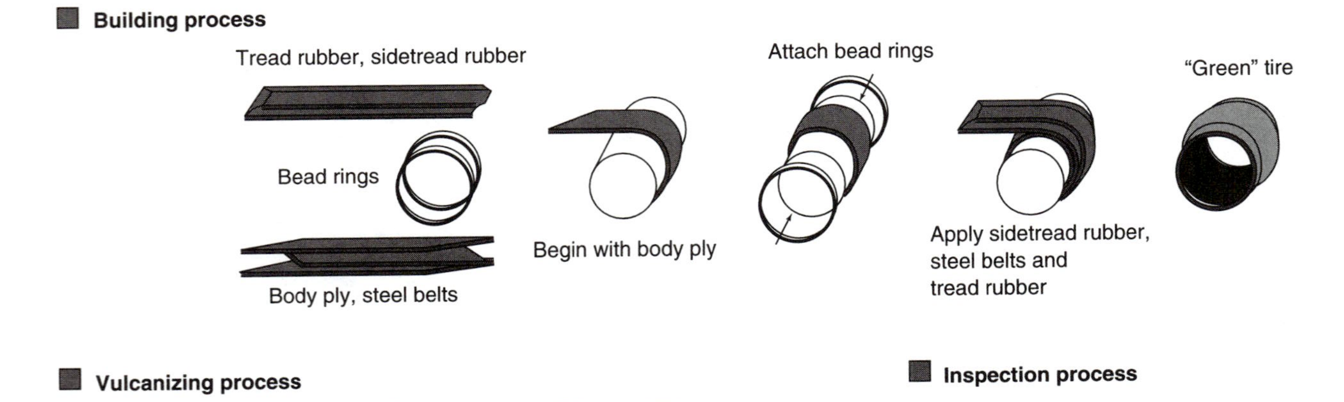

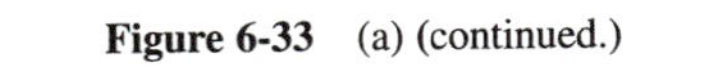

Figure 6-33 (a) (continued.)

Figure 6-33 (b) Runflat tire and wheel system. (Bridgestone Firestone.)

processing and, upon cooling, become solid yet maintain their elastic behavior. Vulcanization is eliminated. The thermoplastic behavior develops as a result of the structure, in which the chains of one monomer form links in the center of block molecules of other monomers. Figure 6-34 shows that the domains in the middle have a hard linking effect, while the other end of the segments has elastomeric behavior. Upon heating through T_g, the hard centers soften to plasticize the rubber for processing. When cooled, the hard blocks re-form to again link the copolymers. This linking is especially effective at low temperatures.

Figure 6-34 Polystyrene domains of thermoplastic styrene butadiene rubber.

Two-block, copolymer, thermoplastic elastomers manufactured under the trade name Kraton include styrene–butadiene–styrene (SBS) and styrene–isoprene–styrene (SIS). The styrene (S) occurs in two thermoplastic blocks, as shown in Figure 6-35(a), on either end of a rubber block of butadiene (B) or isoprene (I). Rather than having randomly distributed monomers, as do most copolymers, block copolymers form uniformly distributed sections as blocks. Figure 6-35(b) shows that a network of block chains has formed. It consists of a highly uniform distribution of ***polystyrene domains*** and the separate butadiene rubber chains. The ***rubber network,*** which is physically joined, links the domains and cannot move because the domains are immobile, thus providing a physical rather than a chemical cross-linking. The physical cross-linking can be broken when heated, which makes the compound a thermoplastic. Kraton is therefore capable of continuous molding and remolding. The polystyrene domain consists of actual particles that become hard and glasslike at room temperature; they take on cylindrical, spheroidal, and platal configurations.

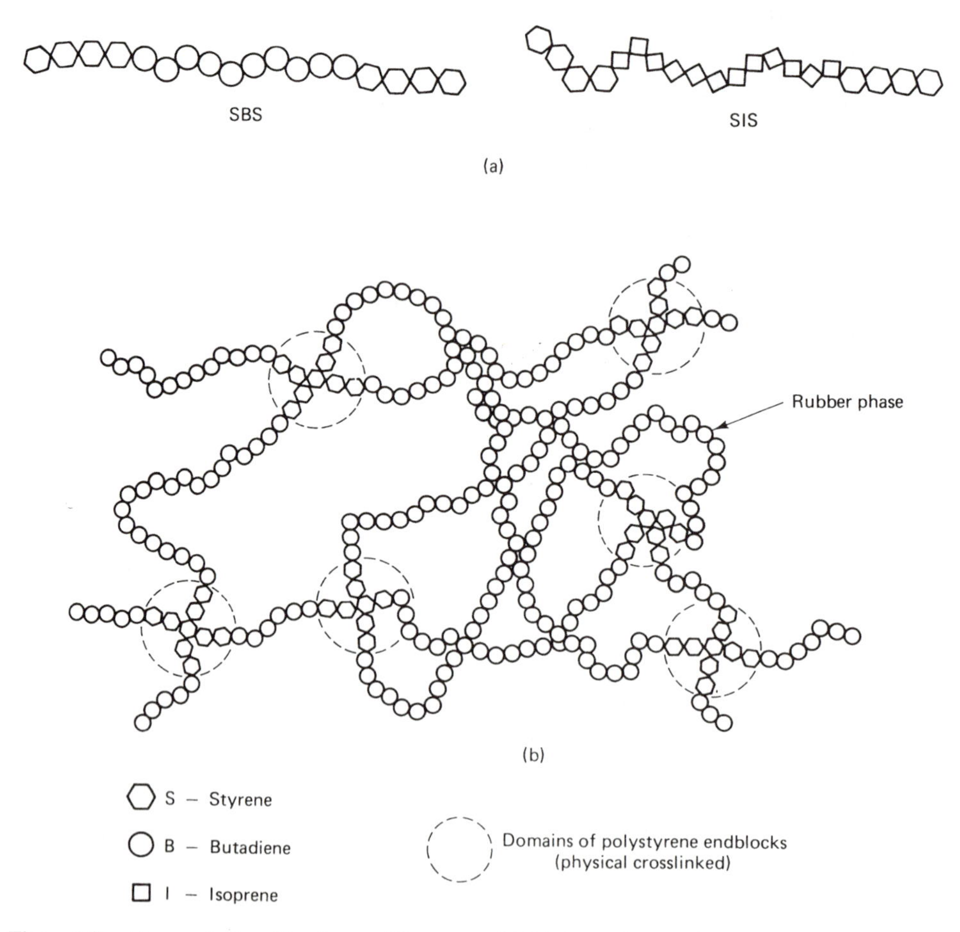

Figure 6-35 Thermoplastic rubber "network" structure. (a) Block copolymers. (b) SBS.

The major elastoplastics are polyester copolymers, olefinics, polyurethanes, and styrene copolymers. The wide service temperature ranges, low-temperature flexibility, and ease of painting place thermoplastic rubbers in contention as a substitute for thermosetting rubbers. Their ease of processibility has given them wide acceptance in the auto industry as flexible bumpers and other exterior panels, steering wheels, hose covers, housings for seat belts and horns, wire covering, and oil seals. Other applications include housewares, toys, sporting equipment, adhesives, rainwear, footwear, and skate wheels. As with plastics, scrap thermoplastic elastomers can be ground and remolded. Because of the complex nature of the modern automotive tire, their "after-life" has been a real problem. Many methods have been employed to recycle the millions of discarded tires, including grinding them into materials for road paving, as tiles for patios and tennis courts, as fence posts, and for burning to produce electricity.

Shape-memory polymers are a class of "smart materials" that return to their original fabricated shape under the influence of temperature. ***Bioelastics*** are synthetic fibers that behave much like muscles. Based on the natural protein, elastin, these biomaterials offer medical applications for surgical repairs as well as industrial uses such as for robotic hands and automatic valve controls to replace metal springs and hydraulic and pneumatic devices. Bioelastics are nontoxic, biocompatible with human tissue, and biodegradable. Because bioelastics can convert chemical energy into mechanical energy, they offer potential as sensors or transducers.

6.4.3 Classification and Properties of Elastomers

Designations of elastomers by the American Society for Testing and Materials (ASTM D2000 standard) and the Society of Automotive Engineers (SAE) group them by type and class (Figure 6-36). ***Type*** reveals the maximum service temperature, which ranges from A

Type	Test temp (°C)	Class	Maximum swell (%)
A	70	A	No requirement
B	100	B	140
C	125	C	120
D	150	D	100
E	175	E	80
F	200	F	60
G	225	G	40
H	250	H	30
J	275	J	20
		K	10

(a) ASTM D 2000/SAE J200 type and class

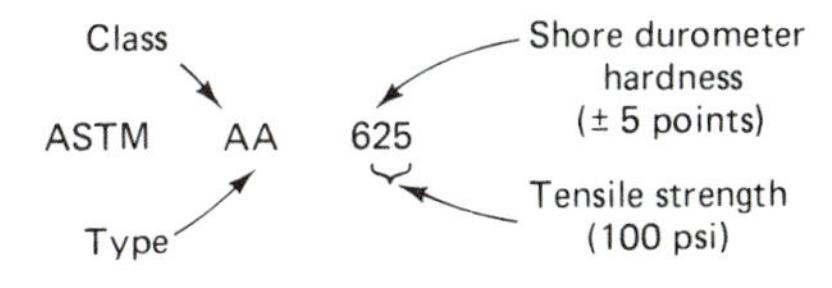

(b) ASTM designation example

Figure 6-36 Elastomer designation. (a) ASTM D2000/SAE J200. (b) ASTM designation example.

(70°C) to J (275°C). ***Class*** indicates the maximum percentage of swell when immersed in oil and ranges from A, with no requirement, to K, at 10% swell. Table 10-10 shows the class and type for selected elastomers. ASTM D1418 establishes abbreviations for elastomers, also shown in Table 10-10. To illustrate, PTR, AK designates a polysulfide elastomer with a maximum service temperature of 70°C that can swell as much as 10%. Thiokol is the trade name for polysulfide. Additional systems of designation indicate hardness and minimum tensile strength (Figure 6-36). ASTM AA625 designates (AA) natural rubber with Shore durometer hardness (6), meaning 60 plus or minus 5 points on the scale and (25) 2500 as the minimum tensile strength.

Table 10-10 compares properties for common elastomers. NR is the standard for comparison of the synthetics, as can be seen by its overall superiority, with excellent resistance to abrasion, tear, impact, electrical current, and water absorption, plus a tensile strength of 3100 Pa and a wide hardness range of 30 to 100 Shore A. Through designed formulation, the synthetic rubbers can surpass NR for special properties. NBR, ACM, FPM, and PTR offer excellent resistance to oil, kerosene, and gasoline, while NR is not recommended for service when exposure to these chemicals is expected. For high-temperature service, SL and VE are excellent. Neoprene resists deterioration from weathering, oxygen, ozone, oil, gasoline, and greases and has good tear and abrasion resistance. It finds applications as soles and heels for work shoes and liners for chemical tanks and pipelines, serves as an adhesive, and is applied like paint as a protective coating. Fluoroelastomers such as FPM are space-age elastomers that meet the demands of the aircraft and aerospace industry because of excellent resistance to heat and fluids, but they carry a high price tag. NR has only fair ***gas permeability,*** which means that it will allow some gas to pass through it. Butyl rubber (IIR) resists gas permeating and serves well as inner tubes, gas hoses, tubing, and diaphragms.

Rubber plays an important role in health care. Silicone elastomers, the most stable group of all elastomers, have polymer chains composed of silicon and oxygen backbones. Silicone rubber is used for various body replacement and enhancement components. Silicone is often used for plastic, reconstructive surgery of body tissue; recent problems with breast implants have caused concern for silicone as an implant. Silicone is a derivative of the second most abundant material on earth—silicon. Silicon composes about 28% of the earth's crust and is usually found in combination with other elements. One combination is silica, a naturally occurring silicon–oxygen compound known as silicon dioxide. Sand is probably the best known silica. Silicones are a family of extraordinary, synthetic materials made from polymers. Of high molecular weight, their molecular structure consists of spines of oxygen alternating with silicon. By attaching various organic, carbon-containing groups of elements to the spines, liquids, pastes, and rubberlike materials can be produced. The liquids and pastes can be cured into solids with application of heat or exposure to air to make adhesives and sealants. In other forms they are used as elastomers, coatings, oils, lubricants, and resins. If the attachment of these compounds to the spines of the SiO_2 polymer are short, the result is a volatile liquid that has the appearance of water and evaporates in air. If the attachment is long and ropelike, the physical properties change and the materials are more viscous, ranging from syrupy to a "solid" that flows if left unattended. You can stretch it, pound it, roll it into a ball and bounce it, and hit it with a hammer so it shatters. This material has the trade name Silly Putty.

Uses for silicone include baby bottle nipples, heat-resistant gaskets, cable insulation, spark-plug boots, encapsulations, bumper gels, brake fluid, adhesives for exterior tile on space shuttles, sealants, and weatherproof caulks. Theoretically, millions of silicone materials can be made just by altering the structure of the silicone chain. Only a few thousand have been produced and studied so far.

Natural-rubber latex has gained considerable attention recently with the concern for the transmission of AIDS and other sexually transmitted diseases. Latex surgical and examination gloves and condoms must meet the highest standards of manufacturing and testing to ensure the highest reliability. ASTM standards D3577, D2578, and D3492 address these procedures.

Hysteresis is an energy loss through heating in elastomers and creates a problem with many synthetic elastomers. If you stretch a rubber band and hold it against your cheek, you can feel the release of heat. The constant flexing of automobile tires generates enough heat in elastomers such as SBR that, if a tire were made wholly of this compound, it would quickly deteriorate. NR, on the other hand, converts the stressing into elasticity and quickly rebounds with less heating. The high hysteresis of SBR has an advantage in preventing slippage on wet surfaces and providing better abrasion resistance than NR, so it is used for the tread of tires while the NR goes into the sidewalls.

Elastomers possess extremely diverse properties, as do plastics. Thousands of additives and fillers are available and as many as 30 may be mixed into a single elastomeric compound. The range of hardness includes soft foam rubbers used for pillows to hard-rubber (ebonite) battery cases. Figure 6-37 provides a comparison of the hardness of selected elastomers and plastics. ***Foam rubbers*** have been formulated to respond to body temperature and pressure. Developed for astronauts, these foams now serve as orthopedic support for bed mattresses and chairs because of their ability to provide even support,

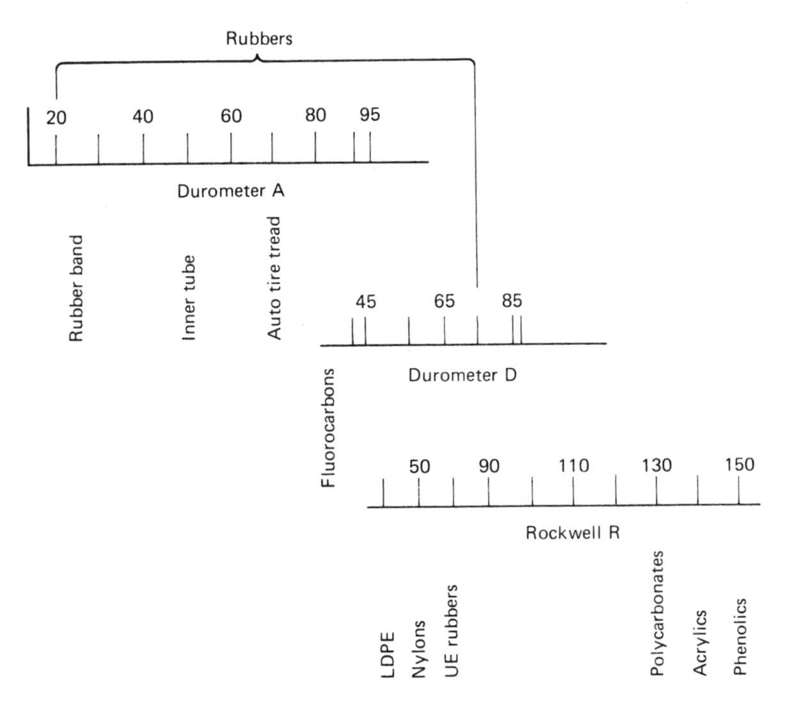

Figure 6-37 Comparisons of polymer hardness.

which reduces fatigue and soreness caused by prolonged body contact. Upon impact the foam firms up, and when heated, it softens. Joggers and other athletes use foam inserts in their shoes to reduce shock from the impact of running and jumping.

EPDM is a class of 13 types of materials that are a miracle of rubber technology because of their low cost and ability to be painted. Through compounding of carbon black, clay, processing oils, and cross-linking agents, the elastomer compounds can be molded or extruded as thermoplastics or thermosets. Temperature insensitivity, impact resistance, high deformation recovery, dimensional stability, and distortion resistance make them good choices for flexible exterior parts on automobiles, including bumpers and fascia.

Heat resistance is an important property in certain elastomer applications. Fluoroelastomers (FPM) can withstand temperatures over 340°C and continuous service temperatures up to nearly 290°C in exposure to oils, steam, and certain fluids. They have been improved to handle low temperatures down to around −35°C, at which point they become brittle. Among the silicon elastomers, room-temperature vulcanizing (RTV) compounds have stable properties through a range of temperatures, from nearly −53° to 250°C. Ethylene acrylics used for belts, seals, rollers, hoses, and gaskets can withstand hydraulic fluids, engine coolants, and hot oils for over a year at temperatures over 200°C. Chlorinated polyethylene elastomer, which is a random compounding of chlorine with high-density polyethylene, has good weather, hydrocarbon fuel, and oil resistance down to −15°C and up to nearly 150°C of continuous service. Grades of thermoplastic elastomers (TP) such as polyurethane provide good weathering as low as −15°C and as high as 120°C. Table 10-10 provides general ranges of maximum service temperatures for the groups of elastomers listed.

6.4.4 Selection of Elastomers

Whether natural or synthetic, rubber offers distinguishing properties. It is flexible, airtight, tough and resilient, and waterproof, and it resists corrosion caused by most common chemicals. Its ability to stretch in any direction and return to its original shape make it valuable in a variety of applications, from tires, to ball bladders, to hoses and gaskets. Reinforcing fibers can make it a nonstretch composite that retains the other valuable attributes of elastomers. Some high-performance elastomers, such as polyurethane (PUR), can be synthesized as thermosets or thermoplastics (TPUs); TS PUR will elongate 650% and has a tensile strength up to 100,000 kPa; TPU can elongate 600% and has a tensile strength of about 40,000 kPa.

Environmental concerns regarding rubber center on disposal of reinforced rubber, especially tires. This concern has been addressed with a variety of approaches. Tires are shredded for use as floor and roof tiles and filler for roads. The shred is burned to produce heat for electric generation or to generate low-cost gas used to incinerate hazardous waste.

The ***cost of elastomers*** varies widely and has a tremendous influence on selection of one over another. The fluoroelastomer mentioned previously that is capable of temperatures over 340°C sells for over $4500 per kilogram on finished parts, whereas the price for SBR would range from $0.60 to $0.80 per kilogram. Natural rubber sells for about $0.90 per kilogram, SI for $3.75 to $11 per kilogram, and heat-resistant ethylene acrylics sell for $3 to $4 per kilogram. While all fluoroelastomers do not cost as much as $4500 per kilogram, they are generally the most expensive elastomers, selling at around $28 per kilogram, and as copolymers with silicones, most fluorosilicones reach prices of $65 to over $90 per kilogram. Of the thermoplastics, the following indicate general prices per kilogram: copolyesters, $3.20 to $3.65; olefins, $1.37 to $1.75; and styrenes, down to $0.90 to $1.37.

6.5 ADHESIVES

Until the mid-1940s, the selection of adhesives for joining metallic materials together normally included either mechanical fasteners, such as bolts, rivets, and pins, or a thermal bonding method, such as welding, brazing, or soldering. Today, the knowledgeable designer learns that adhesive bonding must receive equal consideration over the traditional methods of joining parts together. The technician and crafter involved in maintenance and repairs can choose adhesives as an alternative technique for fastening broken or unjoined components. Generally, welding, riveting, and adhesive joining are classified as permanent joining methods, while mechanical fasteners such as bolts, screws, nuts, and pins allow disassembly.

The ideal material described in Module 1 does not exist, and neither does the ideal adhesive. If it did exist, it would stick to any material, need no surface preparation, and maintain its strength and other properties forever under all types of conditions. An adhesive bond has a great advantage over a mechanical joint, such as a nailed, screwed, bolted, or riveted joint. The strength of these mechanical joints depends on the combined strengths of the materials joined together and the fasteners involved. This interaction takes place where these fasteners come in contact with the material to be joined, causing a concentration of stress that may cause one of the materials to exceed its ultimate strength and cause the connection to fail. Not so with an adhesive joint, which permits an even distribution of stress to be applied over the complete bond area.

Adhesives provide the benefit of joining dissimilar materials. Adhesives also provide light weight; joint sealing; sound and vibration dampening; thermal and electrical insulation; uniform strength; and low-cost, low-skill techniques. These advantages mean that adhesives find many uses in joining for building construction, including the replacement of nails with mastic adhesives to bond plywood or particleboard to flooring joists and drywall panels to wall studs. ***Elastomeric adhesives*** are good for joining wooden furring strips to concrete. Various other adhesives are used to fasten vinyl and ceramic tiles (see Figure 6-38).

The wide use of adhesives by the aerospace industry results from its need to join thin sheets of aluminum, plastics, or composites to framework and still maintain smooth aerodynamic surfaces, light weight, uniform strength, and good fatigue resistance. ***Epoxy*** adhesives stick together structural members formerly joined by rivets, which caused problems due to their nonuniform stress distribution that created stress concentrations; also, rivets are heavy and prone to corrosion.

To help the automotive industry achieve more fuel-efficient vehicles through reduction of weight, adhesive-joining technology provides many unique joining methods. Certain ***structural acrylic*** and ***plastisol*** adhesives offer the auto industry and other such

Figure 6-38 Adhesives in construction. Low-cost, lightweight panel resulting from strong honeycomb paper core bonded with adhesives between plywood sheeting. (U.S. Forest Products Laboratory.)

assembly lines an advantage because of their compatibility with oily metal surfaces, and they bond many dissimilar metals and plastics in addition to wood, glass, hardboard, and asbestos board. These acrylics resist moderate temperatures, have high impact and peel strength, and cure rapidly with minimal shrinkage. ***Urethane adhesives*** bond together the steel and glass sunroofs. Elastomeric and plastic bumper systems and plastic body side molding use adhesives that dissolve the contaminating oils and films found on production lines, thus providing a dependable joint. The successful use of ***cyanoacrylates,*** or ***"super-glues,"*** on electronic, electrical, appliance, and instrument assembly lines represents the widespread acceptance of these instant bonding adhesives (Figure 6-39).

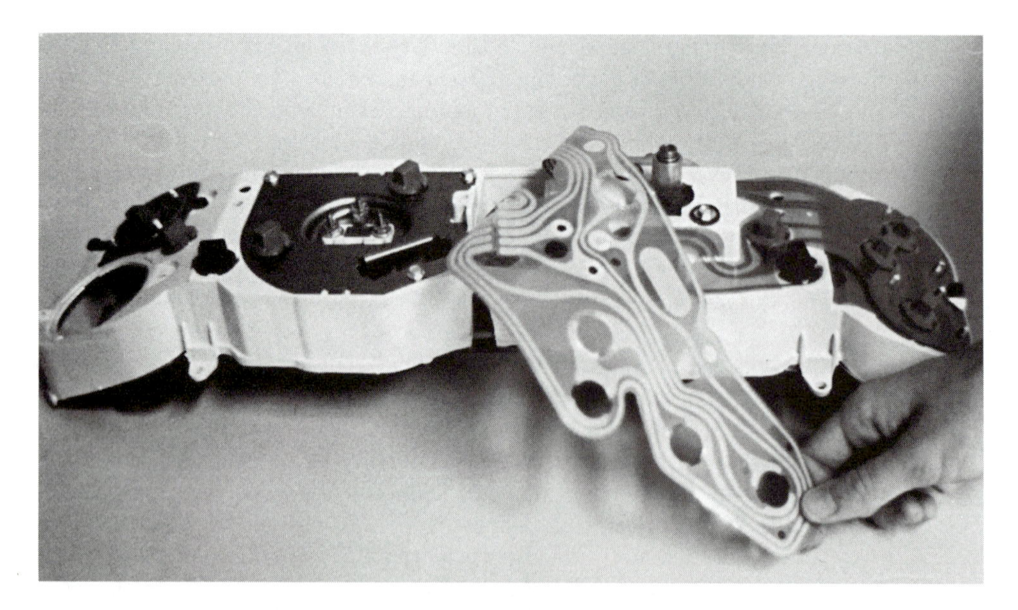

Figure 6-39 Cyanoacrylics set rapidly for electronic-component assembly. (General Motors.)

Among the major reasons for rejecting adhesive bonding over mechanical or thermal joining is heat sensitivity, with a normal service temperature ranging from 160°C to over 500°C. However, most common adhesives withstand a lower range, from −90° to 290°C. The space shuttle vehicle employs an adhesive bonding system for its ceramic heat shield tiles, with design temperatures up to 1260°C. Epoxy adhesives in the form of microcapsules serve as locking systems for mechanical fasteners. With the shearing force of a nut on an epoxy-coated bolt, the adhesive mixes and cures quickly.

6.5.1 Adhesives Systems and the Bond Joint

An adhesively bonded joint results from the adhesive's ability to flow, wet, and set. ASTM defines ***adhesive*** as a substance capable of holding materials together by surface attachment. The materials held together are *adherends* or *substrates*. ***Substrate*** is the broader term that defines a material upon the surface of which an adhesive-containing substance is spread for any purpose such as bonding or coating. As liquids of varying viscosity ranging from low viscosities below that of water to a semisolid (viscoelastic) or syrup consistency, including hot-melted solids, adhesives must wet the surfaces to which they bond. ***Wetting*** involves flowing (1) into large openings such as pores of woods or cells of foam plastics or (2) into the microscopic hills and valleys that exist on the smoothest of surfaces, such as polished metals or glass. A key factor in adhesive selection is the ability of the adhesive to wet all or most of the available bonding sites of the ***substrates*** (surfaces making the bond joint) (Figure 6-40). Refer to Figure 6-44 for a microscopic view of wood to contrast with smoother surfaces. The free spreading of the adhesive on the substrate surface may be impaired by conditions such as air entrapment, moisture or oxide buildups, and contaminants such as oil and dirt. Figure 6-40 illustrates a nonwetting condition, such as water beading up on a waxed surface, and a wetting condition, such as water absorbed in tissue paper. To achieve wetting of an adherend, the adhesive must not have ***cohesive forces*** (attraction of molecules inside the adhesive) that provide a very strong ***surface tension*** (force contracting the liquid into a droplet). The surface tension must be less than the adhesive forces between the adherend and adhesive. A high-surface-energy liquid such as epoxy

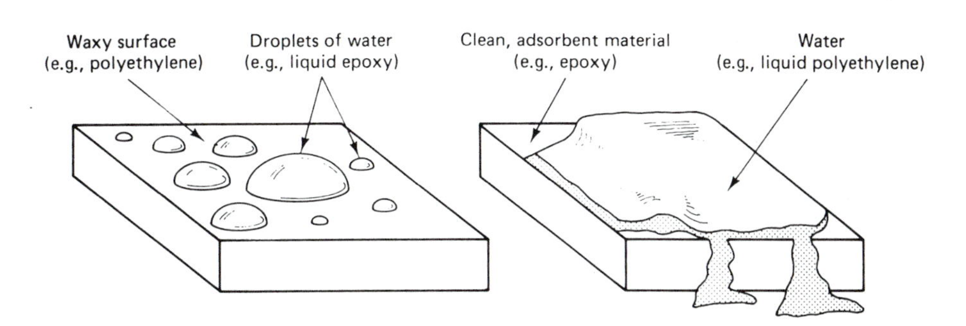

Figure 6-40 Wetting of substrate. Poor wetting: liquid beads up on surface. Good wetting: liquid flows into micro hills and valleys on surface, and porous materials absorb liquid.

adhesive will not wet a low-energy polyethylene solid. But the polyethylene liquid with lower surface energy will wet and form good adhesion to an epoxy solid.

Contact angle, or wetting angle, is a measure of how a liquid will spread on a solid. Many liquids spread easily on solids and have zero or very small contact angles. If the contact angle is greater than 90°, it is considered nonwetting. Water on clean glass would have a contact angle of 0°; on polyethylene the water would form an angle of over 100°. The analysis of contact angle has many applications in interface technology, ranging from the spread of adhesives, lubricants, printing inks, and protective coatings, to achieving maximum coverage with insecticides and herbicides. Once the wetting has occurred, the adhesive must ***cure*** into a solid (occasionally a viscoelastic) or ***set*** so that it develops adhesion, thus bonding materials ***(adherends)*** together into an adhesive joint.

Four theories attempt to explain the forces involved in the adhesion of an adhesive to its substrates. Both ***chemical*** and ***physical bonds*** are involved in adhesive bonding, but the exact nature of all forces is not fully known. Covalent bonds and physical attraction through van der Waals forces seem to operate in adhesives as they do in other polymers, but scientists differ on how the adhesion develops. On porous materials such as wood, ***mechanical interlock*** (penetration of adhesives into openings) provides added surface area for the adhesive to attach, but again some scientists believe that mechanical interlock has a minor role in adhesion. Testing of the adhesive bonding systems must deal with the ***interfacial adhesion*** (attraction between adhesives and substrates) and ***practical adhesion*** (breaking strength of joint).

Synthetic resins and natural adhesives ***cure*** through the various polymerization processes covered earlier in this chapter. Addition and condensation polymerization develop in the thermosetting adhesives that form permanent bonds; thermoplastic adhesives can be softened to reposition parts or for disassembly.

6.5.2 Adhesive Bond Chain Links

To grasp the complexities of adhesive bonding, see the model developed by Alan A. Marra (Figure 6-41), which illustrates the nine links in the adhesive-bond chain. The chain-link model reveals the interrelations of factors in the bonded product that include (1) adhesive composition, (2) adhesive application, (3) the bonded materials' properties, (4) their preparation, and (5) the stress and environment that the product will encounter in actual service. The physical and chemical theories and principles associated with the chain links become quite involved and require adhesive specialists to deal with them. The user of adhesives must recognize that proper adhesive bonding depends on the choice of the proper adhesive for the intended environment and on application according to specifications. The best adhesives will fail if not used properly. Link 1: the adhesive film must have proper ***cohesion*** to stick together when in service. Links 2 and 3: the intra-adhesive boundary layers are strongly influenced by the chemical and physical properties of the adherend; for example, extractives in wood may have an undesirable chemical interaction with the adhesive and change the composition and/or action of the adhesive as it is curing. Links 4 and 5 are critical links in the systems because this is where the adherend and adhesive engage. Problems such as surface contamination can prevent the anchoring of the adhesives at the necessary bonding sites. The compatibility of adhesive to adherend is required here so that proper wetting takes place; several theories attempt to explain the actual forces operating in these

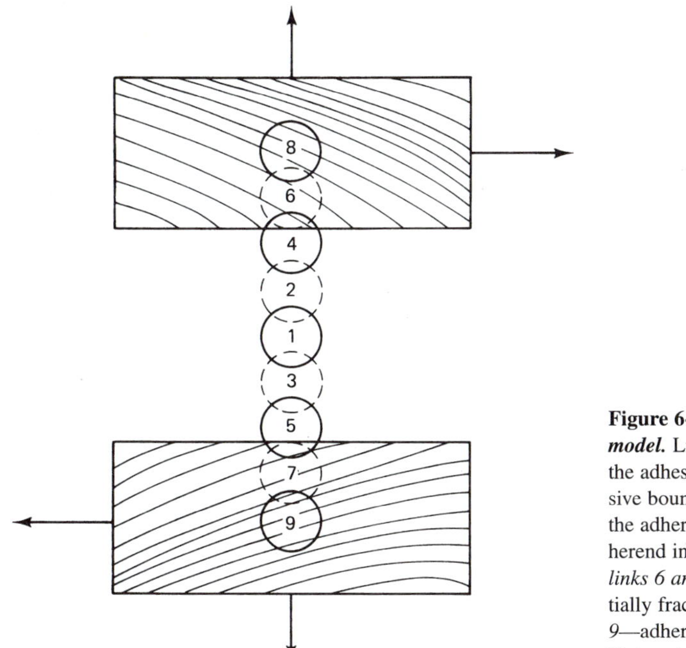

Figure 6-41 ***Marra's adhesive-joint model.*** Links of an adhesive bond: *link 1*—the adhesive film; *links 2 and 3*—intra-adhesive boundary layer, strongly influenced by the adherend; *links 4 and 5*—adhesive-adherend interface, site of adhesion forces; *links 6 and 7*—adherend subsurface, partially fractured in preparation; *links 8 and 9*—adherend proper. (Dr. Alan A. Marra, University of Massachusetts.)

two links. Links 6 and 7: the adherend/substrate is critical in a material such as wood or ceramic, where a damaged surface must be repaired through penetration of the adhesive into the pores. Mechanical interlocking may help with the repair. Finally, the adherend material's links 8 and 9 must possess sufficient cohesive strength and compatibility with the bonded joint. For example, a porous material such as a cellular plastic or elastomer may absorb and transport harmful liquids to the bonded joint, where the chemical could degrade the adhesive.

These problems lead to four ***main factors in adhesive selection:***

1. Materials to be bonded and type of joint
2. Conditions under which the bonding must be done
3. Durability and strength levels required during service of the product
4. Cost of the bonding process, including joint preparation, material cleaning, cost of adhesives, and so on

These four factors will vary considerably and require close attention to specifications by those making adhesive selections. Technical sales representatives, handbooks, and other sources of technical information guide the user, but as with any industrial production system, the manufacturer of the product must evaluate the adhesively bonded product throughout its development and maintain quality control over the adhesives and the application of those adhesives to products.

Advances in microprocessors and computers resulted from improvements in semiconductors and ceramic substrates. Many of those advances were possible because of continually improved ***interface technology,*** which focused on joining materials. Multilayered ceramics (MLCs) are examples of improved integration, one product of interface technology. Very-large-scale integration (VLSI) of semiconductors also presents interface challenges, as solder balls of only 0.002 in. join microthin metal circuitry deposited on ceramics. Printed-circuit boards use ***surface-mount technology (SMT)*** of electronic components (see Figure 8-43).

6.5.3 Types of Adhesives and Their Properties

Adhesives fit into various classifications, depending on factors such as the ***substrate,*** physical form (solid, liquid, viscoelastic), and bonding characteristics. Among the characteristics is the determination of whether the adhesive is structural or nonstructural. ***Structural adhesives,*** by ASTM definition, are those used to transfer required loads between adherends exposed to service environments typical for the structure involved. Some authorities classify structural adhesives as those whose failure could result in a threat to public safety, loss of life, or substantial property damage. These adhesives were formulated to replace nuts, bolts, rivets, and welds. A U.S. government study called the primary aircraft bond structure test (PABST) demonstrated that adhesive bonding is feasible and economical for replacing mechanical fasteners and welds for the main structural components on aircraft. Adhesives must possess ***serviceability*** (the ability to withstand design conditions such as heat, cold, stress, moisture, or chemical attack) and ***durability*** (the ability to maintain serviceability of the bond joint over its expected life).

Structural adhesives for wood include ***phenol–resorcinol–formaldehyde*** polymers, which are thermosetting, all-weather exterior adhesives used on laminated structures. ***Casein*** adhesives are structural water-resistant adhesives (moisture content of wood less than 16%) commonly used on softwood laminated beams and doors. ***Nonstructural adhesives*** include ***urea–formaldehyde,*** which is used on particleboard and furniture; ***poly(vinyl acetate)*** (Elmer's) for furniture and interior woodwork; and *starch* and *dextrins* for paper products. ***Protein*** adhesives include animal glues made from bone, hide, and fish (they are used for woodworking and gummed labels); casein from skimmed milk; and ***blood*** and ***soybean flours*** formulated with phenolic and formaldehyde resins, which cross-link to serve as water-resistant adhesives on interior plywood.

Thermoplastic adhesives are generally serviceable at temperatures below the thermosets. Among the thermoplastics, ***cyanoacrylates*** (Superglue) have a service range of −5° to 80°C; ***polyamides,*** 15° to 100°C; ***poly(vinyl acetate),*** up to 45°C; and ***butadiene styrene,*** 5° to 70°C. These compare with thermosets such as ***epoxy,*** −10° to 140°C; ***phenolic epoxy,*** −25° to 260°C; and ***phenolic neoprene,*** −20° to 95°C. Some synthetic adhesives come as thermosets and thermoplastics, such as ***polyimides*** (TS up to 370°C and TP −160° to 350°C), ***silicones*** (TS −25° to 250°C and TP −75°C), and polyurethanes (TS and TP −60° to 175°C).

Elastomeric adhesives are formulated with natural or synthetic rubber or blended with other resins to provide impact resistance and to withstand wide temperature variations. These adhesives include *neoprene, polysulfides, butyls,* and *nitriles.* A phenolic–

rubber blend adhesive bonds brake linings that withstand severe shock and temperature and saltwater exposure.

Except for pressure-sensitive tapes that remain tacky, adhesives change from the liquid state to the solid state upon curing. Tapes such as adhesive tape, duct tape, and transparent tape do not develop the strength found in solid adhesives; however, double-coated pressure-sensitive tapes serve as hold-down tapes for use on machine tables of milling machines and surface grinders.

Curing of adhesives involves the cooling of hot melts, loss of solvents, anaerobic environments, and two-part mixtures. ***Hot-melt adhesives*** usually begin as a solid that becomes liquid in heat guns or furnaces, or with other heat sources that develop around 150°C temperatures; most cool and set within seconds after removal of the heat source. ***Loss-of-solvent adhesives*** employ volatile liquids, including water and organic solvents, that dissolve the adhesive base material to form the fluid; upon application to the joint, the adhesives harden as the solvents evaporate or penetrate porous adherends. With solvent adhesive, especially, and other adhesives, one must determine whether the adhesive will attack the substrate so severely that it may produce weaknesses in the material being joined. ***Anaerobic adhesives*** retain their fluid state when exposed to oxygen, but when squeezed into thin joints that block the oxygen, they set up into hard, strong adhesives. ***Two-part mix adhesives*** are stored with the resin and catalyst (or hardener) separated; when mixed the catalyst causes cross-linking of the resin.

Loss-of-solvent adhesives often have long setting times; and except for water-solvent types, the solvents present health and fire safety problems. Also, they rely on our short supply of petroleum, which is causing a trend toward water-based adhesives. Anaerobics come in one part, cure quickly, and develop high shear and impact strength, and some have gap-filling abilities. Other adhesives such as cyanoacrylates cure as a result of exposure to the moist alkali environment found on most surfaces. Contact cements are usually rubber-based structural adhesives that form an instant bond when substrates coated with the cured adhesives are brought into contact, thus eliminating the need for clamping. Most other adhesives require pressure long enough for the adhesive joint to cure.

Caulks and *sealants* work like adhesives and protective coatings to fill gaps and seal against weather, liquids, and many commercial applications. ***Polymer caulks*** are classified by their expected service life, known as *range.* Oil-based caulks are low-range and low-priced. Acrylics, butyls, and latexes are medium-range caulks and serve many construction needs. High-range caulks are the most costly; they include silicone (for considerable joint movement, ultraviolet-light exposure, waterproofing, and wide temperature variations), various copolymers, and urethane. Some high-range caulks such as urethane are also available as foam to fill large gaps. Industrial suppliers and home improvement centers have hundreds of caulks and sealants; some even allow patching and joining under water. An advanced material, Gore-Tex joint sealant, is 100% PTFE in expanded, continuous cord form that allows it to be used as an "asbestos-free" sealant. It can serve as gaskets for engine blocks [Figure 6-42(a)] that will never leak or need replacing and can be used on irregular or damaged surfaces because it fills in and conforms to surface imperfections to provide an ideal seal against gas or liquid. ***Sealants,*** applied as liquids or pastes, cure to form seals against gases or liquids. Often, the sealants also act as adhesives, especially in building construction. See the silicone sealants in Figure 6-42(b).

(a)

(b)

Figure 6-42 (a) Gore-Tex® gaskets of PTFE provide excellent seals for rough surfaces where oil, fluids, and temperatures may damage other sealing materials. (W. L. Gore & Associates, Inc.) (b) One part insulating glass, silicone sealant offers exceptional hydrolytic stability, fast cure, high strength, high modulus, and primerless adhesion to glass and many substrates. (G.E. Plastics.)

6.5.4 Adhesive Selection

The wide range of available adhesives and sealants is partially the result of the difficulty of finding suitable matches of adhesive to adherend. Many general guides exist, but they should be used with caution. For important production jobs, it's best to consult suppliers to obtain the best choice of adhesive. For general-purpose jobs, some broad guidelines help.

As discussed earlier, plastics present the most difficulty because of wetting problems. Table 6-7 lists some common adhesives and their typical applications and provides

TABLE 6-7 COMMON ADHESIVES

Adhesive	Typical adherends and applications
Acrylics (two-part)	Plastics (ABS, phenolic-melamine, polyester, polystyrene, polysulfone, PVC) to metal, rubber to metal, and plastics to plastics
Casein	Interior wood components
Cyanoacrylates (anaerobic)	Thread locking, gear and bushing mounting, aluminum, copper, steel, ceramics, glass, rubber, polycarbonate, phenolic, acrylics
Epoxy	Most adherends except some plastics such as acetals, polypropylenes, polyurethanes, and silicones; many modifications including nylon epoxies and epoxy phenolics
Melamine formaldehyde	Paper, textiles, hardwood, plywood for interior use
Natural rubber	Cork, foam rubber, leather, paper-building materials, pressure-sensitive tapes
Neoprene rubber	Many plastics (ABS, fluorocarbons, polystyrene), aluminum, ceramics, copper gasketing and laminating
Nitrile rubber and phenolic	Many plastics (ABS, cellulose, fluorocarbons, PVC, polyester), aluminum, ceramics, glass, magnesium, microwave isolators
Phenol–formaldehyde	Cork, cardboard, softwood, plywood for exterior use
Poly(vinyl acetate)	Textiles, polystyrene, wood, other porous and semiporous materials, interior furniture and construction
Resorcinol–formaldehyde	Asbestos, cork, paper, rubber, wood furniture, laminated exterior wooden beams, wooden boats
Silicone	Aluminum, ceramics, glass, magnesium, phenolics, polyester, acrylics, rubber, steel, textiles, integrated circuits, sealants, gasketing
Urethane elastomer	Many plastics (ABS, polyester, acrylics, PVC), aluminum, ceramics, copper, glass, magnesium, steel-bond solid propellants, cryogenic applications, insulation

an overview of selected adhesives, adherends, and applications. The list below provides a basic guide for adhesive selection in selected materials groups. Read the technical data for each adhesive to learn more specific details on compatibility. For important jobs, conduct a test on small samples before doing the complete bonding. If adhesive tapes won't stick to an adherend, you must abrade the surface to remove coatings or apply another coating that is compatible with the adhesive.

General Material Group	General adhesive groups
Metals	Epoxy, cyanoacrylates, urethane
Wood	Vinyl emulsion, resin, epoxy
Ceramic	Cyanoacrylates, epoxy, urethane
Plastic	Numerous options: read packaging information

Adhesives have been developed to meet the most varied of circumstances. In medicine, methylmethacrylate resins join artificial hip sockets to natural hip bones and repair broken, damaged bones. Refer to Applications & Alternatives in Module 1. Cyanoacrylates bond eye tissue and other skin tissue. Thermoplastic polyester resins bond copper foil to flexible backing on printed-circuit boards and, with epoxy and phenolic adhesives, withstand etchants and the heat of soldering operations. ***Weld bonding*** combines spot welding and adhesives. The weld immobilizes the adherends to allow the adhesives to cure. The increased use of resin-based composites promises even wider use of adhesives.

6.6 OTHER IMPORTANT POLYMERICS

Polymer science and technology continue to generate innovative materials. In the medical field, research and development efforts encompass replicating human tissue such as skin and producing a polymer that will stimulate cartilage regeneration. Polymers that will not burn yet are easy to process, lightweight polymer batteries, water-soluble polymers, and controlled-release polymers are some materials of continuing interest. The following groups of polymers are other significant engineering polymers.

6.6.1 Protective Coatings and Preservatives

Because polymers include synthetics, an endless variety of them can be synthesized. Protective coatings cover another category of polymers not discussed in this module. Protective coatings include natural polymers such as shellac, oil-based paint, and many synthetic polymers, including the same general types used for plastics, elastomers, and adhesives: for example, epoxies, silicones, latex, urethanes, and acrylics. These coatings develop their protective skins in a number of ways including polymerization, cooling of hot melts, and evaporation of solvents or water. Evaporation leaves behind a film of the polymer that was dissolved in solvent or suspended in water.

Environmental concerns about ozone layers, water quality, and overall air quality have required that the petrochemical industry put considerable effort into eliminating the traditional materials-related uses of chlorofluorocarbons (CFCs) and ***volatile organic compounds (VOCs).*** VOCs create air-quality problems at the ground level (troposphere) by forming ozone and photochemical oxidants, the primary ingredients in smog. CFCs create problems in the upper atmosphere (stratosphere) by reacting with the delicately balanced ozone layer that protects the earth from the sun's ultraviolet rays. CFCs deplete the "good" ozone. Much of the current development in protective coatings focuses on eliminating VOC solvents because they create health hazards to workers and harm the atmosphere. By replacing VOCs with water as the vehicle to carry the polymer dispersions, they evaporate with few harmful effects. Much of the paint used today is water-based.

As a part of the ***"green movement,"*** paints have been developed based on natural substances (plants and minerals), without pesticides, preservatives, and antimildew agents. Some paint manufacturers separate the solvents from the solids in excess paint, leaving pigment and resin that is recycled as primer or utility paint.

A polymerization process known as ***group-transfer polymerization*** (GTP) permits unlimited polymer-chain growth in acrylics, with the potential for high-performance auto finishes that will cure at lower temperatures (180°F versus 250°F to 300°F). The process also uses lower concentrations of polluting solvents.

Although wood offers natural beauty and warmth, it is susceptible to weathering, moisture, insect attack, and stress on weaker sections that have suffered during the life of the tree. Many protective coatings and polymer enhancements have been developed, such as diffusion of ***poly(ethylene glycol) (PEG)*** to preserve archeologically significant wooden pieces. Cross-linking chemicals can permanently bond cell walls of wood for stability and resistance to fungi, but their use results in decreased strength. Enhancement of mechanical properties of wood is achieved with vacuum impregnation of liquid monomers such as epoxy and methyl methacrylate into openings that result from sap leaving the tree during curing. The resins are polymerized after impregnation to make harder, stronger modified wood for boat hulls, billiard cues, knife handles, and parquet flooring. Large structural timber and lumber is both pressure and nonpressure treated (brushing, soaking, dipping, and heating) using oil-type preservatives: creosote, pentachlorophenol (penta), copper naphthenate, and waterborne inorganic arsenical compounds.

6.6.2 Oils, Lubricants, and Fluids

Polymers play an important role in many types of lubricating, cutting, and hydraulic applications as solids and as additives or suspensions in liquids. Engine-oil viscosity for multigrade oils, such as SAE 10W-40, changes when the ambient and engine temperatures change because polymers control the oil thickness. New long-lasting engine oils are the result of synthetic polymers.

Electrorheological (ER) fluids (introduced in Section 6.3.5 as smart polymers) congeal from a liquid into a solid mass in a few milliseconds when an electrical current is passed through the liquid. When the current is removed, the solids return to liquids. ER fluid has promise as a hydraulic fluid for robots, automotive clutches, and artificial human limbs. Experimentation with combinations of selected oils with particles (microscopic solids of polymethacrylic acid and water in long, tangled chains) will continue in the quest for the ideal ER fluid. What possibilities might this type of solid lubricant hold for problems encountered with high-temperature ceramic engines or automotive brakes?

6.6.3 Conductive Polymers

A major disadvantage of lead-acid and other metallic batteries is their weight. The development of conductive polymers such as polyacetylene electrodes could open the way to long-sought electric automobiles. Remember that polymers do not normally serve as conductors because their electrons are locked in covalent bonding rather than as free electrons in metals. Oxidation, resulting from doping polyacetylene with iodine, allows free electron movement. Reduction in the same polymer, achieved by reaction with the metal sodium, turns polyacetylene into an anion while the sodium becomes a cation. The polymeric form of sulfur nitride becomes conductive at low temperatures without doping. Poly(*p*-phenylene), or PPP, is made up of single and double bonds that can achieve electron mobility through doping. Polypyrrole also has conductor potential because of bonding similar to that of PPP. Both can be switched back and forth between conductors and insulators. Each of these polymers offer potential as conductors, but major problems exist in converting them to reliable engineering materials. Disulfide polymers, long chains of sulfur–sulfur bonds, provide battery cathodes that act with lithium anodes to produce a depolymerization–polymerization process that yields electrochemical energy.

6.6.4 Geosynthetics

ASTM defines ***geosynthetic*** as:

> a planar [lying in one plane] product manufactured from polymeric material used on soil, rock, earth, or other geotechnical engineering related material as an integral part of a man-made project, structure, or system.

Geotextiles, geogrids, geofoams, geofilters, and geocomposite drains serve many purposes in building roads, dams, landfills, and similar applications where earth erosion, pollution, or containment of liquids is a factor. Civil, building, and ***geotechnical*** (related to the earth's crust) ***engineering*** often call on impermeable membranes to contain water in dams, to keep oil and other liquid chemicals in reservoirs, or to line sanitary landfills to prevent leached liquids from leaving the landfill and polluting groundwater. Porous woven polymer fabrics hold soil on river or lake banks until rocks and vegetation stabilize the banks. Polymer fabrics go into road construction to develop laminar composites that prolong the life of macadam (asphalt-and-rock composite) roads.

A major concern of the geotechnical community is how long a geosynthetic will last. A landfill has many potential pollutants. Once the geomembrane liner is in place, will it prevent pollution of the groundwater indefinitely? ***Geomembranes*** of PVC, PE, and ethylene copolymers undergo extensive testing based on ASTM Committee D35 standards. Geotechnical engineers also use expanded polystyrene (EPS) blocks, also known as ***geofoam,*** to rebuild highway embankments. Increased awareness of the fragile nature of Earth has prompted use of innovative geosynthetics; continued improvements in these synthetic polymers is inevitable.

SELF-ASSESSMENT

6-39. What is the major source of natural rubber? What are other possible sources?

6-40. Define the term *elastomer* and diagram a polymer of natural thermoplastic rubber and a polymer of vulcanized rubber.

6-41. Diagram the polymer-chain structure of rubber at rest and again with tensile stress applied.

6-42. What are the major advantages and disadvantages of natural rubber (NR)? Give examples of applications of NR to show how its advantages cause selection and its disadvantages prevent selection.

6-43. Use Table 10-9 to choose a suitable elastomer for the following containers: **(a)** hydraulic hose, **(b)** hot water bottle, **(c)** diver's wet suit, **(d)** bicycle-tire inner tube.

6-44. What is the major raw material for synthetic rubber?

6-45. At what stage in tire manufacture does vulcanizing occur?

6-46. Why is natural rubber used in automobile tires? Why are synthetic rubbers also used?

6-47. What additives to synthetic polymers accomplish the following: **(a)** retard aging from ultraviolet rays, **(b)** soften raw rubber, **(c)** increase rigidity, **(d)** strengthen?

6-48. Explain TR in terms of structure and applications.

6-49. Explain the following elastomer designation: ASTM BH 530.

6-50. What is the importance of hardness in rubber? Compare the hardness of elastomers to that of other polymers.

6-51. What is the term to describe loss of energy through heating in elastomers? When is it an advantage for this property to be **(a)** high or **(b)** low?

6-52. Cite examples of an application where the cost of an elastomer may be so high that another material would be selected.

6-53. Select elastomers and specify their ASTM abbreviation and their favorable properties for the following applications: **(a)** fire hose, **(b)** lawn mower wheels, **(c)** stopper on chemical test tubes, **(d)** golf-ball winding, **(e)** racket ball, **(f)** tennis-shoe soles, **(g)** rubber band, **(h)** seal on auto-brake master cylinder, **(i)** carpet backing, **(j)** work-shoe soles, **(k)** pad for bicycle caliper brakes, **(l)** wrestling mat.

6-54. What trends in engineering make adhesives so desirable?

6-55. List three advantages of adhesive bonding over other joining methods. Name three applications.

6-56. **(a)** In the adhesive system, what three events must occur to obtain a bonded joint? **(b)** What types of forces are involved in adhesion?

6-57. What is the difference between adhesion and cohesion?

6-58. **(a)** What is the difference between practical adhesion and interfacial adhesion? **(b)** What can prevent wetting?

6-59. Describe structural adhesives and list two examples.

6-60. **(a)** Name the five integrated factors that can affect a bonded product. **(b)** What must be considered in adhesive selection? **(c)** What contributions has interface technology made to new computers?

6-61. **(a)** Why would a thermoplastic adhesive be selected over a thermosetting adhesive? **(b)** Name two protein adhesives and give an application of each.

6-62. What adhesive can be used to bond structural components that will meet service conditions in southern deserts or in arctic regions?

6-63. Describe three methods in which adhesives develop their cured attraction between the substrate and the adherend.

6-64. Name one method of classifying caulking.

6-65. Select an adhesive for the following joining applications: **(a)** aluminum electronic component to fiberglass printed-circuit board; **(b)** PUR-foam bumper fascia to steel bumper, **(c)** ABS control mount to porcelain-enamel dishwasher panel, **(d)** plywood veneers, **(e)** aluminum identification label to PC plastic electric-drill housing; **(f)** aluminum towel rack to ceramic tile wall, **(g)** glass walls to chrome-plated steel aquarium, **(h)** PS-foam pad to wooden seat, **(i)** exterior laminated beam, **(j)** copper foil to flexible printed-circuit board, **(k)** brake liners to steel brake shoe.

6-66. What is a major thrust in the development of protective coatings? List three broad groups of protective coatings.

6-67. Name the term for synthetic polymers used on soil, rock, and earth in construction and cite three examples of these products.

6-68. Why has geotechnical engineering begun to seek greater use of synthetic polymers?

OBJECTIVE QUESTIONS

6-24. The objective of vulcanizing rubber.

a. To improve processibility **b.** To increase hardness and chemical resistance
c. To reduce materials cost **d.** To increase amorphousness

6-25. Vulcanization of rubber uses sulfur to transform it into a thermosetting elastomer.

T (true) **F** (false)

6-26. Before vulcanization, natural rubber has many double covalent bonds known as ______________.

a. Reactive sites **b.** Benzene rings **c.** Isoprenes **d.** Copolymers

6-27. The major source of raw materials for synthetic rubber.

a. Hevea **b.** Oil **c.** Trees **d.** Corn

6-28. The purpose of an antioxidant additive.

a. Prevents aging **b.** Speeds up processing
c. Accelerates curing **d.** Strengthens

6-29. The purpose of carbon black as an additive.

a. Prevents aging **b.** Speeds up processing
c. Accelerates curing **d.** Strengthens

6-30. Thermoplastic rubbers process easier than most thermosetting rubbers.

T (true) **F** (false)

6-31. Application for shape-memory polymers.

a. Robotic hands **b.** ATMs **c.** Tire reinforcement **d.** Netting

6-32. Energy loss through heating in elastomers, which creates problems in applications such as car tires.

a. Ablation **b.** Oligomer **c.** Cohesion **d.** Hysteresis

6-33. Elastomer used most often to help prevent the transmission of AIDS.

a. Neoprene **b.** Fluoroelastomers **c.** Latex **d.** IIR

6-34. A most stable elastomer used for human implants.

a. Latex **b.** NR **c.** Fluoroelastomers **d.** Silicone

6-35. Elastomers offer a wide range of properties from very soft to quite hard, they resist most chemicals, they provide good traction and abrasion resistance, and they are quite flexible.

T (true) **F** (false)

6-36. Which advantage(s) would adhesive bonding have over other joining methods?

a. Ease of assembly and disassembly
b. Light weight, load spreading, joint sealing
c. Withstands very high temperatures normally exceeding the base material
d. Easy selection for joining any type of material

6-37. A key factor in adhesive selection is the ability of the adhesive to ____________ the substrate material.

a. Wet **b.** Cover **c.** Cure **d.** Entrap

6-38. Attraction of molecules within an adhesive or substrate.

a. Adhesion **b.** Contact **c.** Cohesion **d.** Interface

6-39. Penetration of adhesives into the openings of porous materials.

a. Interface **b.** Mechanical interlock
c. Cross-linking **d.** Surface tension

6-40. Which group of materials presents the greatest challenge for adhesive selection?

a. Plastics **b.** Wood **c.** Metal **d.** Ceramics

6-41. How does Marra's adhesive joint model aid in dealing with bonding?

a. Illustrates the relationships between the adherends and adhesives
b. Shows how to design the substrate
c. Diagrams methods to select plastic adhesives
d. Provides data on curing times

6-42. Loss-of-solvent curing of adhesives usually has shorter curing times than cooling of hot melts.
T (true) **F** (false)

6-43. Polymer strands classified as either regenerated cellulosics or noncellulosics.
a. Natural fibers **b.** Jute **c.** Manufactured fibers **d.** MLC

6-44. Materials used to work like adhesives, provide protective coatings, and keep out liquids and gases.
a. Adhesives **b.** Caulks **c.** Sealants **d.** Stains

6-45. Considerable emphasis has been placed on eliminating ______________ in protective coatings to improve air quality and other aspects of our natural environment.
a. Silicone **b.** Water-based coatings **c.** ER **d.** VOCs

6-46. The term for synthetic polymers used on soil, rock, and earth in construction.
a. Aramides **b.** Geosynthetics **c.** Styrofoams **d.** Earth screens

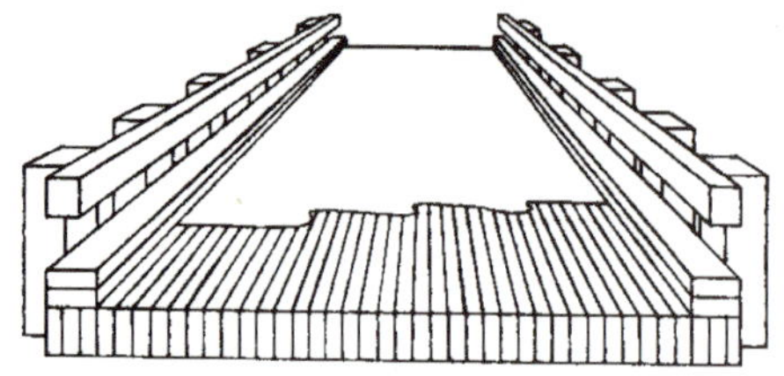

Type A: Stressed Timber Deck

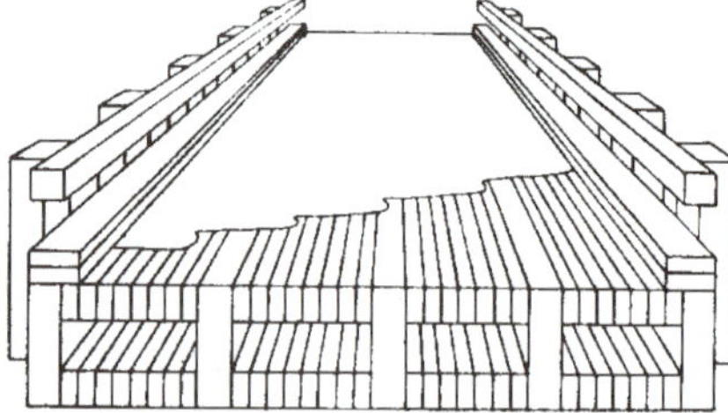

Type B: Stressed Timber Box Section

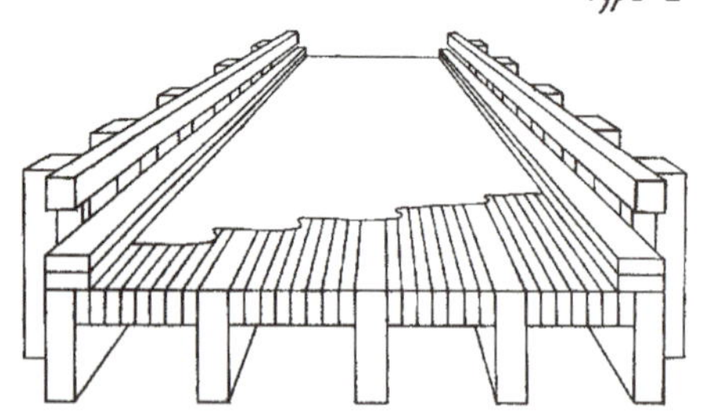

Type C: Stressed Timber Tee Section

Rebirth of wooden bridges. For decades, concrete and steel replaced the old timber bridges that had served humanity since early civilization. Recently, we've learned that the acid rain, road salt treatments, and other environmental conditions have rendered nearly half of the 600,000 short-span bridges in the United States unsafe and/or unusable. New technology applied to wood has resulted in timber bridges that should provide much longer service life than steel and concrete bridges. Timber treated with wood preservatives and laminated into stress timber decks, T-sections, and box sections for bridge construction offer the following advantages: pleasing aesthetics, good shock resistance, high strength-to-weight ratios, sound and thermal insulation, and weather and chemical resistance (see photo). Using wood, a renewable resource, localities should be able to harvest regional materials and labor for economical alternatives to those hundreds of thousands of defective concrete and steel bridges. (Barry Dickson, Construction Facilities Center, West Virginia University.)

6.7 WOOD AND RELATED PRODUCTS

Wood, one of nature's unique, natural composites, and rock were humanity's first materials. As a ready weapon or fuel for the fire, the stick required no major processing. Since those early beginnings, wood has been the target of much technology and is still researched for methods not only of improving the yield of forests, but also for ways of better using the tree and the many by-products of wood.

As with other living organisms, trees are subject to environmental conditions. Acid rain, lack of biodiversity in tree farms, overharvesting, fires, and similar problems present constant challenges to maintaining the world's fragile balance with its forests. Many underdeveloped nations have harvested their trees too heavily to sell the timber for needed income and/or for cooking and heating because of the scarcity of other fuels. Where forests are lost, regions turn to deserts, causing disasters to all living things. As with most other materials in our Family of Materials, all aspects of wood must be closely monitored in the materials cycle.

Wood must be considered a valuable engineering material and should be exploited to its fullest extent because it is an environmentally friendly material. Fossil fuels and minerals required an eon of time to develop, so civilization faces shortages of oil, gas, iron, chromium, and other raw materials used to produce many of our engineering materials. Trees, on the other hand, are renewable resources also crucial for the air we breathe, and they stabilize soil, provide shelter and nutrition for many species, and add to the natural beauty of our landscape. It is possible to grow wood in a variety of ways that yield a full range of construction and consumer products. If grown in plantations using new planting and harvesting techniques, then processed with new technology to maximize all of the plant, and finally recycled efficiently, wood will serve humanity well. In our modern era, wood has been replaced by other materials, such as steel, plastic, and concrete; however, experience with these alternative materials has not always been completely positive.

Many plants possess woody or cellulosic substances, but the tree is our major source of wood. ***Extractives*** from the tree that contribute to properties of wood such as decay resistance, color, odor, and density are also useful in a number of industries. These extractives include tannin (tannic acid), used in processing leather, and polyphenolics, used to make phenolic plastics, plus coloring agents, resins, waxes, gums, starch, and oils. In addition to the fruit and nuts that are harvested, trees serve as the raw material for paper, cellulosic plastics, explosives, rayon fibers, films, lacquers and drilling muds, ethyl alcohol, food flavoring, concrete additives, and rubber additives. Wood was our first fuel, but with the technological development of coal, oil, and gases, wood became only a romantic fuel for most of the technologically advanced people of our world. However, tree products once again are being given serious consideration as a fuel. The ability to renew wood supplies through forestry increases humanity's interest in this material as nonrenewable resources such as oil, coal, and minerals rapidly become depleted.

6.7.1 Structure

Wood develops through ***photosynthesis:*** the chlorophyll in trees uses carbon, hydrogen, and oxygen to produce sugar, starches, and cellulose. The dry wood used as an engineering material has an approximate composition of the following percentages by weight: cellulose (50%), lignin (16% to 33%), hemicelluloses (15% to 30%), extractives (5% to 30%), and ash-forming mineral (0.1% to 3%). ***Cellulose*** ($C_6H_{10}O_5$) [Figure 6-43(a)], a high-molecular-weight linear polymer, forms fibers that make up cell walls of vessels and ducts. ***Lignin*** [Figure 6-43(b)], an amorphous polymer, forms a matrix around the cellulose, much like the plastic matrix in fiberglass. Through removal of lignin, wood can be broken down into fibers that are used in making paper and other synthetics. The cellulose forms cellular networks of ducts, vessels, fiber rays, and pits, which transport and store the extractives and minerals throughout the living tree. The network differs, depending on whether it is a hardwood or a softwood tree. Softwoods are not always softer than hardwoods; Douglas fir, a softwood, is about twice as hard as basswood, a hardwood. ***Softwood*** species (Figure 6-44) include firs, pines, and spruces; they bear cones and have needles or scalelike leaves. ***Hardwoods*** are broadleaved and not normally evergreen; they include maples, oaks, birches, and mahogany. ***Hemicellulose*** is another polymer closely akin to cellulose; like cellulose it breaks down into sugars when chemically treated. The ash-forming minerals include calcium, potassium, phosphate, and silica. Extractives are removed by heating the wood in water, alcohol, or other chemicals; they represent a significant commercial value.

Most common engineering materials, such as steel and concrete, are ***isotropic,*** which means that generally the strength is the same in all directions because the materials are ho-

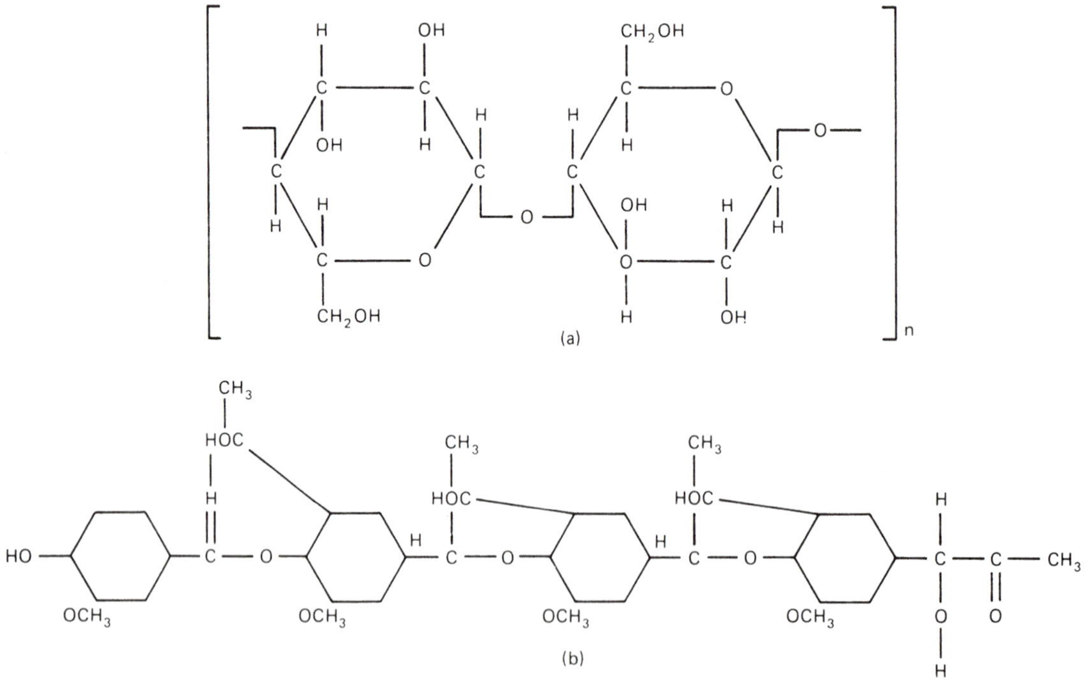

Figure 6-43 (a) Cellulose formula. (b) Lignin formula.

Figure 6-44 Softwood block showing three complete and part of two other growth rings in the cross-sectional view (X). Individual cells can be detected easily in the earlywood (EW), whereas the smaller latewood cells are difficult to distinguish in the latewood (LW). Note the abrupt change from earlywood to latewood. The two longitudinal surfaces (R—radial; T—tangential) are illustrated. Rays, which consist of food-storing cells, are evident on all three surfaces. 70×. (Dr. Wilfred A. Côté, SUNY College of Environmental Science and Forestry.)

mogeneous. Wood is ***anisotropic,*** meaning that it has greater strength in some directions than others. The anisotropy develops as a result of the way a tree grows, with various factors influencing the wood structure: cellular structure, branches, and bending by prevailing wind. A full discussion of all these factors is beyond the limits of this book, but the resulting tensile strength and stiffness of wood are greater in the radial direction than in the tangential direction (Figures 6-45 and 6-46). Rays that radiate from the pith outward on the tree tend to tie together the layers of cells (tracheids) growing longitudinally. This ***directional strength*** results partly from the complex structure of cell walls, with long polymer chains of cellulose that form layers running in varying patterns to reinforce each other. Primary covalent bonding holds together these ***microfibrils,*** which are almost parallel to the cell axis. Weaker secondary bonds operate in the perpendicular axis.

Lumber consists entirely of dead cells. Figure 6-47 illustrates a cross-section of a white oak tree trunk. The ***cambium layer*** produces all the growth of cells (wood and bark). ***Sapwood*** is a transitional wood of both living and dead cells. The sapwood's main role is to conduct sap and store food. ***Heartwood*** evolves from sapwood as the cells die and become inactive. The heartwood cells no longer transport sap but rather become storage cells. Sapwood of all species is highly susceptible to rotting, while heartwood of some species is protected from decay because it has taken on infiltrating materials that serve as preservatives. Additional information on wood structure, plus other information on wood, is available in the excellent source *Wood Handbook: Wood as an Engineering Material.* The handbook defines ***lumber*** as the product of the saw and planing mill and is not manufactured further; ***timber*** is a classification of lumber that is nominally 5 or more in. in its least dimension.

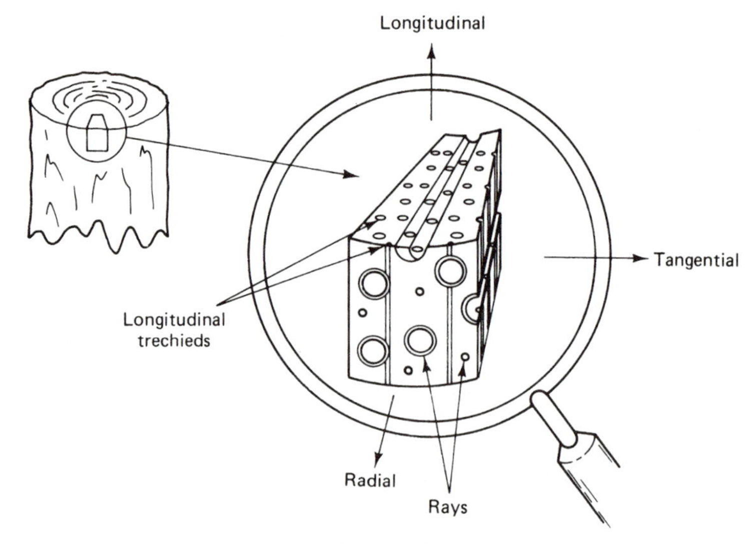

Figure 6-45

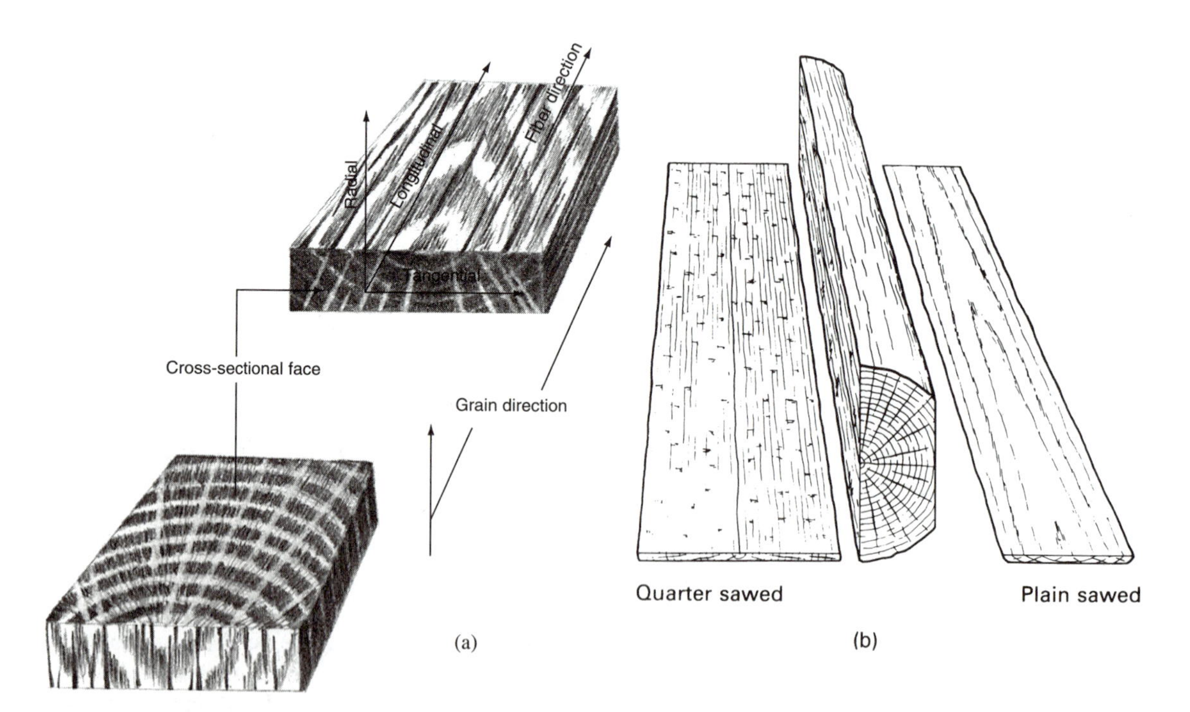

Figure 6-46 (a) Grain direction. (U.S. Forest Products Laboratory, Madison, Wisconsin.) (b) Sawed boards.

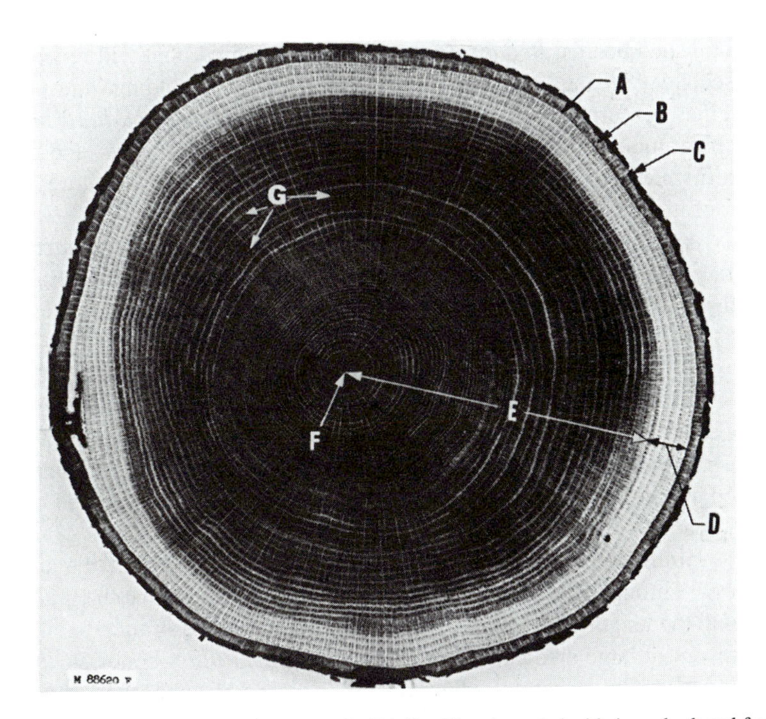

Figure 6-47 Cross-section of white oak tree trunk: (A) Cambium layer is inside inner bark and forms wood and bark cells. (B) Inner bark is moist and soft, and contains living tissue; carries prepared food from leaves to all growing parts of tree. (C) Outer bark containing corky layers is composed of dry dead tissue; gives protection against external injury. (D) Sapwood, which contains both living and dead tissue, is light-colored wood beneath the bark; carries sap from roots to leaves. (E) Heartwood (inactive) is formed by a gradual change in sapwood. (F) Pith is the soft tissue about which the new growth takes place in newly formed twigs. (G) Wood rays connect the various layers from pith to bark for storage and transfer of food. (U.S. Forest Products Laboratory, Madison, Wisconsin.)

6.7.2 Wood Products

6.7.2.1 Commercial lumber. Lumber is graded to provide the buyer with an indication of the strength and appearance of the lumber. Unlike most engineering materials, wood is not homogeneous. The grading thus provides an averaging of the quality of boards in a given grade. Visual grading spots defects such as knots, stains, bark pockets, decay, shakes (cracks), pitch pockets (resin buildup), and checks (splits). ***Softwood lumber*** is a vital material for construction and manufacturing and is graded as construction and remanufactured lumber based on the American Lumber Standard. ***Construction grade*** includes stress-graded, nonstress-graded, and appearance lumber. ***Stress-graded*** provides an indication of engineering properties and covers pieces of 2 to 4 in. nominal size, larger timbers (5 or more in. nominal thickness), posts, beams, decking, stringers, and boards (less than 2 in. nominal thickness). ***Nonstress-graded*** lumber includes a combination of allowable properties and visual defects that is used for boards, lath, batten, and planks. ***Appearance lumber*** that is nonstress-graded is cut to patterns such as trim, siding, flooring, and finished boards. ***Remanufactured*** grade is lumber that will receive further processing or secondary manufacturing, such as cabinet stock, molding, and pencil stock. Grading specifies conformity of the lumber to its end use. ***Hardwood lumber*** is graded as factory lumber (amount of usable lumber in a piece), dimensional parts, and finished market products (maple flooring, stair treads, trim, and molding). The furniture industry is the prime user of most hardwood grading.

A system for designating the degree of dressing or finish ***surfacing*** specifies one side surfaced (S1S), two sides surfaced (S2S), one edge surfaced (S1E), two edges surfaced (S2E), and various combinations, such as one side and one edge surfaced (S1S1E) or all sides and edges surfaced (S4S).

6.7.2.2 Composite wood.

To gain the maximum use of trees and to achieve properties not possible from solid wood, composite woods have been developed for use in construction and manufacture. Among composite woods are laminated timber, impreg-wood, plywoods and veneer, particle board, hardboard, sandwiched materials, and insulation board.

Laminated Timber. Laminated timber is a product of the adhesive-joining technology. Through adhesive bonding of pieces of lumber so that the grain of all pieces is parallel to the length of the timber, it is possible to produce straight to curved structural-wood members of large size and outstanding strength. ***Glulam*** bonds sawn lumber laminations with waterproof structural adhesives, allowing nearly unlimited depth, width, and length in a variety of shapes. One glulam product, Parallam, achieves higher stress ratings and uniform strength than those of conventional materials through a patented process in which thin strands of veneer are glued using waterproof adhesives, then cured with microwaves.

Laminated beams offer the architect flowing lines and the warmth and beauty of wood. Custom-ordered laminated beams have the capability to span more than 100 m of unobstructed space for sports arenas, convention centers, entertainment halls, and churches (Figure 6-48). Seasoning of lumber for laminated beams prior to gluing provides improved strength, dimensional stability, and elimination of unsightly surface cracks and shrinkage. The ablative nature of the large beams provides an advantage in protection from the hazards of fire, since the thin layer of char protects the interior of the beam and the wood does not become plastic at high temperatures, as do metal and plastic members.

Impreg-wood. ***Impreg-wood*** achieves a very stable lumber by the bonding of phenolic resins to the cell wall microstructure. This process is accomplished through saturation of thin veneers with phenolic resin that is polymerized before the veneers are stacked into thick laminates. Applications of impreg-wood include sculptured models for the huge metal dies used to stamp automobile sheet metal parts. ***Compreg-wood*** employs impregnation of veneers with phenolic resin, but the polymerization is accomplished when the veneers are stacked and compressed into the desired final shape. The dimensionally stable, high-density, hard compreg laminate yields a glossy finish throughout the material. Knife handles, bowls, jigs for manufacturing, and parts for textile looms use the compreg laminates.

Plywood. Plywood is another form of laminate that is most common in building construction. It is produced by stacking veneer with the grain direction in each layer at right angles to the next, beginning with the grain running the length of the panel [Figure

Figure 6-48 Custom-laminated beams. (U.S. Forest Products Laboratory.)

6-49(a)]. This cross-laminating takes advantage of the fact that wood is much stronger in the direction of the grain than across the grain. Thus, plywood yields equalized strength in the plane directions, while also being shear, puncture, and split resistant. Numerous grades of plywood are available; Figure 6-49(b) shows the marking applied to three grades of plywood using the American Plywood Association standards. Letters assigned to veneer grades specify the quality of exposed plys [Figure 6-49(c)]. Typing of plywood divides it into ***interior*** plywood, which has its laminates glued with moisture-resistant glue, including phenolformaldehyde, urea, and melamine. Veneers in interior plywood may be of lower quality than the exterior type. In contrast, ***exterior*** plywood is glued only with phenolformaldehyde to provide a waterproof bond. The waterproof glue can withstand aging and boiling water. Marine plywood is the superior-quality exterior plywood. The exterior and interior typing are applied to both appearance- and engineering-grade plywood.

Thin appearance plywood, in the form of paneling, is very popular for walls in houses, mobile homes, recreational vehicles, and boats. There is a nearly unlimited variety of hardwood-faced paneling. An inexpensive grade of Philippine mahogany known as ***lauan*** receives many types of coating and textures for decorative paneling. Thicker appearance-grade plywood is used for furniture and cabinets. Engineering-grade plywood serves as forms for casting concrete, underlayment for flooring and carpets, structural panels, roofing, decking, walls, and cabinets. Plywood siding has gained wide acceptance in contemporary-style buildings. Siding offers warmth, low maintenance, and good thermal and acoustical insulation, among other properties, and is available in various species with tex-

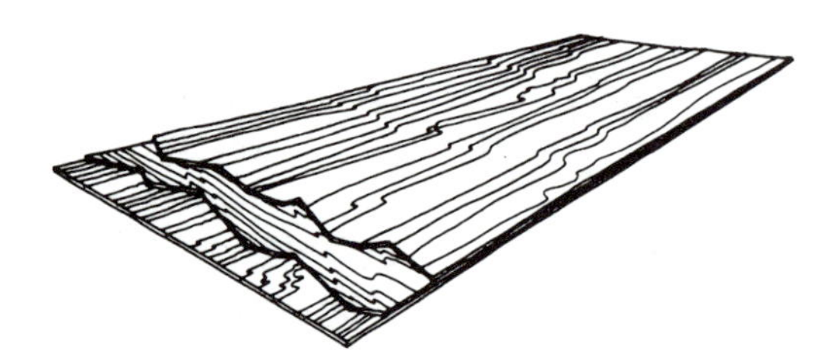

3 ply construction (3 layers of 1 ply each)

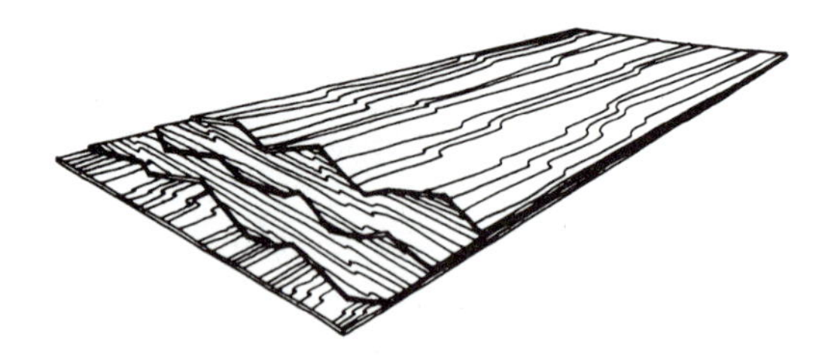

4 ply construction (3 layers: Plies 2 and 3 have grain parallel)

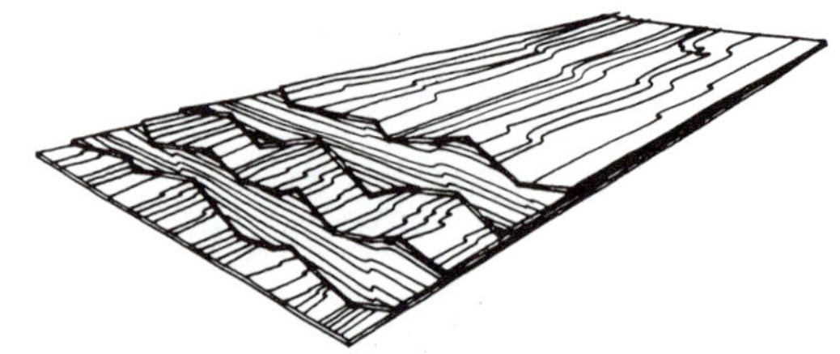

5 ply construction (5 layers of 1 ply each)

(a)

Typical APA Registered Trademarks

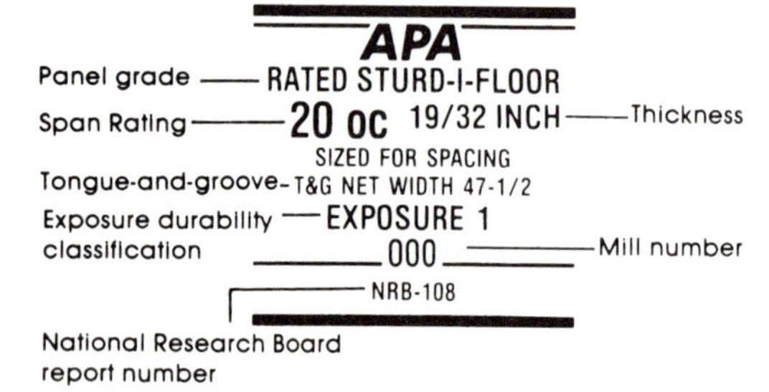

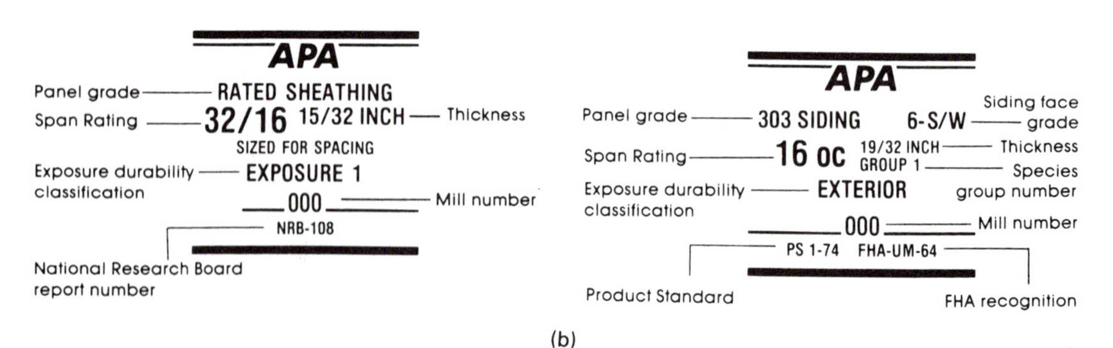

(b)

Figure 6-49 (a) Plywood. (b) Marking used by American Plywood Association.

N	Smooth surface "natural finish" veneer. Select, all heartwood or all sapwood. Free of open defects. Allows not more than 6 repairs, wood only, per 4 × 8 panel, made parallel to grain and well matched for grain and color.
A	Smooth, paintable. Not more than 18 neatly made repairs, boat, sled, or router type, and parallel to grain, permitted. May be used for natural finish in less demanding applications.
B	Solid surface. Shims, circular repair plugs and tight knots to 1 inch across grain permitted. Some minor splits permitted.
C Plugged	Improved C veneer with splits limited to 1/8-inch width and knotholes and borer holes limited to 1/4 × 1/2 inch. Admits some broken grain. Synthetic repairs permitted.
C	Tight knots to 1-1/2 inch. Knotholes to 1 inch across grain and some to 1-1/2 inch if total width of knots and knotholes is within specified limits. Synthetic or wood repairs. Discoloration and sanding defects that do not impair strength permitted. Limited splits allowed. Stitching permitted.
D	Knots and knotholes to 2-1/2 inch width across grain and 1/2 inch larger within specified limits. Limited splits are permitted. Stitching permitted. Limited to interior (Exposure 1 or 2) panels.

(c)

Figure 6-49 (c) Veneer grades. (American Plywood Association.)

tures including smooth, deep grooves, brushed, rough sawn, and overlays. ***Veneer*** in thin sheet form is used in single thickness for baskets, for boxes, and as ornamental inlays.

Wood-Based Fiber and Particle Panel Materials. Wood-based fiber and particle panel materials include a variety of panels and boards used in building construction, packaging, furniture, and other manufactured products. ***Particleboard*** panels, also known as reconstituted panels, are produced through the use of thermosetting resins, such as urea–formaldehyde and phenolformaldehyde, which serve as a matrix to bind together wood residues or shavings in the form of small wood flakes, wood flour, and additives (Figure 6-50). Water resistance is improved through the addition of wax. The resins account for 5% of the dry panel's weight and polymerize as the particles are pressed in flat presses or extruded through thin die presses. Many particle sizes and shapes are manufactured to yield panels with specific properties. Particleboard serves as backing for plastic and wood laminates or as a decorated panel that is normally painted. ***Hardboard*** is produced by processing wood chips with steam and/or pressure, and the lignin bonds the fibers together. Medium-density, high-density, or special densified hardboard offers high strength, wear resistance, moisture resistance, and resistance to cracking and splitting and has good working qualities. ***Tempered hardboard*** has oils added to increase water resistance, hardness, and strength. Hardboard finds use as painted house siding, diestock, and electrical panels; with plastic and other laminates; and as components in furniture such as drawer bottoms, mirror backs, and television sides and backs.

Sandwiched Materials. Sandwiched materials include treated paper honeycomb faced with laminating veneers. Doors and panels employ these materials. Fiberboards

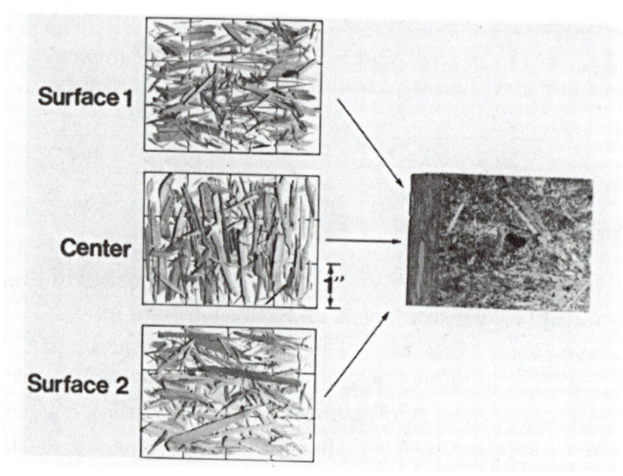

Figure 6-50 Particleboard (also called reconstituted panel) products consist of a variety of compositions and organizations. Flakeboard uses relatively square chips; oriented strandboard uses long, narrow chips that can be aligned for maximum directional strength. (U.S. Forest Products Laboratory.)

made of wood fibers bonded with rosin, asphalt, alum, paraffin, oils, fire-resistant chemicals, and plastic resins serve as insulation panels on walls and roofing, onto which some other material is added, such as siding or tar and gravel.

6.7.3 Wood By-Products

The most common by-product of lumber and plywood processing is *chips* for the production of pulp used in various paper products. Pulp is obtained through chemically defibrating the chips by dissolving the lignin to obtain fibers. Other by-products include naval stores (turpentine and rosin), bark for mulching plants, charcoal briquettes, and various resins from the tree.

6.7.4 Physical Properties

6.7.4.1 Appearance. When selected for furniture, house trim, doors, floors, wood turning, or other applications in which wood is exposed, appearance is paramount because it lends warmth and beauty. Color, figure (patterns produced by grain, texture, and machining), luster, and the manner in which finishes affect wood all become important factors in selection.

6.7.4.2 Sawing effects. The Forest Products Laboratory of the U.S. Department of Agriculture defines grain and texture as follows:

Grain: direction, size, arrangement, appearance, or quality of fibers in wood or lumber

Texture: finer structure of wood rather than annual rings

Twenty grain specifications cited in the *Wood Handbook* provide precise meaning to various grain patterns. The angle at which a log is cut is one factor that determines grain pattern. Figures 6-47(b) and 6-51 show the difference in quarter-sawed and plain-sawed boards.

While lumber is used in many stages, from logs with the bark removed or rough-sawn timbers, finished lumber is the most common form encountered. Standard dimensions include 2 × 4, 1 × 12, and 4 × 4, which are nominal measurements in inches. Nominal measurements indicate the approximate size of the lumber when it is in the rough state. A finished 2 × 4 is actually 1½ × 3½ in. Following rough sawing, lumber is ***seasoned*** to remove the moisture, reduce warping and cracking, reduce weight, and increase strength and many other properties. There are a variety of seasoning methods, including air drying and kiln drying. Air drying is a slow process in which rough timber is stacked in a specified manner and allowed to dry naturally. Kiln drying employs ovens, in which the rough lumber is stacked and hot air is forced over the wood. Research into solar kilns shows promise for low cost and quick seasoning.

Beyond seasoning, many methods are employed to improve the physical, chemical, and mechanical properties of wood. ***Chemical modification*** involves a chemical reaction (covalent bonding) between the cellulose, lignin, and hemicellulose and a chemical that acts as a preservative. Woods are also impregnated with polymers, oils, salts, and other

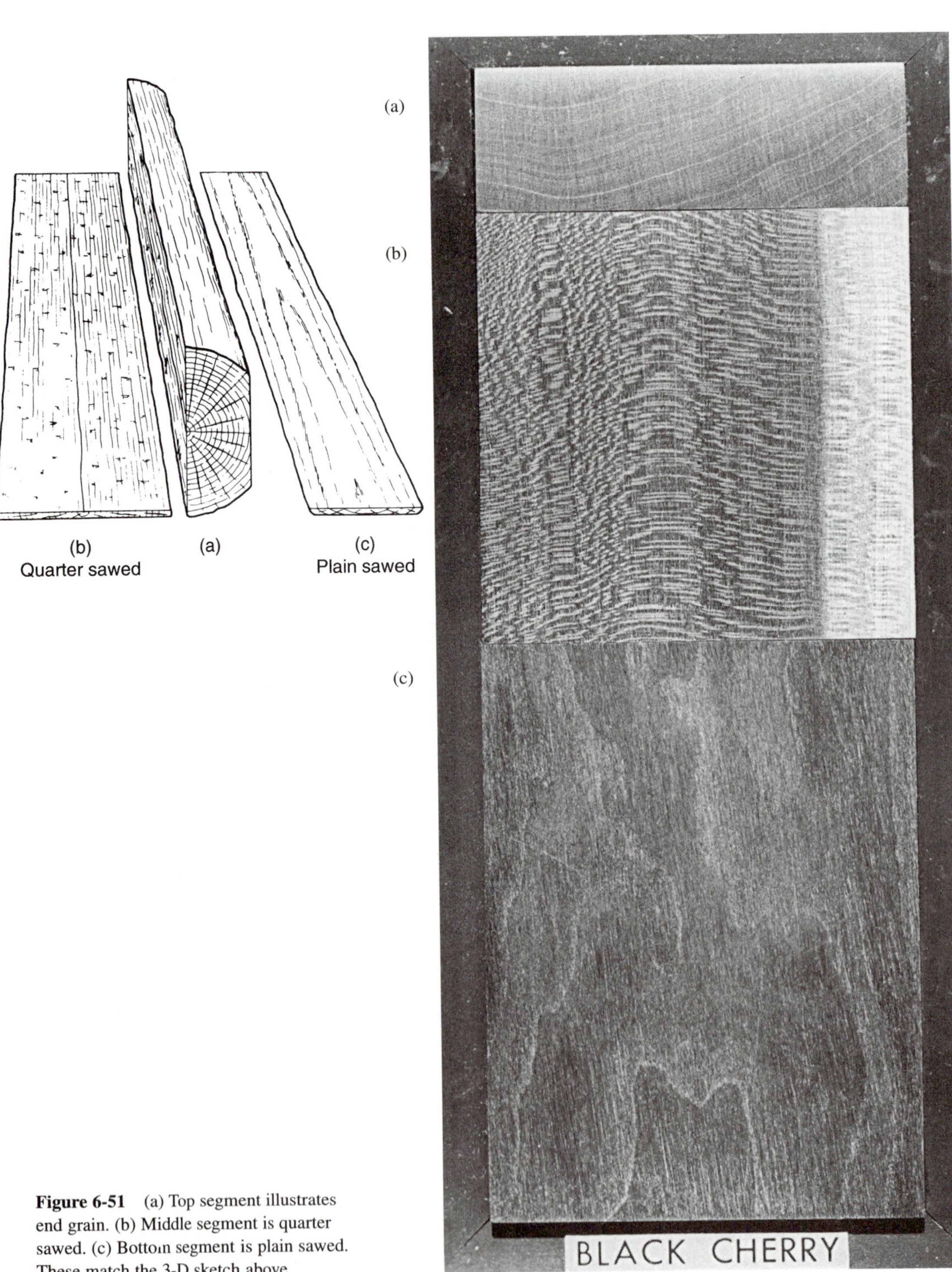

Figure 6-51 (a) Top segment illustrates end grain. (b) Middle segment is quarter sawed. (c) Bottom segment is plain sawed. These match the 3-D sketch above.

solutions. A wide variety of coatings, from natural clear finishes to stains and paints, provide protective and decorative value.

6.7.4.3 Moisture content.

Moisture content is a measure of the water in wood. Normally expressed as a percentage of the amount of water weight to the weight of oven-dry wood, moisture content is an extremely important factor in the mechanical and other physical properties of wood. Trees have moisture content ranging from 30% to 200%. Drying of freshly cut wood removes moisture, but the hygroscopicity of wood substance allows it to pick up moisture from the surrounding atmosphere. For this reason, seasonal and environmental changes in relative humidity cause a constant shift in moisture content. You may recall that, during humid and rainy seasons, doors and drawers become hard to open, while during dry days the same doors or drawers become easy to use. Treatment of wood must include steps to minimize moisture content changes, weathering, chemicals, and biological attack. Expanding and shrinkage or improperly finished wood cause ***warping, splitting, checking*** (lengthwise separation across the annual rings), and other problems. Damp wood also provides opportunity for bacteria and fungi to ***decay*** or ***rot*** wood. Seasoned lumber has been dried in varying degrees. To ship lumber and avoid molding, moisture is reduced to 25% or below. Thoroughly dried lumber has from 12% to 15% moisture. The key to seasoning lumber is to attain a moisture content corresponding to the average atmospheric conditions in which the lumber will serve.

6.7.4.4 Weatherability and decay resistance.

Similar to the reaction of plastics, woods change color and structure due to heat, light, moisture, wind, and chemicals in the environment. Ultraviolet (UV) light causes chemical degradation and graying of color, while metal fasteners (screws, nails, etc.) and hardware (hinges, etc.) also cause color changes. The UV light breaks down cellulose in the fibers. However, the degradation of the surface fibers, which is rapid when wood is first exposed, slows to a negligible rate of about 1 mm per 100 years. Other effects of weathering include warping, checking, abrasion, and surface roughening. While the even gray color is attractive, the appearance of dark gray and blotchy colors indicates attack by biological organisms and probability of rotting. See Table 6-8 for the varying resistances of woods to decaying. The cut of the boards, such as vertical grain rather than flat grain, aids weatherability, but special treatment with preservatives is normally required.

Generally, heartwood provides more resistance to decay than does sapwood. The naturally occurring wood extractives determine the resistance of a particular species of heartwood to attacking fungi. The southern and eastern pines formerly were slower growing and older and thus had more heartwood, which made them more resistant to decay. New forestry methods produce faster growth but more sapwood, which reduces these species' decay resistance. Bald cypress has outstanding natural durability.

6.7.4.5 Density.

The variation in a wood's moisture content depends on the environment and preparation of the wood and will greatly affect the density, weight, and specific gravity. The range of density for most woods falls between 320 and 720 kg/m^3. Calculations of ***specific gravity*** are determined on either kiln-dried wood or wood with a specific percentage of moisture. Twelve percent is frequently used. At a 12% moisture

TABLE 6-8 PHYSICAL AND MECHANICAL PROPERTIES OF IMPORTANT COMMERCIAL WOODS GROWN IN THE UNITED STATES (AVERAGE VALUES OF SMALL, CLEAR, STRAIGHT-GRAINED SPECIMENS)

Common species	Specific gravity (12% moisture)	Hardness[a] (N)	Modulus of elasticity[b] (MPa)	Decay resistance[c]
Hardwoods				
Ash, white	0.60	5,900	12,000	S
Basswood	0.37	1,800	10,000	S
Beech, American	0.64	5,800	11,900	S
Birch, yellow	0.62	5,600	13,900	S
Cherry, black	0.50	4,200	10,300	V
Elm, American	0.50	3,700	9,200	S
Maple, sugar	0.63	6,400	12,600	S
Oak, northern red	0.63	5,700	12,500	S
Oak, white	0.68	6,000	12,300	V
Walnut, black	0.55	4,500	11,600	V
Yellow, poplar	0.42	2,400	10,900	S
Softwoods				
Bald cypress	0.46	2,300	9,900	V
Cedar				
Eastern red cedar	0.47	4,000	6,100	V
Northern white cedar	0.31	1,400	5,500	V
Western red cedar	0.32	1,600	7,700	V
Douglas fir				
Coast	0.48	2,200	13,400	M
Interior west	0.50	2,300	10,300	M
Fir				
Eastern species	0.38	2,200	10,300	S
Western species	0.36	1,800	8,500	S
Pine				
Eastern white	0.35	1,700	8,500	M
Sugar	0.36	1,700	8,200	S
Western white	0.38	1,900	10,100	S
Redwood				
New-growth	0.35	1,900	7,600	V
Old-growth	0.40	2,100	9,200	V

Source: Wood Handbook: Wood as an Engineering Material, by Forest Service, U.S. Department of Agriculture.

[a]Newtons required to embed an 11.26-mm ball one-half its diameter in a direction perpendicular to the grain.

[b]Measured from a simple supported, center-loaded beam, on a span/depth ratio of 14:1.

[c]V, very resistant or resistant; M, moderately resistant; S, slight or nonresistant.

level, most species of trees yield specific gravities in the range of 0.32 to 0.67, as shown in Table 6-8.

6.7.4.6 Working qualities. In hand working, normally the lower-specific-gravity wood works best. Machinability of wood depends on characteristics such as ***interlocked grain, hard deposits*** of minerals such as calcium carbonates and silica, ***tension wood,*** and ***compression wood,*** in addition to density. High density and hard deposits will dull tools. Reaction wood is tension wood and compression wood. Reaction wood and interlocked grain cause binding of boards as they are fed through saws and planers. Other considerations in the workability of wood include nail splitting, screw splitting, and surface results of sanding.

6.7.4.7 Ablation. An advantage offered by the use of wood timbers and boards in building is their ability to char when burned. This ablation property allows woods to retain much of their strength because ***charring*** slows the burning. Conversely, metals act as good thermal conductors and uniformly transmit the heat of a fire throughout the metal. This can cause metal supports to lose much of their strength and bend under their loads. Figure 6-52 shows the ablative value of wood; the steel beams failed (softened) under the heat, while the wood structural members held some of the load.

6.7.5 Mechanical Properties

Mechanical properties of woods can vary widely because of the lack of homogeneity in wood structure. Knots, cross-grain, checks, and growth rings provide varying properties within boards and between boards from the same tree. The properties shown in Table 6-8 reflect values obtained from small, clear, straight-grained specimens. They provide average values for comparing common woods and for comparing wood with other materials on the basis of modulus of elasticity.

Figure 6-52 Ablative nature of wood: steel beams became plastic and draped over wooden beams in a factory fire. (U.S. Forest Products Laboratory.)

SELF-ASSESSMENT

6-69. In light of diminishing supplies of many raw materials, what makes wood such a promising engineering material for future generations?

6-70. What is the relation of lignin to cellulose in a tree? When is lignin removed?

6-71. Explain how anisotropy develops in wood at both the microstructure (cellular) and macrostructure levels.

6-72. How do hardwoods differ from softwoods? Cite examples to illustrate these differences.

6-73. Describe how the various layers in the cross-section of a tree affect the properties of the lumber produced from the tree.

6-74. Name typical applications for the following grades of lumber: **(a)** nonstressed graded, **(b)** construction appearance, **(c)** remanufactured, **(d)** stress graded.

6-75. Name four wood products that are not solid wood, and cite typical uses for each.

6-76. Explain the following APA designation: B-D, Group 3, Interior.

6-77. What properties make wood a desirable material for furniture and paneling?

6-78. What properties can limit the use of wood as an engineering material?

6-79. What effects can moisture have on wood? How can they be altered?

6-80. Name a wood to fit into the following categories: **(a)** high hardness, high decay resistance, high specific gravity; **(b)** very low hardness, low specific gravity, good modulus of elasticity.

OBJECTIVE QUESTIONS

6-47. Which polymer serves as a matrix in natural wood to bond the cellulose fibers?

a. Sulfur **b.** Hemicellulose **c.** Lignin **d.** Isoprene

6-48. As a natural composite, wood's grain structure gives it ______________ properties.

a. Uniform **b.** Anisotropic **c.** Isotropic **d.** Poor

6-49. The main difference in hardwood and softwood is ______________.

a. Polymer structure **b.** Hardness
c. Economics **d.** Growing environment

6-50. The reason for grading lumber is its lack of ______________.

a. Value **b.** Processing **c.** Homogeneity **d.** Visual defects

6-51. A means of overcoming the problem in question 6-50 to ensure uniform strength.

a. Seasoning **b.** Curing **c.** Heating **d.** Laminated timber

6-52. The key difference between interior and exterior plywood.

a. Number of plys **b.** Direction of grain **c.** Glue **d.** Appearance

6-53. A material developed to maximize the yield of wood from forest products.

a. Plywood **b.** Paper **c.** Glulam **d.** Particleboard

6-54. A major reason wood is prized as an engineering material.

a. Natural beauty **b.** Low cost
c. Plentiful supply **d.** Uniform properties

6-55. Wood should be considered as a valuable engineering material because the raw materials are ______________, whereas those for most other materials are not.

a. Recycled **b.** Cheap **c.** Renewable **d.** Limited

6-56. To prevent wood from warping, splitting, and checking, it is important to control ______________.

a. Growth **b.** Weather **c.** Moisture **d.** Lignin

6-57. A problem with wood not found in most other engineering materials.

a. Rotting **b.** Weight **c.** Processing **d.** Supply

6-58. Because of the way that wood grows, most wood of the same species has identical properties.

T (true) **F** (false)

6-59. Wood timber may offer better protection in a fire than steel beams.

T (true) **F** (false)

6-60. Wood's ability to char when burned.

a. Ablation **b.** Isotropy **c.** Homogeneous **d.** Check

6-61. Most wood has very good workability because of the uniformity of its structure.

T (true) **F** (false)

Advances in polymer materials development, much like those discussed here, occur daily. If you understand the basic concepts covered in this module, you will possess the ability to understand these developments. To stay current with polymer innovations, make it a regular habit to read some of the periodicals listed at the end of this module.

6.8 EXPERIMENTS WITH POLYMERS

Check the reference section for experiments and demonstrations useful for exploring the nature and properties of polymers. These resources were developed through the National Educators' Workshop: Standard Experiments in Engineering Materials, Science and Technology.

APPLICATIONS & ALTERNATIVES

Recycling: Myths and Promises

The world is running out of places to bury its garbage. There are concerns about the effect of burning rubbish: Some experts believe that it is safe to turn waste into energy by burning some garbage to generate electricity; others argue against incineration. We cannot expect simple solutions to environmental concerns. Some issues about the environment involve myths that might have sprung from facts coupled with assumptions on how those facts should (but may not) apply to a new situation. It is important for all citizens to develop a base of knowledge about technology to separate fact from myth.

There is general agreement that the best manner to handle ***municipal solid waste (MSW)*** is through the following procedures, listed in order of priority:

1. ***Source reduction.*** Keeping the amount of waste, especially toxic waste that may result from processing of the original product, to a minimum should be the first step.

One example of reducing waste would be the use of ceramic dishes, cups, and glasses instead of disposable items (plastic or paper) when there is no problem with the spread of disease.

2. ***Removing recyclable or reusable materials.*** MSW has already been reduced dramatically by recycling programs prompted by legislation. Not only does recycling reduce MSW, it saves natural resources. For recycling to be effective, there must be regional centers that can make use of the materials economically feasible.
3. ***Burning for volume reduction and/or energy.*** Although many localities have developed waste-incineration programs, there is still concern about global warming as a result of burning hydrocarbons. One example is rubber tires, which have created tremendous disposal problems, yet are a potential source of hydrocarbon energy. Utility companies have built tire-to-energy plants to rid regions of tires while generating electricity. Many planners feel that the existing technology is adequate to safeguard against harmful emission from incinerators. This reasoning follows the facts that gases given off by landfills are composed of about 50% methane, a much more dangerous gas in terms of the ***"greenhouse effect"*** (causes global warming) than the CO_2 (carbon dioxide) given off through incineration. The ***NIMBY (not in my backyard)*** syndrome also presents problems in terms of finding communities willing to permit incinerators for waste-to-energy processes.
4. ***Landfilling.*** This is the least desirable method of dealing with MSW; however, it is still the most common. Figure 6-53 shows the composition of the average landfill in both weight and volume. Volume became the major concern when we recognized the lack of space to bury trash. Weight is a concern in transporting waste.

In Module 1 we said that any product should be considered in terms of its total life cycle. What are the results of the processing methods? Are polluting gases released into the environment, and do toxic metals get flushed into our rivers and streams? During use, does the product safeguard our health? At the end of the product's useful life, how can it be disposed of safely? MSW is a hidden cost of a product that we pay in the form of taxes and quality of life. Recycling depends on existing technology. If the technology does not exist, recycling cannot occur. Recycling technology and availability of the technology to localities is needed to ensure a proper stream of materials flow to recycling centers.

Note in the graph in Figure 6-53 that plastics account for 18% of the volume in our landfills, while paper consumes 38% of the space. By weight, paper and cardboard comprise 40% and plastics 6.5%. These two materials clearly occupy the greatest amount of space compared with metal, glass, and food, while yard waste (grass, limbs, shrubs, etc.) uses almost 11% of the volume. Yard waste can be reduced dramatically by using mulching lawn mowers and ways to turn limbs, leaves, and so on into mulch.

Biodegradable materials break down through the natural action of the environment. Sun, water, heat, and microorganisms will break down waste materials and reduce their effect on the MSW. However, recent digging through garbage dumps unearthed food (corn on the cob, chicken, hot dogs, etc.) that looked like it had been buried yesterday but, in fact, had been in the landfill 10 years. Newspapers buried 60 years ago can be read easily (Figure 6-54). Other items unearthed that show little sign of biodegrading after a decade or more are paper containers and wrappers, steel cans, clothing, grass clippings, and a va-

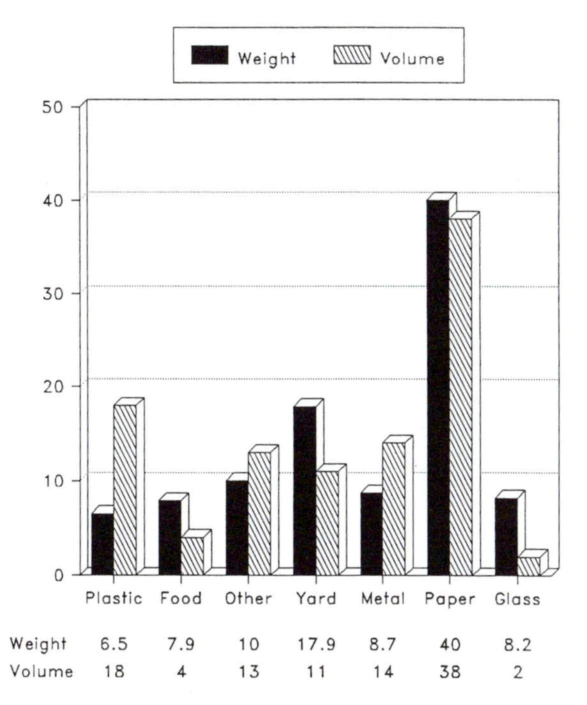

	Plastic	Food	Other	Yard	Metal	Paper	Glass
Weight	6.5	7.9	10	17.9	8.7	40	8.2
Volume	18	4	13	11	14	38	2

Figure 6-53 Weight and volume compositions of typical landfill. (Dow Chemical.)

Figure 6-54 The newspaper shown here was discovered by an archaeological team digging through a landfill. It was buried 12 years. (Dow Chemical.)

riety of plastic products. For biodegrading to occur, there must be 65% moisture, but landfills typically are held to 20% to 25% moisture to prevent washing out chemicals into the groundwater below.

The materials that were dug up did not biodegrade because landfills protect much of the waste from the natural elements. On the other hand, some materials do break down because the proper conditions are present for biodegrading. Rainwater "leaches" out chemicals that enter the natural water reservoirs underneath the landfills, thus introducing toxic chemicals to water supplies that people drink.

If newspaper, packaging, and clothing do biodegrade, they release ink, dye, and paint that can pollute. This is the concern with ***biodegradable plastics.*** These hydrocarbons contain many elements and additives such as chlorine, formaldehyde, dyes, and metals. Like newspapers, some of these materials should not degrade. Newer landfills with liners and gas-collection systems are designed to contain the pollutants to the dump.

Photodegradable plastics are susceptible to the ultraviolet rays of the sun, whereby polymer chains break, causing the material to lose strength and crumble into small bits. These plastics are intended for products that often become litter that might harm wildlife and degrade the landscape. An example of a use for photodegradable plastics is six-pack-beverage carrying rings, which should decompose after several months of exposure to the sun. It should be pointed out that these products do not break down completely, as does rusting steel or most paper products.

Although modern landfills should be constructed to contain toxic chemicals, problems may be avoided by keeping toxic chemicals out. As with newspapers, food, and other materials, the landfill will not have the conditions necessary for biodegrading of plastics, so biodegradable plastics can present more problems than they solve. Photodegradable plastic might encourage people to litter. Also, lacking sunlight, they will not degrade and might harm the marine environment. The development of polymers that will degrade in the absence of light, leaving only carbon dioxide, water, and a small amount of biological matter (e.g., Bispol resin), should make plastics more friendly to the environment and more acceptable as packaging material.

Effective methods for recycling plastics and paper products would address 56% of the volume of MSW. To encourage recycling, a profit must be realized. Aluminum is a good example. It requires 95% more energy to produce virgin aluminum from bauxite ore than it does to melt down and use recycled aluminum. This fact has promoted a concerted effort by the aluminum industry and retailers to promote aluminum recycling. The Aluminum Association reports that, at present, 50.5% of aluminum cans are recycled.

A number of other products, such as iron, steel, and glass, are also in a healthy recycling stream. Other items, such as scrap tires and used engine oil, are a challenge to dispose of in a manner safe to the environment. Having engineering designers consider secondary uses of a product once its useful life is at an end, and designing for disassembly, should help with MSW and environmental problems.

The plastics industry has attempted to make plastic more environmentally compatible. One type of plastic, polystyrene (PS), which is widely used but also criticized because of its effect on the environment, serves as a good example for a study of plastic recycling issues. The National Polystyrene Recycling Council (NPRC) reports that PS products use an average of one-third less energy to produce than is used by comparable paper products. Some communities banned certain PS packaging based on a questionable belief that paper packaging was preferable in terms of the total materials cycle.

Polyethylene terephthalate (PET), used for soft-drink bottles, is a good recycling plastic. Some reclaimed PET goes back into food packaging, while another significant amount finds use in recycled applications such as for carpet backing, fiberfill for sleeping bags and ski jackets, and transformation into polyol, a chemical ingredient used to make rigid urethane foam.

High-density polyethylene (HDPE) is another widely recycled plastic. You encounter virgin HDPE in milk jugs and the base caps of large PET soft-drink bottles. Recycled HDPE finds use in trash cans, flowerpots, piping, and traffic cones. "Plastic lumber" made of recycled HDPE is finding uses as railroad ties, decking for boat docks, and fencing. It will not rot or splinter, nor does it need paint; the color is added during production.

In keeping with the idea of waste reduction, polycarbonate-resin returnable milk cartons provide recycling without a need to scrap the product for reprocessing. The bottles are produced to permit refilling (after sterilization) up to 100 times and should last 8 to 10 years. This keeps them out of the MSW stream for an extended period. Once their life cycle is up, they can be reground and reprocessed for uses such as engineering TP for automobiles and building components.

The National Resources Defense Council (NRDC) presents views on plastic use and recycling from a non-industry-aligned position. They support a ban on the use of polystyrene (PS) based on several factors:

1. There is skepticism as to the cost-effectiveness of recycling.
2. The goal of 25% recycling leaves 75% as MSW.
3. During the manufacture of foam-PS, nonfood products, chloroflurocarbons (CFC), and gases used as blowing agents contribute to depletion of the upper atmospheric ozone layer.
4. As single-use products they create MSW problems because they are composed of small, air-filled beads that do not compress.
5. Pentane and butane, foaming gas for PS food containers, contribute to lower-atmospheric ozone (smog) problems.

The NRDC supports the use of plastics when they can be recycled. PET and HDPE have good potential here. To reduce waste, single-use (disposable) items are discouraged. We can, for example, follow the European custom of shopping with string bags instead of using single-use plastic and paper bags.

Plastic-Container Coding System

Symbols on packaging containers aid in sorting plastics. These symbols, usually found on the bottom of packages and other products, are there as part of the Society of Plastic Industry, Inc. (SPI) voluntary ***plastic container coding system***. Figure 6-55 shows the seven symbols used in this coding system and the legend of the code. Note that "1—PETE" represents polyethylene terephthalate, which has an American Society of Testing and Materials (ASTM) abbreviation of PET; the ASTM acronym for poly(vinyl chloride) or vinyl is PVC, whereas the SPI code is V. These departures from ASTM standards are made to avoid confusion with registered trademarks.

The coding system required for packaging by most U.S. states is scheduled for replacement due to the confusion it has caused. Much of the labeling on packages results in the erroneous impression that the plastic will be recycled, when in fact it will not. Only

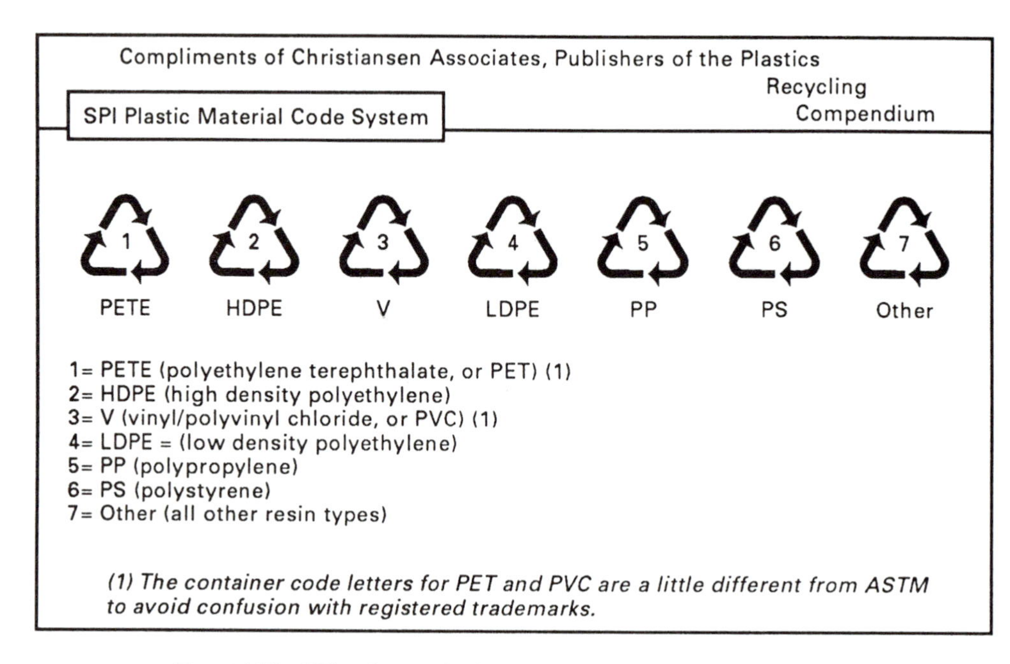

Figure 6-55 SPI's voluntary plastic container coding system. (Dow Chemical.)

about 3% of plastic garbage actually make it back into new products. Recycling centers can use only select materials, such as HDPE for blow-molded milk bottles or PETE for soda bottles. HDPE used for injection medicine bottles has a different composition that should not be mixed with the milk bottle resin.

A new system that conforms to International Standards Organization (ISO) symbols is being developed to replace the current seven symbols code now in use. The new system will require more detailed information about the plastic to allow ease of separation and thus more efficient recycling.

Positive Trend with Recycling

Public awareness, legislative mandates, and improved technology have all helped to move society away from some of the MSW dilemma. At the beginning of the move toward recycling, municipalities established programs to collect waste, but the ability to process it was limited. Many localities had to pay for the removal of good-quality paper, plastic, and glass. However, the potential for profit in recycling spurred improved processing technology and increases in recycling processing centers.

U.S. News & World Report published information showing that, during the short period from 1993 to 1995, commodity prices for scrap materials increased substantially. Prices paid for clear glass rose 78%, old newspaper soared 1,338%, and both clear plastic and ledger paper made solid price gains.

The auto industry also reports large gains in the plastic recycling of junked cars. Because of the savings realized from using recycled rather than virgin plastics, more new cars carry parts with recycled plastics. For example, PET from soda bottles recycles into parts such as luggage rack siding, headliners, and reinforcement panels. Ford estimates that nearly 95% of scrapped cars make it to recycling centers, where more than 75% of the contents are recycled. Research reveals that many plastics can be scrapped, shredded, pelletized, and remolded to new parts up to a dozen times.

Achieving these positive results of recycling required considerable education, discussion, and cooperation. Change is often difficult! Many people complained about the costs and inconvenience of recycling. The effort seems to be paying off, however, with fewer problems with MSW, better use of limited resources, and new opportunities for jobs and business growth.

REFERENCES & RELATED READINGS

Plastics

American Fiber Manufacturers Association. "Manufactured Fibers," video. New York: AFMA, 1989.

American Society for Testing and Materials. *Annual Book of ASTM Standards,* Vol. 08.01, *Plastics* (D-20). Philadelphia: ASTM, 1989.

ASM International. *Engineering Materials Handbook,* Vol. 2, *Engineering Plastics.* Materials Park, Ohio: ASM International, 1988.

Brady, George S., and Henry R. Clauser. *Materials Handbook.* New York: McGraw-Hill, 1977.

Coalition on Resource Recovery and the Environment. *The United States Conference of Mayors.* "Incineration of Municipal Solid Waste: Scientific and Technical Evaluation of State of the Art." Washington, D.C., February 1, 1990.

Constance, J. "Dendrimers: Materials by Design." *Compressed Air Magazine,* June 1992, pp. 12–17.

Consumers Union. "The Right Glues." *Consumer Reports,* July 1995, pp. 470–473.

Crane, F. A. A., and J. A. Charles. *Selection and Use of Engineering Materials,* 2nd ed. Woburn, Mass.: Butterworth, 1989.

Data Business Publishing. *Plastics Materials DATA Digest.* Englewood, Colo.: DATA, 1996.

Fischer, David. "Turning Trash into Cash." *U.S. News & World Report,* July 17, 1995, p. 43.

Gobstein, Saul. "Introduction to Plastics and How to Choose Them." ASM International continuing education short course materials. Materials Park, Ohio: ASM International, 1993.

Hocking, Martin B. "Paper Versus Polystyrene: A Complex Choice." *Science,* June 7, 1991, pp. 504, 506.

Jacobs, James. "Recycling Plastics." *The Technology Teacher,* September/October 1990, pp. 15–22.

Labana, S. S. "Recycling of Automobiles—An Overview." Paper presented to *NEW:Update 93,* NASA-LaRC, November 4, 1993.

Marshall, John A. "Liquids That Take Only Milliseconds to Turn into Solids." National Educators' Workshop: Update 93—Standard Experiments in Engineering, Materials Science, and Technology. NASA Conference Publication 3259, April 1994, pp. 315–322.

Mittal, K. L. *Polyimides: Synthesis, Characteristics, and Applications,* Vols. 1 and 2. New York: Plenum Press, 1984.

PATER, RUTH H. "Interpenetrating Polymer Network Approach to Tough and Microcrack Resistant High Temperature Polymers." *Technical Support Package for Tech Brief,* Part 2, LAR-14338, LaRC-RP41.

PEIFFER, ROBERT W. "Solid Imaging: New Opportunities for Photopolymers." *Radtech Europe '91 Conference,* Edinburgh, September 29–October 2, 1991.

PETERSON, I. "Geometry for Segregating Polymers," *Science News,* Vol. 134, no. 19, p. 151.

ROGERS, CRAIG A. "Intelligent Materials Systems: The Dawn of a New Materials Age." Center for Intelligent Materials and Structures, Virginia Polytechnic Institute and State University, 1992.

SALEM, A. J., et al. "Fabrication of Thermoplastic Matrix Structural Composites by Resin Transfer Molding of Cyclic Bisphenol—A Polycarbonate Oligomer," *SAMPE Journal,* January/February 1991, pp. 17–22.

SEYMOUR, RAYMOND B. *Polymers for Engineering Applications.* Materials Park, Ohio: ASM International, 1987.

SHEER, ROBERT J. "Incorporating Intelligent Materials into Science Education." National Educators' Workshop: Update 94—Standard Experiments in Engineering, Materials Science, and Technology. Gaithersburg, MD, 7–9 Nov. 1994.

SPROW, EUGENE E. "Chrysler's Concurrent Engineering Challenge," *Manufacturing Engineering,* April 1992, pp. 35–42.

WIRKA, GENE. *Wrapped in Plastic.* Environmental Action Federation, 1990.

Elastomers

AMERICAN SOCIETY FOR TESTING AND MATERIALS. *Annual Book of ASTM Standards,* Vol. 09.01, *Rubber, Natural and Synthetic.* Philadelphia: ASTM, 1991.

BHOWMICK, A. K., and H. L. STEPHENS. *Handbook of Elastomers.* Materials Park, Ohio: ASM International, 1988.

BRADY, GEORGE, and H. R. CLAUSER. *Materials Handbook,* 12th ed. New York: McGraw-Hill, 1986.

HARPER, CHARLES A., ed. *Handbook of Plastics and Elastomers.* New York: McGraw-Hill, 1975.

INTERNATIONAL INSTITUTE OF SYNTHETIC RUBBER PRODUCTS. *Synthetic Rubber: The Story of an Industry.* New York: IISRP, 1973.

PUGH, BRADLEY. "The Role of Standards in the Development of the Condom," *ASTM Standardization News,* August 1992, pp. 23–29.

SHELL. *Kraton Thermoplastic Rubber.* Shell Chemical Company, Polymer Division, Houston, 1972.

Wood and Other Polymers

AMERICAN SOCIETY FOR TESTING AND MATERIALS. *Annual Handbook of ASTM Standards,* Vol. 04.09, *Wood* (D-7); Vol. 06.03, *Paints, Related Coatings and Aromatics.* Philadelphia: ASTM, 1992, 1991.

AMERICAN SOCIETY FOR TESTING AND MATERIALS. *Standardization News,* "Geosynthetics," special issue, September 1992.

CASSENS, DANIEL L., and WILLIAM C. FEIST. *Exterior Wood in the South: Selections, Applications and Finishes,* FPL-GTR-69. Forest Products Laboratory, 1991.

FOREST PRODUCTS LABORATORY. *Wood Handbook: Work as an Engineering Material.* Washington, D.C.: U.S. Department of Agriculture, 1974.

INDUSTRIAL FABRICS ASSOCIATION INTERNATIONAL. *A Design Primer: Geotextiles and Related Materials.* St. Paul, MN: IFAI, 1991.

MATERIALS EDUCATION COUNCIL. Vol. I, *Wood: Its Structure and Properties*; Vol. II, *Wood as a Structural Material*; Vol. III, *Adhesive Bonding of Wood and Other Structural Materials*; Vol. IV, *Wood: Design of Structural Elements.* 1983. MEC, 1988.

MITTAL, K. L., ed. *Adhesive Aspects of Polymeric Coatings.* New York: Plenum Press, 1983.

MITTAL, K. L., ed. *Adhesive Joint Formation, Characteristics, and Testing.* New York: Plenum Press, 1984.

RITTER, MICHAEL A. *Timber Bridges: Design, Construction, Inspection and Maintenance.* EM 7700-8. Washington, D.C.: U.S. Forest Service, 1990.

ROWELL, R. M., and P. KONKOL. *Treatments That Enhance Physical Properties of Wood.* FPL-GTR-55. Forest Products Laboratory, 1987.

SARIKAYA, MEHMET, and ILHAN A. AKSAY (eds.). *Biomimetrics: Design and Processing of Materials,* AIP Press, 1995.

Periodicals

Adhesive Age
Advanced Materials & Processes
Automotive Engineering
Journal of Adhesion Science and Technology
Materials Engineering
Modern Plastics
Plastics Technology
Plastics World
Popular Science
Scientific American
SME Composites in Manufacturing

EXPERIMENTS & DEMONSTRATIONS IN POLYMERS

NEW:Update 89 NASA Conference Publication 3074

GREET, RICHARD, and ROBERT COBAUGH, "Rubberlike Elasticity Experiment."

KERN, KRISTEN T., WYNFORD L. HARRIES, and SHEILA ANN T. LONG. "Dynamic Mechanical Analysis of Polymeric Materials."

KUNDU, NIKHIL K. "The Effect of Thermal Damage on the Mechanical Properties of Polymer Regrinds."

KUNDU, NIKHIL K., and MALAY KUNDU. "Piezoelectric and Pyroelectric Effects of a Crystalline Polymer."

OHUNG, WENCHIANG R. "The Assessment of Metal Fiber Reinforced Polymeric Composites."

STIBOLT, KENNETH A. "Tensile and Shear Strength of Adhesives."

WIDENER, EDWARD L. "Industrial Plastics Waste: Identification and Segregation."

WIDENER, EDWARD L. "Recycling Waste-Paper."

NEW:Update 90 NIST Special Publication 822

BROSTOW, WITOLD, and MICHAEL R. KOZAK. "Instruction in Processing as a Part of a Course in Polymer Science and Engineering."

CORNWELL, L. R., R. B. GRIFFIN, and W. A. MASSARWEH. "Effect of Strain Rate on Tensile Properties of Plastics."

GRAY, STEPHANIE L., KRISTEN T. KERN, WYNFORD L. HARRIES, and SHEILA ANN T. LONG. "Improved Technique for Measuring Coefficients of Thermal Extension for Polymer Films."

NEW:Update 91 NASA Conference Publication 3151

ALLEN, DAVID J. "Stress–Strain Characteristics of Rubber-Like Materials: Experiment and Analysis."

CHOWDHURY, MOSTAFIZ R. "An Experiment on the Use of Disposable Plastics as a Reinforcement in Concrete Beams."

GORMAN, THOMAS M. "Designing, Engineering, and Testing Wood Structures."

LLOYD, ISABEL K., et al. "Structure, Processing and Properties of Potatoes."

MCCLELLAND, H. T. "Laboratory Experiments from the Toy Store."

SORENSEN, CARL D. "Measuring the Surface Tension of Soap Bubbles."

WICKMAN, JERRY L., and DAVID PLOCINSKI. "A Senior Manufacturing Laboratory for Determining Injection Molding Process Capability."

NEW:Update 92 NASA Conference Publication 3201

KUNDU, NIKHIL K. "Performance of Thermal Adhesives in Forced Convection."

LIU, PING. "Solving Product Safety Problem on Recycled High Density Polyethylene Container."

WICKMAN, JERRY L. "Thermoforming from a Systems Viewpoint."

NEW:Update 93 NASA Conference Publication 3259

CSERNICA, JEFFREY. "Mechanical Properties of Crosslinked Polymer Coatings."

EDBLOM, ELIZABETH. "Adhesives: The State of the Industry."

EDBLOM, ELIZABETH. "Testing Adhesive Strength."

ELBAN, WAYNE L. "Three-Point Bend Testing of Poly (Methyl Methacrylate) and Balsa Wood."

LABANA, S. S. "Recycling of Automobiles: An Overview."

LIU, PING, and TOMMY L. WASKOM. "Application of Materials Database (MAT.DB>) to Materials Education and Laminated Thermoplastic Composite Material."

MARSHALL, JOHN A. "Liquids That Take Only Milliseconds to Turn into Solids."

QUAAL, KAREN S. "Incorporating Polymeric Materials Topics into the Undergraduate Chemistry Curriculum: NSF-Polyed Scholars Project: Microscale Synthesis and Characterization of Polystyrene."

NEW:Update 94 NASA Conference Publication 3304

FINE, LEONARD W. "Concrete Repair Applications and Polymerization of Butadiene by an 'Alfin' Catalyst."

HALPERIN, KOPL, CHARLES ECCLES, and BRETT LATIMER. "Inexpensive Experiments in Creep and Relaxation of Polymers."

KERN, KRISTEN, and HEIDI R. RIES. "Dielectric Analysis of Polymer Processing."

KUNDU, MUKUL, and NIKHIL K. KUNDU. "Optimizing Wing Design by Using a Piezoelectric Polymer."

KUNDU, NIKHIL K., and JERRY L. WICKMAN. "An Affordable Materials Testing Device."

STIENSTRA, DAVID. "In-Class Experiments: Piano Wire and Polymers."

NEW:Update 95 NASA Conference Publication 3330

FINE, LEONARD W. "Polymerization (Jumping Rubber)."

LIU, PING, and TOMMY L. WASKOM. "Plastic Recycling Experiments in Materials Education."

MARSHALL, JOHN A. "Application Advancements Using Electrorheological Fluids."

WICKMAN, JERRY L. "Cost-Effective Prototyping."

Module 7

Ceramic Materials

After studying this module, you should be able to do the following:

7.1. Name some goals of industry and society that caused a push for greater use of ceramics. Describe attributes of ceramics that make them competitive with plastics and metals. Contrast traditional and advanced ceramics.

7.2. Relate bonding, structure, and processing of ceramics to their properties (electrical, thermal, optical, and mechanical). Contrast ceramics to ceramics and ceramics to metals in terms of structure and properties. Sketch common ceramic microstructures.

7.3. Recall the major processing and raw materials used to produce ceramics. Differentiate among empirical, deterministic, and probabilistic design. Describe some advanced processes and their benefits in terms of high and low technology.

7.4. List and describe the favorable major properties of ceramics (e.g., wear and heat resistance and low thermal expansion) and the units to express them. Name the common test to measure the properties.

7.5. Use diagrams and explanations to show stress concentrations plus methods to toughen ceramics, including composite matrix ceramics. Define work of fracture and compare work of fracture versus tensile strength in glass, cements, and biomaterials. Make sketches to show methods of improved ceramic structure for greater toughness.

7.6. Express fracture toughness in MPa and psi, and density in SI and U.S. customary units. Compare typical properties such as wear, density, hardness, flexural strength, fracture toughness, and thermal expansion and thermal shock resistance of CMCs with other ceramics, metals, and polymers. Calculate the thermal shock resistance of given materials.

7.7. Explain, with the aid of sketches, methods of strengthening glass and four types of glass, and typical applications and properties allowing those methods.

7.8. Explain the difference between cement and concrete and between setting and curing. Discuss the advantages offered by cement and concrete for developing nations. Explain the advantages accrued from improved cements.

7.9. Explain the process of making polycrystalline glass and applications of this material. Draw the heating curve for glass ceramics.

(a)

(b)

Figure 7-1 (a) Examples of traditional ceramics from Colonial Williamsburg Museum Exhibit. Left to right: glazed clay pitcher, split mold made of plaster of paris for slip casting, samples of slip cast/glazed pots, potter's wheel with sample bowls and platters. Ancient ceramic processes seen here for traditional ceramics are still used today with some automated aspects. (b) Examples of newer (advanced ceramics) for engines. Top, left to right: silicon nitride (Si_3N_4) turbocharger, Si_3N_4 valve, cast steel diesel engine rocker arm with partially stabilized zirconia (PSZ) cam follower and wear button, Si_3N_4 cam follower, and valve. Bottom row, left to right: valve; silicon carbide (SiC) water pump seal; piston pin, valve spring retainer, and valve guide of Si_3N_4; and PSZ diesel head plate with integrated valve seats. (ORNL.)

PAUSE & PONDER

What comes to mind when you think of products made of ceramics? Do you think about teapots, dishes, vases, toilets, bathtubs, or perhaps even cement? Do you think about ceramics for bearings, valves, and rotors in car engines? Most of us tend to think of ceramics in the first group of applications, which are ***traditional ceramics*** [Figure 7-1(a)]. They are made from naturally occurring materials such as clays and silicas. For example, silica slurry (powder in water) is slip cast, as described later, and clay can be shaped on a potter's wheel; other processes for traditional ceramics include press molding and extruding. Processes for traditional ceramics have been automated to provide many useful products, such as for the construction industry (buildings and roads) and in the home (plumbing fixtures and dishware). Traditional ceramics are usually heavy, easily broken, and not considered for use as structural (load-bearing) materials. Because they are fragile and heavy, traditional ceramics have not been candidates for engineering applications in engines. They were not competitive with metals. New technology is changing the nature of ceramics such that they must now be added to the designer's list of engineered materials. Greater demand for engineering properties in materials has caused the emergence of ***industrial*** or ***technical ceramics.*** Pictured in Figure 7-1(b) are a few of the newer ceramics that you will learn about in this module. The silicon nitride (Si_3N_4) turbocharger rotor makes possible more efficient engines that have smaller displacement and operate at higher temperatures. Si_3N_4 valves are 50% lighter than steel valves and therefore reduce friction in the valve train; they also run faster and hotter. Si_3N_4 roller bearings replace the complex steel needle bearings and cage while saving about 40% in cost; refer back to Figure 8-24(c). They also operate at higher temperatures, which means reduced cooling needs. Partially stabilized zirconia (PSZ) has also found applications in heat engines, as shown in Figure 7-1(b) in the diesel-engine rocker arm, head plate, piston pins, and valve guides.

Most of these new ***advanced ceramics*** were developed within the past 60 years, which is quite recent in materials technology. Much of that development resulted from aerospace ***research and development (R&D),*** but gradually, the costs associated with advanced ceramics have made them feasible for more earthly applications, such as in the automobile, sports, and machine-tool industries. Most are actually synthetic ceramics that are produced from fine, relatively poor powders using new technology that includes microwaves, electron beams, and polymer chemistry. Figure 7-2 lists some differences between traditional and advanced ceramics, including some applications of each category of ceramic. Advanced ceramics are receiving attention for wide-ranging scitech development because they offer the designer properties such as light weight, good strength at elevated temperatures, and wear resistance. Many of these newer ceramics are composites reinforced with whiskers and fibers to improve their fracture toughness, so they do not break catastrophically as we generally expect dishes and coffee cups to do when we drop them. The wear resistance of ceramics makes them valuable as coatings for cutting tools, surgical instruments, punches, and dies. For a modest cost, it is possible to apply a TiN coating only a few micrometers thick to tools, which can extend their life 7 to 15 times. You may have seen golden TiN coatings on twist drills in the hardware store.

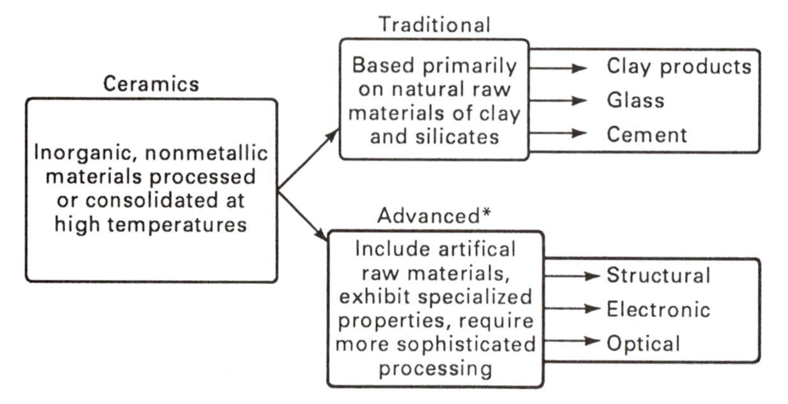

Figure 7-2 Broad classifications of ceramics. Traditional ceramics are contrasted with the newer advanced ceramics, which are also referred to as *fine ceramics, engineered ceramics, new ceramics,* and *value-added ceramics.* (ORNL.)

In the ongoing search to produce superior materials with a minimum of energy (low temperature, low pressure, and minimum number of steps), researchers have turned to nature to replicate its processing techniques. ***Biomimickry*** or ***biomimetics*** is the attempt to mimic or replicate natural processes. For example, seashells are intricate, lamellar ceramic composites made with low energy yet at a nanometric scale, yielding complex patterns that hold promise for ceramic electronic devices and similar advanced applications.

This module should help you to gain a perspective on the structure, properties, selection, and basic processes in ceramics, both traditional and advanced. In your study of ceramics, ponder their nature and their ability to compete with metals, polymers, and composites. Be alert to news articles and advertisements that describe emerging applications of advanced ceramics. Also, imagine where there may be new opportunities for the use of ceramics. This information will serve you well in our advancing age of materials and prepare you for the dramatic changes in this era.

7.1 NATURE OF CERAMICS

Ceramics are loosely defined as crystalline solids composed of metallic and nonmetallic materials. They are inorganic materials; that is, they do not contain carbon and are derived from mineral sources. Ceramics are crystal structures made of metallic ions. Bonding is either partially or completely ionic. Variables involved are (1) the magnitude of the electrical charge on the ions, and (2) the relative size of the ions. The crystal must be neutral; that

TABLE 7-1 COMPARISON OF METALS WITH CERAMIC MATERIALS

Metals	Ceramics
Crystal structure	Crystal structure
Large number of free electrons	Captive electrons
Metallic bond	Ionic/covalent bonds
Good electrical conductivity	Poor conductivity
Opaque	Transparent (in thin sections)
Uniform atoms	Different-size atoms
High tensile strength	Poor tensile strength[a]
Low shear strength	High shear strength
Good ductility	Poor ductility (brittle)
Plastic flow	None
Impact strength	Poor impact strength
Relatively high weight	Lower weight
Moderate hardness	Extreme hardness
Nonporous	Initial high porosity
High density	Initial low density

[a]Small flaws; pores act as stress concentrators that are not reduced by ductility/plastic deformation.

is, all the cation positive charges must be balanced by an equal number of anion negative charges. The ***oxidation state*** or ***oxidation number*** designates the number of electrons an atom loses or gains or otherwise uses in joining with other atoms in compounds. The total of the oxidation states of all atoms in a molecule or formula unit is zero. Na has an oxidation state of + 1; Cl has an oxidation state of −1. When they join to form the compound expressed by the formula unit NaCl, their oxidation states must add up to 0.

Some of these ceramics are intermetallic compounds (see Section 5.2). Originally, ceramics were clay-based materials. The word *ceramic* comes from the Greek ***keramos,*** meaning "burnt earth." Glass, a rigid liquid with an amorphous molecular structure, is sometimes classified as a ceramic. Silicon carbide, a very popular ceramic, does not meet the definition of a ceramic material given above. Ceramics are inexpensive compared with competing materials. Consisting primarily of forms of silicon, aluminum, and oxygen, the most abundant elements in the earth's crust, ceramics can be produced at less cost than that of competing metal-alloy components. Further, many, if not all, of the strategic minerals, including cobalt, that are needed to produce sophisticated metal alloys, such as the superalloys presently used in high-temperature applications, are not found in the United States. AerMet 100, one such recently developed superalloy steel, was shown in Figure 5-53(a) and described in Section 5.10.2.3. What sets ceramics apart is their set of properties common to nearly all forms: (1) extreme hardness, (2) heat resistance, (3) corrosion resistance, (4) low electrical and thermal conductivity, and (5) low ductility, or *brittleness.* Table 7-1 is a comparison of these and other properties of ceramics with those of metals. Table 7-2 provides an additional comparison of the properties of ceramics and those of competing materials.

TABLE 7-2 TYPICAL PROPERTIES OF SELECTED ENGINEERING AND TECHNICAL CERAMICS

	Ceramics				Comparison materials		
Properties	Alumina	Beryllia	Silicon carbide	Zirconia	Mild steel	Aluminum	Nylon
Melting point (approximately °C)	2050	2550	2800	2660	1370	660.2	215
Coefficient of thermal expansion (m/m/K) $\times 10^{-6}$	8.1	10.4	4.3	6.6	14.9	24	90
Specific gravity	3.8	—	3.2	—	7.9	2.7	1.15
Density (kg/m^3)	3875	2989	3210	9965	7833	2923	1163
Dielectric strength (V/m) $\times 10^6$	11.8	—	—	9.8	—	—	18.5
Modulus of elasticity (MPa) $\times 10^4$	34.5	39.9	65.5	24.1	17.2	6.9	0.33
Hardness (Mohs)	9	9	9	8	5	3	2
Maximum service temperature (K)	2222	2672	2589	2672	—	—	422

The several examples that follow add emphasis to the properties of ceramics listed above. The Knoop hardness cubic boron nitride (CBN) is 5000 kg/mm^2; that of diamond, the hardest known material, is about 7500 kg/mm^2. Partially stabilized zirconia (PSZ), to be discussed in more detail later, can withstand a pressure of 78,000 psi for about 2000 hours without any appreciable effect. Its ***coefficient of friction*** α with steel is only 0.17 at room temperature, and its coefficient of thermal expansion has very nearly the same value between 0° and 1000°C.

Applications for advanced ceramics have received major media attention in recent years, particularly for use as parts in a future ceramic heat engine. However, corrosion resistance, chemical inertness, thermal shock resistance, and other properties that materials scientists and engineers can design into ceramic materials make both traditional and advanced ceramics highly attractive in a large number of applications. The combination of properties mentioned above make ceramics good candidates for wear-resistance applications. Electrical properties place ceramics in great demand as solid electrolytes in experimental batteries and fuel cells. Other uses include automotive sensors, packaging for integrated circuits, electronic/optical devices, fiber optics, microchips, and magnetic heads. In the marriage of the computer and communications technologies, ceramics play a major role. The chemical inertness of ceramics is finding many uses in the medical field, where contact with body fluids is less of a problem than with most other materials. Finally, ceramics play a big role in the machine-tool industry. Their thermal and mechanical stability allows them to retain their smooth, accurate cutting surfaces longer than metals do. Coated cutting tools and inserts, some with as many as 12 extremely thin coatings, each designed to serve a special function, can run productively at faster cutting speeds and at faster feed rates than can any metal-alloy tool in the machining of hard steels, superalloys, and ceramics. What is remarkable about these multicoated carbide inserts is the fact that, with slightly over a dozen thin ceramic layers, the total thickness of the coatings is only 10 to 12 μm.

7.1.1 Porosity and Density

When referring to a solid material such as a part made from copper or stainless steel, the word ***density*** takes into consideration a microstructure that contains no porosity. We are not speaking about voids or vacancies in the atomic structure when we use the term ***porosity.*** Density in such a situation can be termed the material's theoretical density. Mass density, which uses the mass of a material divided by its volume, as in Section 3.4.6, refers to this theoretical density. Such a density may be expressed in terms of its weight instead of its mass, as discussed in Section 7.5.2. Atomic weight is a major factor in determining the density of a material. Low-atomic-weight elements have low densities. A second factor is how the atoms/ions are stacked in the microstructure. Close-packed metals are more dense than open-structured materials. Our study of crystalline materials in Module 3 revealed that a bcc unit cell with a maximum coordination number of 8 is only 68% of full density, whereas fcc and cph unit cells with a coordination number of 12 can have a density of 74% (of a maximum of 100%). Ceramics are composed primarily of different phases, each with its own density. As an example, a ceramic that has a porosity of 20% is described as being 80% of theoretical density. Green density is the bulk density of a compact prior to its densification. In preparing particles for the production of powdered metals or ceramics, the packing of spherical shaped, monosized (0.1 μm) particles, even when vibrated to settle the particles, does not achieve a density level much higher than 60%. Despite tapping or vibrating, these particles do not rearrange themselves completely to eliminate pores. The optimum density of spherical particles can be reached by varying the size distribution of particles to permit smaller particles (50 nm or less) to locate in the interstices of the larger ones. One disadvantage with this procedure is that the larger particles tend to grow excessively during high-temperature sintering, which takes place later in the process. After particle sizing is complete, the pores remaining must be reduced by additional steps, such as mixing and kneading of the initial green mixture prior to the making of a green compact.

Ceramics are, by nature, generally porous materials with varying degrees of porosity. The term ***bulk density*** is used in this instance to refer to a ceramic's density, and it includes the material's porosity and the fact that most ceramics contain both a crystalline and a noncrystalline phase. ***Open porosity*** refers to the network of pores in a material that is open to the surface and into which a liquid such as water can penetrate if the part were submerged in it. ***Closed porosity*** refers to those pores that have become sealed within the grain structure. New technologies are under development to reduce, if not eliminate, pores in ceramic materials. Pores affect the strength of ceramics in two ways. First, they produce stress concentrations. Once the stress reaches a critical level (see Section 7.1.4), a crack will form and propagate. Because ceramics possess no plastic-deformation attributes to absorb any energy transferred to these materials once a crack is initiated (started), it propagates (grows) until fracture occurs. Second, pores (i.e., their size, shape, and amount) reduce the strength of ceramics because they reduce the cross-sectional areas over which a load can be applied and, consequently, lower the stress that these materials can support. The bulk density of complex geometric parts as well as their various porosities can be determined by using ASTM procedure C373, which is based on Archimedes' principle.* This principle states that the weight of an object in a fluid equals its dry weight minus the buoyant force (or the weight of the fluid displaced). Taking only three weight measurements—the object's dry weight, its saturated weight in a fluid, and its net saturated suspended weight—several physical properties relating to the density and porosity of an object can be calculated. For example, the following five relationships expressed in terms of these three weights can be calculated:

$$1.\quad d_B = \frac{W_D}{V_B} = \frac{W_D \cdot d_L}{W_S - W_{SS}}$$

$$2.\quad V_B = \frac{W_S - W_{SS}}{d_L}$$

$$3.\quad V_A = \frac{W_D - W_{SS}}{d_L}$$

$$4.\quad d_A = \frac{W_D}{V_A} = \frac{W_D \cdot d_L}{W_D - W_{SS}}$$

$$5.\ \%P_A = \frac{W_S - W_D}{W_S - W_{SS}} \times 100\%$$

where

W_D = dry weight
W_S = saturated weight
W_{SS} = saturated suspended weight
d_L = density of saturating liquid
d_B = bulk density
d_A = apparent density
V_A and V_B = apparent volume (bulk volume)
P_A = apparent porosity

Notes:

1. If water is the saturating liquid, its density may be assumed to be 1 g/cm³; the ratio *W/d* results in units of volume (cm³).

2. The apparent density is its dry weight divided by the difference between its bulk volume and its open-pore volume.

Illustrative Problem

A specimen of sintered silicon carbide has a dry weight of 3.2 g, a saturated weight of 3.5 g, and a saturated suspended weight of 2.3 g after being suspended and soaked in water. Calculate the bulk volume, bulk density, apparent density, and apparent porosity.

Solution

$$V_B = \frac{3.5 - 2.3}{1} = 1.2\ \text{cm}^3$$

*Refer to the experiment by Gail Jordan at the end of this module, NEW:Update 90.

$$d_B = \frac{3.5}{1.2} = 2.6 \text{ g/cm}^3$$

$$d_A = \frac{3.1 \cdot 1}{3.1 - 2.3} = 3.9 \text{ g/cm}^3$$

$$\%P_A = \frac{3.5 - 3.1}{3.5 - 2.3} \times 100\% = 33.3\%$$

Note: The saturated weight is the weight of the object after it is removed from the water.

7.1.2 Structure, Bonding, and Properties

The structure of most ceramics varies from relatively simple to very complex. Being compounds, ceramics are made of different types of atoms of varying sizes. ***Silicates,*** compounds of silicon and oxygen that make up many ceramics, are the most common minerals on earth. They include sand; clay; feldspar ($K \cdot Al \cdot Si_3O_8$); quartz (SiO_2); and the semiprecious stone, garnet. Pure silica (SiO_2) has three common polymorphs: cristobolite, tridymite (high-temperature), and quartz (low-temperature). The silica structure is the basic structure for glasses and many ceramics. It has an internal arrangement consisting of pyramid (tetrahedral or four-sided) units, as shown in Figure 7-3(a). Four large oxygen (O) atoms surround each smaller silicon (Si) atom. The silicon atoms occupy the openings (interstitials) between the oxygen atoms and share four valence electrons with the oxygen atoms through covalent bonding. The silicate structures [Figure 7-3(b)] can link together by sharing the atoms in two corners of the SiO_4 tetrahedrons, forming chain or ring structures. By sharing three corner atoms, they produce layered silicates (talc, kaolinite clay, mica); or by sharing four corner atoms, they produce framework silicates (quartz, tridymite). The alternating Si and O covalent bonding is most common in ceramics and compares to the carbon to carbon (C—C) bonding in organic materials. The unique atomic structures of ceramics leads to the properties possessed by most ceramics. Figure 7-4 (a) and (b) indicates some similarities and differences between ceramic and metal unit cells. See Section 3.4.1 for metal unit cells.

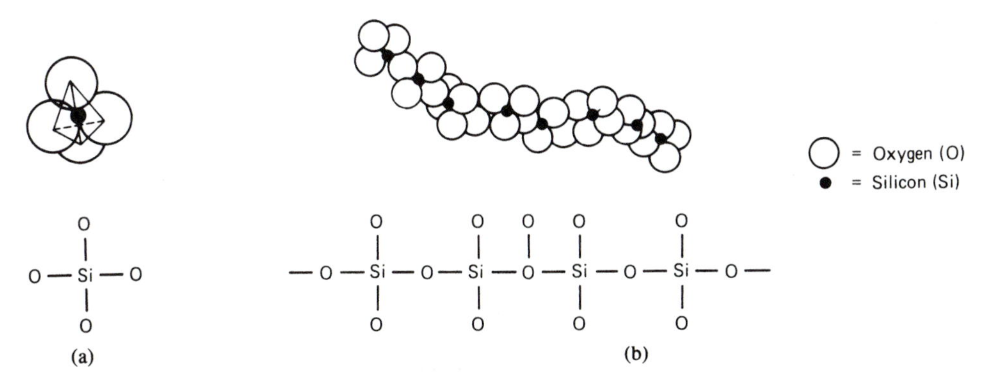

Figure 7-3 Silica structure: (a) silicate tetrahedral, (b) single chain of silicate.

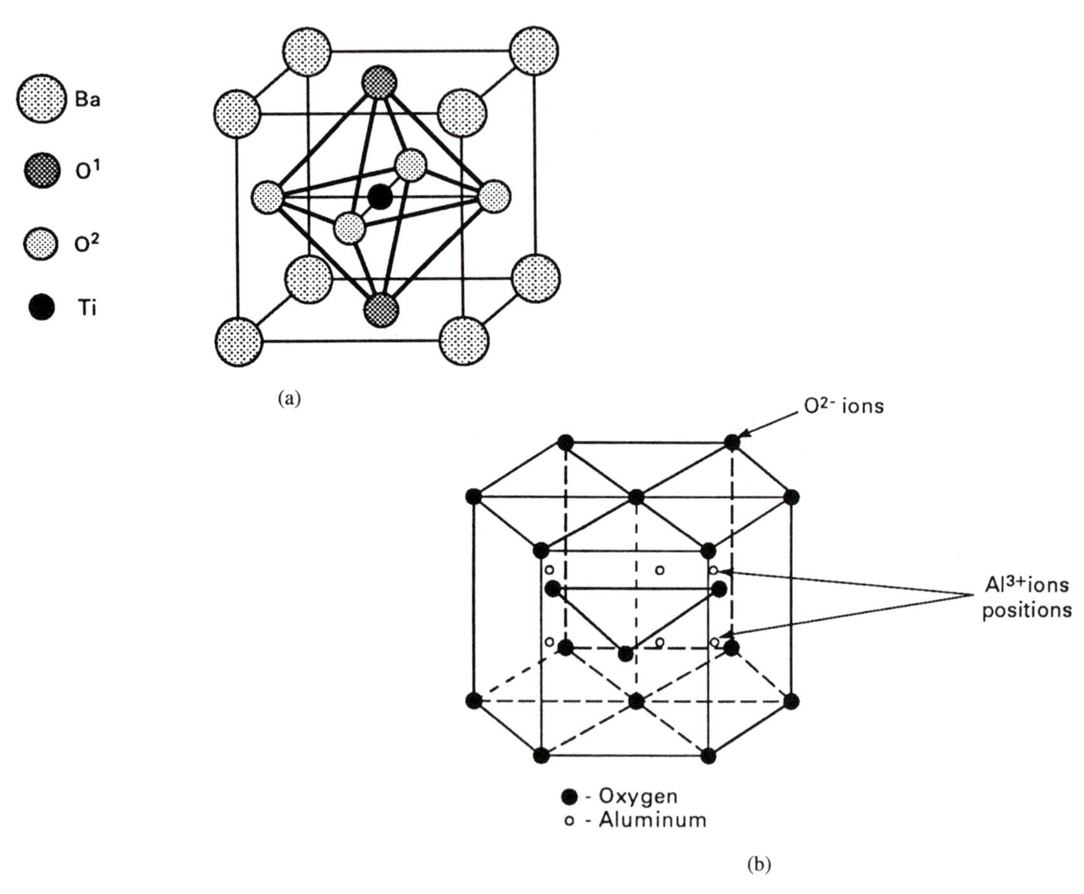

Figure 7-4 (a) Cubic cell of ideal perovskite structure of $BaTiO_3$ above 130°C with the origin on a barium ion. The ideal cubic perovskite structure, abbreviated ABX_3, has an "A," a large cation; "B," a smaller cation; and "X," an ion. Various arrangements or views of this structure might show the "B" cation in the center of the cube or the "A" cation in the center. This figure depicts this structure using $BaTiO_3$. The "A" ions are barium (Ba); the "B" ions are two oxygen ions, 01 and 02; and the "X" ion is titanium (Ti). Superconductivity occurs in oxides with this same structure, and the new high-temperature, superconducting copper oxides closely resemble it also. Such a ceramic possesses large values of dielectric constant as well ferroelectric properties. [*Journal of Materials Education* (JME 13 (3))]. (b) Crystal structures of Al_2O_3 showing O^{2-} ions occupying the hexagonal close-packed (hcp) unit cell sites and Al^{3+} ions occupying about ⅔ of the interstitial sites to maintain electrical neutrality.

The bonding mechanisms for ceramics produce the primary bonds: ionic and covalent. The crystal structure of bulk ceramic compounds is determined by the amount and type of bonds. The percentage of ionic bonds can be estimated by using electronegativity determinations (refer to Figure 2-8).

Being compounds, ceramics have different types of atoms. Hence, their resistance to shear and high-energy slip is extremely high. Because the atoms are held (bonded) so strongly compared to metals, there are fewer ways for atoms to move or slip in relation to each other. Thus, the ductility of ceramic compounds is very low and these materials act in a brittle fashion. Fracture stresses that initiate a crack build up before there is any plastic deformation and, once started, a crack will grow spontaneously. The combination of high shear stresses and reduced ductility produces high compressive strength but low tensile strength. The maximum bending stress in tension at failure is called the ***bend strength*** for ceramic materials and is often referred to as the ***modulus of rupture*** (MOR). At room temperature, metals and ceramics are often competitive, but at temperatures above 1500°F, metals weaken while ceramics retain much of their strength. Ceramics are noted for their heat resistance. The maximum service temperature for alumina is 3450°F and for silicon carbide, it is 3000°F. Heat-resistant nickel alloys are considered unserviceable above 1500°F. (Table 7-1 contrasts some characteristics of metals with those of ceramic materials.)

7.1.3 Wear Resistance

Wear can be defined as the removal of surface material. Wear is a facet of tribology that was introduced in Module 4. Several mechanisms are involved in wear. One is adhesion, or the bonding of two surfaces with subsequent removal of material. Inertial welding is an example of how this mechanism is used in the welding of metals. The roughness of the surface results in removal of part of the protrusions and depressions. When there is a large difference between the hardness of two materials in contact, gouging can occur. Gouging is how an abrasive grinding stone removes material from a workpiece. ***Impact,*** also known as ***erosion,*** occurs when abrasive particles in a fluid impinge on a surface. Other variables at play are abrasion, oxidation, and contact stress. Wear is responsible for many equipment breakdowns and attendant downtime. With wear comes friction, and with friction comes a large expenditure of financial resources to overcome friction. Without considering the loss of material from wear, the dollar value of energy lost each year due to friction is in the billions of dollars. Ceramics are very likely candidates to help reduce this exorbitant expenditure of energy and funds. A quick perusal of their properties should convince people that greater selection of ceramic materials for use in their area is indeed warranted.

Materials can be ranked according to their wear resistance by using a variety of wear tests. Many are single-purpose tests with a particular application in mind. Some concern themselves with only one type of wear factor. A recently developed test method uses flat rectangular specimens ($100 \times 25 \times 6$ mm) mounted on a cylinder that rotates inside a container filled with an abrasive medium tailored to simulate different types of wear. This test, called the rotating-cylinder wear test, is supposed to be a better indicator of a material's performance under service conditions. Specimens are weighed before and after the test, and the percent weight difference is an indicator of the material's wear resistance. In reporting, the test conditions must accompany the results. Figure 7-5 illustrates the properties of a ceramic metal composite produced by Alanx using a new process called Dimox, discussed later in this module. This composite is designed specifically to provide maximum wear resistance. Alanx CG 896 consists of densely packed silicon-carbide particles within a matrix of aluminum oxide and a metal alloy. The particles as well as the alumina resist wear, while the metal alloy provides the bonding between the two ceramic phases. The accompanying graph relates this wear-resistance composite with other materials by use of a wear-life factor. This factor compares the service lives of competitive materials. The larger the number, the longer the wear life. A slurry erosion test is used to measure the

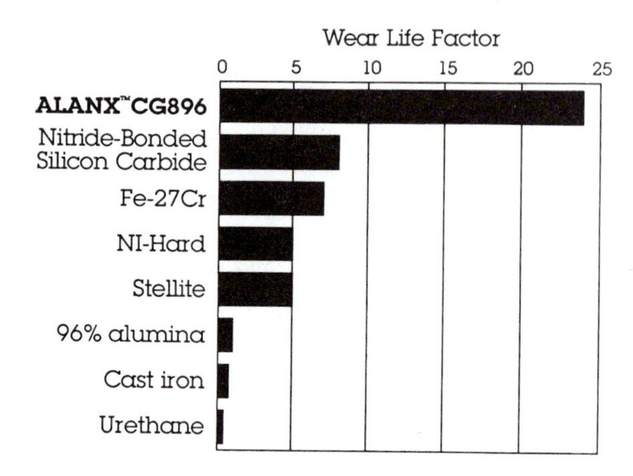

Figure 7-5 Typical properties of ALANX ceramic/metal composite.

TYPICAL PROPERTIES OF ALANX™ CERAMIC/METAL COMPOSITES

	ALANX™ CG896	ALANX™ CG273	ALANX™ FGS	Developmental metal matrix composite (MHC)
Wear life factor	24	24	12	9.2
pH range	5–8	1–14	5–8	5–8
Density (g/ml)	3.32	3.42	3.32	2.95
Hardness (R_A)	80	80	80	57
Flexural strength (MPa)	137	100	216	216
Fracture toughness K_{1c} (MPa · m½)	6.0	4.4	5.5	11.3
Shear modulus (GPa)	131	144	144	79
Young's modulus (GPa)	313	357	366	204
Poisson ratio	0.19	0.23	0.27	0.29
Thermal expansion (10^{-6}/K)	5.43	5.20	5.89	10.24
Thermal conductivity (W/m · K)	160	160	160	160

wear resistance of a test specimen. The material loss is compared to a standard aluminum test specimen and the wear life factor is determined. Refer back to Figure 4-38 for more wear comparisons.

7.1.4 Stress Concentrations and Fracture Toughness

When selecting a material that might be subjected to ***cyclic loading*** or where the load cannot be distributed uniformly over an area, the subjects of stress concentration and fracture toughness come into play. For example, when a hole is drilled in a flat plate, such as that sketched in Figure 7-6(a), and is loaded with a force *P*, the stress in the plate is assumed to be uniformly distributed [Figure 7-6(b)] across the cross-section of the plate everywhere but at the cross-section containing the hole. Figure 7-6(c) shows the distribution of stress as it finds its way around the hole, thus producing a concentrated stress at the edges of the hole, as shown. Such a buildup of stress may cause the material to crack. It is important to point out that stress concentrations depend on the geometry of the flaw (hole, notch, or crack) and the geometry of the specimen or component (shape and thickness). They have nothing to do with the properties of the material. In materials that may contain cracks due to flaws, voids, inclusions, or oxides in their microstructure, the stress concentration is greatest at the tip of the crack. Flaws that cause fracture of ceramic materials usually are smaller than 50 μm. This size of flaw is not readily detectable by most nondestructive-testing (NDT) methods.

Fracture toughness, often perceived as the limiting factor for ceramics applications, is being improved to make ceramics competitive as engineering materials. A stress–strain diagram [Figure 7-7(a)] illustrates the effects of ceramic brittleness compared to those of relatively ductile metals. Figure 7-7(b) contains a graph contrasting the fracture strength of metals with that of ceramics; the graph also shows the increase in this property using the toughening techniques discussed in this module, to the point where advanced ceramics are approaching metals in their fracture toughness.

Using fracture mechanics techniques, the critical stress–intensity factor (K_{1c}), or fracture toughness of the material, is determined independent of the nature of the material. The "1" in the symbol K_{1c} refers to the mode of loading of the specimen. In this instance, the load is parallel to the "1" or x material axis, producing a normal stress. When used with the symbol for fracture toughness, the numerals "2" and "3" refer to the shear loads acting on the y or z axes, respectively. Figure 7-8 is a sketch of a fiber-reinforced part being stressed parallel to the "2" axis. There are several mechanical tests, including indentation methods, used to measure the fracture toughness of ceramics. Which one is chosen depends in part on the type of information needed. One popular test used for testing ceramics

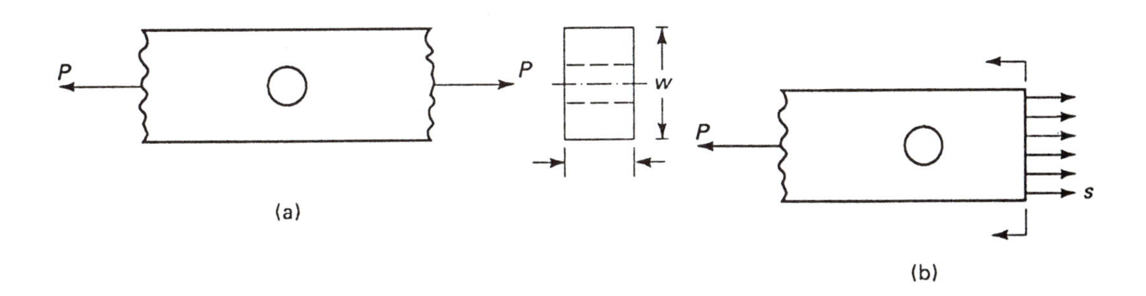

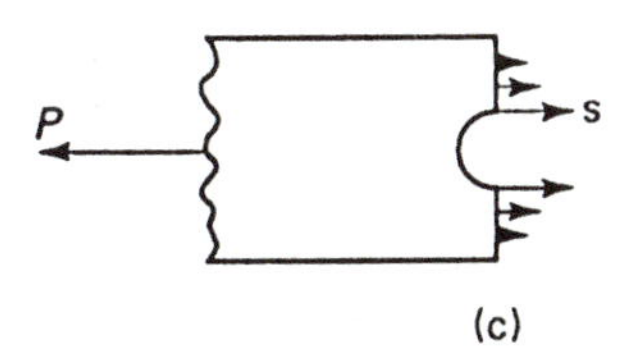

Figure 7-6 A stress concentration resulting from a through hole in a flat plate. (a) Sketch of the plate loaded with a tensile force P, showing a side view of plate. (b) Cross-section of the plate at some distance from the hole, with uniform distribution of tensile stress indicated by small vectors. (c) Cross-section through the center of the hole, with the largest vectors portraying the maximum stress concentrations at either edge (side) of the hole.

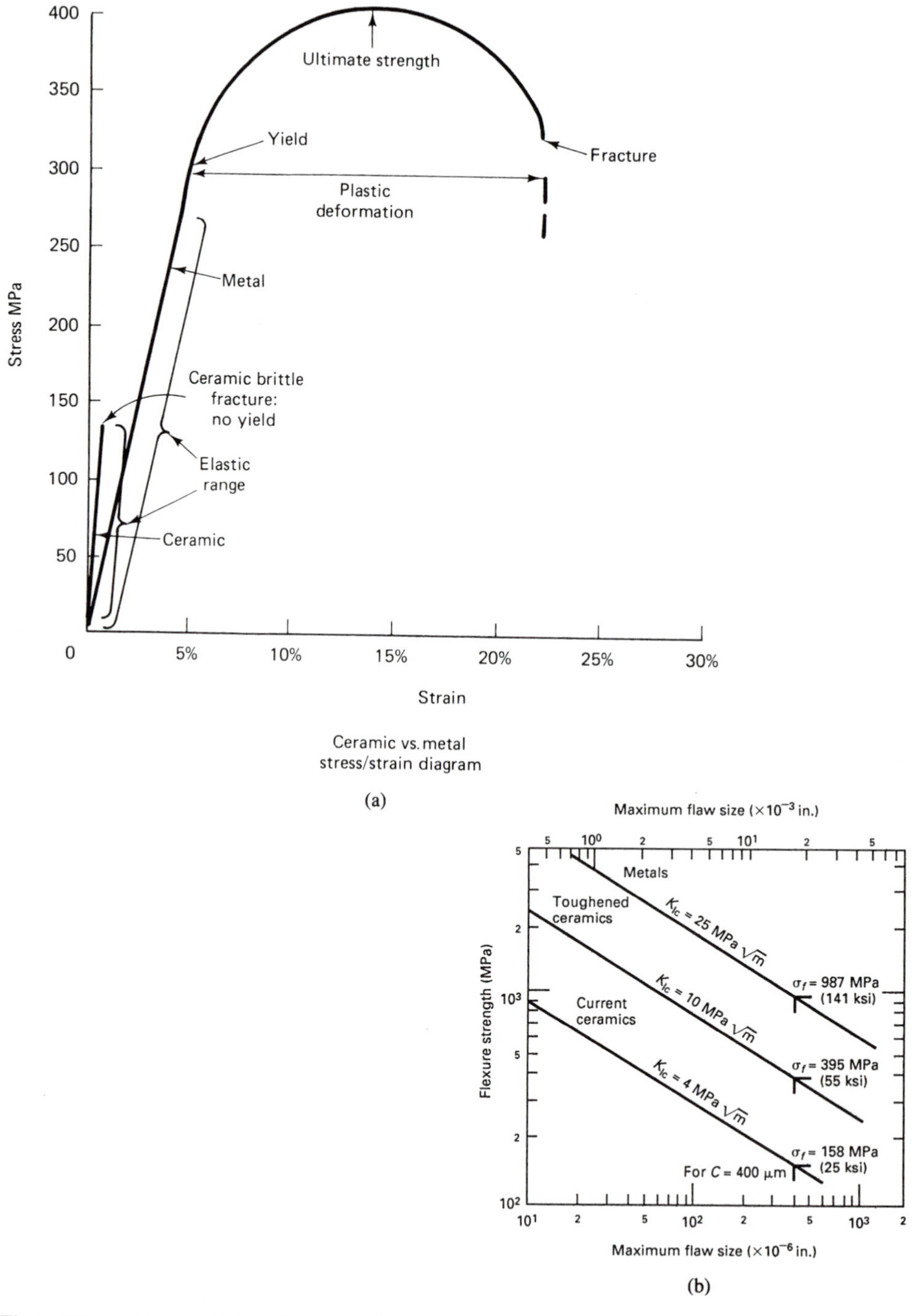

Figure 7-7 (a) Ceramics' brittle fracture results from their inability to yield like metals do. This diagram shows a large percentage of plastic deformation in metals that occurs because localized stresses are relieved by localized plastic deformation. No yielding point shows on the ceramic curve, so localized stress eventually causes catastrophic failure when the elastic range is exceeded. Percentage of strain before failure on ceramics can be measured in hundredths of a percent compared to tenths of a percent for metals. (b) A plot of flexural strength versus maximum flaw size for metals, toughened ceramics, and traditional ceramics, which translates into (represents) toughness values ranging into the 20s. (ORNL.)

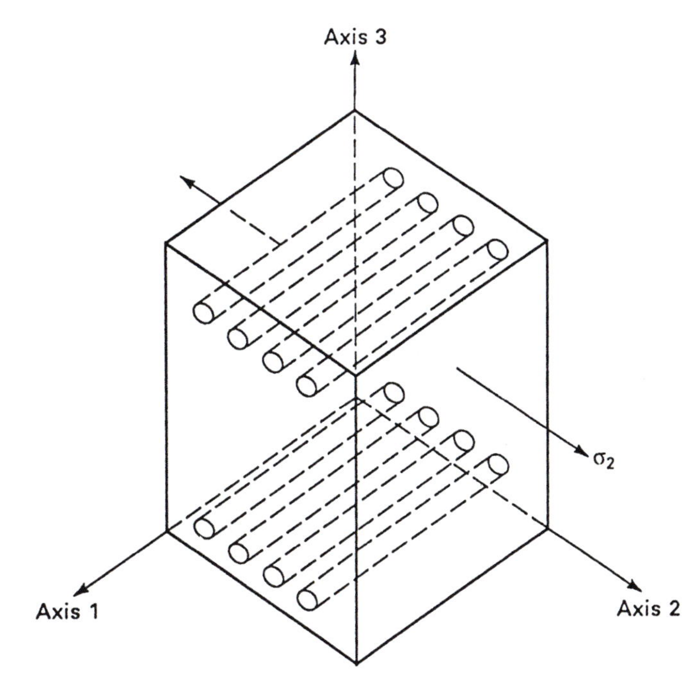

Figure 7-8 A composite material with fiber reinforcement. The axes (1, 2, and 3) are shown with a normal tensile stress acting in the direction of the 2 axis.

is the single-edge, notched beam (SENB) test [Figure 7-9(a)]. It is similar to a four-point bend test except that an artificial crack is placed in the specimen before testing. As Figure 7-9(b) illustrates, the chevron-beam test specimen is similar in loading to the SENB test but differs because it has a chevron notch rather than a straight-through notch machined or cut into the surface. The reason for this type of notch is to force the crack to propagate slowly through the specimen. It does so because it is always passing into a zone of greater

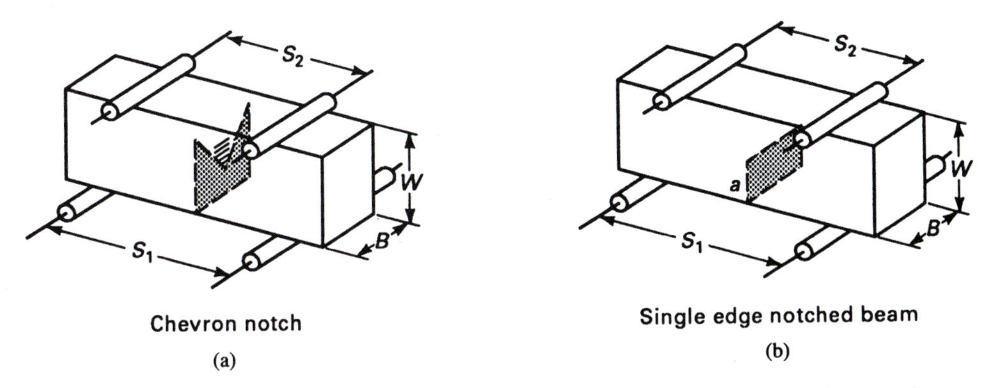

Figure 7-9 Fracture toughness is determined using single-edge notched and chevron notch beam specimens. An artificial cut is placed in both specimens. (a) A straight-through notch in the SENB is cut with a diamond-impregnated blade and (b) a chevron is machined into the chevron notch (CN) specimen.

load-bearing area. Both tests give a direct measure of the fracture toughness of a material. The SENB test uses the following equation:

$$K_{1c} = \frac{3P(S_1 - S_2)a^{1/2}Y}{2BW^2}$$

where

P = applied load
a = crack length
S_1, S_2 = dimensions between the outer and inner roller spans
B = specimen thickness
W = specimen height
Y = dimensionless calibration factor

The units for expressing fracture toughness are either psi $\times \sqrt{\text{inches}}$ or MPa $\times \sqrt{\text{meters}}$. It is of interest to note that once this critical stress–intensity factor is exceeded, the crack will propagate and the material will fracture, producing a catastrophic failure. The failure can be caused by an increase in the applied stress (load) or in the crack length. Once the fracture toughness is known, the fracture stress or the stress at which the crack will propagate can be calculated. Also, the stress at failure may be less than the yield strength. It is also noted that flaws in almost all silicate and oxide-based ceramics and glasses will grow at stress levels lower than the calculated stress that produces a catastrophic or fast fracture. This environmentally assisted crack growth is produced by a reaction between the strained bonds at the tip of the crack with water or other molecules. Figure 7-10 is a sketch of a crack zone toughened as a result of the transformation of ZrO_2. Transformation toughening is covered in greater detail in Section 7.4.

Flaws in ceramic materials that produce failure are usually smaller than 50 μm. Most NDT procedures do not detect flaws smaller than 50 μm. In summary, the fracture toughness, a property of a material, is an indicator of the ability of a stress concentrator (crack or flaw) to produce a catastrophic failure. To predict the ability of a material to resist fracture (fracture strength), knowledge of both fracture toughness and the presence of ***stress concentrators*** in a material is vital.

7.1.5 Thermal Shock Resistance

Thermal shock is a mechanism that can produce mechanical failure in materials. It involves the buildup of thermal stress in a material as a result of exposure to a temperature difference between the surface and the interior of a material. If the temperature difference occurs rapidly, the possibility of failure increases. Testing of ceramic turbine vanes for thermal shock subjects them to repeated recycling from 340° to 1230°C in less than 2 s. At 1230°C, these vanes glow as a result of the heating. Future automobile gas-turbine engines will require materials that can withstand a rapid temperature change, from room temperature to 1200°C or more in a few seconds, and can be exposed to stresses in excess of 207 MPa. Some engine designers are contemplating future engines capable of operating for 3500 h with internal operating temperatures as high as 1745°C. Some of the material properties involved in producing thermal shock are the elastic modulus *(E)*, linear coeffi-

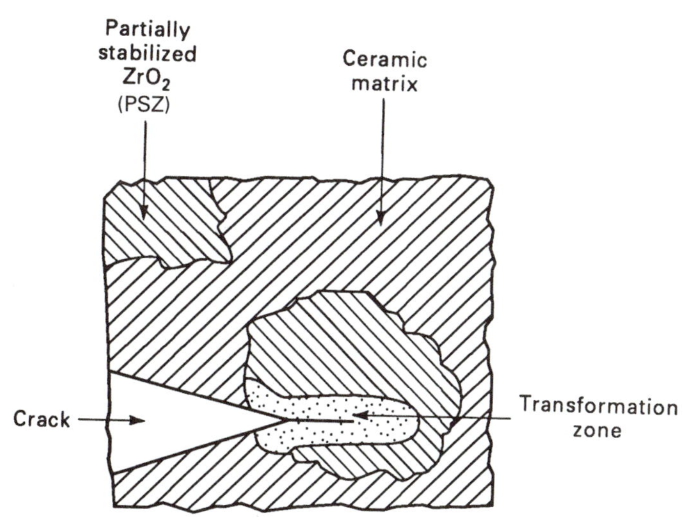

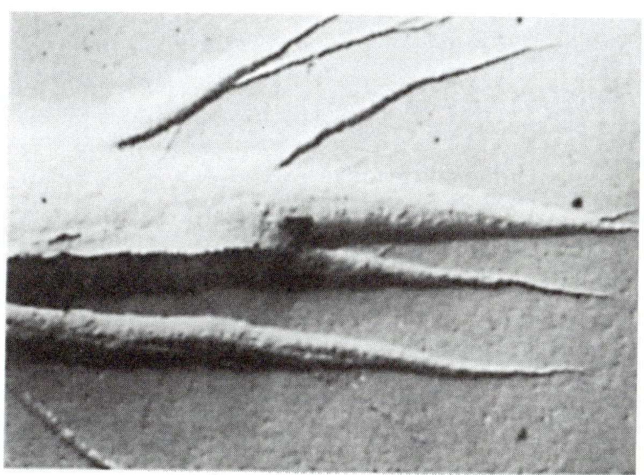

Figure 7-10 A diagram showing the formation of a transformation zone surrounding the tip of a crack due to the polymorphic transformation of ZrO_2, including a micrograph of the same area. (ORNL.)

cient of thermal expansion (α), thermal conductivity *(k),* tensile strength (σ), and fracture toughness (K_{1c}). Also, if a material is allotropic and undergoes a phase transformation as a consequence of the temperature change, this factor must be considered. In the selection of materials for good resistance to thermal shock, each of the variables noted above must be analyzed to determine how it affects the material's ability to limit and/or reduce the thermal shock produced. For example, a material with high thermal conductivity, high strength, a low coefficient of thermal expansion, and a low elastic modulus would be a good candidate for resisting thermal shock. Lithium aluminum silicate (LAS) has an extremely low coefficient of expansion along its hexagonal structure ($-2.0 \times 10^{-6}K^{-1}$) and, despite its low tensile strength, low thermal conductivity, and fracture toughness, is a good candidate for thermal-shock-resistance applications.

Due to their ability to deform both elastically and plastically, most metals have few problems with thermal shock. The presence of free electrons, which conduct heat as well as electricity in metals, results in good thermal conductivities. Organic materials have low thermal conductivities, due to their covalent bonding, which ties up their electrons. Ceramic materials exhibit a diverse range of thermal conductivities due to their complex microstructures. Those ceramics with the least complex structures, such as silicon carbide (SiC), composed of atoms of similar size and shape, achieve the highest thermal conductivities. Although not a ceramic material, diamond has the simple crystal structure of pure carbon, with a thermal conductivity at room temperature (RT) of 900 W/m · K. Copper's conductivity is half that of diamond's. Graphite, another polymorph of carbon, has a layered microstructure, which makes it isotropic. In addition, graphite undergoes different fabrication techniques that result in different orientations of the crystal structure, which only compounds the problem of assessing its ability to conduct heat or electricity. Finally, porosity, dispersions, noncrystallinity, grain size, and grain boundaries all affect the thermal shock resistance of materials. Pores and microcracks are probably the most important of these.

To assist in the selection of likely candidate materials for thermal shock resistance, several parameters have been developed that take into consideration the factors mentioned above. One such parameter is the ***thermal shock index (TSI),*** expressed in equation form as follows:

$$\text{TSI} = \frac{\sigma \times k}{\alpha \times E}$$

where

σ = tensile strength
k = thermal conductivity
α = linear coefficient of thermal expansion
E = modulus of elasticity (Young's modulus)

TSI is a ratio of the products of four variables. High values for the variables in the numerator and low values for the variables in the denominator render large or increased values for TSI. When comparing different materials based on the value of TSI, it is essential to use the same units of measure for each of the variables. For example, if you use MPa to express tensile strength for one material, you must also use MPa (not psi) for the material you are comparing it with. Table 7-3 lists the TSI and some representative property values for selected materials.

TABLE 7-3 THERMAL SHOCK INDEX (TSI) FOR SELECTED MATERIALS

Material	σ (MPa)	k (W/cm · °C)	α (°C^{-1} $\times 10^{-6}$)	E (GPa)	TSI[a] (W/cm)
Fused SiO_2	68	6×10^{-2}	0.6	72	94
Al_2O_3	204	3×10^{-1}	5.4	344	33
Graphite	8.7	1.4	3.8	7.7	416
Soda-lime silica glass	69	2×10^{-2}	9.2	68	2.1

[a]TSI units in the U.S. customary system are Btu · in/hr · ft^2 · F°; in the SI system, they are W/cm.

Illustrative Problem

Determine the TSI for aramid fiber and E-glass fiber, and indicate which has the best thermal shock resistance based on these calculations. Selected properties for each of these materials are provided in the table that follows:

Aramid fiber	E-glass fiber
$E = 18 \times 10^6$ psi	$E = 10.5 \times 10^6$ psi
$\sigma = 400 \times 10^3$ psi	$\sigma = 500 \times 10^3$ psi
$k = 3.5$ Btu · in./hr · ft² · °F	$k = 7.9$ Btu · in./hr · ft² · °F
$\alpha = 0.8 \times 10^{-6}$ °F^{-1} (long.)	$\alpha = 1.6 \times 10^{-6}$ °F^{-1}

Solution

$$\text{TSI}_{\text{aramid}} = \frac{4 \times 10^5 \times 3.5}{0.8 \times 10^{-6} \times 18 \times 10^6} = 9.7 \times 10^4 \text{ Btu} \cdot \text{in./hr} \cdot \text{ft}^2$$

$$\text{TSI}_{E\text{-glass}} = \frac{5 \times 10^5 \times 7}{1.6 \times 10^{-6} \times 10.5 \times 10^6} = 2.1 \times 10^5 \text{ Btu} \cdot \text{in./hr} \cdot \text{ft}^2$$

E-glass fiber was the greater and therefore has the best shock resistance.

7.1.6 Emissivity

Emissivity is an important property, particularly in high-temperature ceramics. When a surface is exposed to radiation, it may absorb part or all of the incident radiation (energy). The fraction of the energy absorbed is the emissivity *(e)* of the surface. A blackbody, the perfect absorber, has a value of $e = 1$; for the perfect reflector, $e = 0$.

Thermal barrier coatings (TBCs) on the metal components of gas turbines have low thermal conductivities and low emissivity. Having a coefficient of thermal expansion similar to that of the base metal allows the coating to adhere to the metal as the metal undergoes changes in temperature. The newer, energy-efficient window glass, which is more costly to the customer, has low emissivity, which means that a much greater percentage of the sun's radiation (heat) will be reflected than from older types of window glass. Another technique to reduce solar heat gain in glazing applications is shown in Figure 7-11(a).

Ceramists probably use phase diagrams more than do metallurgists. Figure 7-11(b) is a phase diagram of a rather simple ceramic system of silicon oxide and alumina showing the formation of ***mullite,*** an important refractory. This diagram can be contrasted with phase diagrams of metals in Module 5. Many ceramics are the product of more than two materials, and thus their phase diagrams are more complex. ***Ternary,*** or three-component, systems, such as $MgO–Al_2O_3–SiO_3$, require a three-dimensional model with a base in the shape of an equilateral triangle, with each vertex representing one of the three pure compositions. The temperature axis usually points or sticks out of the paper on which the diagram is drawn. It is important to remember that a phase diagram represents equilibrium conditions, which for ceramics would require substantial time to be reached.

(a)

Figure 7-11 (a) Reducing heat gain through its reflective nature. Lexan® polycarbonate sheet material (either MR® or XL-1®) is laminated with a proprietary, metallized reflective film for commercial glazing applications. The plastic also provides natural light and the high impact strength of polycarbonate sheet. It is a competitor to glass glazing. As a materials selection issue, what are the pros and cons for each type of glazing, beyond their emissivity properties? (G.E. Plastics.) (b) Rough sketch of SiO_2-Al_2O_2 binary system used to produce many refractories. Mullite (3 Al_2O_3 $2SiO_2$), an intermetallic compound with a small range of compositions, is a ceramic with many high-temperature applications.

Liquid
Mullite ($Al_2Si_2O_{13}$)
Temperature (°C)
2000
1820
1700
1600
1400
I
II
0 80 100 Al_2O_3
SiO_2 100 20 0
Composition (%)

(b)

7.2 DESIGNING CERAMICS

Because of their properties—their inherent ***brittleness;*** inability to yield, change shape, or plastically deform; sensitivity to defects; tendency to break at any point of high stress concentration; microstructures, which possess flaws that can cause failure without being detected beforehand—ceramics require a design model that calls for greater degrees of accuracy in the design and manufacture of ceramic parts. Anyone who has experienced breaking a piece of glass will know about the sensitivity of brittle materials to defects. A slight scratch on the surface of the glass and a small bending force to produce a tensile stress are all that is needed to break a relatively thick piece of glass. This flaw sensitivity means that a ceramic cannot be considered as having only one strength but rather, a distribution of strengths, depending on the flaws in the material that were added during processing, final machining, or testing, or while in service. The three general design models used by engineers are the empirical, the deterministic, and the probabilistic. ***Empirical design*** can be defined as the trial-and-error method of selecting a material for a particular application. ***Deterministic design*** is based on mathematical analysis of average properties of ductile materials, together with a factor of safety to accommodate any errors. ***Probabilistic design*** is required for designing ceramics, particularly when the parts are complex in form and when there is a complex set of applied stresses to contend with. Using the tools of fracture-mechanics theory, highly refined stress-analysis techniques, and statistical representations of material properties, this probabilistic design model allows the engineer, aided by numerous computer programs, to determine the stresses statistically, and parts are then designed not to exceed a specified probability of failure. The density, size, and severity of flaws are also determined statistically. Because ceramic parts have slight variations in their structure, which can have a severe effect on their properties, each part must be designed individually. The mathematical means for representing the statistical nature of brittle failure is the Weibull equation, which relates the cumulative probability of failure to the failure stresses of all the elements of the sample. The graphical plot of the data along with a best-fit curve is known as a ***Weibull plot*** of strength for a material. The slope of the best-fit curve, known as the ***Weibull modulus,*** indicates the degree of variance in the strength data. A high value for the modulus (m) is desired because it indicates low variance or a narrow range of failure stresses. Many critical parts are ***proof tested*** at a predetermined proof-stress level. Those parts with large flaws will fail the test. Several assumptions are made at this point. One is that parts that pass the test will perform satisfactorily in service. It is also assumed that the testing did not initiate cracks or cause cracks to propagate that will cause failure later in service.

7.3 CERAMIC PROCESSING

A major problem in ceramic processing is achieving the same properties every time a process is repeated. ***Intelligent processing*** of materials is an attempt to solve the problem of making complex materials more uniform in their structure and properties. As materials become more complex in composition, structure, and processing, it becomes more difficult to make a uniform product. The National Institute of Standards and Technology (NIST), in cooperation with industry, has an ongoing program to automate ***quality control*** that allows quality to be built in during processing rather than attempting to obtain it by inspection after the fact. Sensors analyze a material as it is processed to warn computers controlling the process to correct errors beginning to occur. ***Adaptive control*** is the name of the method to improve the quality of manufacturing processes. The key to tougher, stronger, more reliable ceramics lies in ceramic processing. By improving existing processes and developing more efficient ones that reduce the size and number of flaws and even out the distribution and size of particles, attaining such ceramics is being realized rapidly.

7.3.1 Thermal Processing-Sintering/Densification/Firing

Densification, also referred to as ***firing,*** is a process by which a particulate compact is transformed into a ceramic part that has adequate properties to satisfy the needs of a particular application. The powder compact, with its relatively small powder size, is heated at a temperature below the melting temperature of its components. The melting temperatures of ceramics make it impractical to actually melt the materials and thus bond the grains of powder together. The bonding between the grains during sintering is due to the action of many complex mechanisms, including diffusion. In summation, ***sintering*** is the removal, by a variety of mechanisms, of the initial pores, combined with the growth of grains and strong bonding forces to provide strength to a powder compact. The main driving force involved is the reduction in surface area (i.e., the surface energy of the item). In general, this results in a part with the same shape as the original or ***"green"*** shape but reduced in size. The density of the part has increased as well as the particle size. The increase in particle size must be controlled because the final strength of the part is determined partly by the size of the final grains. Relatively large particles have a greater number of interstices. Interstices are filled with air, which itself is a good insulating material. This may be satisfactory if the ceramic is designed for its heat resistance. Usually, a ceramic requires strength, and strength comes with a homogeneous microstructure that has a fine grain structure. The problem is that the initial pores between the original particles have not been completely eliminated. Further, the final density reaches a limit of only some 95% of the maximum *theoretical density* (see Section 7.1). Figure 7-12 shows graphs of the sintering of alumina and zirconia using microwave sintering versus standard furnace-heating temperatures. *Sintering furnaces* usually heat ceramic compacts externally by radiant and convection thermal transfer. Ceramics that have rather high dielectric-loss factors (a measure of electric energy lost as heat energy by a capacitor in an ac circuit) can take advantage of this property, as Al_2O_3 and ZrO_2 illustrate in Figure 7-12. The significant benefits of, as well as a few applications for, these new ceramic materials also accompany the graphs, which also show the benefit of using microwaves to lower temperatures for sintering and to achieve finer microstructures and better mechanical properties. In addition to the conventional sintering process discussed above, other sintering processes, such as liquid-phase and solid-state sintering, are used to densify compacted powders.

Many processes employed by industry improve on the pressureless sintering process described above. Some that involve the application of pressure and heat simultaneously are (1) hot pressing, (2) overpressure sintering, and (3) hot isostatic pressing (HIP). In ***hot***

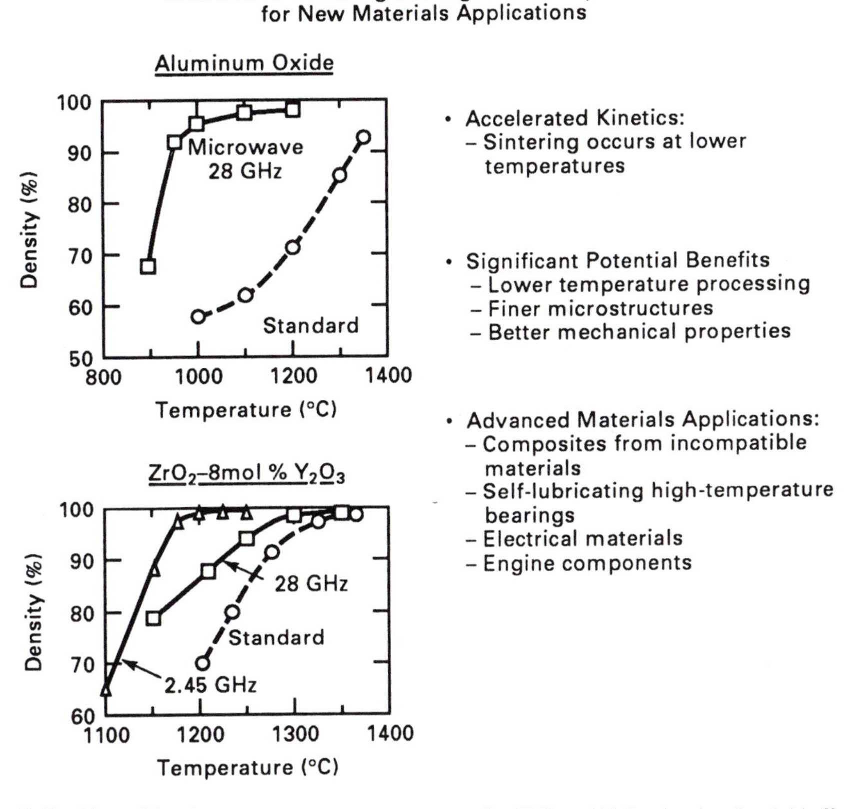

Figure 7-12 Plots of density percentage versus temperature for Al_2O_3 and ZrO_2, showing the vivid effects of sintering these materials with microwaves versus standard heating procedures. Note the significant increases in attained density as a result of microwave heating. (ORNL.)

pressing, also known as ***pressure sintering,*** powders are compacted using a die within an evacuated chamber. Temperatures are about one-half of the absolute melting temperature (T_m + 273°)* of the powdered material. Due to the uniaxial techniques used to apply the pressure in the die, hot-pressed materials may have a preferred orientation of their grain structure, which produces anisotrophy. ***Hot isostatic pressing*** (HIP) combines compounding (compressing and densifying loose powders into a desired shape) and sintering using temperatures as high as 2500°C and pressures as high as 200 MPa in a furnace within a pressure vessel. The process relies on plastic deformation to achieve densification. Using argon or helium gas to exert pressure from all sides, HIP results in near-shape forming of ceramic parts, particularly with cemented-carbide particulates. As with the other mentioned densification processes, HIP can be used in combination with other processes to achieve highly densified parts.

*T_m in °C; absolute (T_m + 273°) is in kelvin.

Sintering is a critical factor in the production of advanced ceramics. A green, or unsintered, ceramic material is softer than chalk, consisting of millions of powder particles. When sintered at high temperatures, usually between 1000° to 2000°C, the particles fuse to become a solid material with properties superior to many metals. ***Microwave sintering*** produces uniformly heated ceramic parts of large volume and irregular shape rapidly in excess of 1600°C in a vacuum or under atmospheric pressure. Enhanced densification rates have brought on reduced sintering times and temperatures, which means that this innovative technology not only produces a higher-quality product, it does so at a lower cost. Microwave sintering causes the molecules throughout the workpiece to vibrate, which produces uniform heating without gradients in temperature. In contrast, in radiant or convection heating, energy is absorbed only at the surface of the workpiece and must be transferred into the bulk of the material by conduction.

In another novel process, graphite–carbon particles with controlled electrical properties are used as both the pressure-transmitting medium and as an electrical resistor for simultaneous heating and consolidation of the powder preforms. Cyclic times are very short compared to other methods, such as HIP (minutes versus hours), and both high temperatures (about 1000°C) and relatively low pressures are involved. This process has the ability to densify complex shapes without "canning." The term ***canning*** refers to the need to encapsulate preforms for HIP with glass to seal the preform in a gas-impermeable envelope. With such rapid densification, this new process has the potential for inhibiting grain growth, resulting in fine-grained microstructures with improved mechanical properties.

7.3.2 Traditional Processing

The first ceramic materials produced by processing were bricks and lime plaster. Limestone was roasted to produce lime that was mixed with water. To this mixture was added clay, straw, or fibers to provide building materials that were used widely and are still in use today. When added to the clay materials, water provides the clay with plasticity or workability, in addition to providing the bonding mechanism that gives the material its strength. Traditional ceramics are still produced with basically the same materials. ***Slip casting*** [Figure 7-13(a); see also Figure 7-1(a)], an age-old process for making vases and other hollowware, is still used as a production process for many ceramics. Many other traditional ceramics are produced by blending, pressing, and firing [Figure 7-13(b) to (d)]. Often, traditional ceramics develop a vitreous or glassy phase upon cooling after firing, which accomplishes densification (Figure 7-14). Glass phases must be avoided in advanced ceramics produced from ceramic particulates. Ceramic materials processed to produce refractories, ***structural clay*** products (brick, pipe), or ***whiteware*** (dinnerware, tile pottery) can be considered traditional ceramics. In early times whiteware, such as pots or vases, was formed by pressing one's hands against material supported on a rotating wheel. This very old process, called ***jiggering,*** is still practiced by people engaged in making hollowware with the aid of a potter's wheel. In industry the same process is highly automated in the fabrication of hollowware.

7.3.3 Advanced Ceramic Processing

More modern ceramics with unique and more sophisticated properties needed in this highly technological age are referred to using a variety of terms. ***High-tech, high-performance,***

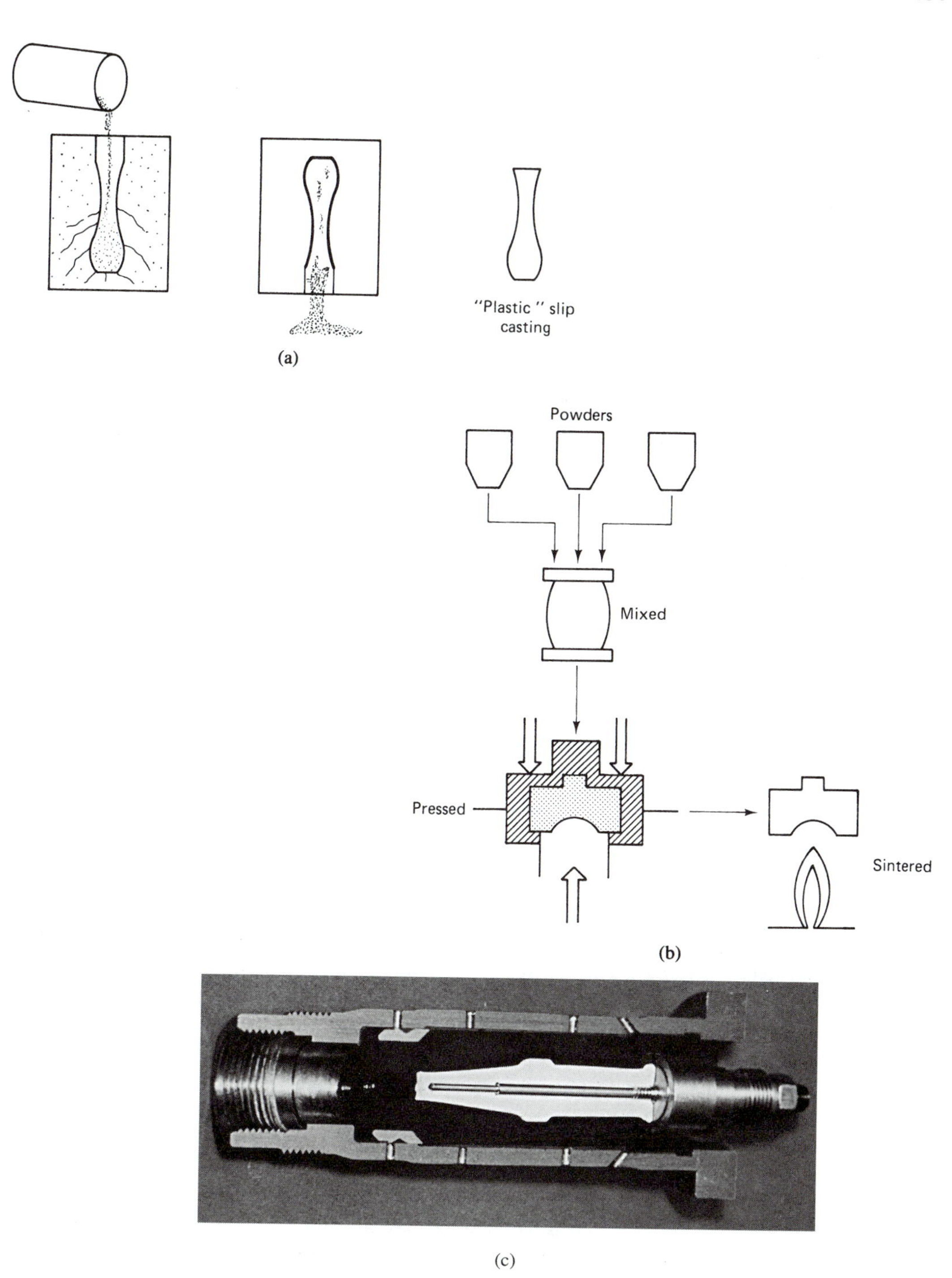

Figure 7-13 Basic processes for ceramics. (a) Casting. (b) Ceramic blending, pressing, and heating. (c) Isostatic pressing mold in cross-section showing insulator blank. (American Ceramic Society.)

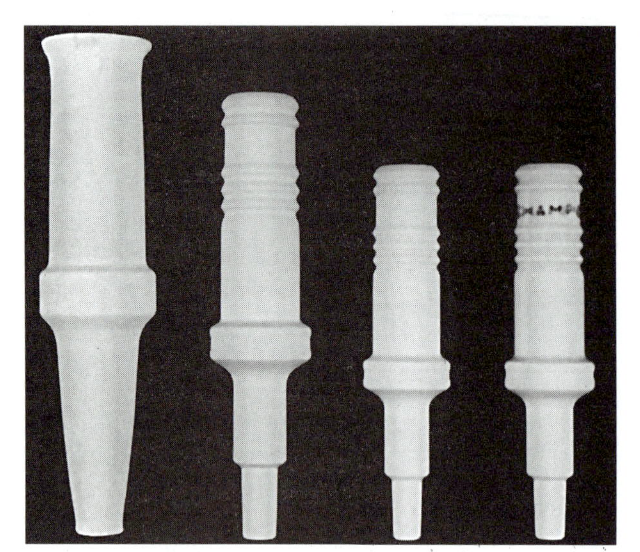

(d)

Figure 7-13 (d) Stages of insulator body development (left to right): pressed "green" blank, turned "green" insulator, bisque-fired insulator, glass glazed and decorated refired to finished insulator. (American Ceramic Society.)

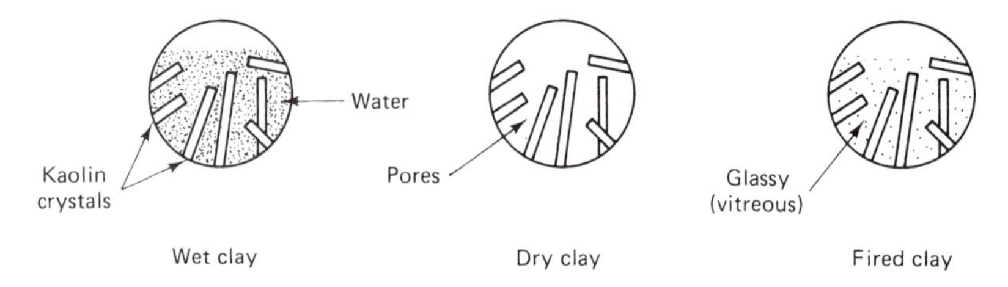

Figure 7-14 Stages of sintering (firing) clay.

fine, new, value-added, engineered, engineering, and ***advanced*** are some of the terms seen in the literature. In this book we use the word *advanced* to describe such ceramics. Sources for advanced ceramics are not new or revolutionary. What is new are the many new processes developed in recent years to produce them. For example, Figure 7-15 shows a hybrid sol-gel process for fiber optics.

The multistage fabrication process for ceramics includes powder production, powder conditioning, shaping (forming), drying, and densification (also called *firing* or *sintering*). Note in Figure 5-63 the similarity of ceramic processing to the processing of powder metals. ***Cermets,*** one product of powder metallurgy, are combinations of metals and ceramics with one or more ceramic phases. Examples include TiC; TiC, N; ZrB_2; SiC; SiO_2; and serve in such applications as cutting, tool inserts, wear parts, and die inserts. The source minerals for these advanced ceramics are metallic elements combined chemically to form oxides, carbides, nitrides, or borides. It is common practice to refer to oxides by a special name ending in *-ia* or *-a*. For example, Al_2O_3 is alumin***a***, MgO is magnes***ia***, Cr_2O_3 is chrom***ia***, ThO_2 is thor***ia***, and ZrO_2 is zircon***ia***.

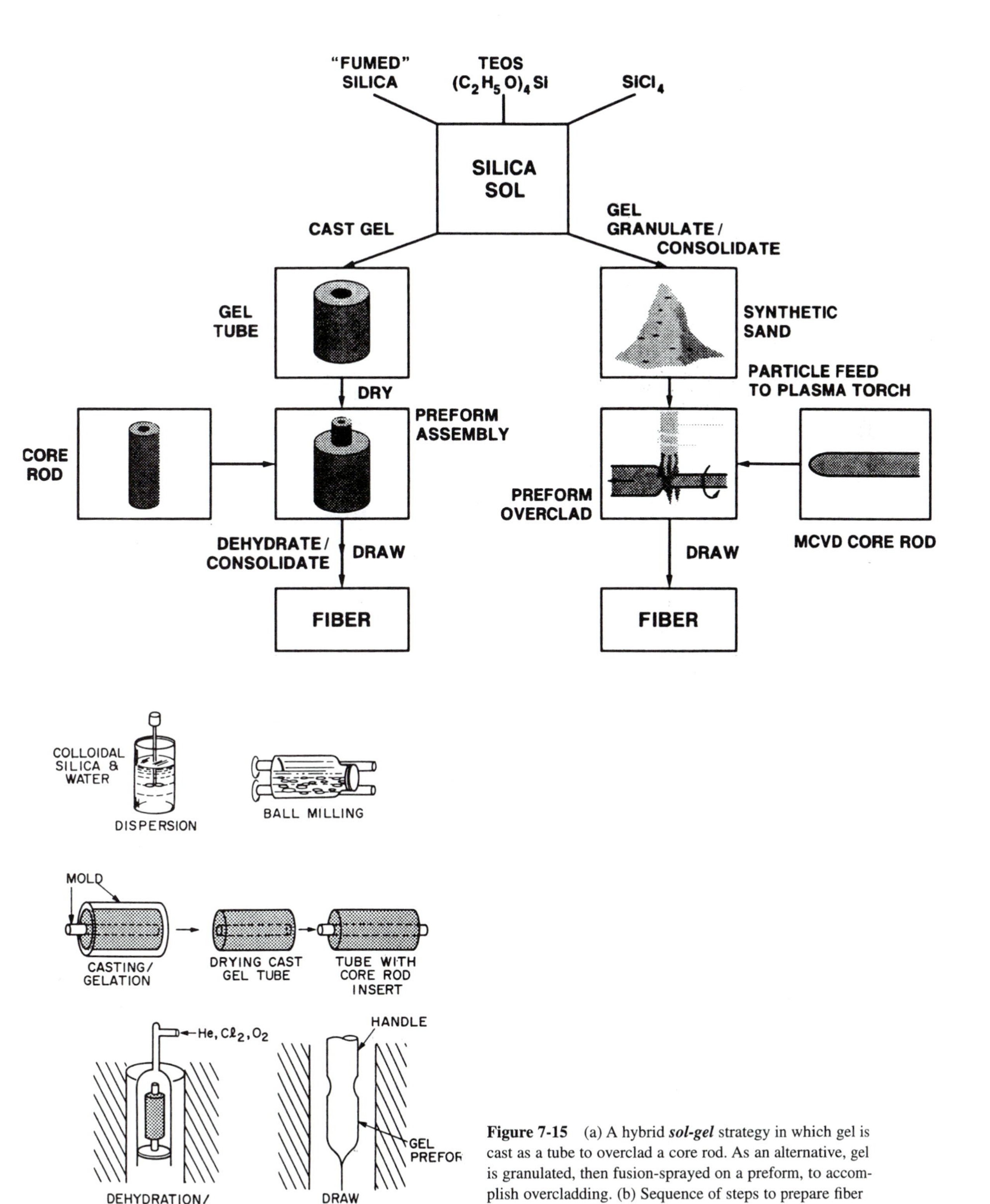

Figure 7-15 (a) A hybrid ***sol-gel*** strategy in which gel is cast as a tube to overclad a core rod. As an alternative, gel is granulated, then fusion-sprayed on a preform, to accomplish overcladding. (b) Sequence of steps to prepare fiber from core-rod and cast gel body.

Ceramic processing can be described simply as consisting of two steps: (1) a cold step, in which the ceramic part is formed or shaped into a "green" part or preform, and (2) a hot step, in which the green compact is first subjected to heat to dry up any liquid phase formed during the processing and then subjected to higher heating, known as firing, sintering, or densification. The firing consolidates the particles and bonds the ceramic particles together permanently to give the ceramic its strength and other final properties. Some processes do not require drying, nor do some require the final step in most processing—that of machining to final specifications.

Some but not all engineering-ceramics fabrication processes begin with minerals ground into fine powders before they are transformed into useful products. The processing of the powder is itself a major undertaking, involving steps such as milling and sizing to produce a powder with the desired particle size, shape, and particle-size distribution. Either physical flaws (i.e., porosity or inclusions) or chemical flaws (i.e., unwanted or unknown chemical elements that combine to produce second phases such as liquids) can occur at any of these stages and may be difficult, if not impossible, to correct or eliminate. It must be stressed that the microstructure of the final ceramic product is dependent on the range of particle sizes and the inclusion of impurities in the starting materials. Clean rooms are evidence of the importance of cleanliness in the processing of advanced ceramics. We also know that the microstructure of any material determines the material's properties. Thus, much importance has been placed on developing low-technology techniques that make the conversion of mineral ores into industrial ceramic products more efficient with the expenditure of less energy and with ever-lower temperatures. Some advanced ceramics require that the powders used in the process be produced by chemical means, which allows greater control over both impurities and the final grain structure.

7.3.3.1 Chemical processes. ***Sol-gel*** processing of ceramics and glass is a chemical process for producing powders for use in advanced ceramics. This process and chemical-vapor-deposition techniques (discussed later) provide the means for designing and controlling the composition and structure of a ceramic at the molecular level. Their benefits include increased purity, lower processing temperatures, finer grain size, and more homogeneous microstructures. To gain further understanding of these techniques, a brief description of some terms is mandated. A ***sol*** is a stable dispersion in a liquid of particles less than 0.1 μm in diameter. In more general terms, it refers to a mixture of solid colloidal particles in a liquid. A ***colloidal*** substance is made up of very small particles that remain in suspension in a liquid for some time, without settling. When a sol loses liquid, it becomes a gel. ***Gels*** are noncrystalline solids formed by chemical reaction rather than by melting. Described in other terms, gels are a jellylike substance formed by the coagulation of a sol (causing a liquid to become a soft, semisolid mass). Through a chemical reaction, the sol-gel process turns the solution of organometallic compounds (sol) into a gel by the formation of polymerlike bonds. The final step is raising the temperature to convert the dehydrated gel into a ceramic material. Such temperatures are much lower (less than 900°C) than those used in the conventional processes of pressing, casting, or plastic forming, which involve high-temperature sintering. Applications for sol-gel processing include preparation of glasses for coatings, optical fibers, fabrication of glass ceramics, forming of oxide matrices for fiber composites, coatings on fibers to lessen the interfacial bonding between the fibers and the matrix to increase the toughness of reinforced fiber composites, and monolithic ceramics. Sol-gel and reaction-bonded ceramic processes are low-temperature, relatively high-energy processes that produce many advanced ceramic materials without the many difficulties (i.e., chemical interactions and residual stresses) associated with the use of high-temperature processes. Figure 7-16 contrasts them with chemically bonded ceramics (CBCs) such as cement and concrete, discussed later in this module.

As mentioned above, ***gelcasting*** is a new ceramic-forming process in which a slurry (a fluid suspension) of ceramic powders in a solution of organic monomers is cast in a mold. The monomer mixture is polymerized in the mold to form gelled parts. Both monolithic and composite ceramic parts, and complex-shaped and near-net-shaped parts, that require minimal final finishing are produced. Figure 7-17(a) shows a gelcasting process flowchart, and Figure 7-17(b) shows an application as well as a list of some materials that are gelcast.

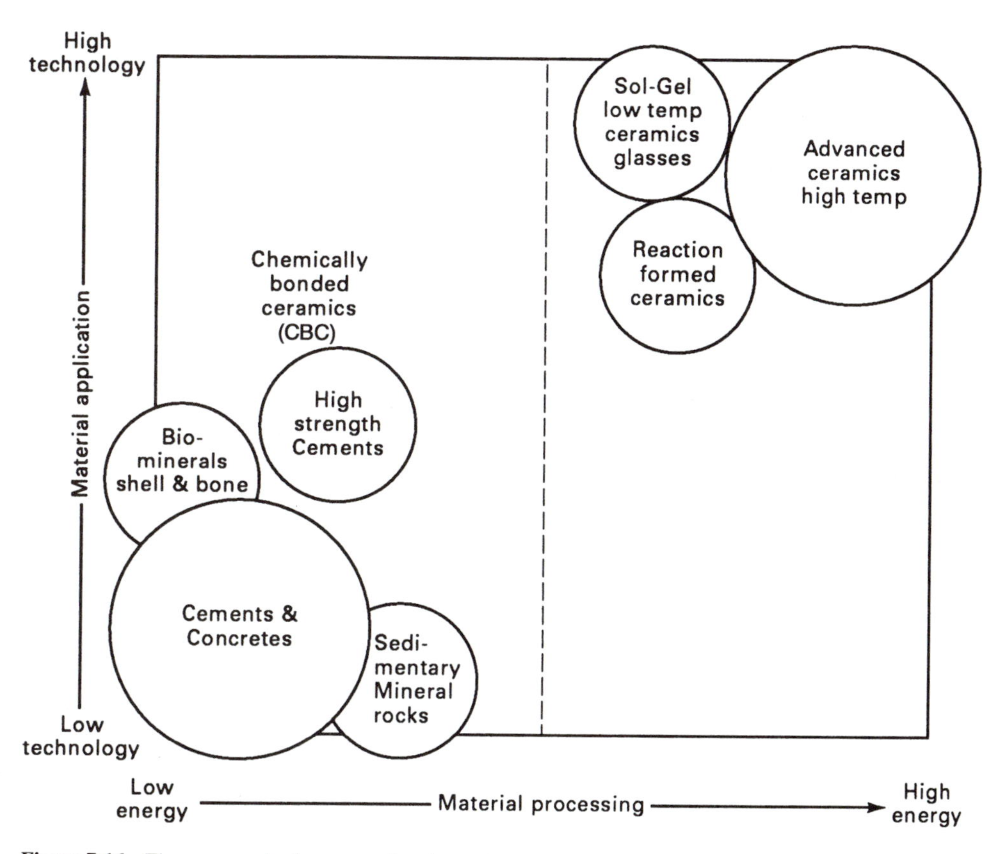

Figure 7-16 The compromise between using the world's materials for an ever-increasing number of technological materials applications and the need to reduce the energy required to process them. The higher the need for complex technology in the processing, the more energy is consumed. An underlying theme in this graphic is humankind's never-ending search to find the methods that nature uses to accomplish many of the technological goals that humans attempt to duplicate. (JME.)

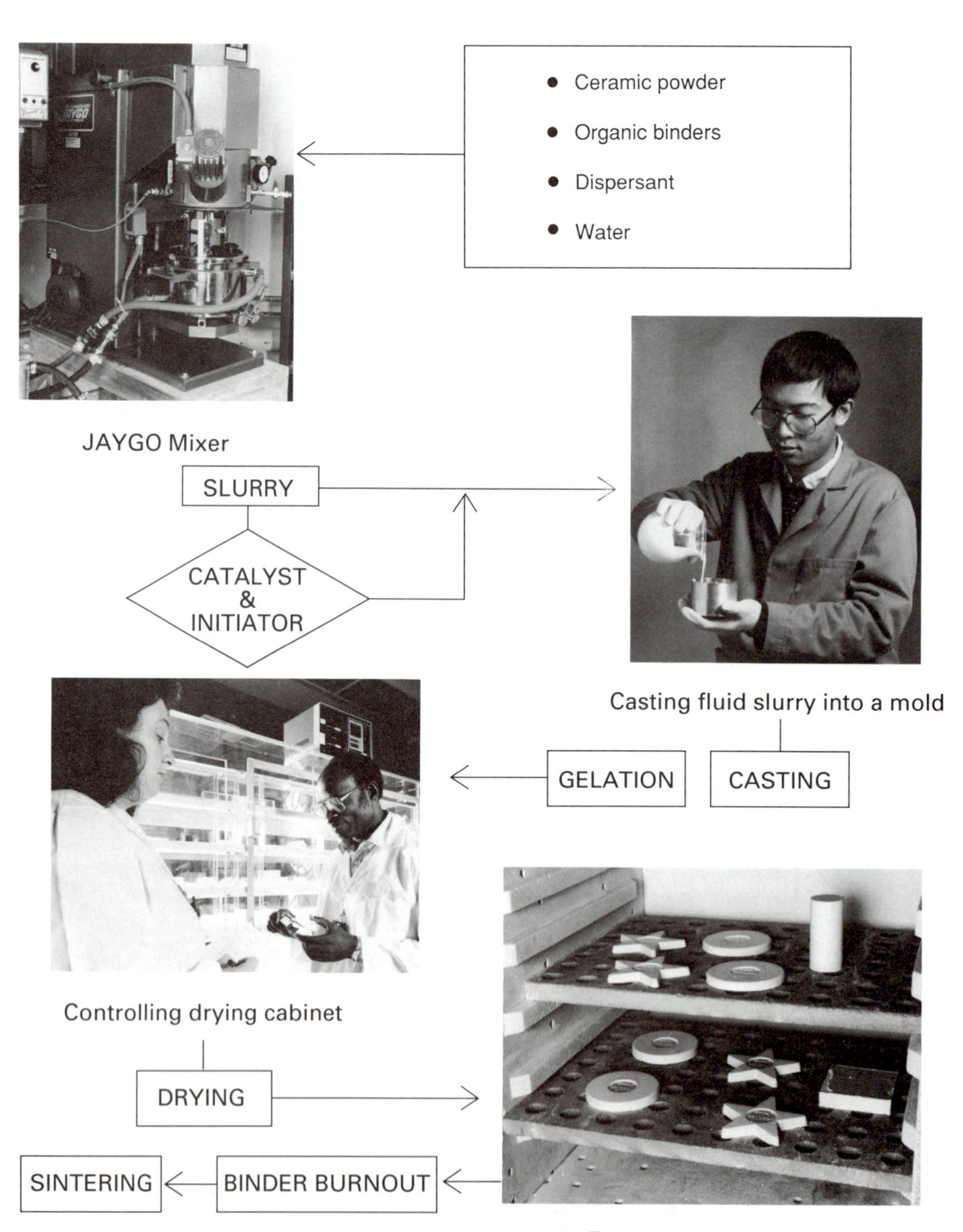
Ceramic powder
Organic binders
Dispersant
Water
JAYGO Mixer
SLURRY
CATALYST & INITIATOR
Casting fluid slurry into a mold
GELATION
CASTING
Controlling drying cabinet
DRYING
SINTERING
BINDER BURNOUT
Furnace with parts in it

(a)

Gelcasting Process Applications

Ceramic gears (Alumina)

Compound curves with varied thickness

List of Materials gelcast

Monoliths

- Alumina
- Silicon
- Silicon carbide
- Zirconia
- Fused silica
- Sialon

Composites

- Nicalon fiber-reinforced reaction-bonded silicon nitride
- Alumina-zirconia

Figure 7-17 Gelcasting. (a) is a graphical flow chart for the Gelcasting process and (b) contains pictures of some spplications for the gelcasting process along with a partial listing of materials that could be selected for their manufacture. (ORNL)

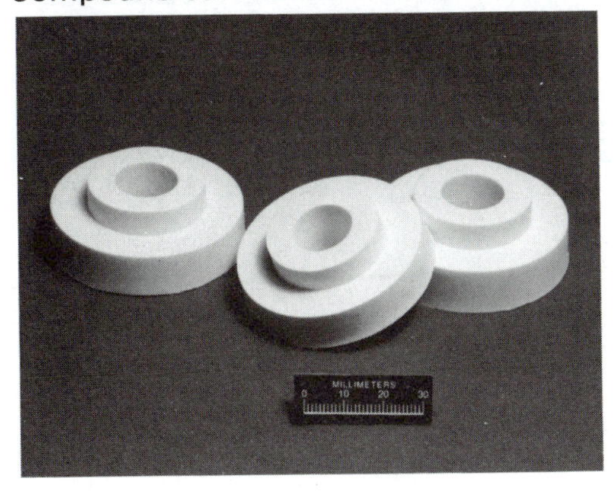

Tapered two-step doughnut

Ceramic rotor fired (Alumina)

(b)

The formation of the microstructures of ceramic materials is the result of chemical changes, interactions, or reactions. Chemically bonded ceramics (CBCs) can be distinguished by the manner in which consolidation of the components is brought about. In one process, consolidation results from sintering (densification) or fusion. In another, consolidation results from cementation by chemical reaction and bond formation. Figure 7-18 is a graphical representation of these two consolidation processes. Traditional sintering, depicted in Figure 7-18(a), is a solid-state diffusion process that results in shrinkage; Figure 7-18(b) illustrates that the volume between reactants and products can be conserved, resulting in net- or near-net-shape forming. Generally, these chemical processes take place when the compacted powders are heated to their sintering temperature. On other occasions the compacted powders react with a gas or liquid (see Section 7.3.4). ***Reaction-bonded silicon carbide*** (SiC), or reaction-sintered SiC, is one such process. A powder mixture of Si and C is formed into the desired shape and exposed at high temperature to Si vapor or molten Si. The reaction with Si produces SiC, which bonds to the original SiC compact, and the excess Si infiltrates the compact's pores at a controlled rate. The result is a nonporous composite (SiC/C) with varying degrees of strength and elasticity. By using carbon fibers to form laminae of fibers woven within the molten Si, a range of fiber composites can be developed having the characteristics desired.

A similar process is the ***reaction-bonded*** $\boldsymbol{Si_3N_4}$ (RBSN) process, which starts with a compacted shape of Si in a furnace under an atmosphere of N, N/He, or N/H gas. Heated initially to about 1250°C, the N permeates the compact, forming Si_3N_4. The temperature is then raised to just below the melting temperature of Si to increase the reaction rate. After about one week in the furnace, the process is completed, producing a ceramic that has

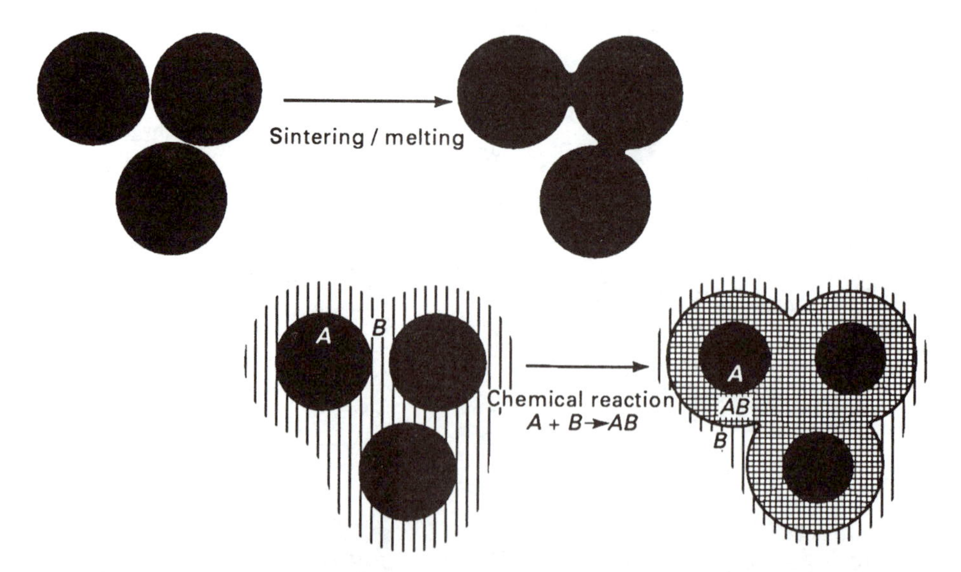

Figure 7-18 How the densification of particles is accomplished under conventional sintering (a) and reaction bonding sintering (b) processes. Conventional sintering depends, in the main, on high temperature (over 2000°C) to cause solid-state diffusion, whereas reaction bonding sintering requires temperatures around 1400°C. (JME.)

good dimensional stability, thermal-shock resistance, and thermal conductivity. Its one disadvantage is its low oxidation resistance due to its interconnected porosity (open pores).

Oxidation describes the chemical reaction of oxygen when it combines with another element. If sufficient oxygen is present, it will react with the surface of the silicon nitride, producing silicon dioxide (SiO_2), which forms a protective layer at the surface that prevents further reaction. This is known as ***passive oxidation.*** If not enough oxygen is present, the gas silicon monoxide (SiO) is formed. Known as ***active oxidation,*** this type of oxidation can be continuous, resulting in complete destruction of the ceramic. Any oxygen-poor environment, such as outer space, would be conducive to such attack. Other factors, such as temperature and porosity, influence the degree of reaction. By using oxidation initially, the surface may be coated with a layer of SiO_2, which will protect the interior from further oxidation attack. This is known as *flash oxidation,* which is useful for protecting $MoSi_2$ heating elements.

7.3.3.2 Melt processing. Most glasses and many ceramics are melted before being processed. Abrasives, glass shapes, and bricks of Al_2O_3 are made by casting from a melt. Glass fibers for fiberglass composites are drawn or spun from a molten state. Molten droplets made from ceramic particles are sprayed onto surfaces through a variety of spray techniques that provide corrosion, wear, and thermal protection to a wide variety of metal, plastic, and ceramic products.

Thermal spray is a generic term for a group of coating processes used to apply metallic or nonmetallic coatings. ***Plasma-arc spray, flame spray,*** and ***electric-arc spray*** all heat the coating material (in powder, wire, or rod form) to a molten or semimolten state. The heated particles are accelerated and propelled toward a prepared surface by process gases or by atomizing jets. On impact, a bond forms with the surface and subsequent particles cause a controlled thickness buildup. Plasma, a high-temperature, ionized, electrically conducting gas, is used for spraying ceramic coatings. Flame and plasma-arc spray guns melt ceramic particles and spray the molten droplets at high velocities. One disadvantage of such a process is that even ceramics that are to be coated often require preheating to avoid thermal shock due to the high temperatures involved. Numerous techniques are used to melt ceramic particles and spray the melt onto a surface. Newer methods are used to produce uniform coatings, with "all porosity effectively sealed," that can withstand continuous exposure to temperatures of 540°C. Such coatings are referred to as ***thermal-barrier coatings*** (TBCs). (See Section 7.10.) TBCs on metal components of gas turbines have low thermal conductivity and low emissivity. Materials such as ZrO_2 applied as a spray are used for such coatings because, in addition to possessing the properties noted above, such materials have a coefficient of thermal expansion similar to that of the substrate metal, which allows the coating to adhere to the metal as the metal undergoes changes in temperature. Designers working on ways to increase gas-turbine operating temperatures have concentrated their efforts on ways to improve superalloy-turbine-blade cooling with TBCs. Their goal is to develop an oxide coating that has thermal stability at temperatures up to 1650°C. Competing developmental efforts are devoted to perfecting a ceramic-matrix-reinforced composite that will be capable of replacing superalloys in such turbine applications. In the meantime, the metal components must be protected from higher and higher operating temperatures by application of various thermal coatings using several coating processes, which are themselves under constant refinement. A ***high-velocity oxyfuel*** (HVOF) system can be used to apply denser and thicker coatings than is possible using a plasma spray. Cored wires made by extruded powder-filled metal tubes are used with arc-spray applications to deposit coatings of alloys and material combinations heretofore considered unsprayable. Computer-controlled spray systems now permit greater control over the process and provide greater safety to personnel in an inherently hazardous environment. The major thermal-spray coating processes are *plasma spray, wire flame spray, electric-arc spray, powder flame spray,* and *high-velocity oxyfuel* (HVOF). Without devoting time to a description of their operation, a few advantages of each process will be noted. The ***plasma spray*** provides a coating that is dense and bonds strongly to the substrate. The ***wire flame*** process is possibly the best overall for applying a coating rapidly at lowest cost. ***Wire-arc spraying*** is used to deposit corrosion-inhibiting zinc on metal structures and aluminum coating on computer components to suppress electromagnetic interference. A new application is the spraying of metal on foam without damaging the foam. The metal to be sprayed is placed in an electric arc drawn between the tips of two wires of the metal to be deposited. The metal is melted and carried toward the substrate by a high-pressure flow of inert gas, such as argon. In the past, thermally insulating foams were electroplated with metallic coats that protected them from damage by handling, seepage of air, or other reasons. ***Electroplating*** is more expensive. It is slow and involves toxic and polluting chemicals. Wire-spraying, on the other hand, exposes the substrate to less severe temperatures, ranging from about 38° to 149°C, than do other spray techniques. The ***electric-arc*** process provides flexible tailoring of the coating characteristics and is a good choice for coating large areas. ***Powder flame*** allows for the widest choice of coating materials and provides a good combination of wear and thermal properties. ***HVOF*** provides the highest bonding strength and the lowest porosity. Thick layers and a high-quality surface finish are characteristic of this method. It is well suited for the production of good-quality tungsten- and chromium-carbide coatings.

7.3.3.3 Producing single-crystal materials. *Single-crystal materials* solidified from a melt are developed to obtain materials with special properties not possessed by polycrystalline materials. Such materials are processed by several different techniques, of which two processes for forming such crystals are mentioned here. Figure 3-58 showed a typical application of this technique as it relates to metals. In the semiconductor industry, great advances have been made in growing silicon single crystals, which are now produced with few, if any, structure dislocations. An *edged-defined, film-fed growth method* (EFG) uses alumina with a molybdenum die that is immersed in molten alumina. The molten alumina is wicked to the top of the die, where it comes in contact with a seed crystal that is slowly pulled away to start the growth process. The growing crystal assumes the shape of the cross-section of the die. Plates, tubes, and various-sized filaments are produced using this method. For example, some continuous filaments are produced at a rate of 62 m/h. A second technique is the *heat exchanger* or *gradient-furnace,* which produces larger single-crystal components. A seed crystal is kept cool by helium flow in a crucible filled with molten alumina. The temperature of both the melt and the seed crystal is controlled such that the contents of the crucible grow into a single crystal.

7.3.3.4 Vapor processing.

Vapor processing involves heating a solid substance to a temperature that transforms the solid into a vapor, which is then deposited onto a surface. When the vapor contacts a cold substrate or when a gas reacts on contact with a hot substrate, the solid produced on that surface solidifies so quickly that the atoms fail to form a crystalline structure. Such noncrystalline, fine-grained, nonporous coatings are difficult, if not impossible, to produce by other means. ***Physical vapor deposition*** (PVD), ***chemical vapor deposition*** (CVD), ***chemical vapor infiltration*** (CVI), ***metalloorganic chemical vapor deposition, molecular beam epitaxy*** (MBE), ***liquid-phase epitaxy*** (LPE), ***sputtering,*** and ***ion plating*** are some of the methods that use this technology. As a group they represent a wide range of ***surface modification*** techniques to achieve surface properties in a controlled manner for ***engineered materials*** needed by today's advanced technologies. In addition to providing high-temperature materials for turbine blades or metallic coatings for plastic parts, these techniques are now involved in producing high-quality optical glass for fiber-optics communications as well as bulk forms in single shapes with structures developed on an atomic scale. These new materials that have designed structures are known as ***artificially structured materials.*** Until recently most developments in this field were somewhat restricted to use in the semiconductor industry, but as each day passes, new applications are found for such materials.

PVD vaporizes a target material by using a high-energy source such as ions or electrons. The vapor is deposited atom by atom on a substrate, building up the required layers of atoms. ***Laser ablation and deposition*** uses a pulsed laser process to produce high-temperature-superconducting thin films. This new process is also based on PVD principles. Using multiple target pellets, each containing the required chemical composition, a thin film is reproduced with the exact composition. Wear problems are constantly being solved by using thin-film coatings made of zirconium, chromium, and carbon. Before a particular coating is used, a complete tribological profile of the conditions in which the part is to operate must be determined. Recently, titanium carbonitride (TiCN) has been replacing titanium nitride (TiN) and chromium nitride (CrN) coatings on ***cutting*** and ***forming tools.*** Such coatings are applied in thicknesses of 1 to 4 μm by a low-temperature, reactive ion plating PVD process. Figure 7-19(a) shows a schematic of the process and Figure 7-19(b) shows a photomicrograph of the edge view of the coating and substrate. Such coatings [Figure 7-19(c)] combat abrasive wear and adhesive welding when metals such as stainless steel, cast iron, brass, titanium, and aluminum alloys are machined. The TiCN coating, for example, is stable to temperatures of 400°C with a hardness of 3000 HV (25-g load). Another PVD-based process is the electron-beam vacuum evaporation process, developed primarily to coat turbine blades. This method generally produces a higher-quality coating than that produced by its competitor—the plasma spray technique—with PVD's cost advantage. Two other versions of the electron-beam evaporation method—the arc method and the plasma-assisted evaporation method—are finding many applications throughout history.

Chemical vapor deposition (CVD) is a process in which a mixture of gases is passed across a heated surface. The temperature causes the gases to react or decompose to form a solid on contacting the surface. In an example taken from the literature, a TiN coating that protects a surface from wear is produced by CVD with a mixture of gases, $TiCl_4$–N_2–H_2, heated to a deposition temperature between 900° and 1000°C. CVD can be used to

(a)

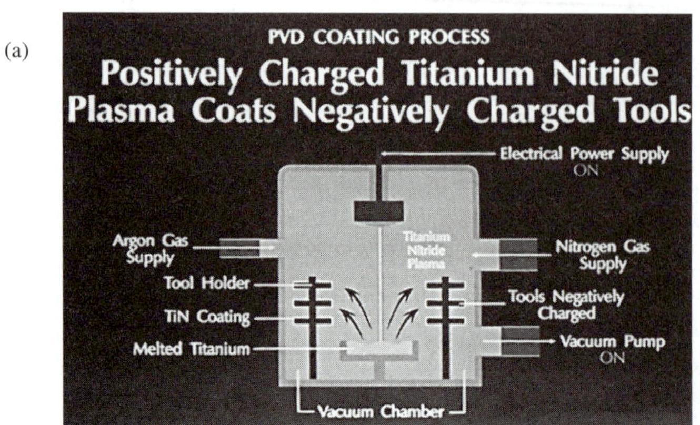

(b)

(c)

Figure 7-19 (a) Physical vapor deposition (PVD) process used for thin film ceramic coatings. (b) Photomicrograph showing edge view with uniformity of 1×10^{-4} in. thin coating of titanium nitride (TiN) on high-speed steel substrate. (c) TiN and other thin film coatings, sometimes as many as 13 layers, are used on cutting tools and tool inserts.

manufacture parts such as a boron-nitride (BN) crucible with wall thicknesses of several millimeters. Figure 7-20 is a micrograph of a boron-nitride layer deposited on a substrate of magnesium oxide (MgO). In the composites field, CVD is used to coat fibers that will be embedded in a metal or ceramic matrix to increase the toughness of the fiber composite. This toughening mechanism is discussed in Section 7.4. For example, silicon fibers are coated with carbon before they are used in a matrix of silicon carbide. This coating increases the toughness of the composite by reducing the interfacial bonding strength between the fibers and the matrix material. Doing so allows the fibers to debond from the matrix materials and pull out before breaking. All of these displacements require energy, decreasing the probability of catastrophic failure from a single mechanical overload. Figure 7-21 shows a toughness model sketch (a) and a micrograph (b) of fiber bridging and pullout as fibers resist the propagation of a crack.

Synthetic diamonds were first made in 1950 by General Electric using extremely high pressures and temperatures. The same technology is in use today to make diamonds.

Figure 7-20 A micrograph of a layer of boron nitride (BN) deposited on a substrate of magnesium oxide (MgO)—magnification 7,200,000×. (ORNL.)

Toughness Model

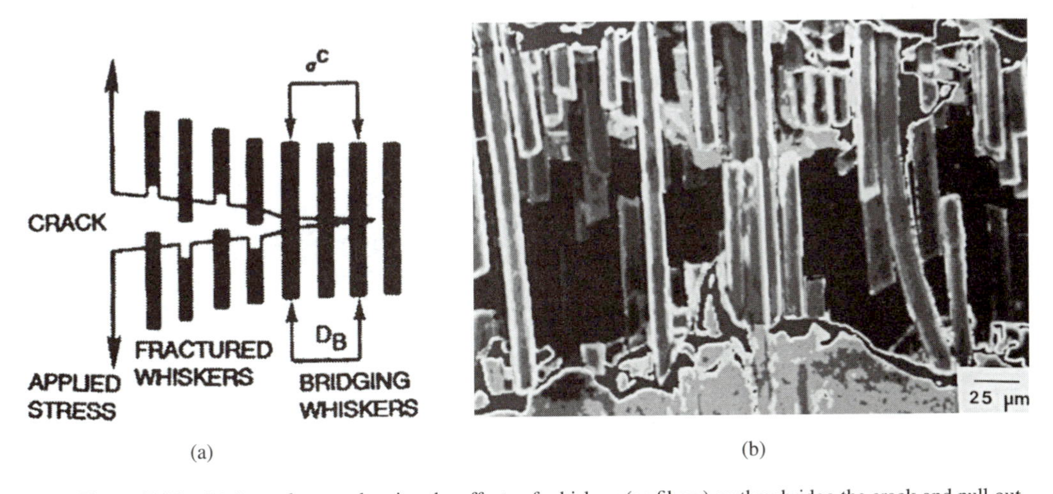

Figure 7-21 (a) A crack zone showing the effects of whiskers (or fibers) as they bridge the crack and pull out of the surrounding matrix material before fracturing. (b) Photomicrograph of a fractured, fiber-reinforced Lanxide ceramic composite showing the fibers spanning the fracture to resist crack propagation. By designing an interfacial bond with the correct strength between the fibers and their matrix, the fibers can expend energy overcoming friction and pulling out of the matrix, thus bridging the fracture instead of fracturing and assisting in crack propagation.

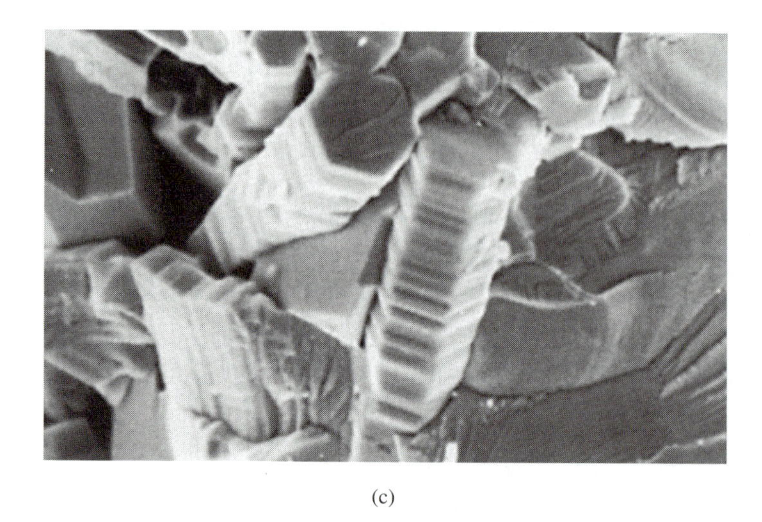

(c)

Figure 7-21 (c) Ceramic composite fracture surface. (ORNL and Lanxide.)

A more recent development is the making of ***synthetic diamond films*** and coatings using low-pressure processes. One of many vapor deposition techniques is CVD, which deposits the film from a gaseous mixture (CH_4–H_2) onto a substrate. Numerous substrates, such as molybdenum, silicon, ceramics, and natural diamond, have been used to grow the films. The largest obstacle to successful film growth is the need to heat any substrate to temperatures above 600°C. Hydrocarbon gases, of which methane is most common, are used as the precursor gas. Various methods for forming the vapor are in use, such as microwave activation and thermal-activation and plasma-jet processes. Figure 7-22 shows a simple sketch of the process, a photograph of a typical reactor, and several grades of diamond film. Raman spectroscopy is used for detection of diamond; it consists of looking for a single, well-defined peak that is very distinctive for the diamond bond. ***Natural diamond*** has a set of unique properties. It has the highest hardness reading on the Mohs scale. The elastic modulus is about five times as large as most steels, at 160×10^6 psi. Its tensile strength is higher than that of steel but lower than that of ceramic fibers, which can go as high as 10×10^6 psi. Natural diamond's 0.1 coefficient of friction ranks about the lowest for a solid material. The coefficient of thermal expansion for synthetic diamond is only about 0.8×10^{-6} °C^{-1}, a consequence of its strong covalent bonds. Electrical resistivity is about $10^{14}\ \Omega \cdot$ cm. If synthetic diamond could be produced in sufficient quantities at an affordable price, it would replace silicon as a semiconductor material.

A comparison of the various forms of diamond is interesting. Natural diamond is a single crystal made up of 99% carbon 12, with 1% carbon 13 isotope. General Electric's synthetic diamond contains more atoms per cubic centimeter than does natural diamond and can be made to any desired C_{12}/C_{13} ratio. CVD polycrystalline natural diamond has no binder, whereas polycrystalline sintered diamond contains up to 10% binder. ***Polycrystalline diamond (PCD)*** and ***polycrystalline cubic boron-nitride (PCBN),*** referred to as ***superabrasives,*** are finding more applications in the turning of hard materials, once the sole province of the grinding field. Using tipped turning inserts from the materials noted

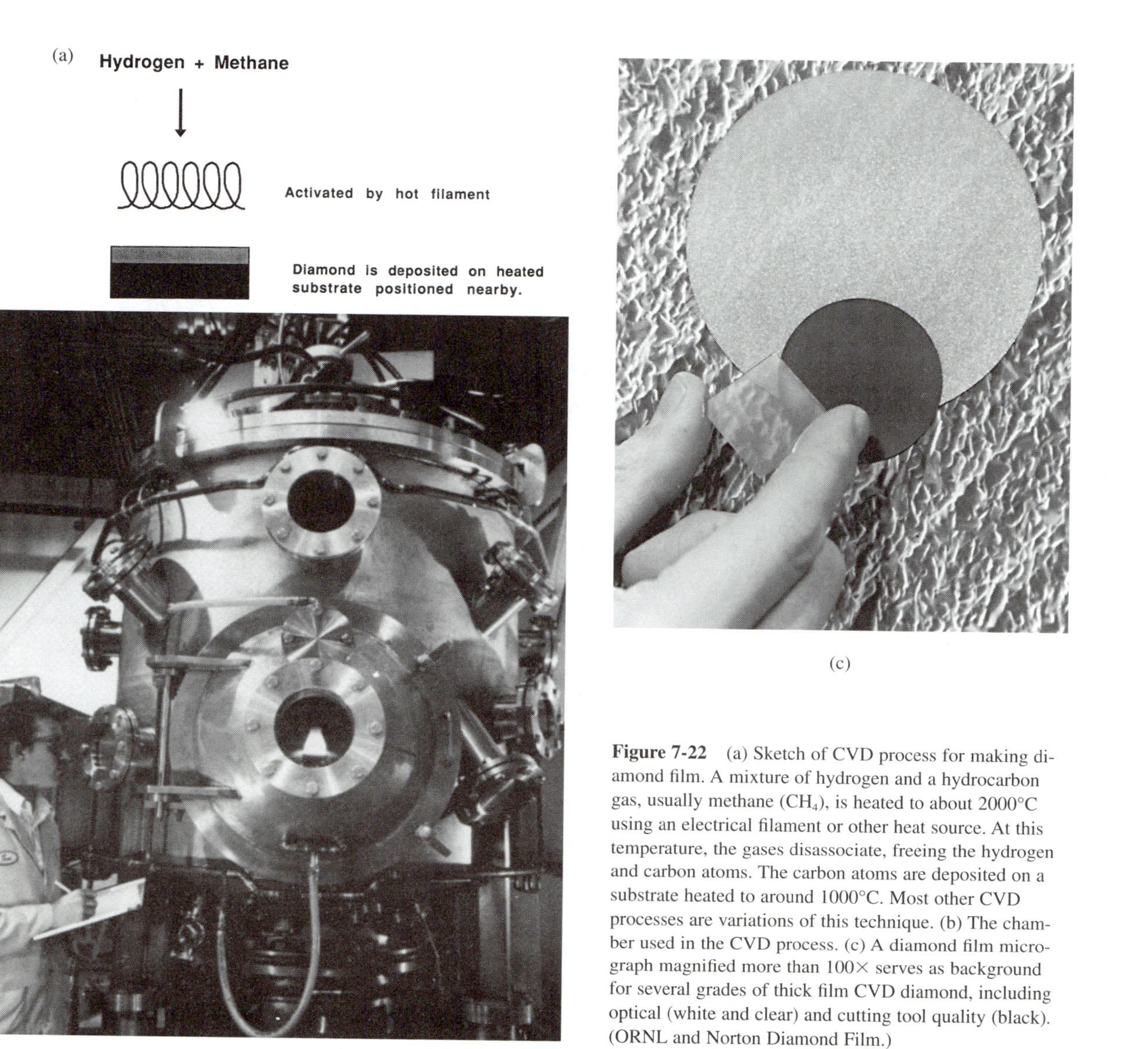

Figure 7-22 (a) Sketch of CVD process for making diamond film. A mixture of hydrogen and a hydrocarbon gas, usually methane (CH_4), is heated to about 2000°C using an electrical filament or other heat source. At this temperature, the gases disassociate, freeing the hydrogen and carbon atoms. The carbon atoms are deposited on a substrate heated to around 1000°C. Most other CVD processes are variations of this technique. (b) The chamber used in the CVD process. (c) A diamond film micrograph magnified more than 100× serves as background for several grades of thick film CVD diamond, including optical (white and clear) and cutting tool quality (black). (ORNL and Norton Diamond Film.)

above, material removal rates were 4 to 10 times greater than rates for grinding with cleaner surfaces. As with all materials, diamond-film properties can vary significantly, not only with deposition techniques but with substrate temperature, precursor-gas composition, and other factors that can or cannot be controlled in the process. Current research is focused on producing diamond films at lower cost and lower deposition temperatures, with larger film areas (present limits are 100 mm square and about 1 mm thick) and better adhesion between the film and the substrate.* Applications for diamond films are as cutting-tool inserts, wear surfaces for dies, heat sinks for electronics, and infrared (IR) and microwave windows (diamond is transparent to both radiations).

Chemical vapor infiltration (CVI) is somewhat related to CVD, the major difference being that the deposition occurs inside a porous preform rather than only on a surface. CVI originated in efforts to densify graphite bodies by infiltration with carbon. Infiltration is the densification of a porous preform by filling the pores with liquid, vapor, polymer, or a sol. The CVD-type coating grows with continued deposition to form the composite matrix. A wide variety of matrix materials are used, including borides, carbides, nitrides, and oxides. A great advantage of this process is that it allows the formation of high-melting-point materials such as TiB_2 ($T_M = 3200°C$) at relatively low temperatures, which reduces the risk of damage to the fiber reinforcement. Also, CVI does not use high pressures, which translates into less residual stress in the final composite materials. CVI is now used to produce carbon–carbon composites as well as ceramic–ceramic composites (see Module 8 for additional details). When fabricating fiber-reinforced ceramic composites such as SiC high-modulus ceramic fibers with diameters of only 10 μm or less in a matrix of dense alumina, the fragile fibers cannot withstand the high stresses and handling of more conventional processes. With its low stress and gaseous precursors, CVI can infiltrate around the bundles of continuous fibers and deposit the matrix material around the fibers, filling in all pores and thus encapsulating the fibers with the matrix material and knitting them together in a strong interfacial bond. The coating of the fibers is a slow molecular process that builds up molecule by molecule. Figure 7-23 shows (a) a sketch of the CVI furnace, (b) a micrograph of SiC fibers reinforcing a SiC matrix composite material, and (c) a dramatic demonstration of the increased toughness of reinforced CVI ceramic materials. (Liquid-phase epitaxy and molecular-beam epitaxy methods are discussed in Section 9.9.)

Reactive melt infiltration is a new process that produces silicon-carbide–based ceramics and composites faster and more economically than do processes such as CVI or hot isostatic pressing (HIP). In comparison to CVI, this process takes minutes instead of days and costs about half as much. In the reactive melt infiltration process, a microporous carbon preform is infiltrated with molten silicon or a molten silicon alloy. The liquid and solid react to form a solid ceramic or composite. If the end product is to be a composite, the preform contains fibers. Products can be formed into complex shapes with full density, controlled microstructures, and tailored thermal and mechanical properties. A composite material consisting of a silicon-carbide matrix reinforced by silicon-carbide fibers has been produced. This ability to produce a composite material of complex shape and nearly final dimensions makes way for the production of silicon-carbide cutting tools.

*See Experiments & Demonstrations in Ceramics at the end of this module.

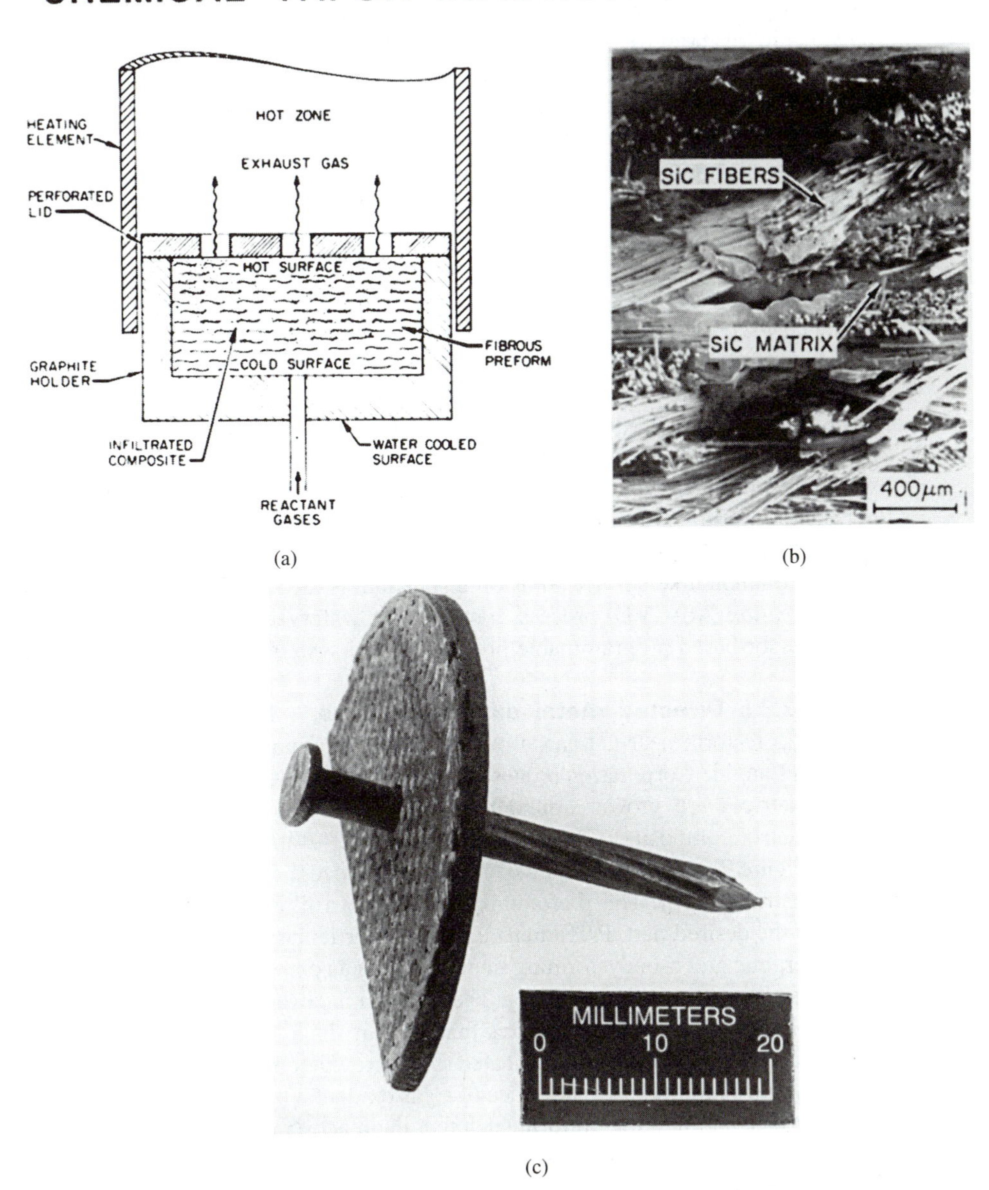

Figure 7-23 (a) Forced chemical vapor infiltration (CVI) process. (b) Photomicrograph of SiC ceramic produced by CVI. (c) Dramatic demonstration of the toughness obtained in ceramics as a result of the reinforcement achieved in CVI. (ORNL.)

Ion plating and *sputtering* involve plasmas. A ***plasma*** is a cloud of gas at high temperature made up of ions and electrons. Extremely high temperatures in the plasma can be reached because the cloud of gas is electrically conductive. This allows electrical heating (i.e., radio-frequency induction or direct contact with a metal electrode that supplies electrical current). The various factors involved in maintaining a plasma are somewhat complex, but in this brief treatment of the subject, they may be said to involve the temperatures of the ions in the plasma, energy input (heat), and the gas pressure. At high pressures, the plasma temperature can reach several thousands of degrees Celsius. Under these conditions, a powdered ceramic can be fed in the plasma, producing tiny droplets that can splatter onto a substrate as in plasma-arc spraying. At lower pressures, the gas atoms in the plasma remain cold and permit high electron temperatures to produce chemical reactions in the cold gas, the products of which can be deposited on a substrate. ***Sputtering*** is a basic coating process that uses a plasma, a target positioned above the substrate that is to be coated, and a sputtering gas. The target acts as the negative electrode, and the substrate acts as the positive electrode. The positive ions (usually argon ions) in the plasma are accelerated toward the target at high speed, resulting in the knocking off of desired atoms from the target, which are subsequently deposited on the substrate. This is known as the sputtering action. There are numerous variations to this basic process that depend, in part, on the type of substrates involved, how the coating atoms are introduced, the manner in which the plasma is maintained, or the differences in the applied voltages between the target and the substrate. One final comment is a reminder that the temperature of the substrate is extremely important to the final coating achieved. A recent innovative process for depositing diamondlike surface films on a substrate is the *microwave-assisted chemical vapor deposition* (MACVD) process. It can handle wafers up to 4 in. in diameter with a microprocessor-based programmable process control system.

7.3.3.5 Directed metal oxygen process. The reaction of a molten metal with a gas to form near-net-shape metal and ceramic-matrix composite parts is exemplified by a Lanxide Corporation process. In the directed metal oxidation process, Dimox, ceramic matrices are grown around preplaced fibers (reinforcements) to produce a ceramic–matrix composite (CMC) to net- or near-net-shape without limitations on size or shape. Figure 7-24 is a schematic of the Dimox process showing an example of unidirectional matrix growth. The preform is shaped into a filler preform of the same size and shape as the desired part. Preforms are made of particulate, which can be produced by any of the ceramic green-body-forming methods, such as pressing, isostatic pressing, slip casting, extrusion, or injection molding. The preform is covered with a special barrier material that limits matrix growth, ensuring that the part shape matches that of the preform. The parent alloy and preform are then heated to the growth temperature, at which the metal begins to oxidize rapidly in the presence of a gas oxidant. Oxidation occurs outward from the metal surface and into the preform such that the reaction product becomes the matrix surrounding the reinforcement in the preform.

Oxidation of a metal generally forms a solid ceramic layer that limits further reaction. However, in one method, additives such as Mg or Si promote wetting of the ceramic by the molten aluminum alloy and reduce the stability of grain boundaries in the ceramic,

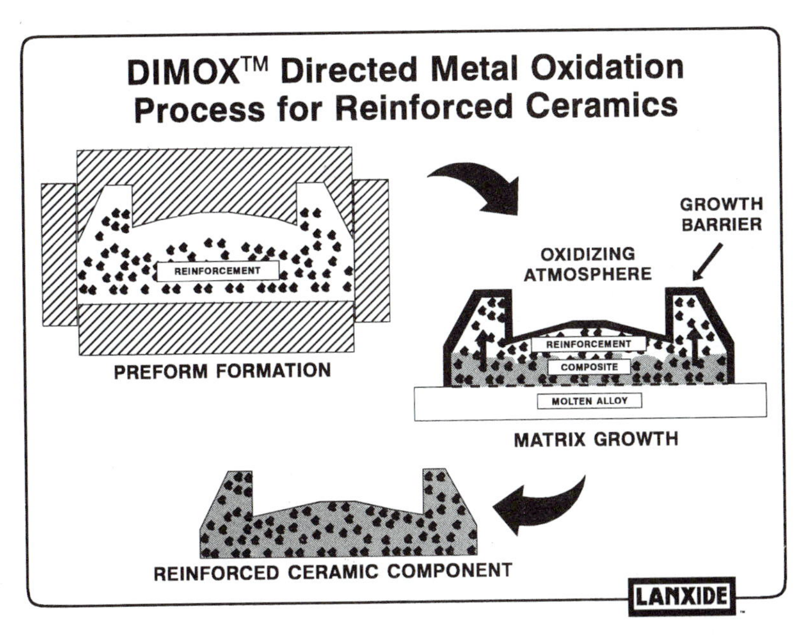

Figure 7-24 DIMOX® process, demonstrating preform infiltration technology with unidirectional growth. A filler preform coated with a growth barrier is placed on top of a solid metal ingot in a refractory container. Temperatures (850°–1300°C) initiate growth, which proceeds through the filler until either the molten metal is consumed or the growth reaches the barrier. No shrinkage occurs.

which allows sustained reaction of the metal with the gas oxidant (oxygen). The metal is continuously made available to react with the gas oxidant, providing a growth rate that is independent of the thickness. In other words, the molten metal is drawn through its own oxidation product. The growth of the matrix into the preform occurs with little, if any, change in dimensions. This characteristic sets this process apart from traditional ceramic processing, with its inherent densification shrinkage.

Figure 7-25 shows a photomicrograph of the microstructure of a CMC made by the Dimox process, composed of a matrix of Al_2O_3 reinforced by woven SiC fibers with some unreacted metal still visible. In low-temperature applications, the residual metal provides added toughness. For high-temperature use, that amount of metal can be removed, leaving a small amount of porosity, or treated to prevent melting or further oxidation during service. Figure 7-26 is a photograph of a Lanxide CMC specimen bar, consisting of an alumina matrix reinforced with CVD-coated SiC fibers, undergoing a flexure test. Table 7-4 lists some mechanical properties of SiC_p/Al_2O_3 CMC for use at high temperatures. The subscript "p" in the notation for the composite material refers to platelets or particles. Other abbreviations, "W" for whiskers and "f" for fibers, are used here.

Shape forming can be done by metal-shape replication and/or infiltration into a preform of the desired filler (reinforcement). Metal-shape replication used in making screw threads is shown in Figure 7-27. The shaped parent-metal cast or machined part is inserted

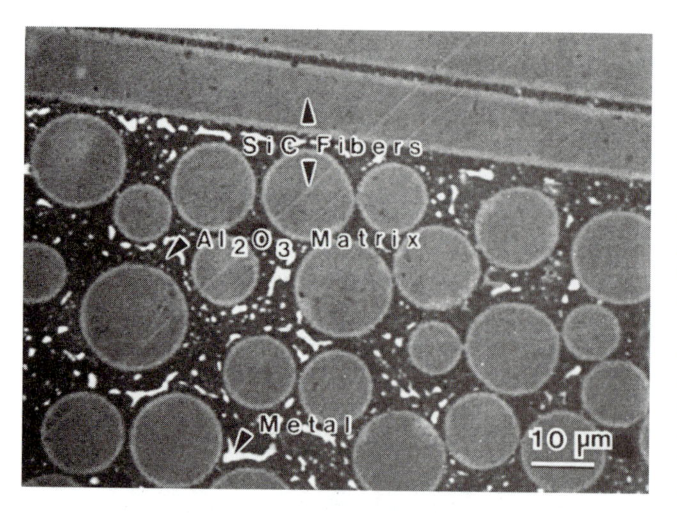

Figure 7-25 A micrograph of a typical microstructure of CMC and MMCs made by Lanxide's DIMOX process. Al_2O_3 reinforced by woven SiC fibers are shown with some unreacted metal also visible. Impurities in the uncoated fibers reacted with the molten metal to form a reaction layer on the fibers. In practice, fibers are given a protective coating prior to matrix formation, as found in continuous fiber ceramic composites (CFCC). (LANXIDE.)

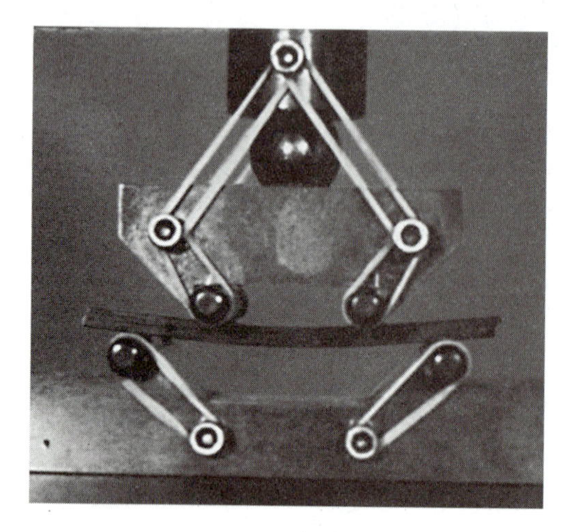

Figure 7-26 A photograph of an alumina composite containing CVD-coated Nicalon® fibers produced by the Lanxide DIMOX® process undergoing a four-point bend test. Despite a deflection greater than 1 mm, sufficient to fail an unreinforced ceramic catastrophically, the bar continues to support the load. (LANXIDE.)

into the central portion of the figure surrounded by the silicon-carbide bedding material. The growth will extend outward from the metal-shape preform toward the barrier bedding.

In the DIMOX™ process, the silicon-carbide bedding materials are incorporated into the ceramic matrix by the directed oxidation of molten aluminum alloy. The matrix reaction product occupies the spaces between the filler without displacing it, so that densification is achieved without shrinkage. Of course, for growth to occur, the filler used, be it fiber, whisker, particle, or platelet, must be compatible with the gas atmosphere and the parent metal. Coating of the filler materials is used to improve compatibility. Lanxide's PRIMEX™ process, a pressureless molten-metal-infiltration technique, is discussed in the composite module, Module 8. Both of these Lanxide processes employ bulk processing techniques that have been traditional in foundries and ceramic processing for many years.

TABLE 7-4 MECHANICAL PROPERTIES OF SiC_p/Al_2O_3 COMPOSITES PRODUCED VIA DIRECTED METAL OXIDATION (DIMOX™) PROCESS

Property	Degrees Celsius	90-X-008
Flex strength MPa (ksi)	25	179(26)
	1000	141(21)
	1400	155(22)
	1550	130(19)
Toughness (MPa · $m^{1/2}$)	25	5.5
	1000	4.0
	1400	3.0
	1550	3.8
Young's modulus (GPa)	25	391
Shear modulus (GPa)	25	153
Poisson's ratio		0.28
CTE (ppm/ · C)	25–1400	6.0
Thermal conductivity (W/m-K)	25	116
	500	54
	1000	31
Bulk density (g/cc)		3.30
Carbide loading (v/o)		73
Residual metal (v/o)		6

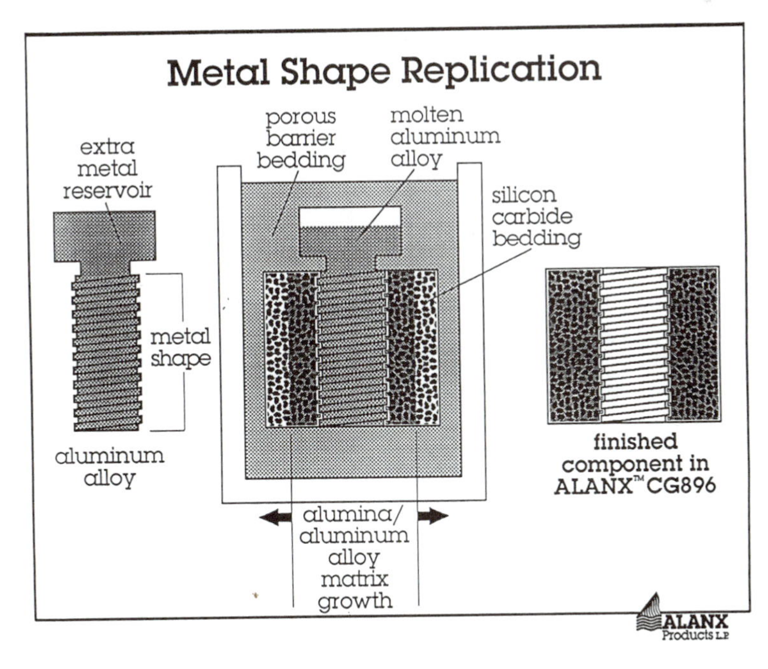

Figure 7-27 Using the Lanxide DIMOX process to demonstrate the making of screw threads with a ceramic composite material through metal shape replication. (ALANX.)

This is in contrast with the expensive equipment and materials encountered in the production of many other advanced ceramics.

7.3.3.6 Finishing (machining). In some instances, the shaped compact (preform) has not acquired the final shape or met the tolerances specified. Machining of the surfaces of ceramics is done to meet dimensional tolerances, improve surface finish, or remove surface flaws. It is to be assumed that all machining will introduce flaws of some size on the surface of ceramic parts. The machining of fine-grained ceramics such as Si_3N_4 significantly reduces their strength—more than that of ceramics of large grain size. After forming, the compact can be "green" machined (i.e., machined before densification). In other instances, final machining is required after densification.

Regardless of the type of machining (i.e., turning, milling, or wheel or profile grinding) or when the machining is used, tool materials must be made from hard, wear-resistant materials to avoid damaging or lessening the toughness of the relatively fragile ceramic workpiece. Internal flaws of less than 100 μm or surface flaws of less than 50 μm will produce fractures due to stresses in heat-engine ceramic parts. Although more costly initially, ***diamond inserts*** in machining tools usually result in significant cost savings in terms of reduced risk of damage to the workpiece, reduced time expended in tool changes, greater cutting speeds, and increased tool life. Figure 7-28 illustrates the phenomenal rise in cutting-tool speed over the past 200 years as a result of the research and development of metals and ceramics. Figure 7-29 is provided to emphasize the advantage of synthetic diamond as a cutting tool, particularly for ceramics.

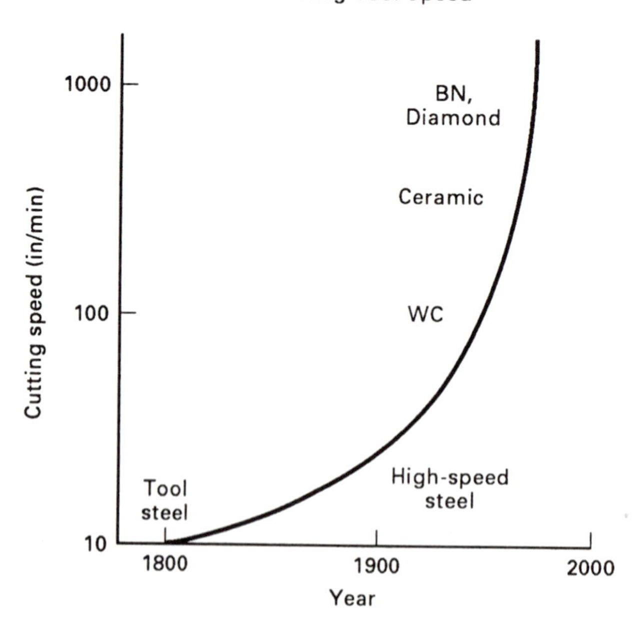

Figure 7-28 Plot of ***cutting speed*** versus time (years), showing the exponential rise in speed as a result of the advances in materials development.

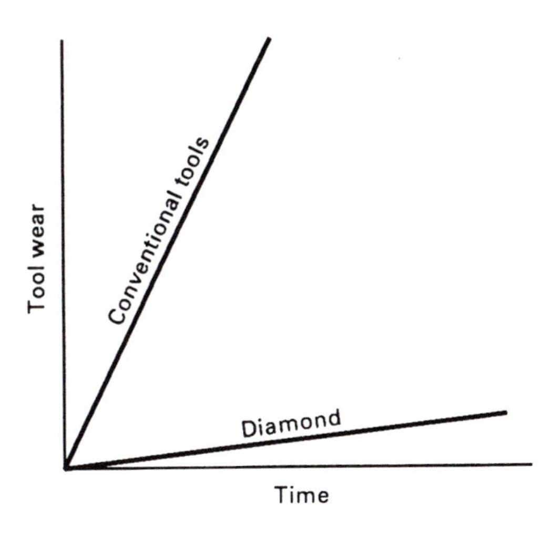

Figure 7-29 Rough plot of tool wear versus time, emphasizing the drastic reduction in tool wear as a result of the introduction of diamond-coated tools and inserts.

7.4 IMPROVING TOUGHNESS

Toughness is an indication of how much resistance a material offers to a crack starting to grow through it. Most materials either contain cracks already or have structural flaws that, under stress, can readily turn into cracks. The work to be done in propagating a crack, also known as the ***work of fracture,*** is measured by the energy absorbed in creating a unit area of new crack ahead of an existing crack tip. This work is a measure of toughness (G_C) and is expressed in joules/m^2. Figure 7-30 shows a graph of specific strength versus work of fracture for some typical ceramics. Work of fracture is the preferred measurement of toughness for fiber-reinforced ceramics.

Another procedure for assessing toughness is to measure the area under the stress–strain curve up to the point of fracture. This procedure is more suitable for a brittle material due to the difficulty in identifying the area under the curve that represents the energy attributable to plastic deformation [see Figure 7-31(b)]. The units for expressing toughness by this procedure are joule/m^2. ***Impact*** is defined as the sudden application of a load confined to a localized area of a material. Exemplified by the striking of a material with a hammer, this relatively quick application of a force (as opposed to a slow or static loading of a material) can cause considerable damage to a material that cannot adequately redistribute the stresses caused by the impact. Figure 4-11 showed a picture of an impact machine and some fractured specimens. ***Ductile materials*** usually survive impact due to their microstructure, which allows slip to take place. Metals are thus considered tough, or having good impact resistance. Due to their inherent nature as compounds of metals and nonmetals, ceramics do not possess the ability to redistribute stresses and deform plastically. Consequently, they have poor toughness, impact resistance, and fracture toughness. To improve the impact resistance of ceramics, measures are taken to increase their fracture toughness. One such procedure is ***fiber and particulate reinforcement.*** Common examples are the use of glass fibers to reinforce plastics, carbon fibers to reinforce carbon-matrix composites, zirconium-diboride platelets to reinforce a zirconium-carbide matrix, and

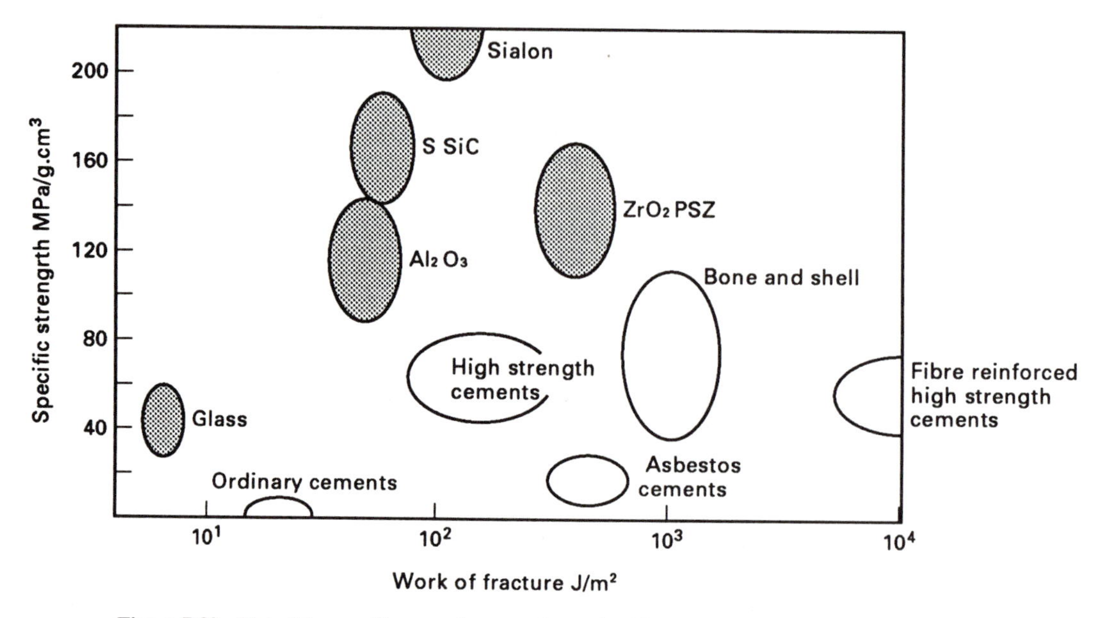

Figure 7-30 Plot of the specific strength versus the work of fracture for some typical ceramic materials—glass, cements, and biominerals. (JME.)

Al_2O_3 fibers in an aluminum matrix. Figure 7-31(a) is a micrograph of precoated SiC fibers reinforcing a SiC matrix by using the forced CVI process illustrated in Figure 7-23. Figure 7-31(b) is an accompanying stress–strain diagram, showing the effect on toughness of the composite by modifying the interfacial bond between fibers and the matrix material. In ceramic-matrix composites, the bond between the fibers and the matrix must be weakened to allow the fiber to absorb more of the energy that is trying to initiate or propagate a crack. This is the opposite strategy to fiber bonding with a polymeric matrix, which calls for an increase in the interfacial bonding. For example, silicon fibers are coated with carbon before they are used in a matrix of silicon carbide [Figure 7-31(a)]. This coating increases the toughness of the composite by reducing the interfacial bonding (friction) between the fibers and the matrix material [Figure 7-31(b)]. Doing so allows the fibers to debond from the matrix material and pull out before breaking. All of these displacements require energy expenditures and thus reduce the energy that can propagate as a crack until there is a failure of the composite. Figure 7-20 was a sketch of fiber bridging (fibers span across the crack) and pullout (fibers pull out of the matrix material) as the fibers resist the propagation of a crack; whereas Figure 7-21(b) was a micrograph of this same phenomenon. Figure 7-32 is a stress–strain plot showing the point of fiber pullout and matrix cracking as well as the progressive fracturing of the fibers. Each of these individual events takes energy away from that required to propagate a crack, leading to total failure of the ceramic. In addition to precoating the fibers, the toughness of the materials is increased partly because a crack that tries to propagate in the matrix encounters fibers that block its movement. The crack is diverted to run along the fiber-matrix interface. The stress applied in this direction is much lower, so the crack might stop propagating. Figure 7-33 is a

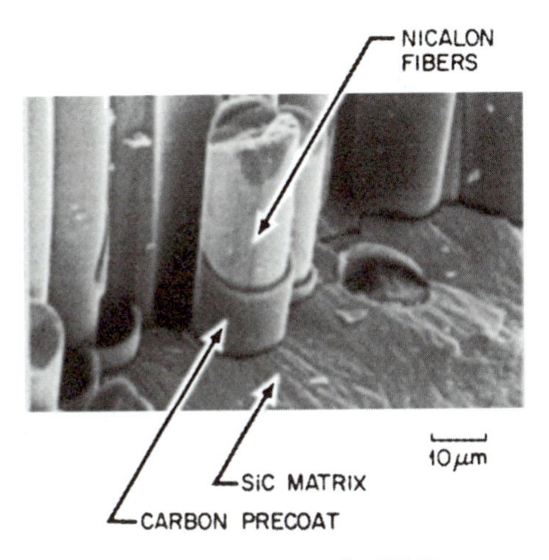

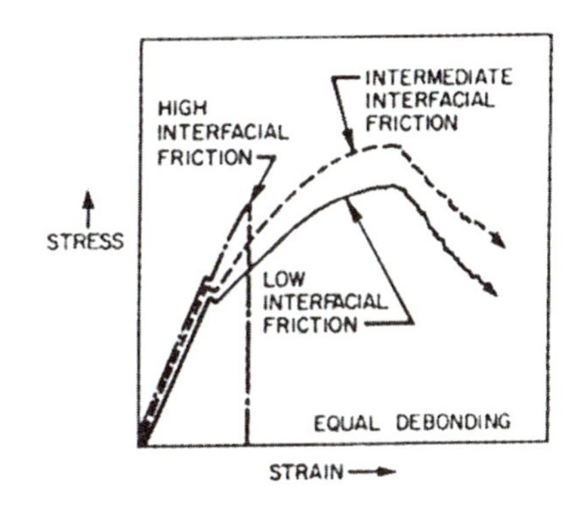

Figure 7-31 (a) Micrograph of a SiC fiber precoated with carbon reinforcing a SiC matrix in a ***continuous fiber ceramic composite*** (CFCC). (b) Stress–strain curve showing the dramatic results of such a coating on the toughness of the composite. Note the differing results of reducing the friction or interfacial bonding between the fiber and its matrix material. (ORNL.)

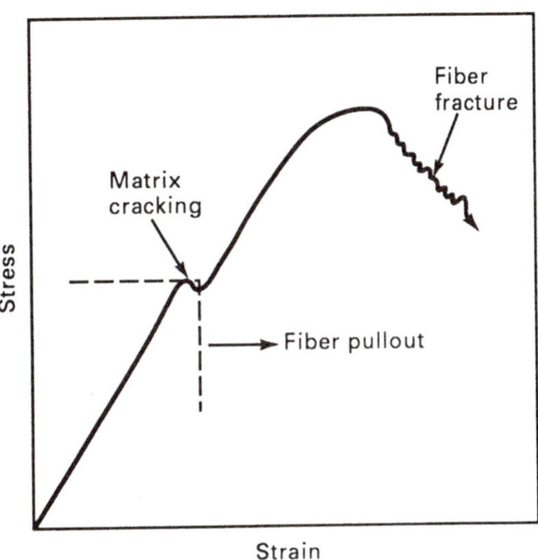

Figure 7-32 A fiber-reinforced ceramic composite has increased toughness due to the added strain energy of the fibers and the work of one in debonding the fibers from the matrix.

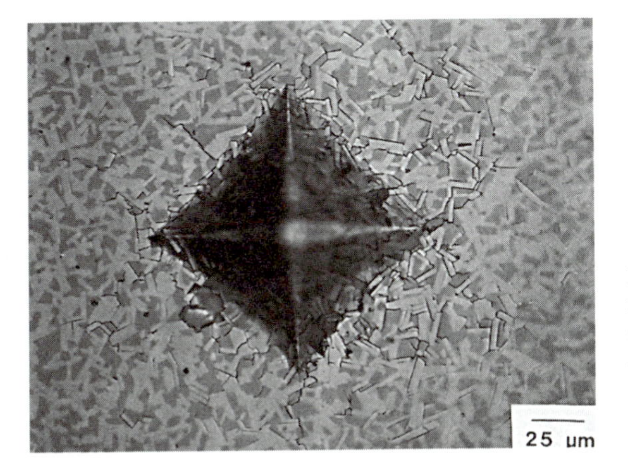

Figure 7-33 A micrograph of a platelet-reinforced Lanxide® ceramic composite showing the crack pattern produced by a diamond stylus indentation that effectively stops the propagation of a critical crack, which could have led to failure. (Lanxide.)

micrograph of the crack pattern surrounding a diamond-hardness indentation in a polished surface of a Lanxide™ zirconium-diboride platelet, zirconium-carbide/zirconium composite. The matrix material of the composite, written or abbreviated $ZrB_{2\ p}/ZrC/Zr$, contains up to 20% metallic zirconium in addition to ZrC. These composites have fracture toughness as high as 23 $MPa \cdot m^{1/2}$ and flexural strength exceeding 1 GPa. These properties are the result of a combination of crack deflection and pullout of the platelets, as shown in the micrograph. Additionally, the excess zirconium provides some ductile yielding (plastic deformation) in the crack zone. These reinforcing platelets have successfully localized the damage, as shown by the numerous small cracks surrounding the indentation.

Another method for toughening materials is the use of ***second-phase reinforcement.*** Examples are the addition of small quantities (about 10%) of cobalt (Co) to tungsten carbide (WC), a ceramic (a hard refractory material), while both are in the particle stage. After additional steps in processing (pressing and sintering), a cermet is produced that has the toughness needed by cutting tools. A third method for toughening materials is called ***transformation toughening.*** In the study of steels, the formation of martensite was discussed in Section 5.6. The martensite reaction was once believed to be confined only to steels, but now a number of other alloy systems, such as iron–nickel and copper–aluminum, as well as titanium produce similar phase transformations. In the field of ceramics, ***zirconia*** or ***zirconium oxide*** (ZrO_2) has been found to possess some unique properties. Zirconia possesses the lowest thermal conductivity of any ceramic material. Pure ZrO_2 has a stable, tetragonal crystal structure of temperatures above 1000°C but changes to a monoclinic crystal structure below 1000°C. This displacing transformation produces an abrupt change in the dimensions of the crystal structure, accompanied by stress buildup and possible cracking. A similar transformation occurs when silica allotropically changes from α-quartz to β-quartz, and from tridymite to cristobalite. When zirconia is cooled through 1000°C, the accompanying volume change (3.25% expansion) causes the material to fall apart. Figure 7-34 is a rough sketch of the effects of temperature on the volume of ZrO_2. However, if a small amount of yttria is added to the ZrO_2, the tetragonal phase can be stabilized down to room temperature. The stabilizer size is critical. In addition to yttria, manganese and calcium can be used. The yttria must be uniformly distributed throughout the microstructure

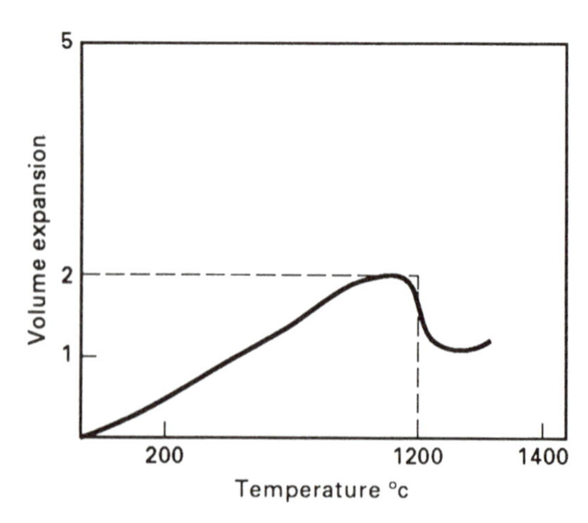

Figure 7-34 Volume expansion curve for ZrO_2 showing the effects of the displacing polymorphic transformation.

of the ZrO_2. This is accomplished by mixing the two materials in powder form, with particle sizes varying from 0.05 to 1 μm. Note that this does not require a phase transformation. As stated, specific microstructures are crucial for optimum properties. High-temperature sintering (firing) is essential for the formation of tetragonal zirconia precipitates within a matrix of cubic stabilized zirconia. The morphology (shape) of these tetragonal precipitates can be controlled by various stabilizers. The ability of these precipitates to be transformed to monoclinic zirconia (with a 4% volumetric dilation on application of a stress or passage of a crack) imparts increased fracture toughness to these materials. The zirconia is now referred to as stabilized zirconia. By not adding enough yttria to stabilize the zirconia and with adjustments in the particle sizes and processing controls, mixtures of the stabilized cubic phase and the unstable monoclinic phase that possess high fracture toughness are obtained. This metastable mixture, known as ***partially stabilized zirconia (PSZ),*** will undergo transformation if it is disturbed sufficiently. If a part is made of this ceramic material and the part is subjected to impact, vibration, or another tensile stress-producing condition, the stress produced will cause the material to absorb energy sufficient to cause transformation of the monoclinic phase. The transformation takes place ahead of the tip of any crack growing in the material, resulting in a diminishing of the stress and in turn causing the crack to propagate, and hence an increase in the toughness of the ceramic material. Figure 7-10 was a sketch of the PSZ transformation zone surrounding a crack. It has been shown that Mg-PSZ, one of many PSZ systems, has the highest fracture toughness. As with other materials, it can be heat treated in a variety of ways to modify its microstructure for particular applications.

The last toughening method to be mentioned is ***surface compression.*** If the surface of a material is placed under compression, any applied tensile stress must exceed the compressive stress before a stress concentration can begin to build up and lead to a crack propagation. Surface compression or prestressing can be accomplished by ion exchange, quenching, or layering. ***Ion exchange*** can be brought about by exchanging ions of greater size with those of the parent material (see Figure 7-35). The larger ions substitute for some of the smaller ions when the material is heated, and after the material is cooled, the larger

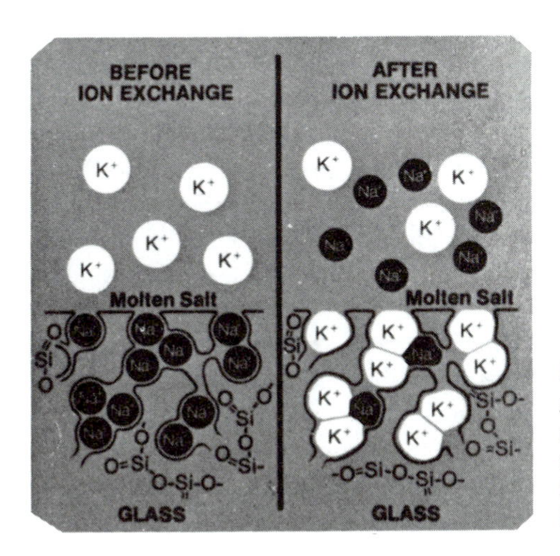

Figure 7-35 ***Chemically strengthened glass.*** Ion exchange causes relatively large potassium ions (K+) to replace smaller sodium ions (Na+), thus crowding the surface and producing compression (prestressing).

ions near the surface cause a compressive stress. This phenomenon comes about because, at the higher temperature, the unit cells of a crystalline material or the atoms in an amorphous material have moved outward, making room for the larger ions that still find room with the smaller ions squeezed, producing the compressive stress.

When a material is heated sufficiently to a temperature that produces some viscous flow and then quickly cooled, the surface will cool more quickly than the interior, forming a solid, hard case around the still-somewhat-viscous interior. By producing a surface compression with a different coefficient of thermal expansion (layering), heating to a temperature to produce some plasticity, and then cooling, the surface layers experience a compressive ***prestress.*** Figure 7-36 is a graphic step-by-step portrayal of this process. An example of this technique is the toughening of tempered glass and composite materials, including ***prestressed concrete.*** With fiber-reinforcing composites and reinforced concrete, this technique places the matrix materials under compressive stress to prevent a tensile stress from acting to cause crack propagation in the matrix. A multistep process is used to accomplish this prestressing. First, the fibers are tensioned mechanically. Second, they are encapsulated by the matrix material, which is allowed to solidify. Third, the tensioning force is released, causing the fibers to contract elastically toward their original length, placing the matrix under a compressive stress. Once an external tensile force (or load) is applied on the composite, the compressive stress in the matrix must be exceeded before the matrix feels the presence of the tensile stress.

7.5 REFRACTORIES

Refractory metals are among the few materials with significant mechanical strength above 2500°F, but they do not have sufficient strength-to-weight ratios for many sought-after applications. See refractory composites and explosive welding/bonding in Module 8 for new applications of refractories.

A ***refractory material*** is one with a very high melting point and other properties that make it suitable for uses such as furnace linings and kiln construction. A broad range of ceramic materials qualify as refractories. In the traditional ceramic group are products of sintered clay, consisting of crystals in a glassy matrix. These materials find use in ovens, kilns, furnaces, and melting pots; in welding and cutting; and as engine parts (see Figure 7-1). Some common refractory ceramics are alumina, alumina-silica, chromite, bauxite, zirconia, kaolin, silicon carbide, magnesite, and graphite. In addition to heat resistance, chemical properties bear heavily on the selection of refractories that come into contact with molten metals and glasses. Ceramics with weak ionic bonds have relatively low resistance to chemical attack, while those possessing strong covalent/ionic bonding are stable as linings for furnaces used for melting metal alloys and glasses at temperatures above 1200°C. The acidic or basic nature of a ceramic will determine how it reacts with a given molten material. The reaction between materials can change the properties of the metal or glass being processed. Operating temperatures of some furnaces are extremely high. For example, the basic oxygen furnace (BOF) for making steel subjects its linings for about 1 hour to a temperature of about 1700°C. Sintering furnaces may reach 2500°C. Blocks of alumina are cast after being melted in an arc furnace at temperatures well above 2000°C

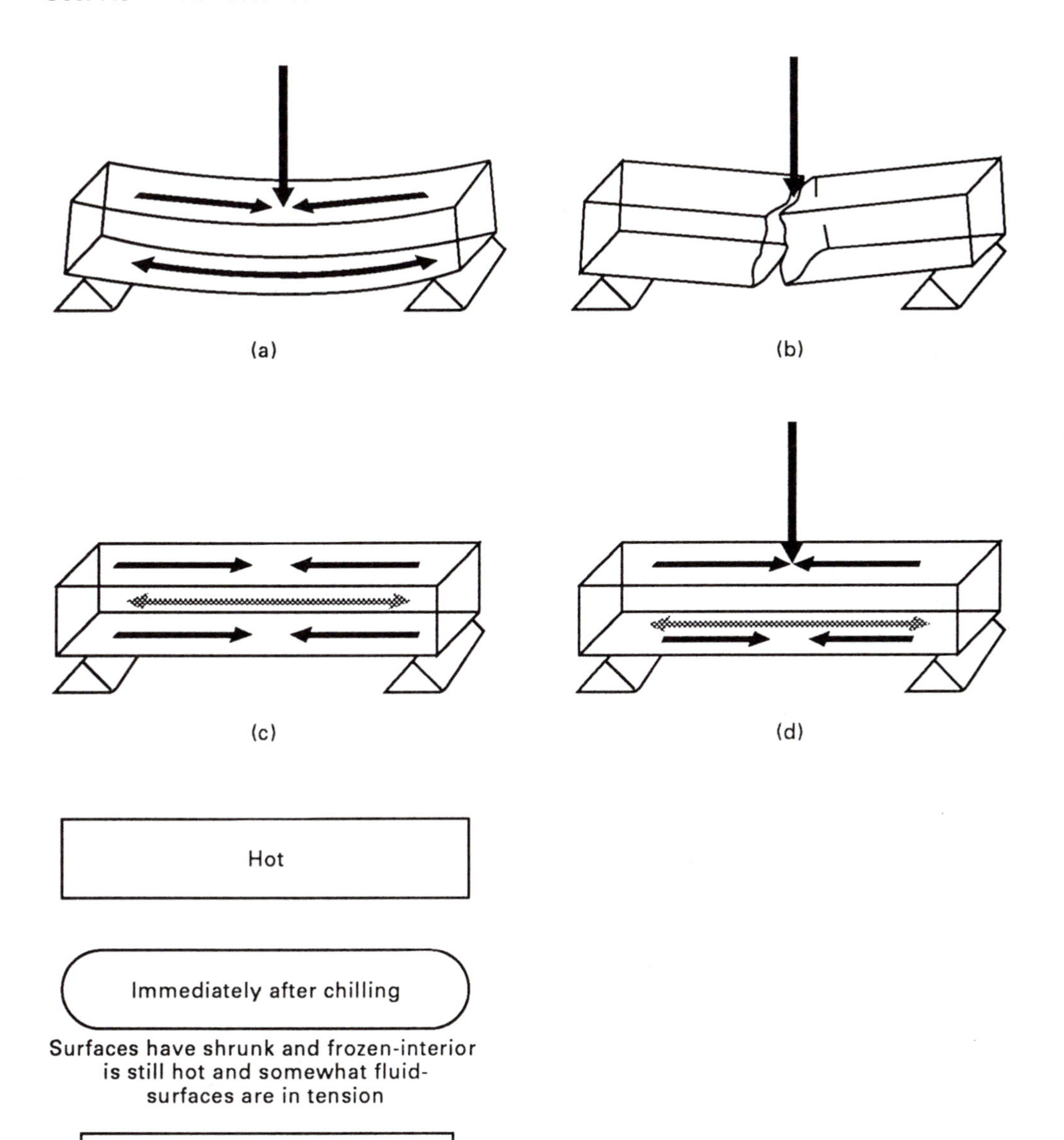

Figure 7-36 (a) Bending load produces tension in the lower surface of the test bar and compression in the upper surface. (b) The bar breaks when the tension on the lower surface exceeds the ultimate strength of the glass. (c) ***Prestressed*** bar shows compression in all surfaces. Reactive tension is buried within the bar. (d) Bending load applied to the prestressed bar must first overcome built-in compression before the surface can be put in tension. (e) Prestressing by physical tempering consists of heating glass until it begins to soften. An abrupt chill then shrinks and freezes surfaces. When the interior of the glass cools and shrinks, the surfaces are compressed. (Corning Glass Works.)

for use as liners in glass-melting furnaces because they are highly resistant to high-temperature corrosion.

These high-temperature ceramic materials must possess a varied assortment of properties. For example, the electrodes mentioned above must have (1) resistance against corrosion and erosion from high-velocity gases, (2) good thermal-shock resistance, (3) high bulk density for good chemical resistance and erosion resistance, (4) good resistance against electrical-discharge arcing, and (5) good thermionic emission. Notice that strength properties were not mentioned, nor were those properties principally related to insulation against high temperatures. Many of our present engineered materials are designed for specific purposes. The fact that ceramic cutting tools can have more than a dozen thin coatings added to their surfaces, each with its own function, suggests that this technique may be useful for refractory ceramics as well. In the case of furnace linings, more than one liner may be installed that has special functions to perform in addition to its basic high-temperature resistance. Such functions could be structural or specialized insulating properties with low thermal conductivity in addition to high-temperature melting resistance. Finally, the possibility of using additives to these ceramics to increase their specific properties must be addressed. Fibers added to refractories serve to reduce thermal conductivities as well as to provide more open space within the material to decrease the material's ability to store up heat as a heat sink.

7.6 CEMENT AND CONCRETE

Traditional cements developed many years ago, as well as those produced in recent years, are actually a group of chemical materials in which consolidation occurs by means of chemical reactions at low temperatures rather than by firing or sintering at high temperatures. When mixed with water in suitable proportions, portland cement sets and hardens. Consequently, it is used in concrete to bind the sand and coarse aggregate into a solid mass. Setting and hardening are caused by the chemical reactions between the cement and water. ***Setting*** (stiffening) occurs within hours, while with most normal varieties of cement, hardening (the subsequent development of strength) is largely complete in a month. If the water–cement mixture (paste) is kept moist, the strengthening process can continue for many years. If high strength is required in less than 1 month, alumina cements are used, which can achieve high strength in 48 hours. Figure 7-37 is a schematic of the setting and hardening process. Setting occurs when the gel coatings formed around individual grains join, and hardening results from densification of the contact areas in the gel.

7.6.1 Chemically Bonded Ceramics

More recent chemically bonded ceramics (CBCs) find uses as tooling mediums, for military armor, and in containers for storing hazardous and radioactive materials. Metal aggregates and fiber reinforcements are added to CBCs to improve their tensile strength and fracture toughness. These low-temperature CBCs are improving to the point where they have nearly 20 times the strength of concrete. As for cost, when raw materials and energy

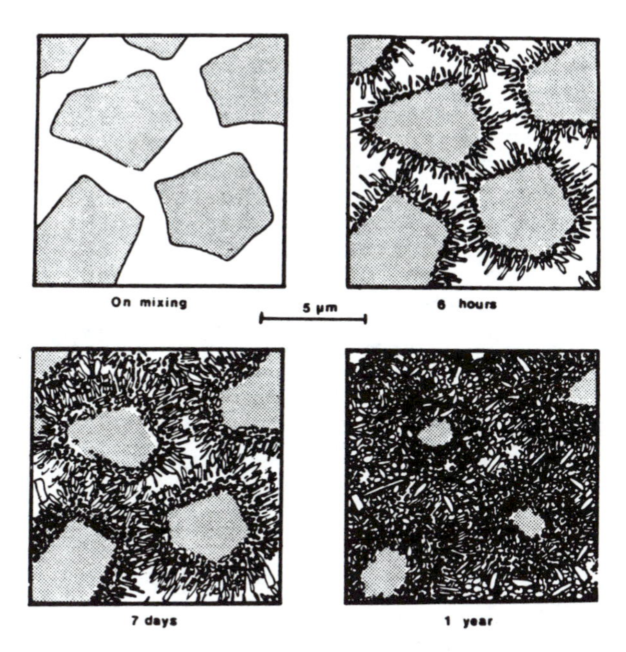

Figure 7-37 Schematic picture of the setting and hardening process. (JME.)

are considered, CBCs are nearly six times less expensive than polymer materials, 20 times less expensive than steel, and 25 times less expensive than aluminum.

Traditional hydraulic cements, as well as reaction and precipitation cements and the concretes made by combining them with various aggregates, are among the most used human-made materials. The term ***cement*** can include any materials that act as an adhesive to bond components together. ***Cementitious*** materials include pozzolans (containing natural hydrating or water-incorporating material), such as shale, lime, gypsum, asphalt, tar, synthetic plastics, and cement. Ceramic cements are most familiar to us in the form of ***concrete*** and ***mortar.*** Concrete and mortar are similar because both use cement to bond together the ***aggregate*** (rocks and sand, which combine only through adhesion with cement), which adds bulk to the material. Concrete is more correctly classified as a composite, especially when rocks and reinforcing metal rods and wires are added to it. ***Masonry mortar*** consists of masonry cement plus a fine aggregate sand. Masonry cement is composed of portland cement with hydrated (water-incorporating) lime, silica slag, and other additives to improve plasticity and slow setting time. Masonry mortar is applied by brick masons between cement blocks, brick, and tile.

Cements can be classified according to their bonding mechanism: hydraulic bonding, reaction bonding, and precipitation bonding. ***Hydraulic cements*** cure or set by interaction with water. ***Portland cement,*** the most common hydraulic cement, is composed primarily of anhydrous (without water) calcium silicate. An example is tricalcium silicate ($3CaO \cdot SiO_2$). Other ingredients are alumina, gypsum, and iron oxides. ASTM C150 recognizes eight types of portland cement, each with different chemical compositions and properties. Figure 7-38(a) depicts the raw materials used to prepare this cement. Portland

(a)

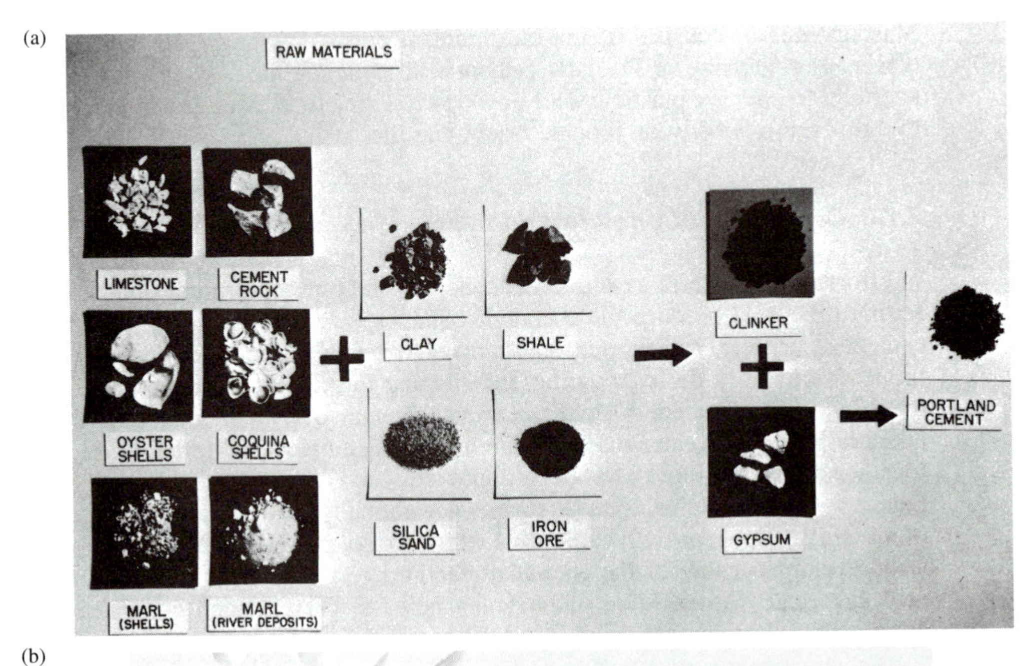

(b)

Figure 7-38 (a) Raw materials to make portland cement. [From Design and Control of Concrete Mixtures (EB001.11T). Portland Cement Association, Skokie, IL.] (b) Porous concrete allows runoff water to seep through, disperses water naturally, and traps some pollutants. (John Corbitt, *The Virginian-Pilot.*)

cement sets by a complex reaction with water to form a hydrated composition. This reaction is ***exothermic*** (gives off heat), which means the heat can dry out the cement, thus stopping the reaction prematurely. Keeping the cement damp until setting is almost complete precludes this from happening. A compromise is needed here because the lower the ratio of water to cement when mixing, the higher the final strength of the mix that will be attained. When hardened, portland cement can contain a porosity of over 30%, which results in flexural strengths in the range of 4 to 10 MPa. Two factors that affect the strength of cement are the size of the pores arising from the entrapment of air and the incomplete packing of cement particles. Another hydraulic cement is gypsum cement, such as plaster of paris. These cements set using the same hydraulic reaction but recrystallize to a higher crystalline structure that has a low degree of adhesion. This property makes them good candidates for components such as wallboard and plaster molds made for slip casting of metals and ceramics. To give wallboard more strength, both sides are covered with paper, producing another example of a stressed-skin composite material.

The "greening" of concrete has been achieved with a "new age," pervious concrete that can trap contaminants as it filters runoff water in parking lots, driveways, and other large concrete pads. ***Pervious concrete,*** as seen in Figure 7-38(b), was developed by Tarmac America, Inc. It is a porous (pervious) material that permits water, oil, sediments, and other solids and liquids to flow through the pores and slowly seep into the ground. With normal impervious concrete and asphalt pads, water and contaminants carried by the water must be directed to catch basins and storm drains, where they are piped through sewer systems to streams, rivers, lakes, and oceans. Any waterborne contaminants flow into the bodies of water and contribute to pollution. Pads made of pervious concrete provide for the slow, natural filtration of water that works its way back into groundwater reservoirs deep in the earth. To keep the pervious concrete clear, it is flushed by water.

New legislation to control storm water runoff has prompted new concrete technology. When introduced, pervious concrete cost more than double the price of impervious concrete. As with any new technology, this "green" concrete must prove its ability to withstand stresses from changing weather conditions, traffic loads, and aging. Its higher cost may be offset by savings in storm systems and improved water quality.

7.6.2 Improved Cements

Two recent advances in hydraulic cements have improved their tensile strength primarily by decreasing their porosity. In one technique, known as DSP (densified with small particles), SiO_2 particles of very fine size (ultrafine) are added. Because they are smaller than the cement particles, they fill spaces between the cement particles, decreasing the overall porosity and pore size. In addition, less water is required to produce plasticity of the mix. These factors reduce the permeability of the mix, which translates into longer service life for the many structures built with reinforced concrete that contain metal bars. However, these concrete mixes suffer corrosive attack from water seeping through the concrete.

A second technique is referred to as MDF (macrodefect-free) cement, in which additives such as small amounts of polymers and other ingredients are added to achieve better particle dispersion and packing. The absence of pores, whether due to initial particle size, particle distribution, or the elimination of entrapped air by kneading, results in the reduction of stress concentrators and a dramatic increase in tensile strength (over 200 MPa in some instances) and fracture toughness, with a corresponding decrease in permeability, all with little increase in overall cost.

High-temperature cements are produced by chemical reaction not involving water. One common reaction cement is monoaluminum phosphate, produced by the reaction between aluminum oxide and phosphoric acid. One example of these cements is dental cement based on zinc oxide. Another type of cement is precipitation cement, which is acid-resistant and has good abrasion resistance. Sodium silicate, which is formed by gels that are precipitated from colloidal solutions, is a very common, inexpensive cement in this category of cements.

Improved cements, such as the CBCs, will no doubt be produced now and in the near future using techniques that call for low temperatures and low energy consumption. As a consequence, high-technology materials and processes will be called upon less often. The significance of this new, low technology is the fact that the resources needed to produce this new class of engineering materials for the twenty-first century are universally available. Third-world nations will benefit greatly, as will more developed countries that now rely on petroleum-based raw materials. As introduced in the Module 1 discussion of the materials cycle, there is much to be said for utilizing ***low technology*** as the world struggles to reduce consumption of limited natural resources and energy. Reduction of environmentally threatening waste (by-products of materials processing and energy consumption) and in the disposal of used products will be some of the results of the new technology. While we have studied high-technology materials and processes, we must continue to reexamine the viability of older materials such as wood (a renewable resource) and cement (from readily available resources) where it is possible to apply new processes.

7.7 CLAY

Clay is classified as either residual or sedimentary. ***Residual clay*** results from the wearing down of rock from the mechanical and chemical action of wind, water, earth movements, and chemicals in the soil. Such clays are found where they were formed. ***Sedimentary clays*** also form through mechanical and chemical erosion of rock, but they are moved by wind or water to places other than where they have formed. Common clays include kaolin ($Al_2O_3 \cdot 2SiO_2 \cdot 2H_2O$), ball clay, fire clays, stoneware clay, and slip and flint clays. In various forms, clay finds many applications: as ingredients in most traditional ceramics, structural products of brick and tile, raw materials for sculptured art, china dishes, and electronic components. Clay is an important ingredient in paper making. Silicon and aluminum are prominent in clay, comprising 28% and 8%, respectively, of the earth's upper crust. Most other elements are present in only small percentages, such as carbon, with nine thousandths of a percent.

7.8 NUCLEAR FUEL

Uranium dioxide, a ceramic material, is used as a fuel in the current generation of nuclear reactors for the production of electric power. This oxide was chosen for its chemical stability and high melting temperature (2750°C). Figure 7-39 shows a cross-sectional view of a typical reactor fuel rod. The uranium oxide is made into pellet form by pressing and sintering; these pellets are then inserted into tubes made of a corrosion-resistant alloy of zirconia called zircaloy, which are then welded shut. A nuclear reactor contains thousands of these tubes, held into frameworks that allow water to flow among them. The water acts as both coolant and moderator. It slows the neutrons produced by splitting the uranium atoms so that they can then split other uranium atoms in a controlled fission process. The fuel inside each tube produces an amazingly large amount of energy, each foot of length producing about 2 kW, enough electric power to run an average household.

Because the ceramic is not a very good conductor of heat, this energy produces high temperatures, 600° to 1400°C, at the center of the oxide pellets. The water outside the fuel is at a fairly low temperature, about 300°C, so the temperature gradient is very steep. As the nuclear-fission reactions proceed, elements of lower atomic weight are produced, and they accumulate inside the solid oxide fuel. Eventually, these elements, called fission products, accumulate to the point that the fuel must be removed and new fuel inserted. The fuel produces a lot of energy before this is necessary; the spent fuel that produced all the electrical power for a household for a period of 3 years weighs about 4 oz. In a typical power reactor, a third of the fuel will be exchanged for fresh fuel every year during a planned outage. Refueling is sometimes done on a 2-year schedule, so fuel usually produces power for 3 to 6 years.

A ***radioactive isotope*** consists of a vast number of radioactive nuclei that do not decay all at once. They decay randomly over a period of time. The rate of decay of any isotope is often specified by giving its ***half-life,*** which is defined as the time it takes for half

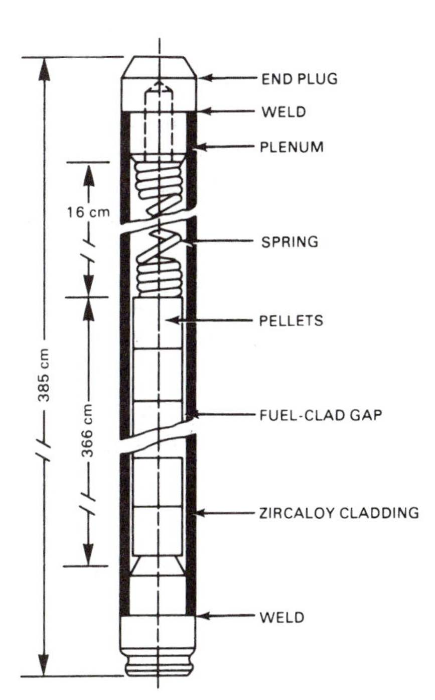

Figure 7-39 Light water reactor (LWR) fuel rod. (U.S. Dept. of Energy—Pacific Northwest Laboratory.)

the original amount of the isotope in a given sample to decay. The longer the half-life of an isotope, the more slowly it decays. Spent fuel from nuclear reactors is extremely radioactive, so it must be kept isolated from people, animals, and the food chain. A radioactive fuel such as plutonium (Pu, at. no. 94), for example, has a half-life of about 24,000 years. Starting with a pound of plutonium, after the passage of about 24,000 years, one-half pound of this element will decay to a stable condition. After another 24,000 years, an additional one-quarter of the original pound will become stable, leaving one-quarter of the original pound still radioactive.

The ***radioactive decay*** of the fission products also produces heat, so cooling must be provided for the first few years that the spent fuel is stored. The usual way to provide cooling is by storing the spent fuel in water. Because of the current lack of a permanent place to dispose of spent fuel, the material is now stored on-site at power reactors. It is estimated that by the year 2020, nearly 100,000 tons of uranium-dioxide fuel will be awaiting a more permanent way of isolating it from the biosphere. Generating the electrical power to operate the average household for 3 years by burning 22.5 tons of coal would generate about 82.5 tons of carbon dioxide and about 2.3 tons of ash. The ash represents its own disposal problem, and the carbon dioxide would be released irretrievably into the atmosphere, where it accumulates. Carbon dioxide is one of the greenhouse effect gases, which impede the Earth from radiating energy into space. Many scientists believe that these gases could cause global warming, changing the climate worldwide.

7.8.1 Disposing of Nuclear and Chemical Waste*

Nuclear materials provide an excellent example of the importance of understanding the total Materials Cycle. While these materials represent some of the highest levels of technological development, they also present huge hurdles in terms of their safe processing, use, maintenance, and disposal. As with other chemicals, biological organisms, and materials, materials science and engineering is often called on to immobilize the end products.

When in a molten state, glasses tend to be very corrosive and dissolve other oxides readily. Glasses are nonselective in this dissolution because their noncrystalline nature permits them to incorporate other atoms without much regard for size, valence, or crystal form. One way to take advantage of this characteristic is to use particularly durable glasses as a matrix material for disposal of high-level nuclear wastes. Development of this concept started in the mid-1960s, and operating nuclear-waste glassmelters exist in several countries at present. Startup of several plants in the United States is scheduled before the year 2000. The glass type used for nuclear-waste disposal is roughly similar in composition to Pyrex, a borosilicate glass developed by Corning Glass Company. This glass is very durable but can be processed at moderate temperatures. The wide variety of elements present in a typical nuclear waste is illustrated in Figure 7-40(a). All of these elements can be dissolved by nuclear-waste glass, although solubility limits are occasionally reached for noble metals and elements such as ruthenium.

Processing of glass at temperatures of about 1100°C inside a shielded, remote-control-operated enclosure called a ***hot cell*** is a substantial engineering challenge. The most common way of heating nuclear-waste glass is to pass electricity directly through it, heating the glass by its own resistance. Glass behaves like a semiconductor, meaning that

*Written for this text by L. Roy Bunnell, Battelle—Pacific Northwest Laboratory.

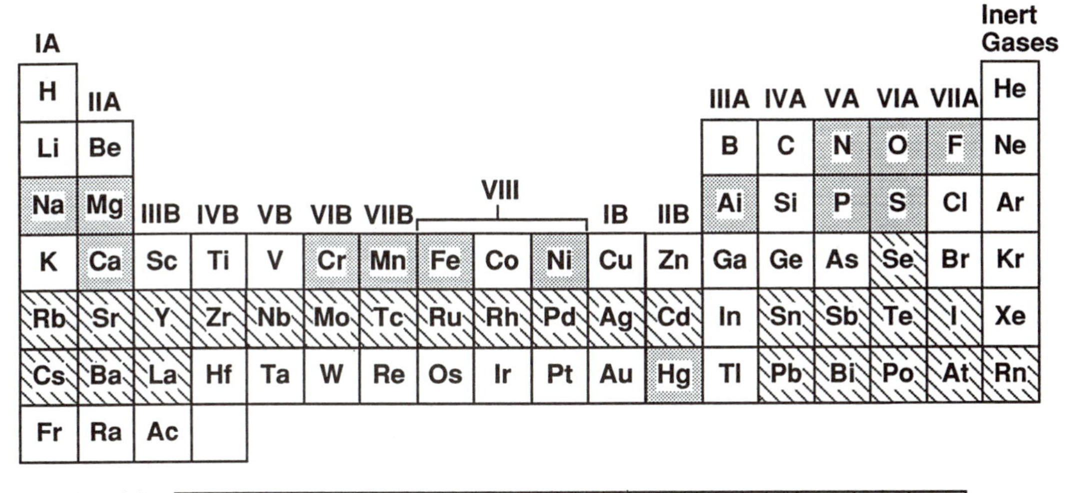

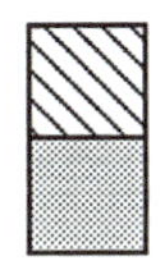

(a)

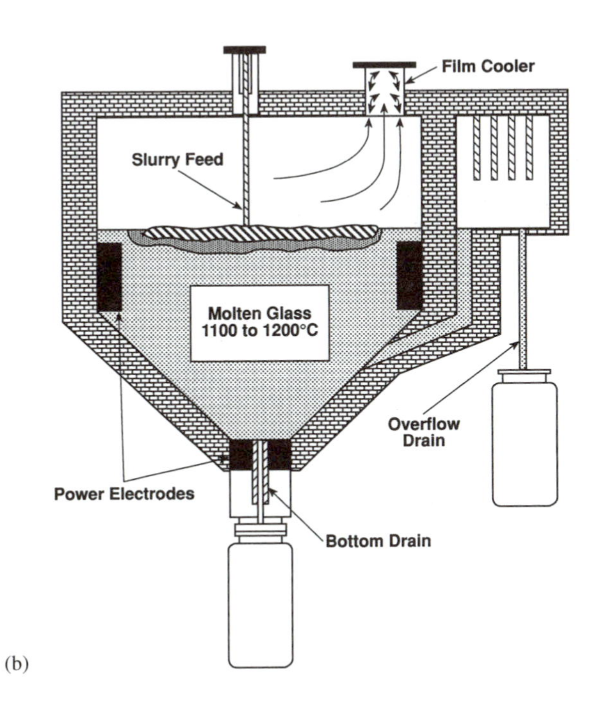

(b)

Figure 7-40 (a) Elements typical in nuclear waste. (U.S. Dept. of Energy—Pacific Northwest Laboratory.) (b) Nuclear-waste glassmelter. (U.S. Dept. of Energy—Pacific Northwest Laboratory.)

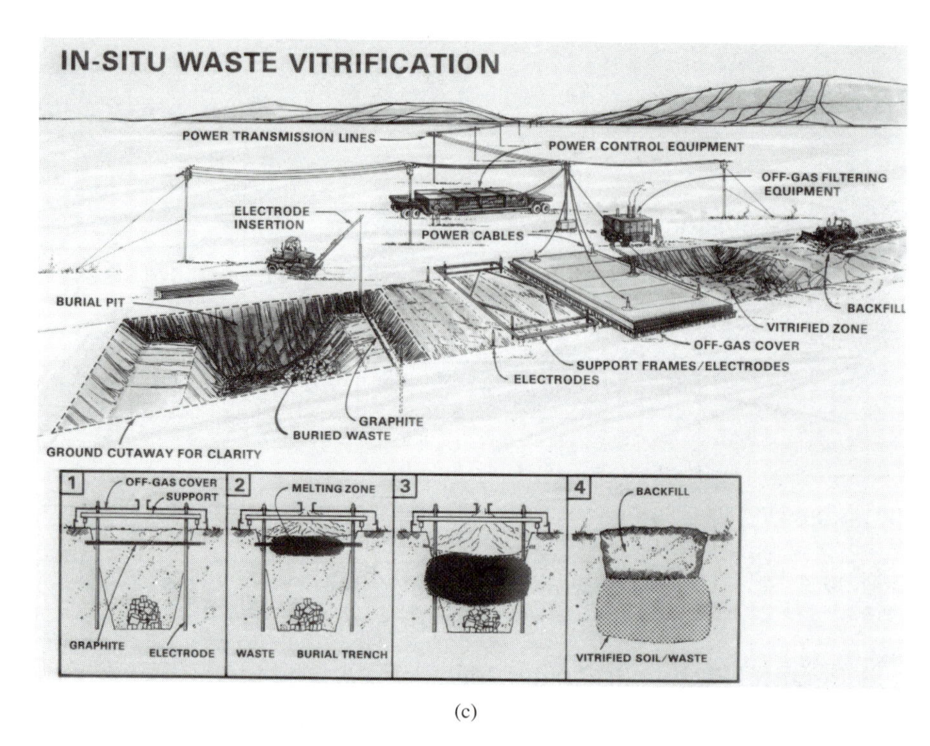

(c)

Figure 7-40 (c) In-situ waste vitrification. (U.S. Dept. of Energy—Pacific Northwest Laboratory.)

the electrical resistance drops as temperature increases, opposite to the behavior of a metal. Figure 7-40(b) shows a typical nuclear-waste glassmelter. Nuclear waste is combined with glassmaking ingredients and fed to the top of the melter as an aqueous slurry. The water evaporates and the ingredients combine to form a homogeneous black glass, in which the radioactive elements are dissolved at the molecular level. After a period to allow a homogeneous glass to form, the product is poured into a stainless-steel canister approximately 2 ft in diameter, where it cools and hardens to form a "log" about 10 ft long and weighing 4500 lb. After a lid is welded on and the canister is cleaned of external contamination, the waste glass is ready for transport to a permanent repository.

This concept of glass as a host for harmful materials has been extended to instances where these materials have been spilled or dumped into the soil as liquids. Eventual penetration of the materials into groundwater can occur. In a process called ***in situ vitrification,*** these harmful materials are either removed from the soil thermally and treated, or trapped within the soil when that soil is melted to form a glass. Figure 7-40(c) shows the process schematically. Electrodes are placed into the soil at the outer margins of the spill zone, and the electrodes are connected to a controlled source of electrical power. After heating a small volume of soil to melt it, the glass resulting from that melting becomes the heater, and the electrical power continues to melt more soil as the process continues. Very high temperatures (1500° to 2000°C) are reached in the molten glass, and the heat drives volatile materials such as organics out of the soil for collection and processing via a hood placed above the molten soil. Nonvolatile materials such as chromium plating wastes are dissolved in the glass. In a test, a glass melt weighing 25 tons was produced using an average power of 263 kW for a period of 110 h. The electrical-power consumption in this test was actually quite small, amounting to about 0.6 kW per pound of glass produced. The black glass resulting from this process has proven to be quite durable, comparable to naturally occurring glasses such as obsidian. The end-product glass has sufficient mechanical strength to be left in the ground after processing is completed.

7.9 ABRASIVES

Abrasives are very hard particles used for grinding, sanding, and polishing. ***Abrasive materials*** include particles of flint, garnet, diamond, aluminum oxide, silicon carbide, emery, pumice (pulverized volcanic lava), rottenstone (shale rock), and corundum. Aluminum oxide and silicon carbide find the greatest industrial application, although garnet is often used as sandpaper—an abrasive paper or belt used for sanding wood. Adhesive papers include paper and cloth sheets, belts, disks, and drums onto which an abrasive is cemented. Wet or dry abrasive paper employs waterproof adhesives. Grinding wheels consist of abrasive particles (grains) held together by a tough bonding material such as rubber or organic resins. The particles constantly fracture on the wheel to expose new, sharp cutting edges, or they break off as the wheel becomes smaller in size. Abrasive particles are also used in loose form for sand blasting, hand rubbing, or abrasive drilling. Selection of the ***grit*** (particle size) is based on a screening method that determines the number of particles per inch, such as 60 grit or 600 grit. A 120 grit would have 120 abrasive grains per linear inch or 14,400 (120^2) per square inch. Sialon (silicon–aluminum–oxygen–nitrogen) is a hot-pressed cermet that works well on nickel-based superalloys.

Superabrasives are synthetic diamond and cubic boron nitride (CBN) in the form of grinding wheels or sintered, polycrystalline cutting tools for milling and turning. Diamond abrasives are four times as hard, and have three and one-half times the wear resistance of common abrasives such as aluminum oxide. CBN is two and one-half times as hard and has two and one-half times more wear resistance than aluminum oxide. In addition, superabrasives are excellent thermal conductors. The combined superior properties of hardness, wear resistance, and thermal conductivity provide sharper, longer-lasting tools that can operate at higher temperatures and speeds. When grinding and machining high-speed steel, cemented carbide, high-silicon aluminum alloys, cast iron, and superalloys, superabrasives offer increased productivity through high material-removal rates, achieving better quality (tolerances and surface finishes), less residual stress, reduced scrap, and lower overall manufacturing costs.

7.10 PROTECTIVE COATINGS

With the emphasis on materials that can withstand high temperatures (greater than about 320°C), whether in space applications or down on earth in turbines or heat engines, new materials developments are reported almost daily. The search for such materials leads in many cases to the field of ***surface engineering.*** One of the first applications was the use of zirconia in the form of a surface layer applied to a metal substrate. This thermal barrier coating (TBC) is very effective in resisting high temperatures in heat engine moving parts. However, it failed to provide the necessary wear characteristics needed to protect materials in sliding contact under high temperatures. To add lubrication properties to this TBC, a new, self-lubricating material consisting of 80% silicon carbide, 10% silver, and 10% barium/fluoride eutectic has been developed that provides lubricity at high temperatures to sliding contact bearing surfaces of internal combustion engines made with conventional iron and aluminum. Many of these engines are now operating at such high temperatures that oil-based lubricants cannot be used. Both TBCs and self-lubricating coatings can be applied by plasma-arc spraying. This new development is another example of a true laminar composite (MMC) material consisting of an aluminum or iron substrate, with laminates consisting of a TBC and a self-lubricating wear coating superimposed on the surface of the TBC.

The cost of holemaking is one of the largest machining costs in automotive engine production. Much research is devoted, therefore, to making it more efficient. Ecological factors as well as cost factors are looked at. Coolant disposal and recycling costs are significant. Some estimates state that the costs associated with coolant and coolant management are 16% of the machining costs in the high-production industries, compared to only 4% for the cost of the cutting tools. And of the total coolant-related costs, some 22% are charged to coolant disposal alone. The optimum objective of such research is stated in terms of eliminating all coolants.

Near-term objectives talk of near-dry drilling as the next step in reaching zero coolant drilling. Past attempts in this direction were the development of cermets and special carbide coatings on cutting tools. One such coating was titanium aluminum nitride (TiAlN). A combination coating of TiAlN and Al_2O_3 provides superior heat resistance in dry milling of cast iron. A more recent coating to reduce friction is a soft coating, somewhat like Teflon, that adheres well to the tool surface but provides a gliding or low-friction surface. Made of ***magnesium disulfide,*** it is called ***MOVIC.*** A second new coating is a variation of MOVIC. The soft coating is placed over a hard coating, except at the point of the tool where hardness is needed. Instead, the flutes of the drill are lubricated with the soft coating to aid chip flow. Another result of research in this area is the development of an improved titanium diboride composite material produced by the self-propagating, high-temperature synthesis (SHS) process. (See Module 8 for applications of this new composite material.)

Ceramic particles are used to coat and decorate pottery and metals when melted in to a ***glassy vitreous state. Glazes*** and ***enamels*** impart color, hardness, and corrosion resistance when they are suspended in a slurry and fired to a vitreous state; as the *substrate* (base material) cools, the coating bonds tightly to the substrate surface. *Flame spraying* (see Section 7.3.3.2) uses a high-temperature heat source, which melts the ceramic oxide powder, and then sprays or blasts the coating onto the surface of a metal. The oxide material cools and bonds to increase its surface hardness while offering protection against heat and oxidation.

7.11 ELECTRONIC AND MAGNETIC CERAMICS

Typical electronic ceramics include some representative compounds and mixtures, such as ferrites (ZrFeO), silicon (SiO_2), zirconia (ZrO_3), steatite (SiO_2MgO), porcelains (Al_6SiO_{13}), and alumina (Al_2O_3). Module 9 deals with some of these and other electronic materials.

7.11.1 Ceramic Magnets

Ceramic magnets that contain a sufficient number of ***dipoles*** (atoms with electron spin in the same direction, discussed in Section 9.2.2) are ferromagnetic (see Section 9.3). Ceramic ferromagnetics can be either soft or hard magnets. The high electrical resistivity of ceramic magnets gives them an advantage over metal magnets for use in high-frequency devices. Their permanent magnetic behavior finds application in microwave devices. The group known as ferrites are iron oxides containing ions of elements such as zinc (Zn^{2+}), iron (Fe^{2+}), magnesium (Mg^{2+}), and nickel (Ni^{2+}) that combine with oxygen. Ferrites are commonly found in the computer industry in memory cores. Sheets of rubber and ferrite are employed for ***"stealth"*** coatings on some steel suspension bridges that cross navigational waterways. The coatings partially absorb radar and minimize clutter on ships' radar screens as they navigate under the bridges.

7.11.2 Thick- and Thin-Film Ceramic Devices

Thick- and thin-film ceramic devices are of major significance to the computer industry and form a group of technical ceramic substrates (mostly alumina) onto which semiconductor integrated circuits are mounted. These substrates generally fit into the categories of single-layer or multilayer ceramic (MLC) devices that employ thick- or thin-film metal layers to serve as circuitry. A substrate is usually a base material, but in these devices it includes the composite of ceramic with the metal layer.

For ***thick-film*** devices, the Al_2O_3 substrate receives a layer of metal paste such as gold, platinum, palladium, or silver. The paste is printed by silk screen in a circuitry pattern (designed to carry the electron signal) onto either greenware or bisque. With the bisque substrate, the paste is sintered at about 850°C, which changes the paste structure into resistors and conductors; greenware and paste are fired together (cofiring). Computer-controlled lasers may trim resistors to exact resistance values before the chip is soldered onto the composite substrate. The layers begin as continuous slip-cast sheets of greenware, which are blanked to size, then silk screened. The ceramic substrates consist of 92% alumina and frit made up of alumina, calcia, magnesia, and silica. Frit is a form of glass that has been melted, cooled, then broken into small particles. The substrates with metal paste are stacked into multilayers (5 to 35 layers), then sintered at about 1600°C.

Thin-film circuits composed of microthin layers (about 150 nm) of material, such as gold or nickel–chromium alloys, are produced through vacuum deposition onto ceramic substrates such as alumina, glass, or beryllia. Study of the references listed at the end of the module can focus on specific techniques such as sputtering, evaporation, and electron-beam deposition. Also see Section 9.7 for more detail.

7.12 GLASS

The American Society for Testing and Materials (ASTM) classifies glasses as inorganic products of fusion that have cooled to a rigid condition without crystallizing. They differ from glass ceramics in their lack of polycrystalline structure. Some authorities prefer not to consider glass as a ceramic because of its structure, and thus give it a separate classification.

7.12.1 Nature of Glass

The base raw material of glass is the very pure, white silica sand found in abundant supply in the central United States and other parts of the world. Although there are approximately 750 different glasses and glass ceramics, most can fit into six groups: soda-lime, lead-alkali, borosilicate, aluminosilicate, 96% silica, and fused silica.

The ***vitreous state*** of glass is mechanically rigid, like crystalline materials, yet it has an atomic space lattice that is amorphous, much like a liquid. Even though raw materials such as quartz sand (SiO_2) have a naturally crystalline structure [Figure 7-41(a) and Figure 7-3], when they melt, the lattice breaks up [Figure 7-41(b)]. Slow cooling will allow some crystals to form, but normal cooling produces an amorphous vitreous or glassy structure [Figure 7-41(c)] such as the SiO_4^{4-} tetrahedron (silicon oxide ion with a net charge of -4).

As discussed under phase diagrams, amorphous materials do not have clearly defined freezing or melting points, as do crystalline materials; rather, they harden or liquefy over a range of temperatures. By cooling molten glass, it becomes more viscous (Figure 7-42) and thickens to a working or softening point. Figure 7-43 plots the heating of crystalline raw materials. Curve *a–b* shows an increase in volume until it becomes molten (liquidus temperature); increased temperature yields a corresponding increase in volume as

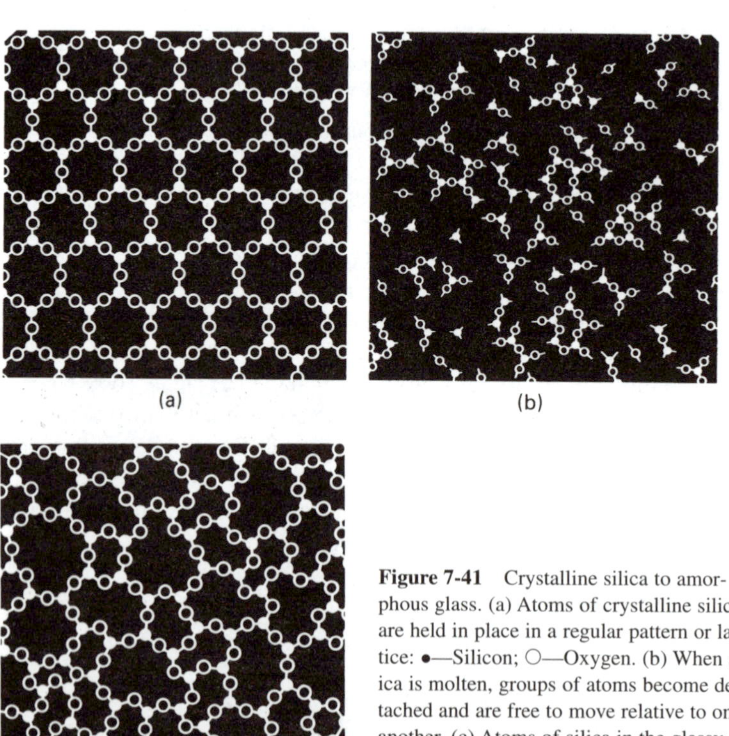

Figure 7-41 Crystalline silica to amorphous glass. (a) Atoms of crystalline silica are held in place in a regular pattern or lattice: •—Silicon; ○—Oxygen. (b) When silica is molten, groups of atoms become detached and are free to move relative to one another. (c) Atoms of silica in the glassy state are frozen in a random, or disordered, manner. (Corning Glass Works.)

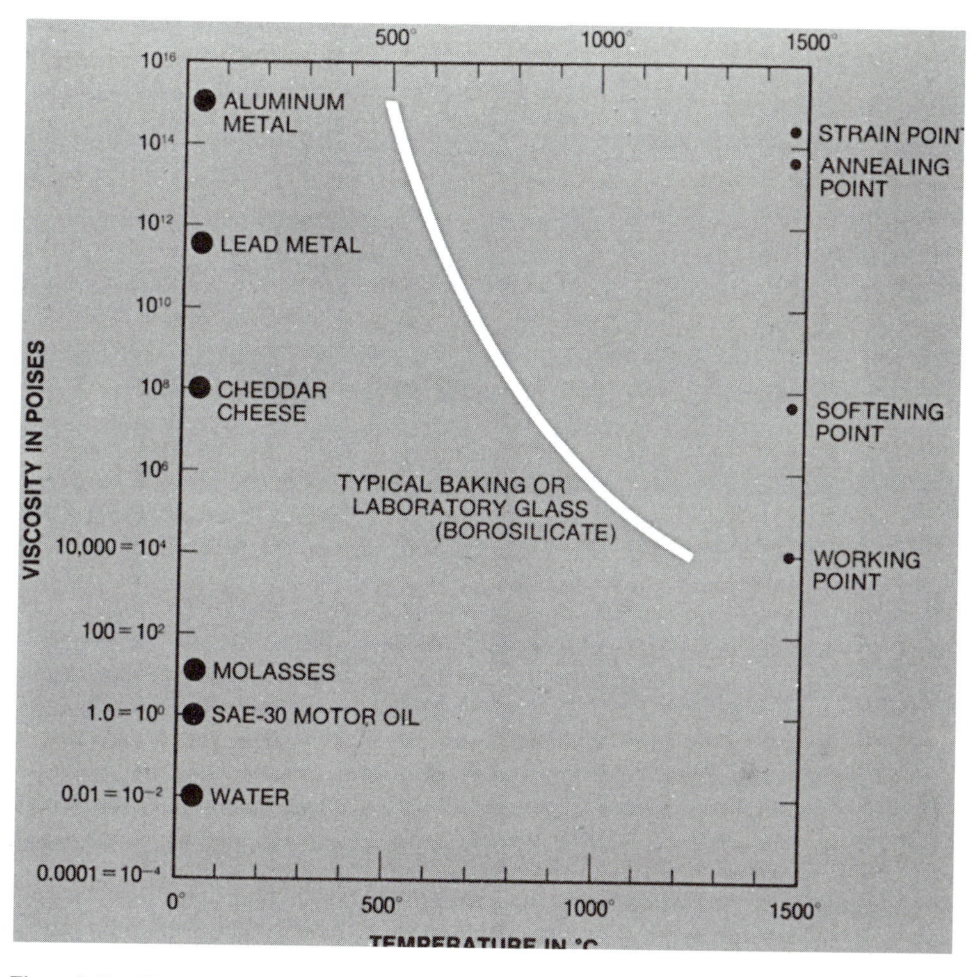

Figure 7-42 ***Glass viscosity.*** As glass is heated it becomes gradually less viscous. (Poise—absolute viscosity in Pascal seconds [Pa · s]. The measure of force required to overcome resistance to flow). (Corning Glass Works.)

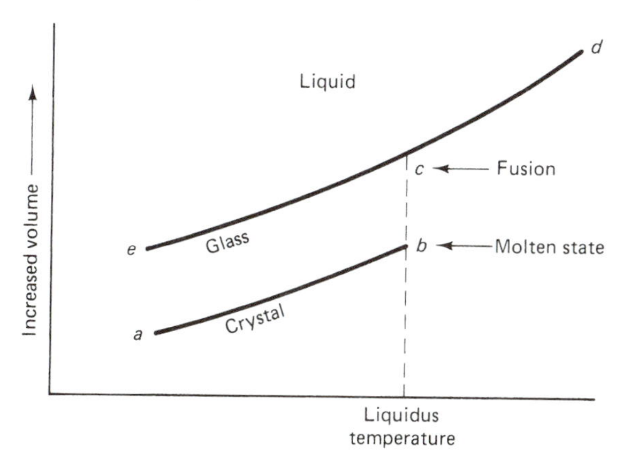

Figure 7-43 Glass formation.

the atoms expand. Glass is formed as the raw materials cool through curve *d–c–e,* but the volume remains higher at point *e* than when the raw materials were in a closely packed crystalline state. This curve represents a normally cooled glass.

Slower cooling permits some crystallization in the process known as devitrification. ***Devitrification*** is the conversion of glass or other noncrystalline solids into polycrystalline solids. This chemical reaction is opposite to that of *vitrification*: producing amorphous structures. Used commercially, devitrification produces high-quality, homogeneous, fine-grained, nonporous *glass ceramics.* Devitrification can occur only below the liquidus. To control the glass structure, it is important to know the location of the liquidus line (point), as with metals. High resistance to devitrification is desirable and prominent in commercial glasses.

A variety of raw materials yield glass compositions of assorted oxides. Oxides such as SiO_2 (silicon dioxide), B_2O_3 (boron trioxide), GeO_2 (germanium dioxide), and V_2O_5 (vanadium pentoxide) are glass formers. *Glass formers* or *network formers* are those oxides that promote the ionic linking or polymerization of oxide molecules in the glass compound. Network modifiers or fluxes such as lead, zinc, and alkali ions lower the liquidus temperature, improve workability, and change thermal and optical properties. Other stabilizers, such as CaO (calcium oxide), improve chemical properties.

Heat treatment of glass can reduce internal stress or create high internal stress. High stresses could lead to breakage from relatively minor forces. ***Annealing*** of glass through slow cooling provides a homogeneous structure by reducing internal stresses to give isotropic (equal in all directions) properties. ***Tempering*** involves rapid cooling of the outer surface of glass while still in the plastic state. The tempering results in compressive stress on the surface and tensile stresses in the core (Figure 7-36). This condition occurs because the slower-cooling core tries to contract but is restrained by the rigid outer glass. The nonequilibrium condition causes tempered glass to fracture into small pieces rather than large splinters. Tempered glass is now required on doors in buildings and homes because of this safety feature.

Glass is a nearly perfect elastic material. At any stress below breaking stress, glass will return to its original shape when the stress is removed. Some plastic flow can be achieved, but the amount of stress must nearly equal its ultimate strength. Except when internal flaws are present, glass will fail in tension on the surface opposite the compressive stress. Whereas crystalline materials fracture along planes of slippage that may not be normal with the tensile stress, the amorphous structure of glass causes fracture to be normal to the stress. Tempering glass provides a compressive stress throughout the material, with the exception of a small amount of tensile stress in the center, which must be overcome before the tensile stresses can act on the surfaces (Figure 7-36). This is similar to prestressing concrete, which is also weak in tension.

7.12.2 Types and Properties of Glass

The six glass categories are diagrammed in Figure 7-44 and broken into soft glasses and hard glasses. The classification of hard or soft does not denote mechanical hardness but, rather, the ability to resist heat. ***Soft glasses,*** those that soften or fuse at relatively low temperatures, include soda-lime and lead-alkali. Soft glasses have lower heat resistance and a

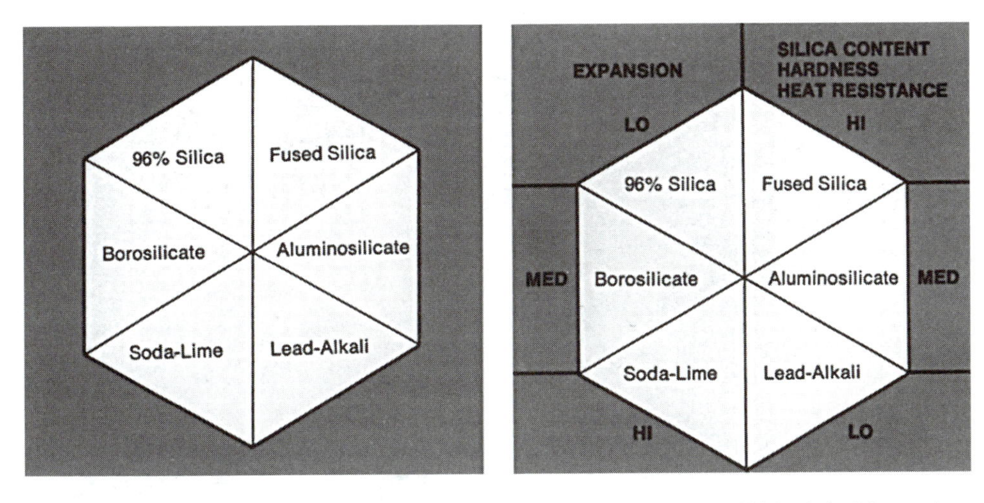

Figure 7-44 Categories of glass. *Hard glasses* are lower in thermal expansion and higher in both heat resistance and silica content than *soft glasses.* (Corning Glass Works.)

higher coefficient of thermal expansion than those of hard glasses. ***Hard glasses*** are borosilicate and aluminosilicate, and the hardest high-silica glasses are 96% silica and fused silica.

Soda-lime silica glass is the oldest glass; it dates back 4000 years and is most familiar to you in the form of window panes and bottles. It accounts for more than 90% of the glass produced. The typical composition is 74% silica (SiO_2), 15% soda (Na_2O), 10% lime (CaO), and 1% alumina (Al_2O_3). The least expensive glass, soda-lime glass, has a composition of oxides of silicon (from silica sand), calcium, and sodium. Although easy to form and cut into many designs, soda-lime glass has fair chemical resistance but cannot endure high temperatures or rapid thermal change (thermal shock). Because it shatters so easily into dangerous splinters, it is not recommended for doors or other glazings where persons might damage it. Thinner soda-lime glass, such as that used for small window panes, is *single-strength glass.* Increased thickness is referred to by strength: for example, *double-strength glass.*

Lead-alkali silica glass is slightly more expensive than soda-lime, but both are considered soft glass; neither will endure high temperatures or thermal shock. ***Borosilicate glass*** is considered a hard (thermally) glass; it is the oldest type of heat-resistant glass. Pyrex is the familiar Corning trade name for this glass. Although three times as costly, ***aluminosilicate glass*** can sustain higher service temperatures than borosilicates and is similar in its ability to handle thermal shock. The ***96% silica glass,*** a valuable industrial glass, is capable of setting on a block of ice and having molten metal poured over it without breaking, attesting to its superior thermal hardness (Figure 7-45). ***Fused silica glass*** is composed of at least 99% silicon dioxide (SiO_2). This pure glass is the most costly and most resistant to heat. It is capable of service temperatures from 900° to 1200°C and has the highest corrosive resistance. It is superior for the transmission of ultraviolet (UV) rays.

Colors are achieved in soda-lime, lead, borosilicate, and 96% silica glasses by the addition of elements that change the structure to absorb certain bands of the light spectra.

Figure 7-45 Bowl of 96% silica glass sits on a block of ice as molten bronze (2000°F) is poured. Failure to break attests to the ability of this *hard* glass to withstand thermal shock. (Corning Glass Works.)

For example, manganese ions (Mg^{+}) can produce purple glass, and iron yields green and brown bottle glass. Colorants such as iron are very undesirable in clear glasses and are kept at low levels by selecting clean sand deposits as a silica source.

7.12.2.1 Specialty glasses. Using the six basic groups just covered, it is possible to develop glasses with unique properties to meet special needs. Some specialty glasses include tempered, optical, colored, sintered, glazing, fibrous, laminated, cellular, photosensitive, and light-sensitive glasses. Advances in optical glass technology have resulted from several societal and economic demands, including the need for greater transmittance (light transmission), safety (impact resistance), styling and convenience, and competition from the plastics industry (refer to the Pause & Ponder in Module 6). Consumer groups and the federal government are requiring improvements for energy efficiency in heating and cooling buildings. Specialty glass and glazing manufacturers are responding with a host of new materials and products. The hard, stable nature of glass permits grinding and polishing of optical lenses for telescopes to 1/6000 of 1%. For styling and convenience, photochromic lenses provide eyeglasses that darken in bright light but lighten with subdued light. This ability to change the transmittance of light results from the use of silver ions. When photons of light strike these silver particles, they change from silver ions to metallic silver and absorb more light. When the light source is removed, they revert back to silver ions (Ag^+) and the lenses lighten in color.

7.12.2.2 Chemically toughened glasses. This type of ***tempered glass,*** also known as chemically tempered glass, is produced by using a molten salt bath in which eyeglass lenses are soaked for hours to achieve appropriate impact strength. During the long-term soaking, larger potassium ions from the salt bath replace sodium ions, as shown in Figures 7-35 and 7-36. This crowds the surface layer and causes compressive stresses. The impact strength of the tempered glass is then a product of its untempered strength and the amount of surface compression developed in chemical toughening. A chemically tempered glass is stronger than a heat-tempered glass.

7.12.2.3 Glazing. Glass is an important architectural material that adds beauty, and it can improve energy efficiency if properly used in design. A variety of types of glass serve as glazing, or glass windows. Most older buildings utilize single soda-lime sheet glass. Because costs for energy to supply heating and cooling have increased, improved glass technology has resulted in ***insulating glasses*** that are constructed of two or three panes, which are often fused together on the edges. Spaces between the sheets of glass are sometimes filled with gases to provide low thermal conductivity and improved sound dampening. Additives to glazing glass and reflective metallic/plastic films are also techniques to reduce the transmission of ultraviolet (UV) rays [see Section 9.4 and Figure 7-11(a)].

7.12.2.4 Glass fibers. These threads of glass are produced by various techniques, including drawing and blowing operations. Fibers can be produced as thin, continuous filaments or discontinuous fiber segments. These thin fibers serve as reinforcers in standard, woven, and other forms for plastic resins, such as fiberglass. (Fiberglass is discussed extensively in Module 8.) Segmented or discontinuous glass fiber is also used as a very effective thermal insulation in buildings, on refrigeration and heating units, and in land and air vehicles. Its light weight and good insulation properties, coupled with low cost, make glass fiber very popular. It is used in both sheet and loose form. Glass fiber is also woven and matted into sheets for filters in heating and air-conditioning equipment.

Optical fibers, used to transmit light, are gaining wide acceptance in the communications field because they can be produced from the plentiful silica sand to replace heavier, bulkier, and more expensive copper and aluminum conductors. Optical fibers of glass and plastic have been used for many years in the medical profession and by engineers as inspection tubes. Flexible fibers on probes or endoscopes are inserted into the human body or into a motor and attached to a television system or magnifying lens to allow viewing of these otherwise inaccessible places. A single optical fiber the size of a hair has the potential to transmit several thousand voice signals, compared to the fewer than 50 voice signals that can be carried on a copper wire of the same size. For this reason, telephone, television, computer, and other communications systems are moving to the smaller coaxial fiber-optics cable to replace copper-stranded cables. Several forms of optical fibers exist, including glass cores clad with silica and coated with plastic (Figure 7-46). This high-purity glass fiber can transmit more than 95% of the light beamed into it for over 1 km and has a tensile strength of over 4137 MPa and good flexural strength.

Table 7-5 provides a comparison of the electrical properties of various insulating materials. Most of the glasses and ceramics have higher dielectric constants than those of plastics and rubber. Except for soda-lime glass, they offer greater volume resistivity than that of organic polymers. Table 10-13 provides a comparison of the mechanical, physical, and chemical properties of selected glasses. With petroleum getting scarcer, it is conceivable that glass and other ceramics will replace plastics in many applications.

7.12.3 Glass Ceramics

The glass ceramics group includes β-cordierite ($2MgO \cdot 2Al_2O_3 \cdot 5SO_2$) and β-spodumene ($Li_2O \cdot Al_2O_3 \cdot 4SiO$), which are specific phases of glass that possess very low expansion coefficients and superior resistance to oxidation when subjected to high heat. As discussed earlier (Section 7.12.1), glass will crystallize under slower cooling processes than are normally used in glass manufacture. Crystals form on the surfaces of glass with

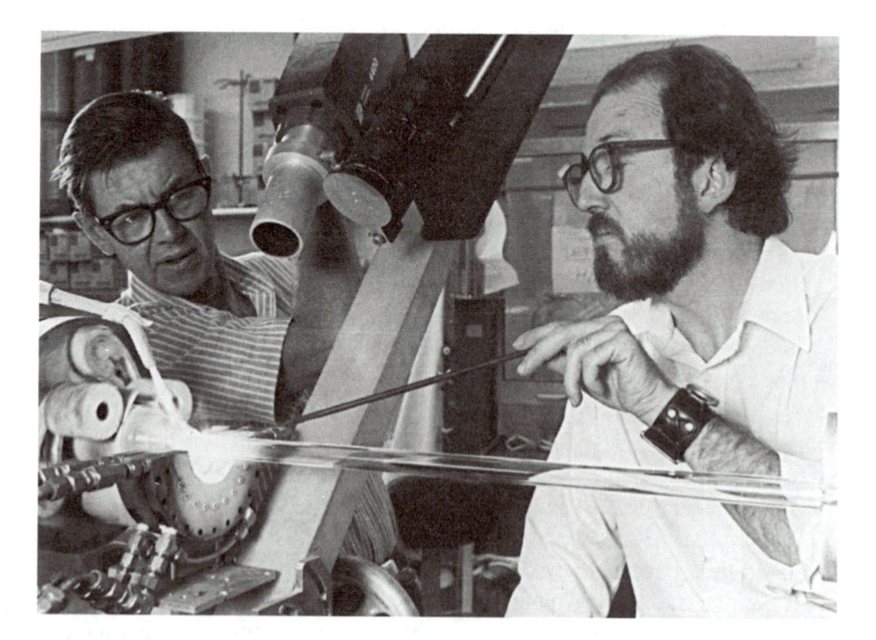

Figure 7-46 Optical fibers. The fibermaking process uses highly purified glass layers that form as chemical vapors react within the silica tube; the tube and its contents are heated until they collapse into a solid glass rod, from which nearly 10 miles of fiber are drawn in the modified chemical vapor deposition process. Single-strand, optical waveguide has the potential, through fiber optic technology, to transmit 10,000 simultaneous telephone conversations when used in pairs. (Corning Glass Works.)

TABLE 7-5 COMPARISON OF ELECTRICAL PROPERTIES OF INSULATING MATERIALS AT ROOM TEMPERATURE

Thickness material	Intrinsic dielectric strength[a] mm	in. × 10^{-3}	kV/cm	kV/in.	Dielectric constant	Volume resistivity (Ω · cm)
Cellulose acetate	0.025–0.12	0.98–4.7	2,300[b]	5,840	5.5	10^{12}
Glass						
Borosilicate code 7740	0.10	3.9	4,800[c]	12,200	4.8	10^{16}
Soda-lime	0.10	3.9	4,500[c]	11,400	7.0	10^{12}
Soda-lead	0.10	3.9	3,100[c]	7,880	8.2	10^{14}
Mica, muscovite clear ruby	0.020–0.10	0.79–3.9	3,000–8,200[b]	7,620–20,850	7.3	10^{17}
Phenolic resin	0.012–0.04	0.47–1.6	2,600–3,300[b]	6,600–8,380	7.5	10^{11}
Porcelain, electrical			380[b]	965	4.4–6.8	10^{14}
Silica, fused			5,000[c]	12,700	3.5	10^{18}
Rubber, hard	0.10–0.30	3.9–11.8	2,150[b]	5,460	2.8	10^{13}
Porcelain, steatite—low loss			500[b]	1,270	6.0–6.5	10^{15}

Source: C. J. Phillips, *Glass, the Miracle Maker.* Pitman Publishing Co., New York, 1987.

[a]Intrinsic dielectric strength can be realized only under special test conditions and is very much higher than the working dielectric strength attainable in ordinary service. These data are listed for purposes of comparison.

[b]Values of S. Whitehead, *World Power,* p. 72, Sept. 1936.

[c]Values of P. H. Moon and A. S. Norcross, *Trans. AIEE* 49, 755 (1930).

impurities. These surface crystals will cause failure of the glass part. However, controlled heat treatment can produce a fully crystallized material known as glass ceramic.

Glass ceramics are polycrystalline glass that has four to five times the strength of glass, with mechanical hardness about equal to that of tool steel. This polycrystalline glass is capable of developing structures that resist extreme thermal shock (rapid change from cold to hot, or vice versa). Cookware is a familiar form of glass ceramic. The fine crystalline structure of glass ceramics is achieved by the introduction of a nucleating agent into the ceramic compound. Typical nucleating agents include phosphorus oxide (P_2O_5), fluorides, titanium oxide (TiO_2), platinum, and zirconium oxide (ZrO_2). These nucleating agents are barely soluble in glass and remain in solution at high temperatures. At lower temperatures the agents precipitate out of solution to become seeds or nuclei around which crystals grow. Therefore, the process shown in Figure 7-47(a) involves allowing a ceramic composition to cool down as a normal glass *(b–c),* raise slightly to nucleation temperature and hold *(d–e–f),* and then raise and hold the temperature for crystal growth *(f–g–h).* Figure 7-47 shows the formation of a glass crystal from an amorphous opal glass with fluorine-rich droplets dispersed throughout. Through heat treatment, it is transformed to a fully crystalline structure. Crystallization of glass ceramics exceeds 95% and forms grains between 0.1 and 1 μm, which are smaller than in typical ceramics. Formation of glass

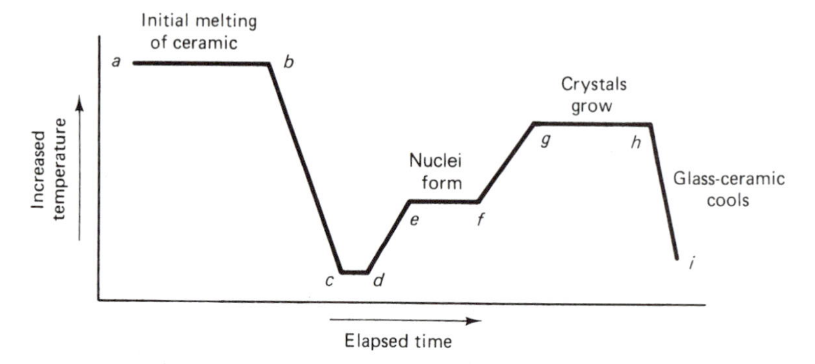

Figure 7-47 (a) Heating curve for crystal formation in glass ceramics.

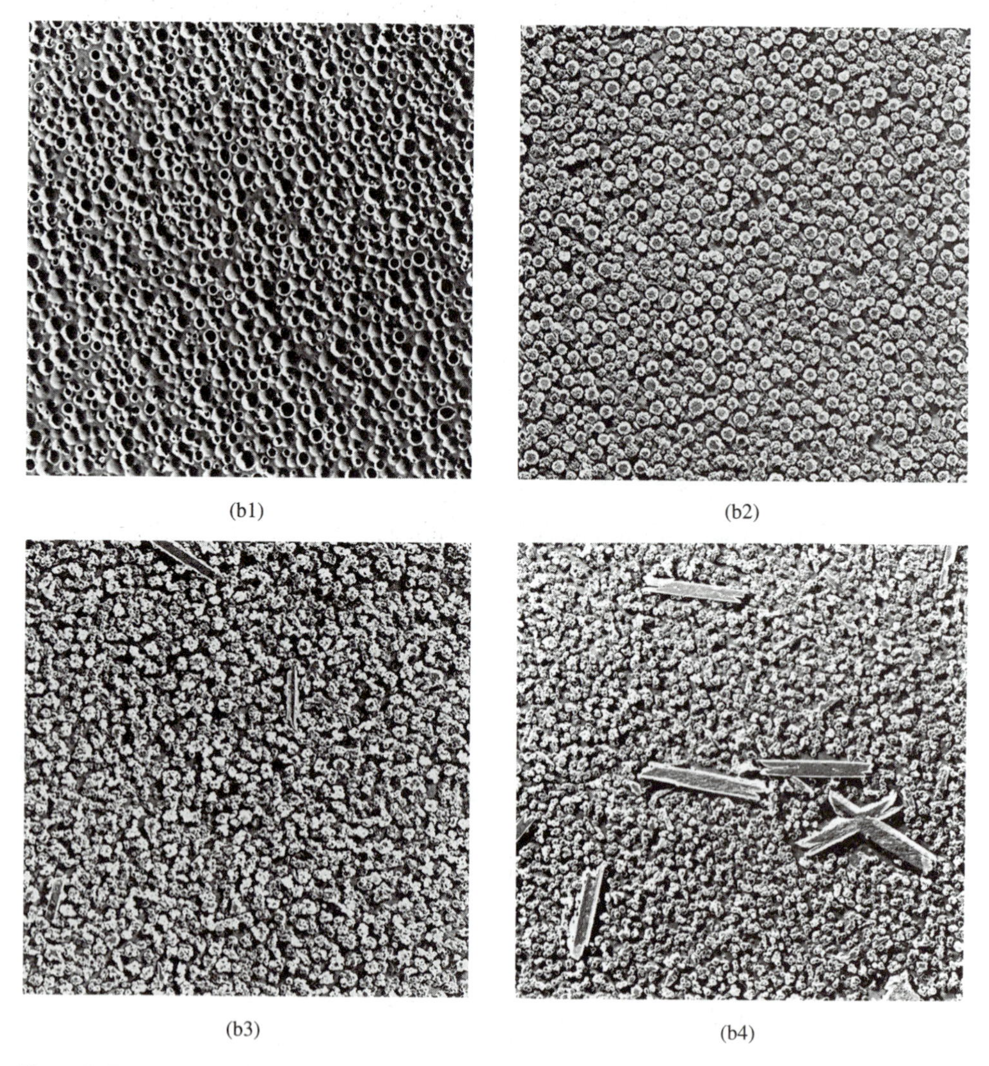

Figure 7-47 (b) Formation of glass ceramics. (b1) Droplet-imbedded parent-glass material before heat treatment. (b2) First intermediate crystal phase seen after heating to 750°C. (b3) Second intermediate crystal phase seen after heating to 825°C. (b4) Beginning of mica-crystal formation seen after heating to 850°C.

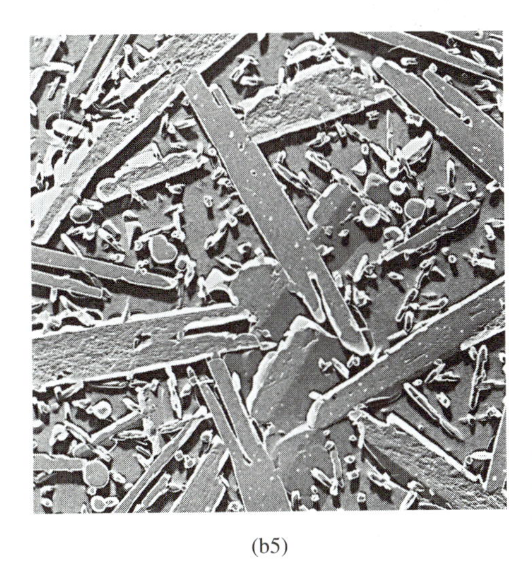

(b5)

Figure 7-47 (b5) Fully crystallized Macor glass ceramic seen after heating to 950°C. (Five micrographs from Corning Glass Works.)

ceramic parts follows the processes used for glassmaking, including pressing, blowing, drawing, and centrifugal casting. One polycrystalline glass ceramic consists of mica crystals in an opal-glass matrix. Known by the trade name ***Macor,*** it can be machined with standard metalworking tools (Figure 7-48). Randomly oriented crystals about 20 μm in length and width keep fracturing locally instead of propagating (spreading), as would a crack in an amorphous glass or plastic.

Figure 7-48 Machinable glass ceramic. (Corning Glass Works.)

7.13 SMART CERAMICS MATERIALS

Among the smart ceramics materials, ***electrorheological*** (ER) and ***magnetorheological (MR)*** fluid use inorganic particles and liquid crystal displays that change their structure with the application of an electrical current. They are covered in other parts of this book. ***Artificial nerves*** and ***sensors*** are constructed of optical fibers, and are part of fiber optic technology. The glass optical fibers can be imbedded in a host of materials or bonded to surfaces. The fibers are connected to instruments to receive signals. As a smart materials system, movement of embedded fibers breaks a light path to reveal material movement or rupture due to various external forces, such as earthquakes shaking a building or bridge.

Optical fibers embedded in composites with "smart skins" employ various light-sensing systems to indicate material damage due to variation in pressure, such as may be experienced by the surface of a model aircraft in a wind tunnel. A similar approach can be used to test designs of automobiles. ***Piezoelectric materials,*** explained in Module 9, include lead zirconate titanate ***(PZT),*** which is the most popular piezoceramic. PZT acts as a control device in intelligent materials systems for active acoustic attenuation (reducing sound), active structural damping (reducing vibration), and active damage control.

Another of the smart ceramics, ***magnetostrictive actuator materials,*** respond to magnetic fields in a manner different than PZT, which changes its microstructure from the introduction of electrical currents. This group of intelligent materials includes Terfenol-D, a compound with the rare earth element terbium. Terfenol-D is controlled with magnetic fields that bring on an alignment of magnetic domains, thereby causing controlled expansion of the magnetostrictive material. It can be used in vibration damping systems; high-torque, low-speed rotary motors; and hydraulic actuators.

7.14 MATERIALS SELECTION APPLICATION

A material(s) is needed to provide good thermal-shock resistance to serve as a liner for a high-temperature vacuum furnace. The liner will be in contact with molten aluminum. Strength requirements for the liner are minimal, but resistance to crack propagation is desired. The requirements are as follows:

1. Based on your knowledge of structures, bonding, and properties of ceramics, make a list of properties that the selected material(s) should have, based on the information given above.
2. For each property listed in (1), explain why each property was chosen (i.e., justify your choice in terms of how the property adds to the overall characteristics of the material selected).
3. List suitable candidate materials that possess all or some of the properties listed in (2).
4. Select the material(s) and justify your selection. Consult the textbook tables of data or outside sources for appropriate data to include with your justification.

Hint: Multicoatings and additives such as fibers can be considered in making your selection.

APPLICATIONS & ALTERNATIVES

The Use of Glass-Ceramics in Dentistry*

Glass-ceramics are highly crystalline ceramics with some residual glass matrix prepared by controlled crystallization of glasses (1). The unique properties of glass-ceramics allows them to be formed into complex shapes in the glassy state and to retain their finished dimensions during firing. Along with their aesthetic qualities, their high strength and toughness, insensitivity to abrasion damage, chemical durability, thermal-shock resistance, and polishability make them appropriate materials for use in dental restorations.

The use of glass-ceramics in dentistry was first proposed by W. T. MacCulloch in 1968 to provide an alternative to the dental porcelains used at that time. He produced denture teeth using a continuous glassmolding process. He also proposed the fabrication of dental crowns and inlays by centrifugal casting of molten glass.

Glass-ceramics are used in the fabrication of crowns, inlays, onlays, veneers, and inserts for composite restorations. The glass-ceramics used in these applications are based on three different crystalline phases: (1) tetrasilicic fluoromica, (2) β-quartz, and (3) leucite (2,3,4).

One of the first commercially available castable glass-ceramic materials used in dental restorations was a fluoromica containing glass-ceramic. This glass-ceramic is used in crowns, inlays, onlays, and veneers. The microstructure of this glass-ceramic consists of small, interlocking fluoromica crystals that have a platelike morphology. The interlocking of these crystals results in greater strength and reinforcement. These crystals also help to divert fractures on a microscopic level and permit the glass-ceramic to be tooled with rotary instruments without physical deterioration. This material is supplied in the form of silica-based glass ingots containing magnesium fluoride, which serves as a nucleating agent for devitrification. The refractive index of the crystals is similar to that of glass surrounding them; this condition helps to maintain translucency in the glass-ceramic casting

*This section is a portion of a paper entitled "Dental Applications of Ceramics," written by Laurie A. George, ADA, Health Foundation, and published in the book entitled *Bioceramics: Materials and Applications Ceramic Transactions,* Vol. 48, Westerville, OH: American Ceramic Society, 1995.

References

1. P. W. McMillan, Chapter 1, 2nd ed., Academic Press, London, 1979.
2. Piddock and A. J. E. Qualtrough, *Glass Ceramics,* "Dental Ceramics—An Update," *J. Dent.,* 18, 5, 227–235, 1990.
3. L. A. George, F. C. Eichmiller, and R. L. Bowen, "An intrinsically colored microcrystalline glass-ceramic for use in dental restorations," *Ceramic Bulletin,* 71, 7, 1073–1076, 1992.
4. J. K. Dong, H. Lüthy, A. Wohlwend, and P. Schärer, "Heat-Pressed Ceramics: Technology and Strength," *Int. J Prosth.,* 5, 1, 9–16, 1992.
5. D. G. Grossman, "Cast Glass Ceramics," *Dent. Clin. North Am.,* 29, 4, 725–739, 1985.
6. DICOR Technique Manual, 14, Dentsply International, York, 1988.
7. J. K. Dong, H. Lüthy, A. Wohlwend, and P. Schärer, "Heat-Pressed Ceramics: Technology and Strength," *Int. J. Prosth.,* 5, 1, 9–16, 1992.
8. G. Beham, "IPS-Empress: A New Ceramic Technology," Ivoclar-Vivadent Report, 6, 1, 1990.

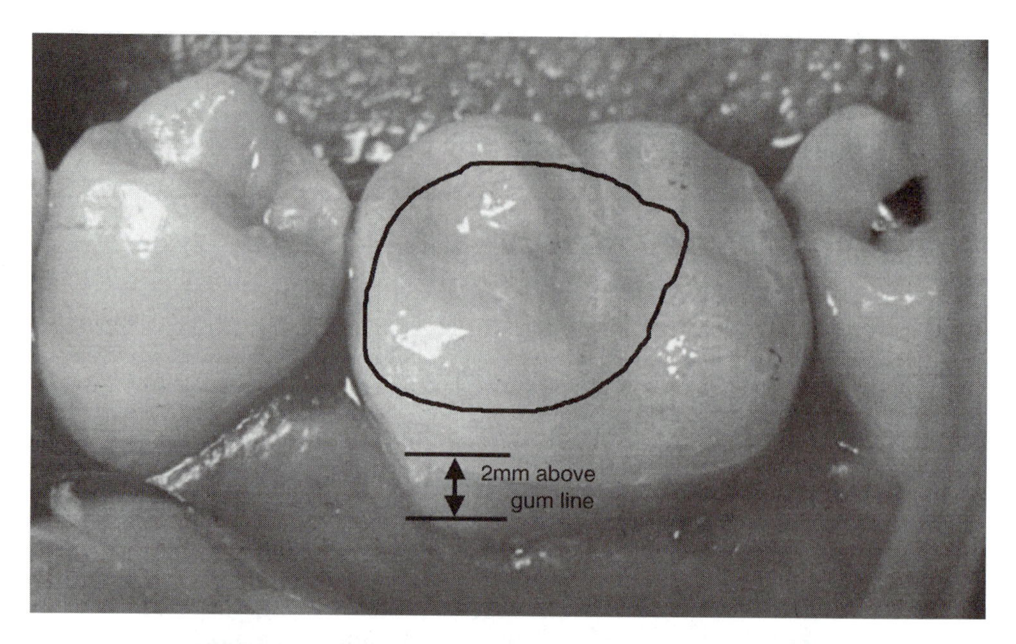

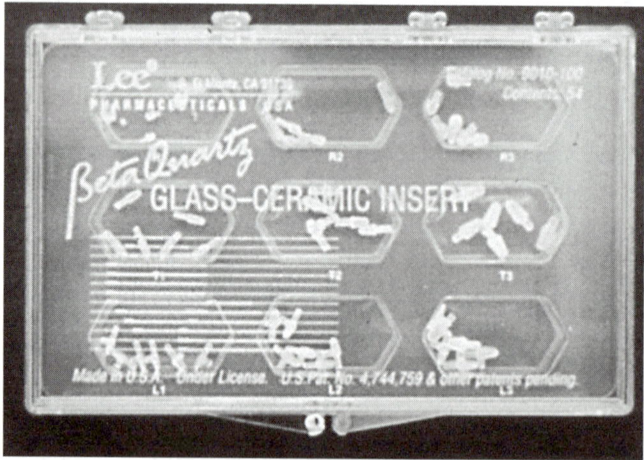

(5). Several important properties of dental restorative materials, including density, hardness, wear resistance, thermal conductivity, and radiographic density, are similar for this glass ceramic and tooth enamel.

Restorations are formed from this glass ceramic by using laboratory techniques similar to the lost-wax casting process used in cast-metal restorations. A wax pattern is invested in a phosphate-bonded refractory. The glass ingot is melted in a crucible in the muffle of the specially designed casting machine. The resultant molten glass, which is heated to 1357°C, is centrifugally thrown into the prepared mold. Following formation, the casting is ***cerammed*** (heat treatment where structure changes from amorphous, to part crystalline and part amorphous) at a temperature of 1075°C for 6 hours (6). The aesthetics of

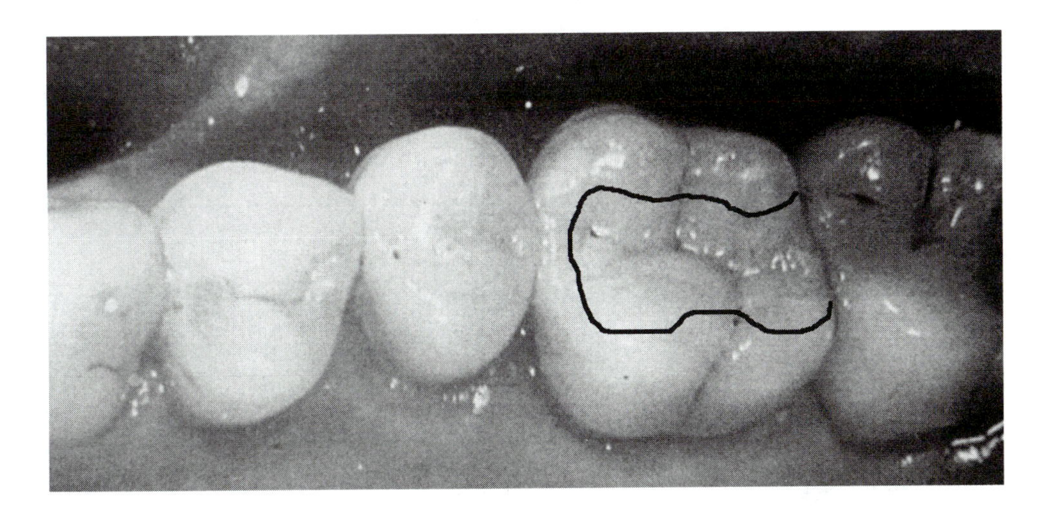

the restoration is enhanced with thin layers of tinted porcelain fused to the surface through sintering.

A second glass-ceramic used for the fabrication of crowns, inlays, and veneers is a leucite-containing glass-ceramic (i.e., the crystalline portion of this glass-ceramic is composed of leucite crystals). This glass-ceramic is formed from a glass composed of SiO_2, Al_2O_3, K_2O, Na_2O, B_2O_3, CeO_2, CaO, BaO, and TiO_2 (7). This material is supplied in preformed ceramic tablets.

The initial steps in the fabrication of restorations from this glass-ceramic are similar to the techniques used to produce porcelain fused to metal restorations. These steps include fabrication of the impression, waxing of the object and spruing it with a wax cylinder, and investing with a phosphate-bonded investment. The set investment, preformed ceramic tablets, and a small alumina press cylinder are placed in a burn-out oven heated to 850°C at 3–6°C/min and held for a minimum of 90 minutes. The investment ring, ceramic tablets, and the heat-saturated plunger are placed in the specially designed furnace at 700°C. Automatic pressing, vacuum, and heating processes are activated. The furnace is heated to 1100°C at 60°C/min and held at this temperature for 20 minutes. The pressing process then begins and continues until the press cylinder movement is less than 0.3 mm in 3 minutes. Once the pressing process is complete, the investment is bench-cooled to room temperature and the object is removed from the mold. Shading of the restoration is achieved with a surface staining technique or a layer technique (8).

Improvements in the physical properties of dental composite restorations, including reduced microleakage and increased stiffness and durability, have resulted from the incorporation of a glass-ceramic insert into the restoration. ***Dental composites*** are a two-phase system that uses polymer as a matrix for the dispersion of ceramic particles.

The glass-ceramic used to form these inserts is largely composed of beta-quartz solid solution. This glass-ceramic is formed from a lithium aluminosilicate glass system. The inserts are formed by using standard pressing techniques. After formation, the inserts are heat-treated to convert the glass to a glass-ceramic. Following conversion to a glass-ceramic, the inserts are coated with an organofunctional silane that provides for chemical

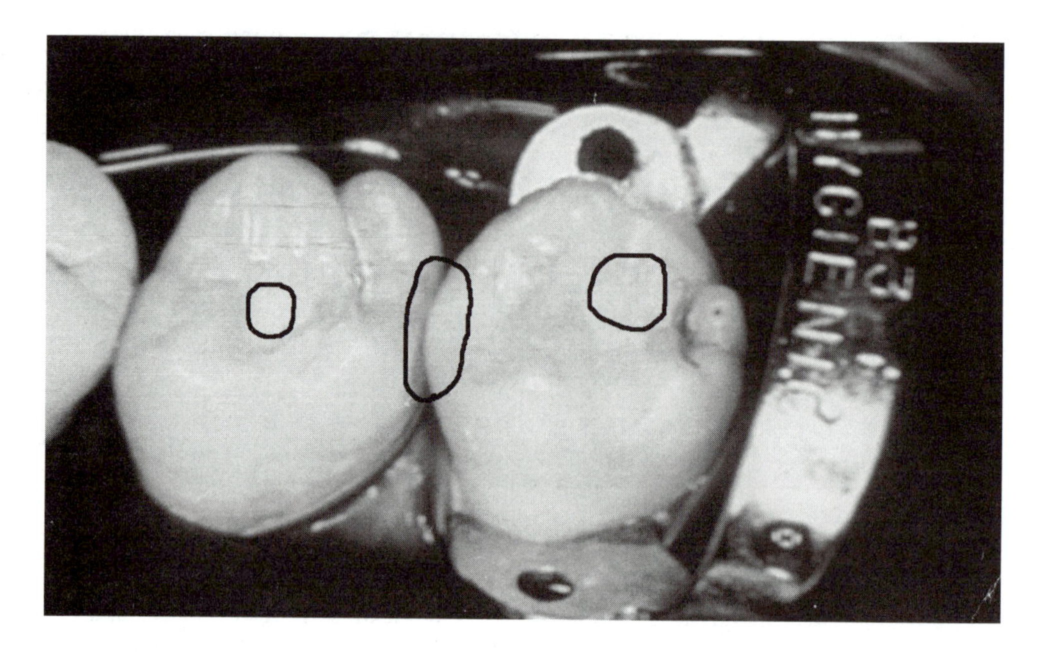

coupling between the insert and the dental composite used for the restoration. The inserts are available in three shapes and sizes, corresponding to the most frequently encountered cavity configurations. These shapes include a round cylinder, a tapered cylinder, and an L-shaped inlay. The insert fills approximately 50% to 75% of the cavity; thus the overall restoration takes on more of the advantageous properties of the insert. The inserts are intended primarily for use in posterior dental restorations.

Using standard composite filling techniques, the dentist places the inserts in a cavity preparation prefilled with composite. The excess composite extruded from the preparation as a result of insert placement is removed, and the restoration is cured with a dental curing light. The handle portion (sprue) of the insert is removed, and the restoration is finished with a series of diamond dental burs and polishing paste.

The refractive index of this glass-ceramic is similar to that of the composite resin, thus providing restorations with good aesthetic properties. The x-ray capacity of the glass-ceramic is nearly equal to that of enamel, providing for accurate interpretation in dental diagnosis. The inherent color of the inserts is similar to that of composite to allow for appropriate aesthetics.

SELF-ASSESSMENT

7-1. From the section Pause & Ponder, what are some of the goals of industry and society that have caused a push for greater uses of ceramics?

7-2. How does biomimetics represent the kind of appeal offered by ceramics for materials selection? What other attributes make ceramics competitive with metals and polymers?

7-3. Ceramics are bonded by what types of bonding mechanisms?

7-4. A material's resistance to the removal of its surface material is known by what terms?

7-5. Name one test used to measure the wear resistance of a material.

7-6. What is a suitable set of units to express thermal conductivity?

7-7. At what location does stress concentrate when a material cracks under load?

7-8. Name one test used to measure or test a material's toughness.

7-9. Explain why the machining of fine-grained ceramic parts reduces their strength and toughness more than that of large-grained ceramic parts.

7-10. What material is added to brittle ceramics to increase their resistance to crack propagation?

7-11. List two properties that a material should have to be a candidate for resisting thermal shock.

7-12. Describe one technique for reducing the emissivity of a roof covering on a two-story house located along a treeless section of an oceanfront property.

7-13. The trial-and-error method of selecting a material for a particular application is also known by what other name?

7-14. In ceramic processing the term *green* is used to describe a particular part. What condition does this term describe?

7-15. At least three ceramic processes involve the application of pressure as well as heat in the sintering of ceramic materials. Name one such process.

7-16. What symbol is used to denote absolute temperatures based on the Celsius scale?

7-17. What are the chemical name and chemical formula for alumina?

7-18. Explain the difference between cement and concrete.

7-19. Surface modification techniques are developed and employed to accomplish what particular goals?

7-20. What is the main difference between CVD and CVI processes?

7-21. Explain how a material is affected by machining with either dull or improper tools.

7-22. Why do ductile materials withstand impact better than brittle materials?

7-23. There are many reasons why fibers add toughness to a composite material. Discuss one such reason.

7-24. The word *cermet* stands for what two words?

7-25. What happens to the volume of ZrO_2 as it is heated past 1200°C?

7-26. Hydraulic cements require what addition for them to set or cure?

7-27. A radioactive material has a half-life of 10,000 years. Using the pound as a unit of mass, how much of the material is stable after 20,000 years?

7-28. Superabrasives possess high hardness and wear resistance. What other property do they have? Name a material from which they are made.

7-29. In terms of microstructure, why do crystalline ceramics have different properties from metals that are also crystalline?

7-30. Match the name of the ceramic to its chemical formula or composition.

Portland cement	Al_2O_3
Soda-lime silica glass	SiO_2
Alumina	99% SiO_2
Quartz	$Al_2(Si_2O_5)(OH)_4$
Fused silica glass	74% SiO_2, 15% Na_2O, 10% CaO, 10% Al_2O_3
Kaolinite	C_3S, C_2S, C_3A, C_4AF

7-31. Explain **(a)** how crystals are formed in glass, **(b)** when crystals are undesirable, and **(c)** when crystals are desirable.

7-32. **(a)** What effect does tempering have on glass?

(b) How does light cause photochromic lenses to darken?

(c) How does toughened glass obtain improved impact strength?

7-33. Match the ceramic to the application for which it is best suited:

(a) Tempered glass	(1) Grinding and polishing
(b) Fused silica	(2) Furnace windows
(c) Ferrite	(3) Protect metal from oxidation
(d) Glaze (porcelain)	(4) Computer memory discs
(e) 96% silica glass	(5) Storm door glazing
(f) Diamond	(6) Contain hot acids

7-34. Consider the characteristics of the general classes of ceramics, plastics, and metals. List the class that is generally superior in the following characteristics; then write down a specific material and value to illustrate its superiority: **(a)** stiffness, **(b)** chemical resistance, **(c)** abundance of raw materials, **(d)** tensile strength, **(e)** lowest coefficient of thermal expansion, **(f)** thermal insulation, **(g)** hardness, **(h)** impact strength, **(i)** flexural strength, **(j)** density.

7-35. Express a cutting tool speed of 1000 in./min in units of meters per second.

7-36. List at least 10 applications for ceramic materials within a home.

7-37. In prestressing a composite material by the layering technique, explain how the surface or coating material's coefficient of thermal expansion should differ, if at all, from that of the substrate or matrix material.

7-38. Using the data from Table 7-3, determine the SI units of TSI for graphite.

7-39. The terms *soft* and *hard* used to describe glass refer to what property of glass?

7-40. What must be done to introduce color into glass? Does the structure of glass affect its color?

7-41. What is the purpose of soaking eyeglass lenses in a molten salt bath?

7-42. Explain the difference between a glass ceramic material and the ceramic glass.

7-43. In studying the relationship of a finished product in service with its properties and the processes used to produce the product from its basic material(s), one interesting observation is that the final properties desired in a product are usually the cause of the problems that need to be overcome in its processing. An example taken from ceramics deals with the desired properties of high hardness and temperature resistance. These are the very properties that must be overcome in processing the product. Using metals and/or polymers, cite a general statement that supports the foregoing observation.

7-44. When selecting a ceramic material to serve as a heat insulator, would a material with a high density qualify? Discuss your answer.

7-45. Name one property that is most desired of a ceramic material for use as an abrasive for a grinding wheel.

7-46. When choosing a ceramic material for a cutting-tool insert, would a ceramic-composite material using SiC fibers be suitable? List the properties required of this material.

7-47. Ceramic armor, initially developed for the Vietnam war and also used in the Gulf war, was a composite made up of boron carbide (B_4C) and fiberglass plus aramid fiber matting. The ceramic has a density of 2.4 g/cm^3. Is this a low- or a high-density ceramic? Explain why you think this ceramic was chosen for this application.

7-48. Match the correct ceramic to its favorable characteristics. Some will have the same characteristics.

(a) Boron nitride	(1) Low cost
(b) Graphite	(2) Machinable
(c) Al_2O_3	(3) Best lubricity
(d) Glass ceramic	(4) Maximum hardness

7-49. Express a fracture toughness of 186 MPa · $\sqrt{m}$ in terms of psi · $\sqrt{in.}$

7-50. Write an abbreviated notation for a CMC that consists of SiC whiskers in an alumina matrix.

7-51. Explain the difference between setting and hardening of cement.

7-52. In terms of their atomic bonding, explain why metals are good conductors and most ceramics are good insulators.

7-53. Explain some of the goals sought when designing advanced, high-tech, high-performance or engineering ceramics.

7-54. Discuss the advantages of improved low-tech cements, such as CBCs, now and in the near future for developing nations and a greener materials cycle.

7-55. Name two advantages of metals over ceramics.

7-56. The text shows several photomicrographs and sketches of methods used to toughen ceramics against cracks and impact. Sketch one of those techniques.

7-57. Discuss the possibilities of zero coolant machining. Mention P/M, surface engineering, protective coatings, and environmental concerns.

OBJECTIVE QUESTIONS

7-1. A broad classification of ceramic materials that use artificial raw materials, exhibit specialized properties, and require sophisticated processing.

a. Traditional ceramics **b.** Advanced ceramics **c.** Cements **d.** Glass

7-2. Ceramics based primarily on natural raw materials of clay and silicates.

a. Traditional ceramics **b.** Advanced ceramics **c.** Cements **d.** Glass

7-3. A major objective in designing new ceramics is to overcome _____________.

a. Hardness **b.** Brittleness **c.** Reactivity **d.** Conductivity

7-4. Generally, ceramics compete well with metals when there is a need for lower weight, lower density, high shear strength, and good tensile strength.

T (true)

F (false)

7-5. Metals are usually superior to ceramics in areas such as impact strength, ductility, and plastic flow.

T (true)

F (false)

7-6. The term for ceramic's density that includes its porosity and both crystalline and noncrystalline phases.

a. Bulk density **b.** Open density **c.** Closed porosity **d.** Green density

7-7. A polymer with a backbone of SiO; the ASTM designation is SI.

a. Silicon **b.** Silica **c.** Silicone **d.** Silicates

7-8. The element Si.

a. Silicon **b.** Silica **c.** Silicone **d.** Silicates

7-9. The ceramic SiO.

a. Silicon **b.** Silica **c.** Silicone **d.** Silicates

7-10. Because of their bonding, different atoms, and unit cell structure, ceramics offer high resistance to shear and are good insulators.

T (true) **F** (false)

7-11. Using Figure 7-5, place in order from good to poor the following materials in terms of wear resistance.

a. Cast iron **b.** Silicon carbide **c.** Urethane **d.** 96% alumina

7-12. Transformation zones such as ZrO_2 serve what purpose in ceramics?

a. Thermal shock resistance **b.** Fracture toughness

c. Improved conductivity **d.** Tensile strength

7-13. Ceramic that has been molded but is still in a plastic condition.

a. Frit **b.** Bisque **c.** Glaze **d.** Greenware

7-14. Ceramic that has been fired but is porous and will absorb liquids.

a. Frit **b.** Bisque **c.** Glaze **d.** Greenware

7-15. The glassy finish such as porcelain fired onto porous ceramic.

a. Frit **b.** Bisque **c.** Glaze **d.** Greenware

7-16. An environmentally friendly concrete known as pervious concrete offers which major advantage?

a. Uses cementitious materials **b.** Porous structure allows natural seepage of water

c. Offers superior toughness **d.** Relies on glass-fiber reinforcement

7-17. A very hard glass that withstands extremes of thermal shock.

a. Soda-lime glass **b.** Polycrystalline glass

c. Lead glass **d.** 96% silica glass

7-18. A ceramic-structured material produced by controlled heat-treating of glass that has nucleating agents added to the mixture.

a. Soda-lime glass **b.** Polycrystalline glass

c. Lead glass **d.** 96% silica glass

7-19. Glass that was rapidly cooled while in the plastic state to place the surface in compression and the core in tension.

a. Annealed glass **b.** Tempered glass

c. Soft glass **d.** Chemically toughened glass

7-20. A method to immobilize hazardous waste by encasing it in glass at the site of the waste.

a. Biomimetics **b.** Densification **c.** In situ vitrification **d.** CVD

7-21. Bonding in ceramic materials is ______________.

a. Ionic **b.** Covalent **c.** Secondary

7-22. Oxidation numbers designate the number of ______________ taking part in the bonding of elements during a chemical reaction.

a. Protons **b.** Atoms **c.** Electrons **d.** Ions

7-23. Mullite, an intermetallic compound, is a ceramic with many applications that take advantage of its properties dealing with _____________.

a. High strength **b.** High conductivity
c. High temperature **d.** Porosity

7-24. Normal sintering depends on convection heating as well as heating by _____________.

a. Contact **b.** Radiance **c.** Conduction **d.** Microwave

7-25. Spraying a protective metal on foam uses _____________.

a. Plasma arc **b.** HVOF **c.** Wire arc

7-26. The ceramic with the lowest conductivity.

a. Thoria **b.** Al_2O_3 **c.** ZrO_2

7-27. The limit on the number of coatings that can be applied to a substrate.

a. Two **b.** One **c.** No limit

7-28. In machining, the tool costs are _____________.

a. Greater than the coolant-related costs
b. Less than the coolant-related costs
c. The same as the coolant-related costs

7-29. The process in which high temperatures cause some ceramics to melt into a glassy matrix.

a. Sintering **b.** Atomization **c.** Vitrification **d.** Annealing

7-30. The process in which clay is heated to change a soft material to a hard thermoset.

a. Sintering **b.** Atomization **c.** Vitrification **d.** Annealing

7-31. Most often the structure of this material is amorphous.

a. Diamond **b.** Cement **c.** Ceramic **d.** Glass

7-32. The measure of energy absorbed in creating a unit area of a new crack ahead of an existing crack.

a. Impact **b.** SENB **c.** Flaw **d.** Work of fracture

7-33. A ceramic metal-matrix composite (densely packed SiC particles in Al_2O_3) produced by the Dimox process that gives maximum wear resistance.

a. CVI **b.** RHM **c.** ALANX **d.** CFCC

7-34. Extremely hard metals such as tungsten carbide or molybdenum that are used for cutting tools.

a. Cement **b.** RHM **c.** Cermet **d.** Abrasives

7-35. A ceramic, such as masonry mortar, used to bond bricks together.

a. Cement **b.** RHM **c.** Cermet **d.** Abrasives

7-36. Ceramic particles floating in a liquid such as water.

a. Slurry **b.** Cement **c.** Amorphous **d.** Composition

REFERENCES & RELATED READINGS

AMERICAN SOCIETY FOR TESTING AND MATERIALS. *Annual Book of ASTM Standards* Part 17. *Refractories, Glass and Other Ceramic Materials; Carbon and Graphite Products.* Philadelphia: ASTM, 1991, or current edition.

ANDERSON, RICHARD M. "Testing Advanced Ceramics," *Advanced Materials and Processes,* Vol. 135, No. 3, 1989.

BEHRENDT, DONALD R., and MRITYUNJAY SINGH. *Melt-Infiltration Process for SiC Ceramics and Composites.* Lewis Research Center (LEW-15767), *NASA Tech Briefs,* August 1994, p. 59.

BESMANN T. M., B. W. SHELDON, R. A. LOWDEN, and D. P. STINTON. "Vapor-Phase Fabrication and Properties of Continuous-Filament Ceramic Composites," *Science,* Vol. 253, September 6, 1991.

BONDS, JAMES W., et al. "Wire-Arc Spraying of Metal onto Insulating Foam." Rockwell International Corp. for Marshall Space Flight Center, *NASA Tech Briefs,* May 1995, p. 84.

BUNNELL, L. ROY. "Tempered Glass and Thermal Shock of Ceramic Materials," *NEW:Update 91:* Oak Ridge National Laboratory, Oak Ridge, Tenn., November 12–14, 1991.

CORNING GLASS WORKS. *All about Glass.* Corning, N.Y.: CGW, 1968.

CORNING GLASS WORKS. *Properties of Glasses and Glass-Ceramics.* Corning, N.Y.: CGW, 1973.

CRAIG, DOUGLAS F. "Structural Ceramics," *NEW:Update 91:* Oak Ridge National Laboratory, Oak Ridge, Tenn., November 12–14, 1991.

CSELLE, TIBOR. "New Directions in Drilling." Applications Research & Development, Guhring, Inc., Brookfield, WI, *Manufacturing Engineering,* August 1995, pp. 77–80.

DOUBLE, DAVID D. "Chemically Bonded Ceramics; Taking the Heat Out of Making Ceramics," *Journal of Materials Education,* Vol. 12, No. 5/6, 1990, p. 353.

EVANS, JAMES W., and LUTGARD C. DEJONGHE. *The Production of Inorganic Materials.* New York: Macmillan, 1991.

GEORGE, LAURIE A. "Dental Applications of Ceramics," in *Bioceramics: Materials and Applications Ceramic Transactions,* Vol. 48, Westerville, OH: American Ceramic Society, 1995.

HARPER, SCOTT. "New-Age Concrete Acts Like a Big Filter." *The Virginian-Pilot,* Sept. 7, 1995, p. B3.

JORDAN, GAIL W. "Adapting Archimedes' Method for Defining Densities and Porosities of Small Ceramic Samples," *NEW:Update 90,* NIST, Gaithersburg, Md., November 12–14, 1990.

LEHMANN, RICHARD L. "Primer on Engineering Ceramics," *Advanced Materials and Processes,* Vol. 141, No. 6, 1992.

MACCHESNEY, J. B. "The Materials Development of Optical Fiber: A Case History," *Journal of Materials Education,* Vol. 11, No. 4, 1989, p. 321.

MCCOLM, I. J. *Ceramic Hardness.* New York: Plenum Press, 1990.

MPR. "Net and Near-Net Shaped Ceramic Composites Formed by Lanxide Technology," *Materials and Processing Report,* Vol. 3, No. 3, 1988.

OMETITE, OGBENI O., MARK A. JANNAYA, and RICHARD A. STREHLOW. "Gelcasting: A New Ceramic Forming Process," *American Ceramics Society,* Vol. 64, No. 12, 1985.

PFAENDER, HEINZ G. *Schott Guide to Glass.* New York: Van Nostrand Reinhold, 1983.

REED, JAMES S. *Introduction to the Principles of Ceramic Processing.* New York: Wiley, 1988.

RICHERSON, DAVID W. *Modern Ceramic Engineering: Properties, Processing and Use in Design,* 2nd ed. New York: Marcel Dekker, 1992.

ROGERS, CRAIG A. "Intelligent Materials Systems: The Dawn of a New Materials Age." Center for Intelligent Materials and Structures, Virginia Polytechnic Institute and State University, 1992.

ROY, DELLA M., ed. "Instructional Module in Cement Science," *Journal of Materials Education.* Materials Education Council, 1985.

SWAIN, BERTA. "Microwave Processing of Ceramics," *Advanced Materials and Processes,* Vol. 134, No. 3, 1988, p. 76.

URQUHART, A. W., G. H. SCHIROKY, and B. W. SORENSON, "Ceramic Composites for Gas Turbine Engines via a New Process," ASME 89-GT-316, *Gas Turbine and Aeroengine Congress and Exposition,* Toronto, Canada, June 4–8, 1989.

Periodicals

Advanced Materials & Processes
Ceramic Bulletin
Ceramic Monthly
Journal of Adhesion Science and Technology
Manufacturing Engineering

EXPERIMENTS & DEMONSTRATIONS IN CERAMICS

NEW:Update 88 NASA Conference Publication 3074

COLORADO SUPERCONDUCTORS, INC. "High-Temperature Superconductors: A Technological Revolution."

NELSON, JAMES A. "Glasses and Ceramics: Making and Testing Superconductors."

SCHULL, ROBERT D. "High T_c Superconductors: Are They Magnetic?"

NEW:Update 89 NASA Conference Publication 3074

BEARDMORE, PETER. "Future Automotive Materials: Evolution or Revolution."

BUNNELL, L. ROY. "Hands-On Thermal Conductivity and Work-Hardening and Annealing in Metals."

LINK, BRUCE. "Ceramic Fibers."

NAGY, JAMES P. "Austempering."

RIES, HEIDI R. "Dielectric Determination of the Glass Transition Temperature."

NEW:Update 90 NIST Special Publication 822

DAHIYA, J. N. "Dielectric Behavior of Superconductors at Microwave Frequencies."

JORDAN, GAIL W. "Adapting Archimedes' Method for Determining Densities and Porosities of Small Ceramic Samples."

SNAIL, KEITH A., LEONARD M. HANSSEN, DAVID B. OAKES, and JAMES E. BUTLER. "Diamond Synthesis with a Commercial Oxygen-Acetylene Torch."

NEW:Update 91 NASA Conference Publication 3151

BUNNELL, L. ROY. "Tempered Glass and Thermal Shock of Ceramic Materials."

CRAIG, DOUGLAS F. "Structural Ceramics."

DAHIYA, J. N. "Dielectric Behavior of Semiconductors at Microwave Frequencies."

WEISER, MARTIN W., DAVID N. LAUBEN, and PHILIP MADRID. "Ceramic Processing: Experimental Design and Optimization."

NEW:Update 92 NASA Conference Publication 3201

BUNNELL, L. ROY. "Temperature-Dependent Electrical Conductivity of Soda-Lime Glass."

HENSHAW, JOHN M. "Fracture of Glass."

STEPHAN, PATRICK M. "High Thermal Conductivity of Diamond."

NEW:Update 93 NASA Conference Publication 3259

BUNNELL, L. ROY, and STEPHEN PIIPPO. "Property Changes During Firing of a Typical Porcelain Ceramic."

BURCHELL, TIMOTHY D. "Developments in Carbon Materials."

DAHIYA, J. N. "Dielectric Measurements of Selected Ceramics at Microwave Frequencies."

KETRON, L. A. "Preparation of Simple Plaster Mold for Slip Casting."

MASI, JAMES V. "Experiments in Diamond Film Fabrication in Table Top Plasma Apparatus."

WERSTLER, DAVID E. "Microwave Sintering of Machining Inserts."

NEW:Update 94 NASA Conference Publication 3304

BUNNELL, L. ROY, and STEVEN PIIPPO. "The Development of Mechanical Strength in a Ceramic Material During Firing."

LONG, WILLIAM G. "Introduction to Continuous Fiber Ceramic Composites."

REIFSNIDER, KENNETH L. "Designing with Continuous Fiber Ceramic Composites."

WEST, HARVEY A., and F. XAVIER SPIEGEL. "Crystal Models for the Beginning Student: An Extension to Diamond Cubic."

NEW:Update 95 NASA Conference Publication 3330

LOUDEN, RICHARD A. "Testing and Characterizing of Continuous Fiber Ceramic Composites."

ROY, RUSTUM. "Diamond Research Overview and a Model for Lab Experiments Using Oxyacetylene Torch."

Module 8
Composite Materials

After studying this module, you should be able to do the following:

8.1. Sketch the Venn diagram that shows how the sets of three main Family of Materials groups intersect with reinforcers to form subsets of polymer matrix composites (PMCs), ceramic matrix composites (CMCs), and metal matrix composites (MMCs). Also diagram how the PMC subset expands into a wide array of composites materials.

8.2. Select examples of composites and advanced composites, sketch their macrostructure with labels of constituents (matrix, interface zone, etc.) and justify their use over monolithic materials. Recall reasons to use composites and cite the applications of advanced composites.

8.3. Discuss carbon and graphite fibers in terms of their definitions; production; and important properties including coefficient of thermal expansion (CTE), specific strength, and wear (erosion) rates.

8.4. Use diagrams to explain isostress and isostrain, and use the rule of mixtures (ROM) to express the elastic modulus in tension for fiber-reinforced plastic (FRP). Explain the value of CMCs and MMCs using examples such as achieved by the Sullivan, pressureless, molten-metal infiltration (PRIMEX) processes and microinfiltrated microlaminated composite (MIMLC).

8.5. Describe the relationship among methods of fabricating composites and the properties that develop from the processes. Explain the factors in machining and repairing advanced composites.

8.6. Calculate the specific strength and specific stiffness of given materials and determine when these properties are major reasons for material selection.

8.7. Explain the uses of advanced composites in automobiles. Also recall the challenge of recycling composites and how hydroglycolysis may help.

8.8. Use composite databases to determine the applications of select composites. Describe NDE techniques to evaluate composite designs and manufacture.

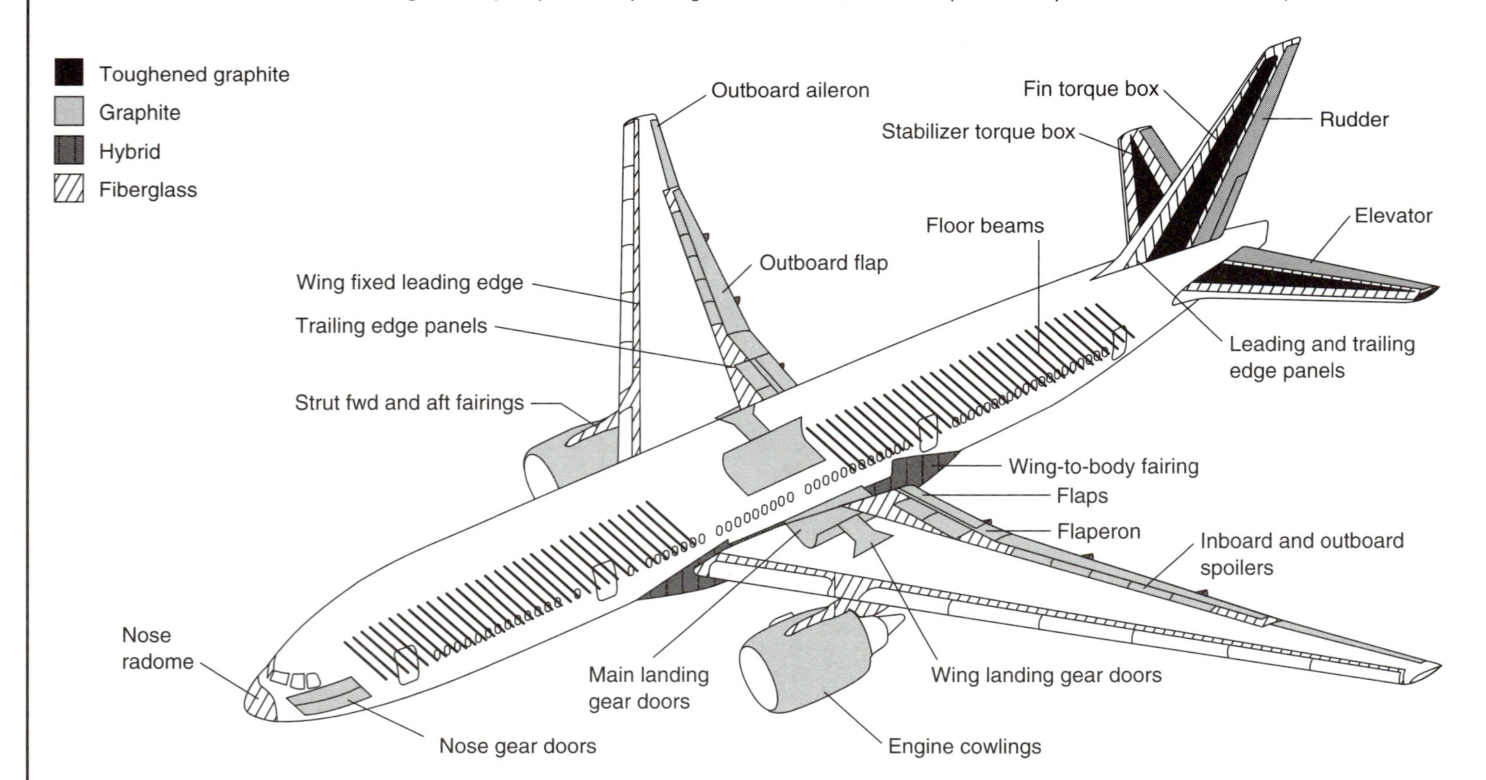

Figure 8-1 (a) The application of composites showing the advantages of reduced-weight, corrosion-free, and durable structures. (Courtesy of The Boeing Company.) The Public Broadcasting System (PBS) series "21st Century Jet" provides an excellent in-depth case study of the product development process, from design through testing, manufacturing, and delivery. The series includes many examples of engineering materials technology. Call PBS at 1-800-344-3337 to obtain further information on the video series. Also consult the book *21st Century Jet: The Making and Marketing of the Boeing 777,* Karl Sabbagh, Scribner, 1995.

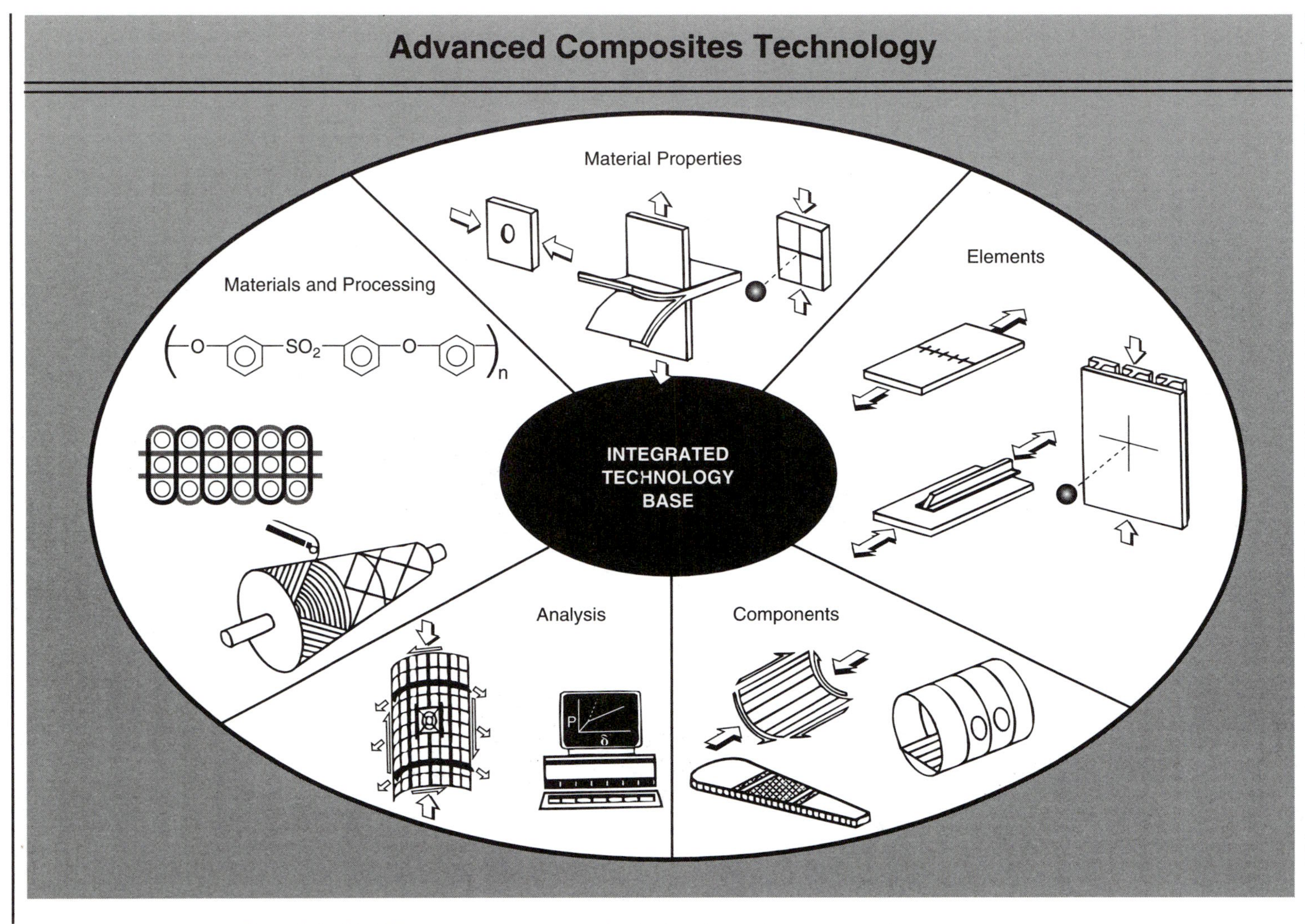

Figure 8-1 (b) ACT program uses an integrated technology base and concurrent engineering to promote reduced costs, improved manufacturing techniques, and high-quality products. (NASA-LaRC.)

PAUSE & PONDER

Economics, often the overriding factor in product and system design, takes into account all aspects of the Materials Cycle outlined in Module 1. To review some considerations in a case study of the use of advanced composites, we turn to two programs: the advanced composites technology (ACT) program and the Boeing 777 program. ACT, a scitech partnership between NASA and the U.S. aeronautical industry, works with a broad objective to develop appropriate technology to design and manufacture the primary structure for commercial aircraft. Cost-effective use of lightweight advanced composites is the emphasis. The goal of ACT is to use advanced composites so that they will provide a 25% reduction in costs below aluminum structures. The Boeing 777 is the first of a new family of wide-body airplanes that makes extensive use of carbon fiber reinforced plastics (CFRP) advanced composites [Figure 8-1(a)].

While weight savings, and corresponding fuel savings, from the use of composite materials would provide large economic benefits over the life of the aircraft, the poor economic health of the airline industry demands that the initial cost of the finished airplane be reduced if composites are to be selected. New design and innovative manufacturing techniques using polymer matrix composites are required to make them competitive because the cost per pound of composites is higher than aluminum, and current production costs of composites is twice that of aluminum aircraft structures.

Figure 8-1(b) depicts the integrated technology base to drive the ACT program to its goal. ***Concurrent engineering*** of the industrial partners involves parallel developments in materials and their processing, determination of the properties for new materials, design and manufacture of individual elements and the structural components they form, and continual analysis of elements and components. This integrated approach to advanced composite technology for primary aircraft structures, which will carry through the year 2002, should provide an answer to the viability of advanced composites for the next generation of subsonic commercial transport aircraft.

Boeing's decision to use advanced materials for its totally redesigned 777 resulted in the application of various lightweight composites, which amount to ten times more mass of composites than in their 757. These composites are approximately 9% of the new plane's structural weight. The 777 success, or lack thereof, may well predict the future of advanced composites in transport aircraft. Uses of fiberglass, as seen in Figure 8-1(a), range from the wing fixed leading edge to the strut fairings. Graphite composites' use continued, as on the Boeing 767 and 757, on rudders, elevators, and flaps. Hybrid composites served as wing-to-body fairing.

The ***toughened graphite*** [Figure 8-1(a)] is a new, toughened epoxy composite used on the fin torque box, stabilizer torque box, and floor beams. Toughened graphite addresses a major concern for composites because of its outstanding resistance to impact: nearly seven times tougher than the previous CFRP.

The costs of all aspects of design, tooling, materials, and manufacturing for the 777's wing were greater than those of a competing aluminum wing. Why? For example, in design and its related testing, more costs from additional analysis spring from a lack of experience and the databases that come with the use of aluminum. Of course, more experience with CFRP should bring reduced costs. The ACT program faces a formidable challenge to reduce cost for advanced composites. The benefits of using advanced composites, as fully developed in this module, can be enjoyed by the entire aerospace industry worldwide and will provide spinoffs for the auto industry, building construction, road construction, and other fields. The 777 represents the global nature of modern manufacturing. Components of the airplane are produced around the world and shipped to the United States and on to Everett, Washington, for assembly.

8.1 INTRODUCTION

A ***composite material*** or ***composite*** is a complex solid material composed of two or more materials that, on a macroscopic scale, form a useful material. The composite is designed to exhibit the best properties or qualities of its constituents or some properties possessed by neither. The combining of the two or more existing materials is done by physical means, as opposed to the chemical bonding that takes place in the alloys of monolithic solid materials. A true composite might be considered to have a matrix material completely surrounding its reinforcing material in which the two phases act together to produce characteristics not attainable by either constituent acting alone. Note that the insoluble phases or main constituents in a composite do not lose their identity. This is not true of solids, which are metallic alloys or copolymers whose phases are lost to the naked eye because these phases are formed as a result of natural phenomena. In other words, steel is a multiphase metal alloy but is not considered a composite. A metallic alloy of Al and Cu, with the addition of Al_2O_3, is considered a dispersion-strengthened copper composite in which a dilute copper/aluminum alloy powder is exposed to an oxidizing environment and the aluminum is oxidized within the copper matrix. The aluminum oxide particles are inert and act as inhibitors of dislocation movement and thus strengthen the alloy at high temperatures. By adding niobium (10%) to this same extruded composite in the form of a uniformly distributed powder, even greater hardness and strength are produced with only a minimal reduction in electrical conductivity. Two phase alloys are not considered composites by some because their phases are not formed in separate processes but originate in a single manufacturing process.

The above discussion reveals that the definition of a composite is still somewhat arbitrary (see Section 2.2.5). A broad definition of composite materials includes the naturally occurring composites, such as wood, as well as the synthetic or human-made composites. A recent example of the development of composites in the medical field is cited. A mineral paste composed of monocalcium phosphate monohydrate, α-tricalcium phosphate, and calcium carbonate, dry mixed, to which a sodium phosphate solution is added to form a paste, is surgically implanted into acute bone fractures and hardens within ten minutes after injection. This paste holds fractured bones in place while the native bone remodeling process replaces the implant with living bone. This process causes the in-situ formation of the mineral phase of bone. Attaining a compressive strength greater than that of long bones, the paste's tensile strength of about 2.1 MPa is comparable to long bones. This new material eliminates the need for heavy and uncomfortable casts.

In answer to the question, ***Why use composites?*** one can reply, in part, as follows:

1. To increase stiffness, strength, or dimensional stability
2. To increase toughness (impact strength)
3. To increase heat-deflection temperature
4. To increase mechanical damping
5. To reduce permeability to gases and liquids
6. To modify electrical properties (e.g., increase electrical resistivity)
7. To reduce costs
8. To decrease water absorption
9. To decrease thermal expansion
10. To increase chemical wear and corrosion resistance
11. To reduce weight
12. To maintain strength/stiffness at high temperatures while under strain conditions in a corrosive environment
13. To increase secondary uses and recyclability, and to reduce any negative impact on the environment
14. To improve design flexibility

The objective of this module is to present a fundamental knowledge of composites; the terms used in this specialized field; the nature, structure, and properties of composites; and an arbitrary classification of the major types. Examples of the applications of some composites in a limited number of industries are included with a view toward providing the readers, many of whom occupy positions in industry that require a degree of expertise in the selection of materials, with a base of knowledge of composites from which to make intelligent selection decisions.

8.1.1 Development of Composites

Composite materials have been used from earliest time. Mongol bows of the thirteenth century utilized a materials system consisting of animal tendons, wood, silk, and adhesives. The ancient Israelites used straw to reinforce mud bricks. The early Egyptians fabricated a type of plywood. Medieval swords and armor are examples of our present-day laminated metal materials. Nature has provided composite materials in living things such as seaweed, bamboo, wood, and human bone.

Such materials were recognized for their strength and light weight when used in the construction of the first all-composite, radar-proof airplane—the Mosquito—around 1940. Composite materials in the form of sandwich construction showed that primary aircraft structures could be fabricated from these materials. World War II saw the birth of glass–fiber–polyester composites for radomes and secondary aircraft structures, such as doors and fairings, which were designed and placed into production. Glass–fiber composites were recognized as valid space materials when they were picked for fabrication and production of Polaris-submarine missile casings.

In the 1950s, fiber technology identified the need for fibers that could compete in strength and stiffness when the state-of-the-art development led to high-performance glass fibers such as the S-type. In the late 1950s, research efforts focused on lightweight elements in the search for fibers of even greater strength that could compete successfully in the marketplace with aluminum and titanium. Boron fibers were the first result of this effort (1963), followed by carbon, beryllium oxide, and graphite. These filaments are surrounded by a material such as aluminum that serves as a matrix. Beryllium-oxide fiber technology developed short fibers, whereas boron research developed continuous filaments with greater strength properties. These developments, made through the collective efforts of government, NASA, industry, and universities, gave rise to ***advanced composites.*** Figure 8-2(a), a Venn diagram, shows how ceramics, metals, and polymers are integrated into sets and subsets of advanced composites. The sets of CMC, MMC, and PMC multiply further into subsets based on the form of reinforcement and the general structural arrangement, as illustrated in Figure 8-2(b) for fiber-reinforced composites, a subset of PMC. The other major ingredients needed for successful production—engineering design and manufacturing—kept pace with developments in materials and resulted in the continuous introduction of new composite materials for use throughout industry.

Figure 8-2(a) also represents the continual search for materials systems that can withstand even greater loads under hostile environments. These composite materials, with ceramic and metal matrices reinforced with similar materials, are being developed and tested for use in many high-temperature applications such as the ***High-Speed Civil Transport*** (***HSCT;*** see Figure 8-3). The National Aerospace Plane (NASP) program is developing and beginning production of advanced materials for the X-30 hypersonic research vehicle (the forerunner of the NASP) by bringing together materials expertise in U.S. aerospace companies and research laboratories.

The Advanced Composites Technology (ACT) program, introduced in the Pause & Ponder, has the broad objective of developing the technology to design and manufacture a

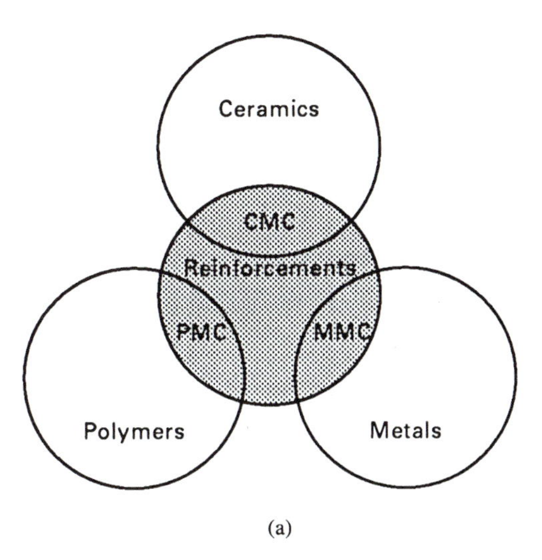

Figure 8-2 (a) Composite materials. The design process for engineering materials involves the selection of components from major members of the family of materials, which, when combined into a composite material, will possess a mix of desired material properties that will satisfy the requirements of a particular application. A ceramic selected as a matrix material combined with a reinforcement in any form and quantity produces a ceramic matrix composite (CMC). A similar combination produces a metal matrix composite (MMC) or a polymer matrix composite (PMC).

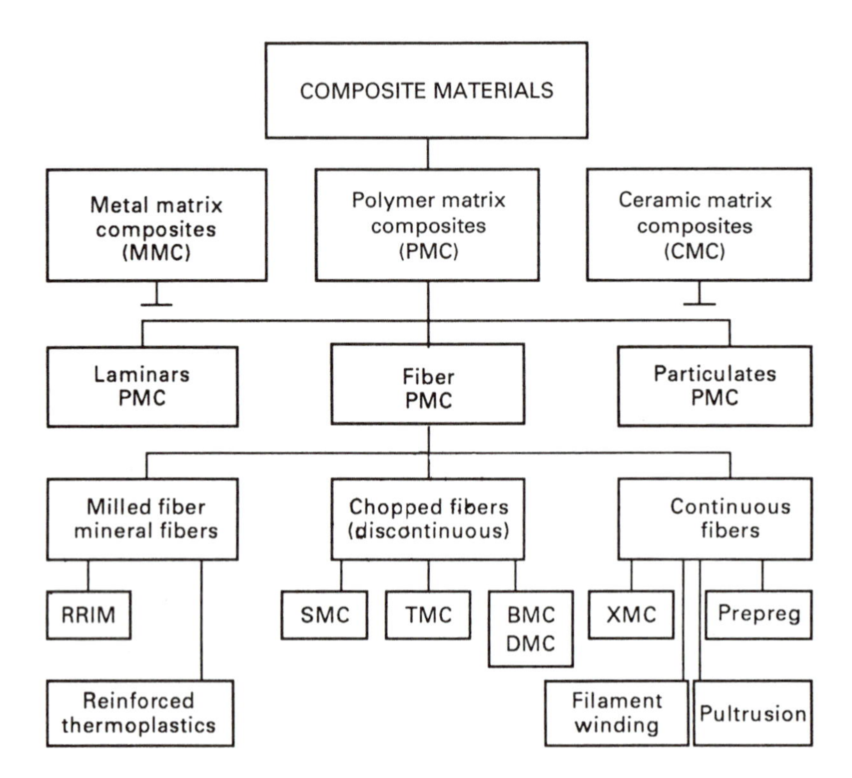

RRIM: reinforced reaction injection molding
SMC: sheet molding compound
TMC: thick molding compound
BMC: bulk molding compound
XMC: x-pattern molding compound (trademark of PPG ind.)

(b)

Figure 8-2 (b) Composite materials—their matrices and reinforcements produce an infinite array of materials.

cost-effective and structurally optimized, lightweight, composite airframe primary structure. In its first phase, the program developed three major manufacturing methods for fabricating these cost-effective composite structures, namely, ***stitched dry preform, textile preform,*** and ***automated tow placement.*** Production costs of a composite structure are now twice the costs of a corresponding aluminum airframe structure. The cost goal for the third phase of this program (1995–2002) is a 25% reduction in costs for a composite primary structure.

A second partnership effort was initiated in 1994 and will be completed in 2002. The objective of the High Speed Research (HSR) program is to develop the technology required to build a large supersonic transport aircraft that can be operated economically by the airlines without requiring high ticket prices. Current technical specifications for this

Figure 8-3 High Speed Civil Transport (HSCT), under development by Boeing for certification in the year 2000, will carry 250 passengers 5000 miles at a speed of mach 2.4. New materials and improvements in conventional aluminum, titanium, and steel alloys and composites will find use in its construction only if they meet stringent cost-per-pound-saved targets or solve a current corrosion or structural design problem. (NASA.)

High Speed Civil Transport (HSCT) call for a speed regime of mach 2.0 to 2.4 (M2.0 to 2.4), seat capacity for 250–300 passengers, flight range of 6000 to 7000 miles, and a design life of 20 years and 72,000 flight hours. (See Figure 8-3 for the HSCT concept being developed by Boeing.) There are no current commercially available composite materials, suitable for an airframe primary structure, that have an acceptable mechanical property retention for 72,000 hours at elevated temperatures under aircraft flight service conditions. New materials are one of the two most important technology development challenges for the next generation of supersonic transport.

It is reasonable to speculate that materials designers will eventually use materials that do not yet exist. This materials effort calls for a "synergistic advancement" of technology where a partnership between materials scientists and materials users will, for the first time, be put into practice. Boeing Aircraft Company is conducting feasibility studies and technology development for a long-range, high-speed civil transport with a year 2000 target certification. Payload requirements call for approximately 250 passengers and a range of 5000 miles (transpacific) at a speed range of mach 2.0 to 2.4. Development of advanced materials is essential to the success of this program.

Down on the ground, designers of the full range of products, which include clothing, athletic gear, appliances, furniture, office supplies, home entertainment systems, packaging, computer systems, boating, camping gear, and housing, have called on materials technology for new composite systems that make better products. Shoes are one example of a line of products making use of new composite systems. Look at inflatable basketball and tennis shoes and cut open a pair of jogging shoes to find the air chambers and wide variety of polymers in the form of rigid plastics, soft foams, dense foams, nylon mesh, and hard rubber. Designers used CAD technology coupled with materials technology to keep the marketplace humming with ever-improving products. The customer should be the winner.

A composite consisting of 80% advanced polymer and 20% aluminum has been developed for automotive applications by XCORP, Beverly Hills, California, at a cost said to be comparable to that of stamped steel. Designated "Composite X," it is reportedly resistant to abrasion and wear, and offers superior crashworthiness. The material is one of a group XCORP calls ***environmental composites*** because they are easily recycled and are made using low energy and a clean manufacturing system. The company has applied for more than 50 patents covering the composition of the materials, proprietary fastening techniques, molding systems, production, and recycling methods. XCORP has combined these technologies to produce its Series 2 SuperCar, which has an all-composite driveshaft, aluminum/polymer sandwich bumpers, and a variety of other composite parts. Known as the 444 Bolero Coupe, its anticipated mileage is about 90 miles per gallon and its sticker price is about $14,000.

Due to the many advances in recent years in composite systems, not only as spinoffs from the nation's space program but as a consequence of the growing energy conservation and concern about the depletion of the world's nonrenewable materials resources by more and more citizens, the emphasis placed on the group of composite materials in this module is deemed appropriate. This is not an attempt to downgrade the importance of the many other composite materials, without which our economic well-being would suffer adverse consequences. In industries' new thrust for "design for manufacturability" and "design for automation," designers seek to reduce the number of components in a product, eliminate mechanical fasteners, and design to facilitate assembly by robots. Composites offer great promise to help meet at least some of these goals.

8.2 CONSTITUENTS OF FIBER-REINFORCED COMPOSITES

A ***fiber-reinforced composite*** is a material system made primarily of varying amounts of particular fiber reinforcement embedded in a protective material called a matrix, with a coupling agent applied to the fiber to improve the adhesion of the fiber to the matrix material. Many of the terms in this modern materials technology field originated in the textile industry. A few such terms are selected for brief definitions to facilitate an understanding of these highly important engineering materials. See the glossary at the end of this module, Section 8.13, for other definitions.

High-performance fiber-reinforced composites exhibit higher structural performance than that of commercial E-glass and polyester composites. The degree of performance over a fiberglass composite depends on the fiber, its orientation and loading, and the matrix. A significant processing aspect that affects performance is the degree to which the fibers are saturated with wet matrix, leaving few or no voids. The matrix's ability to adhere to the fibers, as well as its inherent toughness, also make a contribution. As for fibers, their loading (density); continuous lengths (high aspect ratios); and orientation, including straightness, degree of knitting, and interlacing, all play a role in contributing to the properties.

In the study of atomic structure, we learned that a material's strength in a certain direction or plane is directly proportional to the type of atomic density and the atomic bonding existing in that particular direction or plane. If the atoms are tightly packed and are bonded with strong primary bonds, the material will possess a high degree of strength. We also learned that properties, particularly in brittle materials, are adversely affected by defects or flaws. The larger the flaw, the less the ability to withstand any degree of load. As an example of the application of this knowledge to the production of carbon fibers (to be discussed), we recall that carbon, an element, is polymorphic. In Figure 2-14, the structure of graphite was sketched showing two basal planes, consisting of atoms covalently bonded, which were connected in a lateral direction by weak secondary bonds. In the production of carbon fiber, for example, the original precursor material is a similar state. As part of the process, the basal planes are oriented along the axis of the fiber, thus placing these atoms with the strongest bonding in the best position to withstand a longitudinal force. Defects in the atomic structure or macrodefects on the surfaces of the fibers are dealt with by reducing the amount of cross-sectional area in the fiber. The chance of a defect remaining when the diameter of a fiber decreases is minimized considerably. This last statement explains the ongoing program to produce very fine fibers, down to the nanometer range.

8.2.1 Fiber

A ***fiber*** is the basic individual filament of raw material from which threads or fabrics are made. Fibers constitute one of the oldest engineering materials. Jute, hemp, flax, cotton, and animal fibers have been used from the earliest days of history. Today, fibers can be organic (plant, animal, and mineral), synthetic (human-made from a polymer or ceramic material), or metallic. ASTM further defines a fiber as having a length at least 100 times its diameter, with a minimum length of at least 5 mm. A fiber can be a filament or a staple. Manufactured polymeric fibers were discussed in Module 6. ***Filaments*** are long, continuous fibers, whereas ***staple*** fibers are less than 150 mm in length. Most natural fibers are in staple form, whereas synthetic fibers may exist in both forms.

A ***whisker*** is a human-made, nearly perfect, single crystal with a diameter ranging from about 1 to 10 μm and lengths up to 3 cm. Because it contains few crystalline defects in the direction of the applied load, its strength approaches a theoretical value of more than 6.9×10^3 MN/m^2. Human-made whiskers of many ceramic materials, such as aluminum oxide (Al_2O_3), silicon carbide (SiC), and silicon nitride (Si_3N_4), with hardness values approaching that of diamond, are used as reinforcements in composite materials.

Continuous fibers are essentially infinite in length and extend continuously throughout the matrix in a composite material.

Discontinuous fibers are not less than 3 mm in length and tend to orient themselves in the direction of resin flow as a composite material is processed.

Other terms relating to fibers and fiber composites can be found in Section 8.13.

8.2.2 Matrix

The functions of a ***matrix,*** the binder material, whether organic, ceramic, or metallic, are to support and protect the fibers, the principal load-carrying agent, and to provide a means of distributing the load among and transmitting it between the fibers without itself fracturing. When a fiber breaks, the load from one side of the broken fiber is first transferred to the matrix and subsequently to the other side of the broken fiber and to adjacent fibers. The load-transferring mechanism is the shearing stress in the matrix. Typically, the matrix has a considerably lower density, stiffness (modulus), and strength than those of the reinforcing fiber material, but the combination of the two main constituents (matrix and fiber) produces high strength and stiffness, while still possessing a relatively low density.

The matrix serves as the structure, keeping the reinforcing fibers at the same distance from the bending axis, while the filler provides the desired properties. There is no significant chemical reaction between the two phases except for the bonding action at their interface. If this were not true, any reaction between the two materials would have a varying negative effect on the inherent properties of the filled composite.

The structure or matrix of a filled composite can take a porous or spongelike structural form. This network of open pores can be impregnated with a variety of materials, from plastics to metals to ceramics, and the final shape of the filled open-pore composite is formed during the processing. Finally, the matrix can have a predetermined shape and size formed by an open honeycomb core made of metal impregnated with a ceramic filler for high-temperature applications.

Maximum operating temperatures for resins as matrices in fiber-reinforced composites are listed in Table 8-1 to give some idea of their limitations. No attempt is made to introduce the element of time in these figures, other than to state that the time a material is subjected to these extremes in temperature is a critical factor that must be taken into consideration in any design. An example taken from the literature illustrates how this heat-resistant property is described for a series of polyimide molding compounds, Tribolo PI-600, available from Tribol Industries. Such compounds perform at 316°C continually and 427°C intermittently. A typical molding retains 56% of its mechanical strength after 100 h at 316°C. Weight loss is 2.1% after 500 h.

Both thermosetting (TS) plastics and thermoplastic (TP) plastics are used with hybrid composites, defined later. Epoxies are the chief TS, but thermoset polyester is a strong contender, particularly for auto applications. Epoxies have a wide range of properties. Intermediate modulus epoxies are used in hybrids. They have good high-temperature properties but suffer from moisture absorption when exposed to temperatures near 180°C. Poly(ether ether ketone) (PEEK), a thermoplastic that is tougher and potentially less ex-

TABLE 8-1 OPERATING TEMPERATURES FOR COMMON POLYMER RESINS

Resin	Maximum temperature (°C)
Polyester	Room temperature (RT)
High-performance thermoplastic polyesters	150
Epoxies	200
Phenolics	260
Polyimides	300
Polybenimidazole	Above 300

pensive to manufacture than conventional resins, is used as doors of F-15E fighters and T-38 trainer aircraft. Polyimides find use for extended operations near 260°C and above. TS polyester, formulated to be hard, brittle, or tough, is used with chopped glass in many applications.

TP resins are the latest matrix materials, and an example of TP use for automotive components is seen later in Figure 8-4(d). Nylon; polysufone (PSU); poly(phenylene sulfide) (PPS); and poly(phenylene terephthalate) (PET), better known as thermoplastic polyester, are used in hybrids and advanced composites. Also, bismaleimide (BMI), which is manufactured like conventional composites, is able to withstand higher operational temperatures and has been tested similar to the manner described above for PEEK. Improved resin-transfer molding (RTM) resins have a longer pot life and lower viscosity. User-friendly prepreg resins have better controlled reactivity and an extended out-time.

Resins can be divided into two main groups: structural pastes and structural film adhesives. Pastes can be stored for one year at room temperature (RT) and cure at RT. Four versions are available: one is aluminum filled for easy handling and high compressive strength, another is a thixotropic nonmetallic for composite surfaces and random structures, another is an unfilled version designed for wet-layup composite repairs, and the final version has high shear modulus at elevated temperatures and can be stored 6 months at RT. Structural film adhesives resist moisture in the bond line and are extremely resistant to severe environmental exposure. They can be cocured with the composite structure and features at 250° or 350°F.

Polyimide molding compounds capable of withstanding continuous-use temperatures ranging from 480° to 600°F and intermittent service at 800°F are now available. A typical molding retains 56% of mechanical strength after 100 h at 600°F. Weight loss is 2.1% after 500 h. All are compatible with reinforcing fibers.

Glass ceramics, polycrystalline solids produced by the controlled crystallization of glass, have highly desirable properties for service as matrix materials. Dimensional stability, thermal stability, heat resistance, corrosion resistance, and the ability to maintain strength at high temperatures make these materials attractive despite their inherent brittleness. Their coefficient of thermal expansion is zero up to about 450°C. Glass ceramic matrices act just the opposite to polymer– and metal–matrix materials because the matrix fails first, having the lowest strain to failure. Failure occurs when the matrix viscosity becomes too low to transfer the load to the reinforcement. Thermal compatibility between the matrix and fibers must be good. If the coefficients of thermal expansion are not similar, the brittle matrix will not comply with any differences. Also, the need for high forming temperatures to bring about matrix flow among the reinforcing fibers requires thermochemical compatibility between the fibers and matrix material to prevent reactions at the fiber–matrix interface that produce unwanted interphases that degrade the fibers and reduce their strength.

Some of the most common glass-ceramic matrices are litha–alumina–silicate (LAH; $Li_2O_2–Al_2O_3–SiO_2$); silica, SiO_2; and $BaO–SiO_2–Al_2O_3–Si_3N_4$ with SiC (Nicalon), the most commonly used reinforcing fiber for glass-ceramic matrices.

A ***coupling agent,*** also known as a bonding agent or binder, provides a flexible layer at the interface between fiber and matrix that will improve their adhesion and reduce the number of voids trapped in the material. ***Voids,*** air pockets in the matrix, are harmful because the fiber passing through the void is not supported by the matrix. The fiber under load may buckle and transfer the stresses to the weaker matrix, which could crack.

Interface and interphase. By definition, composites contain a combination of mutually insoluble components (constituents). With fiber-reinforced plastics (FRPs), the fibers and the polymer resin are the main constituents. A wide variety of other material additives, such as coupling agents, fillers, pigments, catalysts, and fire retardants, are also used to satisfy specific needs. Consequently, there exists an ***interface***(s), a surface that forms a common area or boundary similar in many respects to grain boundaries between any two constituents in a monolithic material. An ***interphase*** is the region formed between two interfaces. It is therefore a distinct phase of itself with its own identity, forming a region contiguous to two phases. Figure 8-4 includes sketches showing these main constituents of an FRP.

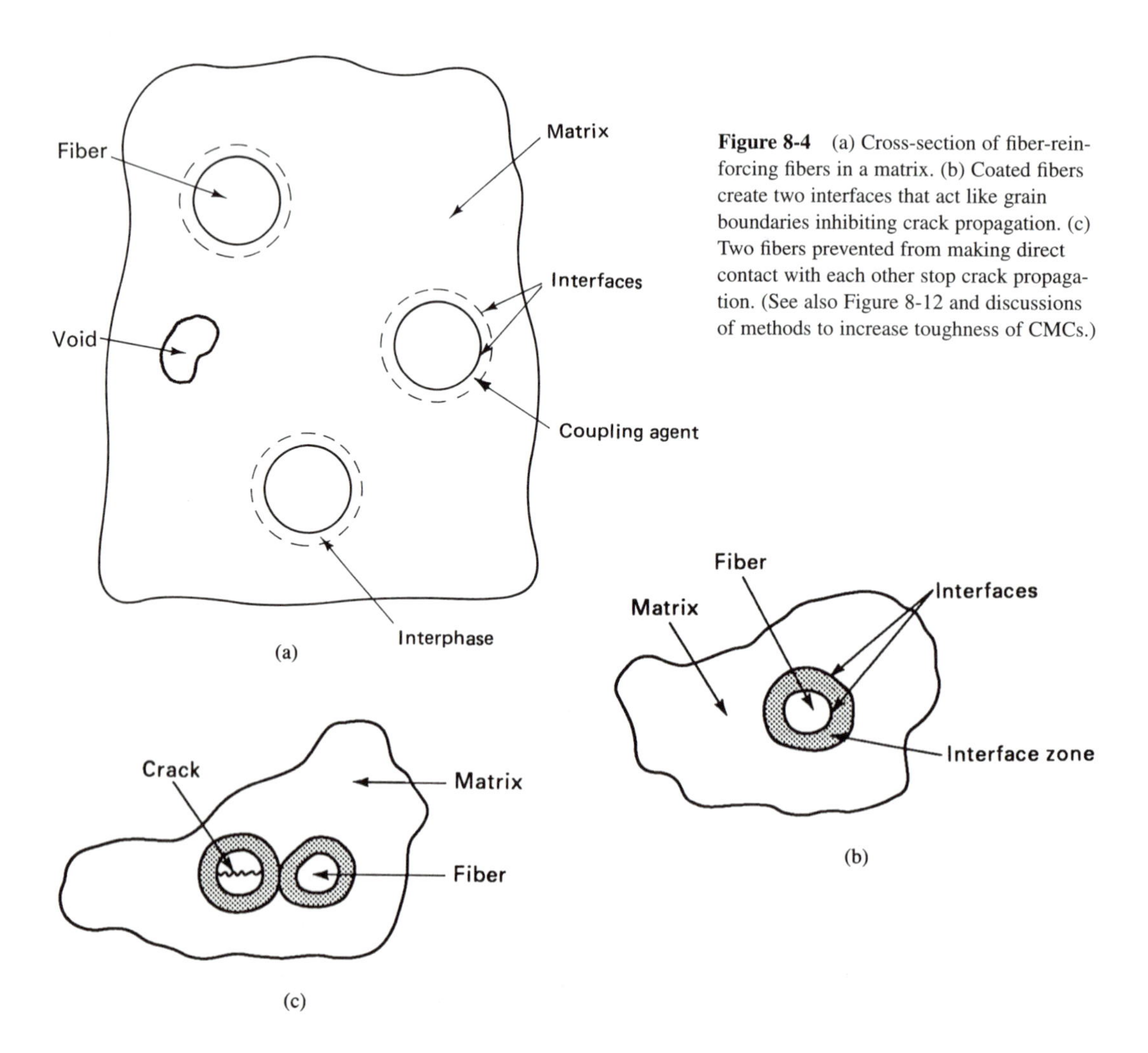

Figure 8-4 (a) Cross-section of fiber-reinforcing fibers in a matrix. (b) Coated fibers create two interfaces that act like grain boundaries inhibiting crack propagation. (c) Two fibers prevented from making direct contact with each other stop crack propagation. (See also Figure 8-12 and discussions of methods to increase toughness of CMCs.)

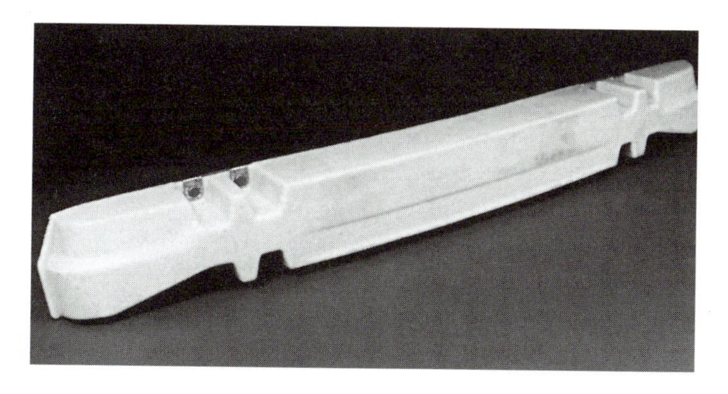

Figure 8-4 (d) TP resins serve as a matrix for glass fibers in the new thermoplastic composite used in Ford's Lincoln Continental bumper beams. AZDEL®Plus is a high-strength, lightweight, impact-resistant composite available in a layered structure of glass mat or chopped glass impregnated with thermoplastic resins. Its superior properties to polypropylene-based glass mat help meet higher bumper collision standards (5 mph versus 2.5 mph). Demonstrating up to 90% higher peak loads, these new bumpers weigh five pounds less than their steel counterparts. This joint development by PPG Industries and AZDEL, Inc. has led to a 40% improvement in the performance of molded parts. (G.E. Plastics.)

8.2.3 Fiber/Matrix Consolidation

Fibers can be saturated with resinous material through a process called ***preimpregnation.*** The resinous material is subsequently used as a matrix material. The preimpregnated fibers are called ***prepregs.*** Unidirectional preimpregnated fibers with a removable backing that prevents the fibers from sticking together in a roll are known as *prepreg tape. Prepreg cloth* or *mats* are also available. Prepregs are ready-to-be-molded resin and fiber laminates [see Figure 8-5(a)]. The fibers impregnated with resin are partially cured (B-staged) and ready to be softened and formed into a permanent shape with the application of heat and/or pressure. Some important advantages of using prepregs are the elimination of the handling of liquid resins, the simplification of the manufacturing of reinforced plastic forms and shapes, and the savings in cost with small production lots. A few disadvantages are the elimination of room-temperature cures, the need for refrigerated storage for the prepregs, and the additional cost of the prepregs over the dry blended or undispersed materials. The latter disadvantage must be weighed against the use of higher mold pressures and increased molding cycle times for undispersed material to effect the required dispersion of the reinforcement throughout the matrix material for the attainment of the optimum properties of the composite.

Polyimides are attractive to the aerospace industry because of their toughness, thermal and thermo-oxidative stability, resistance to solvents, and excellent mechanical and electrical properties over a wide range of temperatures. One disadvantage is that they have poor melt-flow characteristics during processing, which leads to voids in the prepreg laminates. By blending linear, high-molecular-weight polyamic acid solutions with semicrystalline thermoplastic polyimide powders, a slurry is formed that can make void-free prepregs. The slurry can be applied to film, fabric, metal, polymer, or composite surfaces. As discussed in Module 6, cyclic thermoplastic resins developed recently have also improved wet-out of reinforcing fibers.

MAGNAMITE GRAPHITE PREPREG
X-AS4/1904

Magnamite graphite prepreg X-AS4/1904 is a 250°F curable epoxy resin reinforced with unidirectional graphite fibers. The reinforcement is Magnamite continuous AS4 graphite fiber that has been surface-treated to increase the composite-shear and transverse-tensile strength. The 1904 resin matrix was developed to cure under tape-wrap or vacuum-bag conditions at 250°F and to operate at temperatures up to 180°F.

	At 77°F	
Typical Composite Properties	Without Glass Scrim	With Glass Scrim
0° flexural strength, psi × 10^3	225	215
0° flexural modulus, psi × 10^6	18.4	16.1
0° tensile strength, psi × 10^3	240	—
0° tensile modulus, psi × 10^6	20.1	—
Shear strength, psi × 10^3	14.5	15.0
Cured ply thickness, mils	5.6	6.4
Fiber volume, %	58	58
Typical Prepreg Characteristics		
Tape width, in.	12	12
Resin content, %	40 ± 3	40 ± 3
Fiber content, [(a)] g/m^2	145 ± 4	145 ± 4
Glass scrim[(b)]	—	104
Yield, ft/lb	19	18

(a) Various fiber contents are available between 95 and 195 g/m^2.

(b) 104, 108, and 120 glass scrim have been used.

(a)

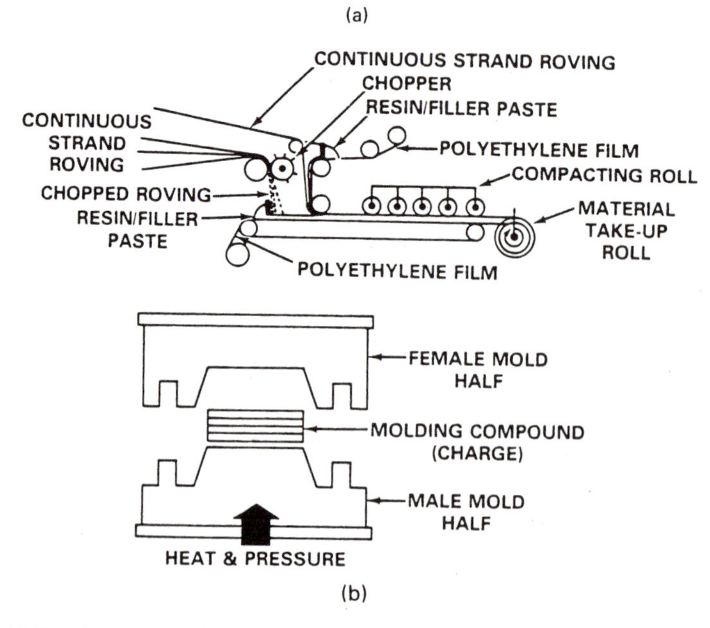

(b)

Figure 8-5 (a) Hercules prepregs. (Hercules, Inc.) (b) Schematic of machines used to make SMC sheet (top) and mold SMC plaques (bottom).

An ***intermetallic compound*** or ***phase,*** formed from two or more metals, has a mixture of metallic and ionic bonds. Cementite, Fe_3C, is a compound or alloy that serves as the backbone for strengthening steel. Titanium aluminides, high-strength, low-density intermetallics, are superseding nickel-based superalloys. The density of $Ti_2 \cdot Al \cdot Nb$ (titanium–aluminum–niobium) alloys is less than two-thirds that of nickel-based superalloys such as UNS N-07718. High specific strength at high temperatures, improved ductility, and fracture toughness are some of the improved properties of such intermetallics that are making these materials feasible for gas-turbine applications at temperatures approaching 1000°C.

Much development work is involved in using intermetallics as matrix materials for metal matrix composites. Intermetallics such as Ni_3Al, TiAl, and $NbAl_3$ reinforced with Al_2O_3 or ZrO_2 fibers form composites that are finding use at temperatures from 1000° to 1400°C. Room-temperature ductility of titanium aluminides containing niobium is improved for these composites by the addition of small quantities of magnesium. A special "pressure-casting" technique, in which the molten intermetallic is poured onto preheated fibers, is used to fabricate these composites. Argon gas pressure of 3 MPa ensures complete infiltration of the ceramic fibers. Titanium or yttrium is added to promote fiber wetting.

Preforms are custom-shaped, resin-bonded mats for reinforcement of molded parts with complex shapes. In other words, a preform is a bundle of fibers having the desired volume and architecture to fill a mold cavity for a one-shot molding.

A ***lamina*** is a flat arrangement of unidirectional fibers or woven fibers in a matrix.

A ***laminate*** is a stack of lamina with various orientations of principal materials directions in the lamina. Laminates can be built up with plates or plies of different materials or of the same material such as glass fibers. Shear stresses are always present between the layers of a laminate because each layer tends to deform independently of its neighboring layers due to each layer having different properties. These shear stresses, including the transverse normal stresses, are a cause of delamination.

Bulk molding compounds (BMCs) are a premixed material of short fibers (chopped-glass strands) preimpregnated with resin and various additives.

Sheet molding compounds (SMCs) are impregnated, continuous sheets of composite material. SMCs cut to proper size and stacked to provide the required thickness before heat curing in matched metal molds are used in applications such as automotive bodies and large structural parts [see Figure 8-5(b)]. They come in three types, depending on the length of the glass fibers used in the reinforcement. Fibers can be as short as 25 mm in length and arranged in random fashion in the resin. Longer fibers (200 to 300 mm) can be oriented in one direction (directional fibers). Continuous fibers laid in only one direction make up another designation. Various combinations of these SMC types are used with fibrous glass reinforcements, reaching 65% by weight of the composite. A recent molding compound designated XMC can contain up to 80% glass fiber by weight and uses continuous fibers running in an X pattern (see Figure 8-16).

The more common fibers used in composites are described below. As research continues, more materials (organic, polymeric, ceramic, and metallic) are becoming sources of fibers for different composites designed for ever-increasing applications in an energy-conscious age.

8.3 REINFORCING FIBERS

8.3.1 Glass Fiber

Glass fibers are made by letting molten glass drop through minute orifices and then attenuating (lengthening) them by air jet. The standard glass fiber used in glass-reinforced composite materials is E-glass, a borosilicate type of glass. The glass fibers produced, with diameters from 5 to 25 μm, are formed into strands having a tensile strength of 5 GPa. Chopped glass used as a filler material in polymeric resins for molding consists of glass fibers chopped into very short lengths.

E-glass is the first glass developed for use as continuous fibers. It is composed of 55% silica, 20% calcium oxide, 15% aluminum oxide, and 10% boron oxide. It is the standard grade of glass used in fiber glass and has a tensile strength of about 3.45 GPa and high resistivity.

S-glass was developed for high-tensile-strength applications in the aerospace industry. It is about one-third stronger than E-glass and is composed of 65% silicon dioxide, 25% aluminum oxide, and 10% magnesium oxide.

8.3.2 Boron Fiber

Advanced composites with boron fibers were developed in the early 1960s by the U.S. Air Force. The year 1963 marked the birth of boron technology. Boron fibers are composites with tungsten, silica coated with graphite, or carbon filaments as a substrate upon which boron is deposited by a vapor-deposition process. The final boron fiber has a specific gravity of about 2.6, a diameter between 0.01 and 0.15 mm, a tensile strength of about 3.45 GPa, and a modulus around 413 GPa. Figure 8-6 is a representation of a boron fiber.

Another boron-fiber composite with the addition of a silicon carbide coating deposited over the boron surface provides a fiber that is more compatible with metal–matrix materials, particularly at high operating temperatures. Figures 8-7 and 8-8 show a sketch of a BORSIC™ filament and an aluminum matrix in which boron is deposited on a graphite-coated tungsten wire, which makes the fiber itself a composite.

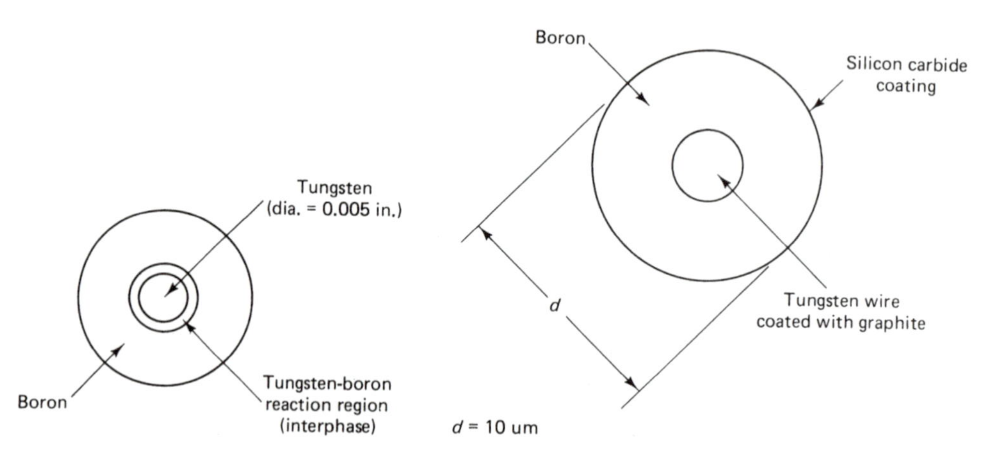

Figure 8-6 Structure of boron fiber.

Figure 8-7 Structure of BORSIC™ fiber.

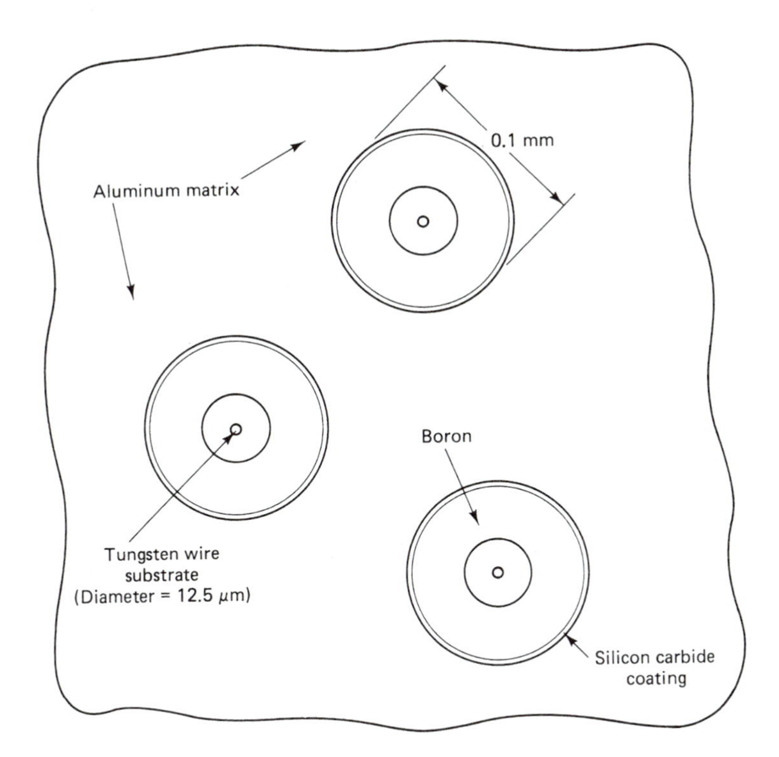

Figure 8-8 BORSIC™ fibers in an aluminum matrix would produce a MMC. The boron fiber itself is a composite, having been deposited on a Ti substrate and coated with SiC.

Boron is more expensive than graphite and requires expensive equipment to place the fibers in a resin matrix with a high degree of precision. Problems with chemical reactions between boron and other metals continue to be a source of concern to materials technologists who are exploring ways to overcome this limitation. One approach to this problem was the coating of the BORSIC fiber with silicon carbide (see Figure 8-8).

8.3.3 Carbon and Graphite Fibers

Carbon is a nonmetallic element. Black crystalline carbon, known as graphite, has a specific gravity of 2.25. Transparent carbon, diamond, has a specific gravity of 3.5. Graphite and diamond are allotropes of carbon, as are amorphous forms of carbon such as coke and charcoal. Both graphite and diamond have very high melting points of over 3732 K, which is explained upon examining the crystal structure and seeing the nearly infinite network of carbon–carbon covalent bonds that must be broken to melt these materials.

The terms ***carbon*** and ***graphite*** are often used interchangeably. However, a line of demarcation has been established in terms of modulus and carbon content. Carbon fiber usually has a modulus of less than 344 GPa and a carbon content between 92% and 95%. Graphite fibers have a modulus of over 344 GPa and a carbon content of 99% or greater. Another distinguishing feature is the pyrolyzing temperatures. For carbon, this temperature is around 1315°C, and for graphite it is around 1900° to 2480°C.

One authority classifies carbon as a ceramic. Another says the pyrolized organic material is a polymer, even though it is mostly carbon. They admit that carbon itself is an element, but if you believe that the word ***carbon*** really means "graphite," then carbon is a ceramic. Another authority uses the word ***carbon*** to describe fibers because the fibers are composed of crystalline graphite regions.

Pyrolysis is the thermal decomposition of a polymer. A carbon-containing polymer can decompose to carbon in an inert or reducing atmosphere and thus produce carbon fibers and/or carbon–carbon composite materials. If elements other than carbon are present in the original polymer, these elements will form a residue as a result of pyrolysis of ceramic compounds. An example of pyrolysis is the production of SiC fibers from polycarbosilane polymers having tensile strengths of 2.7 GPa.

As with composite materials, graphite fibers are not a new development. They were first produced in small quantities in the nineteenth century for use in incandescent-lamp filaments.

Several methods are used today to make carbon fibers of varying length and diameter and having the versatility of glass fibers. The oldest method of producing these fibers, used in the late 1960s, was ***graphitization*** of organic fibers, such as a rayon, at temperatures up to 3250 K. An acrylic fiber, polyacrylonitrile, abbreviated PAN, is also used as a source of such fibers, which the Royal Air Force (RAF) in England produced in 1961. Heated (pyrolized) under tension to stabilize the molecular structure at temperatures between 920 and 1140 K, the noncarbon elements (O_2, N_2, H_2) are driven off, leaving a fiber high in carbon content. A more recent process for producing high-modulus carbon fibers uses low-cost pitch; the pitch is converted to a liquid crystal, or mesophase, state before it is spun into fibers.

Twenty different carbon fibers with cross-sectional configurations, varying from circular to kidney shaped, in diameters of 0.008 to 0.01 mm, with strengths ranging from 1.72 to 3.1 GPa, and modulus values from 193 to 517 GPa, are now available from six manufacturers. One carbon yarn comes in plies from 2 to 30, with each ply composed of 720 continuous filaments about 8 μm in diameter. This fiber is 99.5% carbon and maintains dimensional stability to 3420 K. Another grade of fiber consists of 2000 filaments gathered together into a uniform strand that provides a modulus of 379 GPa and a tensile strength of 2.0 GPa. Thornel 300 carbon fiber, produced by Union Carbide, consists of 6000 filaments in a one-ply construction. The surface of the fiber has been treated with sizing to increase the interlaminar shear strength in a polymer–matrix composite in excess of 89.6 MPa. The tensile strength is listed as 2.69 GPa, with an accompanying modulus (tensile) of 229 GPa. Carbon and graphite fibers and yarns are used to produce various fabrics to meet the ever-increasing demands of industry.

Carbon fibers, first introduced in 1959 at a price of over $500 per pound, have steadily dropped in price. By 1984, the price had declined to $13 to $15 a pound for lower-grade graphite fiber and about $19 a pound for aerospace-grade fibers. Generally, the decline in prices of materials can be expected to continue with an increase in use, as long as there is a good availability of raw materials and improved production technology. Graphite fibers are linear elastic and, like other carbon materials, are anisotropic. Graphite possesses excellent creep and fatigue resistance, as well as dimensional stability at high temperatures. It is one of the few truly efficient structural reinforcements, with both high specific

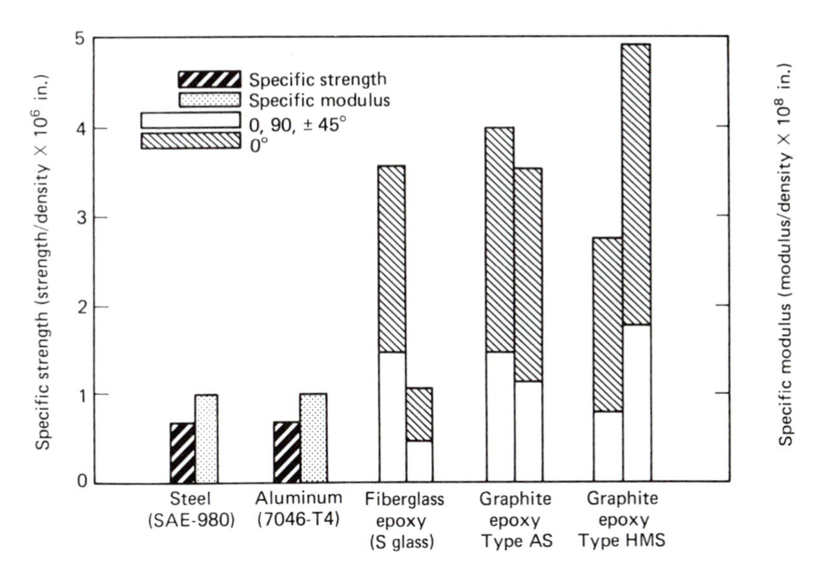

Figure 8-9 Comparison of specific strength and specific modulus (specific stiffness) of commonly used fiber materials. (DuPont.)

modulus and strength. Figure 8-9 compares the strength and modulus of two grades of graphite with some other fiber materials.

Thermal properties also must be considered unusual. The most important properties are low *coefficient of thermal expansion* (CTE or COT), high thermal conductivity, and dimensional stability at temperatures of 2000°C and above. The CTE of graphite is negative at 0°C, but increases with temperature (see Table 8-2). However, its generally negative CTE, coupled with the ability to tailor the orientation of the fibers, permits the designer to adjust the thermal coefficient over a wide range of values. Its thermal conductivity in the same temperature range is similar to that of many metals, including steel and nickel. Aluminum is the only metal that has a significant advantage in thermal conductivity over

TABLE 8-2 DIMENSIONAL STABILITY OF SOME FIBER MATERIALS (CTE)

	Thermal Expansion (in./in./°F) × 10^{-6}	
Material	0°	0°, ± 45°, 90°
Graphite AS	− 0.20	1.01
Graphite HTS	− 0.25	0.80
Graphite HMS	− 0.30	0.45
Fiber glass	3.50	6.00
Aluminum	13.0	—
Steel	7.0	—
Nylon (6/6)	45.0	—

graphite. However, aluminum's high CTE and low-temperature capability make it useful only in low-temperature composite curing applications. Finally, graphite is essentially free from residual stresses and never needs stress-relief annealing, as is common with many metals.

Graphite is commercially available in three different types: high modulus, intermediate modulus and tensile strength, and high stiffness and strength. All three have low density but also low impact resistance. The strength of graphite fiber/polycarbonate matrix composites is increased by precoating (sizing) the fibers with a thin layer of polycarbonate. Sizing can promote the wetting of fibers by matrix polymers, thereby enhancing fiber/matrix adhesion. The COT of graphite increases with temperature.

Graphite fibers can be coated with other materials to extend the usefulness of their properties to higher temperatures. Titanium carbide (TiC) is used to coat graphite or carbon/carbon composites with usual thicknesses (about 150 μm) to provide exceptional resistance to erosive wear and to corrosion by chemical agents. In addition, this coating has demonstrated very high thermal-shock resistance as well. Table 8-3 provides a list of a few properties of TiC-coated graphite and a graph showing the erosion rates of various materials, including TiC-coated graphite.

Electrically conducting composites, made from bromine-intercalated graphite fibers in an epoxy material, are used for many applications, such as grounding planes. The electrical conductivity of graphite fibers (derived from pitch blend) has been increased from about 4,000 to about 50,000 $\Omega^{-1}cm^{-1}$ through ***intercalation*** (insertion between something). Common metals have conductivities that range from 10,000 to 550,000 $\Omega^{-1}cm^{-1}$. The

TABLE 8-3 ENGINEERING PROPERTIES OF TiC-COATED GRAPHITE

Property	Value	
Flexural strength [MPa (ksi)]	90	(13)
Elastic modulus [GPa (Msi)]	11	(1.6)
CTE [ppm/K (ppm/°F)]	8.4	(4.7)
Thermal conductivity [W/m · K (Btu/hr · ft · °F)]	121	(70)
Density [g/cm³ (lb/ft³)]	1.81	(113)

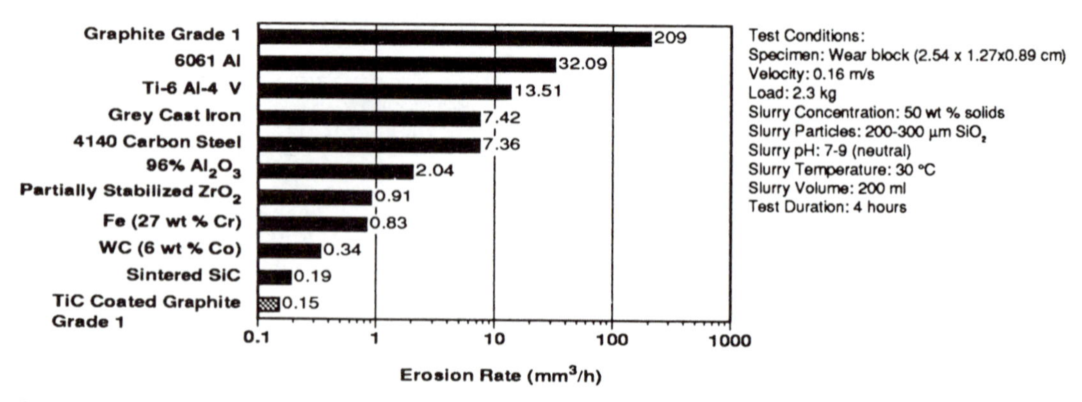

Source: Courtesy of Lanxide.

process of intercalation introduces donor/acceptor atoms, such as bromine, between the layers of graphite atoms in their characteristic layer arrangement. A cloth made of graphite fibers immersed in bromine is stacked between epoxy films making up the composite, which is then pressed and cured.

One disadvantage of graphite is its poor oxygen resistance. It begins to oxidize in air at about 700 K. Unlike metals, graphite does not form a protective film. Instead, the graphite oxide is volatile. To protect graphite from oxidation, several coatings have been developed, one of which, silicon carbide, protects graphite for a limited time at temperatures approaching 1920 K. Another limitation of carbon fiber is that it is a brittle, strain-sensitive material that cannot be depended on to offer much impact resistance.

Graphite fibers are available to the fabricator in the following forms:

1. Continuous fiber
2. Unidirectional prepreg as tow or tape in widths from 3 to 36 in. to form laminations
3. Chopped fiber
4. Pultruded shapes
5. Woven fabrics as dry cloth or prepreg for laminations

8.3.4 Kevlar Fiber

Kevlar, an organic fiber introduced in 1972 by DuPont for use in radial tires, is an aramid, or aromatic, polyamide fiber. The aromatic ring structure (see Figure 2-13) results in high thermal stability. The rodlike nature of the molecules classifies ***Kevlar*** as a liquid-crystalline polymer characterized by its ability to form ordered domains in which the stiff, rodlike molecules line up in parallel arrays. These domains orient and align themselves in the direction of flow during processing, which results in a high degree of alignment parallel to the fiber axis. This alignment results in anisotropy with high strength and tensile modulus in the fiber-longitudinal direction. Refer to Figure 6-7 and the accompanying text for additional discussion of polymer crystallinity. The overall properties above, plus a density of 1.44 g/cm^3 (about one-half that of glass), are the key to Kevlar's use in weight-limited applications. There are three grades of Kevlar. Kevlar 29 provides high toughness with a tensile strength of about 3.4 GPa for use where resistance to stretch and penetration are important. Kevlar 49 is a high-tensile-strength modulus of 130 GPa for use with structural composites. Kevlar 149 is an ultrahigh-tensile-strength modulus of 180 GPa. Figure 8-10(a) is a stress–strain diagram showing the relation of Kevlar 49 to other common reinforcing fibers. The luxury yacht seen in Figure 8-10(b) makes use of Kevlar aramid fibers, which were selected over competing materials shown in Figure 8-10(a). The craft's plastic superstructure is reinforced with Kevlar and E-glass, which provides nearly a 20% weight savings while maintaining structural integrity equivalent to more conventionally built vessels.

8.3.5 Ceramic Fibers

Both continuous and discontinuous ceramic fibers that are based on oxide, carbide, and nitride compositions make up most of the production of these fibers. Their development was started by the need for high-temperature reinforcing fibers in composites for the aerospace industry. Most oxide fibers are compositions of Al_2O_3 and SiO_2, although a few are almost

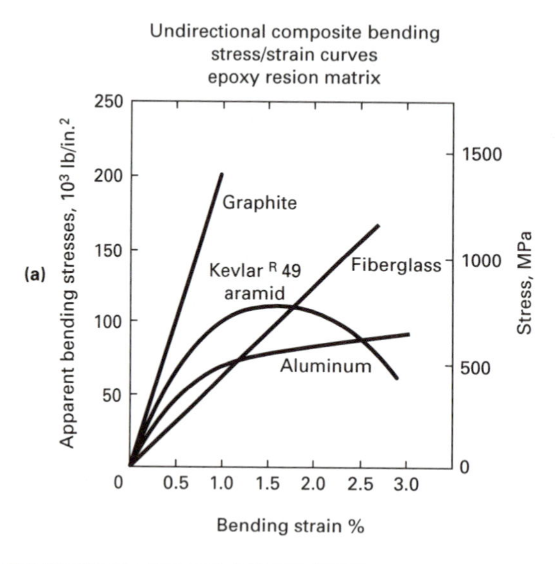

Figure 8-10 (a) Bending stress–strain curves of some fiber materials. (DuPont.)

Figure 8-10 (b) An artist's concept drawing of the luxury yacht *Evviva* shows a 49-m (160-ft) custom motor yacht constructed from a plastic superstructure reinforced with a hybrid combination of Kevlar aramid fibers and E-glass. Nomex aramid structural sheet, fabricated into honeycomb core panels, is used for the infrastructure and accessories such as furniture. The advanced composites allowed the designer to engineer the 180-ton displacement yacht for luxury and also service as a high-speed craft that will cruise at 37 km/h (20 knots) over a 9260-km (5000-nautical-mile) range. *(Advanced Materials & Processes.)*

pure oxides of aluminum and silicon (alumina and silica). Average properties of continuous oxide fibers are as follows: density, 3 g/cm^3; diameter, 12 μm; tensile strength, 2 GPa; tensile modulus, 200 GPa; and use temperature, around 1300°C.

Carbide fiber properties differ significantly from those of the oxides. Representative of continuous carbide fibers is SiC. Approximations of SiC's properties are as follows: density, 2.5 g/cm^3; diameter, 10 to 20 μm; tensile strength, 3.0 GPa; tensile modulus, 100 to 400 GPa; and use temperature, 1200°C. Both SiC and Si_3N_4 are available as discontinuous (whisker) forms. The principal advantage of carbide fibers over oxide fibers is the great increase in the modulus of elasticity.

8.3.6 High-Performance Manufactured Polymeric Fibers

Included in this category are fibers based on polyester, nylon, aramid, and polyolefin. As discussed in Module 6, fibers such as aramid (Kevlar and Nomex), polybenzimidazole

(PBI), Sulfar, and Spectra have increased the range of choices for materials engineers in designing materials with tailor-made properties. However, most do not possess the thermal properties found in ceramic- and metal-based fibers. Kevlar, a trademarked material, is discussed separately in this module. A polyolefin fiber, ***Spectra*** (ultrahigh-molecular-weight polyethylene) has a low specific gravity (0.9), which allows 40% more fiber per pound than aramid (Kevlar) fibers. It also has higher specific strength and specific stiffness, and has 7 to 10 times more abrasive resistance than aramid, with low moisture absorption and good chemical resistance.

8.4 FIBER PROPERTIES

8.4.1 Fiber Strength

A material's brittleness greatly affects its strength. Brittle materials have little resistance to the propagation of cracks. Pure metals possess varying degrees of ductility (the opposite of brittleness), which allows for some yielding in the face of longitudinal stress concentrations, which in turn prevents fracture. Brittle fractures, on the other hand, are usually catastrophic and come without advance warning. Alloying of metals and thermal treatments are two ways of improving the resistance of a material to plastic deformation (strength) by providing for the dispersion of harder particles within a matrix of softer material, which tends to limit the motion of dislocations through the matrix. Another way to increase the strength of a material is to add another material that has greater load-carrying capacity. The most prevalent form of this added material is fiber. How well fibers strengthen a material depends on the efficiency with which the relatively soft matrix material interacts with the fibers to transfer the load between them. The strength of an individual fiber is dependent on the absence of surface defects. The presence of microcracks anywhere in the small fiber causes localized stress and load concentrations and eventual failure of the individual fiber. The properties of a fiber are mainly dependent on the fiber's length, diameter, and orientation. The optimum reinforcing fiber for a composite material has a ratio of length (L) to diameter (D) of about 150. As this ratio, known as the ***fiber-aspect ratio,*** increases, the strength of the composite increases (see Figure 8-11). A single polymer chain has the highest aspect ratio of any "fiber." Incorporating molecular fibers into composite polymers is a technique with great potential for the future. With fibers of small diameter, 1 to 25 μm, aspect ratios range from 100 to 15,000.

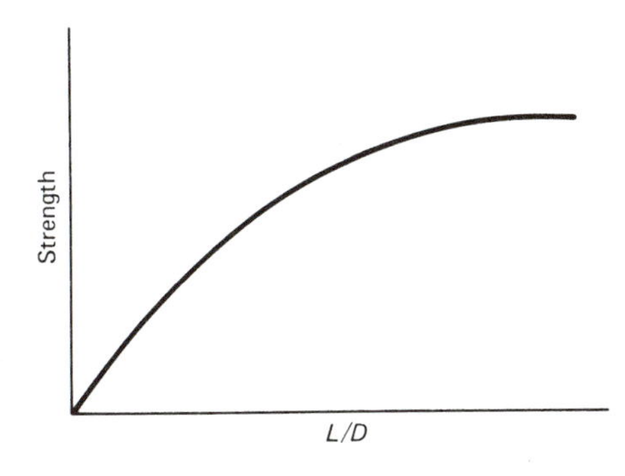

Figure 8-11 Strength versus fiber aspect ratio (*L/D*).

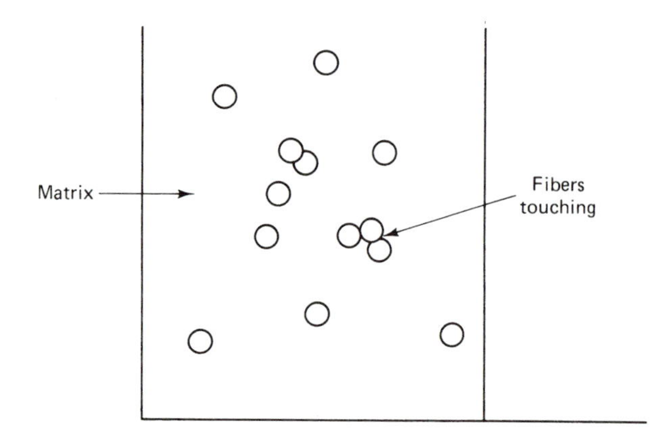

Figure 8-12 Cross-section of a fiber-reinforced material showing a poor spatial arrangement of fibers within the matrix material. See also Figure 8-4(b) and (c).

Illustrative Problem

Given a fiber that is 25 μm in diameter and has an aspect ratio of 15,000, find the length of the fiber expressed in meters.

Solution

$$\text{Length } (L) = \text{diameter } (D) \text{ of fiber} \times \text{aspect ratio (AR)}$$

$$L = 15 \times 10^3 \times 25 \times 10^{-6}$$

$$L = 0.375 \text{ m}$$

The shape of the reinforcing phase, rodlike for fibers, plays an important role in determining the performance of a composite, as does the shape of the unit cell in a homogeneous solid. Their size and distribution control the texture of the material and determine, in part, the interfacial area between the fibers and the surrounding matrix. The topology of the fibers (i.e., their spatial relation to each other) is important because certain properties of the composite may be affected by how much the individual fibers or filaments touch each other (see Figure 8-12).

8.4.2 Specific Tensile Strength

Fiber composites find use in many weight-sensitive applications in space exploration, general aviation, and the automobile and sporting-goods industries. One indicator used to portray the effectiveness of the strength of a fiber is specific strength.

Specific tensile strength, the ratio of the tensile strength of a fiber material to its weight density, is an indicator of structural efficiency, giving the relative load-carrying ability of equal weights of material. Typical units of specific strength are millimeters (mm), inches (in.), or kilometers (km). Table 8-4 and Figures 8-9 and 8-13 list the specific strength for some typical materials from which fibers are formed for use in composite materials.

8.4.3 Specific Stiffness—Specific Modulus

Another indicator of the special properties of fiber composites, in particular, the effectiveness of a fiber, is ***specific stiffness.*** It is a ratio of the modulus of elasticity (or tensile stiff-

TABLE 8-4 SPECIFIC STRENGTH

Materials	Weight density, ρ (kN/m³)	Tensile strength, *S* (GN/m²)	Specific strength, *S*/ρ (km)
S-glass	24.4	4.8	197
E-glass	25.0	3.4	137
Boron	25.2	3.4	137
Carbon and graphite	13.8	1.7	123
Beryllium	18.2	1.7	93
Steel	77	4.1	54
Titanium	46	1.9	40
Aluminum	26.2	0.62	24

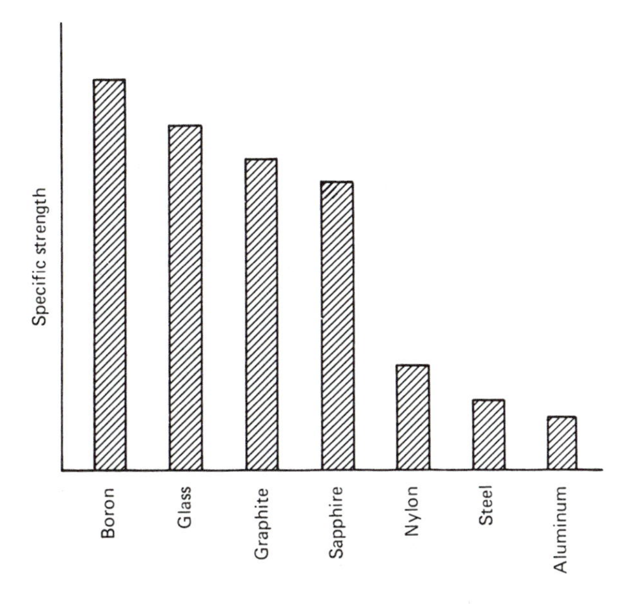

Figure 8-13 Relative specific strengths of typical materials used in composites.

ness) to the weight density of the fiber. Table 8-5, using the same materials as in Table 8-4, and Figures 8-9 and 8-14 show the average specific stiffness of such fiber materials, starting with materials with the highest values. Note that graphite, boron, and carbon have values almost six times that of steel. Graphite is known to have six times the strength of steel with six times less weight. These values of both strength and stiffness of materials in fiber form over the same materials in bulk form are most significant. As an example, the strength of glass in bulk form, such as plate glass, is only a few megapascal, yet in fiber form this figure rises to around 3.5 GPa. Structural steel (plain-carbon steel) in bulk form has a tensile strength around 0.5 GPa, but in fiber form the value is 4 GPa. This contrast is due to the fact that the structure of the material differs between the two forms of the material. In fiber form, the microstructure approaches the nearly perfect structure, with the crystals aligning themselves along the fiber axis in ordered fashion. As a result, there are fewer

TABLE 8-5 SPECIFIC STIFFNESS/MODULUS

Materials	Weight density, ρ (kN/m³)	Tensile elasticity, E (GN/m²)	Specific stiffness, E/ρ (Mm)
Graphite	13.8	250	18
Beryllium	18.2	300	16
Boron	25.2	400	16
Carbon	13.8	190	14
S-glass	24.4	86	3.5
E-glass	25.0	72	3.0
Steel	77	207	2.7
Titanium	46	115	2.6
Aluminum	26.2	73	2.8

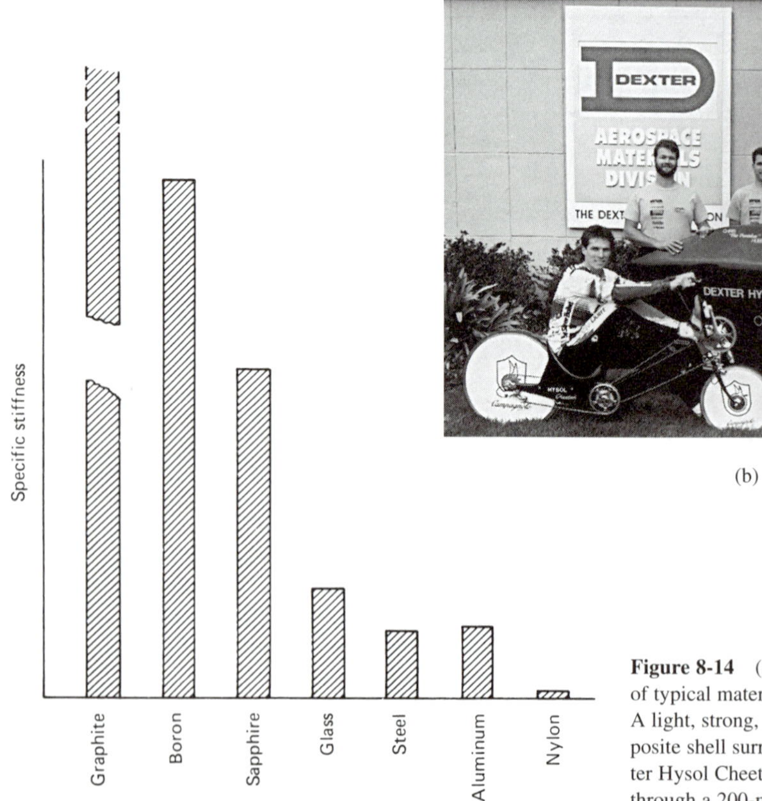

Note: Graphite has a specific stiffness of 5000 versus steel with 25 in units of 10^6N 'm/kg (using mass density in kg/m³)

(a)

(b)

Figure 8-14 (a) Relative specific stiffness of typical materials used in composites. (b) A light, strong, and rigid aerodynamic composite shell surrounded the rider of the Dexter Hysol Cheetah bicycle as it traveled through a 200-meter speed trap at San Luis Valley, Colorado, going 68.73 mph (110.6 km/h) to capture the new human-powered vehicle speed record. *(Advanced Materials & Processes.)*

internal defects or dislocations than would be present in bulk form. As we recall from our previous discussions, the presence of dislocations explains many properties of a material. In this situation, the movement of dislocations permits the material to yield, changing the internal structure accordingly and accounting for ductility and accompanying decreases in strength and stiffness. Figure 4-5 showed the increase in stiffness gained by reinforcing materials (aluminum and titanium) with boron fibers.

Specific stiffness and specific strength can be expressed in terms of the mass density rather than the weight density. Such a practice is more compatible with the SI. Typically, units of mass density are expressed in $kg \cdot m^{-3}$. Modulus of elasticity and tensile strength can be expressed either in $N \cdot m^{-2}$ or Pa. The specific ratio will have units of $m \cdot N \cdot kg^{-1}$. For example, a wood's longitudinal modulus *(E)* is $1.2 \times 10^{10}\ N \cdot m^{-2}$ and its density measured at 12% moisture content is $0.68 \times 10^3\ kg \cdot m^{-3}$. The specific modulus of this wood is

$$\frac{E}{\rho} = \frac{\dfrac{1.2 \times 10^{10}\ N}{m^2}}{\dfrac{0.68 \times 10^3\ kg}{m^3}} = 17.6 \times 10^6\ N \cdot m \cdot kg^{-1}$$

Values of the stiffness–mass density parameter (stiffness modulus) for some other materials are:

Graphite	$5000 \times 10^6\ N \cdot m \cdot kg^{-1}$
Boron	$190 \times 10^6\ N \cdot m \cdot kg^{-1}$
Steel	$25 \times 10^6\ N \cdot m \cdot kg^{-1}$
Nylon	$3 \times 10^6\ N \cdot m \cdot kg^{-1}$

Figure 8-14(b) provides a striking example of how the specific tensile strength and specific stiffness values found in composites provide designers with the ability to select and design engineered materials to form a materials system for specific applications. The human-powered, record-setting vehicle used aerospace-grade adhesive bonding to join the lightweight, rigid structure. The bicycle system used paste adhesives that bond the frame, seat mold, fairing mounts, aluminum inserts, steering assemblies, and composite parts. *Syntactic core material* (reinforced plastic foams made with microspheres instead of entrapped gas voids) was used for the bucket seat and as a laminar core stiffener. The windshield and fairing were attached by epoxy patch kits.

8.4.4 Fiber Loading

Fiber loading refers to the amount of reinforcement in a composite material. The strength of the composite is directly proportional to the volume of fiber ***(volume fraction)*** present. In addition to the fiber loading, the arrangement or orientation of the fibers plays a major role in the strength of the given product. Using the analogy of filling a shoebox with pipe cleaners, the maximum number of fibers that can be placed in a given volume (shoebox) is determined by the arrangement of the fibers. If all fibers are placed parallel to each other, a maximum number can be attained. In fiber loadings, a maximum of about 85% can be achieved. Between 50% and 75% load range can be reached if half the fibers are arranged at right angles to each other (fabrics), while a random arrangement (chopped strands) permits a range of only 15% to 50% (see Section 8.4.5).

8.4.5 Fiber Orientation

As the direction of the applied load moves 90° to the fiber orientation, the strength of a directionally oriented fiber composite decreases to about 20% to 30% of the longitudinal direction. Several techniques are employed to orient the fibers in a fiber composite. If continuous fibers are oriented such that their length is in the direction (longitudinal) of the loading, this type of arrangement is known as ***directionally oriented.*** If the applied loads and/or their directions are not known, a random oriented arrangement with continuous fibers running at various directions may be used. Alternating layers of continuous fibers at various angles provide full strength at these various directions. However, this calls for additional layup time, with its increased costs. Discontinuous fibers less than 30 cm long, which tend to orient themselves in the direction of the resin–matrix flow, provide another partial solution to providing a balanced strength in several directions. This technique is a compromise between the continuous and random orientation of continuous fibers. By sandwiching random short fibers between continuous fibers, a multiple-configuration fiber composite results that is superior to a fully randomly reinforced material. The effect of adding some continuous fibers to a random, chopped-fiber reinforced SMC is highlighted in Figure 8-15(a), which compares, among other things, SMC-R50, a 50% (by weight) of 25-mm chopped E-glass fibers, 16% calcium–carbonate filler, and 34% polyester–resin–matrix formulation, to SMC-C20/R30, a hybrid of chopped (30%) and continuous (20%) E-glass fibers in a polyester–resin matrix. Figure 8-15(b) shows the effect of a change in the fiber orientation on the fatigue failure of XMC, an X-pattern molding compound produced by PPG Industries. All the preceding arrangements are available in SMC form. Figure 8-16 contains sketches of some of these arrangements of fibers in a matrix.

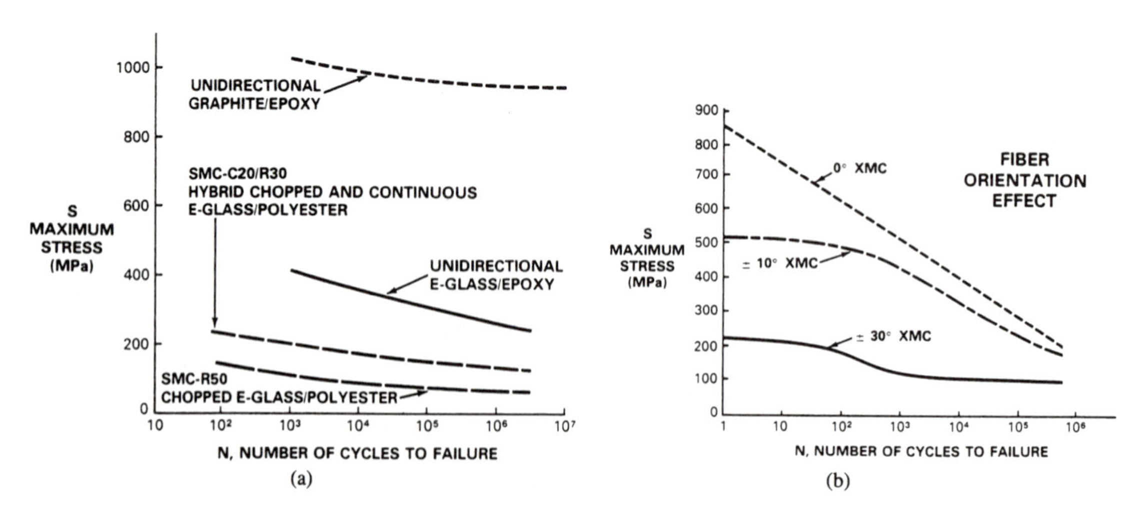

Figure 8-15 sN Diagrams (a) Range of fiber-composite tensile fatigue performance. (General Motors.) (b) Effect of fiber orientation on fatigue of XMC. (General Motors.)

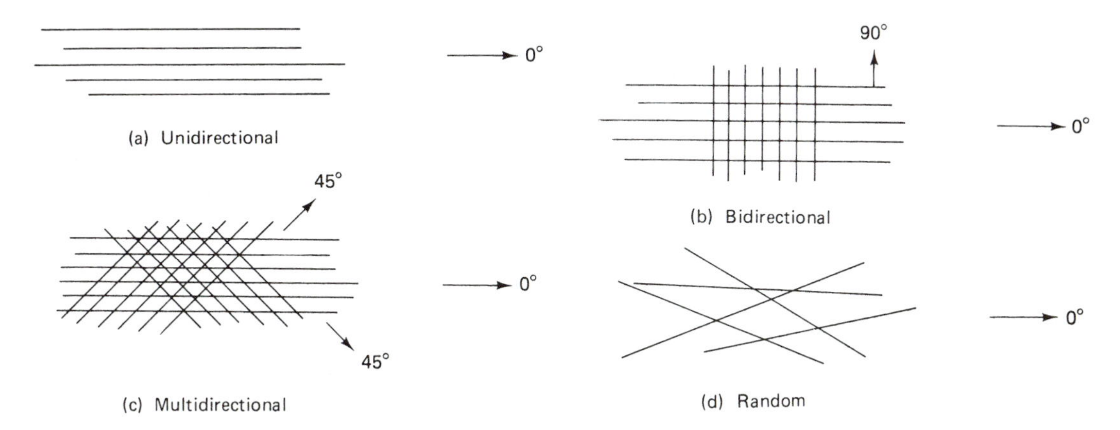

Figure 8-16 Fiber orientations.

8.5 STRUCTURE AND PROPERTIES OF COMPOSITES

Throughout this book we have stressed that the properties of a material are dependent on the material's structure. In monolithic materials such as steel or polymers, the structure commonly referred to is that of the crystalline- or molecular-material microstructure. Many times we peer beyond this atomic or molecular level to the subatomic level for further understanding of the behavior of such materials. With composite materials, we can identify with the naked eye (macroscopic) the major ingredients or constituents, such as metal particles or glass fibers and different matrix materials involved in understanding the collective performance of these components in the composite material. Just as with steel alloys, where we can change the microstructure of these multiphase materials to obtain different properties by varying the amount of the basic ingredients (elements) or by changing the number, size, and shape of the ingredients through appropriate thermal processing, we can vary the properties of composite materials by varying their composition and their structure. We know also that the final behavior of materials such as steel depends on how well the various phases interact with each other. This is vitally important with composites. As human-made materials, composites can be designed to have isotropic properties and, if so designed, they probably would be homogeneous materials with their components distributed and arranged in a uniform pattern. Or the components can be distributed in a nonuniform manner and, if so, the composites would have anisotropic properties. Third, similar to forged metals, a composite material can have its components distributed and arranged in a particular orientation that would result in directional properties. These last statements help explain some of the interest in composites in recent years by materials scientists who can, with the new technologies of materials, design their own material to fit the specific requirements of a particular application that cannot be met by existing technology or conventional homogeneous materials.

The characteristic properties of individual constituents of a composite interact in various ways to produce the collective properties or behavior of the composite. Some properties obey the ***rule of mixtures*** (ROM) because the composite properties are the weighted sums of the values of the individual constituents. In other words, properties are a function of the amounts and the distribution of the contributing material. Density, specific heat, thermal and electrical conductivities, and some mechanical properties, such as modulus of elasticity, follow this rule (see Section 8.5.1). In some composites the properties of the components are somewhat independent and supplement each other to produce a collective performance by the composite. Other composites have resultant properties that are the net results of the interaction or interdependence of the components with each other. This later type of composite behavior is by far the most important because the end result of combining materials into a materials system is a performance or set of properties that far exceeds the individual properties possessed by the components acting alone. An example of this type of composite can be taken from the fiber-reinforced plastics (FRP) composites, such as glass fiber embedded in plastic resin. The glass fibers possess extremely high tensile strength, but being very brittle they cannot be used alone. A plastic resin, on the other hand, is relatively weak but very ductile. Once a sufficient number of glass fibers are embedded in a plastic–resin–matrix in a unidirectional manner parallel to the direction of the load, the two components act together as a unit to withstand the load by deforming equal amounts and sharing the load proportionately so that the composite or materials system achieves higher tensile strengths than otherwise is possible by the individual components.

8.5.1 Rule of Mixtures

Using fiber–glass filler and a polyester–resin–matrix as our composite, we can sketch the stress–strain curves for each material as shown in Figure 8-17. The modulus of elasticity *(E)* for glass is about 69 GPa and for polyester, around 69 MPa. Tensile strength for glass (single fiber) is about 3.45 GPa.

The curves show that there are great differences in the stiffness (slope of the curves) between the two materials. Glass is extremely brittle and shows little evidence of any elastic strain up to the point of fracture. As indicated by its values of modulus and the slope of its stress–strain curve, polyester is a very ductile material. If each material were deformed an equal amount, the stress carried by the glass would be over 1000 times greater than the stress on the polyester.

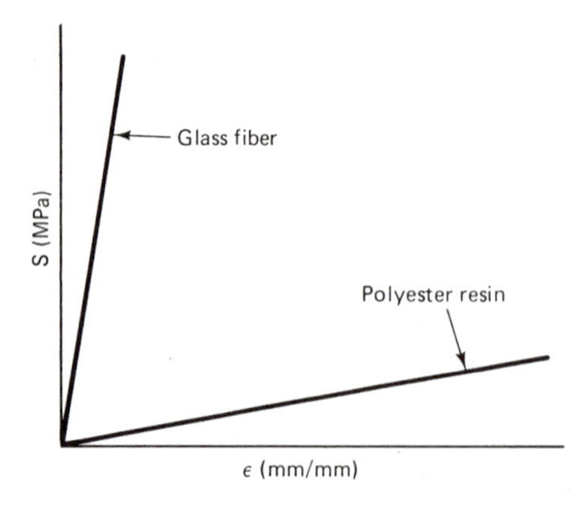

Figure 8-17 Stress–strain curves for glass fiber and polyester–resin.

If these two materials were combined in one ideal material, the glass would act as continuous fibers completely surrounded by a matrix of polyester, which forms a bond between the fibers and the matrix. The bond would allow both to deform (elongate) the same amount under a uniaxial load.

Research shows that the modulus of a unidirectional composite material in which (1) the load is applied parallel to the continuous fiber and (2) the bond between the fiber and the matrix material is strong enough to permit both materials to deform equally is the weighted sum of the modulus of the fiber and the modulus of the matrix. This last statement is an example of a ***simple mixture law.*** Mixture laws for physical properties such as specific heat, density, and conductivity for a composite material can also be developed. For such composite materials, most of the flow of thermal energy or electrical current will be through the component that is the best conductor.

The preceding discussion assumed that the chemical bond between the fibers and the matrix was greater than the strength of the matrix materials. This is not always the situation. Much research goes into determining the conditions necessary to improve the bonding so that the stress transfer between the matrix and the fibers can be made without fracture. Studies of metal–matrix-composite materials have led to the development of composite materials that can withstand temperatures in excess of 670 K and retain great stiffness. Such materials are represented by tungsten, boron, graphite, or silicon carbide fibers in aluminum, cobalt, or nickel matrices.

Certain properties of a composite depend on the relative amounts and properties of the components of the composite material. Using the law or rule of mixtures (ROM), an estimate of properties such as mass, density, modulus of elasticity, thermal conductivity, or electrical conductivity can be determined provided that the reinforcing fibers are continuous, unidirectional, and have the same length as the matrix. Also, the load on the composite is a tensile load parallel to the longitudinal axis of the fibers. It is to be emphasized that not all properties of a composite follow the rule (i.e., some composite properties are in simple proportion or are weighted averages of their components).

The ROM will be developed to express the elastic modulus in tension for a fiber-reinforced matrix material conforming to the constraints stated above. Figure 8-18 is a

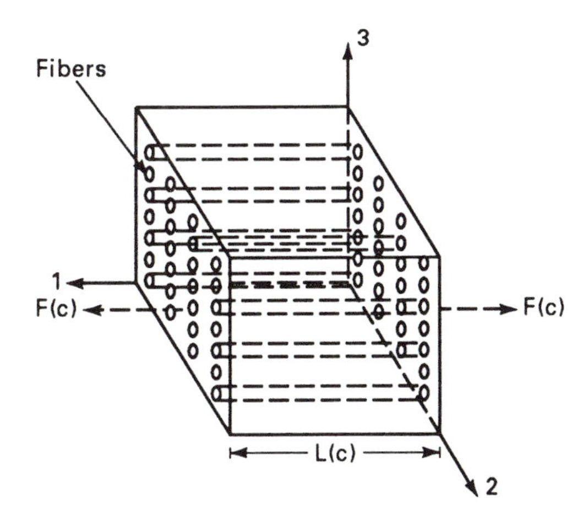

Figure 8-18 Fiber-reinforced-composite material undergoing isostrain loading conditions.

sketch of the composite loaded with a tensile load, $F(\mathrm{c})$. The following abbreviations represent the quantities involved in this development and in further discussions of the ROM to follow:

V = volume fraction
s = stress
e = unit deformation, strain
E = modulus of elasticity in tension
A = area
(f) = refers to fibers
(m) = refers to matrix
(c) = refers to the composite material
S = tensile strength
F = load or force acting on a material
L = length
p = density
m = mass

Referring to Figure 8-18, the load *(F)* is carried by both the fibers and the matrix acting to share the burden:

$$F(\mathrm{c}) = F(\mathrm{f}) + F(\mathrm{m}) \quad (1)$$

Assuming that the bond between the fibers and the matrix is good, we can say that both the fibers and the matrix will stretch or deform the same amount in withstanding the tensile load.

$$e(\mathrm{c}) = e(\mathrm{f}) = e(\mathrm{m}) \quad (2)$$

Using the definition of stress (Section 4.2.1), $s = F/A$, and Hooke's law (Section 4.2.3), $s = E/e$, equation (1) can be written

$$s(\mathrm{c})A(\mathrm{c}) = s(\mathrm{f})A(\mathrm{f}) + s(\mathrm{m})A(\mathrm{m}) \quad (3)$$

Dividing equation (3) by $A(\mathrm{c})$ yields

$$s(\mathrm{c}) = \frac{s(\mathrm{f})A(\mathrm{f})}{A(\mathrm{c})} + \frac{s(\mathrm{m})A(\mathrm{m})}{A(\mathrm{c})} \quad (4)$$

The ratios $A(\mathrm{f})/A(\mathrm{c})$ and $A(\mathrm{m})/A(\mathrm{c})$ are area fractions. Because the composite and its components are arranged in parallel, areas $A(\mathrm{f})$ and $A(\mathrm{m})$ are proportional to their volumes. Thus, volume fractions can be substituted for area fractions:

$$s(\mathrm{c}) = s(\mathrm{f})V(\mathrm{f}) + s(\mathrm{m})V(\mathrm{m}) \quad (5)$$

Returning to equation (2) with Hooke's law and substituting for s yields

$$E(\mathrm{c}) = E(\mathrm{f})V(\mathrm{f}) + E(\mathrm{m})V(\mathrm{m}) \quad (6)$$

Equation (6) expresses the ROM in terms of the elastic modulus for the ***isostrain*** situation described in Figure 8-18. Note that if one volume fraction is known, the other volume fraction can be found using the relationship $V(\mathrm{f}) + V(\mathrm{m}) = 1$.

Illustrative Problem

A laminar composite of epoxy resin reinforced with graphite fibers is loaded similarly to the model in Figure 8-18. Assume that there are equal volumes of graphite and epoxy sustaining the load. The modulus of elasticity for the graphite fibers is 58×10^6 psi and for epoxy it is 0.6×10^6 psi. Find the modulus for the composite.

Solution

$$\begin{aligned} E(\text{c}) &= E(\text{f})V(\text{f}) + E(\text{m})V(\text{m}) \\ &= 58 \times 10^6 \times \tfrac{1}{2} + 0.6 \times 10^6 \times \tfrac{1}{2} \\ &= 29.3 \times 10^6 \text{ psi} \end{aligned}$$

Figure 8-19 illustrates a model of the ***isostress*** loading of a composite material. The load, $F(\text{c})$, acts transversely as opposed to longitudinally or parallel to the long axis of the fibers. The fibers and the matrix each resist the load equally. Stated in equation form, $F(\text{c}) = F(\text{f}) = F(\text{m})$. This also means that the stress in the fibers and in the matrix is equal. However, the strain or deformation in each material is different. The weaker material will deform more than the stronger component. In equation form,

$$e(\text{c}) = e(\text{f}) + e(\text{m})$$

The equation that expresses this type of loading, known as the ***inverse rule of mixtures,*** reveals that this isostress loading condition does not follow the rule of mixtures. In equation form, the inverse rule of mixtures is expressed arbitrarily in terms of modulus:

$$\frac{1}{E(\text{c})} = \frac{V(\text{f})}{E(\text{f})} + \frac{V(\text{m})}{E(\text{m})}$$

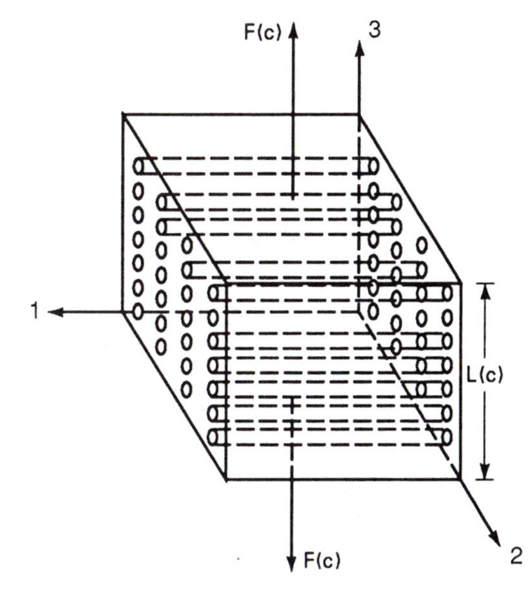

Figure 8-19 Fiber-reinforced-composite material undergoing isostress loading conditions.

Figure 8-20 contrasts these two types of loading and helps illustrate that, for a given volume fraction of fiber in a composite, the composite loaded under isostrain conditions is more efficient, having the higher modulus of elasticity for the composite. In fact, when looking at the curve in Figure 8-20, you can see that the modulus for the composite is not increased much over the modulus for the weaker component in the composite. On the contrary, using the same materials components and with the same volume fractions arranged in the isostress model, the composite modulus is about 10% less than that obtained in the isostrain model.

When determining how to arrange the components in a composite, it is important to recognize how a composite material is to be loaded. If the direction and approximate size of the loading are known beforehand, this simplifies the problem. If these conditions are not known, the correct components must be chosen and best arranged to support the load regardless of its magnitude (within limits) and direction.

8.5.2 Fatigue

The design of a new or improved product for an industrial and/or technical application entails the detailed analyses of many factors. The design process links the scientific knowledge of materials with modern industrial methods to reach an overall objective of producing a better product at a lower cost. The interrelationship among materials, process, and design is nowhere more pronounced than in the use of composite materials to replace traditional materials such as steel in both structural and nonstructural components. Figure 8-21(a) is a flowchart showing the interaction of design, materials, and manufacturing processes that must exist to ensure the success of any design project.

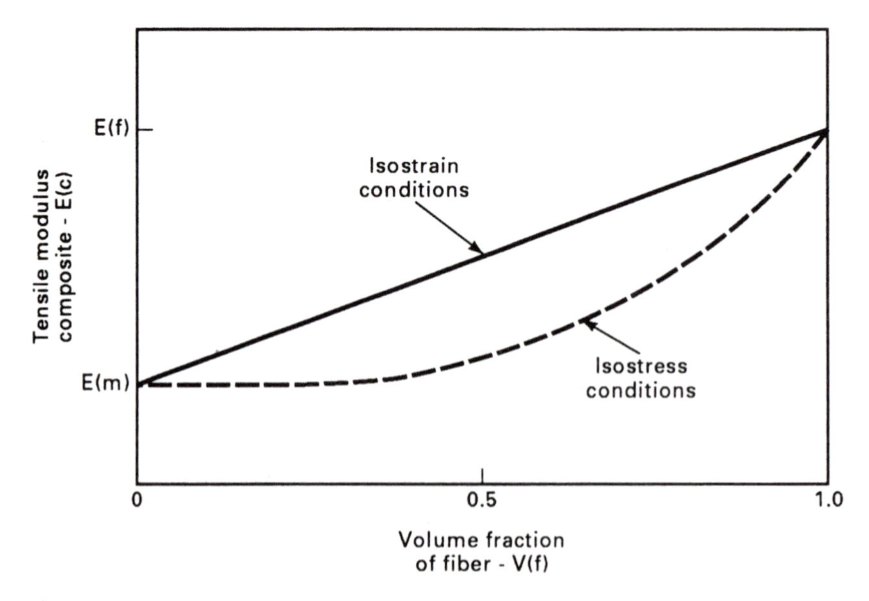

Figure 8-20 Graphical plot of the tensile modulus of a unidirectional composite material loaded under isostrain and isostress conditions as a function of volume fraction of reinforcing fiber.

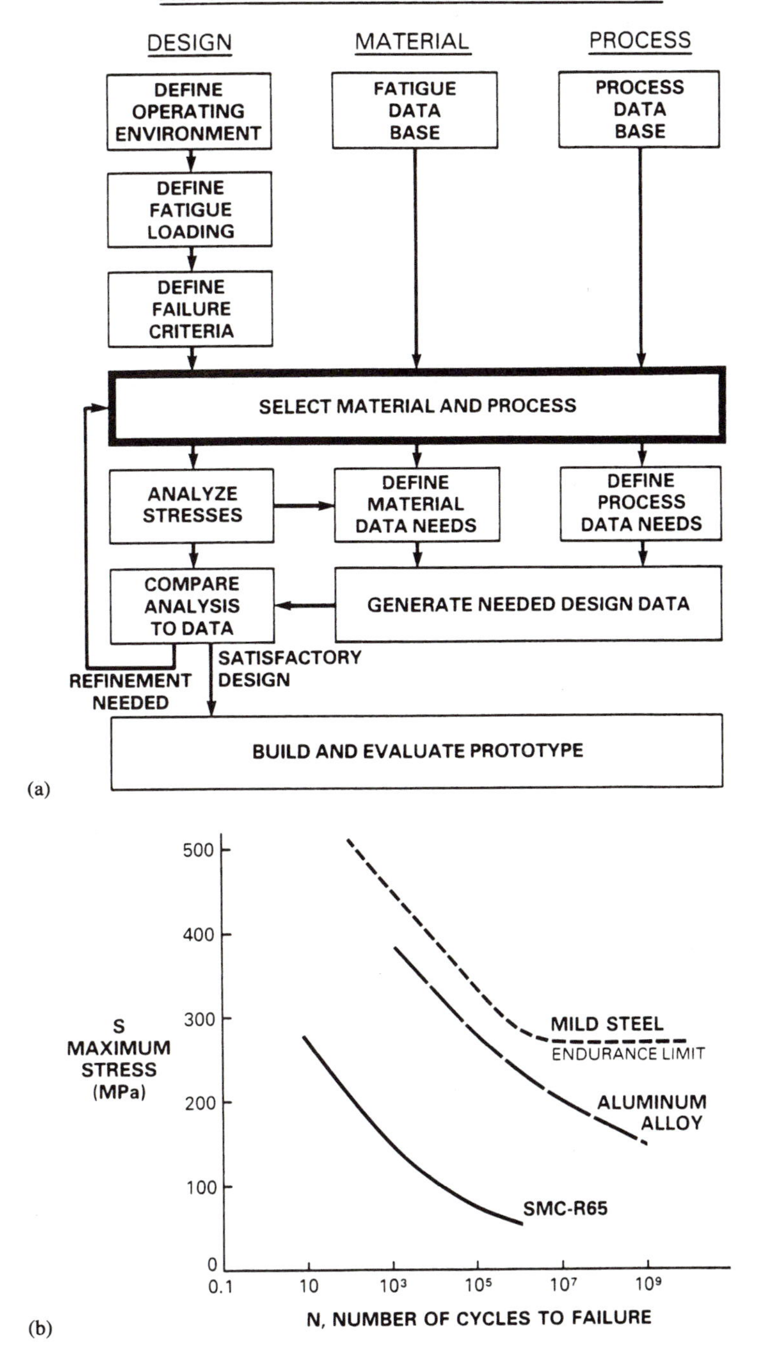

Figure 8-21 (a) Flowchart for fatigue design with fiber-reinforced PMCs showing need for databases. (General Motors.) (b) Comparison of metals and PMC typical flexural fatigue sN diagram curves. (General Motors.)

Composite materials owe their success in large part to the fact that they provide the engineer with considerable design latitude compared to metals. This fact is in addition to their many other desirable properties. Fatigue in metals, discussed in Module 5, differs from fatigue in composites [see Figure 8-21(b) and Figure 8-24(a)]. The mechanisms of crack initiation and crack propagation result in the failure of metals under cyclic, repeated loads at stress levels well below their ultimate strengths. Crack initiation is caused by stress concentrations, inclusions, or voids, which are then acted on by the mechanism of crack propagation, reducing the net load-bearing area to the point where the fatigue stress exceeds the metal's ultimate strength. Final failure in metals usually results in a relatively clean fracture surface. In many instances the failure is of catastrophic proportions. Finally, metals exhibit an ***endurance limit,*** a stress level below which fatigue failure does not occur, regardless of the number of fatigue cycles applied.

In composites, crack initiation and propagation produce simultaneous growth of cracks that may (1) extend through the matrix, (2) be stopped at a fiber, or (3) propagate along a fiber–matrix interface. Cracks are initiated by factors such as filler or fiber debonding, voids, or fiber discontinuities. The crack propagation results in cracks joining each other to the extent that the matrix is unable to perform its basic function of transferring the load from one fiber to the next in fiber composites. The fracture surface usually shows evidence of a complex assortment of matrix failure, fiber failure, and fiber pullout.

Other ***fatigue–failure characteristics*** of FRP composites are as follows:

1. A gradual decrease in the slope of the stress–strain curve (modulus of elasticity, E) occurs during cyclic loading.
2. Elevated temperatures normally reduce the performance of matrix-dominated composites (stress carried mainly by the matrix material). Fiber fatigue properties are better than those of the polymer matrix.
3. Moisture and chemical exposure can greatly affect the polymer matrix.
4. Notch sensitivity in high-cycle tensile-fatigue conditions is less than for metals.
5. There are considerably fewer matrix-dominated properties than fiber-dominated properties. Hence tensile, compressive, and shear stresses in the interlaminar (through-the-thickness) direction, being matrix dominated, should be kept to a minimum.

8.5.3 Toughness and Impact Strength

As measured by the ASTM D256 test on unnotched Izod specimens, the impact strength at room temperature for "tough" plastic molding materials is at least 8 J/cm (15 ft · lb/in.). A value of at least 10 J/cm^2 (50 ft · lb/in.^{2}) using the high-speed tensile impact test (ASTM D1822) also qualifies such materials as being tough. Considerable impact testing of such materials using glass- or carbon-fiber reinforcement in varying amounts with different resin materials has been done. Only recently, however, have such tests been carried out at low temperatures (down to 200 K). From our study of other materials, we know that the impact strength of a notched specimen is less than that of an unnotched specimen because notches produce stress concentrations. We also know that it takes energy not only to start a crack but also to propagate one. With notched specimens the crack is already provided, and depending on the sharpness of the notch, less energy is usually needed to propagate it. The tremendous effect of surface imperfections such as cuts, tool marks, and scratches on the toughness of materials is well known. For many reinforced thermoplastics, toughness drops off and the notch sensitivity increases at low temperatures. However, there are exceptions. Polypropylene, as well as acetal copolymer, becomes tougher, as evidenced by higher test values for unnotched specimens. Glass-reinforced polycarbonates show little change in toughness, although their notch sensitivity is greater at lower temperatures. With DuPont's "super-tough" nylons (Zytel ST resin), notch sensitivity decreases with increased glass- or carbon-fiber content, while at low temperatures, notched Izod impact-strength readings increase with fiber content. Using unnotched specimens, the toughness of these nylon resins increases with increasing fiber content at both room temperature and low temperatures (Figure 8-22).

Impact testing measures the material's resistance to fracture under certain prescribed test conditions when a standard specimen is struck at high velocity. In addition to the Izod and Charpy notched and unnotched specimens using a swinging pendulum, other impact tests, such as the falling ball, falling dart, high-velocity tensile stress, and fracture-toughness tests, are also used to measure toughness. ***Fracture toughness,*** in units of stress times

Figure 8-22 An application that takes advantage of the low notch sensitivity of glass-reinforced Zytel nylon.

the square root of a crack length (MPa · $m^{1/2}$), is an indicator of a material's resistance to the extension of a preexisting crack. The high-speed tensile test uses unnotched specimens and defined impact strength as the area under the stress–strain curve, or the energy required to break a material (toughness). Impact strength using this test setup is expressed in units such as kilojoule per square meter (kJ/m^2). Those tests using notched Izod or Charpy specimens express impact toughness in terms of the energy per length of notch, or kJ/m, as one example.

8.5.4 Test Values and Data

Test values and other data presented in this module and elsewhere are typical values for the material. They are offered as an aid to understanding. The properties of parts fabricated from a particular material are too contingent on many factors, such as part configuration, molding techniques, or curing times, to guarantee reproducibility of the data.

8.6 TYPES OF COMPOSITES

Products resulting from different manufacturing techniques, different reinforcing components, and the specialized nature of the parts themselves result in many types of composites. We have divided this ever-increasing variety of composites into fiber, laminated sandwich, particulate, flake, and filled composites [see Figure 8-2(b)].

8.6.1 Fiber-Reinforced Composites

8.6.1.1 Polymer matrix composites (PMCs)—Fiberglass-reinforced plastics (GRPs). GRPs represent the earliest and the most widely used (over three-fourths of total fiber–reinforced composite production) fiber–resin composites. With glass fibers in various forms coupled with either a thermosetting or thermoplastic resin, these composites can be produced without the need for high curing temperatures or pressures. The product contains a very good balance of properties, has high corrosion resistance, and is low in cost for a multitude of uses as structural, industrial, and consumer-related products, ranging in size from minute circuit boards to boat hulls. Using 20% to 40% fiber loadings, the composites will, in general, double the strength and stiffness of the plastic resin used alone. Continuous fibers will increase these properties fourfold, with accompanying desirable decreases in thermal expansion and creep rate and with increases in impact strength, heat-deflection temperatures, and dimensional stability. These fibers may take the form of continuous filaments (monofilaments) or yarn. Figure 8-23 shows applications of GRPs.

The disadvantages accompanying these composites arise, in the main, from the fact that they are especially two-phase structures. This leads to a degree of environmental degradation greater than that experienced by either component material alone. Residual stresses and electrochemical effects result from the marriage of two dissimilar materials. Furthermore, the diffusion of fiber materials into the matrix materials, and vice versa, may take place at several stages. Variation in the thermal expansion of these two materials leads

(a)

Figure 8-23 (a) Fiberglass composite wind-turbine blades (layers of glass mat and rovings in a vinyl resin matrix). GRP provides high strength, light weight, low cost, extended service life, and increased weatherability. (Morrison Molded Fiberglass Co.) (b) A redesign of the Ford Taurus chassis. This prototype chassis reflects designers' keen interest in using reinforced-plastic composite technology for improved vehicles. GRP would allow 8 or 10 parts to replace over 400 stamped steel parts. A GRP chassis would save about 30% of the car weight because a lighter chassis means downsizing engines, brakes, and other components. Additional benefits come through more efficient manufacturing techniques. (Courtesy of Ford Motor Company.)

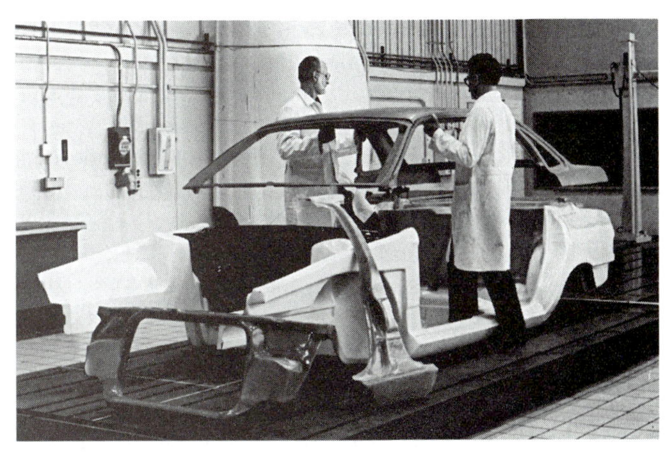

(b)

to thermally induced stresses that result in warping, plastic deformation, cracking, or combinations thereof.

A recently developed FRP composite rebar, called ***C-bar,*** is a likely candidate to replace epoxy-coated steel reinforcing bar (rebar) in many applications. C-bar is composed of unidirectional glass fibers wetted by recycled terephthalic acid resin in the center core. Unidirectional mat and ½-in. chopped fibers with a polyester surfacing coat are added to the exterior. The base resin used only on the outside is a polyurethane-modified vinyl ester. What sets C-bar apart from other forms of FRP rebar is the combined use of both mat and fibers, plus the existence of a mechanical and a chemical bond between the acid resin in the center core and the vinyl ester used on the outside. Tensile strength is reported to be more than twice that of steel rebar, with four times less weight. C-bar is nonmagnetic and noncorrosive, and has a more compatible coefficient of thermal expansion than that of concrete, attributes that can lead to many potential uses. The hybrid process combines pultrusion with compression molding (see Section 8.7), applying the fiber circumferentially to the inner core of the rebar. The result is a consistent, fully cured cross-section at a processing rate of 20 ft/min, which is unattainable with any pultrusion process.

Naturally, the advantages of such composites outweigh the disadvantages. Some of these major advantages can be summarized briefly as follows:

1. Greater matrix strength and stiffness, which permits little or no cross-plying and greater joint and stress concentration load strength
2. Increased operating temperatures and increased durations of time spent at those temperatures
3. Increased resistance to high temperatures and humidity environments
4. Better comformability with existing metal fabrication techniques

8.6.1.2 Metal matrix composites (MMCs) and ceramic matrix composites (CMCs). As with plastic resins, many fibers are used with metal matrices to form composites. The volume fraction of reinforcing materials plus interface–interphase reactions play a leading role in determining the ultimate success of any metal matrix composite in meeting the demanding requirements of a particular design. Figure 8-24(a) shows photomicrographs of metal reinforcing fiber embedded in a metal matrix.

The major rationale for the development of such composite materials is to satisfy the important need for tough, strong materials capable of maintaining their special properties under high-temperature operating conditions. Figure 8-24(b) is a plot showing how the specific strength of some materials varies with temperature.

Originally developed for the aerospace industry, these composite materials are now finding many uses in other industries. Metal matrix composites (MMCs) and ceramic matrix composites (CMCs) consist of a base (matrix) that is reinforced with continuous fibers or discontinuous materials in particulate form. Figure 8-24(c) depicts how the Chrysler Corporation used the Sullivan Process to produce a single-piece, CMC Si_3Ni_4, cam roller follower. As discussed in Module 7, this CMC component replaces 18 small steel bearings, potentially costs less to mass produce than steel parts, and offers longer wear life. ***Continuous reinforcement*** is defined as fibers, usually oriented in one direction, that produce

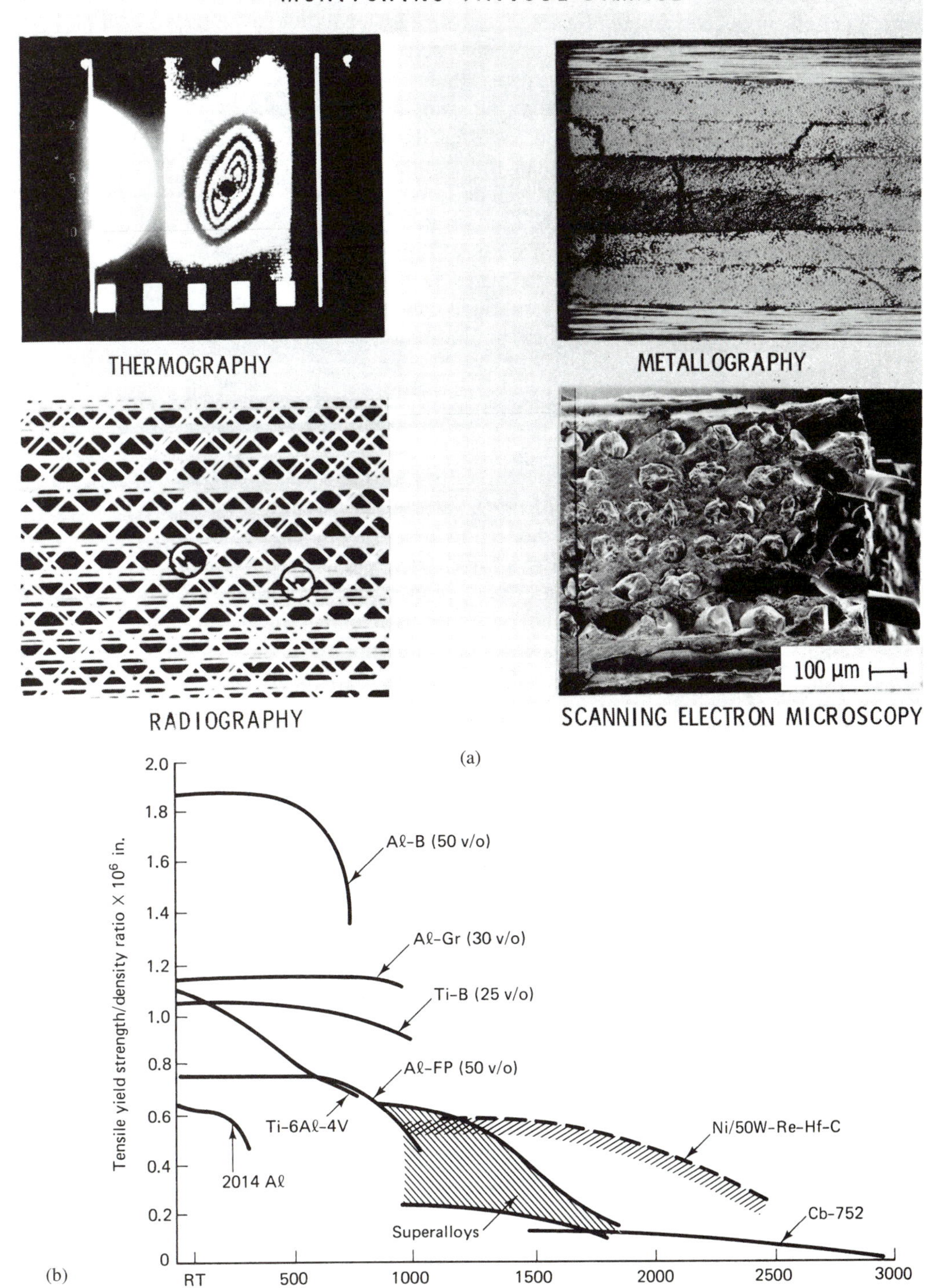

Figure 8-24 (a) Metal matrix composites (MMCs): photomicrographs. (NASA.) (b) Specific strength versus temperature: metal alloys and composites. (NASA.)

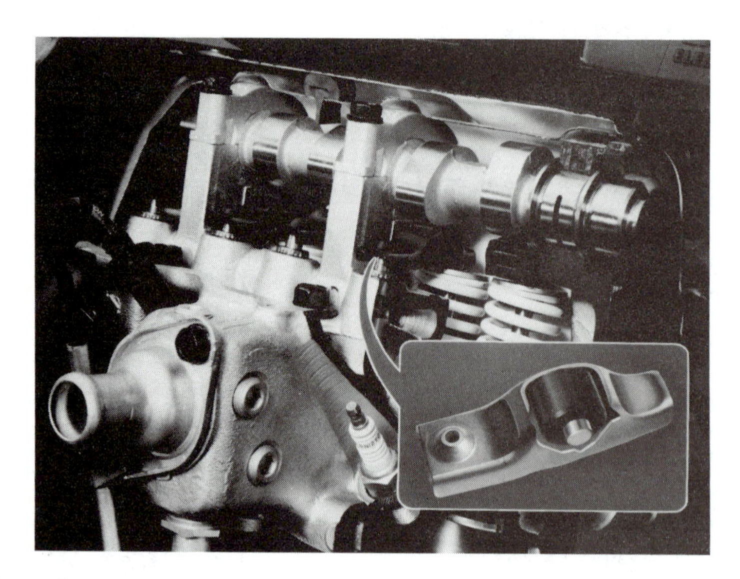

Figure 8-24 (c) Ceramic-matrix-composite (CMC) parts, such as this cam roller follower, are made from a liquid process, developed by Sullivan Mining, that does not use the conventional powder process. The process is less damaging to reinforcing fibers and capable of mass producing complex shapes at about one-tenth the cost of parts made by conventional processing. At least as durable as steel rollers, they may run more quietly. *(Advanced Materials & Processes.)*

an anisotropic composite in which the fibers predominate, with the matrix serving as a vehicle for transmitting the load to the fibers.

To satisfy the need for materials that are lighter, stronger, more corrosion resistant, and capable of performing at elevated temperatures, the U.S. Department of Energy established a comprehensive program to develop a relatively new family of materials known as continuous fiber ceramic composites (CFCCs). Near-term applications are in the mid-temperature (500° to 1000°C), secondary aerospace structures such as exhaust components, hot gas ducting, and leading edges. In the longer term, high-temperature applications in the automotive and industrial fields will have great potential, provided the costs of fabrication can be brought down to a competitive basis. Several methods of fabricating these composites are used. Among them is the use of preceramic polymers for the low-cost fabrication of large, complex-shaped CFCCs. After forming in the polymer condition, the composite is pyrolized to form the ceramic matrix and then densified through repeated reinfiltration/pyrolysis cycles. Regardless of what fabricating process is finally determined to have the most merit, the deciding factor will be the development of a cost-effective, oxidation-stable fiber interface that can withstand high strain fracture.

Continuous fiber ceramic composites (CFCCs) have been developed to overcome the brittle nature of ceramics and to capitalize on their favorable properties (see Figure 7-31). ***Discontinuous reinforcement*** refers to whiskers, short or chopped fiber, or particulates that produce a composite exhibiting a blend of the reinforcement and matrix properties. Table 8-6 compares the physical properties of two MMCs reinforced by particulates, emphasizing the difference in properties due primarily to the type of reinforcement. Most of these com-

TABLE 8-6 PHYSICAL PROPERTIES[a] FOR SILICON-CARBIDE- AND ALUMINA PARTICULATE-REINFORCED (SiC_P/Al AND Al_2O_3/Al) METAL MATRIX COMPOSITES

Property	Reinforcement: Alumina (Al_2O_{3p}/Al)		Silicon carbide (SiC_p/Al)	
Filler content[b](%)	50	—	55	—
Density [g/cm³ (lb/in³)]	3.2	(0.12)	2.9	(0.10)
Tensile strength [MPa (ksi)]	530	(77)	250	(36)
Modulus [GPa (Msi)]	160	(23)	200	(29)
Elongation (%)	0.6	—	0.3	—
Compressive strength				
0.2% yield [MPa (ksi)]	520	(75)	460	(66)
Ultimate [MPa (ksi)]	900	(130)	490	(70)
Fracture toughness[c] [$MPa\sqrt{m}$ ($ksi\sqrt{in.}$)]	18.5	(16.8)	12	(11)
Coefficient of thermal expansion [ppm/°C (ppm/°F)]	13.5	(7.5)	8.5	(4.7)
Thermal conductivity $\left[\frac{W}{m \cdot K}\left(\frac{Btu \cdot in.}{hr \cdot ft^2 \cdot °F}\right)\right]$			160	1100

Source: Courtesy of Lanxide.

[a]Room temperature (25°C) properties. Elevated temperature properties available.

[b]Filler content can be tailored (20% to 80%) to obtain desired properties. Properties as a function of filler content available.

[c]Chevron notch beam.

posites tend to be isotropic. MMCs and CMCs are composed of three constituents: the fibers, having high strength and stiffness; the matrix, which holds the fibers together and distributes the applied load; and the interface zone between the fibers and the matrix, which determines the wetting and bonding (coupling between the fiber and the matrix) as well as the transfer of the load between the fibers and the matrix (see Figures 8-4 and 8-8). The formation of a shallow interface zone is generally desirable to ensure a strong bond.

Most MMCs and CMCs can be synthesized using standard metal-working practices such as extrusion, hot forging, hot molding, superplastic forming, squeeze casting, powder metallurgy, liquid-melt infiltration, and rolling. A recent Mixalloy process uses turbulent eddies and jet impingement to mix two or more liquid metal streams at turbulent velocities in a chamber. The resulting mixture is then solidified fast enough to preserve the required microstructure. This process is unique because it is successful in generating nanoscale particles (e.g., 50-nm refractory boride) for use in a copper matrix. Coarser particles can also be obtained by careful control of the mixing conditions. This new process is replacing older methods of producing a wide variety of dispersion-strengthened alloys (see Sections 8.6.4 and 8.6.4.2 for further information). In addition, brass and bronze can be strengthened by the Mixalloy process using boride particles.

Whisker-reinforced MMCs have distinct advantages over conventional alloys. These MMCs, classified by some as dispersion-strengthened alloys, provide the highest-strength discontinuous-reinforced MMCs. Typical fibers are SiC, Al_2O_3, and Si_3N_4, which are combined with some common metal matrices, such as Al, Mg, Ti, and Cu. Powder metal (P/M) or liquid-melt infiltration methods are used to synthesize the fine-mesh metal-alloy powders and the whiskers by thorough mixing, blending, and consolidating to produce a near-net-shaped component. SiC whiskers reinforcing an Al matrix have transverse tensile properties that are nearly equal to the longitudinal properties, making them capable of outperforming fiber-reinforced composites in a situation requiring multidirectional reinforcement.

Lanxide Corporation fabricates MMCs by using a pressureless, molten-metal-infiltration process (trademarked Primex) that features excellent wetting of the reinforcement by the matrix alloy. (See Module 7 for information on Lanxide's fabrication of CMCs.) Net- and near-net-shaped parts, small to large, having complex shapes can be produced economically. Figure 8-25 is a schematic of the process for making an electronics package. To produce an aluminum-alloy MMC, for example, the infiltration is carried out in a nitrogenous atmosphere at a temperature somewhat above the melting point of aluminum. No external pressure or vacuum is needed. The MMC produced is completely infiltrated with metal alloy and features high and uniform loadings of particles or fibers. Figure 8-26 is a photomicrograph of an aluminum–matrix MMC filled with fused-Al_2O_3 particles, and Figure 8-27 is a similar photograph showing Al_2O_3 agglomerates (clusters of particles). These figures illustrate complete infiltration of the spaces between the particles by the molten aluminum alloy. One application of this SiC/Al MMC is for support structures for electronic components. Typical requirements include a low coefficient of thermal expansion (CTE) to reduce mechanical stresses imposed on the component during attachment and operation, high thermal conductivity for heat dissipation, and low density for minimum weight. Table 8-7 compares the physical properties of SiC_p/Al MMCs to a typical aluminum alloy. In addition, the dimensional stability of this MMC can be more than 3.2 times that of unreinforced steels, aluminum alloys, and titanium alloys. Combined with

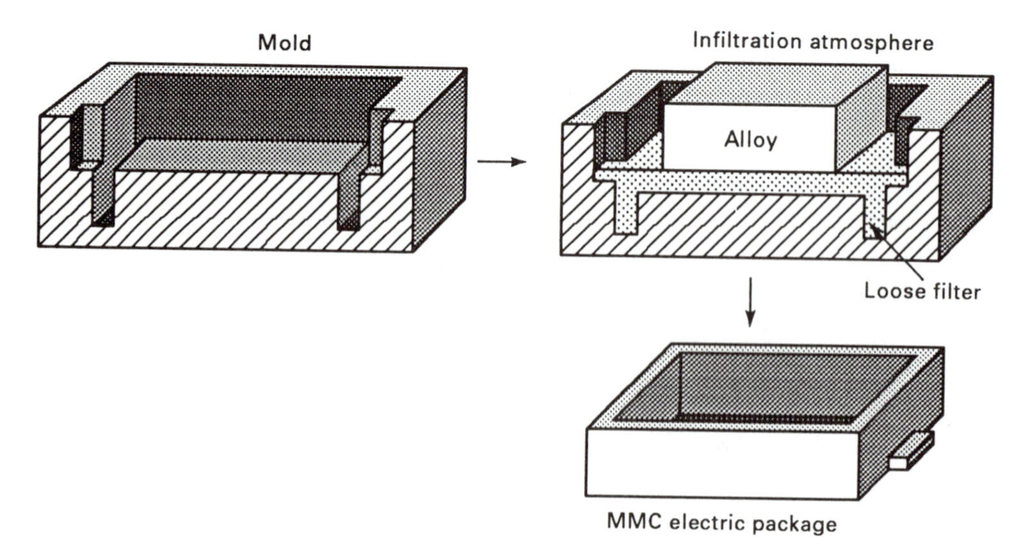

Figure 8-25 Primex® pressureless metal infiltration process by Lanxide for producing MMC parts such as this electronics package to net- and near-net-shape, ranging in size from small to large. (Lanxide Corp.)

Figure 8-26 A photomicrograph of Al_2O_{3p}/Al MMC showing a high and uniform particle loading. (Lanxide Corp.)

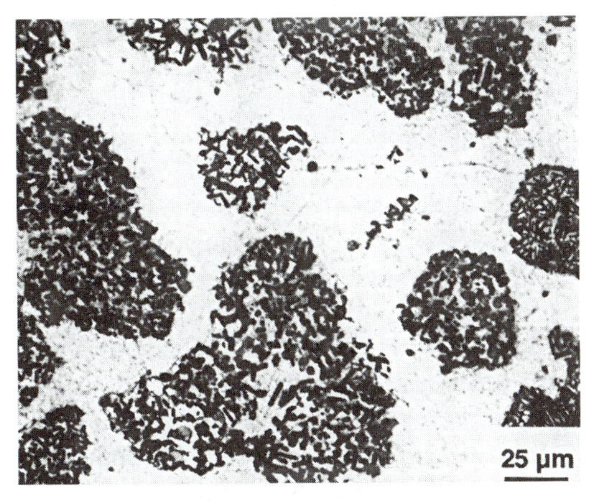

(a)

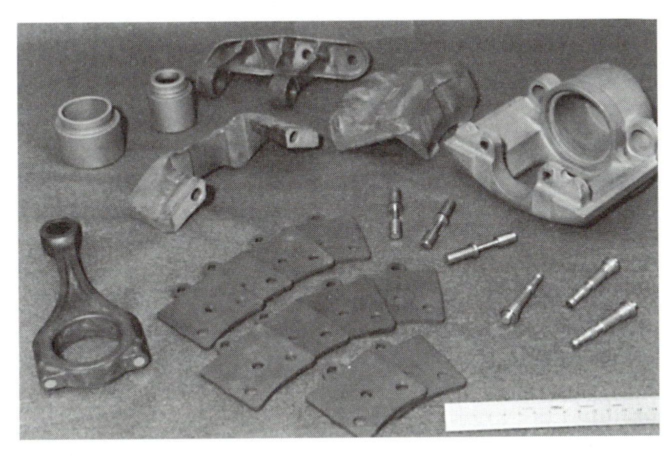

(b)

Figure 8-27 (a) A photomicrograph of an aluminum–matrix MMC filled with Al_2O_3 agglomerates. The molten metal completely infiltrates both the interagglomerate spaces and the minute openings within the agglomerates. (Lanxide Corp.) (b) The Primex process for aluminum MMCs, with 60% to 70% (volume change) particulate reinforcement, can be used in engine components such as connecting rods, piston pins, brake parts, brackets, and stiffeners. (Lanxide Corp.)

TABLE 8-7 PROPERTIES OF SiC-PARTICLE-FILLED Al MMCS FOR ELECTRONIC APPLICATIONS

	Composite examples		Typical Al alloys (for comparison)
SiC loading (vol %)	55	75	—
Coefficient of thermal expansion [$\times 10^{-6}$ ft^2 K ($\times 10^{-6}$ ft^2 °F)]	8.5 (4.7)	6.2 (3.4)	22–24 (12–13)
Thermal conductivity [W/m · K (Btu/hr · ft^2 · °F)]	160 (93)	170 (99)	150–180 (87–104)
Density [g/cm^3 (lb/in.^{3})]	2.95 (0.106)	3.0 (0.108)	2.7 (0.097)
Elastic modulus [GPa (msi)]	200 (29)	270 (39)	70 (10)

Source: Courtesy of Lanxide.

high thermal conductivity and low CTE, this array of properties provides this MMC with excellent rigidity and dimensional stability for use in space stations, automotive-brakes calipers, or optical system components. Table 8-8 lists some properties of $(Al_2O_3)_p/Al$ MMC in which the particles are more coarse, resulting in a composite that has higher strength and fracture toughness than those of the parent alloy. Despite the lower ductility, these MMCs retain good fracture toughness. Table 8-9 provides an opportunity to compare properties of a CMC with an MMC. In particular, note the retention of strength properties at high temperatures for this CMC.

A major inhibition to the use of MMC parts is the difficulty in joining them. A newly developed fabrication process produces MMC parts with integral inserts. Parts consisting of a magnesium matrix and graphite fibers with titanium inserts are formed in final shape in a single casting step using a pressure casting machine. No finish machining is required. Molten magnesium is infiltrated into the graphite fiber preform reinforcement, which is inserted into a mold cavity along with the titanium insert. Pressures, temperatures, and times

TABLE 8-8 PROPERTIES OF AN Al_2O_3-PARTICLE-FILLED Al MMC

	Composite (51% Al_2O_3)	Base alloy (AA 520)
Tensile strength [MPa (ksi)]	531 (77)	331 (48)
Elastic modulus [GPa (msi)]	161 (23)	65 (9)
Fracture toughness[a] [MPa · m$^{1/2}$ (ksi · in.$^{1/2}$)]	18.5 (16.8)	N.A.[b]
Elongation (%)	0.63	16

[a]Chevron notch method.

[b]N.A. = not applicable.

TABLE 8-9 PHYSICAL PROPERTIES FOR SILICON–CARBIDE FIBER-REINFORCED ALUMINA SiC_f/Al_2O_3[a]

Property	Temperature					
	25°C	(73°F)	1000°C	(1832°F)	1200°C	(2192°F)
Fiber content (eight plies) (%)	35	—	35	—	35	—
Density [g/cm^3 (lb/in.^{3})]	2.9	(0.10)	2.9	(0.10)	2.9	(0.10)
Tensile strength [MPa (ksi)]	283	(41)	—	—	—	—
Modulus [GPa (Msi)]	122	(17.8)	—	—	—	—
Poisson ratio	0.10	—	—	—	—	—
Flexural strength[b] [MPa (ksi)]	450	(65)	400	(58)	350	(51)
Modulus [GPa (Msi)]	200	(29)	—	—	—	—
Poisson ratio	0.29	—	—	—	—	—
Fracture toughness[c] [MPa$\sqrt{m}$ (ksi$\sqrt{in.}$)]	28	(26)	23	(21.5)	23	(21.5)
Interlaminar shear strength[d] [MPa (ksi)]	63	(9)	—	—	—	—
Shear modulus[e] [GPa (Msi)]	90	(13.1)	—	—	—	—
Coefficient of thermal expansion [ppm/°C (ppm/°F)]	5.8	(3.2)	—	—	—	—
Thermal conductivity $\left[\frac{W}{m \cdot K}\left(\frac{Btu \cdot in.}{hr \cdot ft^2 \cdot °F}\right)\right]$	8.7	(60)	5.7	(39)	5.5	(38)
Thermal shock (% retained flexural strength after quenching into RT water)	1000 to 25°C 92.7% 1000 to 25°C 85.2% . . . after 5 cycles 1200 to 25°C 83.6%					

Source: Courtesy of Lanxide.

[a]Silicon–carbide fiber (Nicalon); 12-harness satin weave, 0/90 two-dimensional preform layup.

[b]Four-point bend.

[c]Chevron notch beam.

[d]Short-beam method, ASTM D2344-84.

[e]Sonic method.

are calibrated to produce a complete infiltration of the densely packed preform and rapid solidification of the melt. Subsequent tensile testing produced fracture, at around 10,000 N, at a point well away from the insert.

The designation for aluminum MMCs, approved by the Aluminum Association Inc., uses a four-part system to identify the matrix alloy and the composite's amount and type of reinforcement. For example, $6063/Al_2O_3/15_p$-T6, 6063 designates an alloy-6063–matrix composite reinforced by 15 vol % alumina particles and heat treated to T6 temper.

Carbon–carbon composites have been used on nose cones and leading edges of high-performance space vehicles subjected to very high temperatures (3000°C). Another high-temperature application has been as brake discs in racing cars. Such composites are stronger at high temperatures than they are at low temperatures, they have high specific strength and good resistance to thermal shock, and they are self-lubricating. Composed of a carbon fabric and an organic resin, the fabric is placed in a mold that is infiltrated (one method is CVD, described in Module 7) with a resin that is subsequently ***pyrolized*** (heated to a high temperature), driving off the noncarbon atoms in the polymer and thus forming a carbonized matrix. A coating of silicon carbide is added to reduce oxidation, which limits the high-temperature applications of these composite materials.

8.6.2 Laminar Composites

Laminar composites (Figure 8-28) consist of layers (***lamina***) of at least two different solid materials bonded together so the fiber orientation runs at different angles (e.g., 0°, 45°, 90°, etc.). Lamination allows the designer of this composite material to use the best properties of each layer to achieve a more useful material. Properties such as wear resistance, low weight, corrosion resistance, strength, stiffness, and many more can be accented by a wise selection of different constituent layers. A fiber-reinforced composite made up of layers of tape, fabric, or mat can be considered a laminar composite. A well-known example of a laminar composite is plywood. Plywood has isotropic properties in the plane of each sheet due to the layers of wood bonded with a thermoset resin such that the longitudinal direction is at right angles in adjacent plies. Only their orientations differ. ***Clad metals*** such as bimetallic strip or copper-clad stainless steel are composed of two different materials. In the former, the two sheets of metal bonded together have different amounts of thermal expansion, and in the latter example the thermal conductivity property is the rationale for the design of the composite material. Similar treatments with paper produce isotropic properties. Laminated paper with plastic film or metal foil, or using polymers, such as nylon fabric, as a base and laminating the polymer with layers of metal, produces an endless variety of plastic-based laminates. Roofing paper, Formica, and Kevlar fibers laminated

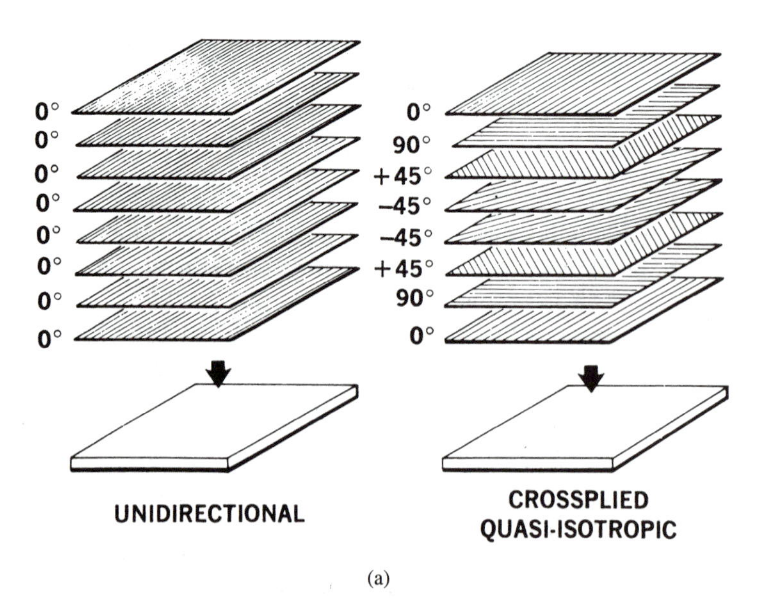

(a)

Figure 8-28 (a) Fiber orientation within laminates. Laminates can be tailored to meet specific requirements, with savings in time and material. (Hercules, Inc.)

(b)

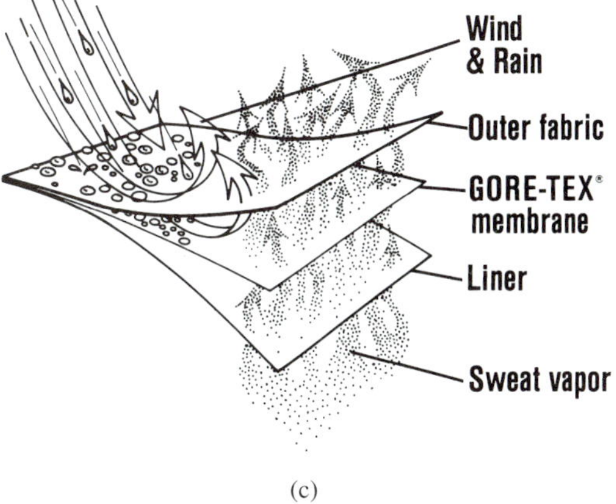

(c)

Figure 8-28 (b) Gore-TEX® laminated membrane used by manufacturers of clothing and footwear for use outdoors in all types of weather. (W. L. Gore & Associates, Inc.) (c) Gore-TEX® lamina between liner and outer fabric allows sweat vapor to escape but keeps large water drops out for foul-weather gear.

with a resin to produce a material for bulletproof vests are additional examples of these composite materials on the market today.

Safety glass is a laminated glass consisting of a layer of poly(vinyl butyral) bonded between two layers of glass. Glass alone is quite brittle and dangerous due to its proclivity to shatter into many sharp-edged pieces. Poly(vinyl butyral) is very tough but weak in scratch resistance. The glass lamination protects the plastic from scratching and gives stiffness, while the plastic contributes its toughness. By protecting the other in different ways, each material allows the other to contribute to the composite material's vastly improved properties over those of its constituents.

A layered composite known as ***Graftskin*** is an artificial human skin replacement tissue that replaces both the outer epidermal and inner epidermal skin layers. It blends dermal cells with collagen—the protein that forms a fibrous framework for skin cells. The collagen is taken from cows. Graftskin is used successfully for treating chronic skin diseases, burns, and deep wounds. The results of such treatment show quick wound closure, accelerated healing, and reduced scarring.

Laminates formed by high pressure and heat are referred to as ***rigid laminates.*** Some examples of rigid laminates are countertops, rods, tubes, and printed-circuit boards (PCBs) for the electronics industry.

Unsolved manufacturing problems associated with the processing of bulk-laminated ceramic or metal composites limit their potential use in structural applications. Design engineers, in their search for solutions to structure problems, have observed how nature produces many examples of laminated composites that possess excellent mechanical properties. Seashells would be one example of this type of structure. Ceramics, with their inherent properties of high-temperature strength, hardness, and wear and oxidation resistance, also possess low toughness. Module 7 contains much information on how materials engineers and scientists are combating this problem of brittleness in ceramics by mechanisms such as crack bridging and transformation of residual stresses, to mention but a few. The field of ***surface coatings*** has had great success in the application of ***ceramic coatings*** to provide other materials with some of ceramic's properties mentioned above. However, the application of coatings is expensive. A processing time of 500 hours is mentioned in the literature to achieve a laminated structure 1 cm thick. Metallic laminated structures made of ultrahigh-carbon steel and iron–silicon alloys using roll bonding would have limitations in producing a bulk ceramic/metal laminated composite due to concern about adhesion between the layers.

Refractory metals are among the few materials with significant mechanical strength above 2500°F, but they do not have sufficient strength-to-weight ratios for many applications. A material with a high strength-to-weight ratio and good mechanical strength at service temperatures above 3000°F is sorely needed for the many high-temperature applications such as space shields, rocket engine fuel turbines, and gas turbines. Combining a dense refractory metal such as tungsten, molybdenum, vanadium, or tantalum with a lighter-weight matrix material creates a unique material—a lightweight ***refractory composite.*** A fabricating process, ***explosive welding,*** is used to join these incompatible metals. Explosive welding or ***explosive bonding*** is a solid-state welding process that uses controlled explosive detonations to force two or more metals together at high pressures. The contaminant surface films are plastically jetted off the base metals as a result of the collision of the two metals. The time duration is so short that the heat-affected zone (HAZ) is microscopic. During the process, the first few layers of each metal become plasma due to the high velocity impact (200 to 500 m/s). The remaining thickness of each metal maintains a near-ambient temperature and acts as a large heat sink, cooling the plasma rapidly. The resultant composite system is joined with a high-quality metallurgical bond. Some explosive bonded materials in industry include niobium/copper, copper/aluminum (for high-voltage busbars), aluminum/stainless steel, titanium/stainless steel (for pacemakers), and zirconium/stainless steel.

Microinfiltrated macrolaminated composites (MIMLCs) offer a new, economically feasible approach to the production of tough composites using cermet/metallic, ceramic/metallic, and ceramic/intermetallic-compound bulk-laminated composites. Figure 8-29 is a schematic of an MIMLC illustrating the conceptual structure. Such materials attempt to duplicate a case-hardened metal with a ductile interior. Such a metal has the surface hardness and strength needed for many applications, such as a cutting tool, along with

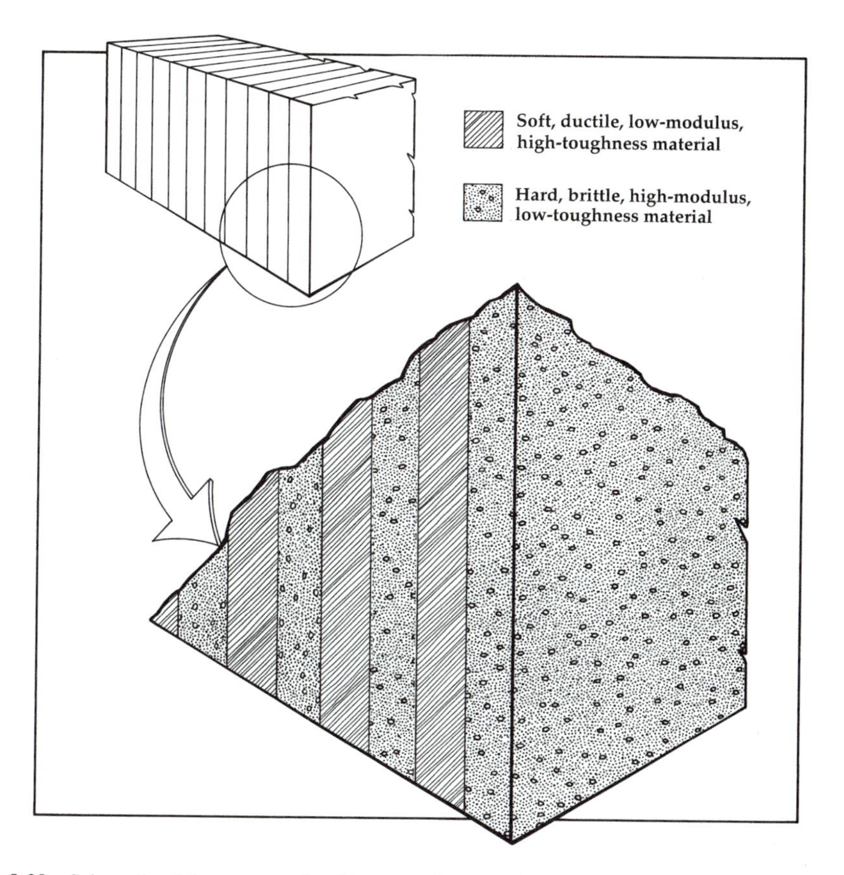

Figure 8-29 Schematic of the conceptual architecture of an MIMLC. The basic double-layer structure, consisting of a soft, ductile material (low modulus and strength and high toughness) and a hard, brittle material (high modulus and strength and low toughness), is repeated as many times as necessary to form a bulk composite material. The brittle material is also infiltrated with the ductile component, tending to increase the brittle material's toughness. (Bose & Lankford, Southwest Research Inst., *Adv. Mats. & Proc.*, Vol. 140, No. 1, July 1991, pub. by ASM Int.)

the ductility to possess great impact resistance (toughness). One problem that must be overcome is the proper adhesion between the layers of the bulk composite. Figures 8-30 and 8-31 illustrate poor and good solid-state bonding with a W–Ni–Fe/Ni MIMLC. W–Ni–Fe is a ceramic alloy that represents a hard material having high strength and modulus but low toughness. It could be compared to the fiber reinforcement in a metal– or ceramic–matrix composite. The nickel (Ni) represents a material having low strength, elastic modulus, and hardness but high toughness, and could be compared to the matrix material in a fiber-reinforced composite. Figure 8-30 shows a cracklike discontinuity between the bottom side of the Ni foil and the W–Ni–Fe heavy sheet when the composite was subjected only to a sintering temperature of 1400°C. Figure 8-31 shows proper solid-state bonding and consolidation of the layers when pressure was added to the process. Many variations of the fabrication process are necessary to accommodate the differences in solubility of the combinations of materials that can be used. Figure 8-32 illustrates the situation where the two materials in a ceramic and a metal that have very little or no solubility in one another are combined. Porous alumina (Al_2O_3) and Ni are tape cast (to be defined) to produce an alumina/nickel MIMLC. Using a temperature just below the melting point of Ni and applying pressure, the Ni is extruded into the fine pores of the alumina structure. With further controls of variables such as time, temperature, and pressure, more adequate bonding can be achieved.

Additional details of the fabrication of the W–Ni–Fe/Ni MIMLC may assist in greater understanding of this process. In this process, both materials are tape cast. ***Tape casting*** consists of mixing powders with a suitable binder to form a slurry, which is then cast as a thin tape (sheet) of material. The material is then cut in the green state when it is

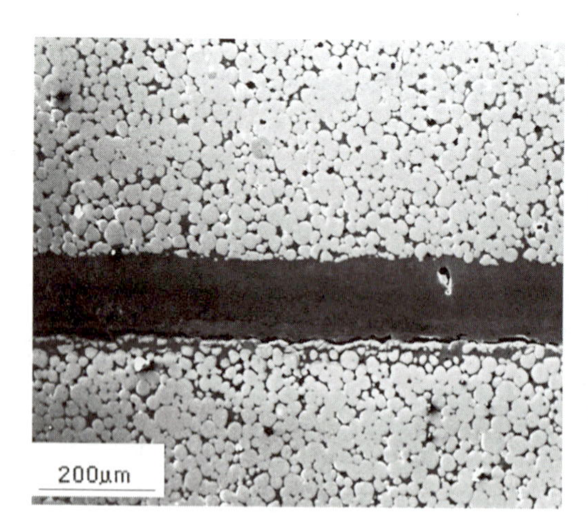

Figure 8-30 W–Ni–Fe/Ni MIMLC composite that was sintered at 1400°C without pressure, showing improper bonding between layers. Note a crack along the bottom side of the nickel foil. (Bose & Lankford, Southwest Research Inst., Adv. Mats. & Proc., Vol. 140, No. 1, July 1991, Pub. by ASM Int.)

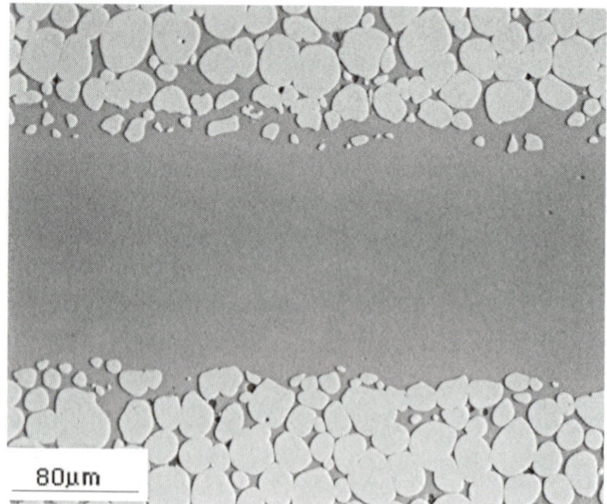

Figure 8-31 W–Ni–Fe/Ni MIMLC composite that was sintered under pressure at 1400°C, showing good bonding between the nickel foil and the heavy alloy sheet, which creates one continuous matrix that interpenetrates the tungsten grains. (Bose & Lankford, Southwest Research Inst., Adv. Mats. & Proc., Vol. 140, No. 1, July 1991, Pub. by ASM Int.)

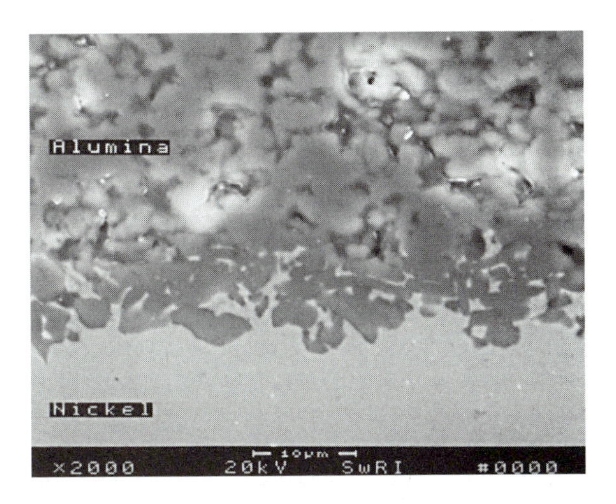

Figure 8-32 Photomicrograph of the microstructure of a hot-pressed alumina/nickel MIMLC where the nickel layer has infiltrated to only a small depth in the alumina sheet, which remains porous. Additional increases in time, temperature, and pressure would be necessary to achieve adequate bonding. (Bose & Lankford, Southwest Research Inst., Adv. Mats. & Proc., Vol. 140, No. 1, July 1991, Pub. by ASM Int.)

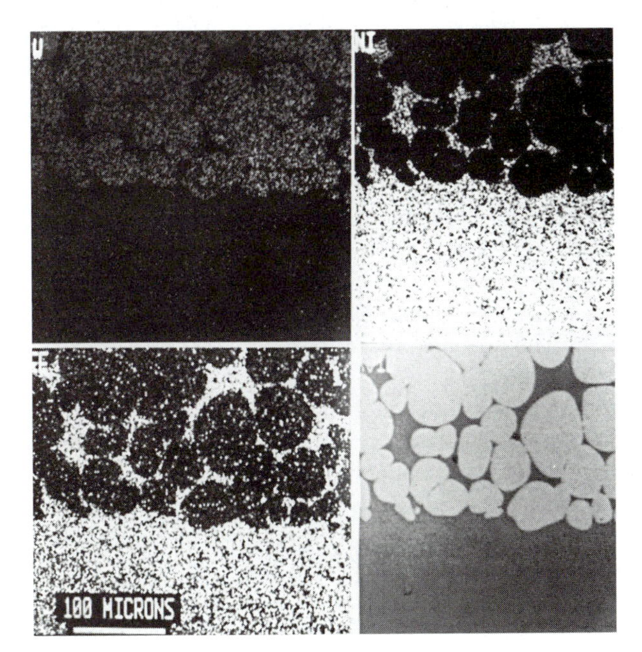

Figure 8-33 An x-ray dot map of a portion of a properly bonded W–Ni–Fe sheet/Ni sheet MIMLC, showing the tungsten spheroids imbedded in a matrix of nickel and iron. (Bose & Lankford, Southwest Research Inst., Adv. Mats. & Proc., Vol. 140, No. 1, July 1991, Pub. by ASM Int.)

relatively soft or after it is sintered. The W–Ni alloy is tape cast and sintered to some 60% of theoretical density to produce an interconnected porosity. An 80:20 ratio of Ni to Fe powders is also tape cast and sintered to full density. Laid up in alternating layers and heated to about 40°C above its melting temperature, the molten Ni–Fe alloy infiltrates the porous W sheet and takes into solution a fraction of the W. In Figure 8-33 the lower white portion (Ni–Fe alloy) shows the infiltration of the porous tungsten sheet. The top, darker portion shows the W spheroids surrounded by the diffusing Ni–Fe matrix, which takes into solution a fraction of the W (darker portion).

This discussion points out the many process control variables, as well as the large number of material combinations of a hard, brittle phase interpenetrated by a soft, ductile material, that can be modified to produce a multitude of MIMLCs. With a plethora of different properties, they will find applications as armor plating; high-temperature composites; tough cutting tools; wear-resistant composites; and components with low density, high strength and high modulus, good fracture toughness, and impact resistance for use in aerospace and automotive structures and components.

8.6.3 Sandwich (Stressed-Skin) Composites

Sandwich composites can also be classified as laminar composites. Their outer surfaces or ***facings*** are made of some material higher in density than the inner material or ***core*** that supports the facings. The primary purpose of sandwich composites is the achievement of high strength with less weight, specifically a high *strength-to-weight ratio* or *specific strength.* A sandwich composite may be compared to a structural I beam. The high-density facings correspond to the I-beam flanges carrying most of the applied load, particularly the bending loads. Like the I-beam web, the low-density core allows the facings to be placed at a relatively large distance from the neutral plane to produce a large section modulus. The core carries the shear stresses. Overall sandwich composites are more efficient than I beams because they possess a combination of high section modulus and low weight.

Typical sandwich facing materials are aluminum, wood, vinyl, paper, glass-reinforced plastics, and stainless steel. Core materials represent all families of materials, are primarily cellular in form, and take on the configurations known as honeycomb, waffle, corrugated, tubes, and cones. These rigid cores provide the greatest strength and stiffness, with honeycomb possessing isotropic properties. Metals that can be made into very thin sheets capable of being diffusion bonded are finding greater use as cellular-core materials. Solid cores made of plywood, as well as foamed cores of polystyrene or ceramics, are examples of other types of core materials.

Ceramic honeycomb panels withstanding temperatures to 1800°C serve as lightweight structural panels. Using CVD, a vapor-depositing ceramic is deposited on a fabric substrate woven in a honeycomb pattern and making a honeycomb structure. The substrate is ultimately eliminated. The fabric can be made of a loosely woven polymer such as polyacrylonitrile (PAN), which is impregnated with an organic binder such as phenolic resin for stiffness. The ceramic materials are not limited to SiC. Others are silicon boride, silicon nitride, or boron nitride. Heated to temperatures between 500° and 1000°C in an atmosphere of 2% to 5% oxygen, the fabric and binder oxidize and pass through the pores of the ceramic as a gas, leaving a microstructure of voids. The voids can be filled with ceramic by additional CVD, as can the hexagonal holes of the honeycomb.

The honeycomb structure or shape, mathematically the strongest possible and the most economical for a mass of adjacent cells, dramatically demonstrates its strength (Figure 8-34). ***Nomex*** honeycomb paneling provides superior specific modulus and finds many applications for structural elements in aircraft, spacecraft, racing vehicles, and watercraft.

8.6.4 Particulate Composites

These composite materials contain reinforcing particles of one or more materials suspended in a matrix of a different material. The particles, either metallic or nonmetallic, by

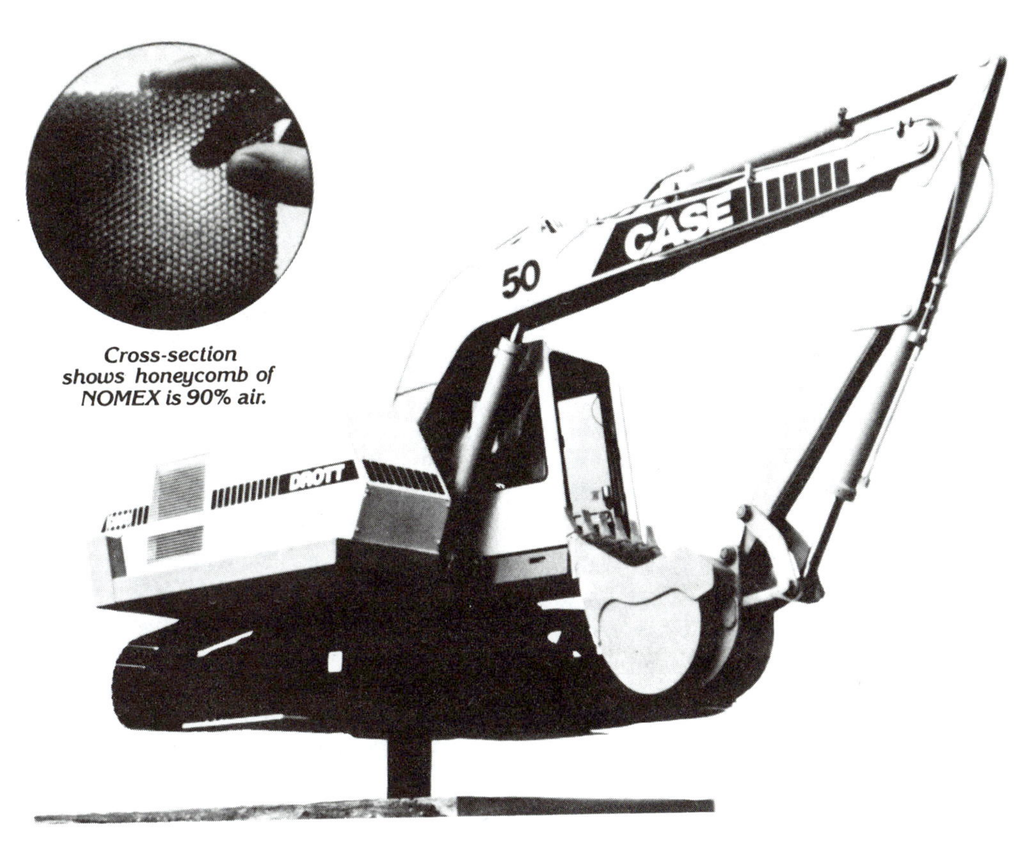

Figure 8-34 A 10-in. square by 22-in. high column of honeycomb of Nomex™ aramid weighing 11 pounds supports a load of 27 tons. (DuPont.)

definition do not chemically combine with the metallic or nonmetallic matrix material. As with nearly all materials, structure determines properties, and so it is with these composite materials. The size, shape, and spacing of particles; their volume fraction; and their distribution all contribute to the properties of these materials. How particles influence the properties of a material are explained in the discussion about dispersed-strengthened alloys (to follow). Particulate composites are many times divided into subclasses by using some characteristic or combination of particular characteristics. This category of composites does not include particulates flat in shape, which possess sufficiently different properties to warrant a special classification. A reinforcing particle can be defined as having all its dimensions nearly equal.

8.6.4.1 Cermets. Cermets provide an excellent example of a particulate composite material. Cermets are produced by sintering a mixture of ceramic and metal powders to form a structure that consists of a dispersion of ceramic particles in a continuous metallic matrix. Carbides of tungsten, chromium, and titanium are widely used in combination with cobalt, nickel, and stainless-steel matrices to provide materials with very high hardness for wire-drawing dies, for very high corrosion resistance for valves, or for very high temperature applications such as turbine parts. Uranium-oxide and boron-carbide cermets suspended in stainless steel find several uses in nuclear-fuel elements and control rods.

8.6.4.2 Dispersion-strengthened alloys/composites. Dispersion-strengthened alloys are similar to cermets and to precipitation-hardened alloys. Differences among these three categories are spelled out in terms of the constituent particles. The particles in dispersion-strengthened alloys are smaller and of a lesser volume fraction (at 3% by volume) than found in cermets. Dispersion-strengthened alloys are formed by the mechanical dispersion of particles, as opposed to precipitation-hardened alloys, in which compounds are precipitated from the matrix by heat treatment. Cold solder is a metal powder suspended in a thermoset to provide a metal that is not only hard and strong, but a good conductor of heat and electricity.

High-temperature structural silicides combine the desired qualities of ceramics with the high-temperature ductility and fracture resistance of metals. New patents have been issued in this field that describe the dispersing of silicon carbide throughout a matrix material of molybdenum disilicide that produces a molybdenum disilicide–alloy matrix composite with eight times the high-temperature strength of molybdenum disilicide by itself. Such a composite has the potential of meeting the demands of advanced high-temperature structural applications in the range of 1200° to 1600°C in oxidizing and aggressive environments. In addition to possessing specialized properties, the composite material is cost-effective in its manufacture compared to other advanced composites, such as silicon carbide ceramics materials.

At this point, it is timely to ask what the differences are between ***precipitation hardening*** (PH; Section 5.7) and ***dispersion strengthening*** (DS). PH metal alloys are strengthened by the in situ precipitation of a hard second-phase particle in a soft, ductile matrix through a multistage heat-treat process. The size and amount of the second phase can be controlled minimally by adjusting the alloy composition and the aging temperature, but certainly not as much as can be produced by the DS process. The second phase particles are *not* stable at high temperatures, which means that, as temperatures rise, these particles, including those under service conditions, will coalesce and agglomerate, reducing the number of particles and thus reducing the hardness and strength of the alloy. DS materials (composites) are produced by mechanically combining a hard and strong oxide particle with soft, ductile, metallic matrix particles. Various combinations of these particles can be used in varying proportions, sizes, and shapes by the powder metallurgy process, yielding a composite material with many desirable properties. The oxide particles chosen for these composite materials are chemically inert. Hence, they do not affect electrical properties. Since they do *not* grow or dissolve in the matrix at high temperatures, they contribute to excellent resistance to softening and to mechanical properties that do not change with changes in service operating temperatures.

The words ***in situ*** mean "in its original position." In relation to composites, the term refers to the formation of components (i.e., the reinforcements, matrices, and coatings of the interfaces) in a material that are produced during processing from the elements that made up the material. Because these reinforcements were produced within the materials, i.e., a metal or intermetallic, they are thermodynamically stable in the matrix (they won't degrade or disappear when heated) and they form as single crystals. These reinforcing structures are different from more conventional composites that have reinforcements in the form of fibers or particles that were deliberately added to the original material. ***In-situ composites*** are proving to be more cost effective and to have higher mechanical strength properties, particularly at high operating temperatures. The cost savings come from the reduction or elimination of the processing steps and the need to add reinforcing components. The many chemical processes that normally occur when a material system is thermally processed, such as phase changes, recrystallization, or nucleation, produce uniform distributions of the reinforcing phases much more efficiently than composites produced by more conventional means. Examples are the processing of conventional aluminum alloys or precipitation-strengthened alloys. Also, it is extremely difficult to handle very fine particles either as reinforcement or matrix contributions smaller than 1 μm. In fact, fine powders, whiskers, or fibers are extremely reactive and present toxicity/health problems. With in-situ composites, reinforcements produced are well within the nanometer range, and they require no handling. With such a scale of reinforcement comes greater mechanical strength.

A good example of in-situ composites is XD (exothermic dispersion) composites, under development since 1983 by Martin Marietta. These XD materials are produced by exothermic (heat given off) reactions of precursors, and they include metals and intermetallics reinforced by dispersions of intermetallic or ceramic particles or whiskers. ***XD technology*** deals with composite fabrication in which reinforcing components such as particulates, short fibers, or whiskers develop in situ within metal or intermetallic matrices. One such composite is a titanium aluminide (Ti_3Al) matrix composite made from three starting materials in powder form: Ti, Ti_3Al (titanium aluminide), and TiAl. To these is added titanium diboride (TiB_2), forming TiB_2 dispersions in a matrix of Ti_3Al, which in turn is formed by reactions of titanium, aluminum, and boron. This composite is replacing nickel-based superalloys in several aerospace applications. With a significantly lower density than stainless steel, titanium aluminide composites have greater strength and stiffness at elevated temperatures. For example, the yield strength at 750°C for 17-PH stainless steel is about 345 MPa; for TiAl (Ti-47Al + 7 vol % TiB_2), it is 415 MPa. The modulus at 920 K for steel is 145 GPa, compared to 162 GPa for the in situ composite.

A titanium diboride–alumina composite material with superior high-temperature wear resistance is produced by a new process called the *self-propagating high-temperature synthesis* (SHS) process developed by the Georgia Institute of Technology. The new chemically based process yields titanium diboride material with smaller (submicron) particles and allows the composite to be produced in final form in molds. The older process using solid-state diffusion left carbides as contaminants in the powder, which accumulated in the grain boundaries. The SHS process mixes powder metal, titanium oxide, and boron oxide; places them in a high-temperature crucible; and ignites them to start a self-sustaining reaction between the components. This produces titanium diboride dispersed within either magnesium oxide or alumina, depending on the choice of metal. Finished products can be hot pressed at 1500°C in less time (2 to 4 h) than the conventional materials. By varying processing conditions for the titanium diboride–alumina composite, either a dense composite material or a porous form (into which materials such as molten metals can be infiltrated) is produced. Some physical properties for this new material are a melting point of 3000°C, hardness superior to tungsten carbide, thermal conductivity better than cubic boron nitride, fracture toughness greater than silicon nitride, and a good stiffness-to-weight ratio. Applications for this new material are cutting tools, dies, and electrodes.

A "particle-dispersed" steel made by mechanical alloying is said to have a Young's modulus of 265 to 285 GPa, the highest ever achieved for steel. Sumitomo Metal Industries, Japan, reports that they have developed the high levels of stiffness through three stages of processing. The first step was mechanical alloying, in which oxides of iron, carbon, and yttrium are comminuted (pulverized or reduced) to ultrafine particles in a ball mill. The second step was the sealing of the powder in a capsule and hot extruding it. The third step was hot recrystallization, followed by heat treatments. The composition of the steel is Fe, 13–15Cr, 1–3 Al, 0.5 Y. Its properties are tensile yield strength, 600 to 900 MPa; ultimate tensile strength, 700 to 900 MPa; and elongation, 10% to 20%. The steel is capable of being fabricated in bar, pipe, or sheet form.

8.6.5 Flake Composites

Flakes of mica or glass in a glass or plastic matrix form a composite material that has a primarily two-dimensional geometry with corresponding strength and stiffness in two dimensions. Flakes tend to pack parallel to and overlap each other, and they provide properties such as decreasing wear, low coefficient of thermal expansion, and increased thermal and electrical conductivity, which are dependent on higher densities in materials. Aluminum in flake form suspended in paint provides good coverage. Silver flakes similarly prepared provide good electrical conductivity.

Metal particles suspended in metal matrices such as lead added to copper and steel alloys lead to increased machinability or reduced bearing wear resistance. These usually brittle metal particles are not dissolved in their metal matrix as in the alloying of metals, but instead, the metal matrix material is infiltrated around the brittle particles in a liquid-sintering process.

8.6.6 Filled Composites

A ***filler*** is a material added to another material to alter its physical and mechanical properties significantly or to decrease its costs. Some fillers, when added to polymer materials, improve their strength by reducing the mobility of the polymer chains, much as an appropriate amount of gravel, as a filler, improves the strength of concrete. Fillers are added to polymer materials for several other reasons (e.g., to improve frictional characteristics, control shrinkage, improve moldability, reduce dielectric properties, lower resistivity, reduce the moisture absorption characteristics, or enhance wear rate). Celluloid and Bakelite are filled polymers, as are electric-circuit boards and countertops made from phenolics, a thermosetting resin. See Module 6 for more details.

Since the accidental discovery early in the twentieth century that natural rubber would accept large amounts of carbon black, thus improving its mechanical properties, great advances in the technology of fillers are continuing. Particles of ***carbon black,*** an inexpensive material, are not only very small (20 to 50 nm) but are spherical in shape. When added to rubber, carbon black enhances its tensile strength, toughness, and wear resistance.

Its reinforcing effect on such properties is the result of its particularly good adhesive bonding with the rubber molecules. Additional reinforcement effects are attributable to the uniformly distributed particles acting as barriers to plastic deformation. Automobile tires (see Sections 6.4.2.1 and 6.4.3) may contain up to 30% vol of carbon black. Just as with multiphase microstructures such as steel, whose structures (micro) can be changed by the relative amounts of the phases present, the phase distribution, and the size of the individual grains making up the phases, fillers are components that are purposely added to other materials to change their properties. Glass fibers are added as fillers to plastics. Not only the type of filler but its size, shape (e.g., spheres), and distribution all play an important role in determining the desired properties of the filled composite.

8.6.7 Hybrid Composites

Hybrid composites, which combine two or more different fibers in a common matrix, greatly expand the range of properties that can be achieved with advanced composites. They can cost less than materials reinforced only with graphite or boron. The common matrix can be either a thermoset or a thermoplastic. Using combinations of continuous as well as chopped fibers also qualifies the material as a hybrid.

There are four basic types of hybrids. One is ***interply,*** which consists of plies from two or more different unidirectional composites stacked in a specific sequence. A second is ***intraply,*** which consists of two or more different fibers mixed in the same ply. A third are ***interply/intraply*** hybrids, in which the plies of interply/intraply hybrids are stacked in a specific sequence. ***Superhybrids*** are resin-matrix-composite plies and metal-matrix-composite plies stacked in a specific sequence. The curing procedure must be compatible with the matrix resin, which places something of a restraint on using plies with different matrix material. In the case of superhybrids, the resin matrix and the prepreg tape must have the same curing cycle as the adhesive used to bond the entire composite. The fabrication of hybrids is the same as for other composites.

The potential number of fiber/resin combinations for hybrids is vast. With two resins and three types of fibers, an unlimited number of different hybrid combinations can be produced. One design problem is how to make the most effective use of each fiber. Graphite fibers have poor impact resistance but high tensile strength and high modulus of elasticity. Aramid fibers have good impact but low modulus. The combination is a natural where each fiber contributes its best properties. A hybrid containing 50% graphite and 5% aramid reportedly shows flex strength on the order of three times that of a straight aramid. Another reason for this good combination is that both have coefficients of thermal expansion that are similar. In fact, the CTEs are slightly negative. This minimizes internal thermal stresses. Another combination is graphite or aramid and glass. Both fibers increase the stiffness, but with aramid there is some loss of compressive strength. Another combination is graphite and glass in polyester. The more glass that is used, the less the cost. The more graphite replaces glass, the higher the mechanical properties. For example, 25% carbon/75% glass fibers gives a modulus of 6.39×10^4 MPa, while the reverse of this proportion gives a modulus (in tension) of 12.3×10^4 MPa. Using SMCs and continuous graphite fibers with a typical loading of 30% graphite fiber and 45% chopped glass, the composite's properties replicated those of the closest all-glass system.

The B-1 bomber stabilizer was made of a hybrid composite—the largest hybrid ever produced, having an area of some 45 m^2—made of aluminum/graphite–epoxy/boron–epoxy hybrid. The saving in weight over the traditional aluminum/titanium structure is about 320%; the stabilizer consists of 108 parts compared to 270 in the metal version. With fewer parts needed, there are fewer holes to drill, countersink, and inspect, and fewer fasteners, all of which reduces the expense of production.

The auto industry will be the biggest user, with hybrid driveshafts. Made of graphite fibers for stiffening and glass fibers to carry the torsional loads, this is another example of the intelligent selection of materials to meet the requirements of the application.

The sports industry uses graphite/aramid and graphite/boron hybrids. Graphite/boron golf shafts are more rigid than an all-graphite shaft, and they are used in golf clubs, for both woods as well as irons.

In the area of high-temperature materials, where composites are beginning to replace metals, a final example illustrates the intelligent design of hybrid composites. It must be remembered that the polymer–resin matrix is the determiner of the high-temperature performance of the composite. Carbon- and glass-fiber–reinforced amorphous composites lose their flexural strength as they approach their glass transitional temperatures (T_g). Any exposure to temperatures above T_g, even for a brief moment, could result in a catastrophic failure. Despite this, amorphous thermoplastic–resin composites are stronger than crystalline–resin composites near their T_g values, while crystalline–resin composites retain measurable strength above these temperatures. By using a hybrid matrix blend of resins (where one resin has a higher T_g value), high-temperature properties can be raised enough to meet the application requirements for a particular service temperature.

8.6.8 Smart Composite Materials

As explained in Module 1, materials engineers and designers seek to mimic biological organisms and to place control and feedback in materials and structures. The infinite variety of structures possible with hybrid composites led to the opportunity for designer ***smart composites*** to fulfill many of the goals sought for smart materials. Such materials are currently available, with more varieties to follow.

Hybrid PMCs incorporating *shape-memory alloys* (SMAs) such as Nitinol give the designer more choice for functionality. As a load-bearing armor material, the PMC with imbedded Nitinol improves low-velocity impact and perforation resistance compared to monolithic PMC. Nitinol's superplastic nature allows it to deform up to a 7% strain with fully reversible recovery when this load is removed, thus allowing greater impacts by the host material. Hybrid composites designed for structural and armor application include various compositions of tough polyethylene, Kevlar, nylon, glass, aluminum, and graphite using the Nitinol SMA to improve strain and impact resistance.

Another smart composite involves ***imbedded particles,*** which allow in-service monitoring of the host composite structure. A polymer or ceramic matrix imbedded in micrometer-size particles can provide active or passive tagging. ***Active tagging*** requires an external source to energize the particles. One external source is a magnetic field to energize ferromagnetic particles. ***Passive tagging*** uses nondestructive testing instruments (e.g., ultraviolet light, x-ray eddy current, or magnetic pickup) to detect the presence of particles and defects in the material. ***Particle signatures*** result from their interaction with their host material and reveal various structural conditions including internal stress, voids, inclusions, degree of curing, and delaminations. The signatures aid in quality control (QC) and quality assurance (QA). Certain QC/QA issues with composites involving defects, such as hidden voids and delamination of composite layers, remain major concerns because composites are becoming more widely accepted in buildings, roads, and other construction as replacements for traditional structural materials.

The 1995 bombing of the Alfred P. Murrah Building in Oklahoma City prompted a call for legislation requiring tagging of explosive materials. An explosive tagging technique patented in 1973 involves mixing a small amount of phosphor with the explosive. Ultraviolet light can easily detect the particles of phosphor before or after the explosion. Other methods use inorganic materials and magnetic particles, including some color coding.

8.7 FABRICATION OF COMPOSITES

Although ordinarily not a part of the study of materials, the numerous manufacturing processes involved in the production of composites are intimately connected with materials because design affects manufacturing and the particular process affects the properties of the final product. Finally and most important, because of the one-piece forming capability of composites, in which the material is made at the same time that the end product is manufactured, a brief mention of the major manufacturing processes as a major factor in evaluating the costs involved is justified.

Two stages in the processing of most fiber composites are the ***layup*** or combining of the reinforcement and matrix materials and the molding or curing stage. ***Curing*** is the drying or polymerization of the resinous matrix to form a permanent bond between fibers and between laminae. It occurs unaided as with contact molding or by the application of heat and/or pressure using vacuum bags, autoclaves, pressure bags, or conventional metal stamping machinery (see Figure 8-35). There are several processing methods for producing fiber-reinforced plastic products.

8.7.1 Contact Molding

Hand layup is the simplest of all the methods. Using a single, inexpensive mold, reinforcing mat or fabric is placed in the mold and saturated with resin by hand. Layers are built up to the desired thickness to form a laminate that is cured at room temperature. This form of curing without the application of heat is called ***contact molding. Spray-up molding*** using a single mold combines short lengths of reinforcing fibers and resin in a spray gun, which deposits them simultaneously on the mold surface. Contact molding follows.

8.7.2 Matched Metal-Die Molding

Matched metal-die molding, for mass producing high-strength parts limited in size by the press equipment, uses pressures of about 1.72 MPa and temperatures of about 120°C. The materials are in the form of SMC or BMC premixes. Most automobile panels are now fabricated from GRP in SMC form by compression molding.

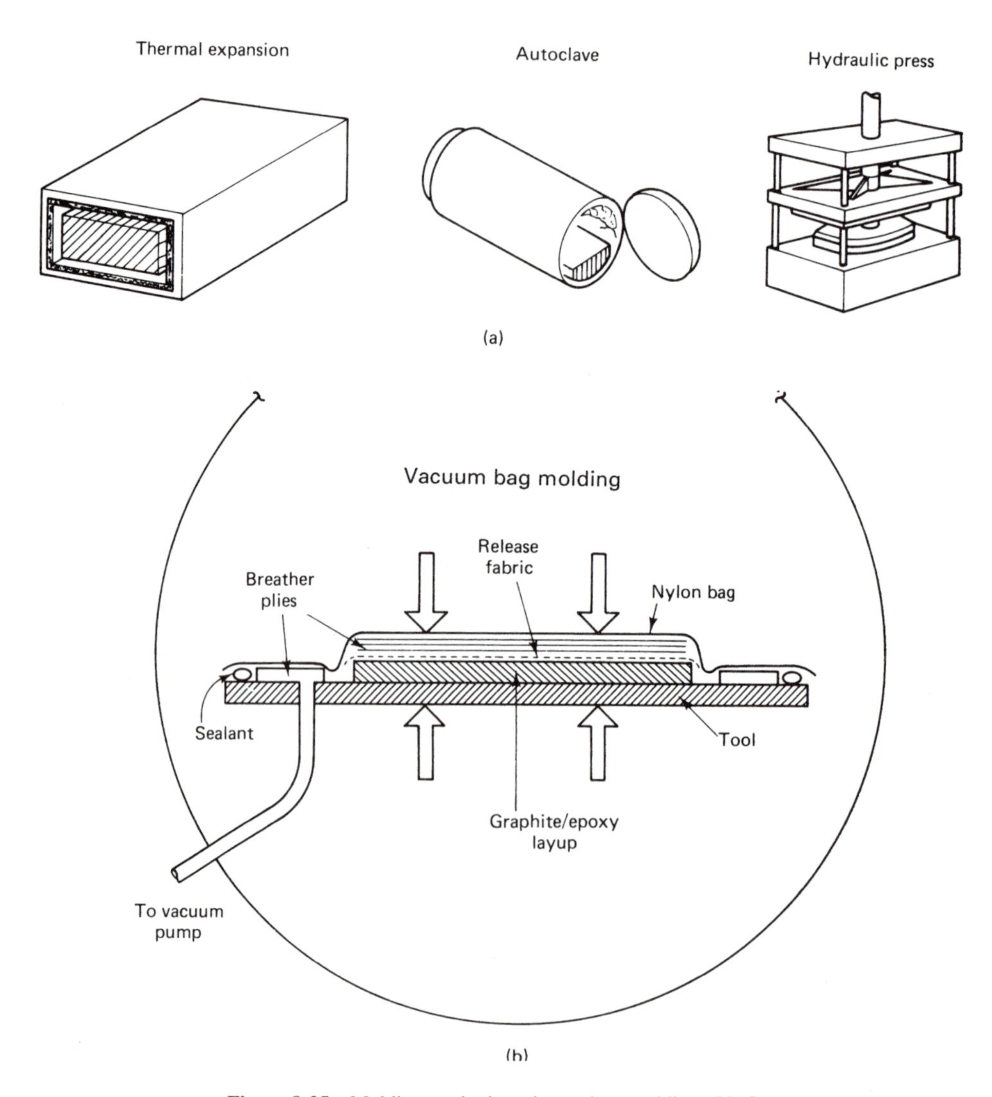

Figure 8-35 Molding methods and autoclave molding. (NASA.)

8.7.3 Injection Molding and Resin Transfer Molding

Injection molding and resin transfer molding (using reinforced plastics) are high-volume molding processes for both thermoplastics and thermosetting plastic resin.

8.7.4 Filament Winding and Tape Winding

Filament winding and tape winding, which produce the highest specific strength and glass content by weight of composite parts (fiber loading up to 85% by weight), is generally limited to parts with round, oval, or tapered inner surfaces. External shapes are unlimited. The continuous glass strand or filament is usually passed through a resin bath prior to winding onto a revolving mandrel. The mechanical as well as other properties of a filament-wound product can be changed by altering the ***wind angle*** (α), shown in Figure 8-36. This angle is measured between the axis of the mandrel and the lay of the filaments. The tangent of this angle equals the ratio of the circumference of the mandrel and the pitch of the filaments. As the angle increases to 90°, the hoop tensile strength increases and the axial tensile strength decreases. Figure 8-36 shows the angle, the pitch, and the axial and hoop directions. The change in modulus of elasticity in the two principal directions is sketched in Figure 8-37. Figure 8-38 shows some internal shapes possible with filament winding, along with a few special design and winding problems.

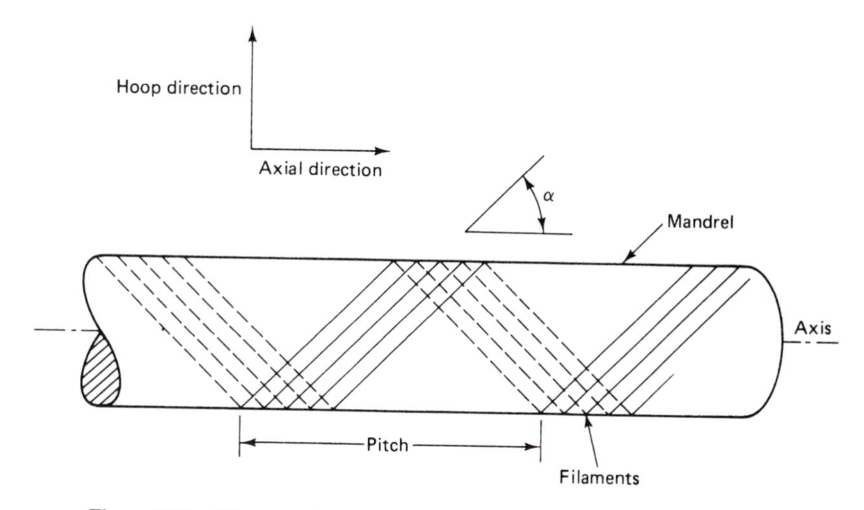

Figure 8-36 Filament winding, showing the wind pitch and the wind angle (α).

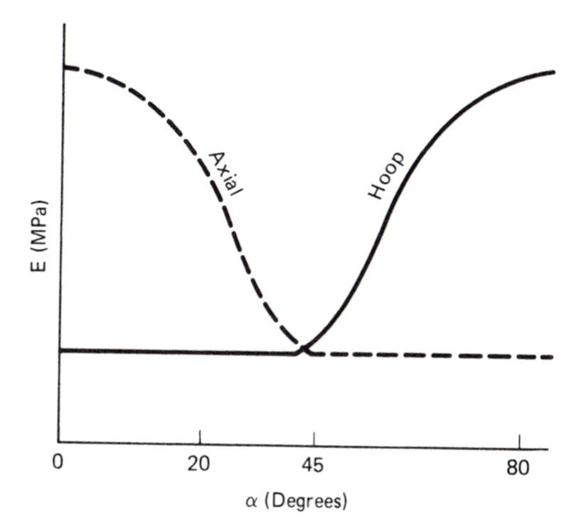

Figure 8-37 Wind angle of a filament-wound product.

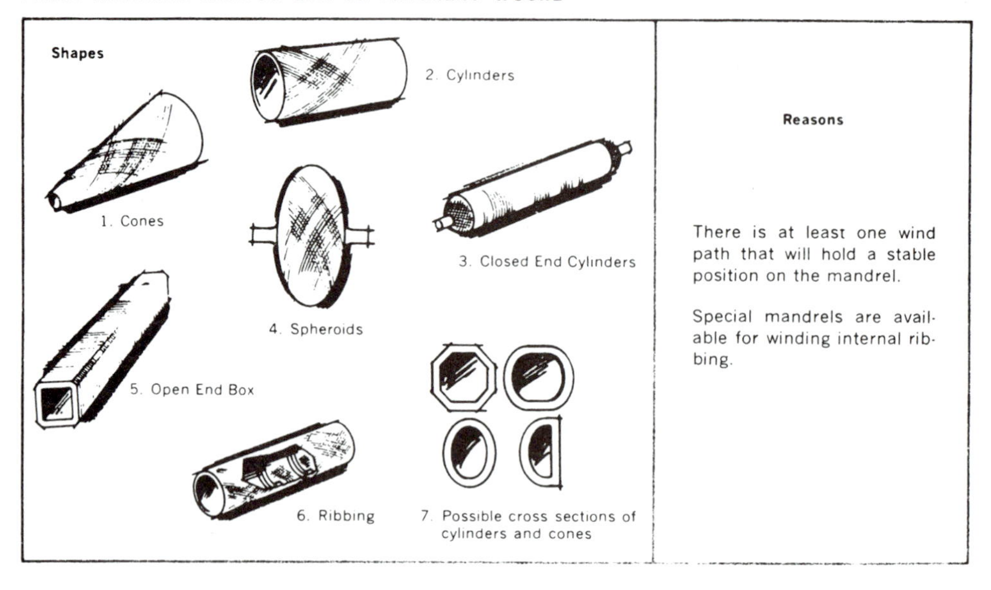

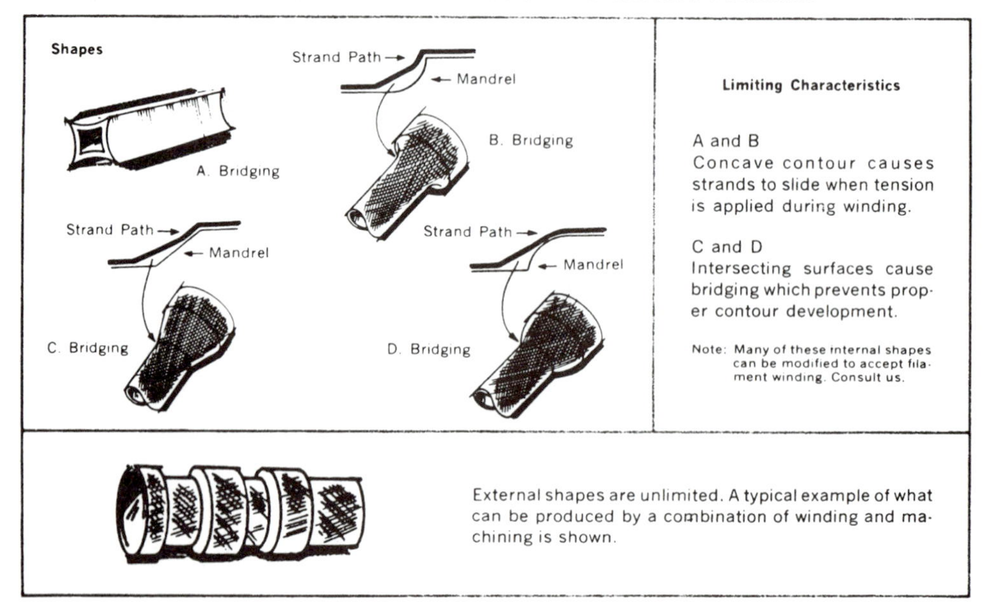

Figure 8-38 Winding problems and some internal shapes of filament-wound products. (Permali, Inc.)

8.7.5 Pultrusion

Pultrusion is a composite-fabrication method by which an extremely long, fiber-reinforced, polymer-matrix material can be produced by pulling a fiber or a bundle of continuous filaments through a resin system for impregnation and then through a heated curing die. As the fiber/resin bundle travels through the curing die, polymerization occurs, and a composite structure emerges from the exit end of the die. Figure 8-39(a) shows a profile view of the pultrusion process used to produce a graphite-reinforced, epoxy composite T-beam for use as a lightweight, high-strength aircraft stiffener. Figure 8-39(b) is a photograph of a section of the T-beam, which can be produced to any desired length. Figure 8-39(c) is a profile view

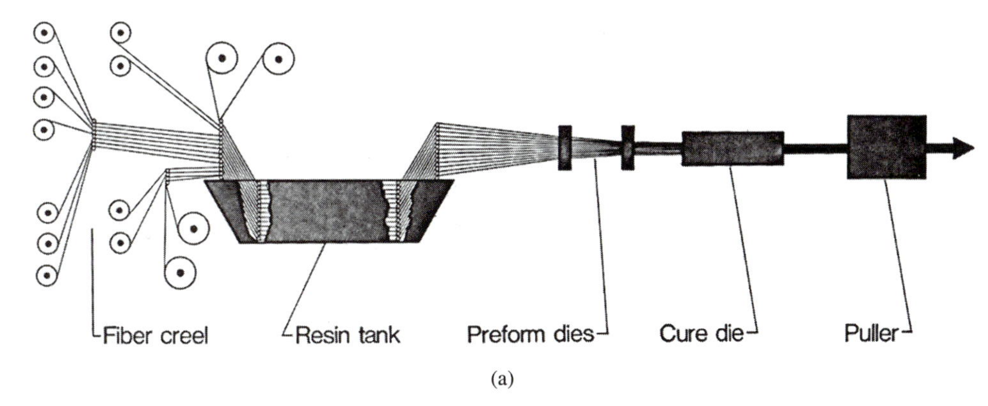

(a)

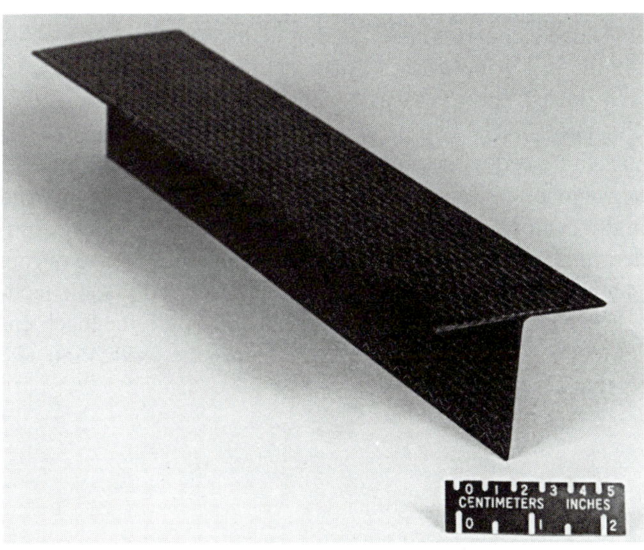

(b)

Figure 8-39 (a) A profile view of the pultrusion process. (NASA.) (b) Photograph of a portion of a T-beam produced by the pultrusion process.

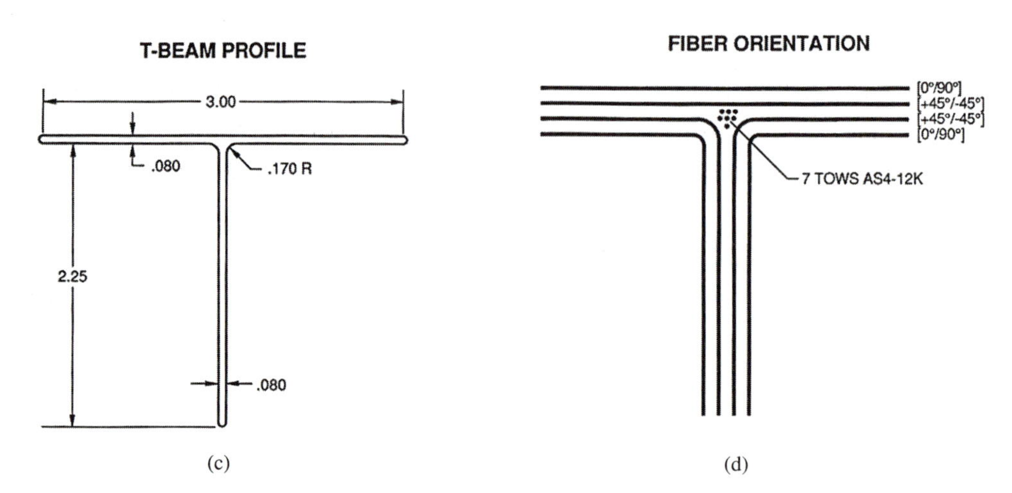

Figure 8-39 (c) T-beam profile showing dimensions in inches. (d) A cross-section of the T-beam showing the placement of the fibers (fiber orientation) in the T-beam. (AS4 is the producer's designation for a graphite fiber.)

giving the dimensions of the T-beam in inches, and Figure 8-39(d) shows the fiber orientation used in the T-beam. The fiber, supplied in 12K tows, was sized with a coupling agent for epoxy-resin compatibility. The fiber layup consisted of eight plies oriented in the 0°, 90°, +45°, −45°, +45°, −45°, 90°, 0° directions (in that order). These tailored fiber orientations were obtained by locking two plies in their positions by stitching with a polyester thread; that is, the 0° (longitudinal) and 90° (transverse) fiber orientations were obtained by knitting the two plies together, thus locking them in their respective positions. The fiber volume was 54%. Preforming, preimpregnation, and layup of the plies to form the T-beam were performed simultaneously in the 47-in. resin tank. Pultrusion produces solid rods, pipe, and many other structural shapes that have a constant cross-section.

This new advancement in the use of pultrusion illustrates the need to minimize, if not eliminate, the labor-intensive manufacturing processes required for any advanced composites. The use of knit-locked, tailor-made reinforcement systems that produce tailor-made properties by pultrusion is contrasted with some older fabrication techniques that require the manufacture and stowage of preimpregnated materials, laborious layup and fixturing methods, vacuum bagging, autoclave curing, and debagging, which produce composite structures that are limited in size and length to the available autoclaves and drying ovens. Parts being pultruded can be inspected as they are fabricated using ultrasonic techniques to detect the presence of porosity and delaminations. This real-time feedback enables changes to be made to correct or discontinue the process before the production of a large number of defective parts that must be rejected.

8.8 FASTENERS

Current fasteners for composites are not meeting all the requirements of structural designers. There is a need to develop fasteners for composites that will eliminate the problems associated with installing metallic fasteners in nonmetallic structures. Material compatibility is a necessity if problems with corrosion, lightning strikes, and low radar signatures are to be avoided.

Fiberlite fasteners (made by the Tiodize Company) have low creep, water absorption, and thermal expansion, and half the weight of aluminum with equivalent strength. Used to bolt composite structures together, these fasteners come in a variety of composite nuts and bolts with shear strengths of about 30,000 psi. They resist temperatures to 600°F. Machined from composite rod stock, the reinforcement in the rod is a three-dimensional, woven graphite-fiber structure with a polyimide-resin matrix. The cost is equivalent to titanium.

8.9 MACHINING OF COMPOSITE MATERIALS

Machining of composites demands special skill and knowledge. Composites cause severe tool wear by abrasion. Dust from the machining of graphite is a health hazard and causes fouling of the machine tools, producing maintenance problems. Graphite and carbon fibers are conductive and pose problems with numerical-controlled equipment. Coolants used with graphite can cause additional problems by combining with graphite dust to produce a slurry that is difficult to clean up. Also, some coolants tend to be absorbed and retained by the composite. Drilling bolt holes to allow joining of composites to other materials has a multitude of special problems, particularly when aligning prefabricated holes with precision tolerances.

Solid ceramic tools boost production rates 5 to 10 times over carbide tools when machining composites. Metal-removal rates are up to 200 times higher than with diamond-coated tools. Tools of fine-grained alumina matrix reinforced with SiC fibers or whiskers are harder and more resistant than carbide. Drills, end mills, and routers can be used to cut graphite/epoxy, fiberglass, Kevlar, polyimide, graphite/graphite, and other advanced composites.

Single-layered diamond tools, comprised of a layer of diamond bonded to a formed metallic preform representing the negative of the shape desired, are manufactured for machining composites. Known as metal-single-layer tools, they are improving precision with longer tool life by remaining cool throughout the cut.

The Suppliers of Advanced Composite Materials Association (SACMA; 1600 Wilson Blvd., Arlington, VA 22209) has a report available that addresses the workplace hazards of advanced composite materials. Many manufacturers are abiding by the Occupational Safety and Health Administration (OSHA) guidelines for the handling of asbestos and confronting the possible hazards of composite particulates released during fabrication. The best remedy for eliminating airborne particles is adequate ventilation provided by a high-efficiency particulate air (HEPA) filtered vacuum. *Conventional machining* poses several problems, including excessive tool wear, frictional-heat generation, crazing effects when using lubricants, and localized fiber breakage and delamination due to tool impact and vibration.

Laser machining has some advantages in cutting and drilling holes in epoxy composites. The laser beam can focus on a very small spot, minimizing the heat-affected zone.

Figure 8-40 Robotic Waternife® waterjet cutting system is used here to finish-cut and trim automotive Class A surfaced rear-end panels with numerous clearance holes for taillights and a license plate mount. (General Electric.)

Its vaporizing of material eliminates fraying. Kevlar and glass/epoxy composites present no laser-machining problems. Because of the high melting temperature and electrical/thermal conductivity of graphite, graphite/epoxy composites do not lend themselves to laser machining.

Waterjet machining, with or without the garnet abrasive added to the stream, is finding use in the machining of composites. Kevlar is cut with a waterjet without an abrasive. Due to safety requirements, only automated waterjet systems are used. Figure 8-40 shows a GE waterjet system used in the automobile industry. Its biggest limitation is the inability to effectively machine material thicknesses greater than 2.54 mm. Otherwise, the abrasive jets cut quickly and leave an excellent edge without tearing, fraying, or delamination. Several companies now specify that composites and other heat-sensitive parts must be cut with abrasive waterjet. Flow International's Paser II waterjet can cut composites at 15 to 30 in./min, in comparison with mechanical cutting of the same materials at 1 in./min. The same waterjet is capable of cutting 0.8-in. Kevlar at 5 in./min. Both waterjet and laser machining exhibit very little tool wear.

8.10 REPAIR OF COMPOSITE MATERIALS

A fast-repair kit produced by Ferro Corporation for fiberglass parts in the aerospace industry consists of an epoxy–resin system—a two-component thermosetting patch and repair material that, when mixed with an activator, produces a hard plastic requiring no external heat to cure. The activated resin saturates glass cloth, mat, or tape for a structurally sound composite with good dimensional stability, impact resistance, and tensile strength. Patches are abrasion and corrosion resistant and not affected by moderately high or subzero temperatures. The resin adheres to most surfaces, including steel, plastics, wood, and ceramics. Figure 8-41 is a line drawing of a typical honeycomb-core repair. Other types of repair patches for laminar composites are illustrated in Figure 8-42. Most composite-repair techniques depend on an adhesive with a certain shelf life. This requires use of the adhesive prior to its shelf-life expiration. Additional restrictions are imposed on its disposal to ensure compliance with existing environmental laws and regulations. Many adhesives require environmentally controlled storage, usually at low temperatures, and their out-times are limited, requiring close monitoring.

A functionally gradient material joint is one in which the materials gradually change in chemical composition and properties across a dissimilar joint, bridging mismatches and reducing thermally induced stress. Traditionally, the joining of dissimilar materials has been a difficult task. A recent exercise demonstrated the joining of silicon carbide and a nickel-base superalloy with a functional gradient material made up of elemental powders of nickel, titanium, aluminum, and carbon. Such a process may lead to the efficient and effective joining of hybrid structures consisting of mixed material systems.

8.11 NONDESTRUCTIVE EVALUATION

Composites have extended the lifetimes of helicopter blades threefold. Older blades made with aluminum sheet and extrusions bonded with adhesives were inspected by x-ray equipment, which took about 8 h. New analysis and characterization techniques, as introduced in Module 3, have drastically reduced the time and improved the reliability of nondestructive evaluation (NDE). Using computerized infrared-thermography, inspecting the blades now takes about an hour. The NDE of composites during the manufacturing process will be the key to 100% reliable inspection. Composites pose a severe challenge due to the variety and complexity of the material's composition and fabricating processes. Also, there is no clear definition of defects. ***Computer tomography*** using x-rays provides high-resolution two- and three-dimensional images of the internal structure of advanced composites for inspecting stress-induced damage in small electronic or mechanical components. Unlike radiography, the system makes measurements from many angles, which a computer program then uses to reconstruct a slice of the object's internal structure. One challenge to

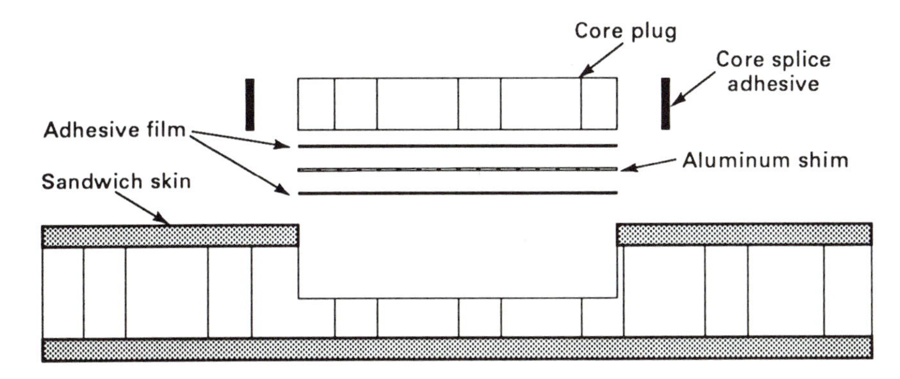

Figure 8-41 Honeycomb-core repair. (USAF.)

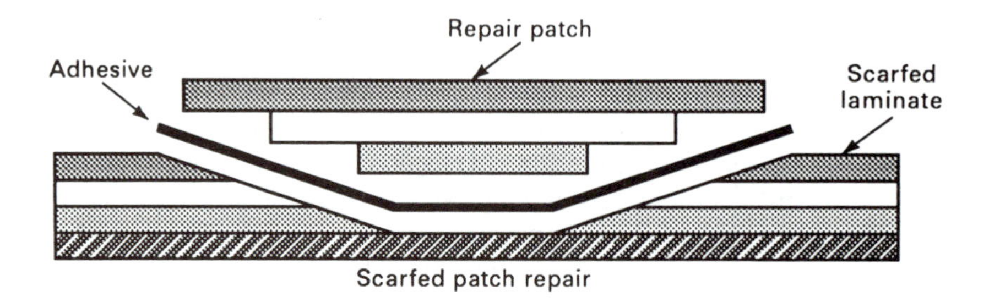

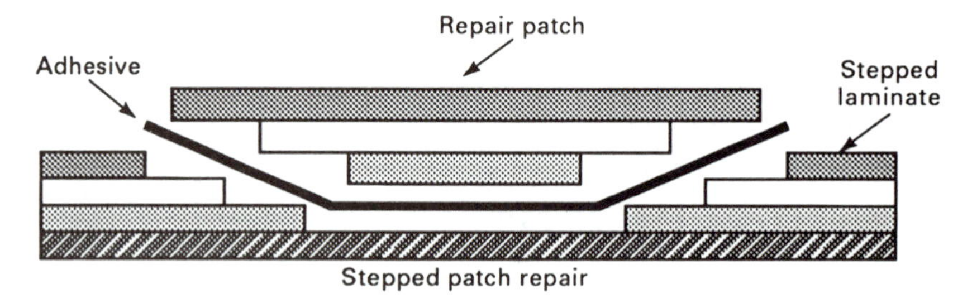

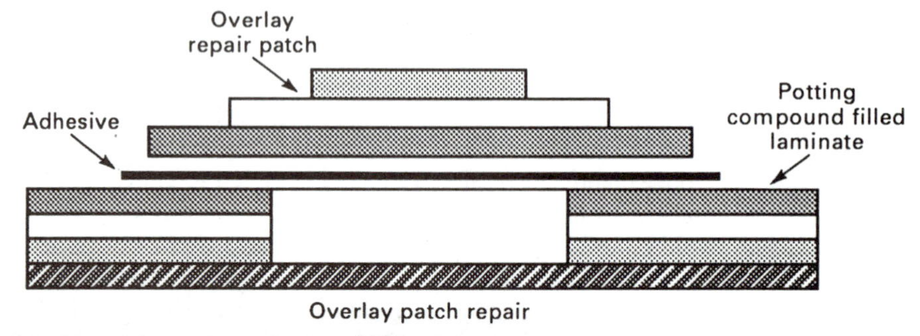

(a)

(b)

Figure 8-42 (a) Scarfed, stepped, and overlay patch repairs. (USAF.) (b) The new material, N-View, is a sheet of Lexan MR5 polycarbonate that receives an abrasive-resistant Margard surface treatment and is then laminated with a thin, abrasive-resistant film of Lexan HPH to avoid damage to plastic glazing. The thin film can be removed if damaged, leaving the MR5 thick sheet and thereby doubling the glazing life. (G.E. Plastics.)

adopting PMCs for use in marine structures, including surface ships, submarines, and marine platforms, is the absence of any effective test methods for electronic or mechanical components.

8.12 COMPOSITE SELECTION: DATABASE FOR COMPOSITES

In the *selection of composite materials* for high-temperature applications, particularly resin-matrix composites, it is imperative to have a detailed knowledge of a resin's glass-transition temperature (T_g) (see Section 6.3.2.4). The need for an up-to-date materials ***database*** is also essential. To replace traditional engineering materials with composite materials requires a database that reflects varying service conditions, including combinations of conditions such as high temperatures with or without a load and/or an adverse environmental condition such as a corrosive atmosphere. Figure 8-21(b) showed a flowchart for designing an FRP composite to withstand fatigue. Note the inclusion of databases for both materials and processes. The lack of good databases inhibits the adoption of new materials by many design personnel who are thus unable to locate sufficient data about materials that would be more efficient in many applications. Using fiber-reinforced thermoplastic composites as materials, the importance of the foregoing two points can be emphasized. The amorphous resin poly(ether sulfone) (PES) and the crystalline resin poly(ether ether ketone) (PEEK), both thermoplastics, are used as matrix materials reinforced with glass or carbon fibers. These reinforcements provide increases in strength and modulus (stiffness) to the polymer resins. In general, carbon-fiber reinforcements provide a 20% improvement in tensile strength and a 30% to 100% improvement in flexural modulus over glass-fiber reinforcement at room temperatures. This is possible due to the ability of the matrix material to transfer the stresses produced by the load(s) from itself to the fibers. Under high-temperature conditions the matrix plays another important role—that of being the determiner of the high-temperature performance of the composite. How well it is able to maintain its strength and stiffness at high service temperatures dictates the service temperature for the composite. There are innumerable high-temperature applications where the critical properties desired are dimensional stability and stiffness. One is the need to reduce mold shrinkage. PES has a T_g of 230°C, whereas PEEK, being crystalline, has a melting temperature (T_m) of 336°C. Testing these materials at various high temperatures shows that the amorphous polymer PES has higher strength at just below its T_g than does the crystalline polymer PEEK, but the crystalline PEEK resin composite maintains its strength above the T_g of the amorphous PES composite. Figure 8-43 illustrates GE Supec poly(phenylene sulfide) (PPS) resin, which has high ductility and good thermal capabilities and finds applications in ***surface-mount technology*** as high-performance connectors with good pin-insertion/retention qualities. This example illustrates that, when selecting thermoplastic composites to replace metals for high-temperature applications, a knowledge of glass transition temperatures and a database containing the results of performance testing at conditions other than ambient room temperatures are essential. Figure 8-44 illustrates a situation where a high-temperature thermoplastic composite, having an array of properties that were required not only to combat high temperatures but a combination of stressful and corrosive environments, has substituted for a metal and performed superior to the metal.

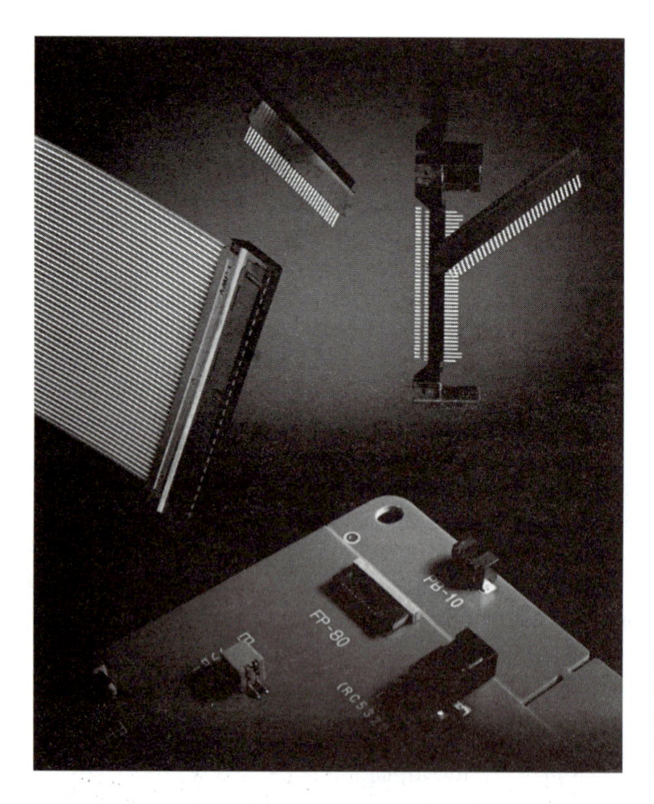

Figure 8-43 High-performance connectors used in the electronics industry's surface mount technology are made of GE Plastics SUPEC® resins with high ductility and low-flash grades. (General Electric Plastics.)

Figure 8-44 Two impellers used in the chemical industry—the metal impeller on the right has an average service life of only two months in an acid environment at 230°C compared to a carbon fiber-reinforced PEEK (the high-temperature thermoplastic composite impeller on the left) with an extended service life of 18 months. (ICI Advanced Materials, Adv. Mats. & Proc., Vol. 140, No. 2, August 1991, Pub. by ASM Int.)

The submersible craft (seen in Figure 8-45) serves as a case study to demonstrate how the many concepts of materials selection presented in this book are applied. Battelle Columbus's two-person, human-powered *Spirit of Columbus* uses varying-density PVC foam in an epoxy-bonded composite hull. Hysol 9460 structural adhesive offers high tensile lap-shear strength (24 MPa or 35,000 psi) when bonded to the aluminum frame. This aluminum-to-composite system is the same one used in Indy 500 race-car bodies and reflects a movement away from mechanical fastening systems that use bolts, rivets, etc. The hull sections consist of PVC foam in different densities, sandwiched between layers of

Figure 8-45 *Spirit of Columbus,* Battelle Columbus's award-winning entry in the International Submarine Race for human-powered submersible craft, allowed their designers to gain experience with emerging composites systems technology. *(Advanced Materials & Processes.)*

fiberglass. High-density, 0.1 g/cm^3 (6.23 lb/ft^3), foam is concentrated on the load-bearing hull bottom, to which air tanks and the drivetrain are permanently bonded. Low-density, 0.06 g/cm^3 (3.75 lb/ft^3), foam is used on the hull's top half, where no other structural support is needed. The propulsion system uses a 455-mm (18-in.) diameter drive sprocket, connected by a chain to a cable takeup pulley. Pulley-cable ends are attached to the driveshaft and connect to two hinged flappers. The flappers are made of 6061-T6 aluminum alloy with lightening holes, polystyrene foam, glass fiber, two layers of carbon fiber, and a top layer of drag-reducing riblet tape. The drag-reducing tape inhibits the transverse movement of turbulent eddies across the flapper surface. The lightweight hull must be both very rigid and very strong because the innovative, articulated linear-thrust propulsion system causes the drivetrain to move back and forth.

8.13 COMPOSITES GLOSSARY

Areal weight is the weight of fiber per unit area (width × length) of tape or fabric.

Braiding is the weaving of fibers into a tubular shape as opposed to a flat fabric.

Commingled yarn is a hybrid yarn made with two types of materials intermingled into a single yarn.

Conductive composites describes electrically conductive composites using metal fibers embedded in a nonconductive matrix to provide materials for uses such as antistatic coatings; fuel-cell catalysts; solar electrical cells; or lightweight, inexpensive batteries.

Cowoven fabric is a reinforced fabric woven with two different types of fibers in individual yarns, such as thermoplastic fibers woven with carbon fibers.

Creel is a device for holding the required number of roving balls (spools) in the desired position for unwinding onto the next step in the process (i.e., weaving, braiding, or filament winding).

Drape is the ability of a fabric or prepreg to conform to a contoured surface.

Fabric is made by interlacing yarn in a variety of ways, including weaving, knitting, and braiding.

Fiber refers to filamentary materials and is synonymous with *filament,* which has an extreme length and very small diameter, usually less than 25 μm. A continuous fiber has an indefinite length, in contrast to a discontinuous fiber represented by whiskers (chopped fiber in short specified lengths) or particulates.

Flash is the overflow or squeeze-out of cured resin from a mold.

Harness satin is a weaving pattern that produces a satin appearance. "Eight-harness" means that the warp tow crosses over seven fill tows and under the eighth in a repeating manner.

To ***impregnate*** is to saturate the voids and interstices of a reinforcement with a resin.

Knitted fabrics are fabrics produced by interlooping chains of yarns.

Macrocomposites are single parts that incorporate more than one composite. An example of the incorporation of more than one composite in a single part would be a CMC and an MMC in a composite-fabricated drive shaft.

Mat is a fibrous reinforcing material comprised of chopped filaments, along with a binder to maintain form, and is available in blankets of various lengths, widths, and weights.

Neat resin is a resin to which nothing (additives, reinforcements, etc.) has been added.

Out-time is the time a prepreg is exposed to ambient temperature (i.e., the total amount of time the prepreg is out of the freezer).

Particulate composites are materials consisting of one or more components (particles) suspended in a matrix of another material. Such particles can be either metallic or nonmetallic (see Section 9.6.4).

Pitch is the residue left from the distillation of coal and petroleum products. It is a base material for the production of some carbon fibers, as well as a matrix precursor for some carbon–carbon composites.

Pot life or ***working life*** is the length of time that a catalyzed thermosetting resin retains a viscosity low enough for it to remain suitable for processing.

Ramping is the gradual, programmed increase or decrease in the temperature or pressure used to control the cure or cooling of composite parts.

Resins are mostly polymers. In reinforced plastics, the matrix is a resin that binds together the reinforcement. Resins are solids or pseudosolid organic materials, usually of high molecular weight, that will flow when subjected to stress.

Roving is a collection of bundles of continuous filaments, either as twisted yarns or as untwisted strands.

Scrim is a low-cost reinforcing fabric made from continuous filament yarn in an open-mesh construction (see Figure 8-4).

Sizing is a compound that binds together and stiffens yarn and fibers, providing resistance to abrasion during weaving and handling. The ingredients, such as gelatin, wax, or oil, provide surface lubricity and binding action. Unlike a *finish,* which contains a *cou-*

pling agent to improve the bond of resin to the fibers, it contains no coupling agent. The size is usually removed by heat cleaning and a finish is applied. A finish can also contain a binder.

Space-grown composites are organic composites grown in space to provide repair materials for a space station. They obviate the need to deliver repair materials from Earth at great expense.

Strand is an untwisted bundle of continuous filaments used as a unit, including tows and yarn. Sometimes a single filament or fiber is referred to as a strand.

Tow is an untwisted bundle of continuous filaments. A tow designated as 100K has 100,000 filaments.

Warp is yarn running lengthwise in a woven fabric.

Weave refers to the particular manner in which a fabric is formed by interlacing yarns.

Weft are the transverse threads or fibers in a woven fabric. They run perpendicularly to the warp. Also called fill, yarn, or woof.

Wet layup is a method of making a reinforced product by applying the resin system as a liquid when the reinforcement is put in place.

Wet-out is the saturation of all interstices between bundles and filaments of reinforcement with resin.

Whiskers are single crystals ranging in size from about 0.5 to 10 μm in diameter, and 3 μm to 3 cm in length. They can be metallic or ceramic. Their aspect ratios are between 100 and 15,000. (See Section 8.4.1 for a definition of aspect ratio.) An example of a popular whisker in use is one made of silicon carbide. It varies in diameter from 3 to 10 μm, 10 to 100 μm in length, and has a Young's modulus of 800 GPa and an ultimate strength of 21,000 MPa. Another popular whisker made of silicon nitride varies in diameter between 0.2 and 0.5 μm, 50 to 300 μm in length, and has a modulus of elasticity in tension of 300 GPa. For example, the size of small defects in silicon carbide, such as voids, inclusions, and surface roughness, is 0.1 to 0.4 μm. This is the secret to the strength of whiskers: the smaller the fiber, the smaller the defect. Adding encapsulation to the fiber, which helps protect it from surface damage, results in fibers (whiskers) maintaining their inherent strength.

Woven roving is a heavy, coarse fabric produced by the weaving of continuous roving bundles.

Yarn is an assembly of twisted filaments, fibers, or strands, either natural or manufactured, to form a continuous length that is suitable for use in weaving or interlacing into textile materials.

APPLICATIONS & ALTERNATIVES

The Composite Automobile: Promise and Challenges. This module has shown how the engineered materials PMC, CMC, and MMC provide many desirable properties. In transportation, fuel efficiency is of paramount importance not only because of the economics of using less fuel, but burning less fuel means that fewer gases and particulates (hydrocarbons) are emitted into the atmosphere, hence less air pollution. Throughout the world, transportation vehicles have diminished the air quality, and legislation is requiring more fuel-efficient vehicles and lowered emissions. CMCs, PMCs, and ceramics offer potential for more efficient engines for air, sea, and ground vehicles. PMCs and MMCs can provide substantial weight savings for both the structures and bodies of those vehicles.

Interest in composites for automobiles dates back to Henry Ford's demonstration of the impact resistance and durability of a trunk lid made of a soybean-based composite. In 1953 the Chevrolet Corvette appeared with an all-FRP production autobody. Since then, numerous applications of composites throughout the automobile industry have appeared, as have concept cars with plastics and composites for complete body assemblies. These developments were helped along by the 1970s fuel crisis and by the demand to increase market share among the world's car and truck manufacturers.

Each year, the pounds of steel and iron in an average car drops and the use of plastic and composites increases. An increase in the use of plastic and composite body panels, from the 8% level in the early 1990s up to a level of about 70% by the year 2000, would bring the following benefits:

1. Improved fuel economy and performance through weight reduction
2. Improved overall product quality and consistency in manufacturing
3. Part consolidation, resulting in lower product and manufacturing costs
4. Improved ride performance due to reduced noise, vibration, and harshness
5. Improved corrosion resistance
6. Greater impact resistance (crash integrity), with a corresponding increase in safety
7. Improved repairability and recyclability

In the United States, the Big Three automakers formed the Automobile Composite Consortium (ACC) to attack the critical area for the future use of high-performance composites—manufacturing techniques. The first two requirements are that they be *rapid* and *economic*. They must be able to combine high rates of production, precise fiber control, and a high degree of part integration. Production vehicles such as GM's Saturn (sedans and station wagons) and all-purpose vehicle (APV) vans and Dodge's Viper have used a combination of SMC and RTM (resin transfer molding) for the entire body, but most other automobile manufacturers have used selected plastic and composite panels. Prototypes of potential production vehicles were seen in concept cars such as the Ford Ghia Connecta [Figure 8-46(b)], with a carbon-fiber and Kevlar body, and GM's Ultralite [Figure 8-46(c)], with a body that sandwiched polyurethane foam between thin layers of carbon-fiber fabric. For auto frames and structural components, Ford's development of a primary PMC structural frame (see Figure 1-6) demonstrates more of the evolution of the composite automobile.

Plastics and PMCs not only save energy with reduced weight and improved aerodynamics on a vehicle, they also provide savings in manufacturing over steel and aluminum. To capitalize on advantages offered by PMCs, the auto industry must meet the major chal-

(a)

(b)

(c)

Figure 8-46 (a) Composite body: the 1953 Chevrolet Corvette was the first fiberglass production body. (Morrison Molded Fiber Glass Co.) (b) Ghia Connecta: Ford Motor Company's electrically powered concept vehicle has a carbon and Kevlar fiber composite body, double-skinned glass roof, and solar panel in the top surface powering the rear roof vent. (Ford Motor Co.) (c) GM Ultralite concept car. (General Motors.)

lenge of determining effective *recycling* programs for PMCs if they expect to head off restrictive legislation.

The industry's successful three-tier (salvage dealers, shredders, and secondary-metal recoverers) recycling program has been quite effective in keeping iron, steel, aluminum, and other metals out of the ***municipal solid waste*** (MSW) stream (see Applications & Alternatives in Module 5). This tiered system recovers more than 80% of the ferrous metals and aluminum and 66% of the copper from scrap automobiles. Most batteries are salvaged to recover lead plates and the polypropylene case. Figure 6-22 showed the increased use (past and projected) of major plastics for automobile use. PMCs are also on a steady incline. There is a declining trend for iron, which is expected to continue, as discussed in Module 6.

Setting up the ***recycling infrastructure*** for PMCs is a major challenge. Currently, much of the polymer content ends up in the landfill. Not only is this unacceptable from an environmental perspective, but from an economic point of view it is unwise to bury polymers at ever-increasing "tipping fees" at landfills when the materials have value as recyclable resources. In the United States, Chrysler, Ford, and GM have established a vehicle recycling partnership to promote and conduct research for improved recycling technology.

Salvaging plastics as whole parts (fenders, fans, etc.) is feasible as long as mechanical fasteners, rather than adhesive bonding, continue to be the primary joining method. Shredders also process plastics in the form of "fluff" (lightweight materials: cloth, foam, upholstery, padding, etc.) and dense plastics after the ferrous metals are removed magnetically. Urethane-foam fluff can be collected for processing through ***hydroglycolysis,*** a method that reverses the polymerization of the foam (depolymerizes it) for use as a starting compound for more plastics (Figure 8-47). Denser plastics such as polypropylene, ABS, and nonrubber, often contaminated with oils, grease, and dirt, present more of a challenge for recycling, although they have use as fillers for low-grade plastics, "plastic wood," and floor tiles.

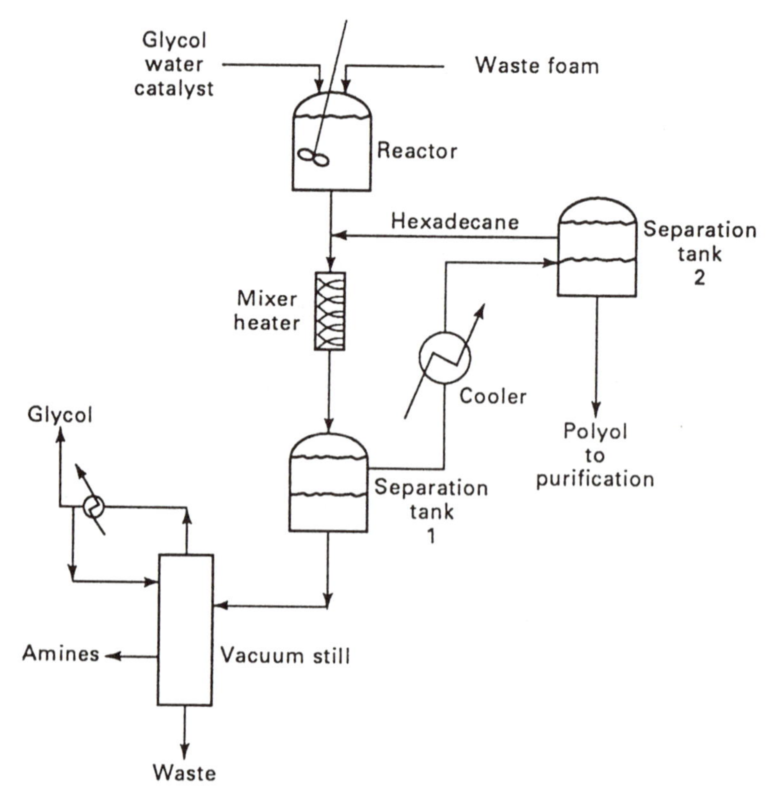

Figure 8-47 Schematic diagram of Ford's hydroglycolysis process. (Ford Motor Co.)

SELF-ASSESSMENT

8-1. From the Pause & Ponder, what advantages were accrued by using engineered polymer composites to replace natural polymers?

8-2. Identify and list the composite materials present in a typical U.S. kitchen. Include furniture, equipment, and tools usually found in the room.

8-3. Reinforced concrete is a composite material. List the constituents that must act in concert with each other to produce this widely used engineering material.

8-4. Name at least three composite materials found in the human body.

8-5. A natural polymeric composite material, wood, has directional properties. Explain what is meant by this, and cite an application of wood that takes this particular property into account.

8-6. The terms *specific strength* and *specific modulus* are often used in describing composites. What does specific modulus refer to, and what does it tell us about a particular material?

8-7. Oilite bearings are an example of what type of composite material?

8-8. When speaking of interface and interphase, which term represents the larger area or volume in a composite material?

8-9. Which fiber material has the greater specific stiffness, graphite or carbon?

8-10. Name two polymorphs of carbon.

8-11. State a major disadvantage of graphite as a fiber material.

8-12. What is the trade name for aramid fibers?

8-13. Compare the tensile strength of Kevlar with nylon. Which material would make better ropes?

8-14. The abbreviation NDE stands for what?

8-15. Using Figure 8-10, which material listed would produce a bending strain of 2% with a bending load that produces a stress of 750 MPa?

8-16. Verify the units of specific strength (inches) using Figure 8-9.

8-17. Describe three tests used to measure toughness.

8-18. The properties of a reinforcing fiber are dependent on their size and length. How are these two qualities related and used in describing the strength of such a fiber?

8-19. Regardless of the name given to the processing method for producing fiber-reinforced plastic products, two stages are included in most of them. Name these two stages.

8-20. Which molding process provides the highest specific strength and glass content by weight?

8-21. Describe the effect of a decreasing wind angle on the hoop stress.

8-22. When a part that is filament wound contains a transition (concave contour), bridging is likely to occur. Explain what *transition bridging* means.

8-23. Laminations in plywood are known by what name?

8-24. List two examples of a rigid laminate.

8-25. Sandwich composites have cores of various configurations. List three such configurations.

8-26. A cermet is basically a composite made from a ceramic and a metal. Which of the two materials is softer?

8-27. What thermal-protection system protected space capsules from the intense heat of reentry into Earth's atmosphere prior to flights of the space shuttle?

8-28. Identify a possible use for a laminated composite consisting of two bonded sheets of different metals, each having a different coefficient of thermal expansion.

8-29. A popular kitchen cooking utensil contains a composite base made of copper clad to stainless steel. What is the purpose of this composite application?

8-30. Discuss the reasons why the emphasis in materials science over the past three decades has been on the research and development of fiber-composite materials as opposed to efforts to develop or refine traditional load-bearing materials such as an alloy steel.

8-31. If present materials technology could produce crystals of practical size free of defects, such high-strength whiskers could be used as reinforcing material in a composite that would outstrip the performance of existing composites. Discuss what problems could arise, if any, in combining this filler material with a suitable matrix material.

8-32. A rod made of polyester under a tensile load that is parallel to the length of the rod complies with the rule of mixtures. If another rod with the same strain under the same load conditions and composed of glass fibers in a polyester matrix is substituted for the rod of polyester, compare the two rods by finding: **(a)** the density, **(b)** the elastic modulus of the composite rod, **(c)** the difference between the cross-sectional areas of the rods, and **(d)** the diameter of the composite rod in millimeters if the area of the polyester rod is 0.7854 in.^{2}. Use the following property values:

E(p), modulus for polyester = 4 GPa
E(f), modulus for glass fibers = 68 GPa
p(p), density of polyester = 1.36 g/cm^3
p(f), density of glass fibers = 2.54 g/cm^3
V(f), volume fraction glass fibers in composite rod = 0.094
V(p), volume fraction for polyester
E(c), modulus of the composite material
F(p), load or force on polyester rod

8-33. Derive the inverse rule of mixtures in terms of modulus and volume fraction as stated in Section 8.5.1.

8-34. Solve for E(c) using the inverse rule of mixtures equation in Section 8.5.1.

8-35. NASA and the Department of Defense (DOD) are both financing an experimental aircraft that will fly from a standing start on a runway into Earth orbit at speeds of mach 25. What are the name of this program and the designation for the aircraft?

8-36. Calculate the specific strength of S-glass that has a tensile strength of 4.8 GN/m^2 and a weight density of 24.4 kN/m^3.

8-37. Sketch a Venn diagram that shows how sets of the three Family of Materials groups intersect with reinforcers to form the wide array of advanced materials.

8-38. Sketch a typical fiber-reinforced composite and label the following: matrix, reinforcing fiber, interfaces, interphases.

OBJECTIVE QUESTIONS

8-1. Composite of plastic reinforced with glass.

a. GPa **b.** FRS **c.** SMC **d.** GRP

8-2. Impregnated, continuous sheets of composite materials such as those used for autobody panels.

a. GPa **b.** FRS **c.** SMC **d.** GRP

8-3. From the Pause & Ponder, list one advantage of using an engineered polymer composite to replace natural polymers.

8-4. List one type of advanced composite used on the Boeing 777, NASP, or HSCT.

8-5. The binding portion of a composite that distributes the load among fibers or particles.

a. Matrix **b.** Reinforcer **c.** Coupling agent **d.** Interface

8-6. Organic, ceramic, synthetic, or metallic materials with a length of 100 times the diameter, with a minimum length of 5 mm.

a. Strand **b.** Fiber **c.** Staple **d.** Particulate

8-7. Bonding agent of binder that provides a flexible layer between the fiber and the matrix.

a. Matrix **b.** Reinforcer **c.** Coupling agent **d.** Interface

8-8. Means of inspecting materials for flaws without adversely affecting them.

a. NDE **b.** SPF/DB **c.** RRIM **d.** RCC

8-9. Method of manufacturing structural shapes, such as channels and angles, in which prepregs are drawn through a die and heated until they set.

a. NDE **b.** SPF/DB **c.** RRIM **d.** Pultrusion

8-10. Composite primary structures in the aircraft industry now cost ______________ as much as corresponding metallic structures.

a. Five times **b.** Just **c.** Two times

8-11. Fibers such as rayon that are pyrolized to drive off organic elements, leaving a very stiff, high-strength carbon fiber.

a. E-glass **b.** Kevlar **c.** Borsic **d.** Graphite

8-12. Trade name for aramid fiber, which is used for bulletproof vests and advanced composites for aerospace applications.

a. E-glass **b.** Kevlar **c.** Borsic **d.** Graphite

8-13. Directional properties in fiber composites are achieved mainly due to ______________.

a. Fiber orientation **b.** Laminate plane **c.** Interphase **d.** Prepregs

8-14. Air pockets that reduce the strength of the composite structure.

a. Voids **b.** Laminates **c.** Particles **d.** Honeycomb

8-15. A hollow core material that provides stiffening but very little weight.

a. Voids **b.** Laminates **c.** Particles **d.** Honeycomb

8-16. One advantage of composites over monolithic cast materials is ______________.

a. Low material cost **b.** Anisotropic properties **c.** Isotropic properties

8-17. An advantage of reinforced plastic composites over sheet steel is ______________.

a. Lower material cost **b.** Common fabrication methods
c. Isotropic properties **d.** Flame resistance

8-18. A molding method for producing FRP, in which parts are placed in a cylinder that is subjected to high heat and pressure.

a. Filament winding **b.** Injection molding **c.** Autoclave **d.** Pultrusion

8-19. A fiber composed of 55% silica, 20% calcium oxide, 15% aluminum oxide, and 10% boron oxide.

a. SMC **b.** GRP **c.** SCAB **d.** E-glass

8-20. The goal of the ACT program for 1995–2002 is to reduce the cost of ______________.

a. Metallic primary structures **b.** Composite primary structures **c.** Aircraft safety

8-21. The HSCT is programmed to fly at mach speeds of about ______________.

a. 1.0 **b.** 2.0 **c.** 3.0

8-22. Name one of the most important technological development challenges for the HSCT.
a. New materials **b.** New runways **c.** New airports

8-23. Environmental composites are so named because they are ______________.
a. Easily recycled **b.** Easily manufactured **c.** Cheap to fabricate

8-24. The chance of a defect remaining in a fiber as the diameter is decreased is ______________.
a. Minimized **b.** Unaffected **c.** Increased

8-25. Electrically conducting composites use components such as ______________.
a. Metal matrix **b.** Polymer fibers **c.** Graphite fibers

8-26. CFCCs for near-term applications will perform in the temperature range of ______________.
a. 250° to 500°C **b.** 500° to 1000°C **c.** 1250° to 2000°C

8-27. Explosive welding is used to ______________.
a. Join incompatible metals **b.** Remove surface contaminants **c.** Weld projectiles

8-28. In precipitation-hardened alloys, the second-phase particles are ______________ at high temperatures.
a. Stable **b.** Unstable **c.** Do not grow or dissolve

8-29. Carbon black is used as an inexpensive reinforcing filler in tires. Its reinforcing effect is the result of the ______________ of its particles.
a. Large size **b.** Irregular shape **c.** Uniform distribution

REFERENCES & RELATED READINGS

AMERICAN SOCIETY FOR TESTING AND MATERIALS. *Annual Book of ASTM Standards,* Part 36. Philadelphia: ASTM, 1993.

AMERICAN SOCIETY FOR TESTING AND MATERIALS. *Composite Materials, Testing and Design,* ASTM STP 617. Philadelphia: ASTM, 1993.

AMERICAN SOCIETY FOR TESTING AND MATERIALS. *Fracture Mechanics of Composites,* ASTM STP 593. Philadelphia: ASTM, 1993.

ARONSON, ROBERT B., ed. "New Materials, New Challenges," *Manufacturing Engineering,* February 1995, pp. 57–59.

ASHBEE, K. H. G. *Fundamental Principals of Fiber Reinforced Composites.* Lancaster, Pa.: Technomic Publishing, 1989.

ASM INTERNATIONAL. *ASM Engineered Materials Reference Book.* Materials Park, Ohio: ASM International, 1990.

ASM INTERNATIONAL. *ASM Handbook,* Vol. 7, *Powder Metallurgy*; Vol. 15, *Casting*; Vol. 17, *NDE and QC.* Materials Park, Ohio: ASM International, 1987, 1988, 1989.

ASM INTERNATIONAL. *Engineered Materials Handbook,* Vol. 1, *Composites.* Materials Park, Ohio: ASM International, 1987.

BITTENCE, J. C., ed. *Engineering Plastics and Composites.* Materials Park, Ohio: ASM International, 1990.

BOSE, ANIMESH, and JAMES LANKFORD. "MIMICs: New Composite Architecture," *Advanced Materials and Processes,* Vol. 140, No. 1, 1991.

BRASHER, DAVID. "Explosive Welding: Principles and Potentials," *Outlook.* Teledyne WahChang, Vol. 15, No. 2, 2nd Quarter 1994, pp. 2–6.

BROSTOW, WITOLD. *Science of Materials.* New York: Wiley, 1979.

CARLSSON, L. A., and R. B. PIPES. *Experimental Characterization of Advanced Composite Materials.* Englewood Cliffs, N.J.: Prentice Hall, 1987.

CATLIN, CHRIS. "Aluminum/Polymer Composite Developed for Automobiles." *Advanced Materials & Processes,* Vol. 147, No. 5, May 1995, p. 8.

CHARLES, J. A., and F. A. A. CRANE. *Selection and Use of Engineering Materials,* 2nd ed. London: Butterworth, 1989.

Compressed Air Magazine. "Get Smart," Vol. 100, No. 3, April/May 1995, pp. 36–41.

EASTERLING, KEN. *Tomorrow's Materials.* Dorchester, England: Henry Ling Limited, Dorset Press, 1988.

GORDON, J. E. *The New Science of Strong Materials.* New York: Penguin, 1976.

HARRIS, C. E. "Industry-University-NASA Partnerships in Aeronautics Research and Technology Development," 4th Technical Conference: New Century Partnerships for Material Systems. Blacksburg, Va., April 23–25, 1995.

KUMAR, K. S. et al. "XD Titanium Aluminide Composites," *Advanced Materials & Processes,* Vol. 147, No. 4, April 1995, pp. 35–38.

LANGLEY RESEARCH CENTER. *Research and Technology,* NASA Technical Memorandum 4243. Hampton, Va.: NASA Langley Research Center, 1993.

LEE, ARTHUR K., LUIS E. SANCHEZ-CALDERA, S. TURKER OKTAY, and NAM P. SUH. "Liquid–Metal Mixing Process Tailors MMC Microstructures," *Advanced Materials and Processes,* Vol. 142, No. 2, 1992.

LEWIS, D. III et al. "In-Situ Composites," *Advanced Materials & Processes,* Vol. 148, No. 1, July 1995, pp. 29–31.

LINDSAY, KAREN. "FRP Rebar: The Next Generation," *CI on Composites,* April/May 1995, p. 7.

LUCE, S. *Introduction to Composite Technology.* Dearborn, Mich.: Society of Manufacturing Engineers, 1988.

Manufacturing Engineering. "Cutting Tool Material Makeover," August 1995, p. 26.

MARC ANALYSIS RESEARCH CORP. "Creep-Forming Simulation Helps Form Al Wing Skins," *Advanced Materials & Processes,* July 1994, p. 15.

Modern Plastics Encyclopedia. New York: McGraw-Hill, 1993.

PAINE, J. S. N. "Enhanced Damage Resistant Composites Incorporating Shape Memory Alloys," 4th Technical Conference: New Century Partnerships for Material Systems. Blacksburg, Va., April 23–25, 1995.

Popular Mechanics. "Synthetic Skin," January 1944, p. 14.

POSTON, IRVIN E. "Recycling of Automotive Polymer Composites," *Society of Plastics Engineers Recycling Meeting,* Detroit, March 9, 1992.

QUINN, KEVIN R., and CARLOS A. CARRENO. "High-Temperature THERMO-PLASTIC Composites," *Advanced Materials and Processes,* August 1991.

RICHARDSON, TERRY. *Composites: A Design Guide.* New York: Industrial Press, 1987.

SCHADE J. P., and R. W. ROSS, JR. "Corrosion Control with New Nickel-Base Alloys," *Advanced Materials & Processes,* Vol. 146, No. 1, July 1994.

SCHAEFER, R. J., and M. LINZER, eds. *Hot Isostatic Pressing: Theory and Applications,* ASM Conference Book. Materials Park, Ohio: ASM International, 1991.

SCHWARTZ, M. *Composite Materials Handbook.* New York: McGraw-Hill, 1983.

SEYMOUR, R. B. *Engineering Polymer Sourcebook.* New York: McGraw-Hill, 1990.

STRONG, A. BRENT. *Fundamentals of Composites Manufacturing.* Dearborn, Mich.: Society of Manufacturing Engineers, 1992.

SUMITOMO METALS INDUSTRIES, LTD. "Particle-Dispersed Steel Has Highest Modulus," *Advanced Materials & Processes,* July 1994, p. 7.

TRACESKI, E. T. *Specifications and Standards for Plastics and Composites.* Materials Park, Ohio: ASM International, 1990.

URQUHART, ANDREW W. "Molten Metals Sire MMCs, CMCs," *Advanced Materials and Processes,* Vol. 140, No. 1, 1991.

WILSON, MAAYWOOD L., GARY S. JOHNSON, and ROBERT MISERENTINO. "Pultrusion Process Development for Multidirectional Graphite/Epoxy T-Beam," *SPI Paper,* NASA Langley Research Center, Hampton, Va, 1992.

ZHOU, SUWEI, ZAFFIR CHAUDRY, and CRAIG ROGERS. "Review of Particle Tagging Methods for NDE of Composite Materials and Structures," SPIE North American Conference on Smart Structures and Materials, February 26, 1995.

Periodicals

Advanced Composites
Advanced Materials & Processes
CI on Composites
Composites in Manufacturing
Compressed Air Magazine
Journal of Composite Materials
Journal of Composite Technology and Research
Journal of Materials Research
Journal of Reinforced Plastics
Machine Design
Manufacturing Engineering
Materials Engineering
Materials Performance
NASA Tech Briefs
Plastics Engineering
Plastics Technology
Plastics World
Popular Mechanics
Popular Science

EXPERIMENTS & DEMONSTRATIONS IN COMPOSITES

NEW:Update 88 NASA Conference Publication 3060

NELSON, JAMES A. "Composites: Fiberglass Hand Laminating Process."

NEW:Update 89 NASA Conference Publication 3074

BEARDMORE, PETER. "Future Automotive Materials: Evolution or Revolution."

CHUNG, WENCHIANG R. "The Assessment of Metal Fiber Reinforced Polymeric Composites."

COLEMAN, J. MARIO. "Using Template/Hotwire Cutting to Demonstrate Moldless Composite Fabrication."

NEW:Update 90 NIST Special Publication 822

BUNNELL, L. R. "Simple Stressed-Skin Composites Using Paper Reinforcement."

SCHMENK, MYRON J. "Fabrication and Evaluation of a Simple Composite Structural Beam."

WEST, HARVEY A., and A. F. SPRECHER. "Fiber Reinforced Composite Materials."

NEW:Update 91 NASA Conference Publication 3151

GREET, RICHARD J. "Composite Column of Common Materials."

NEW:Update 92 NASA Conference Publication 3201

THORNTON, H. RICHARD. "Mechanical Properties of Composite Materials."

NEW:Update 93 NASA Conference Publication 3259

MASTERS, JOHN. "ASTM Methods for Composite Characterization and Evaluation."
WEBBER, M. D., and HARVEY A. WEST. "Continuous Unidirectional Fiber Reinforced Composites: Fabrication and Testing."

NEW:Update 95 NASA Conference Publication 3330

CRAIG, DOUGLAS F. "Role of Processing in Total Materials Cycle."
WILKERSON, AMY LAURIE. "Computerized Testing of Woven Composite Materials."

Module

9

Electronic and Other Important Materials and Materials Systems Technology

After studying this module, you should be able to do the following:

9.1. Use diagrams, explanations, and calculations to determine electrical, magnetic, and optical properties of materials. Recall the nature, processing, and applications of electronic, magnetic, and optical materials.

9.2. Describe the goals of materials science in developing smart materials and intelligent structural systems. Recall some current and potential applications of various smart materials.

9.3. Describe the goals of materials science in research and development with biomimetic materials. Explain the properties and some applications of biomaterials.

9.4. Name some of the techniques, explain the needs, and list some developments in nanotechnology and microtechnology.

9.5. Gain an appreciation for the role of engineering materials technology and its value to you in the future.

Figure 9-1 A CD-ROM leans against the keyboard of a mobile notebook computer. When played in the computer's built-in CD-ROM drive, the CD provides optimum multimedia capabilities as well as mobile computing uses.

PAUSE & PONDER

Starting around 1986 with the adaptation of optical disks to storage media, ***CD-ROMs*** (compact disc—read-only memory) brought a revolution in the storage and retrieval of digital information. Offering a capacity to store 500 times as much data as the ordinary magnetic diskette, CD-ROM technology, with its associated laser technology, has already moved the music business away from audio tapes and now offers both computer and entertainment business arenas new variations of ***multimedia technology.*** CD-ROM provides high-density data storage, durability, and accessibility, and allows for the storage and retrieval of text, graphics (both still and motion), and sound. ***Information technology,*** the term often used to label the combination of electronics, optics, computers, telecommunications, graphics communications, entertainment, and the wide range of other forms of communication systems, owes much of its evolution to materials science and engineering developments, which spurred its progress.

Once the master disc is produced by lasers, replication of a limitless number of disks is accomplished by injection molding of polycarbonate [see Figure 9-2(a)]. With recordings on magnetic media such as cassette tape, a recording head had to pass over the entire surface to transfer data. The 4.75-in. (121-mm) CD-ROMs contain up to 2.8 billions pits, which spiral for approximately 3 miles as measured from the hub to the outer rim. The cross-section [Figure 9-1(c)] shows the structure of a recordable CD. A regular CD uses aluminum instead of gold and has no dye layer.

A study of CD-ROM materials processing and manufacturing techniques provides insight into modern production technology. The processing of a metal master and the subsequent replication of polycarbonate CD-ROMs, shown in Figure 9-2(b), involves eight basic steps:

1. Preparing the coated glass substrate
2. Applying a photoresistant coating to the glass substrate
3. Using a laser beam recorder to "cut" pits in the photoresist layer and then developing the glass master
4. Electroforming the glass master to produce a metal "master," or father
5. Additional electroforming produces the "mother" and "stamper"
6. Injection molding of polycarbonate plastic discs from the "stamper"
7. Aluminizing the disc
8. Adding a protective coating to the aluminized surface and applying a label

With multimedia computers becoming commonplace, the demand for CD-ROMs will continue. Most PC computers use these optical storage media in the form of OROM ***(optical read-only memory);*** however, the demand for recordable CD-ROMs, known as WORM ***(write once, read many),*** has already caused an increase in production and a lower price for these more versatile optical storage disks.

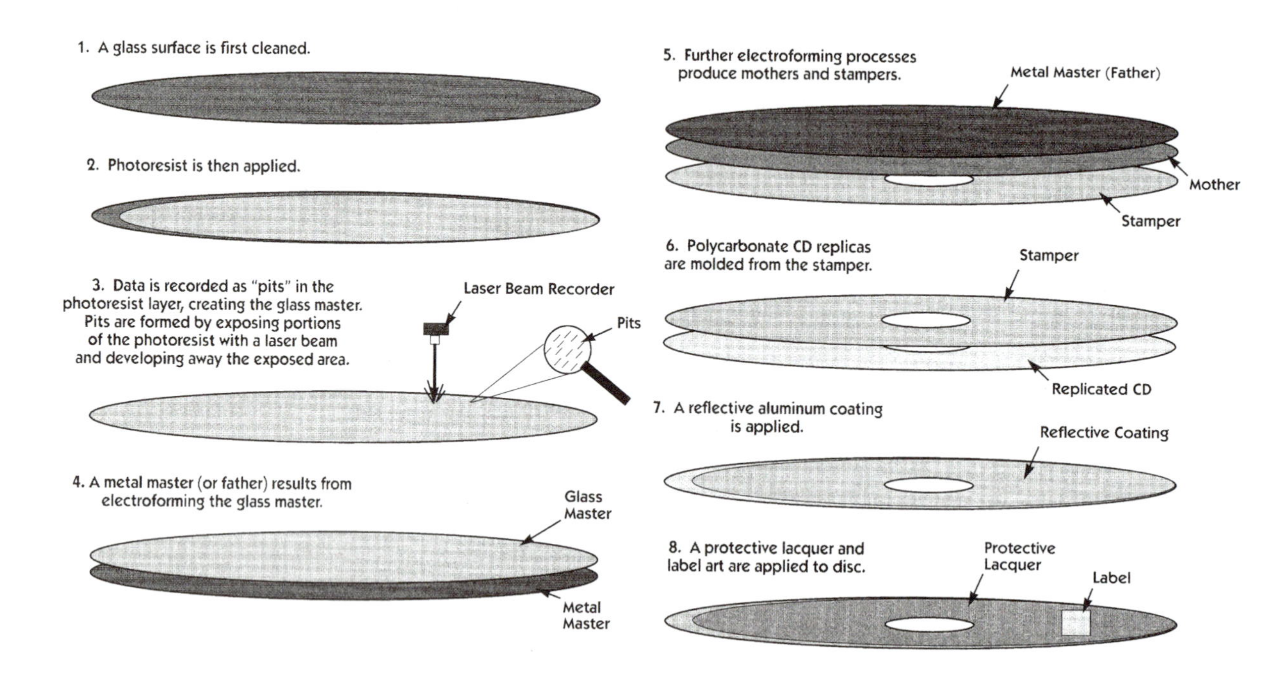

Figure 9-2 (a) How a compact disc (CD) is made. (Nimbus Inc.) (b) Cross-sectional view of a recordable CD.

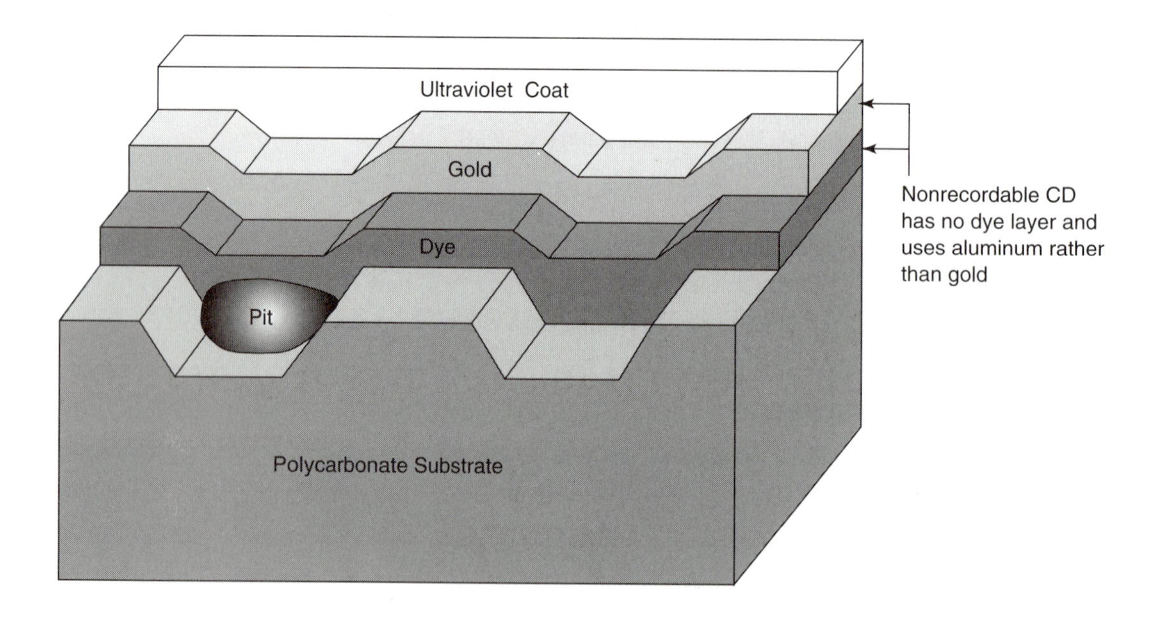

Pause for a minute and consider all the rapid developments in information you have witnessed in your lifetime. As with CD-ROMs, many of those developments were a direct result of advances in engineering materials technology. This module will introduce you to some of that technology.

9.1 ELECTRICAL PROPERTIES

Before beginning our discussion of electrical properties, a review of some common terms used to describe electrical behavior of materials is essential. First it is important to realize that electrical behavior is affected not only by the structure of the material and how it was processed but by the material's environment as well.

Using a simple conducting wire circuit to which a source of electrical potential such as a battery or generator is applied, Ohm's law relates mathematically the electrical properties of current flow, resistance to the flow of current, and the voltage drop or potential difference across a resistance. Stated mathematically, $I = E/R$. Ohm's law states that the current flow (I), measured in amperes (A), is directly proportional to the voltage (E), measured in volts (V), and is inversely proportional to the resistance (R) to the current flow, measured in ohms (Ω). This resistance might be considered as the macroscopic definition of ***electrical resistance*** because it is related to the size, shape, and properties of the materials that make up the electrical circuit. For example, a copper wire, 1000 ft in length with a diameter of 0.1 in., has the same resistance of 1 Ω as a steel rod 1000 m long with a diameter of 1 cm. The unit of current is the ampere, which is equal to a rate of flow of electric charge of 1 coulomb per second.

Illustrative Problem

The current in a circuit is 0.25 A when a potential difference (voltage) of 120 V is applied across its terminals. Find the value of the resistance.

Solution

$$I = \frac{E}{R} \quad \text{and} \quad R = \frac{E}{I}$$

$$= \frac{120\text{ V}}{0.25\text{ A}} = 480\ \Omega$$

9.1.1 Resistivity

Resistivity (or *volume resistivity*) is the term used to describe the relationship between electric current and the applied electric field. It is a measure of the resistance to the flow of current from a microscopic level, that is, as explained in terms of the atoms, the basic building blocks of all solid materials. The resistivity, ρ (Greek letter rho), depends on the behavior and number of free, or conduction, electrons and not on the shape of the conductor, as does resistance. Like density, this inherent property of a material will change as the structure changes, as in the alloying of metals or the doping of semiconductor materials. It depends on the movement of charge carriers, electrons in metallic conductors, or ions in ionic materials. In fact, electrical resistivity is the reciprocal of electrical conductivity (σ, the Greek letter sigma) (i.e., $\rho = 1/\sigma$). This electrical property is discussed later in this module. Microstructure again plays a large role in these properties. Any imperfections in the crystalline structure, whether they are atoms out of their normal positions, dislocations, or grain boundaries, to mention but a few, increase the collisions between electrons. This, in turn, prevents the transfer of energy in the form of electron flow to some intended user.

Figure 9-3(a) and (b) shows a sketch of the effects of an increase in temperature on metallic conductors and semiconductors, respectively. With metallic conductors, the increased vibrational energies of the atoms as a result of an increase in energy make the passage of free electrons through the structure even more difficult. The ***mean free path,*** the average distance an electron can travel as a wave without hitting or deflecting off a positive-ion core (atom) in the lattice structure, is decreased. Consequently, the mobility of the electrons decreases, which produces an increase in the resistivity. For semiconductor materials, the resistivity decreases (conductivity increases) with an increase in temperature because more charge carriers become available to act as conduction electrons. Carbon's resistivity decreases with an increase in temperature.

The doping of semiconductor materials also lowers the resistivity by increasing the number of charge carriers. Additional information on this subject appears in Section 9.6.

In the discussion of resistance, we stated that resistance varies directly with the length L and indirectly with the uniform cross-sectional area A of a conductor. To write this as a mathematical statement, we need a constant to make the units agree on both sides of the equation. Thus,

$$R = \rho \frac{l}{A}$$

where ρ is the proportionality constant or resistivity in ohm-centimeters, assuming that l and A are expressed in centimeters. Note that a conductor, 1 cm in length with an area of 1 cm^2, has a resistance R equal to the resistivity ρ.

In selecting conducting materials for the express purpose of generating heat from the flow of electricity, such as Nichrome, a heat-resistant alloy of nickel and chromium, the

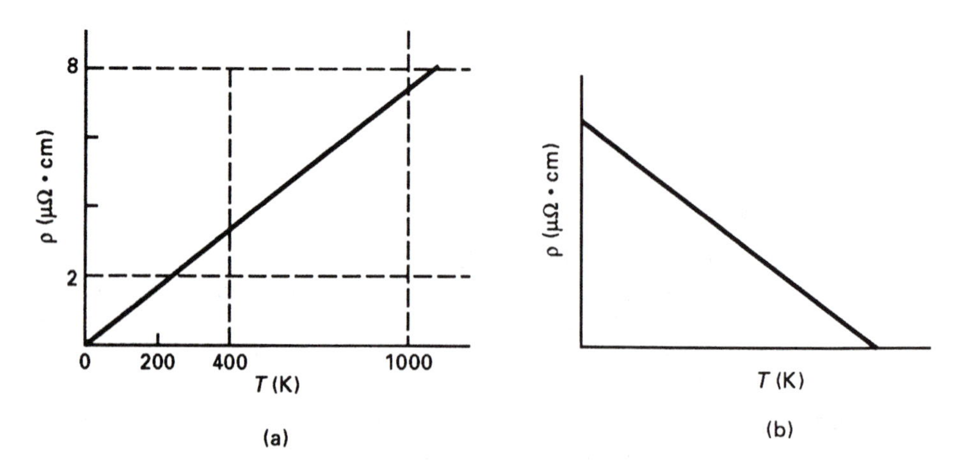

Figure 9-3 (a) Resistivity (ρ) versus temperature (T) for a metallic conductor. (b) Resistivity versus temperature for a semiconductor material.

material must have a carefully selected variety of properties, such as a moderate resistivity, excellent resistance to oxidation, and a capability to operate effectively at high temperatures without failure. Table 9-1 lists the electrical resistivities of some selected solids from three families of materials.

Illustrative Problem

(a) A metallic wire 100 cm long has a diameter of 0.05 cm. If it has a resistance of 0.08 Ω when 10 A of current is running through the circuit at standard temperature, find the wire's resistivity in Ω · cm. (b) Using Table 9-1, identify the material from which the wire was fabricated.

Solution

(a) $R = \rho\,(l/A)$; therefore, $\rho = R/(A/l)$.

$$A = \frac{\pi d^2}{4} = 0.7854 \times (0.05)^2 = 1.96 \times 10^{-3}\ \text{cm}^2$$

$$\rho = 0.08\ \Omega \times 1.96 \times 10^{-3}\ \text{cm}^2/10^2 = 1.7 \times 10^{-6}\ \Omega \cdot \text{cm}$$

(b) Either silver or copper would be adequate answers.

Note in our discussion that, where heat is involved in the flow of electric charge, efficiency is reduced. The incandescent light bulb is a good example. This bulb gets hot when used. A fluorescent bulb remains cool. Consequently, the fluorescent bulb is more efficient in the use of electricity (energy). As a matter of fact, the incandescent light is only about 3% efficient; that is, 97% of the energy needed to produce light is lost mainly to heat.

Summarizing, the resistance R of an electrical circuit is a function of the shape, size, and nature of a solid material. Just like specific heat or density, resistivity ρ is a function of the intrinsic nature of the material itself. Instead of thinking of resistivity, one can think

TABLE 9-1 DIELECTRIC PROPERTIES AND RESISTIVITY FOR SOME COMMON MATERIALS

Material	Resistivity (Ω · m at 20°C)	Dielectric constant, κ (10^6 Hz)	Dielectric strength (10^6 V/m)	Dielectric loss factor, κ tan δ (10^6 Hz)
Mica	10^{11}	7	79	
Nylon 6/6	10^{13}	3.5	17	0.04
$BaTiO_3$	10^{9}	1600		
Al_2O_3	10^{12}	9	98	0.001
Steatite	10^{12}	6	8	
Phenolics—transparent	10^{10}	8	6	0.05
Paraffin	10^{15}	2.3	10	
Water	10^{12}	78		
Polyethylene	10^{15}	2.3	20	0.0001
Soda-lime-silica glass	10^{13}	7	10	0.01
Silver	16×10^{-9}			
Copper	17×10^{-9}			
Tungsten	56×10^{-9}			
Aluminum	29×10^{-9}			
Steel (301)	7×10^{-7}			
Graphite	10^{-5}			
SiC	0.1			

in terms of conductivity, σ. Low resistivity and high conductivity refer to a similar situation in different terms. To further reinforce the concept of resistivity, remember that 1 lb of aluminum has the same resistivity as 1 g of aluminum, whereas the resistance of 1 lb of aluminum is very different from that offered by 1 g. Figure 9-4 illustrates an industrial use for a human-made nonconductive material.

9.1.2 Conductivity

The reciprocal of resistivity is known as *conductivity*. It is a measure of how readily electrons flow through a material. Using the symbol σ (Greek letter sigma) to represent conductivity, this quantity is determined for a particular material by measuring the amount of electric charge that passes through a unit cube of the material per unit of time. Conductivities of common metals range from 10,000 to 550,000 $\Omega^{-1}\text{cm}^{-1}$. (See Section 8.13 for information on increasing conductivities of graphite fibers.) Consequently, σ is dependent on three factors:

1. *n*, the number of charge carriers in a cubic centimeter (cm^3) of material
2. *q*, the charge per carrier (coulombs/carrier)
3. μ, the mobility of each carrier, or the ease of movement of the charge carriers (e.g., electrons in a metal)

$$\mu = \frac{\text{velocity of the carriers (cm/s)}}{\text{voltage gradient (V/cm)}}$$

Figure 9-4 Pultruded fiberglass boom is superior to older metal booms in terms of mechanical, weathering, and electrical properties. The boom exhibits less than 100 microamp leakage over any 2-foot section after a 48-hour full immersion soak while a 100-KV (dc) test voltage is applied. (Morrison Molded Fiber Glass Co.)

The product of these three factors, with units of $1/\Omega \cdot \text{cm}$ or $(\Omega \cdot \text{cm})^{-1}$, is the conductivity. Expressed in mathematical terms,

$$\sigma = nq\mu$$

This calculation agrees with units obtained for conductivity if the basic equation for resistivity is used ($\rho = RA/l$).

Illustrative Problem

A tungsten wire is used in an electrical circuit at a temperature of 20°C. Find its electrical conductivity.

Solution Conductivity is the reciprocal of resistivity. From Table 9-1, tungsten's resistivity is $56 \times 10^{-9}\ \Omega \cdot \text{m}$, so the conductivity is $1/56 \times 10^{-9}\ \Omega \cdot \text{m} = 17.9 \times 10^{6}\ (\Omega \cdot \text{m})^{-1}$ or $\Omega^{-1} \cdot \text{m}^{-1}$.

Illustrative Problem

A rectangular block of aluminum has the dimensions 25 mm × 25 mm × 50 mm.

(a) Make a three-dimensional sketch of the block and place dimensions on it.

(b) What is the resistivity of the block, in $\mu\Omega \cdot \text{cm}$, measured between the two squared ends at 20°C?

(c) Determine the electrical conductivity for aluminum at 20°C.

Solution

(a)

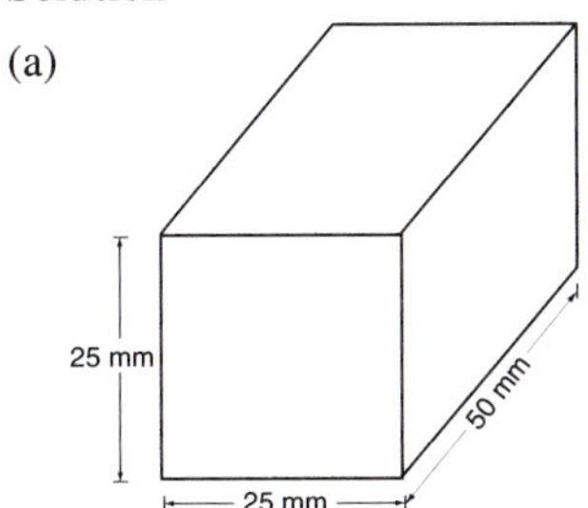

(b) The cross-sectional area is $25\text{mm}^2 = 6.25\ \text{cm}^2$.

$$R = \rho\frac{l}{A} = \frac{2.9 \times 10^{-6}\ \Omega \cdot \text{cm} \times 5\ \text{cm}}{6.25\ \text{cm}^2}$$

$$R = 2.32\ \mu\Omega$$

(c) $\sigma = 1/\rho = 1/2.9\ \mu\Omega \cdot \text{cm} = 3.45 \times 10^5/(\Omega \cdot \text{cm})^{-1}$.

9.2 DIELECTRIC PROPERTIES

9.2.1 Dielectrics

The term ***dielectric*** refers to nonconductive materials whose polarization is caused by an electric field. A material or medium such as a vacuum or gas is called a dielectric because it permits the passage of the lines of force of an electrostatic field but does not conduct the current. Subjecting these materials to a magnetic field also produces similar effects, although the sources are different. A dielectric has two functions to perform. One is to act as an insulator and the second is to increase the capacitance of a condenser (capacitor) beyond that which can be derived from an air gap or vacuum between the plates of the capacitor. As a covering for an electrical wire, the plastic coating is acting as an insulator. Good electrical insulators such as ceramics are referred to as dielectric materials. Even though these materials do not conduct electrical current, they are affected by the applied electric field. This effect, known as ***polarization,*** results in a shift in the distribution of charge within the material forming an electric dipole. ***Dipoles*** are atoms/ions or groups of atoms/ions that have an unbalanced or asymmetrical charge. When subjected to an outside electric field, the dipoles become aligned with the direction of the applied field in the material, causing polarization. It should be remembered that atoms and ions move (i.e., they translate as well as rotate within their structures). Several sources are identified as contributing to polarization. Electrons surrounding the positive nucleus of atoms are one contributor as they shift slightly toward the positive terminal (electrode). Concurrent with this shift is the shifting of the positive nucleus slightly toward the negative terminal. Figure 9-5(a) is a sketch of electronic polarization. A second contributor to polarization, called ***orientation polarization,*** involves the molecules that are already polarized (i.e., they contain an electric dipole). A simple example is the asymmetric water molecule described in Module 2 in conjunction with a discussion about secondary covalent bonding and hydro-

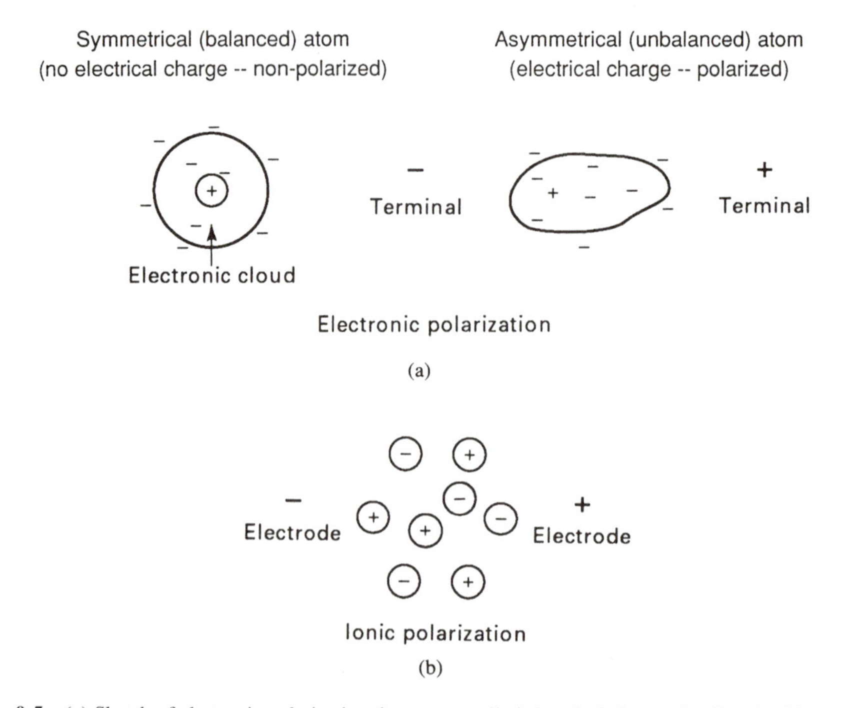

Figure 9-5 (a) Sketch of electronic polarization due to an applied electrical charge. (b) Sketch of ionic polarization showing the effects of attraction/repulsion forces produced by the applied electrical charge.

gen bridging. When an electric field is applied, the molecules will align themselves such that their negative side will align with the positive terminal (or electrode) of the applied field. Other unbalanced-charge molecules exist of varying complexity, such as CH_3Br, which is larger than the water molecule. The result is greater contribution to the electronic polarization. One final contributor to polarization of a material to be discussed is called ***ionic polarization,*** sketched in Figure 9-5(b). The individual ions are displaced in the crystal structure, producing temporary induced dipoles that can possibly change the overall dimensions of the material. Polarization is affected by many structural features. The presence of permanent dipoles, ease of movement of molecules, moisture, cracks, and grain-boundary imperfections are just a few of such features.

Dielectrics and their properties are important in the design and fabrication of both ceramic and polymer composites. *Dielectrometry* uses electrical techniques to measure the changes in dielectric loss factor and capacitance during cure of a resin in a laminate. Ceramics are widely used as insulators and as substrates in MMCs, where the need is for high resistivities and dielectric strength. High thermal conductivity might also be needed if heat must be dissipated.

9.2.2 Dielectric Constant

The ***capacitance*** *(C)* of a condenser or capacitor is a measure of its ability to store electricity. It is determined by measuring the amount of charge *(Q)* that must be placed on a conductor to raise its potential by 1 V. Mathematically, $C = Q/V$, where C is expressed in farads (coulombs per second), Q in coulombs, and V in volts. Solving the expression for $Q = CV$, it can be seen that C is a constant of proportionality with no units. A farad (F), the unit of electric capacitance, is defined as the capacitance of a capacitor, between the plates of which there appears a potential *(E)* of 1 V when it is charged by a quantity of electricity of 1 coulomb. A farad is a fairly large unit, and most capacitors have small values of capacitance. So prefixes such as *pico* or *micro* are used to express smaller units of the farad (1 pF $= 10^{-12}$ F).

Two conductors separated from each other by some insulating material (including air) form a condenser. The conductors make up the plates of the condenser, and the insulation between the plates is the dielectric. The condenser is charged by connecting its plates to a battery or a source of electrical energy. Electrons will flow and collect on one plate until equilibrium is reached and the potential difference between the two plates equals the electromotive force (emf) of the battery. The ratio of the condenser charge *(Q)* to the potential difference *(E)* defines the *capacitance* of the condenser. The capacitance of a parallel-plate condenser with free space as the dielectric is proportional to the area of the plates and inversely proportional to the distance between the plates. Expressed mathematically, we have

$$C = \epsilon_0 \frac{A}{d}$$

where A is the area of the plates, d is the thickness of the dielectric or the plate separation, and ϵ_0 (Greek letter epsilon) is the ***permittivity constant of free space*** with a value of 8.85×10^{-12} F/m. Note that C depends on the geometry of the arrangement.

Illustrative Problem

A condenser is made of two thin sheets of copper with a total area of 1 m^2. The sheets are separated by air at a distance of 0.05 mm. What is the capacitance of the condenser?

Solution

$$C = \epsilon\frac{A}{d} = \frac{8.85 \times 10^{-12}\text{ F/m} \times 1\text{ m}^2}{5 \times 10^{-5}\text{ m}}$$

$$= 0.177\ \mu\text{F}$$

If a dielectric such as mica is placed between the plates of the same parallel-plate condenser using the same potential difference E, the charge Q on the plates will be greater and will produce a corresponding increase in the capacitance C of the condenser. This result can be expressed:

$$C = \kappa\epsilon_0\frac{A}{d}$$

where κ (Greek letter kappa) is the ***dielectric constant*** or ***relative permittivity*** of the dielectric. The dielectric constant is a measure of a material's value toward making a desired capacitance when the material is placed between two conducting plates. The dielectric permits a higher potential difference to be applied than when air is used. It is a dimensionless number that indicates the reduction of the field due to the dielectric. (See Table 9-1 for dielectric constants for some common materials.)

Illustrative Problem

Using the preceding problem with only air between the plates of the condenser ($\kappa = 1$), insert paper between the plates with a dielectric constant of 3.5. Find the capacitance.

Solution

$$0.05\text{ mm} = 5 \times 10^{-5}\text{ m}$$

$$C = \kappa\epsilon_0\frac{A}{d}$$

$$C = 3.5 \times 8.85 \times 10^{-12}\left(\frac{\text{F}}{\text{m}} \times \frac{1\text{ m}^2}{5 \times 10^{-5}\text{ m}}\right)$$

$$C = 6.19 \times 10^{-7}\text{ F} = 0.619\ \mu\text{F}$$

Note: Adding paper increased the capacitance almost fourfold.

When a dielectric material such as a polymer, glass, or ceramic is placed between the plates of a condenser, an induced charge Q appears on the plates without any change in the voltage. This naturally increases the capacitance of the condenser. The increased charge is due to the *polarization* of the entire volume of the dielectric, in turn due to the applied electric field. Polarization is the process of producing electric dipoles. A ***dipole*** is two equal but opposite electric charges ($+Q$) and ($-Q$) that are separated by a distance d whose product (the magnitude of one of the charges times the distance d), Qd, is known as the dipole moment or strength *(p)*. The units of dipole moment are coulomb meters ($C \cdot m$). In dielectric materials, electrons and their atoms are not free to move much under the influence of an external electric field, but they move enough to produce polarization of the material.

Dielectrics are used in alternating-current (ac) circuits with a wide range of frequencies. The higher the frequency, the faster the dipoles must switch directions to allow the device to perform satisfactorily. If the dipoles have difficulty moving, as is the situation with complex organic molecules, dipole friction results, with attendant energy losses. These effects show up in the value of the dielectric constant. Temperatures also affect the dielectric constant. In general, the dielectric constant increases with an increase in temperature. Because the dielectric constant is a ratio of similar quantities, it has no units. Some typical values for κ for some common materials taken from a variety of sources for a frequency of 10^6 Hz and 20°C are listed in Table 9-1.

High values of dielectric constant are needed for capacitors, whereas low values are used as insulating materials. Much research is devoted to the miniaturization of electrical circuits, including capacitors, which means that a continual search is under way to find materials such as ceramics that not only have high dielectric constants, but have such values at higher and higher frequencies and temperatures for use in, for example, dielectric heating applications. A need for materials that have low (< 3) dielectric constants for use in coatings, films, matrix resins, and fibers is in evidence, particularly in the electronics and aerospace industries. A low dielectric constant polyimide fiber with high thermal stability and good tensile properties similar to standard textile fibers has been developed for possible use in printed-circuit boards and in aircraft composites.

9.2.3 Dielectric Strength

Dielectric strength is the voltage gradient (voltage per unit thickness, such as volts per mil), which produces electrical breakdown through the dielectric. It is an expression of the maximum voltage that a dielectric can withstand before electrical discharge occurs through the dielectric. This insulating strength is dependent not only on the usual items, such as bonding type and crystalline structure, but on moisture absorption and the nature of the applied electrical energy. In the latter, the source might be a direct current (dc) or an alternating current (ac), or a combination of the two. If the magnitude of the dielectric field is sufficient, it will overcome the attraction of the electrons to their positive-charged nuclei and produce a *leakage* and eventual rupture of the material. The breakdown voltage has then been reached. With an ac source, the continual reorientation of the material's atoms and electrons, in addition to the deformation of the paths of the electrons, results in a hysteresis loss (to be explained) that produces heat in the dielectric material. Therefore, the breakdown voltage will vary inversely with the frequency of the source voltage. Some typical ranges of values of dielectric strength for some nonmetallic materials, expressed in 10^6 volts per meter (10^6 V/m), are listed in Table 9-1.

9.2.4 Dissipation Factor/Dielectric-Loss Factor

When alternating current provides the voltage (sinusoidal) needed to maintain the charge on a capacitor or any dielectric, the current leads the voltage by $90° - \delta$, where the angle

Figure 9-6 Dielectric heating using microwave assist in the drying/sintering/curing of materials as well as the bonding of composites using adhesives that have different dielectric loss factors than the materials being joined. (a) Microwave ovens are used in ceramic processing to achieve lower temperatures, reducing the possibility of inflicting damage on materials due to high-temperature processing. (b) Contrasting ceramic microstructures between conventional and microwave heating. (ORNL)

(a)

Microstructure of Si_3N_4-6% Y_2O_3-2% Al_2O_3 After Annealing at 1200°C for 20 hours

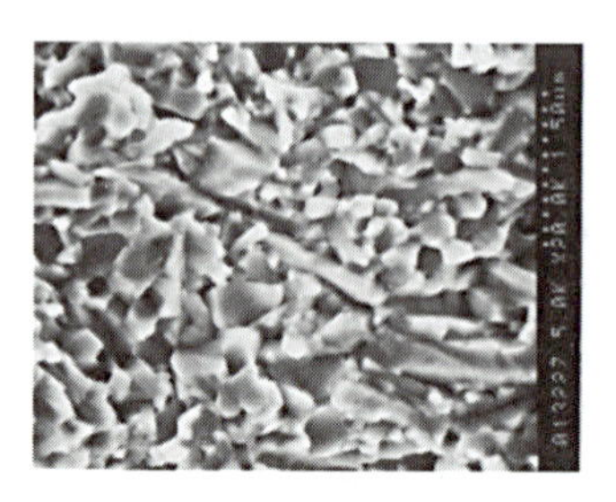

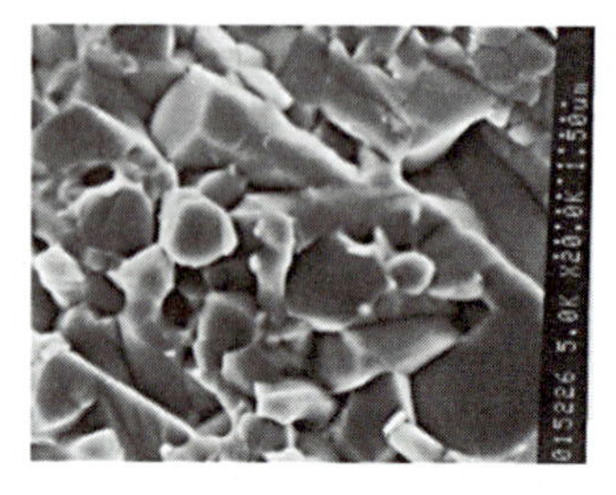

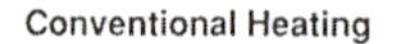

Microwave Heating

(b)

δ is called the dielectric-loss angle. If the dielectric were perfect, the angle δ would be zero and no loss would occur. In this situation, the current would lead the voltage by 90°. Most of the time, however, perfection is not attained and the current and voltage are therefore out of phase, resulting in a loss of electrical energy. This loss of energy, hence power, usually takes the form of heat. The tangent of the angle δ (tan δ) is known as the ***dissipation factor.*** Multiplying tan δ by the dielectric constant (κ) expresses the ***dielectric-loss factor, dielectric heat-loss factor, dielectric coefficient,*** or ***loss tangent.*** Dielectric heating is the heating of dielectric materials by subjecting them to a high-frequency, alternating electric field. This heating assists in the bonding or drying of materials (see Figure 9-6). Dielectric loss is affected by frequency and temperature changes. The loss usually increases at low frequencies and as the temperature increases. Table 9-1 lists the dielectric loss factor for a few nonmetallic materials.

9.3 MAGNETIC PROPERTIES

9.3.1 Magnetic Permeability

In an electrical circuit, the conductivity (σ) of a conductor is a constant because it does not depend on either the potential difference *(E)* across a unit length of a conductor of unit cross-section or on the current *(I)* passing through that section of the conductor. In mathematical terms, with unit value for *(l)* and *(A)*,

$$\sigma = \frac{I}{E} \qquad \text{(a constant)}$$

In a magnetic circuit, the permeability (μ) of a material, a measure of the ease with which a magnetic field can be set up through a material, can be compared to the conductivity in an electrical circuit. The magnetic field, represented by the symbol *(H)* with units of amperes per meter (A/m), corresponds with the voltage or potential drop. The magnetic flux density, magnetic induction, or magnetic field *(B)*, expressed in units of webers per square meter (Wb/m^2), corresponds with the electric current. The ratio of *(B)* to *(H)* (B/H) is known as ***permeability*** (μ) (Greek letter mu), with units of webers per ampere meter ($Wb/A \cdot m$). An important distinction between electrical and magnetic circuits is that permeability is not constant. As *(H)* changes, so does (μ).

Figure 9-7 illustrates the permeability of a vacuum or free space in the core of an electrical coil through which an electrical current is passing. When a core of material replaces the vacuum in the coil, a change in the flux density *(B)* is detected. This change

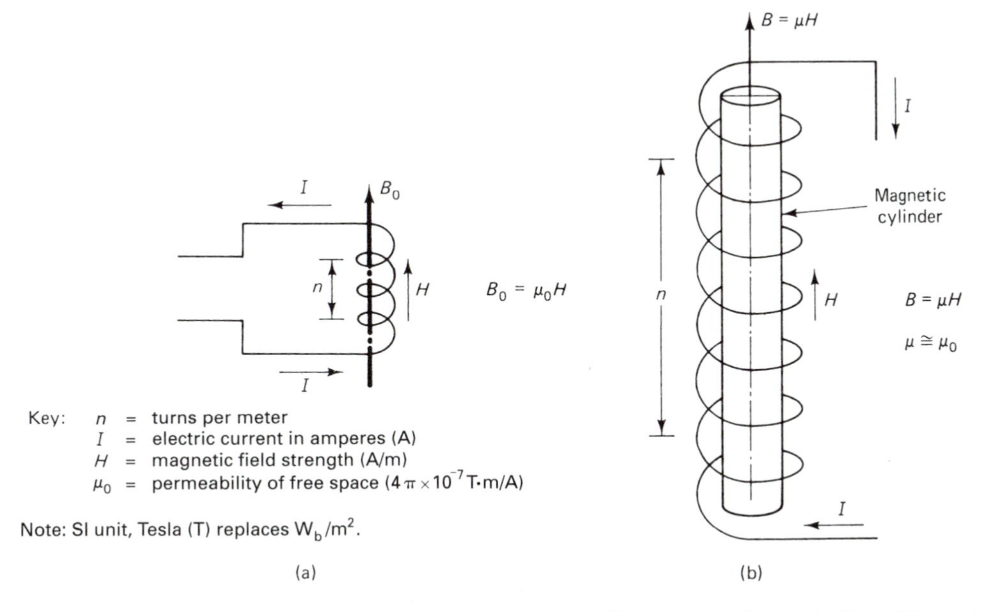

Figure 9-7 (a) Magnetic helical coil with a vacuum core. (b) Magnetic helical coil with a solid material core.

is reflected in a new value of (μ). If the material is made of iron, cobalt, or nickel, an exceedingly large rise in *(B)* is noted.

9.3.2 Relative Permeability

Relative permeability (μ_r) is the ratio of two permeabilities and therefore has no units. The permeability of a material is thus compared mathematically with the permeability of free space (vacuum), $\mu_r = \mu/\mu_0$. μ_0 equals $4\pi \times 10^{-7}$ with SI units $T \cdot m/A$. This ratio may be used to classify magnetic materials into three main categories:

1. Diamagnetic material if $\mu_r < \mu_0$.
2. Paramagnetic material if $\mu_r > \mu_0$.
3. Ferromagnetic material $\mu_r \gg \mu_0$.*

Actually, μ_r for diamagnetic and paramagnetic materials is nearly 1, while for ferromagnetic material it can be up to 10^6 times greater than μ_0. This direction of magnetization can be changed provided that the necessary external energy is provided. This energy is supplied by an external magnetic field. When $B = 0$ (no external magnetic field present), each domain is balanced with a domain of opposite alignment. Therefore, no external magnetic flux is present outside the material. Figure 9-8(a) is a schematic of magnetic domains (defined on page 630) in a crystal.

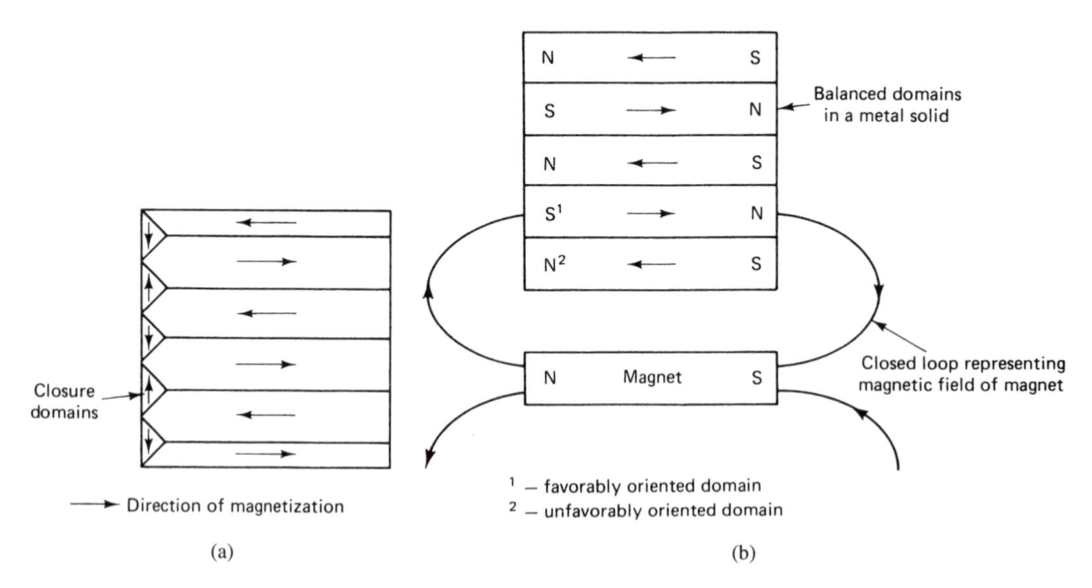

Figure 9-8 (a) Schematic of magnetic domains in a crystal at $B = 0$. (b) Balanced domains in an external magnetic field.

*The symbol $\gg$ means "much greater than."

If a magnetic field is applied, the domains whose magnetization is in the direction of the external magnetic field will grow at the expense of those domains of opposite alignment, as shown in Figure 9-8(b). The end result of this domain-boundary migration, assuming the presence of a large enough magnetic field, is the alignment of all domains in the direction of the magnetic field. Using domains, one can readily understand why an iron nail placed in the vicinity of a magnet will become magnetized in a direction dictated by the location of the north-seeking end of the magnet used to magnetize the nail. Proof of the existence of domains is readily available by using a colloidal solution of iron-oxide particles spread over the polished surface of a piece of magnetic material. When viewed under a microscope, the particles of iron oxide will be concentrated around the domain boundaries. All ferromagnetic materials begin to lose their magnetic properties as their temperatures increase. The thermal oscillations produced tend to destroy the magnetic domains. Above a characteristic temperature known as the ***Curie temperature*** *(T_c)*, ferromagnetism will no longer exist even when subjected to large external magnetic fields.

A spinning electron is a moving electric charge. It induces a magnetic field because of its spin (see Module 2). A pair of electrons with opposing spins has no net magnetic field. Their opposing magnetic fields cancel. A diamagnetic atom such as Zn has all paired electrons. A paramagnetic atom has unpaired electrons that induce a magnetic field, which causes the atom to be attracted into an external magnetic field. Mn is such an atom. Its orbital diagram (Figure 9-9) shows its five unpaired electrons in the 3d orbital. The 1st transition elements, Sc through Zn in period 4 of the periodic table, have unique properties due to the late or delayed filling of the next-to-outermost electron energy levels. The 3*d* sublevel gains from 1 to 10 electrons as the atomic numbers of these atoms increase. All, with the exception of Cr and Cu, have two paired electrons in the 4*s* orbital. The d orbitals in these elements are of primary importance in chemical bond formation, as well as the source of many other distinct properties such as color, catalytic activity, and magnetism. One such property is ferromagnetism, which is unique to Fe, Co, Ni, and some metal alloys. This property cannot be accounted for by considering the paired/unpaired electrons in an atom or ion. Instead, we will leave the world of the individual atom with its electrons and investigate the interactions of atoms with their neighboring atoms. This is necessary to obtain a macro view of the contributions of individual regions of a magnetized material. These regions, visible under powerful electron microscopes, are where the atomic magnets (the magnetic moments resulting from the electronic structure of the individual atoms) are aligned in the same or parallel direction. This macro view, which is a more acceptable one,

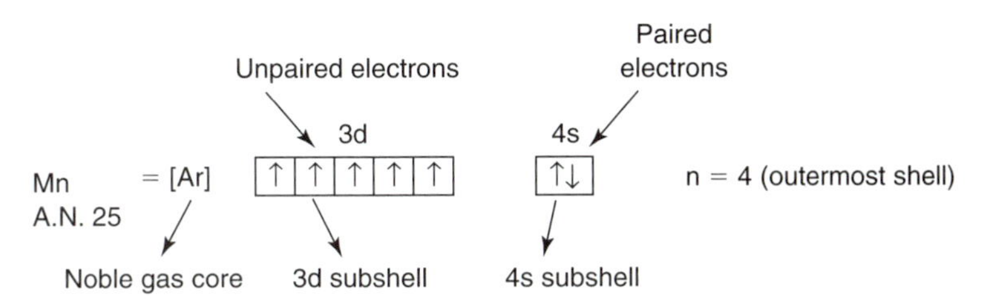

Figure 9-9 Orbital diagram for manganese (atomic number 25). The noble gas core consists of a shorthand notation for the orbital diagram for argon (Ar, atomic number 18).

uses the term ***domain.*** A tiny region, with a volume of about 10^{-12} m^3 containing at most around 10^{23} atoms, behaves like a tiny magnet with a north (N) and a south (S) pole. In a magnet, these domains are preferentially aligned in one direction. This preferred direction is a function of the crystallographic structure of the atoms making up the material. With bcc iron, the <100> direction is the easiest direction to magnetize, which indicates that such materials have a large degree of magnetic anisotropy (different magnetic properties). The interatomic distances must be just the correct magnitude to make possible the ordering of atoms into domains. If the distances are too large, the interactions are too weak to produce ordering. If they are too small, atoms tend to pair and their magnetic moments to cancel. The critical factor in the case of Fe, Co, and Ni is their atomic size, which permits this ordering.

9.3.3 *B–H* Hysteresis (Magnetization) Curve for Ferromagnetic Material

In our earlier discussion of permeability, we used the terms *magnetic flux density (B)* and *magnetic field (H).* If we plot these two quantities, we arrive at a *B–H* or magnetization curve for a ferromagnetic material. Such curves are depicted in Figures 9-10 and 9-11 for soft- and hard-magnetic materials, to be discussed. Referring to Figure 9-10, an "unmag-

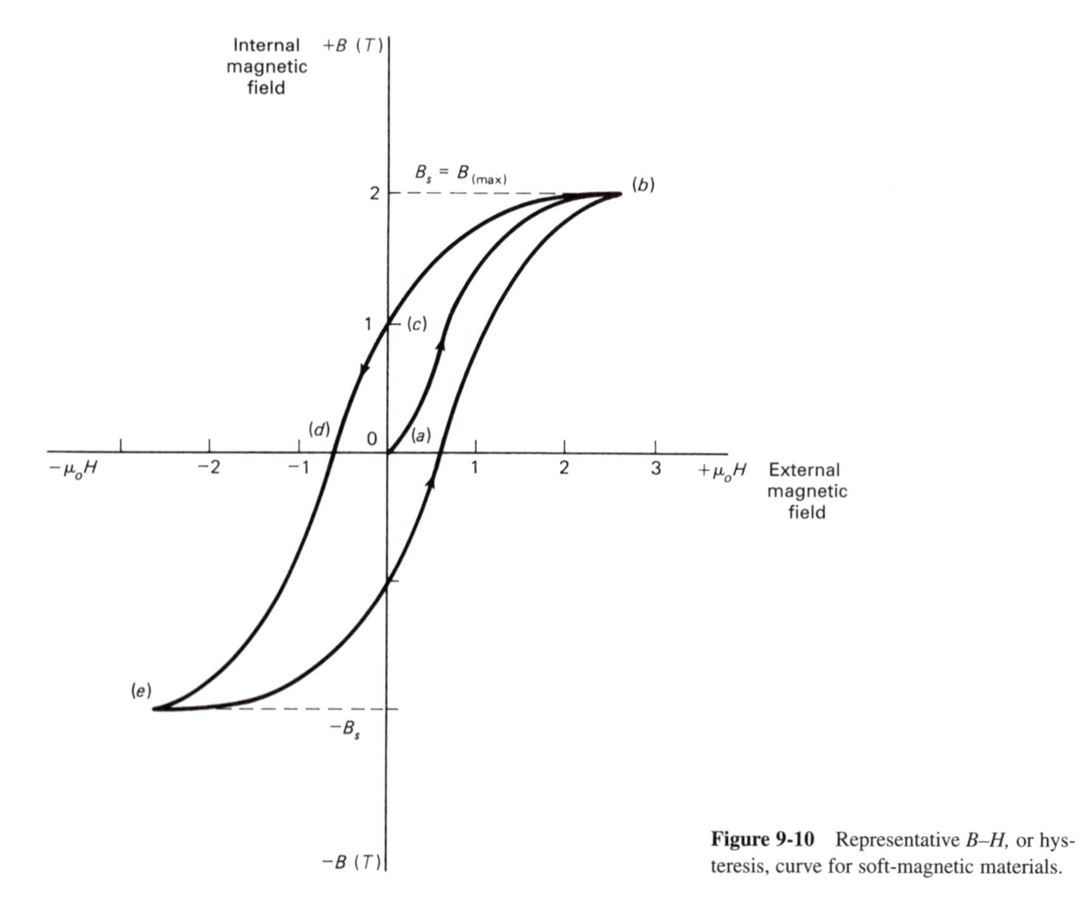

Figure 9-10 Representative *B–H,* or hysteresis, curve for soft-magnetic materials.

netized" material at point a begins to receive an increasing H as the external magnetic field is built up, possibly by use of an electrical coil winding as shown in Figure 9-7. Inside the material the domains are growing, as described previously, and reach point b where the maximum B_s (saturization) or $B_{(max)}$ is reached. At this point the material is essentially one large domain. Upon decreasing the external magnetic field *(H),* the domain tends to return to a balanced domain arrangement; but because of the many structural imperfections, such as crystal defects, grain boundaries, or inclusions, the balanced domain state is not reached. Such a situation can be compared to the plastic region of a stress–strain curve. Point c represents the remanent or remanent magnetization with $H = 0$. As the magnetic field increases in the opposite direction, that is, H increases in the negative direction $(-H)$, point d is reached, $B = 0$. The corresponding value of H at point d is known as the *coercive field* or *force.* This value of H is needed to demagnetize the material (balanced domains). Most domains can be turned around at this point. Continuing to increase the field in the negative direction produces a similar domain growth, as described previously, but in the opposite direction $(-B_s)$. Reversing the direction of the field a second time produces a curve that will plot out a *hysteresis loop.* Notice that the curve did not pass through the origin, point a in this cycle. The fact that these curves did not retrace themselves on the same path is called ***hysteresis.*** In such a cycle, much energy is transformed into thermal energy

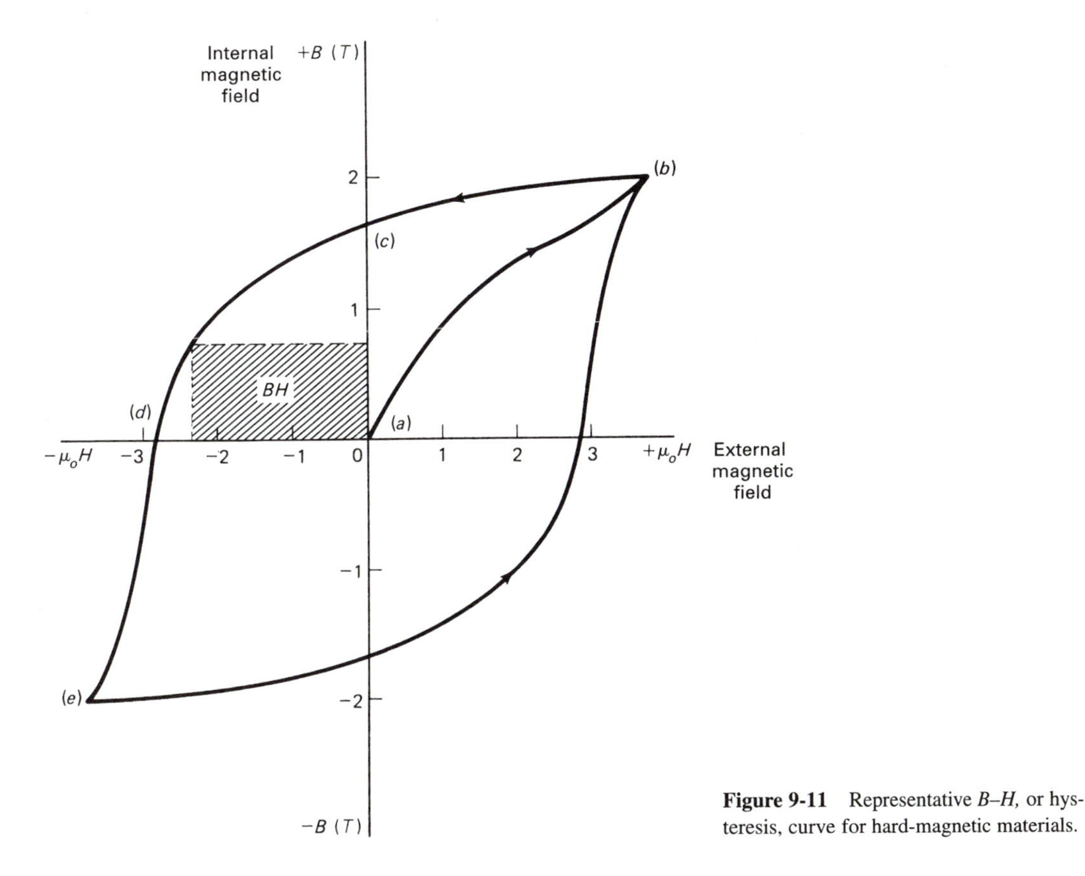

Figure 9-11 Representative B–H, or hysteresis, curve for hard-magnetic materials.

(friction) due to the reordering of the domains. This energy loss is proportional to the area of the hysteresis loop.

9.3.4 Soft-Magnetic Material

Using Figure 9-10 again and treating it as a representative hysteresis curve for a soft-magnetic material, note that for small values of *H,* a relatively large value of *B* is produced. Second, note that relatively little remanent magnetism needs to be canceled. Third, the area enclosed by the loop is relatively small. Such a material would find use as a transformer core. A ferromagnetic material can be demagnetized by reversing the magnetic field repeatedly while decreasing its magnitude. Soft-magnetic material is used for electromagnets since the field can be switched off, reversing the field with little loss of energy. Tape recorder heads are demagnetized in this way.

9.3.5 Hard-Magnetic Material

Figure 9-11 is a typical hysteresis curve for a hard-, or permanent-, magnetic material. Comparing this curve with the similar curve for a soft-magnetic material in Figure 9-10 and assuming that the scales for the axes are the same for both curves, several major differences can be observed. Hard-magnetic materials gain their magnetization in manufacture. Their essential property is their resistance to demagnetization. In other words, they must retain their magnetization when the external magnetic field is removed. The relatively large coercive field then is characteristic of this property, as well as the large hysteresis loop. Note the differences in the values of points *c* and *d* in both figures. The area of the largest rectangle that can be inscribed in the second quadrant of the hysteresis loop in Figure 9-11, called the *B–H product,* is a measure of the power of a permanent magnet. Table 9-2 is a typical listing of magnetic properties for some nickel alloy steels.

Time and space limit discussion of past and ongoing research and development efforts in the field of new materials (metals, ceramics, and composites) to meet the ever-increasing needs of our technological society. This treatment of magnetic materials will permit a greater understanding and appreciation of the breakthroughs in this area as they are made public in the future.

9.3.6 Smart Electronic Materials: Electrorheological and Magnetorheological Materials

The word ***rheology*** literally means "a knowledge of flow." Today rheology is an interdisciplinary science that, in the main, deals with all aspects of deformation of materials under the influence of an external force (load, electrical or magnetic field, etc.) In fact recent attention in the field concentrates on materials that can be classified as neither solid nor liquid. This intermediate state of matter between the solid and fluid states is represented by many rheological materials such as biological fluids, polymer melts, solutions, emulsions, and pastes. Rheology uses terms such as *elasticity, viscosity, solvents, solutes, gels,* and *suspensions,* many of which have been described previously. Two examples of such "smart" materials and their potential for future applications are *electrorheological* and *magnetorheological* materials, which were introduced in Module 6.

TABLE 9-2 MAGNETIC PROPERTIES OF NICKEL ALLOY STEELS[a]

Steel[b]		Temperature		Initial permeability,	Maximum	Remanence, B	Coercivity, H	Hysteresis loss
Type	Condition	°C	°F	μ_i	permeability, μ_{max}	(gauss)	(oersteds)	(ergs/cm^3/cycle)
1015	NT	20	68	190	1590 (H = 4.7[c])	11,400	4.0	2.7×10^4
(0.10 Ni)		93	200	260	1680 (H = 4.2)	11,100	4.0	2.3×10^4
		165	329	260	1850 (H = 3.9)	10,250	3.8	1.9×10^4
		300	572	290	1900 (H = 3.3)	8,500	2.9	1.5×10^4
9 Ni	NNT	23	73	140	700 (H = 13.5)	12,500	10.8	5.7×10^4
(8.56 Ni)		104	219	130	740 (H = 12.7)	12,050	10.2	5.2×10^4
		193	379	130	850 (H = 9.5)	11,700	8.9	4.3×10^4
		308	586	220	1000 (H = 8.6)	10,900	7.2	3.4×10^4
4340	NT	25	77	43	330 (H = 30)	13,700	22.5	11.4×10^4
(1.88 Ni)		114	237	34	440 (H = 25)	12,700	13.9	7.6×10^4
		167	333	40	420 (H = 28)	12,000	18.0	7.9×10^4
		302	576	85	460 (H = 23)	10,250	11.0	4.9×10^4
4340	QT	27	81	64	535 (H = 22)	13,700	18.0	8.6×10^4
(1.90 Ni)		87	189	64	510 (H = 23)	12,700	17.0	8.0×10^4
		157	315	80	490 (H = 24)	12,300	17.5	7.8×10^4
		288	550	82	500 (H = 20)	10,800	16.0	6.1×10^4

Source: Courtesy of Inco.

[a]These data are based on the initial magnetization curve and the magnetic hysteresis loop for applied field strengths (magnetizing forces) up to about 90 oersteds.

[b]N, normalized; NN, double normalized; Q, quenched; T, tempered.

[c]Oersted.

Electrorheological fluids change their flow characteristics when subjected to an electric field. The response, almost instantaneous, to such a field is a progressive gelling of the fluid, which is proportional to the field strength. These materials are composed of two components, a carrier fluid and the suspended particles in the fluid. The fluid does not conduct electricity. The suspended particles are typically polymers, minerals, and ceramics, which may take the form of suspended or dissolved particles or molecules. When an electrical field is applied across the fluid, the particles are polarized and link themselves together. (See Sections 2.1.1.2 and 9.2.2 for discussions of polarized molecules and polarization.) This linking produces a gelling of the carrier fluid, inhibiting its flow. Once the applied field is removed, the polarization of the particles ceases and the fluid can then flow freely (see Figure 6-30). Many future applications are envisioned in the development of automatic transmissions, brakes, clutches, hydraulic valves, and actuator devices.

A second example is ***magnetorheological materials.*** Similar in many respects to electrorheological materials, these materials use magnetic particles suspended in a nonmagnetic fluid (the host fluid). The host fluid could be a light machine oil or vegetable oil. A simple system such as one with iron filings suspended in a vegetable oil and a horseshoe magnet could amply demonstrate the principles of magnetorheological behavior. When a magnetic field is applied, the magnetically susceptible particles in the nonmagnetic fluid align themselves and remain stationary, thus restricting the movement of the suspension fluid. This situation illustrates that the viscosity of these materials is a function of the magnetic field in which they are immersed. The greater the field, the more solidified is the slurry. Applications for such materials are as varied and numerous as they are for electrorheological materials, especially for controlling fluid flow as in hydraulic braking. A recent application for such fluids is in controlling the suspension of truck seats by automatically adjusting the seat to changing driving conditions. During routine driving, a height sensor notes the downward movement of the seat and sends a signal to a microprocessor, which activates an electromagnet. The electromagnet quickly stiffens the fluid in a liquid state in a shock absorber to a jellylike near-solid. The fluid is silicon-based hydraulic oil plus an additive that permits a greater suspension of iron filings.

9.4 OPTICAL PROPERTIES

9.4.1 Absorption, Refraction, Reflection

Light is energy in the form of radiation. It is regarded as being composed of either photons in the energy range (2.5 to 5.6×10^{-19} J) or electromagnetic waves in the wavelength range of 800 nm down to 380 nm (see Figure 9-12). Light, being a form of electromagnetic radiation, interacts with the electronic structure of atoms of a material. The initial interaction is one of absorption; that is, the electrons of atoms on the surface of a material will absorb the energy of the colliding photons of light and move to the higher-energy states. The degree of absorption depends, among other things, on the number of free electrons capable of receiving this photon energy. The electrons can then do several things. They can have sufficient energy to jump into higher-energy states, which permits them to be accelerated within an electric field and thereby conduct electricity (photoconductive ef-

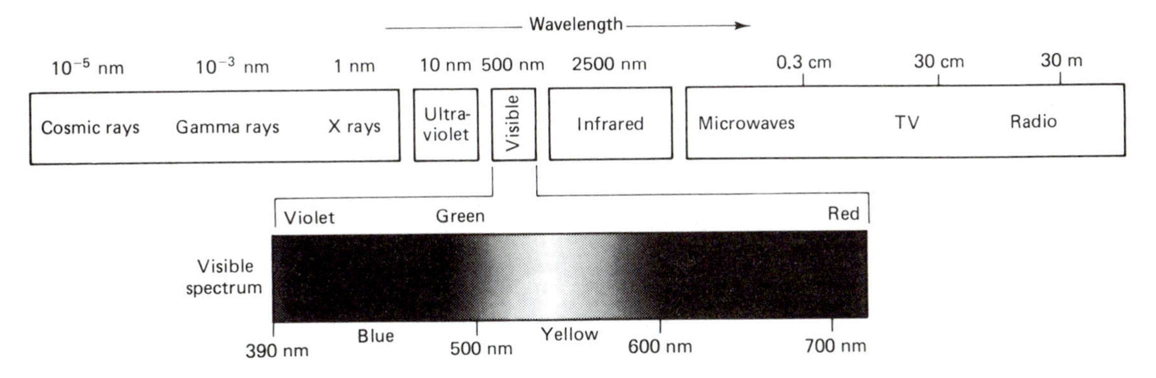

Figure 9-12 Wavelengths of electromagnetic radiation characteristic of various regions of the electromagnetic spectrum. (Theodore L. Brown and H. Eugene LeMay Jr., *Chemistry,* 2nd ed., Prentice-Hall, Inc., Englewood Cliffs, N.J., 1981.)

fect). They can collide with atoms and release their excess energy in some form of electromagnetic radiation, or they can convert their excess energies to atoms in the form of thermal energy.

As light enters a nonmetallic material, it is absorbed (Figure 9-13) by the action of several mechanisms, such as electronic polarization (Figure 9-5) and electron transitions. The electron transitions depend on the electron energy band structure of the material. The interaction process characteristic of photons depends on their energy (E_ρ). Low-energy photons interact principally by ionization or excitation of the outer orbitals in a solid's atoms. Light is composed of low-energy photons (E_ρ is less than 10 eV) represented by infrared (IR), visible light, and ultraviolet (UV) in the electromagnetic spectrum. (See Figure 9-12.) UV light is being tested for curing solvent-free urethane paint on autobodies as a technique to eliminate the release of VOCs into the atmosphere from the spraying as well as the costly baking process newly painted autobodies must undergo. High-energy protons (E_ρ greater than 10^4 eV) are produced by x-rays and gamma rays (Figure 9-14). The minimum energy (E_ρ) required to excite and/or ionize the component atoms of a solid is called the absorption edge or threshold.

When crystalline materials are exposed to radiation, the change in their properties is due to a displacement of atoms from their normal locations in the crystal structure, i.e., the creation of crystal defects. The total energy of the incident radiation is measured in terms of dose, the amount of radiation affecting the material, or the dose rate, which is the rate of energy deposition. Irradiation conditions are often given in terms of the total number of particles incident on the sample, i.e., the ***fluence*** (fluence equals the summation of flux, or the particles per unit time deposited, over irradiation time) [see Figure 9-13(c)]. A greater understanding of this topic requires a study of the mechanisms by which energy is transferred from the radiation (in this case, light) to the atoms of a solid. In the case of metals, only the energy transferred to the nuclei by the beam of energetic, charged particles such as electrons, photons, and fission fragments is effective in creating lattice defects. The crystalline structure can also be converted to an amorphous one by sufficient exposure to

(a)

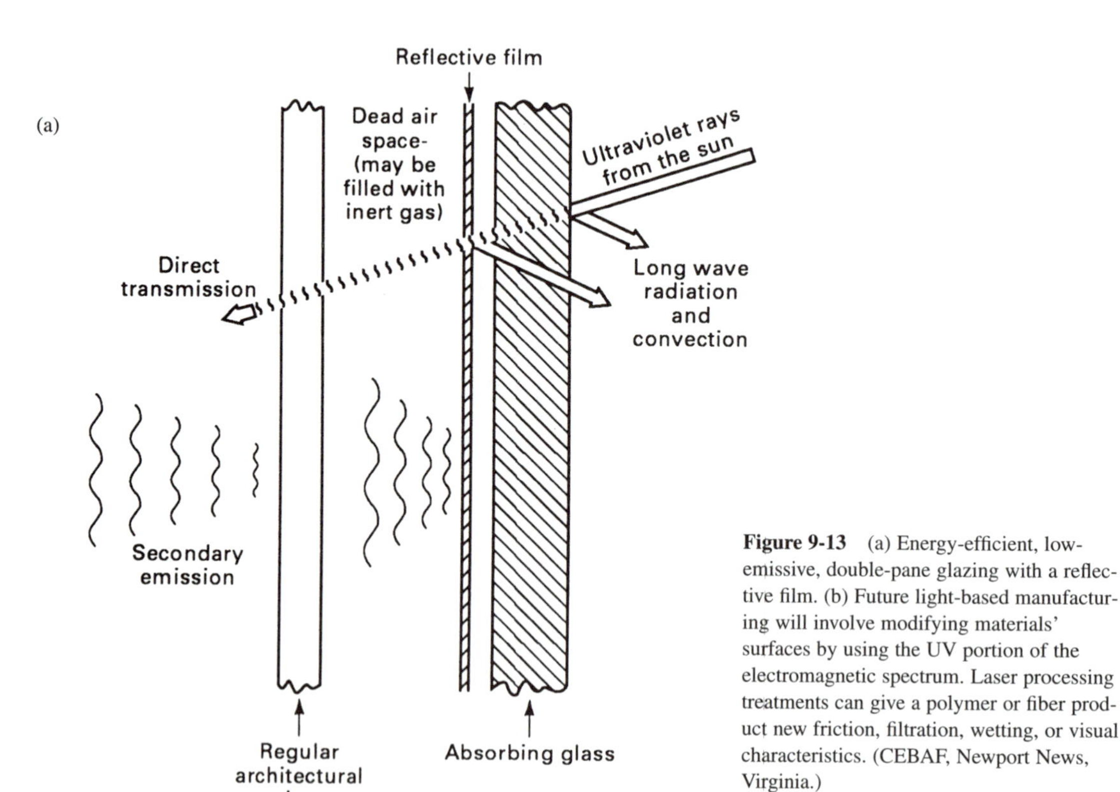

Figure 9-13 (a) Energy-efficient, low-emissive, double-pane glazing with a reflective film. (b) Future light-based manufacturing will involve modifying materials' surfaces by using the UV portion of the electromagnetic spectrum. Laser processing treatments can give a polymer or fiber product new friction, filtration, wetting, or visual characteristics. (CEBAF, Newport News, Virginia.)

(b)

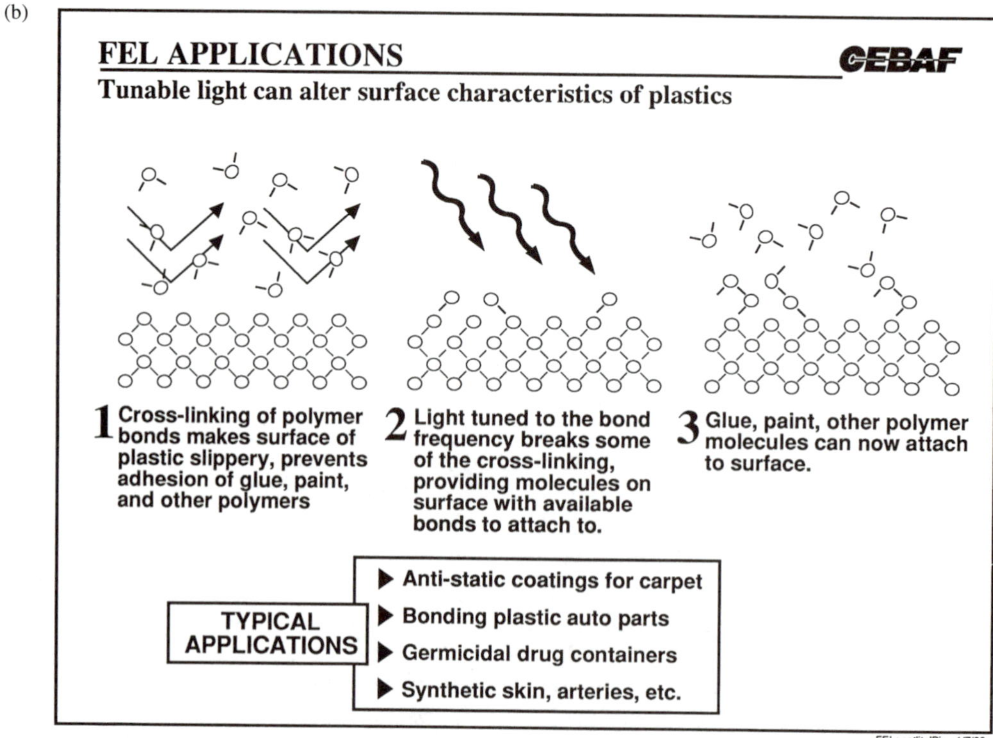

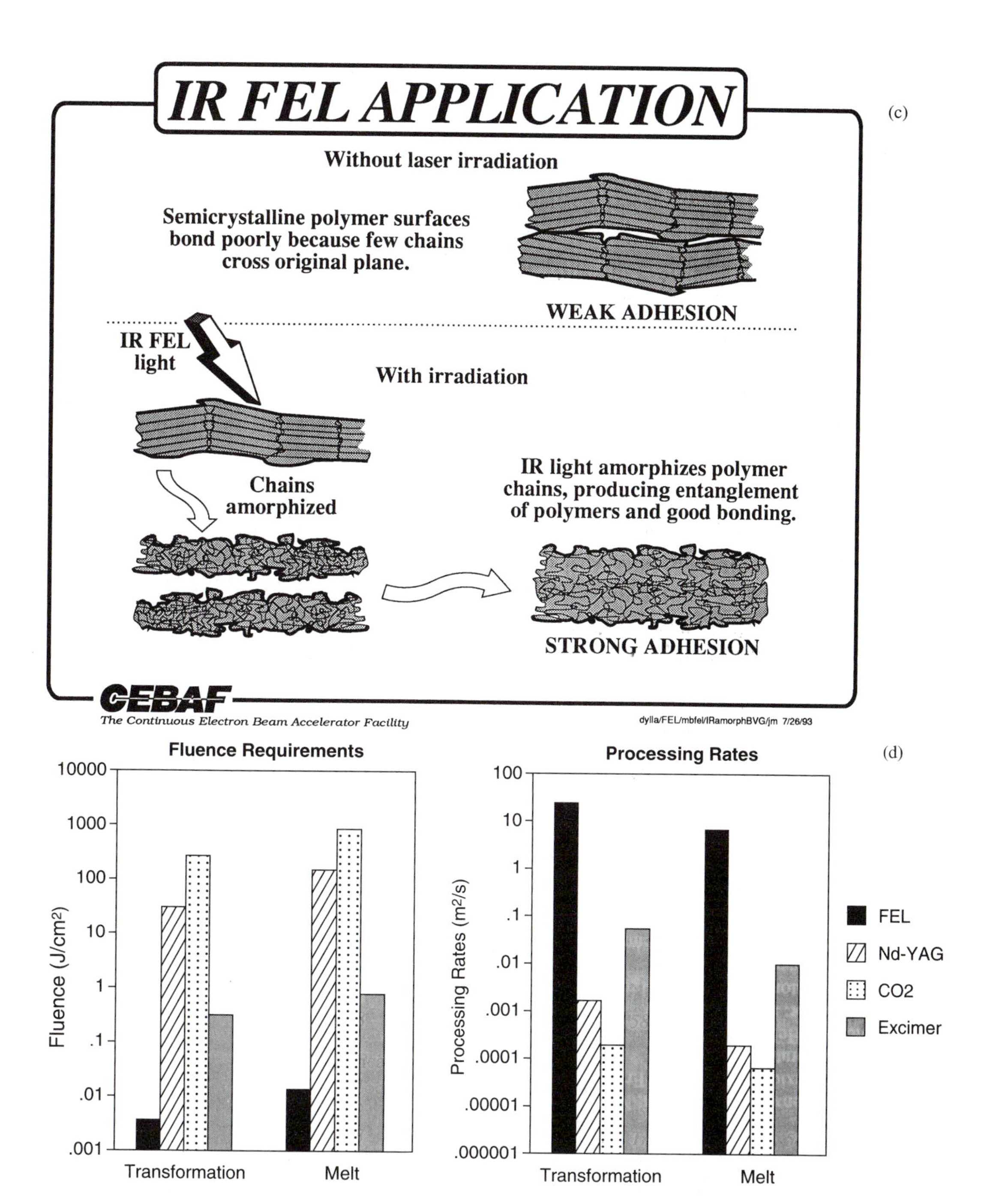

Figure 9-13 (c) Surface modification by light can add value to existing products, serve as the value basis for new products, or be incorporated into special interfaces of packaging materials or structural composites. In this application, surface amorphization (carefully tailored light energy absorption in the transformed region in the surface layers of a polymeric material) produces greater adhesion, with the additional benefit of eliminating the environmental costs of processing with wet chemistry. (CEBAF, Newport News, Virginia.) (d) Calculated laser fluence and processing rates for metal surface processing with conventional lasers and the UV demonstration FEL. Large-scale exploitation of UV surface processing will require sources of at least a few tens of kilowatts, light that costs under a cent per kilojoule, and a full wavelength tunability. (CEBAF, Newport News, Virginia.)

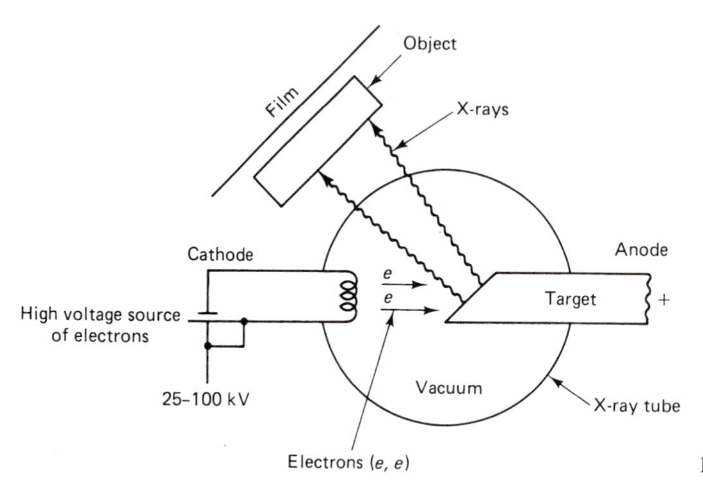

Figure 9-14 Schematic of an x-ray tube.

radiation. In metals, the band gap is zero and photons of all energies can be absorbed to increase the kinetic energy of the conducting electrons and holes. In polymers (covalently bonded materials), photons of suitable energy are capable of exciting vibrations directly in both molecules and crystalline structures. The chemical bonding can be altered by electronic excitation, which can also create cross-links and break primary bonds.

When light is transmitted into the interior of a transparent solid, it experiences a decrease in its velocity. The result is the light bends as shown in Figure 9-13a. This phenomenon is known as ***refraction.*** The ***index of refraction*** (n) of a solid material depends on the wavelength of the incident light. For air, the index is nearly equal to 1, whereas water has a higher index of refraction, 1.33. Other indices of refraction for some materials are as follows: silica glass, 1.46; corundum, 1.76; polyethylene, 1.51; and polypropylene, 1.49. Lucite (Plexiglas) has an index of 1.51 and diamond's is 2.42. When light passes from one medium to another having a different index of refraction, some of the light is scattered at the interface between the two media. Both of these media could be transparent materials. The term ***reflexivity*** (R) is the ratio of the intensity (I) of the reflected light to the intensity of the incident light (I_R/I_0) that is reflected at the interface (see Figure 9-15).

9.4.2 Photoelectric Effect, Photons, Emissivity, and Lasers

Certain metallic materials will become charged positively when exposed to electromagnetic radiation (Figure 9-12). Because no electron charge can be carried by such radiation, the conclusion reached is that the radiation must interact with the electrons attached to the metallic atoms and cause the ejection of electrons from the surface of the metal. Such a phenomenon is known as the ***photoelectric effect.*** The ejected electrons are called ***photoelectrons.*** Furthermore, not all electromagnetic radiation produces this effect. Only when a certain frequency of radiation is reached ***(threshold frequency)*** does this effect occur. The threshold frequencies differ with different metallic elements. As an example, the metal sodium will produce this effect once the frequency of the incident radiation reaches 5.6×10^{14} Hz. The corresponding wavelength of 5.4×10^{-7} m places such radiation in the visi-

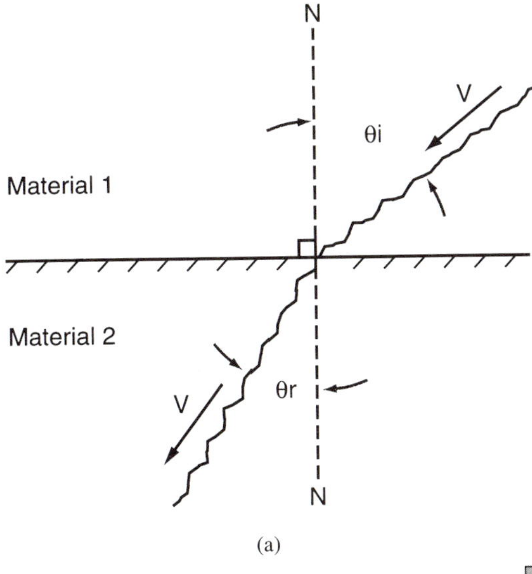

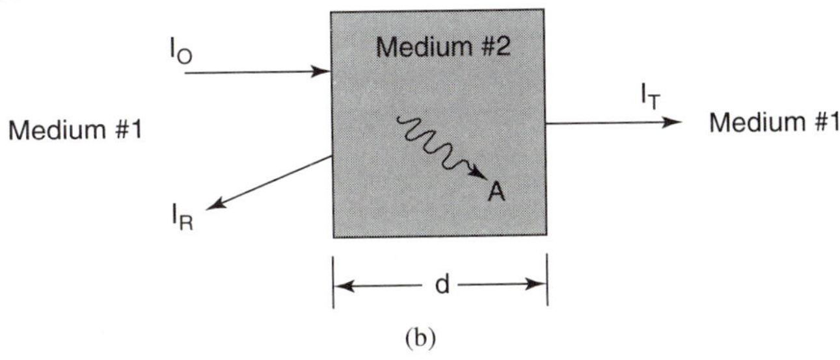

Figure 9-15 (a) Refraction of light, showing the angle of incidence (θ_i) and angle of refraction (θ_r), both measured from a common normal *N*. (b) The transmission of light through a transparent medium of thickness *d* is shown. I_0 is the incident light, I_R is the reflected light, and I_T represents the transmitted light leaving the material. The absorbed light in medium #2 is shown by a wiggly arrow labeled A. Both versions of medium #1 are the same material.

ble-light portion of the electromagnetic spectrum. Zinc requires a threshold frequency of 8.0×10^{14} Hz, which is in the ultraviolet region of the spectrum. The intensity of the radiation plays no part in producing the photoelectric effect.

To explain this phenomenon adequately, Einstein proposed in 1905 that this radiation be treated not as a series of waves (wave phenomenon) but as a series of particles (light photons) that interact with single electrons. Most of the questions dealing with light radiation confine themselves to large-scale bodies such as glass lenses. The films of materials taken by x-rays and the electron microscope are some examples of radiation that exhibits wavelike properties. The pattern of electrons that projects itself onto the x-ray film and the highly magnified images of the internal structure of a material as viewed under an electron microscope are diffraction patterns of waves typical of those exhibited by the diffraction of light through a grating. When we begin to probe into the land of the atom with its subatomic electrons, we use the particle nature of radiation to explain the results of such interactions of radiation, not with a whole surface but with individual electrons.

A ***photon*** of light (a quantum of electromagnetic radiation) exchanges its energy with a single valence or conduction electron. We have learned that such electrons possess the highest kinetic energies and are the outermost electrons in an atom. When a photon of light with sufficient energy ($E = hc/\lambda$) strikes a valence electron, it causes the excited electron to leave a solid metal. Note that the photon reacts with a single electron, not with an entire atom or a solid metal electrode. The kinetic energy (KE) of the ejected electron (now called a photoelectron) can be expressed in two terms. The first term, the work function *(W)*, is that part of a photon's energy (work) expended in digging out or removing the valence electron from the surface of the metal. The second term represents the remainder of the photon's energy appearing in the form of kinetic energy of the escaping electron. The photon disappears, having done its job of transferring its energy, and a photoelectron is emitted from the surface. The electrons in a metal that are eligible to receive energy from colliding photons may be in various energy levels of an atom. Those with the highest energies will require the least energy expenditure to break free from the surface; hence, they will have the greatest kinetic energy. Using electron volts (eV) as units, this energy transfer can be expressed mathematically as

$$\text{KE} = E - W$$

Suppose that violet light is incident on a piece of cesium (Cs). Reference books can be consulted to find the photoelectric properties of metallic elements. Referring to such sources, we find the threshold frequency for Cs to be 4.6×10^{14} Hz, the work function W is 1.9 eV, and the wavelength (λ) of violet light is 4×10^{-7} m. To find the frequency f of violet light, we use

$$f = \frac{c}{\lambda} = \frac{3 \times 10^{8}\ \text{m/s}}{4 \times 10^{-7}\ \text{m}} = 7.5 \times 10^{14}\ \text{Hz}$$

Note this frequency exceeds the threshold frequency for Cs. The photon's energy E is next calculated using

$$E = \hbar f$$

where $\hbar$ is Planck's constant (6.625×10^{-34} J · s) and 1 eV $= 1.60 \times 10^{-19}$ J. Thus,

$$E = \frac{(6.625 \times 10^{-34}\ \text{J} \cdot \text{s})(7.5 \times 10^{14}\ \text{Hz})}{1.6 \times 10^{-19}\ \text{J}} = 3.1\ \text{eV}$$

The kinetic energy (KE) of the photoelectron emitted can then be determined:

$$\begin{aligned} \text{KE} &= E - W \\ &= 3.1 - 1.9 = 1.2\ \text{eV} \end{aligned}$$

When a surface is exposed to radiation, it can absorb part or all of the incident radiation (energy). The fraction of the energy absorbed is the emissivity *(e)* of the surface. For a blackbody, the perfect absorber, $e = 1$; for the perfect reflector, a white body, $e = 0$.

Thermal-barrier coatings on metal components of gas turbines have low thermal conductivities and low emissivity. Materials such as ZrO_2 applied as a spray are used for such coatings. These coatings have the properties noted above and the desired property of having a coefficient of thermal expansion similar to that of the base metal, which allows the coating to adhere to the metal as the metal undergoes changes in temperature. In energy-efficient, low-emissivity window glass, a much greater percentage of the sun's rays (radiation heat) are reflected than from older types of window glass (Figure 9-13a).

By synchronizing the light emission from many electrons, one can produce light that is monochromatic (one frequency) and coherent (photons are in phase with each other). More and more materials are being found whose electromagnetic radiation can be controlled by technicians. These include solids, liquids, and gases. Semiconductor materials can produce laser action that emits photons with wavelengths in the range 7 to 8×10^{-7} m. A ***laser*** (*l*ight *a*mplification by *s*timulated *e*mission of *r*adiation) is a special example of the emission phenomenon we have been discussing. The light produced is further amplified through stimulating the emission. A ***maser*** (*m*icrowave *a*mplification by *s*timulated *e*mission of *r*adiation) differs from lasers mainly in the wavelength of the radiation produced.

The Laser Processing Consortium, a collaboration involving nine U.S. corporations and companies, seven research universities, and the Continuous Electron Beam Accelerator Facility (CEBAF) in Newport News, Virginia, has proposed to develop a profitable, production-scale capability to use laser light in manufacturing. It involves, among other things, developing a cost-effective, high, average-power ***free-electron laser*** (FEL) that would deliver light at wavelengths fully adjustable across the IR (Figure 9-12) and UV portions of the spectrum. This laser, a nonintrusive processing tool, would fundamentally improve industry's abilities to (1) modify polymer film, fiber, and composite surfaces [see Figure 9-13(b) and (c)]; (2) process metal surfaces and electronic materials; (3) micromachine or surface-finish metals, ceramics, semiconductors, and polymers; (4) evaluate materials nondestructively; and (5) monitor manufacturing processes. Conventional lasers suffer from limitations in cost, power, and choice of wavelength. Figure 9-13(d) compares the fluence (a measure of the irradiation on a surface) and the processing rates of commercial lasers with the UV demonstration FEL to be developed in Phase I of this project. Commercially available ***excimer lasers*** are limited to tenths of a kilowatt, tens of cents per kilojoule, and a few isolated specific wavelengths. In contrast, the free electron laser, with its superconducting radio-frequency electron accelerator operating in continuous wave with high average power on a kilowatt scale and full wavelength tunability, will be capable of much greater processing rates. If successful in this endeavor, ***light-based manufacturing*** will find increased opportunities to apply surface modification to products that are presently coated, plated, etched, ground, or burnished by wet-chemical or mechanical processes. Table 9-3 sets forth the status (level of maturity) of some representative industrial applications of free-electron lasers using the categories of commercializable, developable, and prospective.

The FEL represents another development in surface modification technology. ***Surface modification*** techniques are referenced throughout the book, particularly in reference to heat-treated surface techniques such as case hardening and nitriding, electroplating, diamond films, titanium coatings, plating, layered composites, chemical vapor deposition (CVD), sputtering, chemical vapor infiltration (CVI), toughening, surface engineering, and the like. Advances in this new technology will propel many improvements in materials, manufacturing processes, and superior product quality.

Many uses for these well-controlled sources of energy will continue to be discovered for boring holes, welding delicate parts of the human eye, detecting and measuring pollutants in the atmosphere, and carrying human communications through space with little or no interference. Present applications of the optical behavior of materials, including glass and plastics that produce light through the interaction of electromagnetic radiation with the

TABLE 9-3 REPRESENTATIVE INDUSTRIAL APPLICATIONS OF FELs

Level of maturity	Polymer surface processing	Micromachining	Electronic materials processing	Metal surface processing
Commercializable				
Demonstrated major market Cost/capacity barrier	Surface texturing Surface conductivity Surface amorphization	Fuel injectors Spinnerets Slitting coated films	Flat panel displays Large-area photovoltaics Large-area diamond coating	Laser glazing and annealing Surface carburizing and nitriding
Developable				
Demonstrated Development needed	Antimicrobial nylon Pulsed laser deposition	Microtexturing Subthreshold ablation	Pulsed laser deposition	Adhesive bond pretreatment Solvent-free cleaning Removal of corrosives
Prospective				
Preliminary results Research development needed Enabling facilities needed	Surface activation of carbon fibers for composites	Nanometer-scale surface contouring for chemical catalysis	High-density CD-ROMs Embedded layer silicon processing	Thermal barrier coatings (TBCs) Functionally gradient coatings Metglas coatings Shape memory alloys (SMAs)

material's electrons, are numerous. Some common examples are worth mentioning. The coating of TV screens with zinc sulfide (ZnS), a semiconductor material on which electrons act to produce light, is one. A second is the phototube that "reads" the sound track on motion-picture film, ultimately producing an electrical signal that can be amplified and broadcast to an audience. Solar batteries using semiconductor materials and exposure meters carried by photographers for properly setting their cameras are additional examples of the many applications of such devices.

Luminescence is the reemission of photon energy at wavelengths in the visible spectrum as the result of the absorption of electromagnetic radiation from some outside source. If reemission occurs at the same time that the material is absorbing the radiation, the phenomenon is called ***fluorescence.*** In a fluorescent lamp, electrons emitted by the incandescent cathodes collide with electrons of the mercury atoms that fill the tube. The collisions cause the emission of radiation in the invisible ultraviolet range. This radiation, in turn, strikes the fluorescent or phosphor material coating the inner side of the tube and causes this material to emit radiation in the visible range of the spectrum.

Should the light emitted continue after the radiation producing it has been removed, the light is called ***phosphorescence.*** The color TV picture tube uses phosphors that phosphoresce.

9.4.3 X-Rays

X-rays, discovered by Roentgen in 1895, can be produced in two ways. The first process uses a high-energy electron capable of knocking an electron from an inner energy level of an atom completely out of the atom. This removal of an electron, say, from the *K* level, immediately creates a hole into which an *L*-shell electron can fall. As an *L*-shell electron falls back to the lower-energy *K* level, the sudden decrease in potential energy is emitted as a photon of electromagnetic energy characteristic of x-rays (atomic radiation). A hole is produced in the *L* shell, which calls for the transition of another electron from an outer level to the *L* shell, which produces an additional x-ray. Several x-rays are emitted as electrons cascade down to fill the lower-energy-level vacancies, or holes, until eventually the atom captures an electron from the surrounding region and changes from an ion to a neutral atom in its lowest equilibrium condition. The wavelength of the x-ray depends on the particular energy levels involved, as do the colors in visible light emitted from a given jump. X-rays lie in the wavelength range between 1×10^{-12} and 1×10^{-8} m. Another process of generating x-rays (also called *cathode rays*) is through the sudden braking (deceleration) of high-energy electrons by striking them on a metal target, which results in a conversion of part of the electron's kinetic energy into a quantum of electromagnetic radiation or x-ray. Such a technique is shown schematically in Figure 9-14.

9.4.4 Thermionic Emission

Another way to exchange energy with the free electrons in a solid metal is to heat the metal to such an extent that the thermal energy of the electrons is sufficient to emit them from the surface. This is known as ***thermionic emission.*** A typical low-power vacuum diode has a cathode consisting of a nickel tube coated with barium oxide, which is heated by an insulated filament contained within the tube.

9.4.5 Transmission

Optical devices possess many desirable properties, but the most important is their ability to transmit light. One measure of this ability is ***transmittance,*** defined as the percentage of an incident light ray remaining after passing through 10 mm of a material. It can also be expressed as a ratio of transmitted-light intensity to the incident-light intensity (***intensity*** is the measure of the strength of a light source). Some materials will not let light pass through ***(opaque).*** Metals are opaque. When white light is incident on a metal surface, most of the light is reflected, and some is refracted into the surface of the metal where the photons strike the electrons, thus transferring discrete amounts of their energy to the electrons and raising them to higher energy levels. With copper, the 3*d* electrons absorb the energy of the photons corresponding to blue wavelengths through collisions, and reflect the others, which give a reddish color. The energy absorbed causes the atoms concerned to increase their energy levels, which allows them to increase their vibrations, which, in turn, results in an increase in temperature, an indicator of heat.

In covalent and ionic solids, the electrons are bound to the atoms by the very nature of the covalent and ionic bonds. It takes more energy to break such bonds and free the electrons with increased energies sufficient to move them to higher energy levels. Therefore, visible light passes through such solids without reacting with the atom's electrons. Such solids are termed ***transparent.*** If impurities are present in such a solid, the photons of the visible light could react with the electrons of the impurity atoms, which are not involved in covalent or ionic bonding. In a ruby, the chromium atoms absorb the photons of blue and green light, but allow the remaining photons to pass through, which are mostly of wavelengths that correspond to red light.

Many polymer materials are transparent. Others are ***translucent***; that is, these materials, because of their structure, allow some light to pass through, but most of the light is scattered due to the reflection and refraction of the light as it transmits across phase boundaries and interacts with pores in the structures. Research in new ceramic materials allows the production of single-phase, pore-free ceramics that are transparent.

The bending of light rays is due to the fact that the speed of light within a dense medium is less than in air. Many of us have the experience of trying to pick up an object in a shallow pool of water, only to find that the object's actual location is displaced from its apparent location.

The ratio of the speed of light in air (c) to its speed in a medium (v_m) is called the ***index of refraction*** of the medium (n) or

$$n = \frac{c}{v_m}$$

Illustrative Problem

Given the index of refraction in a medium of 1.5, how fast would ordinary light travel in this medium? What material has an index with this value? (See below.)

Solution $n = c/v_m = (3 \times 10^8 \text{ m/s})/1.5 = 2 \times 10^8$ m/s; polyethylene.

The index is related to polarization in a material by the expression $n = \kappa^{1/2}$, where κ is the dielectric constant. This may vary with the change in density of materials as well as in certain crystal directions (i.e., n is anisotropic). The index also varies with frequency. Using a wavelength of 589 nm, some selected indices are: diamond, 2.4; polyethylene, 1.5; water, 1.33; and air, 1.0.

Another way in which the index may be defined is in terms of the angles of incidence (θ_i) and refraction (θ_r). Figure 9-15 is a sketch showing how these angles are measured from a common normal.

$$n = \frac{\sin \theta_i}{\sin \theta_r}$$

Illustrative Problem

Visible light traveling in air enters a medium with an index of refraction of 1.5 at an angle of incidence of 60°. Find the angle of refraction.

Solution

$$n = \frac{\sin \theta_i}{\sin \theta_r} \qquad \sin \theta_r = \frac{\sin \theta_i}{n}$$

$$\sin \theta_r = \frac{0.866}{1.5}$$

$$= 0.577$$

$$\theta_r = 35.3°$$

The refractive index n is for the medium the incident light penetrates into. Solids with dense atomic packing generally have higher indexes of refraction. In crystals the index of refraction is different for different directions, which can result in the incident rays being split into two polarized rays.

In explaining the phenomenon of ***polarized*** light, we use the concept of the electromagnetic wave. All electromagnetic radiation of whatever frequency consists of two fields, one electrical and the other magnetic. Figure 9-16 shows these two fields designated by the vectors **E** and **B.** Each field is separated by 90°; that is, for each **E** there is a corresponding **B** set up in a plane at right angles to each other. For simplicity, only two fields are shown. It must be remembered that any radiation is composed of many, many waves with **E** and **B** vectors representing the electronic and magnetic fields pointing in all directions. Using just the **E** field and Figure 9-17, if you position yourself as standing out on the x axis and looking back at the origin of our x–y–z axes system, the many **E** fields could be represented by the sketch. These **E** vectors could also be represented by their respective components in the y and z directions, thus simplifying the representation of these **E** vectors. A schematic of unpolarized light would then look like the sketch in Figure 9-18(a). This view of *unpolarized* light represents **E** vectors that oscillate back and forth along the y and z axes. If the radiation consists only of **E** vectors along only one axis, it is known as *polarized.* Polarization can be accomplished by the use of a filter that absorbs the electric field oscillations in one particular direction. A representative sketch of polarized light is shown in Figure 9-18(b). The light in this example is polarized in the y direction.

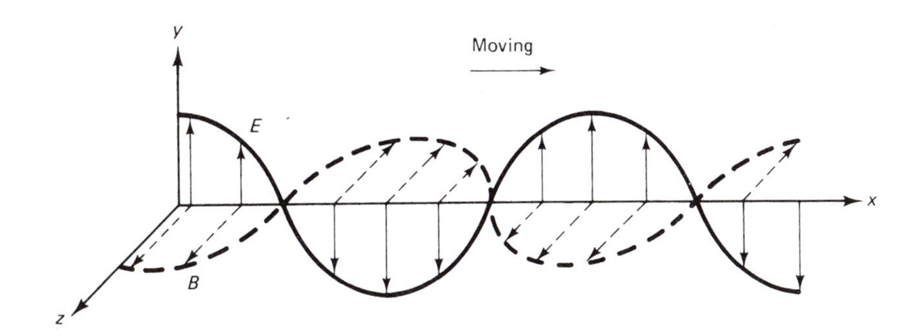

Figure 9-16 Electromagnetic fields in a light wave.

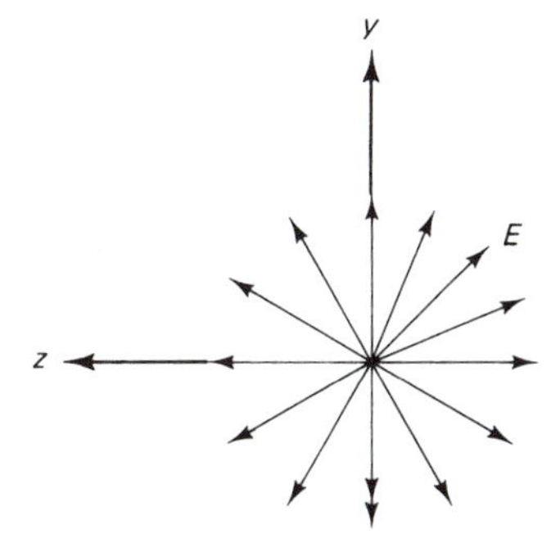

Figure 9-17 A point view of electric field vector **E** in an electromagnetic radiation of unpolarized light.

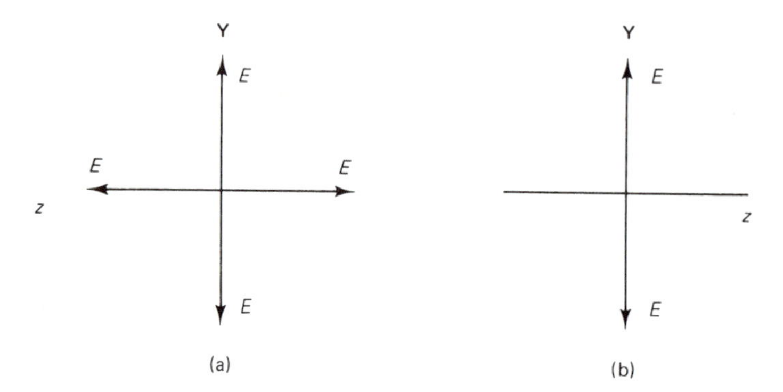

Figure 9-18 (a) Unpolarized light with **E** vector resolved into *y* and *z* axes. (b) Polarized light in *y* direction.

Unpolarized and polarized light can be further understood by comparing them to vibrating a rope. If you fix a rope at one end and vibrate the other end up and down in all directions, this would be a model for ordinary or unpolarized light. If you vibrate the rope in just one direction, say, the vertical direction (up and down), this would be representative of polarized light. Furthermore, if the same rope were pulled through the slots in a picket fence, the rope could be vibrated only in this up and down manner. The picket fence then could serve as a model of an optical filter (a polarizer) that permits light vibrations in only one direction and absorbs (prevents) other vibrations from being transmitted. When light is transmitted through material, the light can become polarized due to the differences in the material structures in different directions. A material whose structure varies with the orientation of the material is not isotropic (anisotropic). As you recall, isotropy is a quality of a material that has identical properties in all directions. The term ***optical anisotropy*** refers to a material having different optical properties in different directions.

Being insulators, most polymeric materials have no free electrons. In addition, their internal structure contains boundaries and pores that interact with light and cause it to be transmitted in different directions at different speeds, hence at different refraction indexes. Polarization of light is used in many applications, such as liquid crystal displays (LCDs), Polaroid sunglasses, and photoelasticity, where polarized light is used to reveal the amount of anisotropy in a transparent-polymer (polystyrene) model of a loaded structure (see Figure 9-19). When subjected to stresses ($\sigma = P/A$), certain transparent materials become ***birefringent*** or double refracting. These substances will divide an incident ray of light into two beams that travel through the material at different speeds. In particular, these two beams are polarized at right angles to each other (see Figure 9-18).

This material property of birefringence is taken advantage of by constructing models of structures such as buildings using transparent materials having double-refraction properties. The models are subjected to forces (strains) and the internal stresses produced are analyzed by translating the stress fringes or isochromatics into stress magnitudes. An ***isochromatic*** appearing as a dark band (see Figure 9-19) represents a region or locus of constant difference in principal stress. A pattern of alternate bright and dark lines represents the variation in refraction produced in the model. A variation of this technique is to coat an actual structure with a photoelastic coating such as Photostress. When the part is

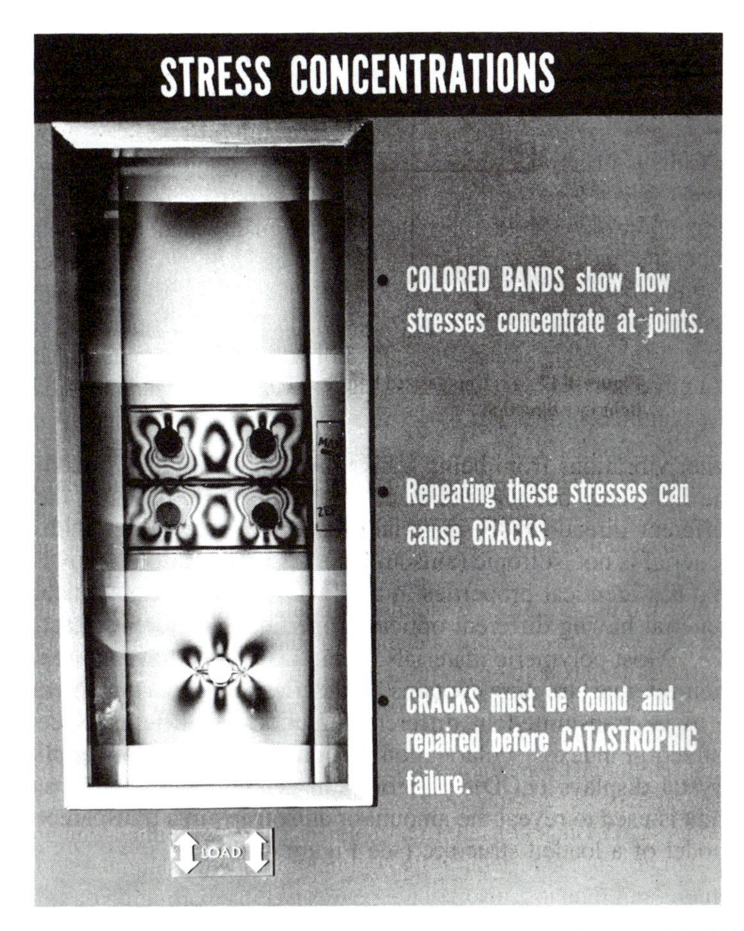

Figure 9-19 Stress concentrations revealed in loaded transparent-polymer models. (NASA.)

subjected to loads, the strains are transmitted to the coating, which then becomes birefringent.

Table 9-4, specifications for Sunglas, lists typical specifications for architectural flat glass for applications such as glazing in windows and doors. This table serves to illustrate concepts discussed under optical properties. In addition to good optical qualities, strength, and weathering resistance, additional requirements imposed on flat glass mandate sophisticated industrial treatments that do not involve changing the composition of the glass. Some of these additional requirements include heat reduction, light and sound protection, and increased resistance to breakage. One example of such treatment is the atoms-thick coating, known as low-emissive coating, which provides an emissivity of approximately 0.14 (83% better than a normal glass surface). This low-emissive coating inhibits the absorptivity of the glass surface, resulting in more of the long-wave heat energy being reflected (see Figure 9-13a). All such glass must meet rigid ASTM test standards E773-83, E774-81, and E6,P-1.

TABLE 9-4 SUNGLAS SPECIFICATIONS[a]

Product	U-Value[b]/R value		Shading coefficient	Relative heat gain[c] (Btu/hr · ft²)	Visible light transmitted (%)	Total solar energy (%)			Ultraviolet transmitted (%)
	Winter nighttime	Summer daytime				Transmitted	Reflected	Blocked[d]	
Single glazed									
Single strength									
$^3/_{32}$ in.	1.16/0.86	1.08/0.93	0.87	189	86	70	7	24	65
Double strength									
$^1/_8$ in.	1.16/0.86	1.09/0.92	0.82	180	84	65	6	29	60
$^3/_{16}$ in.	1.15/0.87	1.10/0.91	0.74	163	79	55	6	36	52
$^1/_4$ in.	1.13/0.88	1.10/0.91	0.69	154	77	49	6	40	45
Double glazed[e]									
Single strength									
$^3/_{32}$ in.	0.58/1.72	0.62/1.61	0.76	161	78	62	10	34	54
Double strength									
$^1/_8$ in.	0.58/1.72	0.62/1.61	0.71	150	76	56	9	38	49
$^3/_{16}$ in.	0.49/2.04	0.56/1.79	0.62	133	71	45	8	46	40
$^1/_4$ in.	0.49/2.04	0.56/1.79	0.56	121	68	39	8	51	33

Source: Ford Motor Company, Glass Division.

[a]All values subject to manufacturing tolerance within applicable provisions of Federal Specification DD-G-451d 1977.

[b]*U*-value is the overall coefficient of heat transmission or thermal transmittance (air to air) in Btu/hr · ft² · °F. Shading coefficient is the ratio of solar-heat gain through a glazing system to solar-heat gain through a single light of double-strength ($^1/_8$ in. thick) glass under the same set of conditions. Shading coefficients and summer daytime *U*-values are calculated for outdoor-air temperature at 89°F, indoor-air temperature at 75°F, outdoor-air velocity at 7.5 mph, indoor-air velocity at 0 mph, and a solar intensity of 248.28 with incident angle of 30°. Winter nighttime *U*-values are calculated for outdoor-air temperature at 0°F, indoor-air temperature at 70°F, outdoor-air velocity at 15 mph, indoor-air velocity at 0 mph, and a solar intensity of 0 Btu/hr · ft².

[c]When solar-heat gain factor from ASHRAE is 200 Btu/hr · ft² and outdoor temperature is 14°F higher than indoor temperature with no indoor shading.

[d]The portion of solar energy that is reflected and reradiated to the outside.

[e]Assumes $^1/_4$-in. air space—consult ASHRAE for calculating storm window.

9.4.6 Photoelasticity

Some materials, however, will transmit light only in a particular direction. If light waves such as those in Figure 9-17 were directed into such material, the resulting emerging beam would have vibrations (oscillations) in one plane only, as shown in Figure 9-18. Other materials rotate the plane of polarized light. If two pieces of such material are arranged so that their polarized planes are perpendicular, no light would emerge. Such polymeric materials experience this rotation as a result of the alignment of their macromolecules due to internal stress. This effect is known as ***photoelasticity.***

9.5 ELECTROCERAMICS

Highly conducting electroceramics are comprised of inorganic single-crystal and polycrystalline compounds as well as glasses that can be used in a variety of electrical, optical, and magnetic applications. Oxide ceramics are the materials of primary interest, but carbides, fluorides, and sulfides are also finding many commercial applications. Additional information on similar ceramic materials can be found in Module 7.

Such ceramics are generally classified as electronic conductors, ionic conductors, mixed (electronic/ionic) conductors, and insulators (Figure 9-20a). The electronic conductors include superconductors, metallic and ionic conductors, and semiconductors. This grouping represents materials with the highest conductivities. The metallic conductors exhibit high conductivities that decrease with increasing temperature while the conductivities of oxide semiconductors increase with temperature. Ionic conductors generally exhibit conductivities in the range 10^{-8} to 100 S/m that increase exponentially with temperature. The units of conductivity used in this discussion and the accompanying figure are expressed using the special SI unit name ***siemens,*** with symbol S (refer to Table 10-6). A siemens is defined as the quotient of amperes divided by volts (A/V). Insulators such as high-purity alumina are at the lower extreme of the conductivity spectrum, with conductivities of 10^{-13} S/m. Figure 9-20a summarizes the electrical conductivity characteristics of these classes of ceramic materials.

9.6 SEMICONDUCTOR MATERIALS

Our discussion of energy levels in Module 2 dealt primarily with the action of electrons in a single atom. When atoms come together to form a solid material, the electrons in one atom come under the influence of other atoms. The energy levels that may be occupied by electrons merge or broaden into bands of energy levels. The ***Pauli exclusion principle*** states that only two electrons in the entire solid have the same energy. According to the energy band theory there are two distinct energy bands in which electrons may exist: 1) the valence band and 2) the conduction band. Separating these two bands is an energy gap in which no electrons can normally exist. The gap is known as the ***forbidden gap.*** A sketch of these bands is shown in Figure 9-20(b).

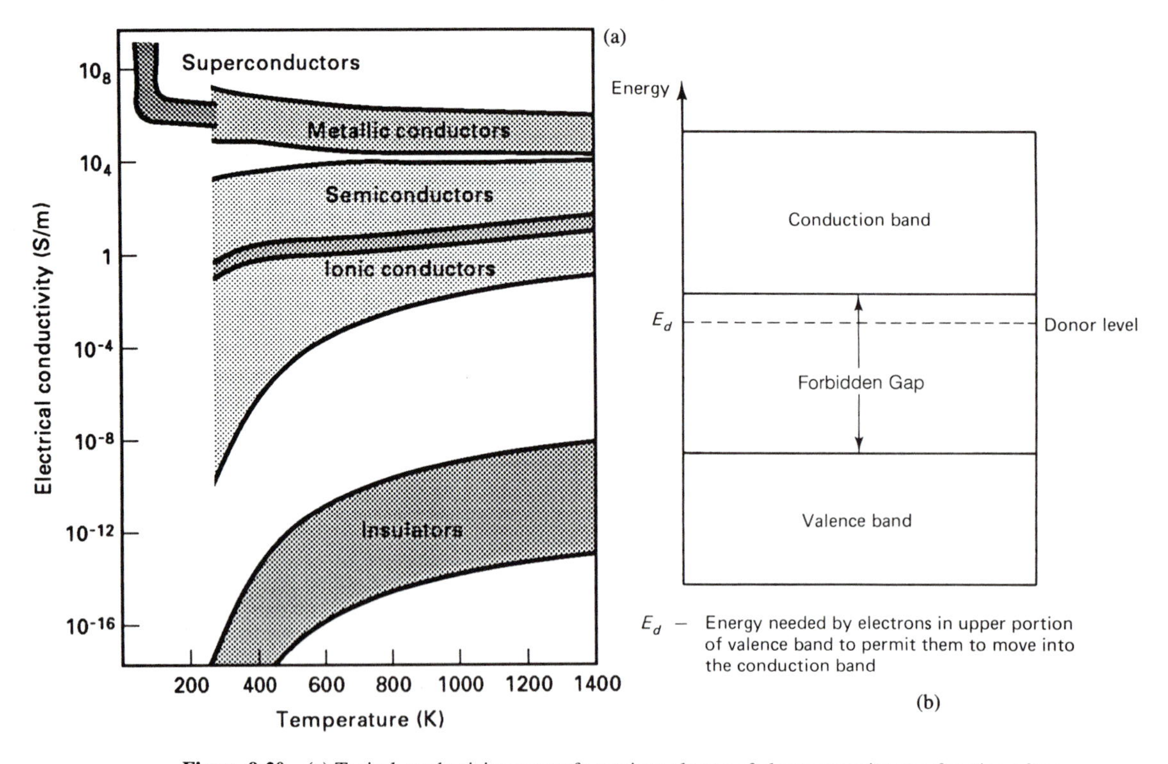

Figure 9-20 (a) Typical conductivity ranges for various classes of electroceramics as a function of temperature. (b) Energy band diagram showing the forbidden gap.

Electrons in the conduction band have escaped from the atomic forces that hold them to their nucleus and therefore they may move about within the material with the application of minimum energy. Electrons in the valence band, in normal orbits around their nucleus, are tightly held or restrained by the attractive forces originating in the nucleus. Much larger amounts of energy must be applied to extract an electron from or move it within this band. The basic differences among insulators, conductors, and semiconductors can be visualized using the energy band diagram sketched in Figure 9-20(b). Conductors have no forbidden gap and the valence and conduction energy bands overlap. This permits large numbers of electrons to move (carry electrical current), even at extremely low temperatures. Insulators have a wide forbidden gap with practically no electrons in the conduction band. Almost all electrons are in the valence band. Large amounts of energy (about 6 eV) are required to cause an electron to cross from the valence band to the conduction band. Finally, a semiconductor material has a narrow forbidden gap, which is more narrow than that of an insulator, with electron distribution similar to that of an insulator. However, since the forbidden gap is narrower, only small amounts of energy (0.785 eV for germanium or 1.2 eV for silicon) can raise electrons from the valence band to the conduction band. At this point it is worthwhile to review Figure 9-20(a) to observe any possible similarities with Figure 9-20(b). The properties of semiconductor materials make possible the transistor, solar cells, tiny lasers, and the multitude of devices using integrated miniature circuits.

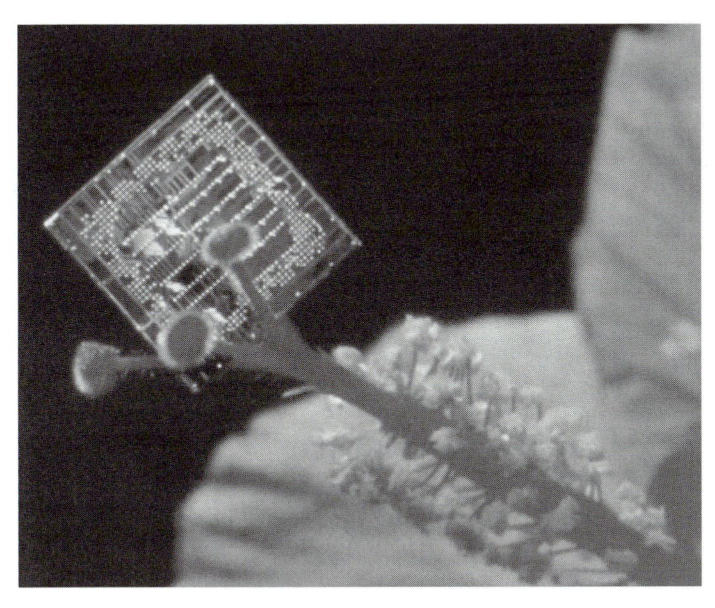

(a)

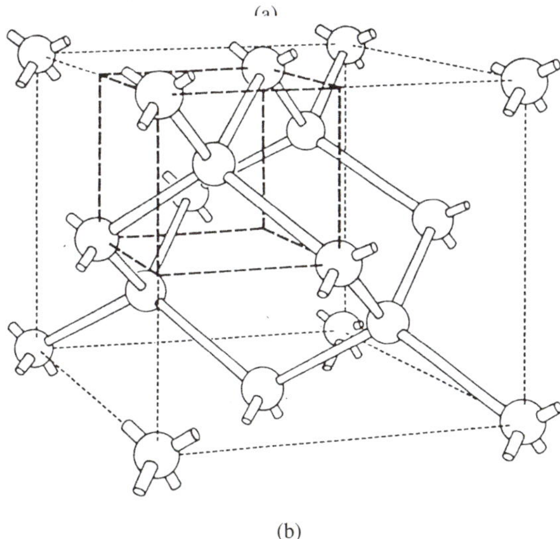

(b)

Figure 9-21 (a) The Power PC 601 microprocessor IC chip, shown behind the pistil of a hibiscus flower, dramatizes its small size. The micro chip, while on 11 mm^2 (compared to 17.6 × 16.6 mm for the Pentium chip), packs in 2.8 million transistors using four 0.65 μm layers made with the CMOS process. (IBM.) (b) Representation of the silicon-crystal lattice arrangement. (U.S. Dept. of Energy.)

This new technology has as its basis the development of human-made materials. Such materials permit the generation and precise control of electron motion within the confines of a tiny crystal of semiconductor material, to which has been purposely added (in most instances) a minute amount of impurity atoms forming a solid solution [Figure 9-21(a)].

Pure semiconductor materials are referred to as ideal or ***intrinsic*** semiconducting materials. They possess a minimum of impurity atoms, certainly none added purposely. We recall that it is a very rare occurrence to find anywhere in a natural state a pure substance. Second, we have learned that impurity atoms can effect profound changes in the properties of materials. So it is with semiconducting materials. If selected impurity atoms in accurately measured amounts are added to certain pure semiconducting materials, highly significant changes in behavior will occur. These materials are now known as ***extrinsic*** semiconducting materials.

Intrinsic semiconducting materials use the elements in group IV in the periodic table, germanium (Ge) and silicon (Si). These elements form crystal structures like that of carbon in diamond [Figure 9-21(b)]. Each atom is bonded to its four neighboring atoms with covalent double bonds. There are no free electrons in the conduction band. How can these bound electrons be excited to act as charge carriers?

9.6.1 Electrically Charged Particles

Electricity is defined in terms of the flow, or movement, of electrically charged particles. This movement to other locations results in increased kinetic energy, which can be converted to some form of work for our benefit. Several forms of charged particles can satisfy these conditions, although some are not as efficient as others. The word *charged* refers to the electrical charge carried by these particles. All particles have this same charge, which is equal to 1.6×10^{-19} coulomb. This is the charge that all electrons carry. Its symbol is e^- and it is the standard unit of charge for measuring electrical charge. The magnitude of the charge in one proton is the same. However, the proton is a positive charge; its symbol is e^+. The measure of electric current *(I)* is the amount of charge *(q)* that passes a given point per unit of time *(t)*:

$$\text{current } (I) = \frac{\text{charge } (q)}{\text{time } (t)} \qquad \text{or} \qquad I = \frac{q}{t}$$

An ampere (A) is defined as an electric current of 1 coulomb (C) of charge passing a point in 1 second (s). Knowing that each electron contains an electric charge of 1.6×10^{-19} C, and using a proportion to determine the number of electrons in 1 C of electric current, the results are 6.25×10^{18} electrons in 1 C. Therefore, 1 A of electric current is the passage of 6.25×10^{18} electrons in 1 s. One microampere (1 μA) equals a current of 6,250,000 million electrons per second (6.25×10^{12}).

A ***hole,*** the absence of an electron, is considered as having the equivalent of an electron's charge only with a positive sign. As you recall, holes in the valence band are produced by supplying electrons with energy sufficient to cause them to be excited to the conduction band. Free electrons have greater mobility than holes. For comparison purposes, the mobility of electrons in pure germanium at 300 K is about 3950 cm^2 per volt second, and for holes it is about 1950.

9.6.2 Donor Doping

If a group-V element such as phosphorus (P), with its five valence electrons, is added to germanium (Ge), only four outer electrons are needed to perfect a covalent bond with its neighboring germanium atoms to form a substitutional solid solution. The extra electron not needed for bonding will be attracted to the impurity atom (phosphorus) by a weak coulomb force of attraction of the nucleus. Because this electron is held much less tightly than the four bonding electrons, it and similar electrons contributed by other impurity atoms can be raised into the conduction band by the absorption of much smaller energies than those required for raising electrons from the valence band into the conduction band. This situation is depicted in Figure 9-22(a) and (b). The extra electron is shown lying just below the conduction band at an extra energy level *(E_d)* called the ***donor level.*** This type of impurity atom makes the material *n*-type (possesses free electrons). The process of adding impurities is called ***doping.*** Figure 9-23 shows a sketch of one phosphorus atom substituting for a germanium atom in the crystal lattice of an *n*-type germanium semiconductor material. Other donor impurities that can be used for donor doping from group V are arsenic (As) and antimony (Sb). Note that these atoms, having five valence electrons, are referred to as *pentavalent* atoms.

9.6.3 Holes

From our study of the electronic structure of atoms, we have learned that any orbit of most atoms, other than the *K* shell or first shell (which is completely filled with two electrons), is stable if it contains eight electrons. Silicon [see Figure 9-24(a)] has only four electrons in its outermost shell (*M* shell). Being its outermost shell, it contains space for four additional electrons. These vacancies or absences of an electron in the valence shell are called *holes.* Holes are created in the valence band when valence electrons receive sufficient energy to move up into the conduction band. With intrinsic semiconductors, the valence electrons with the highest energies (outer-shell electrons have different energies) will be freed

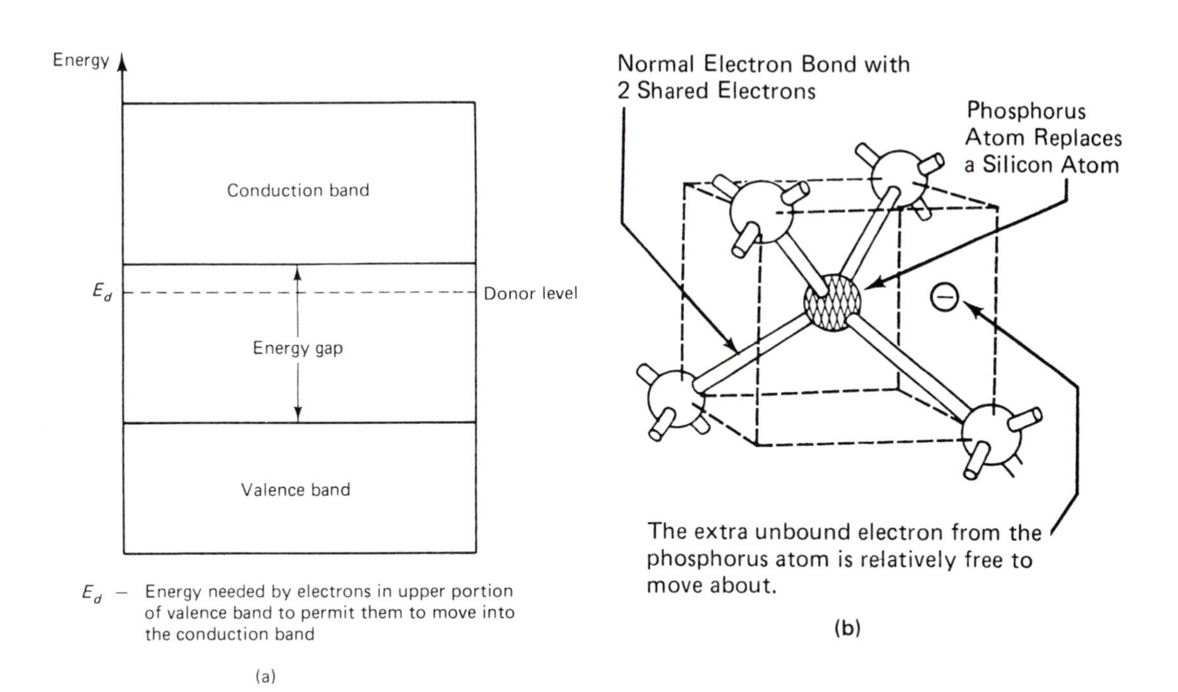

Figure 9-22 (a) Energy-band diagram of an *n*-type semiconductor. (b) Substitution of a phosphorus atom having five valence electrons for a silicon atom leaves an unbonded electron. (U.S. Dept. of Energy.)

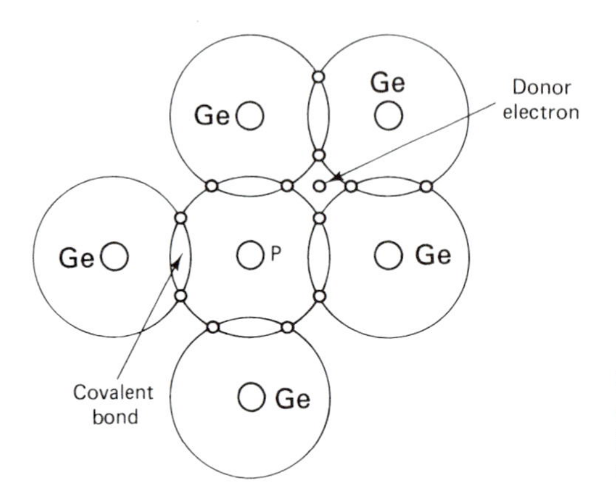

Figure 9-23 Effect of a dopant phosphorus atom substituting for a germanium atom in the crystal structure of an *n*-type semiconductor material.

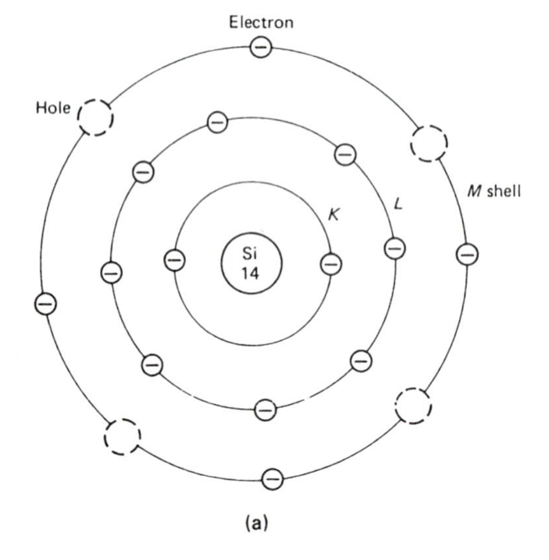

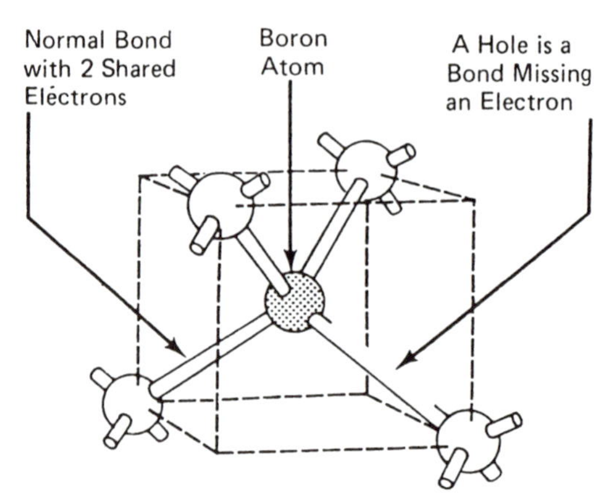

Figure 9-24 (a) Two-dimensional representation of the silicon (Si) atom. (b) Boron, a three-valence-electron atom in a silicon crystal, is normally bonded, except that one of the bonds is missing an electron, creating a hole. (U.S. Dept. of Energy.)

first when energy is added to the material as a result of an increase in temperature. The transfer of these electrons to the upper band creates holes in the lower (valence) band, which can then be occupied by electrons lower down or deeper in the valence band. When an electron receives an input of energy from some outside source, the electron will be loosened from its bonded position in an atom and be free to move about in the crystal structure (in the conduction band). If the energy received is insufficient to loosen the electron, it may just vibrate, which gives off heat. Once the electron breaks loose, it leaves behind a bond missing an electron. The bond missing an electron may also be called a hole [see Figure 9-24(b)]. Electrons and holes freed from their positions in the crystal structure are said to be ***electron–hole pairs.*** A hole, like a free electron, is free to move about in the crystal (Figure 9-25).

9.6.4 Conduction in Solids

Electricity flow (conduction) occurs in a solid when an applied voltage (electric field) causes charge carriers within the solid to move in a desired direction. In the absence of an electric field, the movements of charge carriers are random and result in zero charge transport. If a metal is placed in an electric field (in an electrical circuit), the free electrons in the conduction band will move toward the positive terminal. As they do so, they receive additional momentum (mass × velocity) and hence more energy. Those electrons moving toward the negative terminal lose momentum and reduce their energy.

Hole transfer involving electrons takes place when an electron from one atom jumps to fill the hole in another atom. This electron jump leaves a hole behind it. Or we can describe this movement, not in terms of what the electron does, but in terms of the hole movement. The hole moves in the direction opposite to the electron. Therefore, the flow of electric current is brought about by the movement of free electrons in the conduction band and/or holes in the valence band. Free electrons are easier than holes to move through a semiconductor material, hence they have greater mobility (***drift velocity*** of the carrier).

With the help of Figure 9-25(a), the movement of one hole can be clarified. The hole starts with atom *A*. As an electric field *(E)* is applied, a valence electron breaks free and moves to its left to fill the vacancy (hole) shown by arrow 1. The hole is now at atom *B*. A similar effect is felt by one of the valence electrons at atom *C*, which also moves to its left

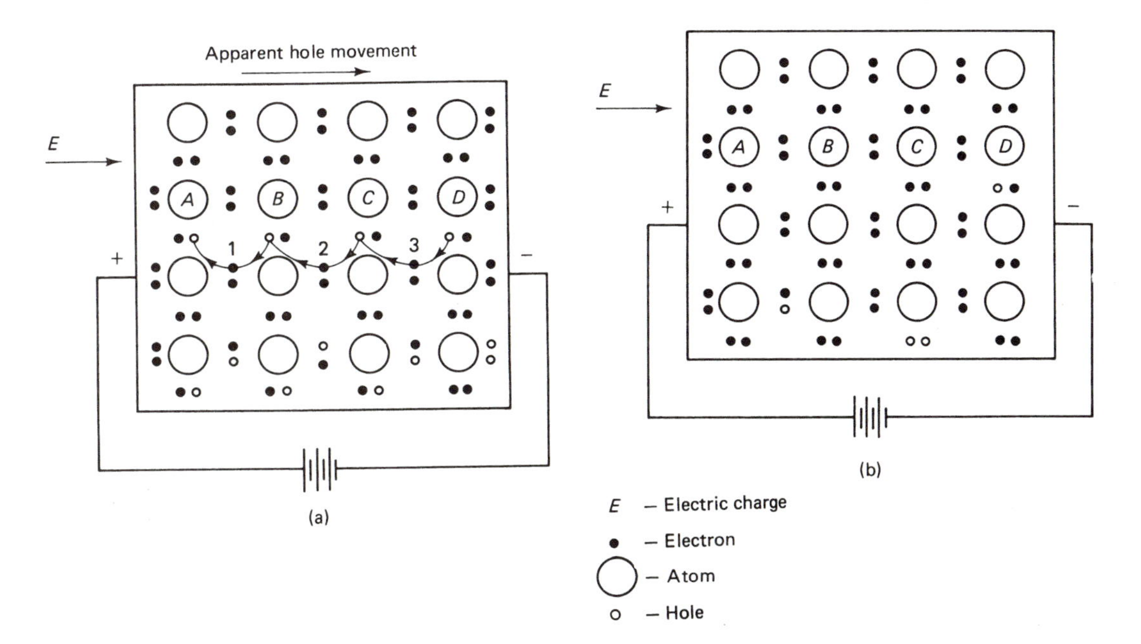

Figure 9-25 Conduction by holes in the valence band.

and fills the hole at atom *B*. Arrow 2 shows this movement. Each time an electron moves, it creates a hole. Figure 9-25(b) shows the final location of the hole at atom *D*. Note that the hole movement, in this instance, from left to right, is toward the negative-charged terminal of the battery, opposite to the flow of electrons as they move toward their left to fill the holes. In other words, this hole movement is a measurement of positive charge in the direction of the applied field *(E)*. The action of the electrons in their moves, labeled 1, 2, and 3, in filling the just-created holes is known as *recombination,* which will be discussed later.

9.6.5 Acceptor Doping

When a group-III, or trivalent, element such as boron (B), aluminum (Al), gallium (Ga), or indium (In) is introduced as an impurity into an intrinsic semiconductor such as silicon (or other group-IV elements), a mismatch in the electronic-bonding structure occurs. Using aluminum as an example of the dopant, this element lacks one electron to satisfy the covalent bonds of the group-IV matrix element such as silicon. This mismatch is illustrated in Figures 9-26 and 9-24(a). One covalent bond near each dopant atom (Al) is incomplete; that is, it is missing an electron, or it contains a hole. If an external electronic field is applied to this solid, one of the neighboring electrons from another covalent bond can acquire sufficient energy to move into the hole. This, in turn, causes the hole to move to the position formerly occupied by the electron. In this manner the hole moves through the solid as a positive-charge carrier. The electric current produced is mainly the result of hole movement (positive-charge carrier). This structure is called *p*-type because of the presence of free positive charges (the moving holes).

In terms of the energy band diagram for this *p*-type structure (Figure 9-27), the aluminum atom has provided an energy level that is only slightly higher than the upper limit of the valence band. This puts it in the forbidden gap. Thus, a nearby valence electron could easily be excited into this intermediate level. This level is called an ***acceptor level.***

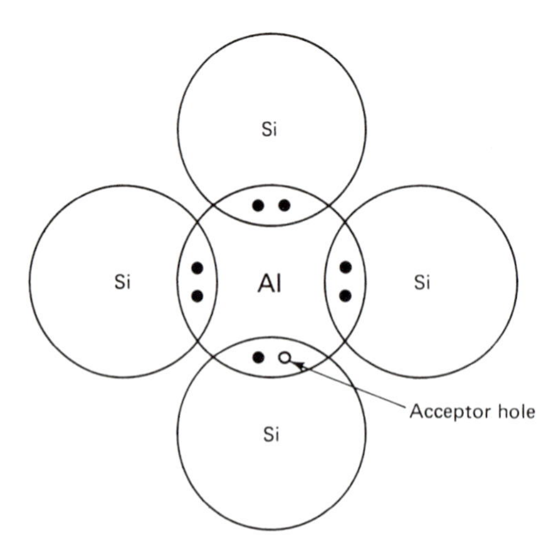

Figure 9-26 Structure of silicon with aluminum atom added as an impurity (dopant).

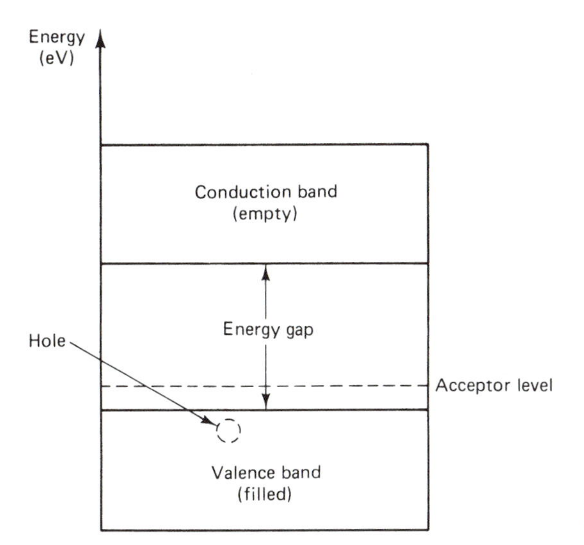

Figure 9-27 Energy-band diagram for a *p*-type semiextrinsic semiconductor material.

As with *n*-type extrinsic semiconductor materials, the extremely small concentration of dopant is measured in parts per million (ppm). In summary, the addition of acceptor atoms to a semiconductor material increases the number of holes in the valence band without an increase in the number of electrons. These positive-charge carriers are termed ***majority carriers,*** and the electrons are called ***minority carriers*** in *p*-type semiconductors, just the reverse of their designations in *n*-type semiconductors.

Figure 9-28 represents an *n*-type semiconductor material in a circuit with a switch in the open position. No electrons will flow in the external circuit. However, within the doped *n*-type semiconductor material, the free electrons contributed by the dopant atoms and the electrons that break away from their parent atoms in the valence band are diffusing throughout the material in a random fashion. In addition, each time a valence electron breaks free of its atom, an electron deficiency in the valence shell of that atom is created, which makes the atom a positive ion with an electric charge of equal magnitude to that of the free electron. The vacancy in the valence shell (energy band) of the electronic structure of the atom is called a hole. The phrase *electron–hole pair* refers to this energy transfer process. When a free electron collides with a hole, it is captured by it, which produces a balanced atom in its lowest equilibrium state. This process is known as ***recombination.*** Taking into consideration the remaining atoms that make up the material, we recognize that overall equilibrium must be maintained, and thus a new electron–hole pair is generated with every recombination (Figure 9-29).

Next we close the switch in the external circuit in Figure 9-28. This causes electron flow to occur from the negative terminal of the source through the semiconductor material and back to the positive terminal of the source. Figure 9-28 attempts to show that only electrons are flowing in the external circuit within the material. The action of the electric field gives direction and movement (drift) to the flow of both free electrons donated by the phosphorus atoms in this instance and the holes (minority carriers) in their movement from right to left. Both the free electrons donated by the donor atoms and the electrons supplied

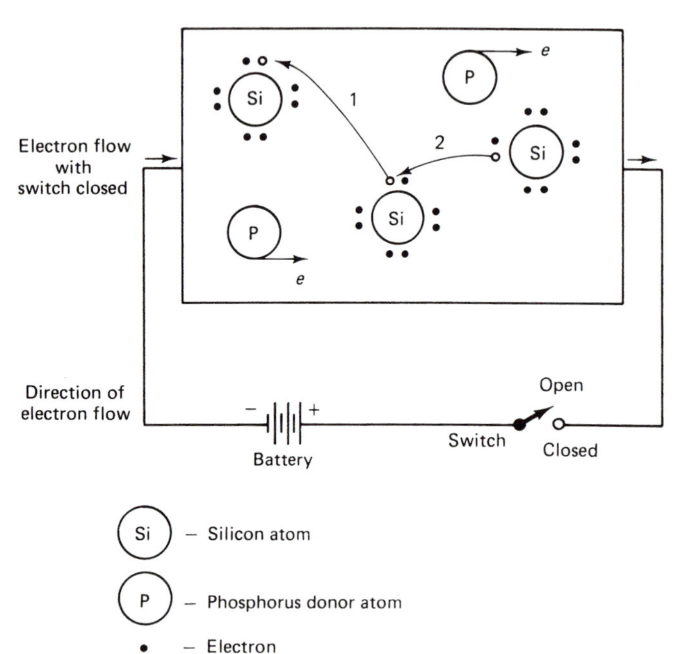

Figure 9-28 Random movement of electrons and holes in an n-type semiconductor material.

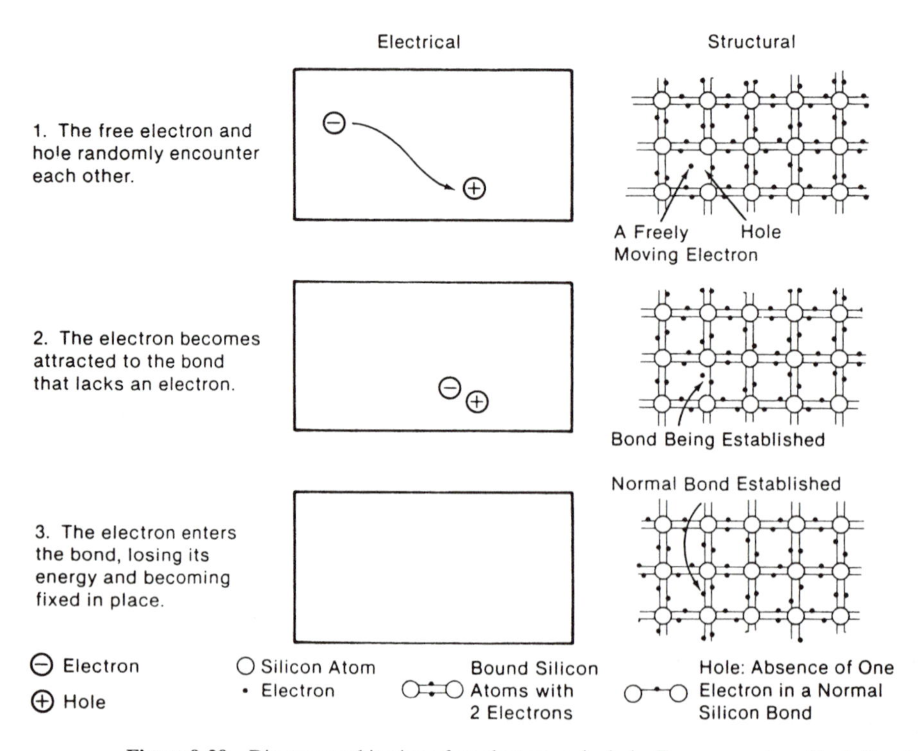

Figure 9-29 Direct recombination of an electron and a hole. Excess energies of both the electron and hole are lost to heat. (U.S. Dept. of Energy.)

through the external circuit combine with the holes to accomplish the conduction process through the processes of generation and recombination.

When a piece of *n*-type material is brought into contact with a piece of *p*-type material, a ***pn junction*** is formed (Figure 9-30). A junction of this type produces a ***diode,*** which is the simplest of the semiconductor-junction devices. Before observing the diode in a simple circuit, it would be helpful to study the interaction of the electrons at the *pn* junction.

The *n*-type material contains an excess of electrons, whereas the *p*-type material has an excess of holes. When the *pn* junction is formed, there is a diffusion of charges between the two materials. The electrons from the *n*-type will move across the junction to the ***p***-type. This will create holes or positively charged particles in the *n*-type. The reason for this diffusion of charges is to counteract the electrical imbalance in the two materials caused by the formation of the junction. The diffusion process will stop once an equilibrium of charge is established between the two materials. This equilibrium is reached when no additional charge carriers have the energy to overcome the electric field that has been built up at the junction. It can be seen in Figure 9-31 that a boundary is formed across the *pn* junction. This boundary is usually referred to as the ***depletion region*** because all mobile charge carriers have been depleted from this area.

The electric field set up by the stationary charges in the depletion region creates a difference of potential between the two materials. This electric field represents the equilibrium potential difference from *n*-type to *p*-type in direction. This voltage is called the contact potential of the junction. For silicon, it is about 0.6 V. All the previous discussion has been aimed at the *pn* junction (a diode) as a closed system. Next the junction will be observed with an externally applied voltage [Figure 9-31(b)].

The *p*-type side of the diode is called the anode and the *n*-type side is called the cathode. Applying a voltage to the diode is called ***biasing.***

If, as in Figure 9-31(b), a battery is connected across the diode so that the positive lead is attached to the anode and the negative lead is attached to the cathode, a large current will flow. This is the forward-biased condition of the diode. The reason for the large current is that the junction potential (contact potential and applied potential) has been lowered due to the forward bias. Electrons from the battery that enter the anode of the device will see little resistance and can flow easily across the *narrowed* depletion region to the cathode.

In Figure 9-31, the leads of the battery have been reversed so that the positive side of the battery is across the cathode and the negative side is across the anode. This is the *reverse-biased* condition of the diode, and very little (only leakage current) will flow. The reason for the extremely small current in this condition is that the battery is acting in the same direction as the electric field across the junction; thus, the junction potential is larger. When the junction potential increases, the width of the depletion region and the amount of stationary charges will increase. This means that the charge carriers must gain an extremely large amount of energy before they can cross from cathode to anode.

The *pn* junction is the basic building block for all semiconductor-junction devices. The preceding discussion can be applied to the transistor, silicon-controlled rectifier (SCR), field-effect transistor (FET), or any other junction device.

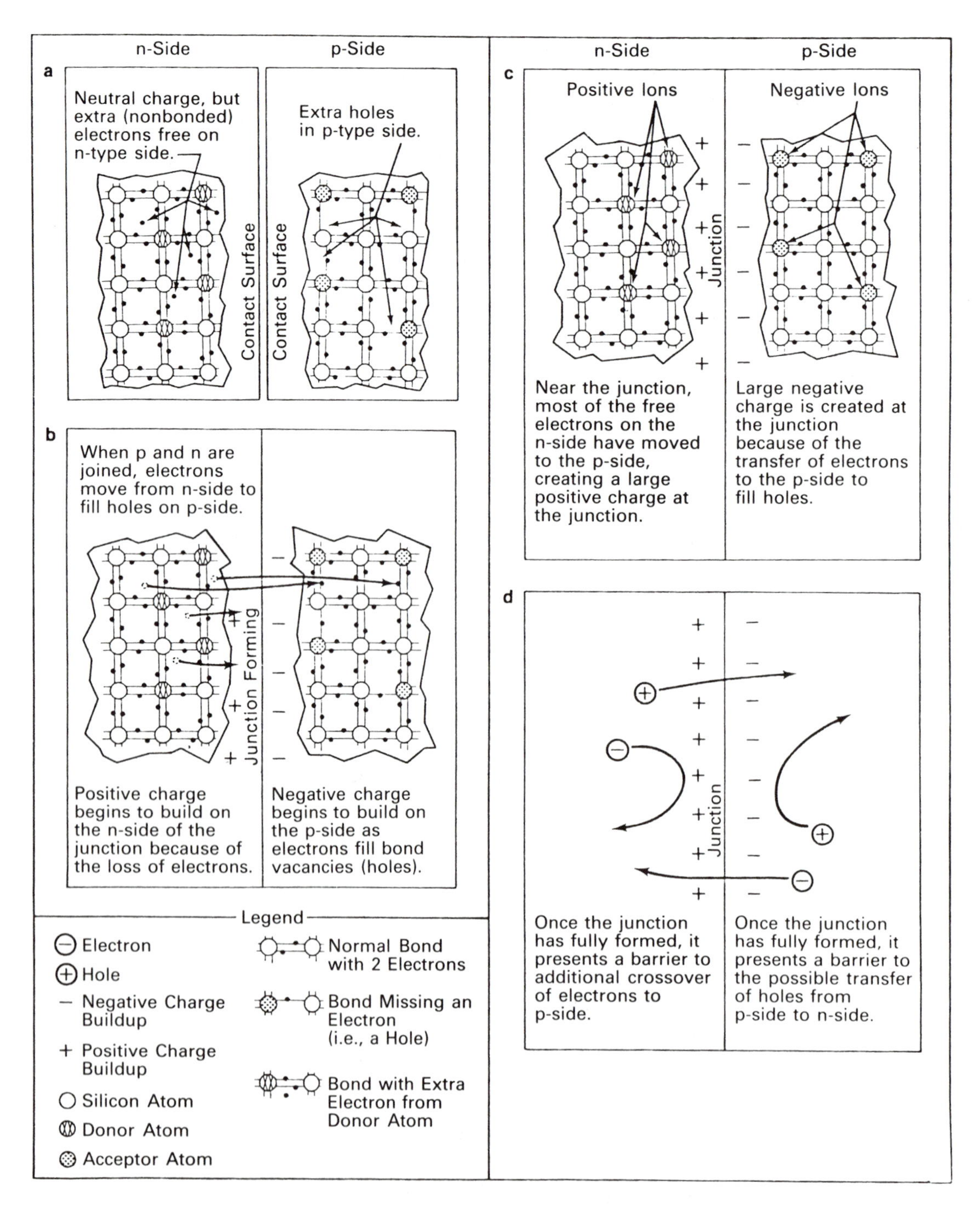

Figure 9-30 Junction formation. Additional movement of charge carriers is stopped and equilibrium is established. (U.S. Dept. of Energy.)

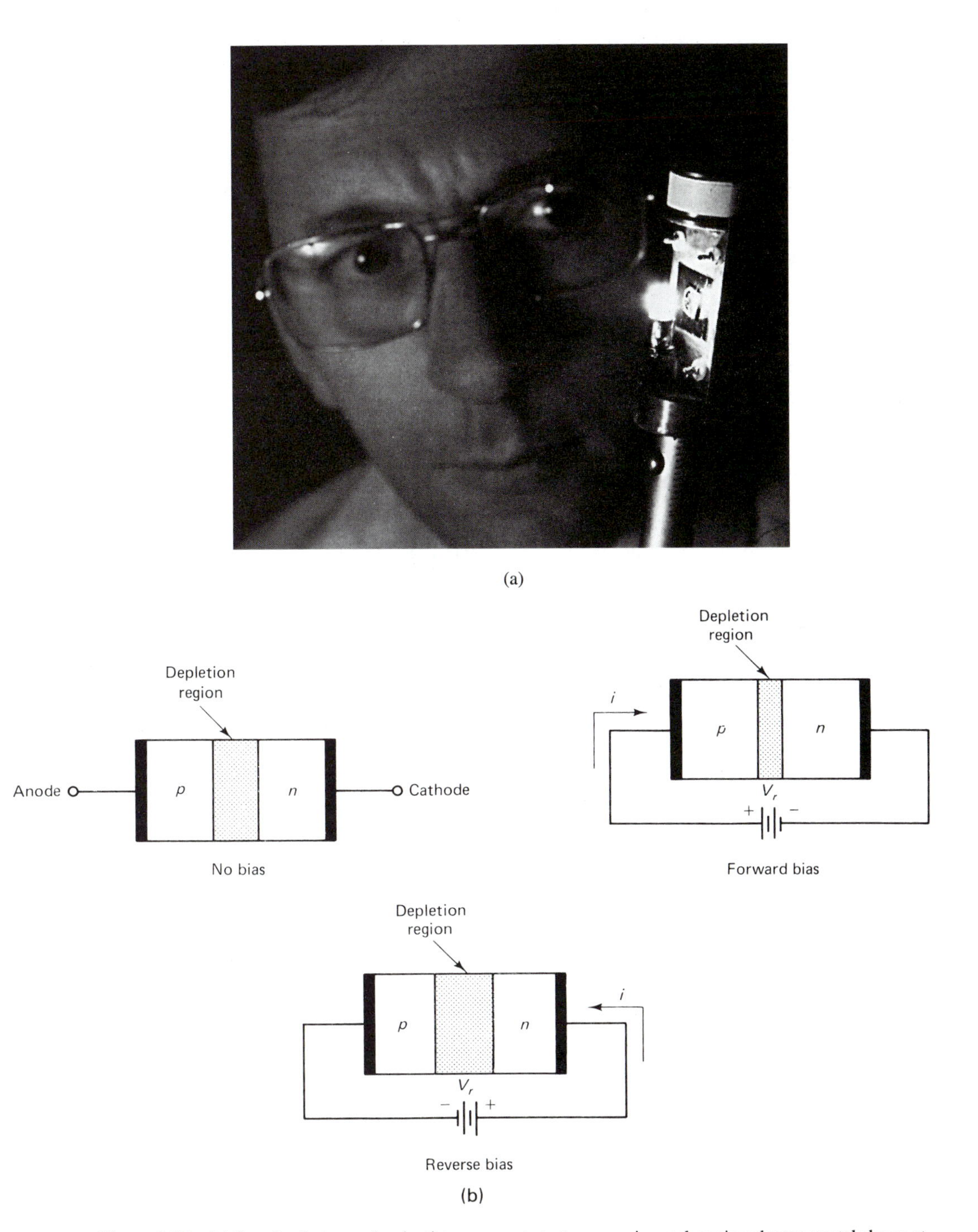

Figure 9-31 (a) Speedy electrons. A scientist prepares to test an experimental semiconductor crystal shown at the right of the light. By increasing the speed at which electrons move through the crystal, a fundamental advance in solid-state technology was achieved. (b) A *pn* junction (diode) showing no-bias, forward-bias, and reverse-bias conditions. (Bell Labs.)

9.7 OPTOELECTRONIC DEVICES

Optoelectronic devices are (1) operated by light (photoelectric), (2) produce or emit light, or (3) modify light. Photoelectric devices, in turn, can be categorized as photoconductive, photovoltaic, or photoemissive. A review of optical properties in Section 9.4 and of semiconductor materials in Section 9.6 is appropriate prior to studying the following treatment of these devices.

9.7.1 Photoconductive Devices

In photoconductive devices, the conductivity of the semiconductor material will vary, provided that the energy supplied by the light (visible, infrared, or ultraviolet) is sufficient to raise the electrons into the conduction band. A photomultiplier is such a device. It can produce an image that is visible when illuminated by a weak light source. A light meter is another example that operates in accordance with the same basic principle.

9.7.2 Photovoltaic Devices

Photovoltaic cells are semiconductor-junction devices that convert electromagnetic energy in the form of light directly into electrical energy. The amount of electrical current (flow of electrons) is directly proportional to the amount of light that is incident on the semiconductor material. A ***solar*** or ***photovoltaic cell*** is a photodiode used to extract electrical energy directly from sunlight. This direct conversion of sunlight to electricity differs from the solar thermal conversion process used in solar heating of homes and offices, which uses panels to absorb sunlight and uses the energy to heat water or other media, including air for heating a building. Photovoltaic cells embedded in the glass of a sunroof for an automobile are a recent application for these devices. They help keep the interior cool. When temperatures inside the vehicle exceed 85°F, the cells power two fans that pull in cooler outside air and exhaust hot interior air through vents in the roof. If it starts to rain, sensors shut the vents.

A typical solar cell [Figure 9-32(a)] contains two extremely thin layers of silicon connected externally by a wire to the load where light energy is converted to work. When light is absorbed in silicon, it creates electron–hole pairs [Figure 9-32(b)]. Electron–hole pairs that reach the junction are separated, holes going to the p side and electrons to the n side. This buildup of charge on either side of the junction creates a voltage that drives current through the external circuit. Reflection of sunlight from an untreated silicon cell can be as great as 30%. By chemical treatments and texturing of the surface, this value can be reduced to around 5%. Light with a certain wavelength (energy) is required to interact with specific materials before the optimum electron–hole generation is achieved [Figure 9-32(c)]. This characteristic energy is called the material's band-energy. (Silicon requires 1.1 eV, gallium arsenide, 1.4 eV.) Overall, these mismatches of light with a solar cell's material waste some 55% of the energy from sunlight. Research is proceeding to find better ways to process sunlight to make it monochromatic (one wavelength) and possess an exact energy match with that required by the cell material to make the conversion of incoming sunlight more efficient.

Theoretically, the silicon solar cell should be able to convert about 25% of the sun's energy into electricity. To achieve greater conversion of sunlight to electricity, two differ-

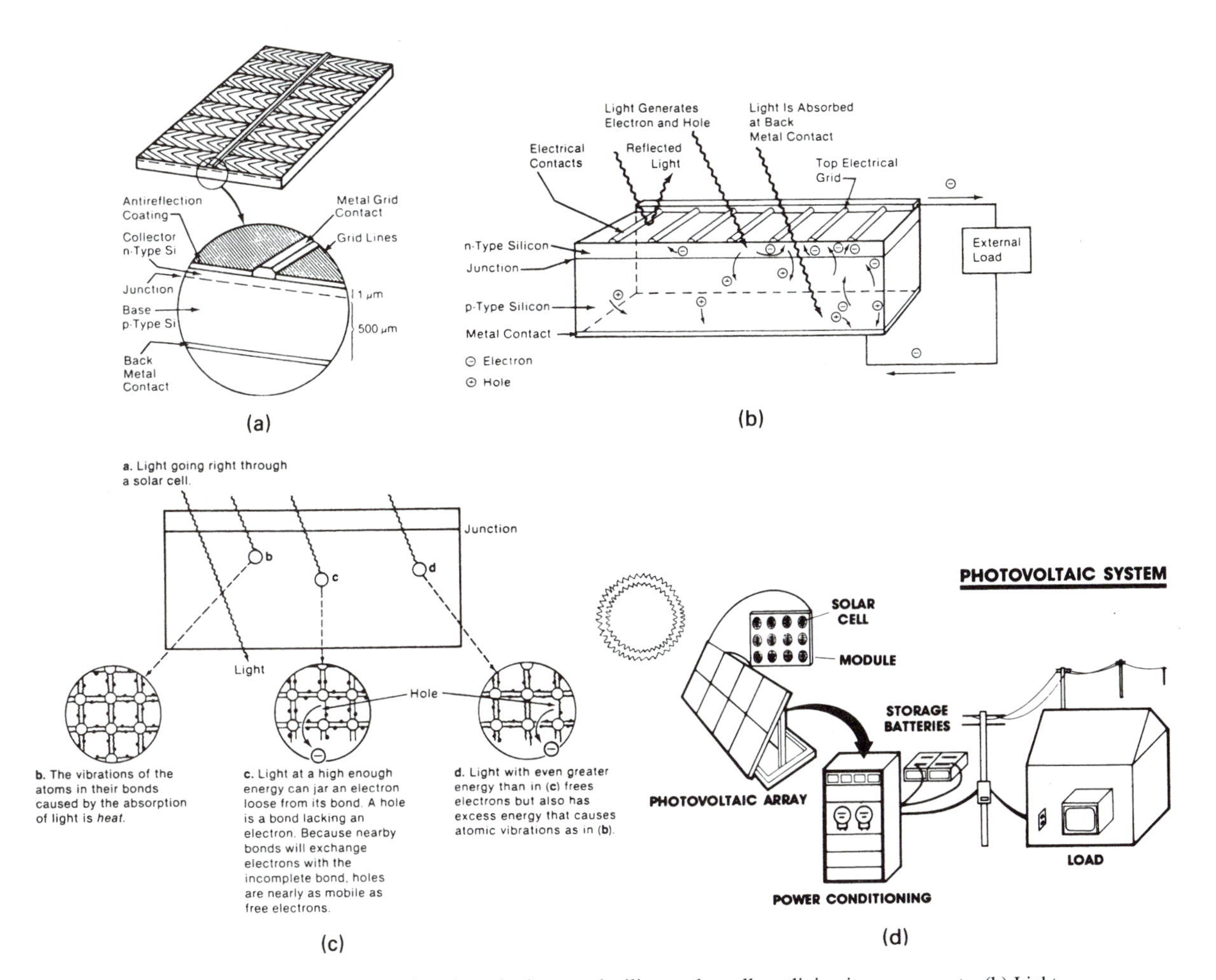

Figure 9-32 (a) A typical *pn* junction, single-crystal, silicon solar cell, outlining its components. (b) Light striking a cell creates electron–hole pairs that are separated by the potential barrier, creating a voltage that drives a current through an external circuit. (c) What happens to light entering a cell? (d) Schematic of a photovoltaic system. (U.S. Dept. of Energy.)

ent solar cells can be used, one of aluminum-gallium-arsenide and the other of silicon. These two cells absorb a wider range of wavelengths from the solar spectrum, but various optical losses reduce the overall system efficiency to about 25%. The term *efficiency* is the percentage of energy in sunlight striking the cell that is converted into electricity. Individual solar cells have limited power. To produce electricity for most applications, they must be joined together electrically to form modules that become building blocks for larger arrays [Figure 9-32(d)].

Solar cells are very expensive primarily because of their fabrication costs. Many breakthroughs in materials technology are needed before solar cells can be developed to compete on the mass market and produce electricity costing about 50 cents per peak watt or about 6 to 8 cents per kilowatthour. Ultrapure (99.999% purity) semiconductor-grade silicon costs around $15 a pound, compared to $0.15 a pound for metallurgical silicon, with a purity of 99.5%. Approximately 30% of the cost per peak watt of solar-produced electricity represents this processing cost. *Peak watt* is a term used to express the amount of power produced by a solar cell in full sunlight at 25°C. A 4-in. cell, for example, can produce 1 peak watt at noon on a sunny day in the U.S. Southwest. This is a description of ideal conditions. A more relevant term is *average wattage,* which is about one-fifth the peak wattage. See Table 9-5 for some current applications.

Invented at Bell Laboratories in 1954 with a 4% efficiency, most common solar cells are made from a single crystal of pure silicon grown artificially in the form of an ingot that is sliced into wafers 0.012 in. thick, which are then polished, trimmed, and doped in an oven. The finished cells are mounted in arrays or panels containing dozens of individual cells whose diameters may range up to 4 in. One manufacturing innovation produces cells continuously in the form of thin ribbons. This thin-film technology, similar to that used in making electronic integrated circuits, could cut manufacturing costs substantially and permit the fabrication of large-area cells that would make solar cells more cost effective.

One solar cell device using thin-film techniques developed by Bell Laboratories consists of a layer of polycrystalline cadmium sulfide deposited on a single-crystal substrate of copper-indium-phosphide. Reportedly, its efficiency is about 12.5%. This new device can also be used quite effectively as a *photodetector* for converting light impulses into electrical signals. Many applications of this device can be found in optical communications systems that detect the presence of infrared- and visible-light transmissions through optical-glass fibers.

9.7.3 Photoemissive Devices

In a photoemissive device, the light (or photons) generated by the recombination of electron–hole pairs is emitted from the surface of the device. Electrons striking the television screen, which is coated with a semiconductor material doped with copper, raise the electrons in the coating into the conduction band, where they recombine and emit energy in the form of light (photons). This is known as ***electroluminescence.***

Fluorescent lamps depend on ultraviolet light striking electrons in the inner coating of the lamps, which emit light in the visible region of the electromagnetic spectrum. To round out our coverage of these devices, a light-emitting diode is treated in more detail below. It is important to repeat at this point that, regardless of the device, the underlying mechanisms involve the mutual interaction of electrons with electromagnetic radiation.

9.7.3.1 Light-emitting diode. Figure 9-33 shows a cross-section of a light-emitting diode (LED). When forward bias is placed across the junction, electrons cross from the *n*-side of the junction to recombine with the holes in the *p*-side, giving off energy in the form of heat and light. The light will be emitted, assuming that the semiconductor material is translucent and the gold-film cathode effectively reflects the light to the surface. Gallium phosphide (GaP) and gallium arsenide phosphide (GaAsP) are typical semiconductor materials used in LEDs to produce red, yellow, or green light. The relatively large amounts of current consumed by LEDs is their primary disadvantage, but their ruggedness and long life tend to make up for this shortcoming.

TABLE 9-5 EXAMPLES OF PRESENT TERRESTRIAL APPLICATIONS OF PHOTOVOLTAIC POWER UNITS

Application	Peak rating (W)
Warning lights	
Airport-light beacon	39
Marine-light beacon	90
Railroad signals	
Highway-barrier flashes	1.2
Tall-structure beacon	
Lighthouse	
Communications systems	
Remote repeater stations for	
Microwaves	50
Radio	109
Television	78
Remote communications stations	3,500
Mobile-telephone communications station	2,400
Portable radio	50
Emergency locator/transmitter	
Water systems	
Pumps in desert regions	400
Water purification	10,800
Scientific instrumentation	
Telemetry—collection and transmission platforms for environmental, geological, hydrological, and seismic data	
Anemometer	100
Remote pollution detectors (H_2S, noise)	3
Industrial	
Remote machinery and processes (e.g., copper-electrolysis installation)	1,500
Cathodic-protection of underground pipeline	30
Electric-fence charger	
Domestic water meter	20
Offshore drilling platforms	
Forest-fire lookout posts	
Battery charging	
Boats, mobile homes and campers, golf carts	6–12
Construction-site equipment	
Ni–Cd-powered military equipment	74
Recreational and educational	
Educational television	35
Vacation home	
Lighting, television	
Refrigerators	200
Sailboats	
Lighting, ship–shore communications	
Automatic pilot	66
Portable television camera	
Camping lighting	
Electronic watches, calculators	
Recreational-center sanitary facility	168
Security systems	
Closed-circuit-television surveillance	150
Intrusion alarms	6

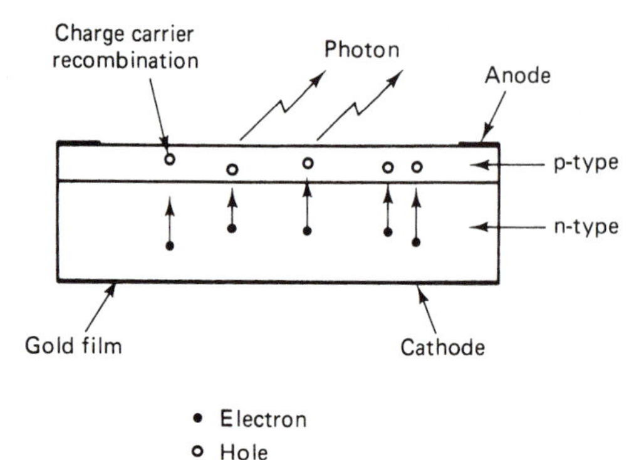

Figure 9-33 Cross-section of an LED.

Silicon, the second most abundant element in the earth's crust, exists as silica or silicon dioxide (SiO_2). The processing of quartzite, a mineral (almost 99% silica) and the source of pure silicon, raises the cost of silicon to about $70 per kilogram. Pure polycrystalline silicon is further processed into single-crystal or large-grain (5-mm) polycrystalline silicon [Figure 9-34(b)]. The single-crystal silicon technology, the oldest and most well established, is slowly giving way to newer technologies using polycrystalline and amorphous silicon.

The latest fabrication techniques form silicon directly into usable wafers (0.5 mm thick) without intermediate steps such as sawing, thus reducing processing costs considerably. More advanced methods produce ribbons (long ultrathin, rectangular sheets). One such method produces ribbons 33 m long, 5 cm wide, and 0.5 mm thick at a linear rate of 55 cm/min, with cell efficiencies of over 10% [Figure 9-34(a)]. ***Polycrystalline silicon*** is cheaper to produce and fabricate than single-crystal silicon, but at present single-crystal cells have maintained their lead in efficiency. In polycrystalline silicon, the grain boundaries impose numerous restraints on the movement of charge carriers [Figure 9-34(b)]. However, with the constant growth of new technologies, the effect of the grain boundaries is being reduced.

A third form of silicon, ***amorphous silicon*** (abbreviated aSi), has been used since 1974 to produce cells with high output voltage (0.8 V), currents greater than 10 mA/cm^2, and efficiencies of 6% using *p-i-n* cells (*p*-type, intrinsic, and *n*-type layers). Figure 9-34(c) is a schematic of this cell, pointing out the ultrathinness of the layers of amorphous silicon. To overcome amorphous silicon's disordered atomic structure [Figure 9-34(d)] with its incomplete bonds, hydrogen is added (doped) to complete this bonding and increase the cells' efficiency. A distinct advantage of amorphous silicon is its ability to absorb light about 40 times more strongly than crystalline silicon. Amorphous silicon cells are used to power hand-held calculators because they are more efficient and cost-effective under fluorescent light than either single-crystal or polycrystalline silicon. This characteristic of amorphous silicon allows such cells to be extremely thin, one-fortieth of the thickness required to absorb the same light. The low mobility of the charge carriers provides relatively rapid recombination, which means that the charge carriers must be separated by the *p-i-n* junction in the short time during illumination. In addition, thinness translates into less material, and depositions of the material tend to be easier.

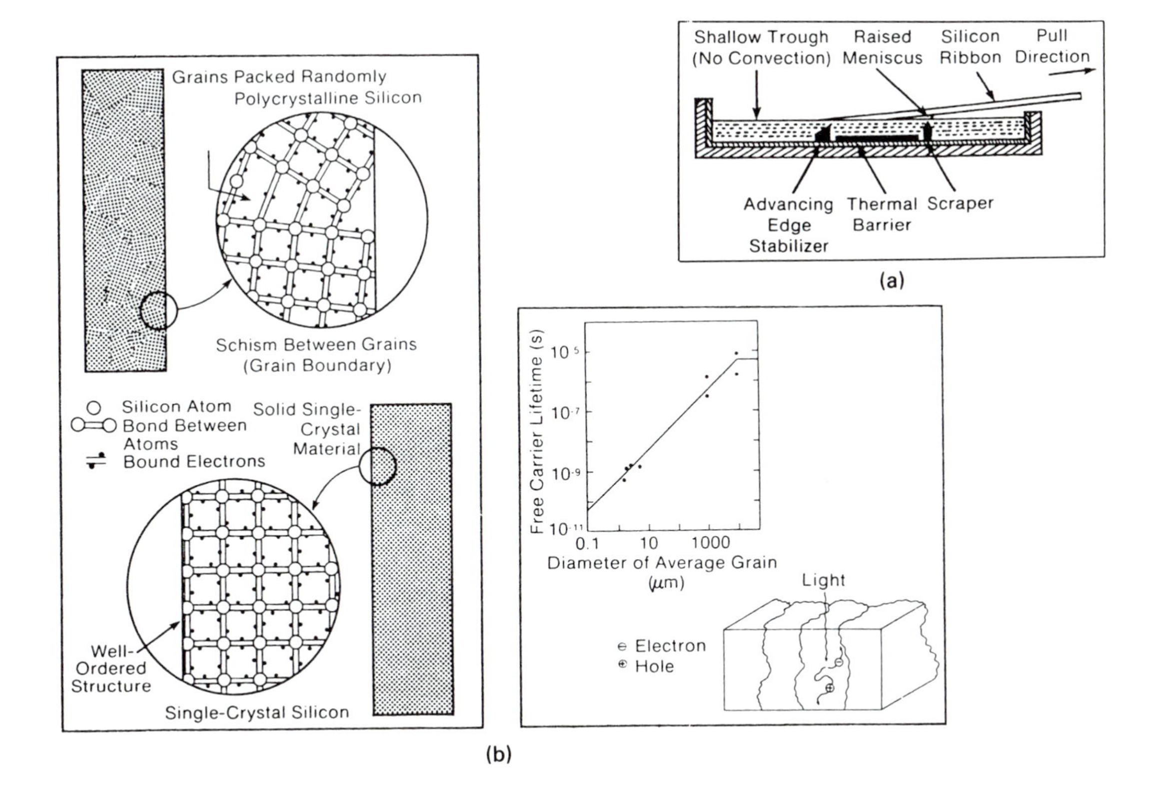

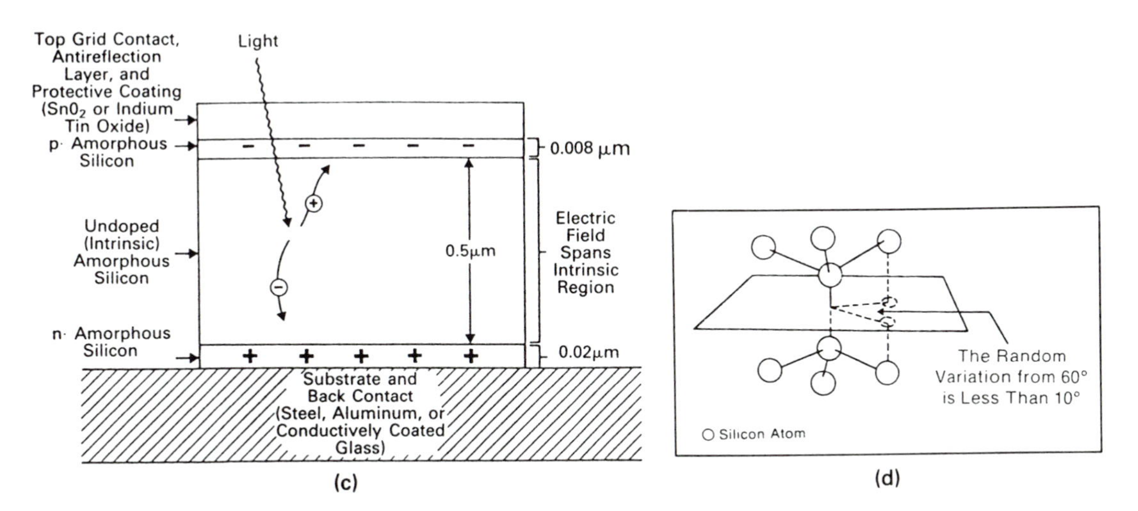

Figure 9-34 (a) Low-angle silicon sheet (LASS) growth is a fast method of drawing a silicon ribbon from a shallow trough of molten silicon. (b) Lifetime of free-charge carriers dependent on grain size and orientation. (c) A *p-i-n* device. An electric field sweeps the charge carriers to opposite ends of the cell. (d) Silicon atoms build a three-dimensional tetrahedral structure that, in amorphous silicon, is rotated randomly, producing a disordered atomic structure. (U.S. Dept. of Energy.)

9.8 PIEZOELECTRIC MATERIALS

Piezoelectric materials can be explained most easily through a discussion of dielectrics with an emphasis on polarization. Piezoelectric crystals are physically uniform solids that are bonded together by ionic bonds. We have learned that ionic bonding is a result of the electrostatic forces of attraction between the ions of opposite charge. Normally, the number of positive-charged ions equals the number of negative-charged ions. Figure 9-35(a) is an attempt to show the symmetric, crystalline structure of a crystal using three representative positive ions located equidistant from each other (at the vertices of an equilateral triangle). Similarly located are three negative-charged ions. The centers of charge both occupy a common center of symmetry. Being of equal magnitude, these centers of charge not only coincide in location, but they cancel each other, leaving the electric charge of these six ions neutralized.

Figure 9-35(b) is a similar representation of the same structure of a crystal with the addition of a mechanical load or force *(P)* that produces a stress assumed to be uniformly distributed over the face of the crystal on which it acts. This force, in this instance, acting

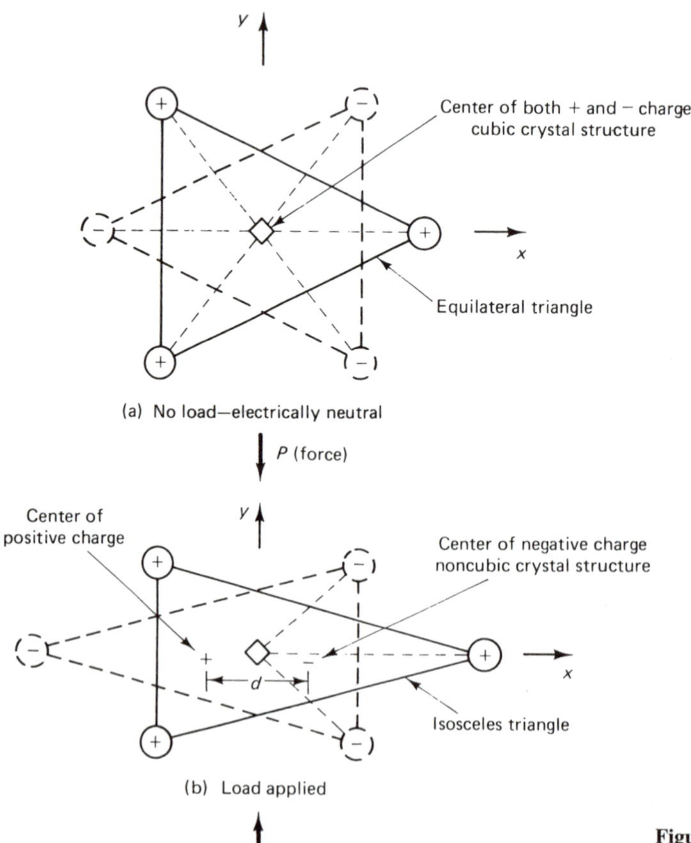

Figure 9-35 Model of piezoelectric crystal structure.

along the y or mechanical axis, tends to compress the crystal, resulting in a compressive, elastic strain or unit deformation. The elastic deformation, in turn, results in the offsetting of the centers of electric charge from each other. In effect, this offsetting of centers of charge creates electric dipoles throughout the material, which combine to produce a measurable electric potential along the x or electric axis of the crystal. Conversely, if an electric potential were to be applied to the x axis, a detectable deformation would be observed along the y axis. The prefix *piezo-* comes from the Greek word meaning "to sit on, or press." The word ***piezoelectricity*** refers to electricity generated by exerting pressure on an ionic crystal.

Figure 9-36 is a model of such a crystal with an electric field impressed along the x axis. This particular arrangement of electric charge acting as the input of energy would produce a mechanical deformation (contraction) similar to that of the force *(P)* in Figure 9-35. Note that the ions of negative charge are attracted to the right toward the positive terminal, with a similar movement of the positive ions and their center of charge to the left, thus increasing the separation of the centers of change (increasing the dipole length). If the polarity of the terminal were reversed, the deformation produced would be distorted in the opposite direction (expansion). An ac voltage, when impressed on the crystal, would cause the crystal to expand and contract (oscillate) at its driven frequency (maximum amplitude at resonance), which could be transmitted as a wave of sound energy into the surrounding medium. In other words, if this potential were produced by an alternating current, it would cause the crystal to vibrate. Each crystal has its own natural, mechanical frequency and the ac potential could be adjusted to match this frequency and produce resonance (maximum amplitude of vibration).

In summary, if a mechanical stress is applied to two opposite ends of a crystal (mechanical energy supplied), the remaining ends of the crystal are charged with electricity (output in electrical energy); if an electric voltage is supplied to one set of faces of a

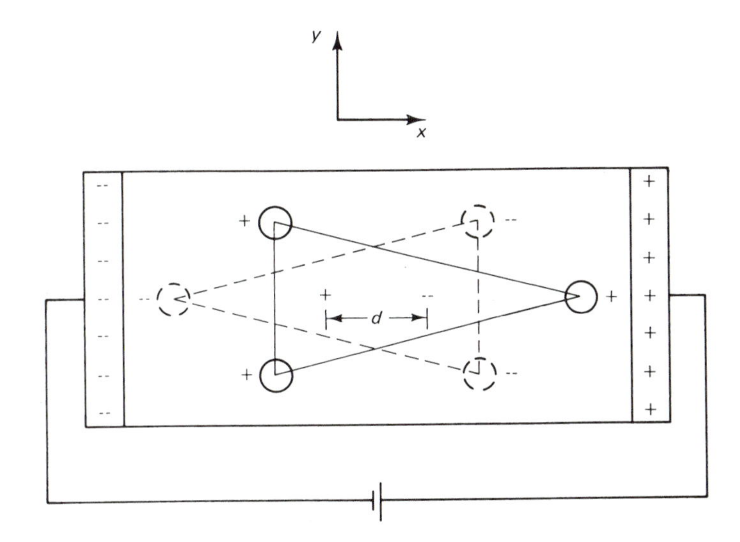

Figure 9-36 An electric field across a piezoelectric crystal.

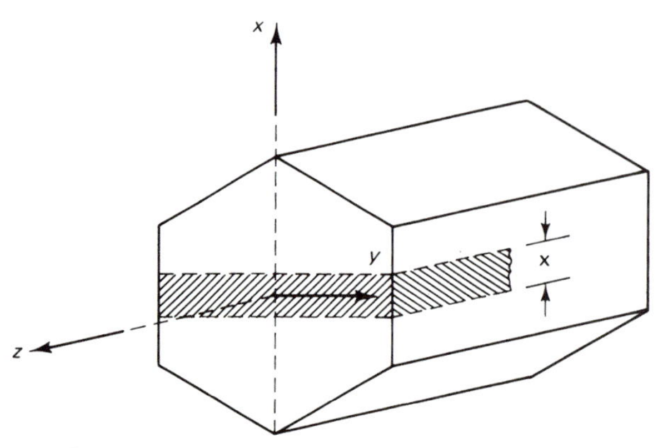

z — Optical (longitudinal) axis
x — Electrical axis
y — Mechanical axis

"x" cut blank shown in section

Figure 9-37 Uncut hexagonal quartz crystal.

crystal, the other two faces of the crystal are deformed (either they contract or they expand); and if an ac potential is supplied, the output is crystal oscillation at its driven frequency, which can be transmitted into a surrounding medium as sound energy at a constant wavelength. A piezoelectronic crystal is a ***transducer*** because it not only transfers energy from one system to another, but it converts energy from one form into another.

Quartz crystals (SiO_2), possessing piezoelectric properties, are both naturally occurring and grown commercially. In its uncut state, the crystal is in the form of a hexagonal prism, as sketched in Figure 9-37. The x axis passing through the corners of the hexagonal cross-section is the *electrical axis,* the y axis perpendicular to the faces of the hexagonal cross-section is the *mechanical axis,* and the z axis is the *optical axis.* The crystal is cut in a variety of ways, depending on the desired characteristics. The sections cut from the crystal are known as *blanks.* One such blank, sketched in Figure 9-37, is called an *x cut* and has a thickness that is parallel to the x axis and a length parallel to the y axis. A mechanical stress applied to its faces along the y axis produces a voltage along the x axis. The dipoles created in quartz crystals do not rotate; thus, there is a permanent orientation. This characteristic permits a quartz crystal, once cut and polished to specified dimensions, to maintain a resonance frequency with an extremely high degree of accuracy. This feature finds great use in controlling the frequency of radio broadcast signals and in chronometers (time pieces of great accuracy).

9.9 LIQUID-PHASE EPITAXY AND MOLECULAR-BEAM EPITAXY

Liquid-phase and molecular-beam epitaxies (MBEs) are methods developed by Bell Laboratories to produce new semiconducting materials not found in nature and having tailor-made atomic structures that provide for an array of new, built-in electronic, mechanical, and optical properties. The word ***epitaxy*** describes the overgrowth in layers of crystalline material deposited in a definite orientation on a base material of different crystal structure and chemical composition. In this world of human-made materials, this scientific research and development effort by Bell Laboratories demonstrates once more our ability to exert precise control over the creation of a new material by methods that ensure the exact composition and structure of the alternating atomic layers that make up a material.

Starting with a substrate of gallium arsenide, a semiconductor material, atomic beams of controlled intensity are aimed at the base material in an ultrahigh vacuum [Figure 9-38(a) and (b)]. Shutters are used to turn the beams on and off as they are directed at the base from a heated oven. A layer of gallium atoms followed by layers of arsenic, aluminum, arsenic, then gallium atoms, repeated many times, produces a crystal resembling a thin, highly polished mirror. Liquid-phase epitaxy, an older technique, produces satisfactory wafers, but MBE appears to be the most promising. Using MBE, ultrathin, multilayer crystals with different structures have been grown. Quantum-well structures with up to 100 alternating layers represent one class of such crystals, with each layer as thin as several hundredths of a micrometer. Monolayer structures with alternating layers of atoms in a stack, each layer two atom planes thick, are being produced. Multilayer structures consisting of several thousands of individual monolayers have been grown with a total thickness of 1 μm. Research has revealed that, because of the thinness of these layers, the electrons and holes normally confined to each layer will interact with their counterparts in adjacent layers and build *superlattice* structures with even more desirable properties [Figure 9-38(c)].

9.10 LIQUID CRYSTALS

Liquid-crystal displays (LCDs), currently used in watches, signs, and similar applications, offer a display system that does not require as much energy as LEDs [light-emitting diodes; Figure 9-38(d)]. The image familiarly seen on the LCD produces a silver display, while the LED is the familiar red display. Images for the LCD that are generated by lasers have potential for multicolored video displays for information systems. The LCD offers many possibilities for easier information retrieval.

Most known substances can exist, given certain conditions, in a gaseous, liquid, or solid state. Very few compounds can exist in a fourth state, between the liquid and solid phases known as the ***liquid-crystal phase.*** This mesomorphic (intermediate) phase or anisotropic liquid was discovered over 100 years ago (1888) but has remained a physical oddity until recent advances in the technologies of electrooptics and thin-film components have permitted several important applications of these materials, particularly in the visual display of information.

Liquid crystals (LCs) are organic compounds that flow like a liquid while maintaining the long-range orderliness of a solid. Normally, crystals of a pure compound, when heated, show a well-defined and characteristic melting point at which the ordered crystalline lattice structure breaks down and the material becomes a liquid. In the liquid phase, the individual molecules show no preferred spatial orientation. The feature of liquid crystals is that, during the melting process, the well-ordered three-dimensional crystalline structure transforms into a one- or two-dimensional state of order. This results in a

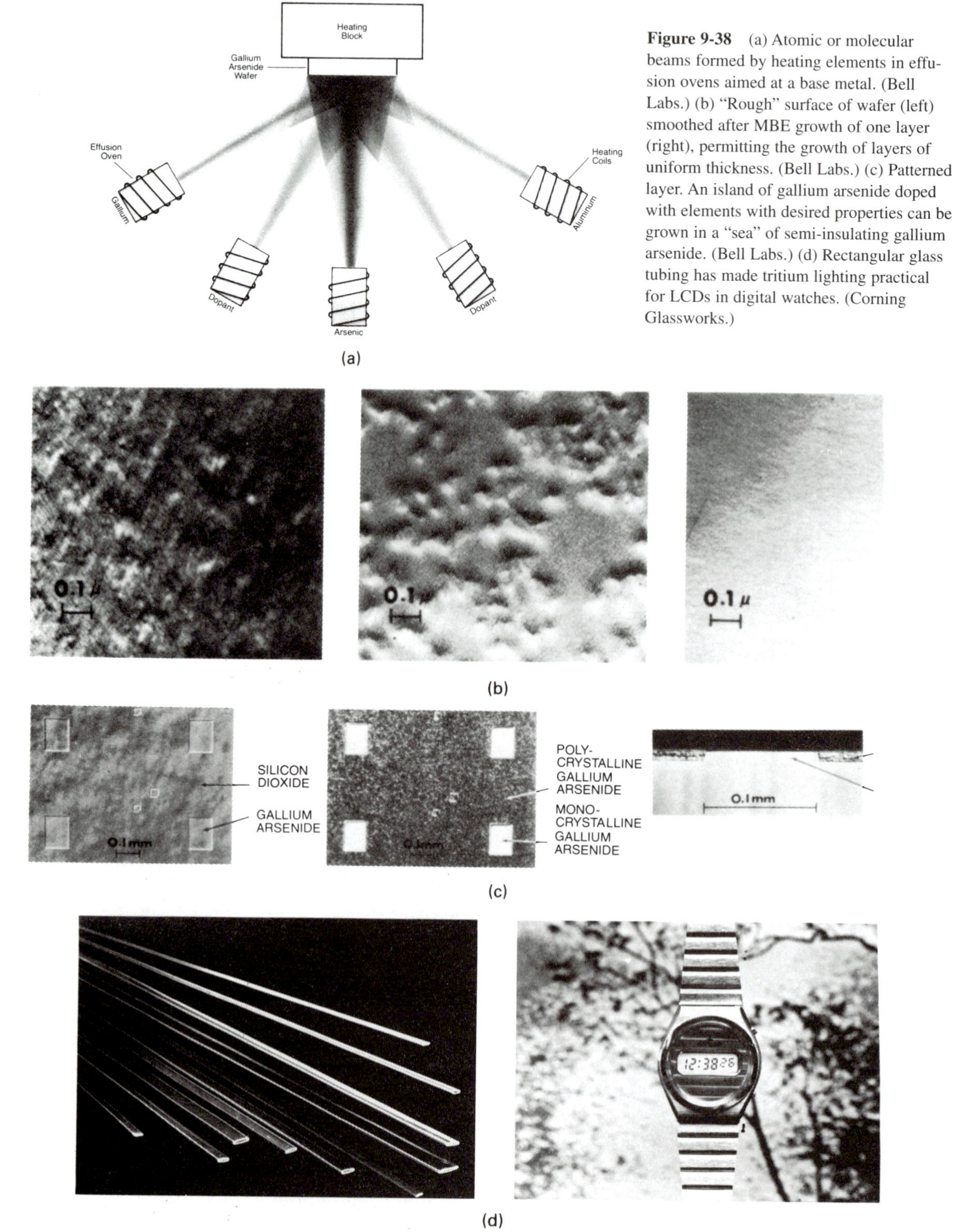

Figure 9-38 (a) Atomic or molecular beams formed by heating elements in effusion ovens aimed at a base metal. (Bell Labs.) (b) "Rough" surface of wafer (left) smoothed after MBE growth of one layer (right), permitting the growth of layers of uniform thickness. (Bell Labs.) (c) Patterned layer. An island of gallium arsenide doped with elements with desired properties can be grown in a "sea" of semi-insulating gallium arsenide. (Bell Labs.) (d) Rectangular glass tubing has made tritium lighting practical for LCDs in digital watches. (Corning Glassworks.)

material that has some optical properties of a solid combined with the fluidity of liquids, resulting in a mix of unique properties.

The molecular aggregates (collection of molecules) of an LC compound are in the form of long cigar-shaped rods. The orientation of these rodlike polar molecules forms the basis for classifying three basic types of liquid crystals: smectic, nematic, and cholesteric. The ***smectic*** phase consists of flat layers of cigar-shaped molecules with their long axes oriented perpendicular to the plane of the layer. This is the most ordered phase. The molecules within each layer remain oriented within each layer and do not move between layers. Figure 9-39 shows a structural model of this smectic mesophase. The molecules lie with their long axes parallel in layers. The molecules can move relative to each other, and consequently several types of smectic structures can be formed, depending on the inclination of the long axes of the molecules to the plane of the layers.

The ***nematic*** phase also has molecules with their long axes parallel, but they are not separated into layers. Rather, their structural arrangement is similar to the ordinary packing of toothpicks in a box. Figure 9-40 depicts such a nematic mesophase structure. While maintaining their orientation, the individual molecules can move freely up and down.

The ***cholesteric*** mesophase (Figure 9-41) can be defined as a special type of the nematic phase in which the thin layers (one molecule thick) of mostly parallel molecules have their longitudinal axes twisted (rotated) in adjacent layers at a defined angle. Each layer is basically a nematic structure. The axes of alignment of contiguous layers differ by a small angle and produce a helix or progressive rotation of many layers in an LC material.

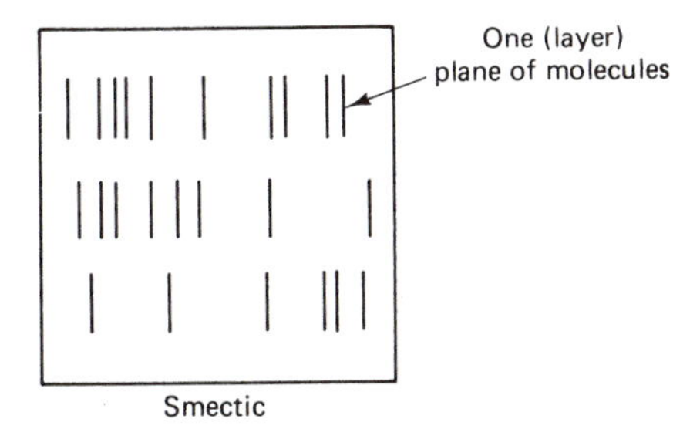

Figure 9-39 Model of the smectic mesophase.

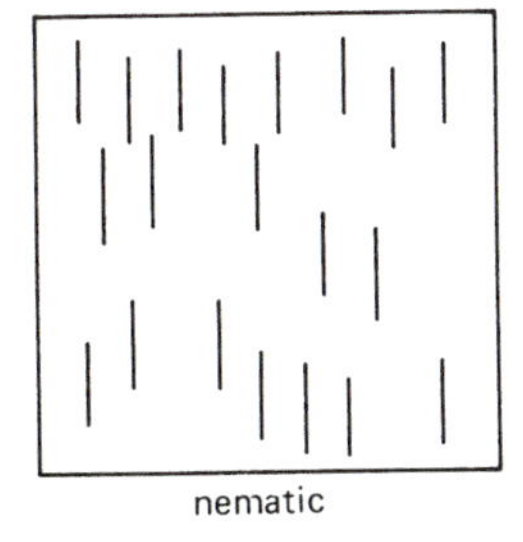

Figure 9-40 Model of the nematic mesophase.

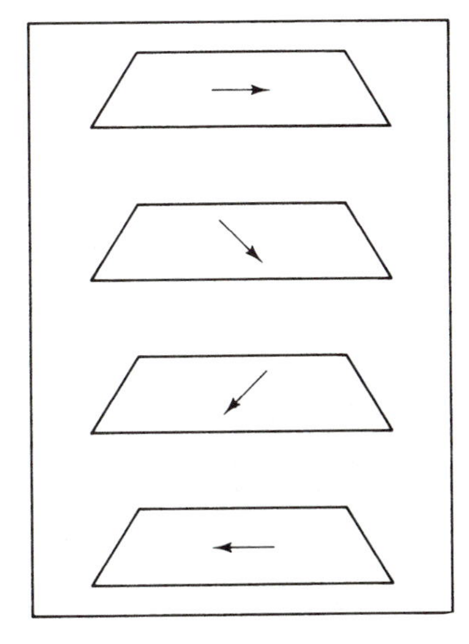

Figure 9-41 Model of the cholesteric mesophase.

Because their molecules can be aligned by electric and magnetic fields to produce changes in their optical properties, the ***nematic*** liquid crystals find increasing use in electrooptical devices. Additionally, several mechanical influences such as pressure, impact, or temperature induce a change in the structure of nematic liquid crystals similar to those brought about by electromagnetic fields. This cholesteric structure exhibits anisotropic optical properties. When such a material is illuminated with white light at certain temperatures, iridescent (having shifting or changing colors) colors brought about by a fraction of the incident light are observed in the material. The much larger portion of light is transmitted by the LC material. This reversible color phenomenon functions over a temperature range of about −20° to 250°C. These properties of LC materials are finding many applications in heat-transfer studies (thermal mapping); nondestructive testing (NDT); toys and games; holography; medical diagnosing of vascular diseases; and flat-panel, full-color display panels.

Switchable privacy glass, produced by Marvin Corporation of Warroad, Minnesota, uses a technology similar to that found in laptop computer screens. The windows contain a thin film of liquid crystals made by 3M, sandwiched between layers of tempered glass. When a small electric charge of about 1 watt per square foot of window is applied, the crystals align, and the glass appears clear. With power off, the crystals arrange themselves in a random fashion, scattering light. The result is that the glass turns a frosty opaque white. See Figure 6-1 and the related text discussion of privacy glass used in the smart house.

9.10.1 Liquid-Crystal (LC) Cell

An LC cell consists of a layer of nematic mesophase LC material between two glass plates that are glued or fused together. The thickness of the LC material is about 10 to 25 μm. The two glass plates have transparent electrodes, made of a transparent and conducting material such as tin or indium oxide, deposited on their inside faces. Figure 9-42 shows a sketch of the construction of an LC cell.

9.10.2 Homeotropic and Homogeneous Orientations

Two preferred orientations of the LC molecules, *homeotropic* and *homogeneous,* are used within LC-display devices (LCDs). Figure 9-43 shows the homeotropic orientation with the long axes of the molecules perpendicular to the surface of the glass plates and elec-

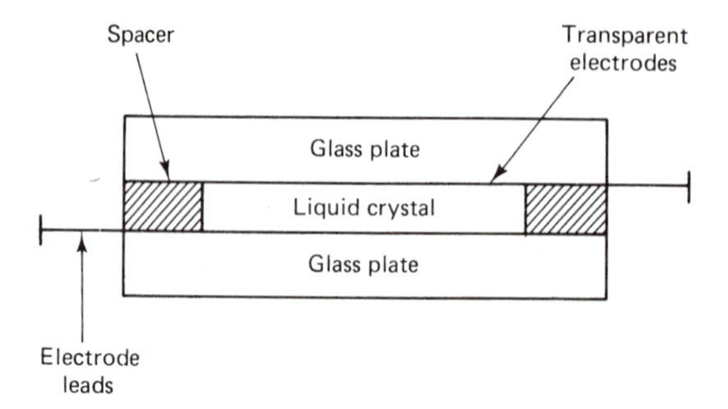

Figure 9-42 Cross-section of an LC cell.

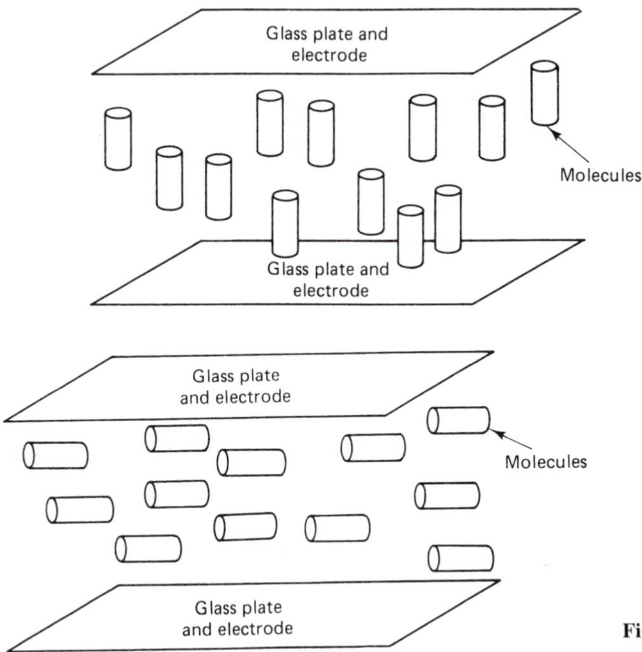

Figure 9-43 Homeotropic molecular orientation.

Figure 9-44 Homogeneous molecular orientation.

trodes. This orientation can be achieved by chemical doping of the nematic phase. The long, cigar-shaped rod orientation of the molecules allows the groups of molecules to act like dipoles in the presence of an electric field.

Figure 9-44 shows the orientation of the molecules to be parallel to the glass plates. This parallel orientation can be produced in the LC material by mechanically rubbing, unidirectionally, the glass plates with a leather cloth prior to assembly of the LCD. Another means of accomplishing this orientation is by the deposition of a layer of dielectric over the transparent electrodes. Each plate is then rotated 90° in relation to each other. In effect, the LC material acts now as a set of polarizers that cause the light passing through the LC cell to be rotated 90°. Another way of altering the alignment of these layers in the LC material is to apply an electric field across the material. The molecules then align themselves in the direction of the electric field, which prevents the 90° twist of the molecular layers. The incident light cannot be transmitted through the cell. By judicious selection of the LC material, polarizers, glass-plate treatment, and the electric field, the incident light can be controlled through the cell; that is, it may be reflected, rotated, transmitted, or extinguished.

9.10.3 Reflective-Type LC Cell Operation

Although there is a transmittive-type LC cell, the reflective-type cell is chosen to explain the operation of a numerical-display LC device. Figure 9-45(a) is a sketch of the components of a reflective-type LCD. The main components of the LC cell consist of the two glass plates, electrodes, and LC material. A vertical polarizer and a horizontal polarizer, plus a

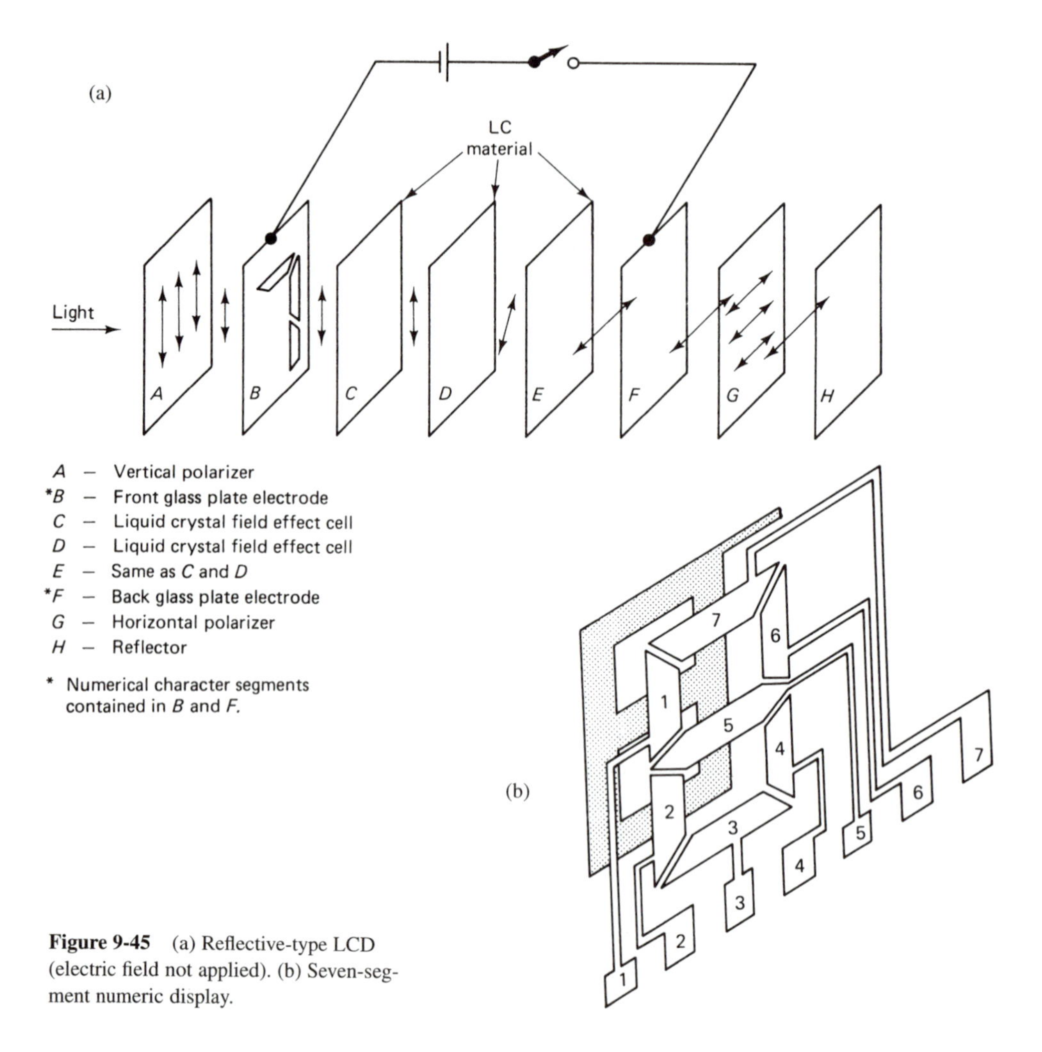

Figure 9-45 (a) Reflective-type LCD (electric field not applied). (b) Seven-segment numeric display.

reflector, are added to the cell. With the plates treated and arranged to produce the 90° twist described previously, unpolarized light enters the vertical polarizer, rotates 90° through the LC material, and passes through the horizontal polarizer and on to the reflector. The reflector reflects the light back through the same path. The face of the reflector appears light all over its surface.

If we wish to show one numeral in the display, then several LC cells are placed together in a particular configuration; each controls one segment of the digital display, as shown in Figure 9-45(b). An electric field activates the selected segments, which act together to form the particular numerical character desired. The digital display can be impressed on the electrode on the front glass plate in Figure 9-45(b), with the electrode on the back plate acting as the base of the common electrode. The molecules in the area of the energized segments will align with the electric field and prevent rotation of the polarized light. Consequently, the vertically polarized light passes through the cell everywhere but in the region of the energized pattern elements. The end result is that the energized display elements appear as black images against a light background. By properly choosing the correct combination of segments to be activated, any digit from 0 to 9 can be displayed.

9.11 SUPERCONDUCTORS

As the world seeks cleaner, more efficient, and readily available energy, many developments, ranging from improved solar collectors to cold fusion, to high-temperature superconductivity, quickly grab the public's attention. High-temperature superconductors [Figure 9-46(a)] achieved just that effect back in 1986. A superconductor is a material that allows an electric current to flow without resistance. Most conductive materials, usually metals, restrict the flow of electricity to some degree; this is known as *resistance.* If metals were compared for their resistance on a scale of 1 to 10, with 1 being the most resistant and 10 being the least, iron would be close to 1 and gold and silver near 10. Copper and aluminum are somewhere in the middle. Resistance of an electric current causes energy losses due primarily to heat generated from the resistance. Superconductors have a theoretical resistance of 0 Ω. This means that there is no energy loss when an electric current passes through the superconductor. Because of this feature, superconductors can be of great benefit in electrical transmission, from very small millivolt signals in computers and other instrumentation to the high-powered, electrical-transmission lines that bring electricity to our homes and businesses. This brings up the question, Why aren't superconductors utilized more? This question is best answered with a little historical perspective.

Superconductivity was discovered in 1911 by a Dutch physicist, Heike Kamerlingh Onnes, while he was observing mercury at liquid-helium temperatures (−452°F, almost absolute zero). Because of the difficulty and expense of working with superconducting materials at such low temperatures, uses for it have been very limited. Research on superconducting materials continued, and by 1973, the temperature had risen to −418°F with the discovery of superconductivity in a niobium–germanium alloy. Because metals are conductive, the most logical choice for a superconductive material would be metal. The discovery in 1986 of superconducting metal oxides by K. Alex Muller and J. Georg Bednorz, physicists at the IBM Zurich Research Laboratory in Switzerland, made quite a stir in the scientific community. Metal oxides are normally nonconductive, but certain oxides, when mixed in the correct ratios, have been found to be superconducting at temperatures that are reasonably achievable.

Since the Muller and Bednorz discovery in 1986, research around the globe has boomed as people seek to find ways to commercialize a product using this so-called high-temperature superconductivity. The number of compositions has increased dramatically. One of the amazing things about this discovery is that it has been the only major discovery in the twentieth century that requires common laboratory equipment and therefore can be duplicated almost anywhere, including the high-school classroom. This means that not only are researchers experimenting with high-temperature superconductivity, but people with an interest—engineers, inventors, industrialists, and tinkers—are trying their hands at making a usable product and furthering the research.

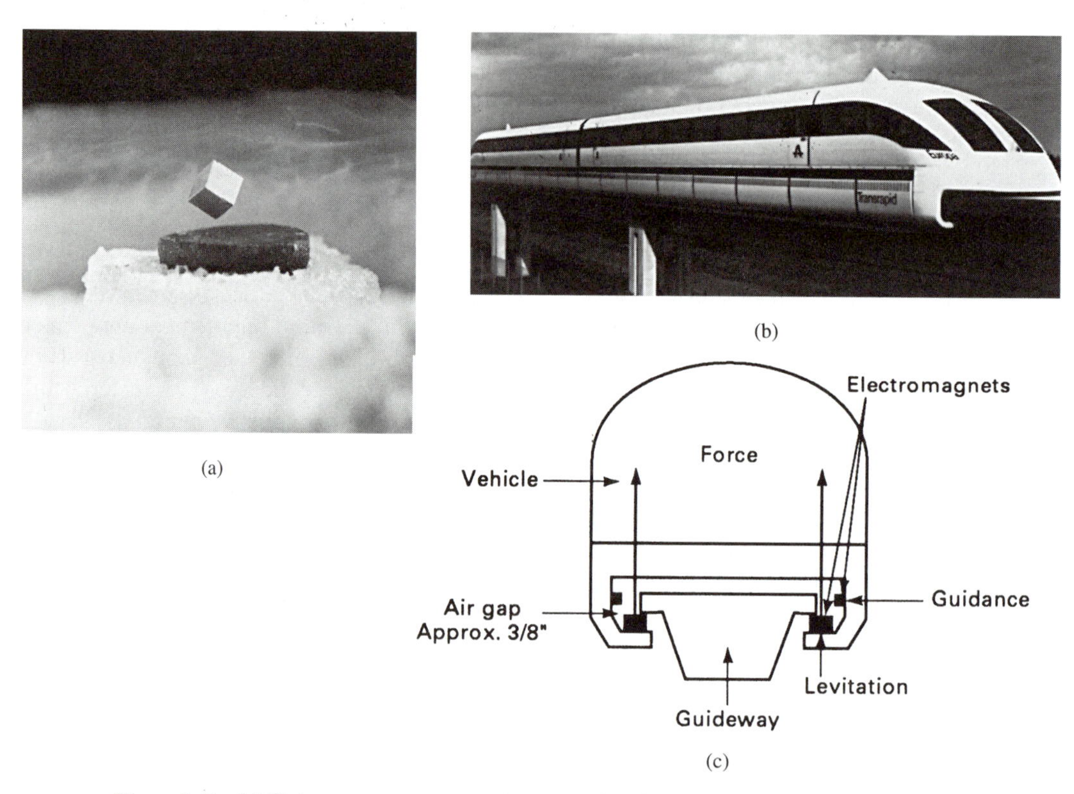

Figure 9-46 (a) High-temperature superconductors resulted from the use of ceramic materials. The Meissner effect, demonstrated here and explained later in this module, shows how superconductivity permits levitation of objects. (NIST.) (b) High-speed travel exceeding 300 mph over long distances with Maglev trains is a desirable alternative in some cases to air and road travel. (Transrapid.) (c) Electromagnetic Maglev, depicted in this simple diagram, is one option to a suspension system whereby electromagnets, perhaps high-temperature superconductors, will levitate the train over the track. (Federal Railroad Administration.)

This discovery will have a tremendous impact on society. Besides wires for conducting electric currents, we are witnessing attempts at revolutionary concepts in medical equipment, trains, batteries, machinery, computers, and so on. Figure 9-46(b) and (c) shows a magnetic-levitation (Maglev) train that uses magnets to suspend the train over, but not touching, the rails. The magnets and rails act together as a long-stator, linear induction motor to propel the high-speed supertrains. The Maglev project in the United States seeks to augment air and road travel with high-speed supertrains. Electromagnetic levitation of trains using superconducting magnets is one application of electronic materials that is sure to receive attention.

Superconducting metals act as normal metals and offer resistance above the ***critical temperature (*** T_c***)*** but have zero electrical resistance below T_c. Figure 9-47(a) depicts electrons flowing through a normal metal conductor [Figure 9-47(a)] and a superconductor [Figure 9-47(b)] from the negative end to the positive end. In the normal conductor, electrons collide with phonons. These quanta, or packets of energy, are caused by the vibration

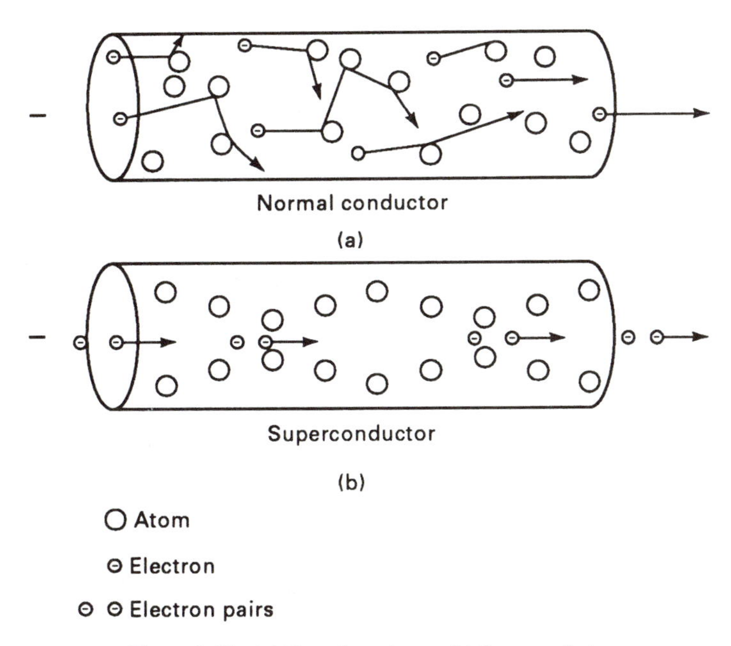

Figure 9-47 (a) Normal conductor. (b) Superconductor.

of the atomic space lattice, in which atoms form into crystalline solids. As materials become hotter, the vibration of their atomic lattices increases, causing a corresponding increase in electrical resistivity. Silver has a resistivity of 1.6 μΩ/cm at 20°C, but that resistance increases to 4.6 at 500°C; copper goes from 1.7 μΩ/cm to 4.5 with the same increase in temperature. The collision of electrons with phonons is resistance, which creates its own heat in the conductor. As long as metals are maintained at moderate temperatures, they are *fairly* good conductors.

A theory known as ***BCS theory*** (after Bardeen, Cooper, and Schrieffer, 1957) holds that, in superconducting metals, pairs of electrons move with ease through the vibrating lattice [Figure 9-47(b)]. As one of the paired electrons passes between two atoms, it causes a drawing together of the positive-charged atoms, which results from their attraction to the negative-charged electron. This is much like a speeding truck sucking cars into its draft. Once the electron passes the two atoms and they are still momentarily close together, they generate a positive attraction for the second electron in the pair and pull it into their region. This seems to set up a pulsing effect that causes masses of valence electrons to move through the superconducting material without resistance. The flow of electrons through the superconductor is perpetual, as long as the critical temperature is maintained.

High-T_c superconductors established some new lines of thought. While ceramics, with their covalent and ionic bonding, are normally insulators, a group of superconducting ceramic oxides possess superconducting properties at temperatures high enough to use the relatively cheap liquid nitrogen as a coolant to achieve superconductivity. Muller and Bednorz's now famous 1:2:3 superconductor was comprised of a ratio of 1 part yttrium to 2 parts barium and 3 parts copper. Copper oxides in these ceramics have positive- and

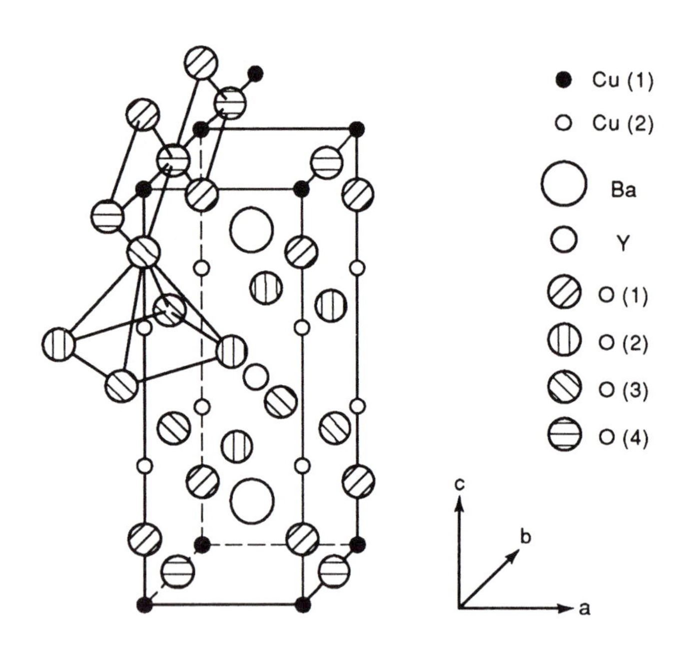

Figure 9-48 A single unit cell from the atomic space lattice of a ceramic–oxide superconductor composed of one atom of rare-earth-element yttrium (Y), two atoms of barium (B), 17 atoms of copper (Cu), and 27 oxygen (O) atoms or vacancies. This unit cell is repeated millions of times in the solid material. The superconductor is abbreviated YBCO.

negative-charged ends (dipole) in the same manner as a bar magnet. It is believed that valence electrons passing through the lattice cause a shifting in the polarity of the charged ends, thereby developing a strong positive charge as they pass (see Section 9.2). Again, as with the superconducting metal, the second electron in the pair is pulled in by the positive attraction. This phenomenon is repeated many times to provide superconducting current. Figure 9-48 provides a diagram of a unit cell for the new $YBa_2Cu_3O_7$ (yttrium–barium–copper oxide).

The ceramic–oxide-superconductor materials provide a major challenge in processing techniques because of their brittle nature and the problems of keeping the composition in proper proportions. A *plasma-spraying technique* can coat a variety of large and small objects, including large copper panels and spherical vessels, with the ceramic–oxide.

Many applications involving superconducting materials call for powerful magnets, as with magnetic-resonance-imaging (MRI) medical-diagnostic devices (Figure 9-49). *Superconducting magnets,* such as the magnets used for these devices, consist of doughnut-shaped field-coil windings made of tightly wound niobium–titanium wire. They are cooled to near 0 K and high amperage is applied. As long as the coil stays cold enough, the magnets continue to conduct the electric current, free of resistance, at a highly stable rate.

Magnetic characteristics of high-T_c superconductors result in their excluding all magnetic flux in the material. This phenomenon, known as the ***Meissner effect,*** is illustrated in Figures 9-46 and 9-50. The figures compare a superconductor to a normal ferromagnetic material. The drawings show the lines of flux *(H)* as arrows penetrating the normal conductor. A critical field *(H_c)* is reached for the superconductor when the temperature is below the critical temperature. Then no lines of flux penetrate the superconductor. When the external field is created by a permanent magnet, it can move the superconductor. Figure 9-46 is the familiar picture showing a permanent magnet levitating over a pallet of

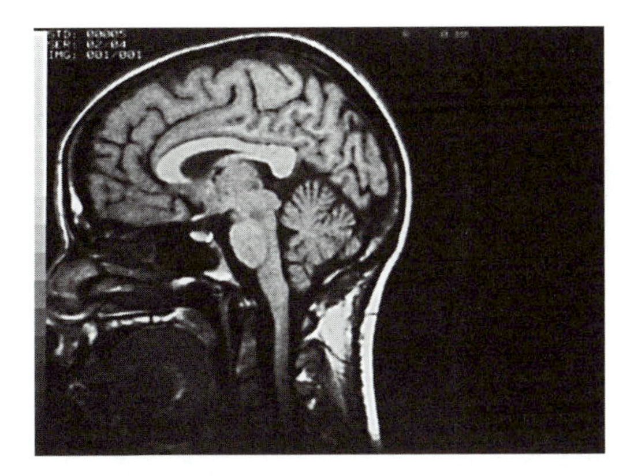

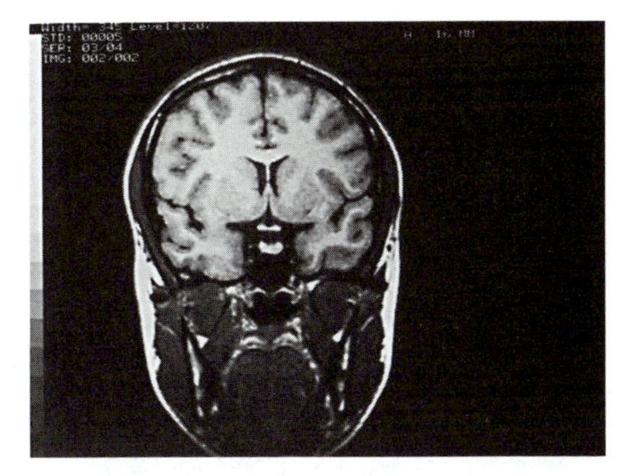

Figure 9-49 Two prints of a set of many slices taken through a human head by a magnetic resonance imaging (MRI) scanner to reveal bone and tissue for diagnosis of suspected cancer. (General Electric.)

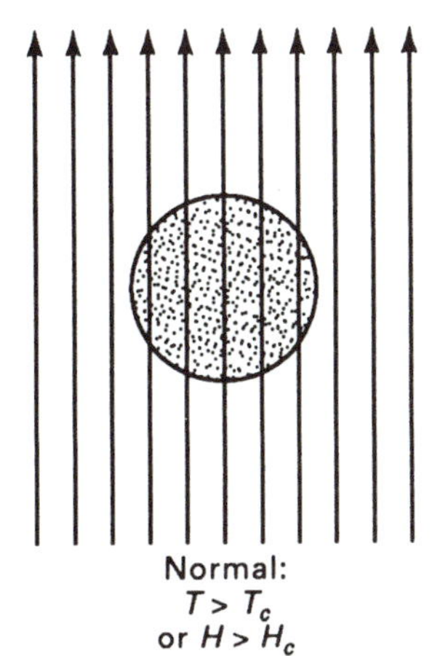

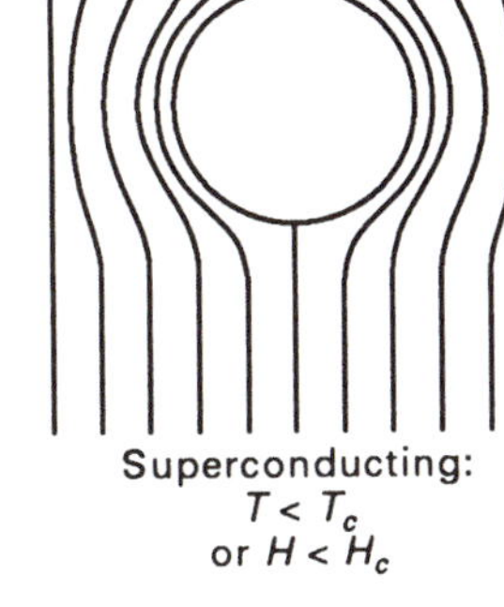

Figure 9-50 Comparison of the effect of the magnetic field *(H)* on normal conductors and superconductors. Critical field (H_c) is reached when the temperature *(T)* falls below the critical temperature.

ceramic–oxide superconductor. The magnet floats in the air above the superconductor because its lines of flux are repulsed by the superconductor and the magnetic field is excluded from the magnet's interiors. In effect, the superconductor acts as a perfect magnetic mirror to the magnet. The superconductor material and magnet are firmly attracted to one another but never touch, as would a normal magnet attracting iron. The cloud beneath the pallet seen in the picture is a result of the liquid nitrogen boiling away at room temperature.

High-T_c superconductors offer many gains to society, such as cost savings in power generation. By applying superconductivity to electric-power generators, it is possible to generate electricity without the losses normally associated with the flow of current in rotor windings in conventional generators, thereby saving millions of dollars (Figure 9-51).

The demand for electricity in the United States is expected to double by the year 2030. Large industrial motors delivering 10,000 HP consume almost 15% of all U.S. electricity. If their coils could be made out of a superconducting wire, the motor could be reduced in size by one half, with a corresponding 60% reduction in weight. The international compe-

Figure 9-51 General Electric Company has built a superconducting generator that produces 20,600 kilovolt-amperes of electricity, or enough for a community of 20,000 people. This is about twice the amount of electricity that would be generated by a conventional generator. Field windings of conventional generators are made of a copper–silver alloy. The low-temperature superconductor generator employs modular field windings consisting of hundreds of filaments of a metal composite of a niobium–titanium alloy in a copper matrix. When available, high-T_c ceramic superconductor materials will provide even more cost savings. But no one can say when those breakthroughs will occur.

tition to produce such a wire is fierce. To date, laboratory samples of such wire are small, consisting of tapes about ½ in. wide, 2 in. long, and a fraction of an inch thick. These samples are extremely brittle and very difficult to make into wire. With a proven high-temperature, superconductor metallic oxide compound abbreviated YBCO (yttium, barium, copper, oxygen) (see Figure 9-48), researchers have used an ion-beam-assisted, deposition spray process to produce a thin film of this material. In the process, one beam of ions lays down a few hundred thousandths of an inch of zirconia on a nickel-alloy tape, while a second beam orients the crystals of YBCO as they form on a nickel substrate. The zirconia acts to prevent the YBCO from reacting with the nickel substrate, and the other ion beam forces the YBCO into a regular crystalline order. In effect, a sandwich is produced that is malleable and has retained much of its superconductivity. Much time and effort will be required to transform the lab process into a low-cost volume manufacturing process.

9.12 SMART MATERIALS AND INTELLIGENT STRUCTURES TECHNOLOGY

Hardly a day goes by that we don't read or hear about some "smart" technology: smart cars, smart highways, smart credit cards, smart bombs, smart explosives, smart pills, smart structures, smart batteries, smart skins, and smart fluids. With each major group of our Family of Materials, we have provided you with examples of how smart materials within each group work and explained some of their uses. This technology shows great promise because it represents the potential for materials systems that often involve electronic and optical materials in conjunction with metals, ceramics, polymers, and composites.

As Figure 9-52 shows, smart materials may find uses such as shape-memory alloys in rather simple products like eyeglass frame temples, or they may involve complex

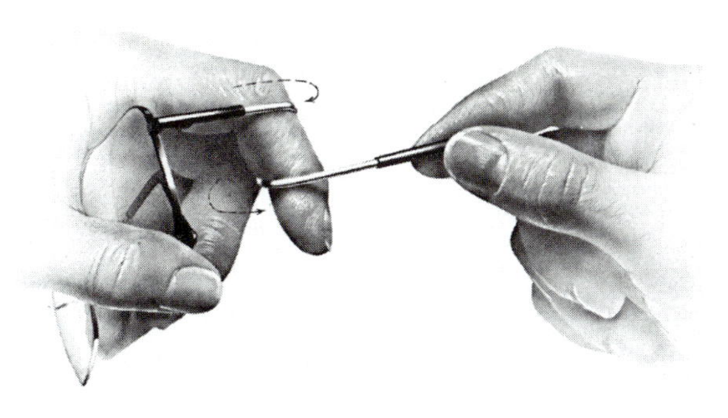

Figure 9-52 Shape memory alloys, used for the Nitinol eyeglass frame temples, "remember" their original shape and immediately return to that shape when the bending force is removed.

intelligent structures such as bridges and aircraft wings. An example of the use of a shape memory alloy is provided by the United States Air Force (USAF) and its use of CryoFit® permanent fittings and Cryolive® dematable end fittings in many of its aircraft hydraulic systems. These fittings are made by Advanced Metal Components, Inc., from nitinol barstock. Nitinol barstock is first machined at room temperature to an inside diameter (ID) that is smaller than the outside diameter (OD) of the tube to be joined. The fitting is then cooled in liquid nitrogen below its martensite transformation temperature. Next it is expanded in the cold condition to an ID larger than the OD of the tube to be joined. The finished fitting is placed over the tube end, where it warms and recovers onto the substrate.

The fastest growing applications for binary NiTi alloys use this shaping memory or superelastic property. These alloys will restore up to 8% strain repeatedly in this pseudoelastic manner. NiTi wires are used to make very flexible eyeglass frames [Figure 9-53(a)], which are essentially unbreakable. Highly flexible antenna wire for cellular phones is produced from these alloys. The service life of these alloys is inversely proportional to the amount of deformation. If deformed close to 8%, a NiTi alloy's life span may be only a few hundred cycles. On the other hand, if the magnitude of the deformations is less than 1%, the alloy's life may be several million cycles.

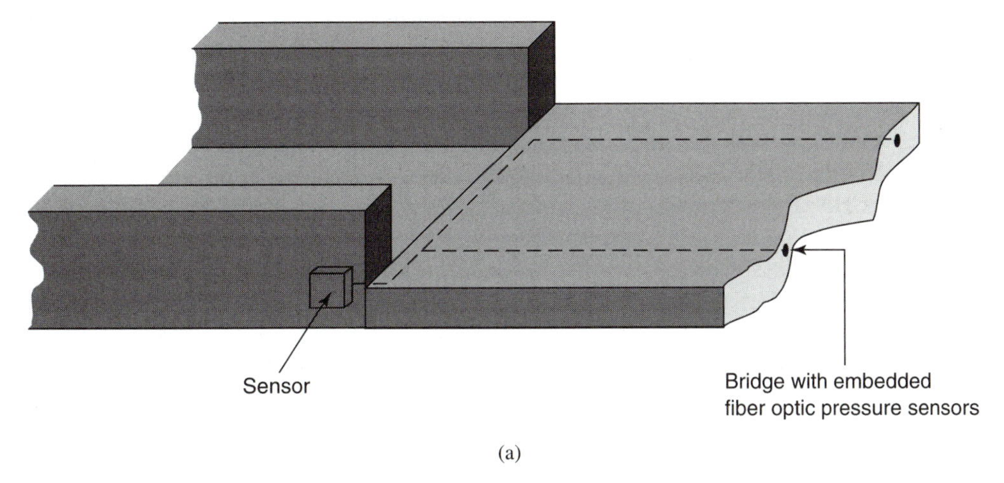

Figure 9-53 (a) Using optical fibers, sensors, and recording devices embedded in concrete or attached to steel structural members, it is possible to monitor stresses on bridges, walls, and other structures.

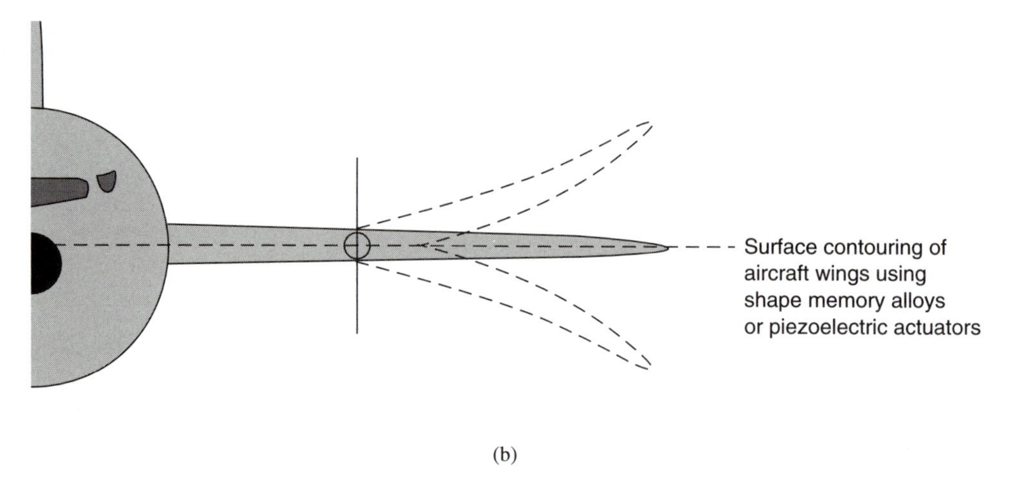

(b)

Figure 9-53 (b) Current research may one day yield aircraft equipped with intelligent structures that can deform adaptive wings in reaction to variations in air pressure and air speed, and thus improve high-speed cruising.

The U.S. Army is developing a smart skin for tanks. This skin uses sensors much like those seen in the bridge in Figure 9-53(a). The armor skin would detect incoming projectiles, such as shells made of depleted uranium, and trigger a repulsive explosion in microseconds to counter the shell's impact. An aerospace intelligent structure is the adaptive wing shown in Figure 9-53(b). It employs piezoelectric materials, which were discussed in this module. Piezoelectrics are being evaluated to see if they can deform wings in response to changes in air speed and air pressure and thus provide for more efficient cruising. Both shape memory alloys and fiber optic systems are also being evaluated by the aerospace industry.

In another development, the "smart battery," an intelligent materials system, integrates battery technology with semiconductor technology. The battery pack designed for laptop computers includes a microprocessor inside each lithium ion battery pack. The system keeps track of battery use and notifies the user of the time remaining on the battery. In the design of these new smart batteries, some major computer manufacturers are teaming up with major battery manufacturers to develop standard sizes and configurations. This would ease the burden on consumers. Just like with the standard AAA and D cells, it would be possible to purchase standard replacement batteries for the new laptop computers in many stores.

In Module 2, we defined *intelligent materials systems, smart materials,* or *smart structures* as materials and materials systems designed to mimic biological organisms and offer the ultimate materials system that can place control and feedback into a material structure. These materials take their cue from biological elements such as muscles, nerves, and bodily control systems, which adapt to environmental changes.

Where will smart materials and intelligent structures go from here? Much of the focus of smart materials research is based on mimicking nature. We will examine that technology next.

9.13 BIOMIMETICS AND BIOMATERIALS

9.13.1 Biomimetics

From the beginning of civilization, technology has reflected human attempts to imitate nature. In the early stages of human flight, the effort to build flying machines that imitated the structure and motion of birds met with failure. On the other hand, imitating the structure used by bees has produced advanced honeycomb composites suitable for aircraft.

Transforming raw materials into finished products usually requires large amounts of energy and many different processes. Because of the concern for the limitations of natural resources, considerable research and development is aimed at reducing the input of both raw materials and energy to process materials. Synthesis and processing determine the structure and properties of a material. How can technology conserve resources and be friendly to the environment in both synthesis and processing, and still obtain the desired structure and properties of materials?

Materials technology has turned to nature in the search for low-energy (low-pressure and low-temperature) and low-environmental-impact synthesis. ***Biomimetics*** is a technology that seeks to duplicate some of the processes that living organisms employ to construct the bones, tissues, shells, spider webs, and other materials in nature. The view of the cross-section of a seashell (Figure 9-54) reveals the intricate structure that mollusks achieve as they construct their hard, durable homes. Some doubt the value of pursuing the biomimetics route to materials synthesis for engineering materials, but others feel that it is useful to study nature to produce biomaterials. Biomaterials for dental and bone transplants have been produced using biomimetics that replicate sea coral. Biomimetic research enhances the development of the smart materials discussed previously.

Figure 9-54 Biomimetics uses techniques that imitate nature to synthesize materials. Mollusks are examples of complex structures at nano scale, which nature produces at low pressure and low temperature, and without harmful effects on the environment. (DOE-PNL.)

Thin-film coatings offer many possibilities if they can be manipulated (1) to achieve unique structures that possess designed properties and (2) for selective application to a variety of surfaces. The ability to orient the crystal structure of coatings offers possibilities for special optical, electronic, magnetic, and mechanical properties. Tough ceramic coatings provide increased durability for many products. The Pacific Northwest Laboratory used biomimetic synthesis to develop an organic interface that controlled crystal growth and formed desired ceramic coatings on polymer, glass, and metal surfaces of complex shapes at lower cost than current vapor deposition techniques. As depicted in Figure 9-55, the coating was deposited at temperatures below 100°C in a water solution that produced no hazardous waste.

Engineers are taught to be curious about the world around them, to observe how Mother Nature does things. Some examples that illustrate this awareness of curiosity are the following. How do honey bees construct their nests? Why doesn't a blade of grass break when stepped on by an animal or human? How does a tree withstand the force of wind? What makes the adhesive produced by barnacles so strong that sand blasting, at considerable expense due to labor and the need for drydocking of vessels, is required to remove them from underwater hulls of ships and pilings? Or how does a bat manage to capture insects as it flies in the dark of night at high speeds? Some twenty years ago NASA researchers were studying how to use networks of strings to fabricate connecting platforms in outer space. Part of their studies involved an experiment with a space-borne spider. As the spider spun a web, it used threads of nonuniform size. NASA researchers used this information recently to help design an improved tennis racket with greater power, feel, and control. Two characteristics of tennis rackets, as well as other sports equipment such as baseball bats, must first be explained. (This discussion builds on the information about testing sports equipment in Module 4.)

The point that reaches the maximum speed in swinging a racket is located at the toe of the racket or at the furthest location from the player's hand. Maximum power is gener-

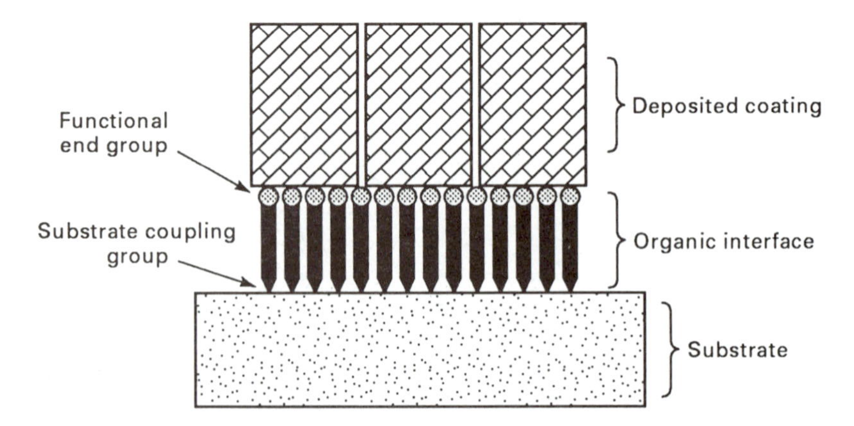

Figure 9-55 Biomimetic synthesis allows thin ceramic films (deposited coating) such as magnetic iron oxide to be deposited on substrates such as polyethylene and polystyrene plastics in a very controlled manner so that the functional end groups covalently bond (discussed in Module 2) to the substrate. (DOE-Pacific Northwest Laboratory.)

ated when a ball hits that point. This point is the desired point of impact when serving a tennis ball. The other characteristic is the point of optimum vibration. Known also as the center of percussion or sweet spot, this location produces the least amount of vibration on impact of a ball. This vibration, kick, or jar is transmitted through the racket to the player's hand and, in some cases, can cause chronic arm problems. Many tennis racket designs have resulted in enlarged sweet spots to reduce this vibration. Sweet spots are usually found at the geometric center of the string area. If the center of percussion could be moved outward toward the location of the position of maximum power, improvements in the power delivered to the ball would be achieved because most of the ball impacts would now be on or close to the sweet spot. Fewer vibrations would be felt by the player, with attendant reduction in stress on the player's arm and hand. Better control would also be experienced because of the increase in the racket's stability. (The racket is less likely to twist in the player's hand when impact occurs at the center of percussion.)

Solving the problem was greatly helped by the knowledge received in the spider experiment. Using a computer and a genetic algorithm, and the laws of physics dealing with vibrating strings, an optimum design that involved distributing the required mass along each string (or tapering the individual strings) was obtained. A ***genetic algorithm*** is defined as one that implements a mathematical evolutionary process that resembles biological evolution in some respects and that arrives at an optimized design by following a sequence of random design changes and preserving those changes that improve performance. Any string material can be used, and the tapering can be achieved by spinning the string thicker or bonding different string materials along a central braid in the string.

9.13.2 Biomaterials

Materials science is not only interested in mimicking nature but in contributing to bioengineering materials and components that work with human and animal systems. ***Biomaterials*** are compatible with human and animal systems; this allows the material to be implanted or manipulated in people and animals. ***Titanium*** is an example of a biomaterial that serves as implants for joint replacements and dental reconstruction (refer back to Figure 5-61). Figure 9-56 shows the many applications of titanium implants used in the human body.

New materials to repair broken and deformed bones that mimic the natural composition of bone have been formulated. This paste is injected into the joint for a speedy recovery without the normal heavy plaster casts. Synthetic polymers have been used as artificial skin to replace burned or otherwise damaged skin. The "plastic" skin, sometimes in conjunction with skin grafts, performs like real skin to protect the wound, allowing new skin to grow naturally.

In Module 1, you were introduced to ***Vitallium,*** a biomaterial used for human prosthetics. The development of this alloy (cobalt, chromium, molybdenum, and nickel), first as a dental material (Figure 9-57) then as a material for orthopedic use (Figure 9-58), is typical of the synergy of engineering materials technology. Once Vitallium was found to be a suitable biomaterial during the 1920s and 1930s, then new methods of molding the alloy were sought. Because of the high melting temperature of the cobalt-based alloy, ethyl silicates were developed to serve as binders for the ceramic refractory to counter the

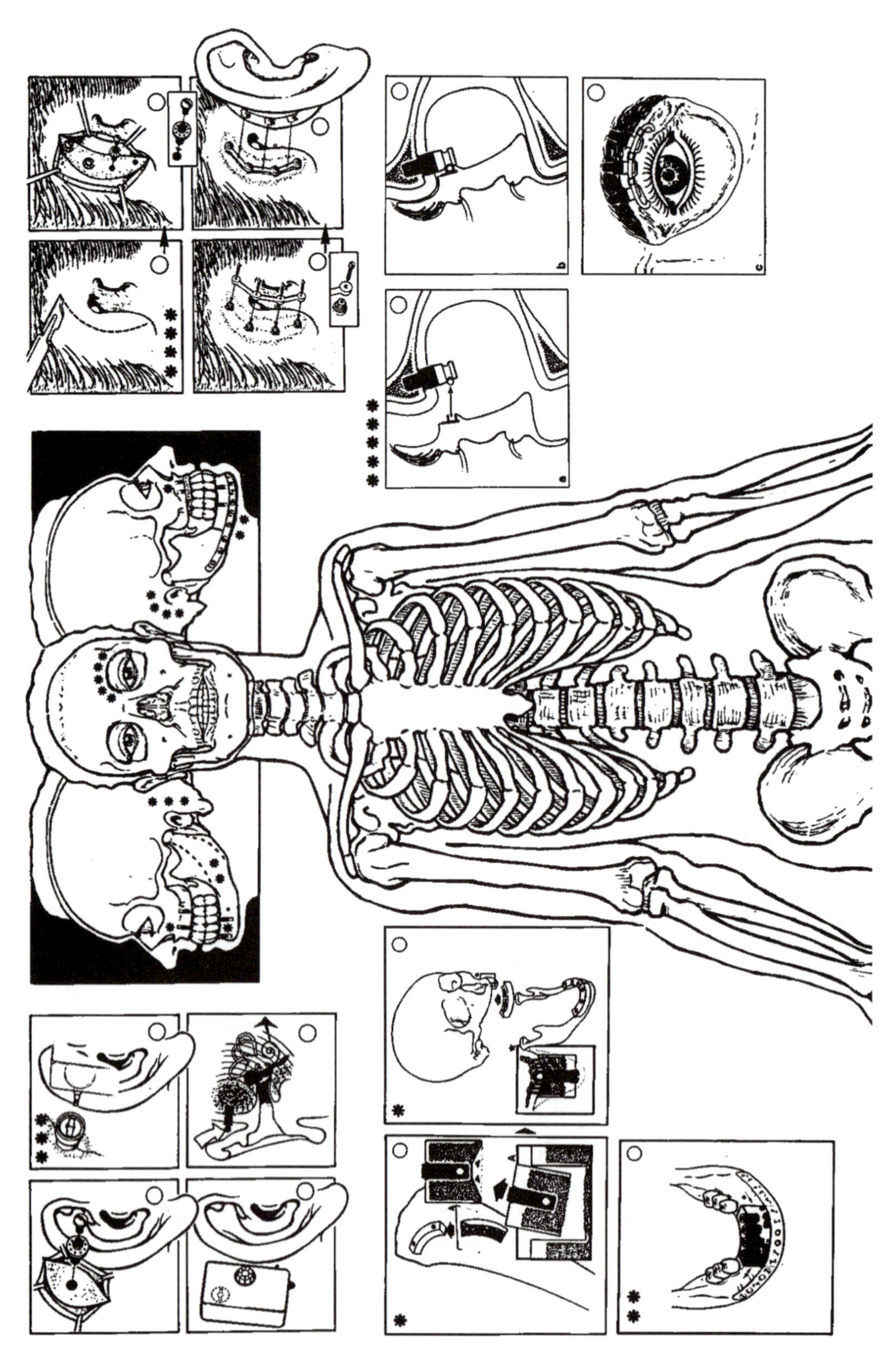

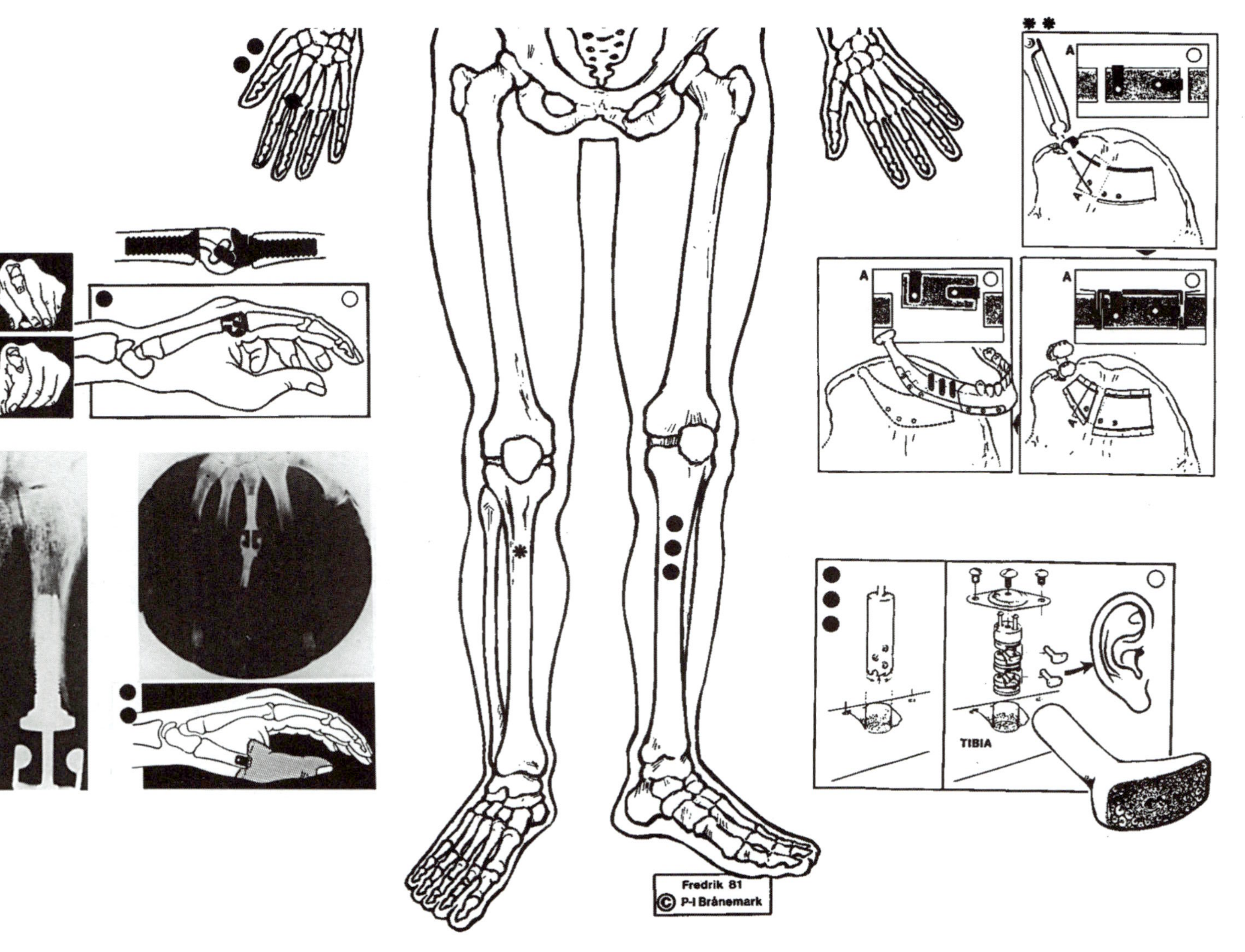

Figure 9-56 Because of its biocompatibility and ability to osseointegrate, titanium is used in many types of prostheses in humans. (Howmedica, Inc.)

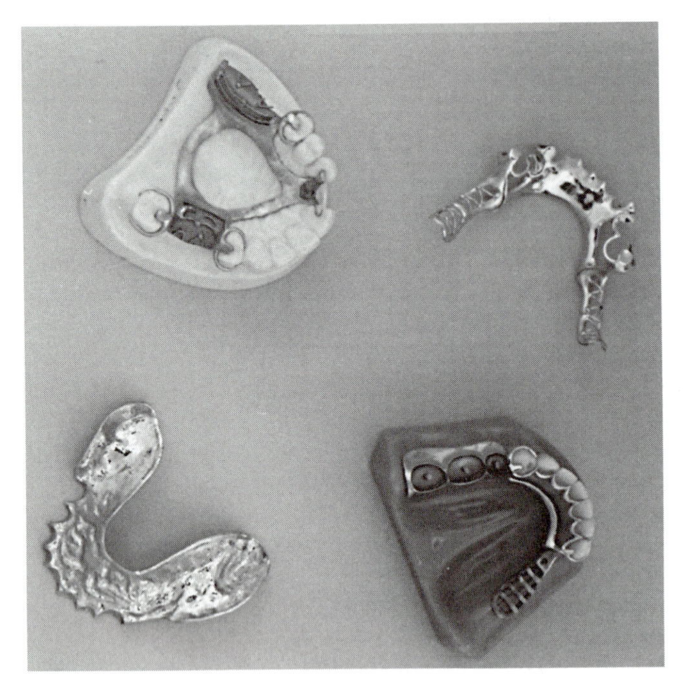

Figure 9-57 Vitallium dental castings. (Howmet.)

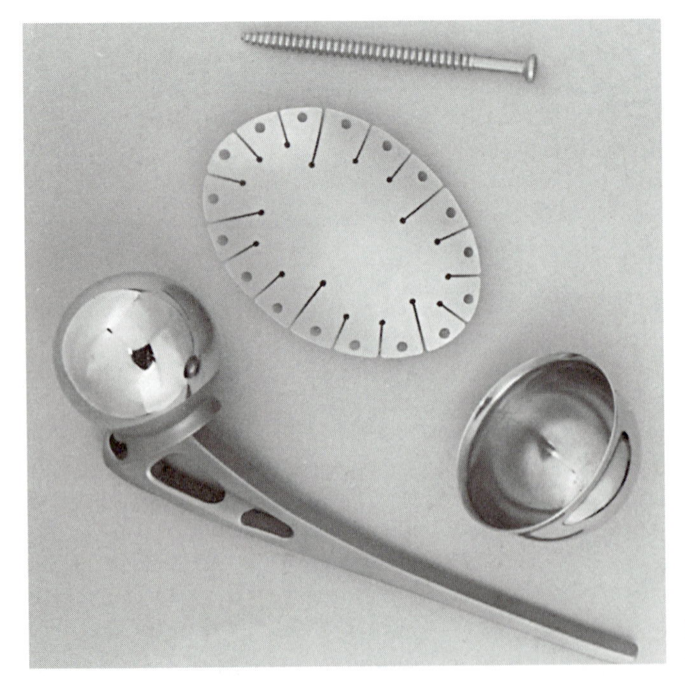

Figure 9-58 Vitallium orthopedic castings. (Howmet.)

problems of mold reaction and expansion. The process that evolved is called investment casting (see Module 5). During the 1930s the experience gained with casting Vitallium dentures and orthopedic prosthetics was refined and used to produce Vitallium and titanium components for jet engines for aircraft. That evolution has resulted in directional and single grained castings.

This synergy typifies the planned progress of materials science, engineering, and manufacturing. Wise and knowledgeable technologists, engineers, and technicians always keep informed about new developments around them. Events of serendipity may occur by chance, but the alert person can recognize their values.

9.14 ON TO THE FUTURE

We set the scene for this book with a high-tech mountain bicycle. When describing the many elements of engineering materials technology, we included customers' desires, designers' efforts to employ new materials and processes to meet the market demands, and manufacturers' methods of manufacturing to produce small lots of products with "agile manufacturing." Virtual reality, artificial intelligence, expert systems, multimedia computers, CAD-CAM, CIM, and robotics are among the evolving technologies and tools that will push engineering materials technology at an ever-increasing rate.

The morning paper, weekly magazines, and evening news bring you a constant barrage of materials evolutions and developments. You might read about the use of supercritical fluids to process plastics and other polymers in a more environmentally friendly method, compared to traditional polymer processing that gives off large volumes of pollutants. Or you may read how the economics of textiles is changed when the world's cotton supply is short and thus pushes up cotton textiles prices. To counter the rising price of fabrics, engineers look to recycled PET soda bottles as an alternative source for polyester and cotton blends. But the new polyester is not the coarse fabric of the 1970s leisure suit; rather, it is composed of "microfibers" that are finer, cooler, and softer than before. Watch for further developments on synthetic textiles as the free-electron laser (FEL, discussed in Section 9.4.2) moves out of the research lab into fabric manufacturing lines.

9.14.1 Molecular Nanotechnology, Molecular Systems Engineering, or Molecular Manufacturing

Much of what has been covered in this module and many of the innovations throughout this book will rely on manipulating the very small nanostructures of materials to achieve the goals of advanced materials, such as smart materials and biomaterials. Electronic components such as transistors, capacitors, and diodes, with dimensions less than 0.25 micrometers, are emerging from the nation's laboratories. Units such as the micrometers are less useful as units of measure. The next smallest unit of measure is the ***nanometer,*** defined as one-billionth of a meter. To emphasize its size, a human hair has a diameter that is 10 times the size of a nanometer. The diameter of most atoms is 0.1 to 0.4 nanometers (nm). This new technology is being investigated for use in manufacturing and is described below in more detail.

Module 1 states that one constant of materials technology is constant change. It reminds one of the use of designer materials versus off-the-shelf materials, with their limited structures and properties. Module 2 recalls that materials will be produced in new ways as materials engineers work with manufacturing engineers and materials scientists to invent new materials processing. The Pause & Ponder for Module 3 asks the reader to recall the synthesis and processing capabilities of contemporary materials technology at the various scales of measurement, i.e., macro, micro, nano, and atomic. Module 3 also contains a reminder that both developed and newly developing countries are involved in improving or maintaining their competitiveness in the race for materials such as those listed in Table 3-1 under the nano and atomic categories.

The heading to this section describes, in present-day terms, the emerging field of engineering that uses nanoscale mechanical systems to guide the placement of reactive ***molecules,*** while building complex structures with atom-by-atom control. K. Eric Drexler has written extensively about the design, performance, and path of development of this system since 1981, when he published this untried concept.* He defines ***molecular manufacturing*** as the construction of objects to complex, atomic specifications using sequences of chemical reactions directed by nonbiological molecular machinery. ***Molecular nanotechnology*** describes the field, as a whole, comprising molecular manufacturing together with its techniques, its products, and their design and analysis. This new field is related to, yet distinct from, mechanical engineering, microtechnology (to be defined), chemistry, and molecular biology. As physics is related to engineering, so chemistry is related to molecular engineering. Molecular manufacturing applies the principles of mechanical engineering to chemistry. Figure 9-59 provides an example of the progress in molecular manufacturing.

Over the next 20 years, the development of this new manufacturing system for the bulk processing of materials and the fabrication of custom products will rely on the use of numerous small manufacturing systems working in parallel with locally available materials. The products will exhibit order-of-magnitude improvements in mechanical properties, and will be of high quality and low cost. Bulk processing will be attained when these machines have sufficient general capabilities to, first of all, manufacture copies of themselves and, second, be reprogrammed to manufacture finished products directly from raw materials, thus bypassing traditional fabricating techniques.

Advances in molecular nanotechnology since the introduction of the concept in 1981 include scanning probe microscopes (Module 3), design of proteins by molecular biologists, design of molecules to trap other molecules and ions, many scientific workstations and molecular modeling software that permit faster testing and construction of new designs, biomimetics (Module 3), and near-net-shape processes (Modules 3 and 5) that have eliminated the traditional secondary processes accompanying manufacturing. The next step is the development of new conceptual and mathematical tools to design and analyze a limited class of molecular manufacturing systems. The total costs of manufacture, ex-

*Readers seeking further information on molecular nanotechnology may wish to contact the Foresight Institute regarding publications, conferences, and relevant research, such as that sponsored by the Institute for Molecular Manufacturing: Foresight Institute, Nanosystems information, PO Box 61058, Palo Alto, CA 94306. Tel: (415)-324-2490. Fax: (415)-324-2497. E-mail: foresight@cup.portal.com.

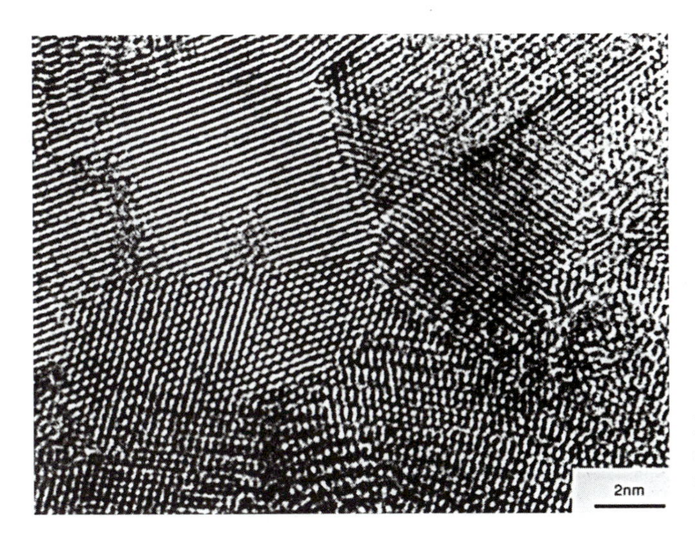

Figure 9-59 High-resolution transmission electron micrograph of a typical area that reveals grain boundaries in a nanophase palladium. (NIST.)

cluding cost of development and distribution, will be almost wholly determined by the cost of materials. The relevant materials presently cost about 0.1$/kg to 0.5$/kg (1992). Most manufactured products fall in the range between about 10$/kg (i.e., automobiles, appliances) and 10^4/kg (i.e., aerospace vehicles).

9.14.2 Microtechnology

Much active research in engineering involves building electronic and mechanical systems on ever-smaller scales using microtechnologies. The micro scale (Module 3 and Tables 3-1 and 10-6) is, in terms of volume, 10^9 times larger than the nanometer scale, and microtechnologies provide no mechanism for gaining precise, molecular control of the surface and interior of a complex, three-dimensional structure. The materials involved can include diamond films, organic films, semiconductors, metal films, dielectrics, ferroelectrics, and piezoelectric films. This technology relies on processing techniques such as photolithographic pattern definition, plasma and wet etchings, physical and chemical vapor deposition, diffusion, and ion implantation. Some examples of the products of such processes are flat panel displays, solid-state sensors, and those related to microelectronics and optoelectronics industries. Refer to Table 9-3 to see how the free electron laser supports these emerging manufacturing processes. These techniques are essentially unrelated to those of molecular manufacturing, so a wide gap exists between ***microtechnology*** using the top-down approach (starting with large complex and irregular structures) and the chemistry of molecular manufacturing (nanotechnology) with its bottom-up path (starting with small, simple, and exact structures). In essence microtechnology's goal is to make imprecise structures smaller, while the goal of molecular manufacturing is to make precise structures larger.

Three micromotors driving a microgear smaller in diameter than a human hair were reported to be fabricated from silicon by researchers at Sandia National Laboratories. This report typifies products being developed using microtechnology (Figures 9-60 and 9-61). These devices were fabricated using both etching processes and silicon materials common

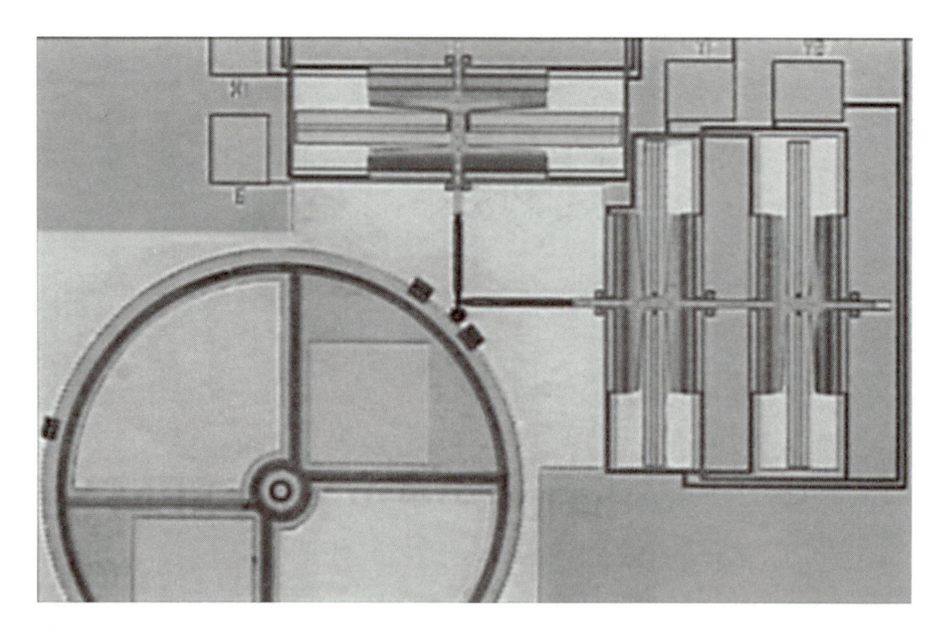

Figure 9-60 Three micromotors (rectangular structures, top and right), when energized by on–off voltages, power a microgear (at center) to drive the larger gear. The motor consists of two silicon combs with a shuttle between them. The edges of the shuttle also have teeth that interdigitate with those of the stationary combs. When energized, the stationary combs alternately pull the shuttle via electrostatic interaction. (Ernest J. Garcia, Sandia National Laboratories.)

Figure 9-61 A closeup of the microgear (drive gear) meshing with the much larger optical-shutter gear. The attached shaft turns a drive gear in a quarter of a circle during the shaft's power stroke. Another micromotor, at right angles to the first, is timed to turn the microgear on the second quarter of its rotation. The two drives, alternating their force, convert reciprocating motion into rotary motion to drive the gear completely around. (Ernest J. Garcia, Sandia National Laboratories.)

to the microelectronics industry to facilitate their mass production. Each micromotor develops 0.5 μW of power rotating at a speed of 30 revolutions per second: sufficient to drive the larger gear (30 times larger than a microgear) at an angular velocity of one revolution per second. Open spaces in the large gear in some positions allow light to pass through, creating an optical shutter.

Figure 9-62 is a photograph of an Africanized, or killer, bee with a chip attached to its thorax. This solar-powered electronic package, with dimensions of 1 mm × 3 mm × 5 mm and weighing less than 50 mg, contains a micro-miniature infrared LED transmitter and tracking system. This product of microtechnology is being developed to allow entomologists to continuously track and study bees in the field to study their mating and foraging habits. When the chip was mounted on the bee, it was unable to fly. This illustrates how research often results in apparent solutions that do not succeed or which require further modifications. These results are normally discovered during field testing.

9.14.3 Challenges in Engineering Materials Technology

Throughout this book we have sought to transport you from the basic concepts of engineering materials technology, dealing with traditional materials, to advanced materials and future materials and processes technology. You have been exposed to new instruments, unique processes, and innovations in materials. At the same time, we have tried to help you

Figure 9-62 A micro-miniature infrared LED transmitter and tracking system is under development that will emit infrared pulses, detectable over a range of 1–2 km. The chip shown is proper size but it is not the actual transmitter. This chip was used to test the bee's ability to fly with the microchip on its back. (Ernest J. Garcia, Sandia National Laboratories.)

develop problem-solving skills, learn to look around your environment, and observe what and how materials are currently being used and abused. We encouraged you to become familiar with the databases outside this book. Now we implore you to join a technical society such as ASM International, Society of Manufacturing Engineers, American Society for Testing and Materials, or any of the other groups who will assist you in keeping up with developments in our ever-expanding field! Student membership rates for these organizations are excellent, as are the benefits.

We encouraged you to consider the Materials Cycle and the consequences of each stage of the cycle. All who venture into technology must carry a concern for the environment. Selection of certain technologies and the associated materials issues to solve environmental problems is sometimes guided by politics and the perceptions of the general public, but the technologies selected may not result from the best use of the facts. An example is the mandating of electric-powered vehicles to relieve air pollution caused by internal combustion and gasoline-powered vehicles.

In Module 5 we discussed the emerging electric vehicles and the new batteries. As we suggested, battery technology has much maturing to do before becoming a strong option for most vehicle needs. Take the issue of weight comparison between gasoline-powered, internal combustion engines versus battery-powered motors. One gallon of gasoline weighs about 6 pounds. To produce the same energy as the 6 pounds of gasoline requires nearly 400 pounds of battery.

With the issue of air pollution, battery-powered vehicles also account for harmful emissions, albeit removed from the location of the electric vehicles. To produce electricity for recharging batteries, generating plants using fossil fuel waste as much as 65% of the energy to transform the fuel into electricity, and then there is another loss of 5 to 10% to wire the electricity down to the battery-charging receptacle. So generating electricity for battery charging not only creates harmful emissions, it is an inefficient use of fossil fuel.

With the issue of batteries, materials science has yet to develop reliable, long-life batteries. Many other issues challenge the viability of battery-powered vehicles, including cost and fire hazards associated with current batteries. Experts feel that we should pursue more viable and cost-effective solutions for dealing with the many environmental issues of the internal combustion engine for vehicles, and wait for battery technology to mature before committing to electric vehicles.

An example of research aimed at cleaner internal combustion engines involves using a new coating on radiators (Figure 9-63) that actually offers the potential to clean the air as the engine runs rather than contributing to pollution. Such a development may buy more time for economical and practical improvements for vehicles powered by electricity, hydrogen, or other alternative fuels.

The technically literate must enter into debates about the impact of technologies on our world and its inhabitants. Those who fear technology need to hear from those who understand it. To parallel the saying, "Those who don't know the lessons of history are doomed to repeat them," we can say, "Those who don't understand the fundamentals of technology may suffer from it and not enjoy its potential benefits."

The knowledge you have gained through this study should serve you well in everyday life and in your career. We would love to hear your comments on this book.

Good luck into the future!

Figure 9-63 "Smart coatings" may buy more time for the auto industry to improve power systems for vehicles that will be easier on our air quality. Smog-eating radiators use PremAir™ coatings to convert ozone (O_3)—the main component of smog—into breathable oxygen (O_2). (Ford Motor Company.)

SELF-ASSESSMENT

9-1. Would a material that has an electrical resistivity of $10^{12}\ \Omega \cdot$ cm make a good conductor or a good insulator?

9-2. Express a dielectric strength of 2000 V per mil in units of (a) V/cm; (b) V/m.

9-3. The parallel plates of an air-filled condenser are 1 mm apart. What is the plate's area (in square meters) if the capacitance is 1 nF?

9-4. Express the permitivity constant of free space in terms of picofarads per millimeter.

9-5. In selecting an optimum polymeric material from a table of dielectric strengths, discuss what effects a lack of knowledge of the thickness of the material selected would have if it were used in a capacitor.

9-6. In designing an electrical circuit, if the resistance is 16 Ω, the current is 7.5 A, and the resistivity is $5.6 \times 10^{-6}\ \Omega \cdot$ cm, what is the conductivity?

9-7. A perfect dielectric has an angle δ of how many degrees?

9-8. What is the product of $\kappa \times \tan \delta$ called?

9-9. Name the three types of polarization discussed in this chapter.

9-10. A material that allows the passage of an electric field but does not conduct electrical current is known by what other name?

9-11. $BaTiO_3$ is chosen as a dielectric for a condenser. The area of the plates is 0.2 cm^2 and the separation between the plates is 0.05 cm. Find the capacitance.

9-12. Using Figure 9-12, the wavelength of red light is about 700 nm. What is the frequency of red light?

9-13. Explain why the electrical conductivity of semiconductor materials increases as temperature increases.

9-14. Iron and steel parts in electrical machinery, by necessity, must possess high permeability and preferably should be made of "soft" iron. What is the significance of this statement?

9-15. Give an example of the use of a metal with high residual magnetism.

9-16. Photoelectric cells are used in a variety of ways. Are solar batteries used in space such an application? If so, how efficient are these devices, in your opinion, in converting solar energy into electrical energy?

9-17. Describe two ways of producing x-rays.

9-18. Distinguish among translucence, transparency, and transmittance.

9-19. Photoelasticity is sometimes defined as a visual, full-field technique for measuring stresses in plastic models of parts, usually before the parts are made. In relation to the material presented in the book, does this definition really define the term? Knowing that this stress analysis of plastic models uses a polariscope, where does photoelasticity play a part in the technique?

9-20. Thermal resistivity is as important as thermal conductivity. Illustrate both properties by citing an example of each from your own experience or knowledge.

9-21. Using magnetic chucks in a machining operation as an example, how does a knowledge of magnetic properties of metals pay big dividends?

9-22. Using Table 9-1 with room temperature of 68°F, determine the resistance of a copper wire with a length of 100 m and radius of 1 mm.

9-23. Monochromatic green light of frequency 6.2×10^{14} Hz produced by a laser consists of 5×10^{15} photons, which pass a point in 1 s. How much energy (joules) does each photon possess? (*Hint:* $E = \hbar f$.)

9-24. The minimum energy needed to extract electrons from inside a metal is known as the work function, *W*. *W* depends on the metal and its surface and is equal to the product of the threshold frequency, f_0, and Planck's constant, $\hbar$. If $W = 1.9$ eV, what is the minimum frequency of the light illuminating a metal surface that will release photoelectricity?

9-25. Yellow light with a wavelength of 5×10^{-7} m traveling in a vacuum enters a glass plate ($n = 1.5$). What is the velocity of yellow light in the glass plate in meters per second (m/s^{-1})?

9-26. Integrated circuits use many types of solid-state devices. Name three such devices.

9-27. What is the purpose of doping a semiconductor material?

9-28. Compare the energy required to raise a donor electron to the conduction band with the energy required to raise a valence electron.

9-29. Name one method by which an ideal semiconductor having no impurity atoms can be made to conduct electrons.

9-30. Once a donor electron has been raised into the conduction band by the receipt of energy from some external source, conduction occurs as a result of the combined actions of what charged bodies?

9-31. A group-VI element, selenium, is a semiconductor element. From what group in the periodic table would you choose a dopant to convert selenium to (a) a *p*-type material; (b) an *n*-type material?

9-32. Identify what are termed the majority carriers in an *n*-type semiconductor.

9-33. Explain what is meant by the term *biasing*.

9-34. The depletion zone in a *pn* junction differs in size when a forward bias is applied from a condition of zero bias. Specify what this difference is in terms of area.

9-35. What device operates in opposite fashion to a solar cell and produces light from electrical energy?

9-36. What creates the voltage in a typical solar cell that drives electrical current through the external circuit?

9-37. Reflection of sunlight from the surface layers of a photovoltaic cell is maximized by treating the surface with some form of antireflective coating. True or false? Explain.

9-38. Down through the ages, humans have dreamed of converting sunlight directly into electricity. How efficient is this process today? Define *efficiency* as used in this process.

9-39. Name four possible things that can happen to sunlight entering a solar cell.

9-40. Explain the terms *solar cell, module,* and *array.*

9-41. What is the major cause of poor efficiency of polycrystalline silicon cells in relation to single-crystal silicon cells?

9-42. Amorphous silicon used in hand-held calculators is more efficient under fluorescent lighting than either single- or polycrystalline silicon. Name one characteristic that amorphous silicon has that explains this advantage.

9-43. Define the word *epitaxy* and explain why it is used in the semiconductor industry.

9-44. Both LED and LCD devices produce light. Which one requires more energy and consequently needs a battery source of power to function?

9-45. Optoelectronic devices all operate in accordance with a basic underlying scientific principle. Explain this principle.

9-46. The three types of liquid crystals are based on what characteristic of liquid crystals?

9-47. Name three influences that can change the structure of liquid crystals.

9-48. A homogeneous orientation of LC molecules can be accomplished in several ways. Name two such methods other than the use of an electric field.

9-49. In a reflective-type LC cell, the light source and the scattered light are on the same side of the cell. One glass plate is given a reflective coating. A transmittive-type cell's light source is on one side of the cell and the scattered light appears on the other. For this to happen, how are the glass plates modified, if at all? Draw a sketch of a transmittive-type cell similar to Figure 9-23.

9-50. Referring to Figure 9-46 to display the numeral "7," which segments need to be activated?

9-51. Referring to Figure 9-36, which axis (x or y) is known as the electric axis?

9-52. Name several applications of piezoelectric crystals.

9-53. The highest frequency of sounds that can be detected by the human ear is about how many cycles per second? Express in units of hertz.

9-54. A certain pickup for a record player converts the movements of a stylus into corresponding electrical signals. Other devices, such as a loudspeaker, use a similar element, a crystal. What material property peculiar to certain materials do these devices depend on to perform their functions?

9-55. Superconductors have great technological potential for many applications in this energy-dependent world. What is one difficulty that limits their development?

OBJECTIVE QUESTIONS

9-1. The dielectric constant for water.

a. 7×10^6 Hz **b.** 12×10^6 Hz **c.** 78×10^6 Hz

9-2. A conductivity in ohm^{-1} cm^{-1} for a common metal would be ______________.

a. 4,000 **b.** 100,000 **c.** 700,000

9-3. A dielectric is a nonconductive material whose polarization is caused by ______________.
a. A magnetic field **b.** An electric field **c.** IR radiation

9-4. Dielectric constants are in demand by the ______________.
a. Machine tool industry **b.** Electronics industry **c.** Oil and gas industry

9-5. If the Mn atom loses two electrons, its paramagnetic properties ______________.
a. Stay the same **b.** Increase **c.** Decrease

9-6. The Na atom has an atomic number of 11. It is ______________.
a. Diamagnetic **b.** Paramagnetic **c.** Neither

9-7. The atomic number of Cl is 17, which tells us that Cl is ______________.
a. Diamagnetic **b.** Paramagnetic **c.** Ferromagnetic

9-8. A pair of electrons with opposing spins has ______________.
a. A large magnetic field **b.** A small magnetic field **c.** No magnetic field

9-9. An example of a transition element.
a. B **b.** C **c.** Sc

9-10. The wavelength of microwaves is about ______________.
a. 10^{-5} nm **b.** 30 m **c.** 0.3 cm

9-11. The hysteresis loss (ergs/cm^3/cycle) for 4340 steel at 25°C is ______________.
a. 2.7×10^4 **b.** 1.5×10^4 **c.** 11.4×10^4

9-12. High-energy protons are produced by ______________.
a. Alpha rays **b.** X-rays **c.** Beta rays

9-13. Electronic polarization deals with the polarity of ______________.
a. Atoms **b.** Ions **c.** Protons

9-14. The bending of light is known as ______________.
a. Reflection **b.** Refraction **c.** Incidence

9-15. The index of refraction for diamond is ______________.
a. 0.67 **b.** 1.36 **c.** 2.42

9-16. Optical anisotropy refers to a material with ______________ optical properties.
a. Good **b.** Poor **c.** Different

9-17. The size of the forbidden gap in conductors is ______________.
a. Large **b.** Zero **c.** Small

9-18. An electric current of one coulomb of charge passing a point in one second is known as a(n) ______________.
a. Micron **b.** Ampere **c.** Angstrom

9-19. The SI prefix μ (mu) stands for ______________.
a. 10^{-6} in. **b.** 10^{-6} **c.** 10^{-6} m

9-20. The *p*-side of a diode is called the ______________.
a. Cathode **b.** Anode **c.** Pentode

9-21. Solar cells are expensive primarily because of the cost of ______________.
a. Material **b.** Fabrication **c.** Tooling

9-22. A YCBO is a ______________.
a. Small yard tug **b.** Superconductor **c.** Computer memory code

9-23. High-T_c superconductors have an advantage over normal conductors because they offer no ______________ to the flow of current.

a. Strain **b.** Resistance **c.** Dielectric **d.** Permeability

9-24. Which smart material might work a robot finger control or an antenna for a cellular phone because of its ability to remember its original shape after being deformed?

a. MMC **b.** Piezoelectric **c.** Titanium **d.** Shape memory alloy

9-25. Biomimetics can help intelligent material systems technology by ______________.

a. Providing feedback and control **b.** Ensuring osseointegration

c. Aiding improved hybrids **d.** Prompting biotransfer

9-26. What is a main concern when selecting a material for bioengineering?

a. Ability to resist heat

b. Ability to provide feedback and control

c. Compatibility with human and animal systems

d. Good formability

REFERENCES & RELATED READINGS

ASM INTERNATIONAL. "Smart Materials Alter Shape of Adaptive Aircraft Wings," *Advanced Materials & Processes,* September 1995, p. 9.

CONSTANTZ, BRENT R. "Mineral Paste Surgically Injected to Repair Bones," Norian Corp., Cupertino, Calif., 95014, *Advanced Materials & Processes,* June 1995, p. 17.

dE NEUFVILLE, RICHARD, et al. "The Electric Car Unplugged," *Technology Review,* January 1996, pp. 30–36.

"Don't Let the Sun Shine In." Webasco Sunroofs of Maumee, Ohio, *Popular Science,* June 1994, p. 12.

DREXLER, K. ERIC. *Nanosystems—Molecular Machinery, Manufacturing, and Computation.* New York: John Wiley and Sons, Inc., 1992.

FISCHER, ARTHUR. "Superconductivity," *Popular Science,* April 1988, pp. 54–58.

FONASH, STEPHEN J. "Micro and Nanofabrication," Surface Engineering and Manufacturing Technology Center (SEMTC), U.S. Navy Manufacturing Science & Technology Program Report, February 1995, p. 3.

"Get Smart," *Compressed Air Magazine,* April/May 1995, pp. 36–41.

HAZEN, ROBERT M. "The New Alchemy," *Technology Review,* MIT, Bldg. W59, Cambridge, Mass., 02139, November/December 1994, pp. 24–26.

HUMMEL, ROLF E. *Electronic Properties of Materials,* 2nd ed. New York: Springer-Verlag, 1992.

KLIENFIELD, SONNY. *A Machine Called Indomitable.* New York: Time Books, 1985, pp. 92–97.

MCWHORTER, PAUL, JEFF SNIEGOWSKI, and ERNEST GARCIA. "Mighty Micrometers Drive Miniature Gears," *Advanced Materials & Processes,* December 1995, p. 6.

MALLARDI, JOSEPH L. *From Teeth to Jet Engines,* Howmet Corporation, 1992.

"Manufacturing Parts Drop by Drop," *Compressed Air Magazine,* March 1995, pp. 38–44.

Moving America: New Directions, New Opportunities, report to Congress, June 1990.

"Nanotechnology," *Compressed Air Magazine,* October/November 1994, pp. 16–23.

National MAGLEV Initiative: Annual Report—November 1992. Washington, D.C.: Federal Railroad Administration, 1992.

"Nimbus Information System." *CD-ROM Replication Guide.* Nimbus Manufacturing, Inc., Charlottesville, Va., 1995.

OLIVER, JOYCE. "Superconductors: The Heat Is On," *The Electron,* November/December 1988, pp. 12–17.

SARIKAYA, MEHMET AND AKSAY, ILHAN A. AKSAY (eds.) Biomimetics: Design and Processing of Materials, AIP Press, 1995.

"Stealth Drapes." Marvin Corp., Warroad, Minn., *Popular Science,* July 1995, p. 10.

TUOMINEN, STEVE, and CRAIG WOJCIK. "Unique Alloys for Aerospace and Beyond," *Outlook Magazine,* Teledyne Wah Chang, Vol. 16, No. 2, 1995, pp. 1–6.

UENOHARA, MICHIYKI. "Electro-Ceramics and Applications in C&C Era," *Journal of Materials Education,* Vol. 10, No. 3, 1988, pp. 259–280.

WEBER, WILLIAM J. "Highly Conducting Electroceramics," Pacific Northwest Laboratory, October 1992.

EXPERIMENTS & DEMONSTRATIONS IN ELECTRONIC MATERIALS

New:Update 88 NASA Conference Publication 3060

SASTRI, SANKAR. "Magnetic Particle Inspection."

New:Update 89 NASA Conference Publication 3074

KUNDU, NIKHIL K., and MALAY KUNDU. "Piezoelectric and Pyroelectric Effects of a Crystalline Polymer."

MOLTON, PETER M., and CLARKE CLAYTON. "Anode Materials for Electrochemical Waste Destruction."

RIES, HEIDI R. "Dielectric Determination of the Glass Transition Temperature."

New:Update 90 NIST Special Publication 822

DAHIYA, J. N. "Dielectric Behavior of Superconductors at Microwave Frequencies."

New:Update 91 NASA Conference Publication 3151

DAHIYA, J. N. "Dielectric Behavior of Semiconductors at Microwave Frequencies."

PATTERSON, JOHN W. "Demonstration of Magnetic Domain Boundary Movement Using an Easily Assembled Videocam–Microscope System."

New:Update 92 NASA Conference Publication 3201

BUNNELL, L. ROY. "Temperature-Dependent Electrical Conductivity of Soda-Lime Glass."

DAHIYA, JAI N. "Phase Transition Studies in Barium and Strontium Titanates at Microwave Frequencies."

New:Update 93 NASA Conference Publication 3259

GRAY, JENNIFER. "Symmetry and Structure Through Optical Diffraction."

PASSEK, THOMAS. "University Outreach Focused Discussion: What Do Educators Want from ASM International?"

WICKMAN, J. L. "Plastic Part Design Analysis Using Polarized Filters and Birefringence."

New:Update 94 NASA Conference Publication 3304

ELBAN, WAYNE L. "Stereographic Projection Analysis of Fracture Plane Traces in Polished Silicon Wafers for Integrated Circuits."

PARMAR, DEVENDRA S., and J. J. SINGH. "Measurement of the Electro-Optic Switching Response in Ferroelectric Liquid Crystals."

NEW:Update 95 NASA Conference Publication 3330

DAHIYA JAI N. "Temperature Dependence of the Microwave Dielectric Behavior of Selected Materials."

MARSHALL JOHN. "Application Advancements Using Electrorheological Fluids."

ONO, KANJI. "Piezoelectric Sensing and Acoustic Emission."

RIES, HEIDI R. "An Integrated Approach to Laser Crystal Development."

EXPERIMENTS & TOPICS IN MATERIALS CURRICULUM

NEW:Update 93 NASA Conference Publication 3259

BRIGHT, VICTOR M. "Simulation of Materials Processing: Fantasy or Reality?"

DIWAN, RAVINDER M. "Manufacturing Processes Laboratory Projects in Mechanical Engineering Curriculum."

KUNDU, NIKHIL K. "Graphing Techniques for Materials Laboratory Using Excel."

MCCLELLAND, H. T. "Process Capability Determination of New and Existing Equipment and Introduction to Usable Statistical Methods."

PASSEK, THOMAS. "University Outreach Focused Discussion: What Do Educators Want from ASM International?"

New:Update 94 NASA Conference Publication 3304

BRIMACOMBE, J. K. "Transferring Knowledge to the Shop Floor."

BURTE, HARRIS M. "Emerging Materials Technology."

CONSTANT, KRISTEN P., and KRISHNA VEDULA. "Development of Course Modules for Materials Experiments."

COYNE, PAUL J., JR., GLENN S. KOHNE, and WAYNE L. ELBAN. "PC Laser Printer–Generated Cubic Stereographic Projections with Accompanying Student Exercise."

MCKENNEY, ALFRED E., EVELYN D. MCKENNEY, and ROBERT BERRETTINI. "CDROM Technology to Strengthen Materials Education."

MASI, JAMES V. "Bubble Rafts, Crystal Structures, and Computer Animation."

OLESAK, PATRICIA J. "Understanding Phase Diagrams."

SCHEER, ROBERT J. "Incorporating 'Intelligent' Materials into Science Education."

SCHWARTZ, LYLE H. "Technology Transfer of NIST Research."

SPIEGEL, F. XAVIER. "Demonstrations in Materials Science from the Candy Shop."

UHL, ROBERT. "ASM Educational Tools Now and into the Future."

NEW:Update 95 NASA Conference Publication 3330

BELANGER, BRIAN C. "NIST Advanced Technology Programs."

BERRETTINI, ROBERT. "The VTLA System of Course Delivery and Faculty in Materials Education."

KOHNE, GLENN S. "An Autograding (Student) Problem Management System for the Computer Literate."

RUSS, JOHN. "Self-Paced Interactive CD-ROMs."

10

Appendix Tables

TABLE 10-1A PERIODIC TABLE OF THE ELEMENTS[a]

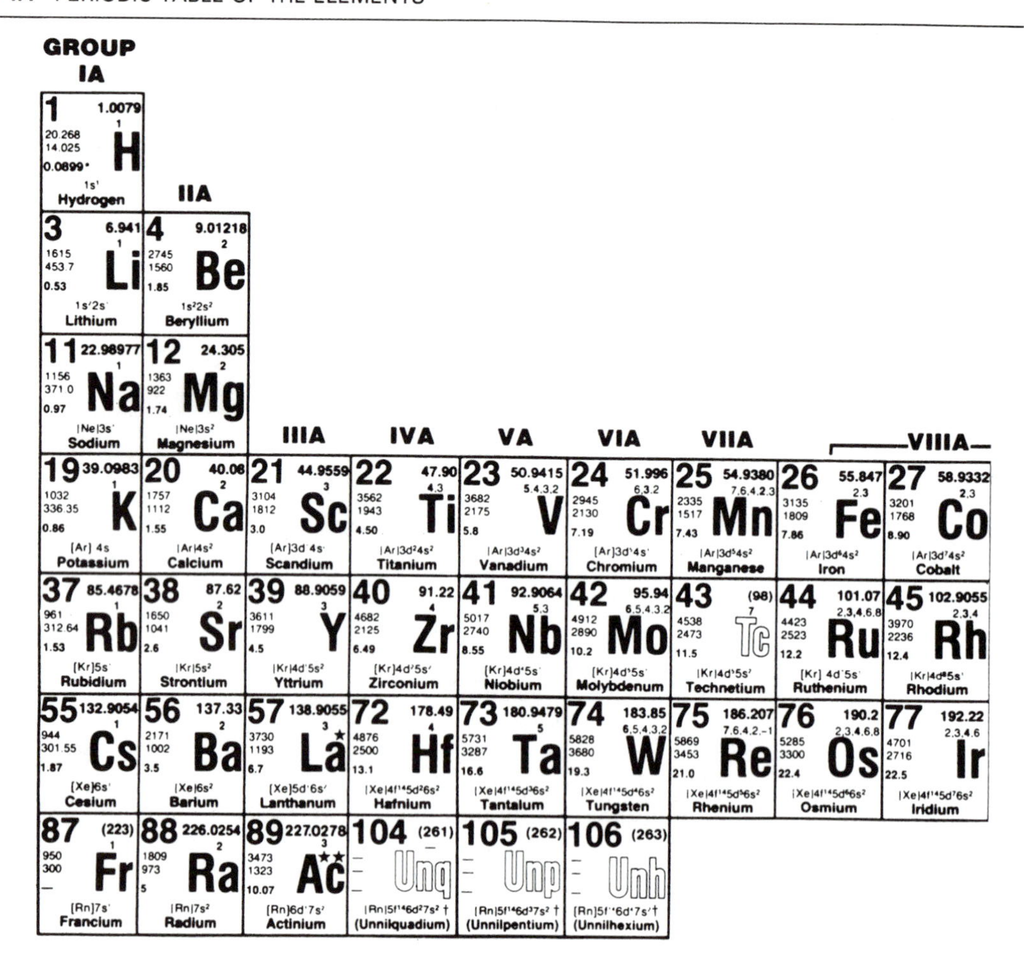

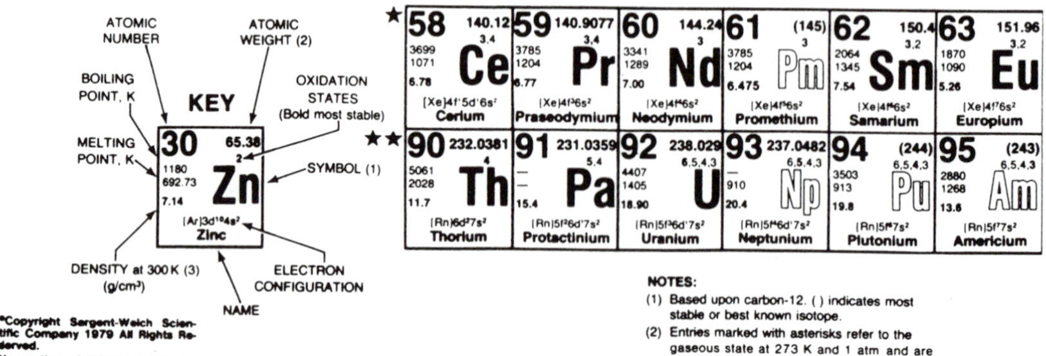

NOTES:

(1) Based upon carbon-12. ( ) indicates most stable or best known isotope.

(2) Entries marked with asterisks refer to the gaseous state at 273 K and 1 atm and are given in units of g/l.

	IB	IIB	IIIB	IVB	VB	VIB	VIIB	VIII
								2 He 4.00260; 4.215; 0.95 (at 26 atm); 0.1787*; $1s^2$; Helium
			5 B 10.81; 3; 4275; 2300; 2.34; $1s^22s^2p^1$; Boron	**6 C** 12.011; ±4,2; 4470*; 4100*; 2.62; $1s^22s^2p^2$; Carbon	**7 N** 14.0067; ±3,5,4,2; 77.35; 63.14; 1.251*; $1s^22s^2p^3$; Nitrogen	**8 O** 15.9994; −2; 90.18; 50.35; 1.429*; $1s^22s^2p^4$; Oxygen	**9 F** 18.998403; −1; 84.95; 53.48; 1.696*; $1s^22s^2p^5$; Fluorine	**10 Ne** 20.179; 27.096; 24.558; 0.901*; $1s^22s^2p^6$; Neon
			13 Al 26.98154; 3; 2793; 933.25; 2.70; [Ne]$3s^2p^1$; Aluminum	**14 Si** 28.0855; 4; 3540; 1685; 2.33; [Ne]$3s^2p^2$; Silicon	**15 P** 30.97376; ±3,5,4; 550; 317.30; 1.82; [Ne]$3s^2p^3$; Phosphorus	**16 S** 32.06; ±2,4,6; 717.75; 388.36; 2.07; [Ne]$3s^2p^4$; Sulfur	**17 Cl** 35.453; ±1,3,5,7; 239.1; 172.16; 3.17*; [Ne]$3s^2p^5$; Chlorine	**18 Ar** 39.948; 87.30; 83.81; 1.784*; [Ne]$3s^2p^6$; Argon
28 Ni 58.70; 2,3; 3187; 1726; 8.90; [Ar]$3d^84s^2$; Nickel	**29 Cu** 63.546; 2,1; 2836; 1357.6; 8.96; [Ar]$3d^{10}4s^1$; Copper	**30 Zn** 65.38; 2; 1180; 692.73; 7.14; [Ar]$3d^{10}4s^2$; Zinc	**31 Ga** 69.72; 3; 2478; 302.90; 5.91; [Ar]$3d^{10}4s^2p^1$; Gallium	**32 Ge** 72.59; 4; 3107; 1210.4; 5.32; [Ar]$3d^{10}4s^2p^2$; Germanium	**33 As** 74.9216; ±3,5; 876 (subl.); 1081 (28 atm); 5.72; [Ar]$3d^{10}4s^2p^3$; Arsenic	**34 Se** 78.96; −2,4,6; 958; 494; 4.80; [Ar]$3d^{10}4s^2p^4$; Selenium	**35 Br** 79.904; ±1,5; 332.25; 265.90; 3.12; [Ar]$3d^{10}4s^2p^5$; Bromine	**36 Kr** 83.80; 119.80; 115.78; 3.74*; [Ar]$3d^{10}4s^2p^6$; Krypton
46 Pd 106.4; 2,4; 3237; 1825; 12.0; [Kr]$4d^{10}$; Palladium	**47 Ag** 107.868; 1; 2436; 1234; 10.5; [Kr]$4d^{10}5s^1$; Silver	**48 Cd** 112.41; 2; 1040; 594.18; 8.65; [Kr]$4d^{10}5s^2$; Cadmium	**49 In** 114.82; 3; 2346; 429.76; 7.31; [Kr]$4d^{10}5s^2p^1$; Indium	**50 Sn** 118.69; 4,2; 2876; 505.06; 7.30; [Kr]$4d^{10}5s^2p^2$; Tin	**51 Sb** 121.75; ±3,5; 1860; 904; 6.68; [Kr]$4d^{10}5s^2p^3$; Antimony	**52 Te** 127.60; −2,4,6; 1261; 722.65; 6.24; [Kr]$4d^{10}5s^2p^4$; Tellurium	**53 I** 126.9045; ±1,5,7; 458.4; 386.7; 4.92; [Kr]$4d^{10}5s^2p^5$; Iodine	**54 Xe** 131.30; 165.03; 161.36; 5.89*; [Kr]$4d^{10}5s^2p^6$; Xenon
78 Pt 195.09; 2,4; 4100; 2045; 21.4; [Xe]$4f^{14}5d^96s^1$; Platinum	**79 Au** 196.9665; 3,1; 3130; 1337.58; 19.3; [Xe] $4f^{14}5d^{10}6s^1$; Gold	**80 Hg** 200.59; 2,1; 630; 234.28; 13.53; [Xe]$4f^{14}5d^{10}6s^2$; Mercury	**81 Tl** 204.37; 3,1; 1746; 577; 11.85; [Xe]$4f^{14}5d^{10}6s^2p^1$; Thallium	**82 Pb** 207.2; 4,2; 2023; 600.6; 11.4; [Xe]$4f^{14}5d^{10}6s^2p^2$; Lead	**83 Bi** 208.9804; 3,5; 1837; 544.52; 9.8; [Xe]$4f^{14}5d^{10}6s^2p^3$; Bismuth	**84 Po** (209); 4,2; 1235; 527; 9.4; [Xe]$4f^{14}5d^{10}6s^2p^4$; Polonium	**85 At** (210); ±1,3,5,7; 610; 575; —; [Xe]$4f^{14}5d^{10}6s^2p^5$; Astatine	**86 Rn** (222); 211; 202; 9.91*; [Xe]$4f^{14}5d^{10}6s^2p^6$; Radon

* Estimated Values

† The names and symbols of elements 104 - 106 are those recommended by IUPAC as systematic alternatives to those suggested by the purported discoverers. Berkeley (USA) researchers have proposed Rutherfordium, Rf, for element 104 and Hahnium, Ha, for element 105. Dubna (USSR) researchers, who also claim the discovery of these elements have proposed different names (and symbols).

The A & B subgroup designations, applicable to elements in rows 4, 5, 6, and 7, are those recommended by the International Union of Pure and Applied Chemistry. It should be noted that some authors and organizations use the opposite convention in distinguishing these subgroups.

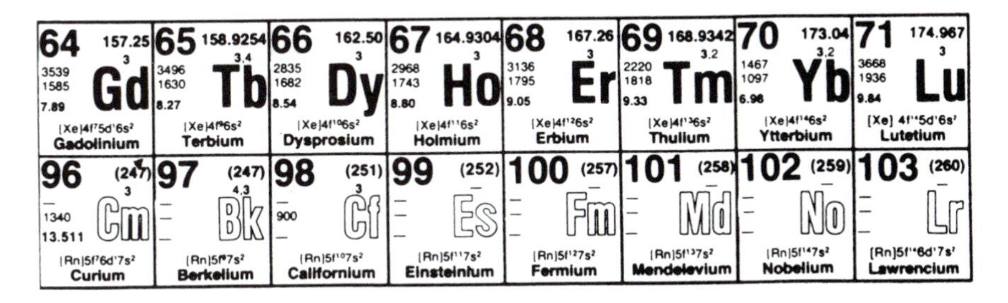

64 Gd 157.25; 3; 3539; 1585; 7.89; [Xe]$4f^75d^16s^2$; Gadolinium	**65 Tb** 158.9254; 3,4; 3496; 1630; 8.27; [Xe]$4f^96s^2$; Terbium	**66 Dy** 162.50; 3; 2835; 1682; 8.54; [Xe]$4f^{10}6s^2$; Dysprosium	**67 Ho** 164.9304; 3; 2968; 1743; 8.80; [Xe]$4f^{11}6s^2$; Holmium	**68 Er** 167.26; 3; 3136; 1795; 9.05; [Xe]$4f^{12}6s^2$; Erbium	**69 Tm** 168.9342; 3,2; 2220; 1818; 9.33; [Xe]$4f^{13}6s^2$; Thulium	**70 Yb** 173.04; 3,2; 1467; 1097; 6.96; [Xe]$4f^{14}6s^2$; Ytterbium	**71 Lu** 174.967; 3; 3668; 1936; 9.84; [Xe] $4f^{14}5d^16s^2$; Lutetium
96 Cm (247); 3; —; 1340; 13.511; [Rn]$5f^76d^17s^2$; Curium	**97 Bk** (247); 4,3; —; —; —; [Rn]$5f^97s^2$; Berkelium	**98 Cf** (251); 3; —; 900; —; [Rn]$5f^{10}7s^2$; Californium	**99 Es** (252); —; —; —; —; [Rn]$5f^{11}7s^2$; Einsteinium	**100 Fm** (257); —; —; —; —; [Rn]$5f^{12}7s^2$; Fermium	**101 Md** (258); —; —; —; —; [Rn]$5f^{13}7s^2$; Mendelevium	**102 No** (259); —; —; —; —; [Rn]$5f^{14}7s^2$; Nobelium	**103 Lr** (260); —; —; —; —; [Rn]$5f^{14}6d^17s^2$; Lawrencium

SARGENT-WELCH SCIENTIFIC COMPANY
7300 NORTH LINDER AVENUE, SKOKIE, ILLINOIS 60077

Catalog Number S-18806

TABLE 10-1B PERIODIC PROPERTIES OF THE ELEMENTS[a]

Percent Ionic Character of a Single Chemical Bond

Difference in electronegativity	0.1	0.2	0.3	0.4	0.5	0.6	0.7	0.8	0.9	1.0	1.1	1.2	1.3	1.4	1.5	1.6	1.7	1.8
Percent ionic character %	0.5	1	2	4	6	9	12	15	19	22	26	30	34	39	43	47	51	55

DATA CONCERNING THE MORE STABLE ELEMENTARY (SUBATOMIC) PARTICLES

	Neutron	Proton	Electron*	Neutrino*	Photon
Symbol	n	p	e (e^-)	ν	γ
Rest mass (kg)	1.67495×10^{-27}	1.67265×10^{-27}	9.1095×10^{-31}	~0	0
Relative atomic mass (^{12}C = 12)	1.008665	1.007276	5.48580×10^{-4}	~0	0
Charge (C)	0	1.60219×10^{-19}	-1.60219×10^{-19}	0	0
Radius (m)	8×10^{-16}	8×10^{-16}	$< 1 \times 10^{-16}$	~0	0
Spin quantum number	1/2	1/2	1/2	1/2	1
Magnetic Moment†	$-1.913\ \mu_N$	$2.793\ \mu_N$	$1.001\ \mu_B$	0	0

Element data

Element	Group	Covalent radius, Å	Atomic radius, Å (7)	Atomic volume, cm³/mol (8)	First ionization potential, V	Specific heat capacity, J g⁻¹K⁻¹ (3)	Electronegativity (Pauling's)	Heat of vaporization, kJ/mol (4)	Heat of fusion, kJ/mol (5)	Electrical conductivity, 10⁶ Ω⁻¹cm⁻¹ (6)	Thermal conductivity, W cm⁻¹K⁻¹ (3)
H	IA	0.32	0.79	14.4	13.598	14.304	2.20	0.44936	0.058.68	—	0.001815
Li	IA	1.23	2.05	13.10	5.392	3.6	0.98	145.920	3.00	[illegible]	0.847
Na	IA	1.54	2.23	23.7	5.139	1.23	0.93	96.960	2.598	[illegible]	1.41
K	IA	2.03	2.77	45.46	4.341	0.75	0.82	79.870	2.334	[illegible]	1.024
Rb	IA	2.16	2.98	55.9	4.177	0.363	0.82	72.216	2.192	[illegible]	0.582
Cs	IA	2.35	3.34	71.07	3.894	0.24	0.79	67.740	2.092	0.0489	0.359
Fr	IA	—	—	—	—	—	0.7	—	—	0.03	0.15*
Be	IIA	0.90	1.40	5.0	9.322	1.82	1.57	292.40	12.20	[illegible]	2.00
Mg	IIA	1.36	1.72	13.97	7.646	1.02	1.31	127.40	8.954	0.226	1.56
Ca	IIA	1.74	2.23	29.9	6.113	0.63	1.00	153.60	8.540	0.298	2.00*
Sr	IIA	1.91	2.45	33.7	5.695	0.30	0.95	144.0	8.30	[illegible]	0.353
Ba	IIA	1.98	2.78	39.24	5.212	0.204	0.89	142.0	7.750	0.030	0.184*
Ra	IIA	—	—	45.20	5.279	—	0.9	—	—	—	0.186*
Sc	IIIA	1.44	2.09	15.0	6.54	0.6	1.36	314.20	14.10	[illegible]	0.158
Y	IIIA	1.62	2.27	19.8	6.38	0.30	1.22	363.0	11.40	[illegible]	0.172
La★	IIIA	1.69	2.74	20.73	5.58	0.19	1.10	414.0	6.20	[illegible]	0.135
Ac★★	IIIA	—	—	22.54	5.17	—	1.1	—	—	—	0.12*
Ti	IVA	1.32	2.00	10.64	6.82	0.52	1.54	421.00	15.450	[illegible]	0.219
Zr	IVA	1.45	2.16	14.1	6.84	0.27	1.33	58.20	16.90	[illegible]	0.227
Hf	IVA	1.44	2.16	13.6	6.65	0.14	1.3	575.0	24.060	[illegible]	0.230
Unq	IVA	—	—	—	—		—	—	—	—	0.23*
V	VA	1.22	1.92	8.78	6.74	0.49	1.63	0.452	20.90	[illegible]	0.307
Nb	VA	1.34	2.08	10.87	6.88	0.26	1.6	682.0	26.40	[illegible]	0.537
Ta	VA	1.34	2.09	10.90	7.89	0.14	1.5	743.0	31.60	[illegible]	0.575
Unp	VA	—	—	—	—	—	—	—	—	—	0.58*
Cr	VIA	1.18	1.85	7.23	6.766	0.45	1.66	344.30	16.90	[illegible]	0.937
Mo	VIA	1.30	2.01	9.4	7.099	0.25	2.16	598.0	32.0	[illegible]	1.38
W	VIA	1.30	2.02	9.53	7.98	0.13	2.36	824.0	35.40	[illegible]	1.74
Unh	VIA	—	—	—	—	—	—	—	—	—	—
Mn	VIIA	1.17	1.79	1.39	7.435	0.48	1.55	226.0	12.050	[illegible]	0.0782
Tc	VIIA	1.27	1.95	8.5	7.28	0.21	1.9	660.0	24.0	[illegible]	0.506
Re	VIIA	1.28	1.97	8.85	7.88	0.13	1.9	715.0	33.20	[illegible]	0.479
Fe	VIIIA	1.17	1.72	7.1	7.870	0.44	1.83	349.60	13.80	[illegible]	0.802
Ru	VIIIA	1.25	1.89	8.3	7.37	0.238	2.2	595.0	24.0	[illegible]	1.17
Os	VIIIA	1.26	1.92	8.49	8.7	0.13	2.2	746.0	31.80	[illegible]	0.876
Co	VIIIA	1.16	1.67	6.7	7.86	0.42	1.88	376.50	16.190	[illegible]	1.00
Rh	VIIIA	1.25	1.83	8.3	7.46	0.242	2.28	493.0	21.50	[illegible]	1.50
Ir	VIIIA	1.27	1.87	8.54	9.1	0.130	2.20	604.0	26.10	[illegible]	1.47
★ Ce		1.65	2.70	20.67	5.54	0.19	1.12	414.0	5.460	0.0115	0.114
★ Pr		1.65	2.67	20.8	5.46	0.19	1.13	296.80	6.890	0.0148	0.125
★ Nd		1.64	2.64	20.6	5.53	0.19	1.14	273.0	7.140	0.0157	0.165
★ Pm		1.63	2.62	22.39	5.554	—	1.13	—	—	—	0.179*
★★ Th		1.65	—	19.9	6.08	0.12	1.3	514.40	16.10	0.0653	0.540
★★ Pa		—	—	15.0	5.89	—	1.5	—	12.30	0.0529	0.47*
★★ U		1.42	—	12.59	6.05	0.12	1.38	477.0	8.520	0.0380	0.276
★★ Np		—	—	11.62	6.19	0.12	1.36	—	5.190	0.00822	0.063

KEY

Zn	
1.25 (Covalent radius, Å)	1.65 (Electronegativity, Pauling's)
1.53 (Atomic radius, Å (7))	115.30 (Heat of vaporization, kJ/mol (4))
9.2 (Atomic volume, cm³/mol (8))	7.322 (Heat of fusion, kJ/mol (5))
9.394 (First ionization potential, V)	[illegible] (Electrical conductivity, $10^6\ \Omega^{-1}cm^{-1}$ (6))
0.39 (Specific heat capacity, $J\ g^{-1}K^{-1}$ (3))	1.16 (Thermal conductivity, $W\ cm^{-1}K^{-1}$ (3))

SYMBOL; CRYSTAL STRUCTURE (2); ACID-BASE PROPERTIES (1); COVALENT RADIUS, Å; ATOMIC RADIUS, Å (7); ATOMIC VOLUME, cm³/mol (8); FIRST IONIZATION POTENTIAL V; SPECIFIC HEAT CAPACITY, $J g^{-1} K^{-1}$ (3); ELECTRONEGATIVITY (Pauling's); HEAT OF VAPORIZATION, kJ/mol (4); HEAT OF FUSION, kJ/mol (5); ELECTRICAL CONDUCTIVITY, $10^6\ \Omega^{-1}cm^{-1}$ (6); THERMAL CONDUCTIVITY, $W\ cm^{-1}K^{-1}$ (3)

[a]The newest elements (rutherfordium, Rf-104, and hahnium, Ha-105) are not included.

1.9	2.0	2.1	2.2	2.3	2.4	2.5	2.6	2.7	2.8	2.9	3.0	3.1	3.2
59	63	67	70	74	76	79	82	84	86	88	89	91	92

* The positron (e^+) has properties similar to those of the (negative) electron or beta particle except that its charge has opposite sign (+). The antineutrino ($\bar{\nu}$) has properties similar to those of the neutrino except that its spin (or rotation) is opposite in relation to its direction of propagation.

An antineutrino accompanies release of an electron in radioactive β (particle) decay, whereas a neutrino accompanies the release of a positron in β^+ decay.

† μ_B = Bohr magneton and μ_N = Nuclear magneton

IB IIB IIIB IVB VB VIB VIIB VIII

Element	Line 1 (left)	Line 1 (right)	Line 2 (left)	Line 2 (right)	Line 3 (left)	Line 3 (right)	Line 4 (left)	Line 4 (right)	Line 5 (left)	Line 5 (right)
He	0.93	—	0.49	0.0845	—	—	24.587	—	5.193	0.00152
B	0.82	2.04	1.17	489.70	4.6	50.20	8.298	[illegible]	1.02	0.270
C	0.77	2.55	0.91	355.80	4.58	—	11.260	[illegible]	0.71	1.29
N	0.75	3.04	0.75	2.7928	17.3	0.3604	14.534	—	1.04	0.0002598
O	0.73	3.44	0.65	3.4099	14.0	0.22259	13.618	—	0.92	0.0002674
F	0.72	3.98	0.57	3.2698	17.1	0.2552	17.422	—	0.82	0.000279
Ne	0.71	—	0.51	1.7326	16.7	0.3317	21.564	—	0.904	0.000493
Al	1.18	1.61	1.82	293.40	10.0	10.790	5.986	[illegible]	0.90	2.37
Si	1.11	1.90	1.46	384.220	12.1	50.550	8.151	[illegible]	0.71	1.48
P	1.06	2.19	1.23	12.129	17.0	0.657	10.486	[illegible]	0.77	0.00235
S	1.02	2.58	1.09	—	15.5	1.7175	10.360	[illegible]	0.71	0.00269
Cl	0.99	3.16	0.97	10.20	22.7	3.203	12.967	—	0.48	0.000089
Ar	0.98	—	0.88	6.447	28.5	1.188	15.759	—	0.520	0.0001772
Ni	1.15	1.91	1.62	370.40	6.59	17.470	7.635	[illegible]	0.44	0.907
Cu	1.17	1.90	1.57	300.30	7.1	13.050	7.726	[illegible]	0.38	4.01
Zn	1.25	1.65	1.53	115.30	9.2	7.322	9.394	[illegible]	0.39	1.16
Ga	1.26	1.81	1.81	258.70	11.8	5.590	5.999	[illegible]	0.37	0.406
Ge	1.22	2.01	1.52	330.90	13.6	36.940	7.899	[illegible]	0.32	0.599
As	1.20	2.18	1.33	34.760	13.1	—	9.81	[illegible]	0.33	0.500
Se	1.16	2.55	1.22	37.70	16.45	6.694	9.752	[illegible]	0.32	0.0204
Br	1.14	2.96	1.12	15.438	23.5	5.286	11.814	—	0.473	0.00122
Kr	1.12	—	1.03	9.029	38.9	1.638	13.999	—	0.248	0.0000949
Pd	1.28	2.20	1.79	357.0	8.9	17.60	8.34	[illegible]	0.24	0.718
Ag	1.34	1.93	1.75	250.580	10.3	11.30	7.576	[illegible]	0.235	4.29
Cd	1.48	1.69	1.71	99.570	13.1	6.192	8.993	[illegible]	0.23	0.968
In	1.44	1.78	2.00	231.50	15.7	3.263	5.786	[illegible]	0.23	0.816
Sn	1.41	1.96	1.72	295.80	16.3	7.029	7.344	[illegible]	0.227	0.666
Sb	1.40	2.05	1.53	77.140	18.23	19.870	8.641	[illegible]	0.21	0.243
Te	1.36	2.1	1.42	52.550	20.5	17.490	9.009	[illegible]	0.20	0.0235
I	1.33	2.66	1.32	20.752	25.74	7.824	10.451	[illegible]	0.214	0.00449
Xe	1.31	—	1.24	12.636	37.3	2.297	12.130	—	0.158	0.0000569
Pt	1.30	2.28	1.83	510.0	9.10	19.60	9.0	[illegible]	0.13	0.716
Au	1.34	2.54	1.79	334.40	10.2	12.550	9.225	[illegible]	0.128	3.17
Hg	1.49	2.00	1.76	59.229	14.82	2.295	10.437	[illegible]	0.139	0.0834
Tl	1.48	2.04	2.08	164.10	17.2	4.142	6.108	[illegible]	0.13	0.461
Pb	1.47	2.33	1.81	177.70	18.17	4.799	7.416	[illegible]	0.13	0.353
Bi	1.46	2.02	1.63	104.80	21.3	11.30	7.289	[illegible]	0.12	0.0787
Po	1.46	2.0	1.53	—	22.23	—	8.42	[illegible]	—	0.20*
At	(1.45)	2.2	1.43	—	—	—	—	—	—	0.017*
Rn	—	—	1.34	16.40	50.5	2.890	10.748	—	0.09*	0.0000364*

Element	Line 1 (left)	Line 1 (right)	Line 2 (left)	Line 2 (right)	Line 3 (left)	Line 3 (right)	Line 4 (left)	Line 4 (right)	Line 5 (left)	Line 5 (right)
Sm	1.62	1.17	2.59	166.40	19.95	8.630	5.64	[illegible]	0.20	0.133
Eu	1.85	1.2	2.56	143.50	28.9	9.210	5.67	[illegible]	0.18	0.139*
Gd	1.61	1.20	2.54	359.40	19.9	10.050	6.15	[illegible]	0.23	0.106
Tb	1.59	1.2	2.51	330.90	19.2	10.80	5.86	[illegible]	0.18	0.111
Dy	1.59	1.22	2.49	230.0	19.0	11.060	5.94	[illegible]	0.17	0.107
Ho	1.58	1.23	2.47	241.0	18.7	12.20	6.018	[illegible]	0.16	0.162
Er	1.57	1.24	2.45	261.0	18.4	19.90	6.101	[illegible]	0.17	0.143
Tm	1.56	1.25	2.42	191.0	18.1	16.840	6.184	[illegible]	0.16	0.168
Yb	1.74	1.1	2.40	128.90	24.79	7.660	6.254	[illegible]	0.15	0.349
Lu	1.56	1.27	2.25	355.90	17.78	18.60	5.43	[illegible]	0.15	0.164
Pu	—	1.28	—	344.0	12.32	2.840	6.06	[illegible]	0.13*	0.0674
Am	—	1.3	—	—	17.86	14.40	5.993	[illegible]	0.11*	0.1*
Cm	—	1.3	—	—	18.28	15.0	6.02	—	—	0.1*
Bk	—	1.3	—	—	—	—	6.23	—	—	0.1*
Cf	—	1.3	—	—	—	—	6.30	—	—	0.1*
Es	—	1.3	—	—	—	—	6.42	—	—	0.1*
Fm	—	1.3	—	—	—	—	6.50	—	—	0.1*
Md	—	1.3	—	—	—	—	6.58	—	—	0.1*
No	—	1.3	—	—	—	—	6.65	—	—	0.1*
Lr	—	—	—	—	—	—	—	—	—	0.1*

NOTES: (1) For representative oxides (higher valence) of group. Oxide is acidic if color is red, basic if color is blue and amphoteric if both colors are shown. Intensity of color indicates relative strength.

(2) Cubic, face centered; cubic, body centered; cubic; hexagonal; rhombohedral; tetragonal; orthorhombic; monoclinic.

(3) At 300 K (27°C)

(4) At boiling point

(5) At melting point

(6) Generally at 293 K (20°C)

(7) Quantum mechanical value for free atom

(8) From density at 300 K (27°C) for liquid and solid elements; values for gaseous elements refer to liquid state at boiling point

TABLE 10-2 SYMBOLS OF THE ELEMENTS AND THEIR ATOMIC NUMBERS AND WEIGHTS[a]

Name	Symbol	Atomic Number	Atomic Weight[b]	Name	Symbol	Atomic Number	Atomic Weight[b]
Actinium	Ac	89	(227)	Lawrencium	Lr	103	(257)
Aluminum	Al	13	27.0	Lead	Pd	82	207.2
Americium	Am	95	(243)	Lithium	Li	3	6.94
Antimony	Sb	51	121.8	Lutetium	Lu	71	175.0
Argon	Ar	18	39.9	Magnesium	Mg	12	24.3
Arsenic	As	33	74.9	Manganese	Mn	25	54.9
Astatine	At	85	(210)	Mendelevium	Md	101	(256)
Barium	Ba	56	137.3	Mercury	Hg	80	200.6
Berkelium	Bk	97	(247)	Molybdenum	Mo	42	95.9
Beryllium	Be	4	9.01	Neodymium	Nd	60	144.2
Bismuth	Bi	83	209.0	Neon	Ne	10	20.2
Boron	B	5	10.8	Neptunium	Np	93	(237)
Bromine	Br	35	79.9	Nickel	Ni	28	58.7
Cadmium	Cd	48	112.4	Niobium	Nb	41	92.9
Calcium	Ca	20	40.1	Nitrogen	N	7	14.01
Californium	Cf	98	(251)	Nobelium	No	102	(254)
Carbon	C	6	12.01	Osmium	Os	76	190.2
Cerium	Ce	58	140.1	Oxygen	O	8	16.00
Cesium	Cs	55	132.9	Palladium	Pd	46	106.4
Chlorine	Cl	17	35.5	Phosphorus	P	15	31.0
Chromium	Cr	24	52.0	Platinum	Pt	78	195.1
Cobalt	Co	27	58.9	Plutonium	Pu	94	(242)
Copper	Cu	29	63.5	Polonium	Po	84	(210)
Curium	Cm	96	(247)	Potassium	K	19	39.1
Dysprosium	Dy	66	162.5	Praseodymium	Pr	59	140.9
Einsteinium	Es	99	(254)	Promethium	Pm	61	(147)
Erbium	Er	68	167.3	Protactinium	Pa	91	(231)
Europium	Eu	63	152.0	Radium	Ra	88	(226)
Fermium	Fm	100	(253)	Radon	Rn	86	(222)
Fluorine	F	9	19.0	Rhenium	Re	75	186.2
Francium	Fr	87	(223)	Rhodium	Rh	45	102.9
Gadolinium	Gd	64	157.3	Rubidium	Rb	37	85.5
Gallium	Ga	31	69.7	Ruthenium	Ru	44	101.1
Germanium	Ge	32	72.6	Samarium	Sm	62	150.4
Gold	Au	79	197.0	Scandium	Sc	21	45.0
Hafnium	Hf	72	178.5	Selenium	Se	34	79.0
Helium	He	2	4.00	Silicon	Si	14	28.1
Holmium	Ho	67	164.9	Silver	Ag	47	107.9
Hydrogen	H	1	1.008	Sodium	Na	11	23.0
Indium	In	49	114.8	Strontium	Sr	38	87.6
Iodine	I	53	126.9	Sulfur	S	16	32.1
Iridium	Ir	77	192.2	Tantalum	Ta	73	180.9
Iron	Fe	26	55.8	Technetium	Te	43	(99)
Krypton	Kr	36	83.8	Tellurium	Te	52	127.6
Lanthanum	La	57	138.9	Terbium	Tb	65	158.9

[a]The newest elements (rutherfordium, Rf-104, and hahnium, Ha-105) are not included.

[b]The values given in parentheses are mass numbers of the principal isotopes of unstable elements.

TABLE 10-2 (continued)

Name	Symbol	Atomic Number	Atomic Weight[b]	Name	Symbol	Atomic Number	Atomic weight[b]
Thallium	Tl	81	204.4	Vanadium	V	23	50.9
Thorium	Th	90	232.0	Xenon	Xe	54	131.3
Thulium	Tm	69	168.9	Ytterbium	Yb	70	173.0
Tin	Sn	50	118.7	Yttrium	Y	39	88.9
Titanium	Ti	22	47.9	Zinc	Zn	30	65.4
Tungsten	W	74	183.9	Zirconium	Zr	40	91.2
Uranium	U	92	238.0				

TABLE 10-3 GREEK SYMBOLS AND THEIR PRONUNCIATIONS

Γ	gamma (cap.)	η	eta (lc)
Δ	delta (cap.)	θ	theta (lc)
Θ	theta (cap.)	ϑ	theta (lc)
Λ	lambda (cap.)	κ	kappa (lc)
Ξ	xi (cap.)	λ	lambda (lc)
Π	pi (cap.)	μ	mu (lc)
Σ	sigma (cap.)	ν	nu (lc)
Υ	upsilon (cap.)	ξ	xi (lc)
Φ	phi (cap.)	π	pi (lc)
Ψ	psi (cap.)	ρ	rho (lc)
Ω	omega (cap.)	σ	sigma (lc)
α	alpha (lc)	τ	tau (lc)
β	beta (lc)	ϕ	phi (lc)
γ	gamma (lc)	φ	phi (lc)
δ	delta (lc)	χ	chi (lc)
ϵ	epsilon (lc)	ψ	psi (lc)
ζ	zeta (lc)	ω	omega (lc)

10.1 INTERNATIONAL SYSTEM OF UNITS AND UNIT CONVERSIONS

SI was created in 1960 by international agreement and represents a worldwide measurement system far superior to earlier measurement systems (gravitational) in expressing scientific and technical data. All gravitational systems using force as a fundamental dimension, including the American engineering system with its pound of mass a fundamental unit, are considered obsolete, and the changeover, though not mandatory, is proceeding at a steady pace, with some industries completing the changeover in record time. For most technology users, this means that they will continue to be confronted with problems that arise when two systems of measurement exist. Consequently, they must be very familiar with both systems and demonstrate ability to convert from one to another upon completion of their calculations.

Units. SI units are grouped into three general classes: base or fundamental units, derived units, and supplementary units. Table 10-4 lists the base units as well as some

TABLE 10-4 NAMES AND SYMBOLS OF SI UNITS

Quantity	Name of Unit	Symbol	Expressed in Base Units Where Applicable
Base units			
Length	meter	m	
Mass	kilogram	kg	
Time	second	s	
Electric current	ampere	A	
Thermodynamic temperature	kelvin	K	
Luminous intensity	candela	cd	
Amount of substance	mole	mol	
Derived units			
Area	square meter	m^2	
Volume	cubic meter	m^3	
Frequency	hertz, cycle per second	Hz	s^{-1}
Density (mass)	kilogram per cubic meter	kg/m^3	
Velocity (linear)	meter per second	m/s	
Velocity (angular)	radian per second	rad/s	
Acceleration (linear)	meter per second squared	m/s^2	
Acceleration (angular)	radian per second squared	rad/s^2	
Force	newton, kilogram-meter per second squared	N	$kg \cdot m \cdot s^{-2}$
Permeability	henry per meter	H/m	$m \cdot kg \cdot s^{-2} \cdot A^{-2}$
Permittivity	farad per meter	F/m	$m^{-3} \cdot kg^{-1} \cdot s^4 \cdot A^2$
Pressure (mechanical stress)	pascal, newton per square meter	Pa	$N \cdot m^{-2}$
Kinematic viscosity	square meter per second	m^2/s	
Dynamic viscosity	newton-second per square meter	$N \cdot s/m^2$	$m^{-1} \cdot kg \cdot s^{-1}$
Work energy, quantity of heat	joule, newton-meter	J, $N \cdot m$	
Power	watt, joule per second	W, J/s	
Quantity of electricity, electric charge	coulomb	C, $A \cdot s$	
Potential difference, electromotive force	volt	V, W/A	
Electric field strength	volt per meter	V/m	$m \cdot kg \cdot s^{-3} \cdot A^{-1}$
Electric resistance	ohm	Ω, V/A	
Capacitance	farad	F, $A \cdot s/V$	
Magnetic flux	weber	Wb, $V \cdot s$	
Inductance	henry	H, $V \cdot s/A$	
Magnetic flux density	tesla	T, Wb/m^2	
Magnetic field strength	ampere per meter	A/m	
Magnetomotive force	ampere	A	
Luminous flux	lumen	lm	
Luminance	candela per square meter	cd/m^2	
Illuminance	lux	lx	
Wave number	1 per meter	m^{-1}	
Entropy	joule per kelvin	J/K	
Specific heat capacity	joule per kilogram kelvin	$J/(kg \cdot K)$	$m^2 \cdot s^{-2} \cdot K^{-1}$
Thermal conductivity	watt per meter kelvin	$W/(m \cdot K)$	$m \cdot kg \cdot s^{-3} \cdot K^{-1}$
Conductance	siemens	S, A/V	
Torque	newton-meter	N/m	$m^2 \cdot kg \cdot s^{-3} \cdot K^{-1}$
Supplementary units			
Plane angle	radian	rad	

TABLE 10-5 SPECIAL NAMES FOR UNITS

Quantity	Unit	Symbol	Formula
Frequency (of a periodic phenomenon)	hertz	Hz	1/s
Force	newton	N	$kg \cdot m/s^2$
Pressure, stress	pascal	Pa	N/m^2
Energy, work, quantity of heat	joule	J	$N \cdot m$
Power, radiant flux	watt	W	J/s
Quantity of electricity, electric charge	coulomb	C	$A \cdot s$
Electric potential, potential difference, electromotive force	volt	V	W/A
Capacitance	farad	F	C/V
Electric resistance	ohm	Ω	V/A
Conductance	siemens	S	A/V
Magnetic flux	weber	Wb	$V \cdot s$
Magnetic flux density	tesla	T	Wb/m^2
Inductance	henry	H	Wb/A
Luminous flux	lumen	lm	$cd \cdot sr$
Illuminance	lux	lx	lm/m^2

of the more common derived units used in this book. Special names are given to some of the derived units (see Table 10-5). For example, hertz (Hz) is the special name given to the SI unit for frequency. Units of force, stress, power, and energy also have special names and therefore need not be expressed in their base units. The amount of force required to accelerate 1 kilogram of mass 1 meter per second squared is given the special name of newton (N). Thus $1\ N = 1\ kg \cdot m/s^2$.

Prefixes. Prefixes corresponding to powers of 10 are attached to the units discussed above in order to form larger or smaller units. In technical work the powers of 10 divisible by 3 are preferred. Table 10-6 contains the authorized prefixes.

TABLE 10-6 PREFIXES

Multiplication factor	Prefix	Symbol
1 000 000 000 000 000 000 = 10^{18}	exa	E
1 000 000 000 000 000 = 10^{15}	peta	P
1 000 000 000 000 = 10^{12}	tera	T
1 000 000 000 = 10^{9}	giga	G
1 000 000 = 10^{6}	mega	M
1 000 = 10^{3}	kilo	k
100 = 10^{2}	hecto	h
10 = 10^{1}	deka	da
0.1 = 10^{-1}	deci	d
0.01 = 10^{-2}	centi	c
0.001 = 10^{-3}	milli	m
0.000 001 = 10^{-6}	micro	μ
0.000 000 001 = 10^{-9}	nano	n
0.000 000 000 001 = 10^{-12}	pico	p
0.000 000 000 000 001 = 10^{-15}	femto	f
0.000 000 000 000 000 001 = 10^{-18}	atto	a

Rules for usage. For standardized usage, the following rules should be observed:

1. Uppercase (capitals) and lowercase letters are never interchanged: kg, not KG.
2. The same symbol is used for plurals: N, not Ns; 14 meters or 14 m.
3. No space is left between the prefix and its unit symbol: GHz, not G Hz.
4. To form products, a raised dot is preferred (or a dot on a line): $kN \cdot m$, or kN.m. The dot may be dispensed with if confusion is not created by its absence.
5. To form quotients, one solidus (an oblique line), a fraction line (horizontal), or a negative power is used to express derived units: m/s, $\frac{m}{s}$, or $m \cdot s^{-1}$.

   Note: The solidus must not be repeated on the same line: m/s^2, not m/s/s. Also, $kg/(m \cdot s)$, $\frac{kg}{m \cdot s}$, or $kg \cdot m^{-1} \cdot s^{-1}$, but not kg/m/s. Note also the use of the parentheses to avoid ambiguity.
6. An exponent affixed to a symbol containing a prefix indicates that the multiple or submultiple of the unit is raised to the power expressed by the exponent.

$$\begin{aligned} 1\ mm^3 &= 10^{-9}\ m^3, \text{ not } 10^{-3}\ m^3 \\ 1\ cm^3 &= 10^{-6}\ m^3, \text{ not } 10^{-2}\ m^3 \\ 1\ cm^{-1} &= 10^{2}\ m^{-1}, \text{ not } 10^{-2}\ m^{-1} \end{aligned}$$

7. A period is used as a decimal marker. It is not used to separate groups of digits. A space is left for this purpose: 5 279 585 J, and 0.000 34 s.
8. Numbers are preferably expressed between the limits 0.1 and 1000, using the appropriate prefix to change the size of the unit: 5.23 GN.
9. For decimal numbers less than 1, the leading zero is never omitted: 0.625, not .625.
10. When units are written in words, they always start with lowercase letters except at the beginning of a sentence. If the unit is derived from the name of a person, the symbol is capitalized. Plurals of special names are written in the usual manner.

    125 watts or 125 W
    0.25 newton or 0.25 N
    58.6 hertz or 58.6 Hz
11. A space or hyphen may be used to form the product expressed in words: newton - meters or newton meters.
12. For quotients, the word *per* may be used: newton per meter squared, kilogram per cubic meter.
13. The kelvin (symbol K) is the standard unit of temperature. In writing this absolute temperature, the word *degree* or its symbol (°) is not used: 472 K. In addition, K may be used to express an interval or a difference in temperature. Celsius temperature is expressed in degrees Celsius with symbol °C. The unit degrees Celsius is equal to the unit kelvin and may also be used to represent an interval or a difference of Celsius temperature: 25°C. Temperature in K = temperature in °C + 273.15.

Computations. The SI system of units makes computations relatively simple because (1) a single unit is used to represent a particular physical quantity, (2) the system is coherent in that the factor of 1.0 replaces many conversion factors, and (3) SI is based on the decimal system.

Prior to computation, all prefixes should be replaced by their respective powers of 10. The final step is to select a suitable prefix to express the answer once the resulting answer is rounded to the appropriate number of significant digits.

Two examples illustrate the solution of typical problems using SI units and the preceding information.

Problem 1

Given: A metal rod under tensile load of 356 kN is allowed to withstand a unit stress of 110 NM/m².
Required: Find the diameter of the rod in millimeters.

Solution

1. Convert SI prefixes to powers of 10:

$$356 \text{ kN} = 356 \times 10^3 \text{ N}$$

$$110 \text{ MN/m}^2 = 110 \times 10^6 \text{ N/m}^2$$

2. Use the direct stress formula ($s = P/A$) and solve for the area (A):

$$A = \frac{P}{s} = \frac{356 \times 10^3 \text{ N}}{110 \times 10^6 \text{ N/m}^2} = 3.24 \times 10^{-3} \text{ m}^2$$

3. Use the formula for circular area $\left(A = \frac{\pi}{4}D^2\right)$ and solve for diameter (D):

$$D = \sqrt{\frac{4}{\pi}(3.24 \times 10^{-3} \text{ m}^2)}$$

or

$$D = \left[\frac{4}{\pi}(3.24 \times 10^{-3} \text{ m}^2)\right]^{1/2}$$

$$= \left[\frac{4}{\pi}(32.4 \times 10^{-4} \text{ m}^2)\right]^{1/2}$$

$$= 6.429 \times 10^{-2} \text{ m}$$

$$= 64.29 \times 10^{-3} \text{ m} = 64.3 \text{ mm (rounding up and using three significant digits)}$$

Problem 2

Given: Modulus of elasticity in tension (E) for steel is 29,120,000 psi.
Required: Express E in GPa.

Solution

1. Express E in terms of powers of 10:

$$E = 29.12 \times 10^6 \text{ psi}$$

TABLE 10-7 CONVERSIONS

Quantity	U.S. customary to SI
Acceleration, linear	1 ft/s^2 = 3.048 × 10^{-1} m/s^2
Area	1 in^2 = 6.452 × 10^2 mm^2
	1 ft^2 = 9.290 × 10^{-2} m^2
Density (mass), ρ	1 lb/in^3 = 2.768 × 10^4 kg/m^3
	1 lb/ft^3 = 1.602 × 10^1 kg/m^3
Electric current, *I*	1 ampere (A) = 1 C/s
Energy, work	1 Btu = 1.055 kJ
	1 in. · lb = 1.129 × 10^{-1} J
	1 ft · lb = 1.356 J
	1 Btu = 0.293 W · h
Force	1 lbf = 4.448 N
	1 kgf = 9.807 N
Impulse	1 lb · s = 4.448 N · s
Length	1 Å = 1 × 10^{-1} nm
	1 microinch = 2.540 × 10^{-2} μm
	1 mil = 2.540 × 10^1 μm
	1 in. = 2.540 × 10^1 mm
	1 ft = 3.048 × 10^{-1} m
Magnetic field strength, *H*	1 oersted (Oe) = 79.58 ampere turns per meter
Magnetic flux	1 maxwell = 10^{-8} Wb
Magnetic flux density, *B*	1 Wb/M^2 = 1 T (tesla)
Modulus of elasticity, *E*	1 lb/in^2 = 6.895 × 10^{-6} GPa
Moment of force, torque	1 lb · in. = 1.130 × 10^{-1} N · m
	1 lb · ft = 1.356 N · m
Moment of inertia, *I* (of area)	1 in^4 = 4.162 × 10^5 mm^4
Momentum, linear	1 lb · ft/s = 1.383 × 10^{-1} kg · m/s
Power	1 Btu/min = 1.758 × 10^{-2} kW
	1 ft · lb/min = 2.259 × 10^{-2} W
	1 hp = 7.457 × 10^{-1} kW
Stress (pressure)	1 lb/in^2 = 6.895 × 10^{-3} MPa
	1 ksi = 6.895 MPa
Temperature	1°F (difference) = 0.555°C
	1.8 °F = 1°C (difference)
Thermal expansion, linear coefficient, α	in./in./°F = 1.8 K^{-1}, K = °C + 273.15
Thermal conductivity, *k*	1 Btu/ft · hr · °F = 1.729 W/m · K
Velocity, linear, *v*	1 in./s = 2.540 × 10^1 mm/s
	1 ft/s = 3.048 × 10^{-1} m/s
	1 in./min = 4.233 × 10^{-1} mm/s
	1 ft/min = 5.080 × 10^{-3} m/s
Velocity, angular, ω	1 rev/min = 1.047 × 10^{-1} rad/s
Viscosity	1 poise (P) = 0.1 Pa · s
Volume	1 in^3 = 1.639 × 10^4 mm^3
	1 ft^3 = 2.832 × 10^{-2} m^3
	1 yd^3 = 7.646 × 10^{-1} m^3

1 joule (J) = 10^7 ergs = 0.625 × 10^{19} eV

1 gauss (G) = 10^{-4} Wb/m^2 = 1 T

1 weber (Wb) = 1 T/m^2

1 Å = 10^{-10} m (obsolete)

1 micrometer (micron) = 10^{-6} m (obsolete)

TABLE 10-7A ADDITIONAL USEFUL CONVERSIONS

$1\ W = 3.413\ Btu/hr$
$1\ W = 1\ N \cdot m \cdot s^{-1}$
$1\ W = 0.7376\ ft \cdot lb \cdot s^{-1}$
$1\ kW = 1.341\ Hp$
$1\ J = 0.2390\ cal$
$1\ cal = 4.186\ J$
$1\ Btu = 252\ cal$
$1\ J = 0.737562\ lb \cdot ft$
$1\ J = 1\ N \cdot m = 1\ W \cdot s$
$1\ Pa = 1\ N/m^2$
$1\ N/m^2 = 1.450\ lb/in^2$ (psi)
$1\ m = 3.28\ ft = 39.37\ in.$
$°C = \frac{5}{9}(F - 32)$
$°F = \frac{9}{5}(F + 32)$
$K = °C + 273$ (K = kelvin)
$°R = °F + 460$ (R = Rankine)
$1\ L = 1.06\ qt$
$1\ Pa \cdot s = 10$ poise

2. Locate conversion ratio (see Table 10-7):

$$1\ \text{psi} = 6.895 \times 10^{-3}\ \text{MPa}$$

3. Express prefixes in terms of powers of 10 (see Table 10-6):

$$1\ \text{psi} = 6.895 \times 10^{3}\ \text{Pa}$$

4. Multiply E by conversion ratio:

$$29.12 \times 10^{6}\ \text{psi}\left(\frac{6.895 \times 10^{3}\ \text{Pa}}{1\ \text{psi}}\right) = 200.8 \times 10^{9}\ \text{Pa}$$

5. Express answer using required SI prefix (see Table 10-6):

200.8 GPa (using four significant digits)

REFERENCES

ASTM E 380: *Standard for Metric Practice.*

ANSI Z 210.1: *American National Standard for Metric Practice.*

ISO 1000: *SI Units and Recommendations for the Use of Their Multiples and of Certain Other Units.*

NBS Special Publication 330: *The International System of Units (SI).*

TABLE 10-8 CONSTANTS

Quantity	Symbol	Value
Acceleration of gravity	g	$9.80\ \mathrm{m \cdot s^{-2}}$
Atomic mass unit	amu	1.66×10^{-27} kg
Avogadro's number	N_A, N_0	6.022×10^{23} molecules/mole
Bohr radius	a_0	5.292×10^{-11} m
Electron charge	$q, -e$	1.60×10^{-19} C
Electron mass	m	9.11×10^{-31} kg
Electron volt	eV	0.160×10^{-18} J
Magnetic permeability of free space	μ_0	$4\pi \times 10^{-7}$ H/m $= 1.257 \times 10^{-6}$ H/m
Permittivity of free space	ϵ_0	$\frac{10^7}{4\pi c^2} = 8.854 \times 10^{-12}\ \mathrm{C/V \cdot m}$
Planck's constant	$\hbar$	$6.63 \times 10^{-34}\ \mathrm{J \cdot s}$
Speed of light (vacuum)	c	$3.00 \times 10^{8}\ \mathrm{m \cdot s^{-1}}$

TABLE 10-9 REPRESENTATIVE PLASTICS

Common Name (Chemical Name)	ASTM Abbre- viations	Trade Names	Common Structure	Grouping[a]	Typical Uses
ABS (acrylonitrile–buta- diene–styrene)	ABS	Absinol Abson Cycolac Royalite	Amorphous terpolymer 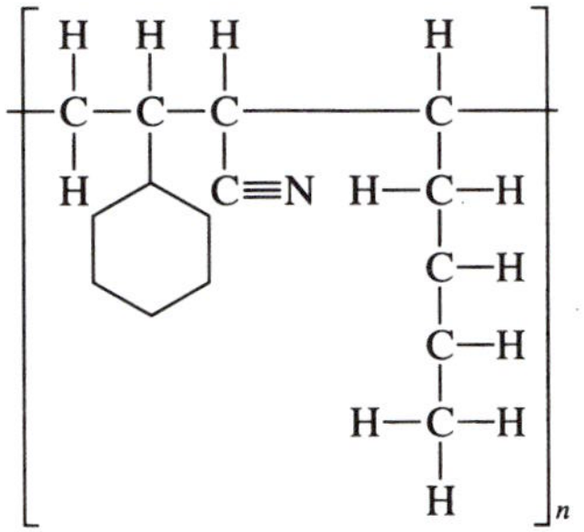 $[-CH_2-CH(C_6H_5)-CH(C{\equiv}N)-CH(CH_2-CH-CH-CH_3)-]_n$	TP, EP	Pipe, toys, luggage, boat hulls, foot- ball helmets, gears, chrome- plated plumbing, and auto parts
Acetal (polyoxymethylene) (polymerized formal- dehyde)	POM	Delrin Celcon Formaldafil	Highly crystalline homopolymer and copolymers 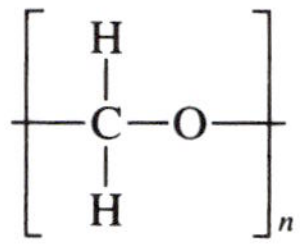 $[-CH_2-O-]_n$	TP, EP	Gears, bearings, fan blades, shower heads, auto parts, and aerosol bot- tles
Acrylics [poly(methyl methac- rylate)]	PMMA	Plexiglas Lucite Acrylite	Amorphous 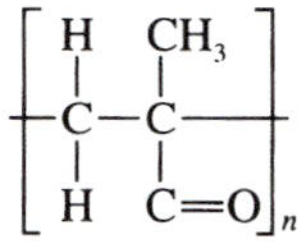 $[-CH_2-C(CH_3)(C{=}O)-]_n$	TP, GP	Lenses, windows, signs, sculpture, light pipes, and skylights
(polyacrylonitrile)	PAN	Sayelle			

[a]TP, thermoplastic; TS, thermosetting; EP, engineering plastic; GP, general-purpose or special. ASTM, American Society for Testing and Materials.
[b]Engineering plastics.

TABLE 10-9 (continued)

Common Name (Chemical Name)	ASTM Abbre-viations	Trade Names	Common Structure	Grouping[a]	Typical Uses
Alkyd plaskon (modified polyester resins)		Premix Dyal Glaskyd	Cross-link network 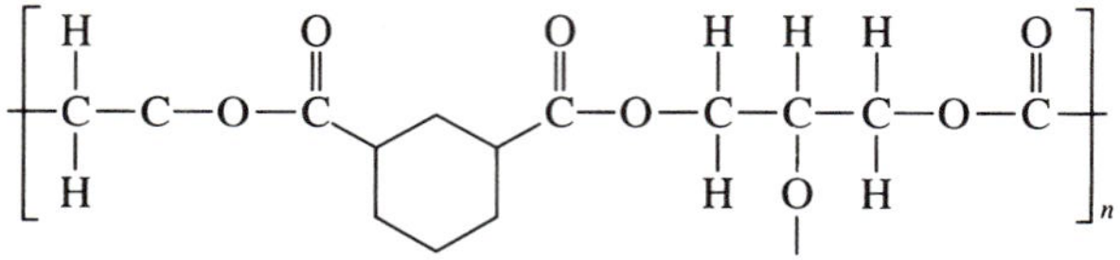	TS, GP	Coatings: enamel, lacquer and paint, molded electrical parts
Allylics					
(diallyl phthalate)	DAP	Diall	Cross-link network	TS, GP	Electronic parts, pump impellers, glass-fiber impregnate, dinnerware, watch crystals
(diallyl isophthalate)	DAIP	Poly-Dap	$\left[CH_2{=}CH{-}CH_2 \right]_n$		
(diethylene glycol bisallyl/carbonate)	CR39-allyl diglycol carbonate			TS	
Aminos					
(urea–formaldehyde)		Plaskon	Cross-link network	TS, GP	Electrical parts, particle-board binders, coatings, dinnerware, paper impregnate
(melamine–formaldehyde)	UF MF	Cymel			
Cellulosics			Highly crystalline 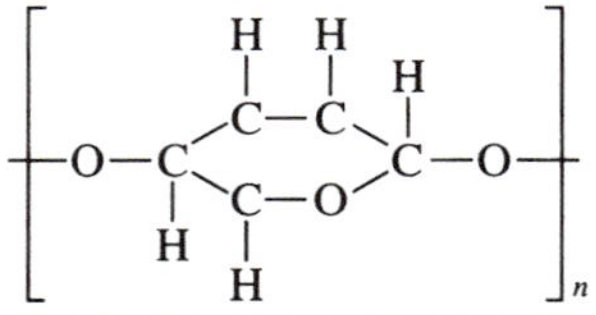	TP, GP	Packaging film, pipe, optical frames, flashbulb shields, helmets, rollers
(cellulose acetate)	CA	Tenite			
(cellulose butyrate)	CAB	Uvex			
(cellulose nitrate)	CN	Nixonite			
(cellulose propionate)	CAP	Forticel			
(ethyl cellulose)	EC	Ethocel Methocel			

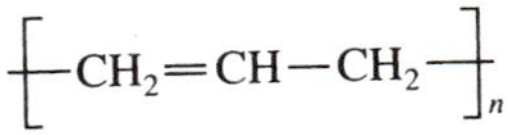

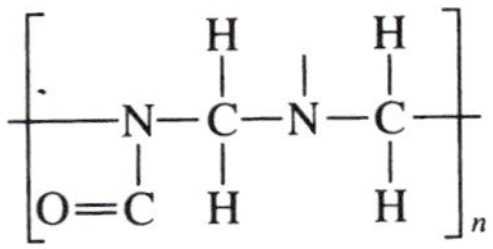

[a] TP, thermoplastic; TS, thermosetting; EP, engineering plastic; GP, general-purpose or special. ASTM, American Society for Testing and Materials.

[b] Engineering plastics.

TABLE 10-9 (continued)

Common Name (Chemical Name)	ASTM Abbreviations	Trade Names	Common Structure	Grouping[a]	Typical Uses
Epoxy	EP	Epi Rez Hysol	Cross-linked network or amorphous $\left[-O-C_6H_4-C(CH_3)_2-C_6H_4-O-CH_2-CH(O)CH_2-\right]_n$	TS, TP, GP	Coatings, fiber reinforcement, potting and encapsulating
Fluorocarbons (polytetrafluoroethylene)[b] (polychlorotrifluoroethylene)[b] [poly(vinylidene fluoride)]	FEP PTFE or TFE CTFE PVDF	Teflon Kelf Kynar	Highly crystalline $\left[-CF_2-CF_2-\right]_n$	TP,[b] EP, GP	Seals, cookware, corrosion liquids hardware, chemical-processing equipment
Nylon (polyamide) (nylon 6) (nylon 6/6) (nylon 6/10) (nylon 6/12) (aramid)	PA	Nylon Celanese Zytel Plaskon Zytrel Nomex Kevlar	Crystalline $\left[-N(H)-C(=O)-(CH_2)_5-\right]_n$ Nylon 6	TP, EP	Bearings, tubing, gears, housings Parts exposed to moisture: nylon 6/10, nylon 6/12 Reinforcing fiber: aramid
Phonelic (phenol–formaldehyde)	PF	Bakelite Durite Resinox	Cross-linked network	TS, GP	Electrical parts, housings, and binders

[a]TP, thermoplastic; TS, thermosetting; EP, engineering plastic; GP, general-purpose or special. ASTM, American Society for Testing and Materials.

[b]Engineering plastics.

TABLE 10-9 (continued)

Common Name (Chemical Name)	ASTM Abbreviations	Trade Names	Common Structure	Grouping[a]	Typical Uses
Phenoxy [poly(hydroxy ethers)]			Amorphous or cross-linked network 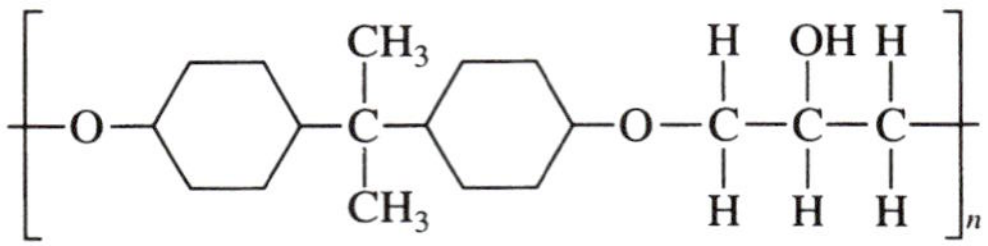	TP, TS, EP	Gas pipe, sports equipment, electrical housing, adhesives and coatings
Phenylene oxide [poly(phenylene oxide)]	PPO	Noryl	Crystalline 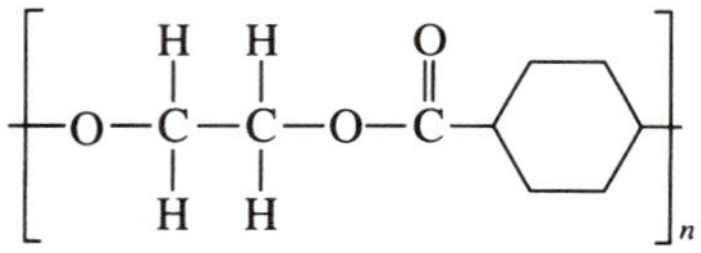	TP, EP	Auto trim, panels electrical housing, TV cabinets, pump parts
Polycarbonate	PC	Lexan Xenoy (blend)	Amorphous 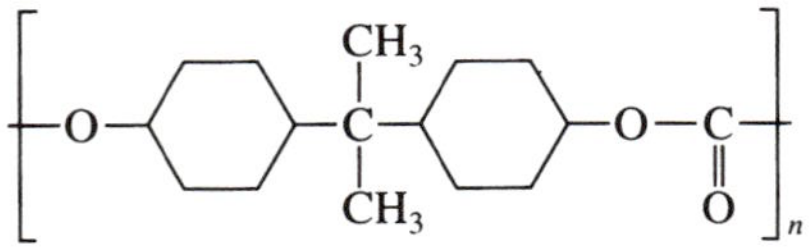	TP, EP	Optical lenses, bullet-resistant windows, housings, cookers
Polyester [poly(ethylene terephthalate)] [poly(butylene terephthalate)] (aromatic polyesters)	 PET (TS) PBT (TP)	Mylar Dacron Kodel Fortrel Laminac Selectron Gafite tp Ekonol	Amorphous or cross-linked network 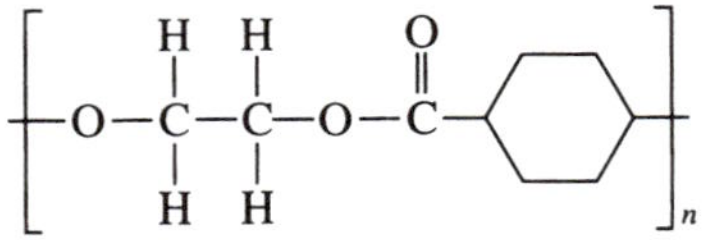	TS, TP, GP, EP	Glass-fiber reinforcer, films, fibers

[a]TP, thermoplastic; TS, thermosetting; EP, engineering plastic; GP, general-purpose or special. ASTM, American Society for Testing and Materials.
[b]Engineering plastics.

TABLE 10-9 (continued)

Common Name (Chemical Name)	ASTM Abbreviations	Trade Names	Common Structure	Grouping[a]	Typical Uses
Polyimide	LaRC-ITP1 LaRC-CP1 UPILEX R_1	Kapton	Cross-linked network or amorphous 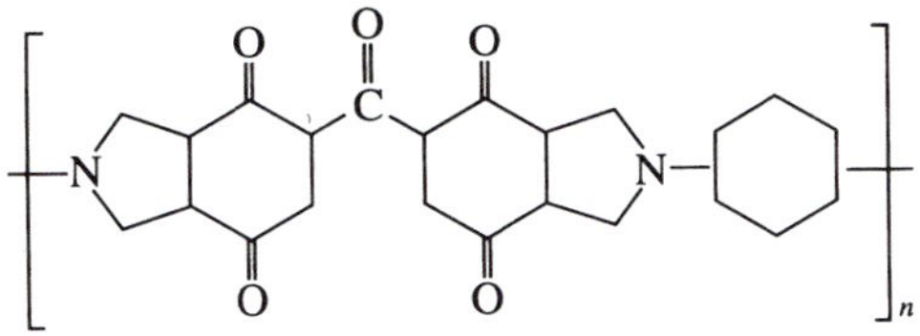	TS, TP, EP	Jet-engine vane bushings, seals, ball-bearing separators, high-temperature film, fiber matrix
Poly(amide/imide)		Torlon Tygon	Amorphous 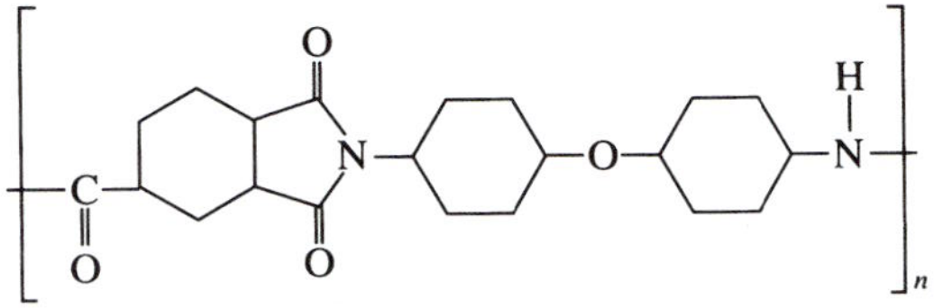	TP, EP	Engineering-plastic gears, structural components, bearings, seals, and valves
Polyolefins			Low to high crystallinity or cross-linked network 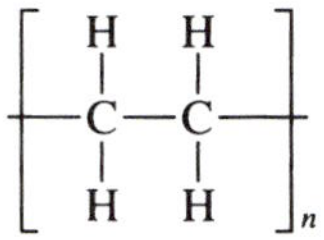	TP, TS, GP, EP	PE: packaging, squeeze bottles, electrical insulation and tubing
(polyethylene, low density)	LDPE	Alathon	polyethylene		
(polyethylene, high density)[b]	HDPE	Dylan			
(polyethylene, ultra-high molecular weight)[b]	UHMWPE	Marlex			PP: packaging, auto-battery cases, housings, electrical components and fan blades
(polypropylene)	PP	Escon			
(polyallomer)		Tenite			
(ethylene–vinyl acetate)	EVA				EVA: shoe soles and hypodermic syringes
[poly(ethylene terephthalate)]	PET	Arnite Dacron Rynite	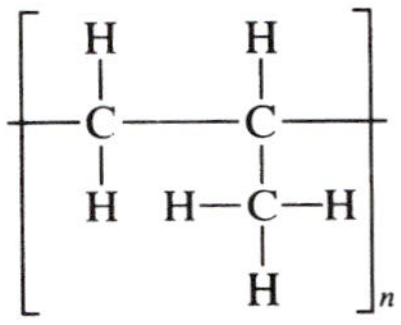 polypropylene		
(polybutylene)					
[poly(methyl pentene)]	TPX				
(ionomer)		Surlyn			

[a]TP, thermoplastic; TS, thermosetting; EP, engineering plastic; GP, general-purpose or special. ASTM, American Society for Testing and Materials.

[b]Engineering plastics.

TABLE 10-9 (continued)

Common Name (Chemical Name)	ASTM Abbre-viations	Trade Names	Common Structure	Grouping[a]	Typical Uses
Poly(ether ether ketone)	PEEK	Victrex	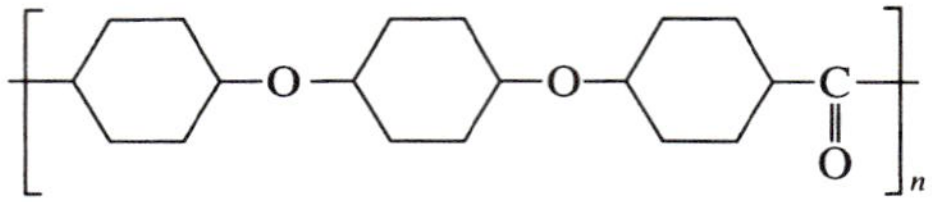	EP	Crystalline polymer for high-tempera-tures and com-posite matrices
Poly(phenylene sulfide)	PPS	Ryton	Amorphous 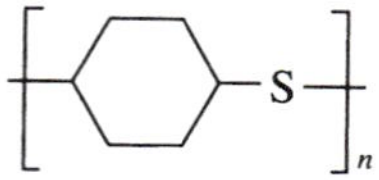	TP, EP	Electrical terminal block and con-nectors, seals, gears
Polystyrene (styrene–acryloni-trile)	PS SAN	Styron Styrofoam Lustrex Dylite Tyril	Amorphous 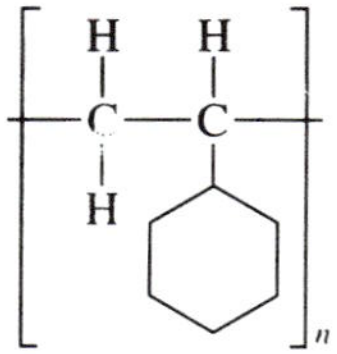	TP, GP	Packaging, control knobs, TV cabi-nets, wood sub-stitute, foam in-sulation
Polysulfone [poly(ether sulfone)] (polyphenylsulfone)		Udel Radel	Amorphous 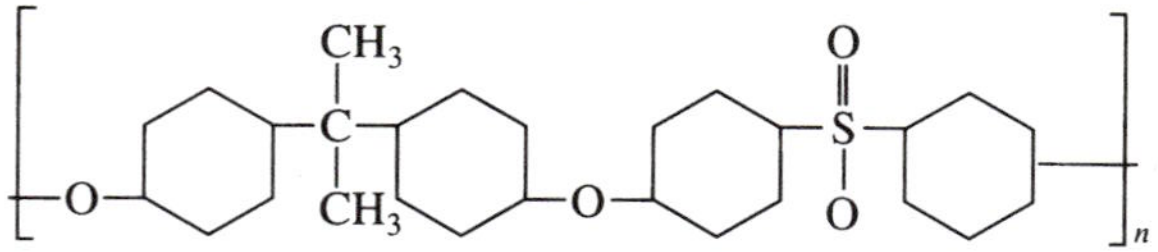	TP, EP	Electrical insula-tors, auto-distrib-utor caps, tubing, and aircraft cabin interiors
Polyurethane (isocyanate polyester or polyether)	PUR	Estane Flexane Texin Calspan	Amorphous or cross-linked network 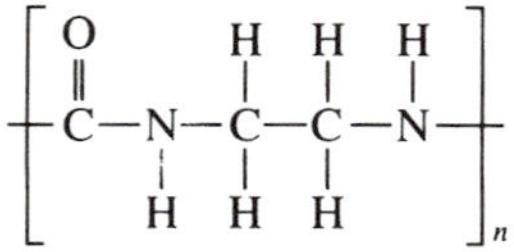	TP, TS, EP	Reaction injection molded (RIM) foamed auto parts, solid tires, auto bumpers, synthetic leather

[a]TP, thermoplastic; TS, thermosetting; EP, engineering plastic; GP, general-purpose or special. ASTM, American Society for Testing and Materials.
[b]Engineering plastics.

TABLE 10-9 (continued)

Common Name (Chemical Name)	ASTM Abbreviations	Trade Names	Common Structure	Grouping[a]	Typical Uses
Silicones	SI	Silastic RTV Silicone RTV	Amorphous $\left[-\mathrm{Si}(\mathrm{CH_3})_2-\mathrm{O}-\right]_n$	TS, GP	Room-temperature-vulcanizing (RTV) molds, fiber-matrix, electronic potting and encapsulating, heat seals
Vinyls [poly(vinyl chloride)] [poly(vinyl acetate)] [poly(vinyl alcohol)] [poly(vinyl butyral)] [poly(vinyl fluoride)] [poly(vinylidene chloride)]	 PVC PVAc PVA PVB PVF PVDC	Vinylite Naugahyde Luxite Elvanol Butrar Tedlar Saran	Cross-linked network $\left[-\mathrm{CH_2}-\mathrm{CHCl}-\right]_n$ PVC $\left[-\mathrm{CH_2}-\mathrm{CH}(\mathrm{C{=}O}\,\mathrm{CH_3})-\right]_n$ PVAC	TS, TP, GP	Plastisol coating, upholstery, pipe, building trim, coatins, film

[a]TP, thermoplastic; TS, thermosetting; EP, engineering plastic; GP, general-purpose or special. ASTM, American Society for Testing and Materials.

[b]Engineering plastics.

TABLE 10-10 REPRESENTATIVE ELASTOMERS[a]

Abbreviation (ASTM):	NR	IR	SBR	IIR	BR	NBR
Common name:	Natural rubber		GRS or buna S	Butyl		Nitrile or buna N
Chemical name:	Natural polyisoprene	Isoprene Polyisoprene	Styrene Butadiene	Isobutene Isoprene	Polybutadiene Butadiene	Nitrile butadiene
(1) Tensile strength (kPa)	31	27.5	24	20.6	20.6	24
(2) Hardness Shore A	30–100	40–80	40–90	40–75	45–80	40–95
(3) Specific gravity	0.93	0.93	0.94	0.92	0.94	1.00
(4) Abrasion resistance	A	A	A	B	A	A
(5) Tear resistance	A	B	C	B	B	B
(6) Flexibility at low temperature	B	B	C	B	B	C
(7) Impact resistance	A	A	A	B	B	C
(8) Resiliency	A	A	B	C	A	B
(9) Creep	A	B	B	C	A	B
(10) ASTM/SAE type class	AA	AA	AA–BA	AA–BA	AA	BF, BG, PK, CH
(11) Maximum service temp. (°C)	70	70	100	100	70	100–125
(12) Heat-aging resistance	B–C	B–C	B	B–A	C	B
(13) Flame resistance	D	D	D	D	D	D
(14) Oil and gasoline resistance	X	X	X	X	C	A
(15) Oxidation resistance	C	B–C	C	C	C	C
(16) Ozone resistance	C	C–D	C–D	A	C–D	C–D
(17) Ultraviolet resistance	C–D	C–D	B–C	B	B–C	B–C
(18) Acid and base resistance	B–C	B–C	B–C	B	B–C	B
(19) Water absorption resistance	A	A	B	A	A	C
(20) Permeability to gases	C	C	C	A	C	B
(21) Electrical resistivity	A	A	A	A	A	D
(22) Adhesion to metals	A	A	A	B	A	A
(23) Trade names	—	Natsyn, Isoprene, Ameripol SN	K-Resin	Enjay Butyl, Petro-Tex Butyl	Diene, Ameripol CB	Paracril
(24) Typical uses	Tires, seals, bearings, couplings, shoe soles and heels	Same as natural rubber	Shock absorbers, belts, heels, sponges, gaskets, belts	Truck and auto tires, shock absorbers, inner tubes	Pneumatic tires, gaskets, seals, abrasion resistance belts	Gasoline, chemical and oil seals, gaskets and O-rings, belting

[a] A, excellent; B, good; C, fair; D, poor; X, not recommended.

TABLE 10-10 (continued)

	ACM Acrylate Polyacrylate	CR Neoprene Chloropene	CSM Hypalom Chlorosulfanyl Polyethylene	FPM Fluorocarbon Fluorinated hydrocarbon	EPDM EPDM Ethylene propylene	SI Silicone rubber Polysilicone	UE Urethane rubber Polyether urethane	PTR Thikol Polysulfide
(1)	17	27.5	27.5	17	20.6	4–10	34–55	4–10
(2)	40–90	40–90	50–95	60–90	30–90	25–80	35–100	20–80
(3)	1.10	1.23	1.12–1.28	1.45	0.86	1.14–2.05	1.06	1.34
(4)	B	B	B	A	B	D	B	D
(5)	C–B	C–B	B	B	C	B	B	D
(6)	D	C	B	D	B	A	C	C
(7)	D	B	C–B	B	B	D–C	B–A	D
(8)	C–B	A	C	C	B	D–A	C–A	C
(9)	C	B	C	B	C–B	C–A	C–A	D
(10)	DF, DH	BC, BE	CE	HK	AA, DA, CA	FC, FE, FK	AD, EC	AK
(11)	125	100	125	250	125	200–225	100	70
(12)	A	A	C	A	A	A	B–A	C–B
(13)	D	B	C	A	D	A	D	D
(14)	A	C	C	A	X	D–C	B	A
(15)	A	A	A	A	A	A	A	B
(16)	B	B	A	A	A	A	A	A
(17)	B	B	B	A	A	A	B	B
(18)	C	B	B	A	A	D	D	X
(19)	A	B	A	A	A	A	A	B
(20)	B	B	B	A	C	D	B	D
(21)	B	C	B	B	B	A	B	C
(22)	B	A	A	C–B	C	A	C	C–B
(23)	Hycar, Acrylon	Neoprene, Perbunanc	Hypalon	Viton, Proflo, Fluorel	Nordel, Epcar, Royalene	Adiprene, RTV, Silastic	Kalrez, Estane, Roylar	Thikol
(24)	Oil hose, colored and white parts, pressure and oil O-rings	Belts, hose, extruded goods, molded sheet, adhesives, chemical-tank liners	Laminated roofing, tarpaulins, reservoir and pond liners, diaphragms, shoe soles and heels, whitewall tires	Brake seals, ducting connectors, carburetors, needle tips, roll coverings	Garden and industrial hoses, belts, bike tires, electrical-wire insulation, paintable auto bumpers	Industrial tires and rolls, mining belts, die pads, gaskets and seals	Chemical O-ring seals, valve seats, gaskets, nuclear, oil, gas, hydraulic, and acid seals	Gasoline hose, printing rolls, caulking, adhesives and binders

TABLE 10-11 REPRESENTATIVE METALS

Material	Nominal Composition (Essential Elements) (%)		Form and Condition	Typical Mechanical Properties			
				Yield Strength (0.2% offset) (1000 psi)	Tensile Strength (1000 psi)	Elongation in 2 in. (%)	Hardness, Brinell or Rockwell
Copper CA 110 Sheet—ASTM B152 Rod—ASTM B124, B133 Wire—ASTM B1, B2, B3	Cu	99.90 min	Strip annealed Spring temper	10[a] 50[a]	32 55	45 4	40 HRF 60 HRB
Commercial bronze CA 220 Plate, sheet, strip, bar—ASTM B36 Wire—ASTM B 134	Cu Zn	90 10	Strip annealed Spring temper	10[a] 62[a]	37 72	45 3	53 HRF 78 HRB
Red brass CA 230 Strip, sheet, plate—ASTM B36 Wire—ASTM B134 Tube—ASTM B135	Cu Zn	85 15	Strip annealed Hard temper	15[a] 60[a]	40 75	50 7	50 135
Copper–nickel CA 715 Sheet—ASTM B122 Plate—ASTM B171 Tube—ASTM B111	Cu Ni Fe Mn	bal 30 0.55 0.5	Tube annealed	25[a]	60	45	45 HRB
Aluminum alloy Alclad 2024 Sheet and plate—ASTM B209	Core: 2024 Al Cu Mn Mg Cladding: Al	 bal 4.5 0.6 1.5 99.3 min	Sheet annealed Heat annealed	11 42	26 64	20 19	— —
Aluminum alloy 3003 Sheet and plate—ASTM B209	Al Mn	bal 1.2	Sheet annealed Cold rolled	6 27	16 29	30 4	28 55
Aluminum alloy 5052 Sheet and plate—ASTM B209	Al Mg Cr	bal 2.5 0.25	Sheet annealed Cold rolled	13 37	28 42	25 7	47 77
Aluminum alloy 6061 Sheet and plate—ASTM B209	Al Si Cu Mg Cr	bal 0.6 0.25 1.0 0.25	Sheet annealed Heat treated	8 40	18 45	25 12	30 95
Aluminum alloy 707S Bar, rod, wire, and shapes—ASTM B221	Al Zn Cu Mg Cr	bal 5.6 1.6 2.5 0.3	Bar annealed Heat treated	15 73	33 83	16 11	60 150
Cast alluminum alloy 13 Castings—ASTM B85 Grade S12A	Al Si	bal 12.0	Die casting As cast	21	43	2.5	—

Typical physical properties								
Density (lb/cu in)	Specific Gravity	Melting Point (°F)	Specific Heat (32 to 212°F) (Btu/lb/°F)	Thermal-expansion Coefficient (32 to 212°F) (in 10^{-6} in./in./°F)	Thermal Conductivity (32 to 212°F) (Btu/sq ft/hr/°F/in.)	Electrical Resistivity (68°F) (ohms/cir mil ft)	Tensile Modulus of Elasticity (10^6 psi)	Torsional Modulus of Elasticity ($\times 10^6$ psi)
0.322	8.91	1980	0.092	9.4	2512	10.3	17	6.4
—	—	—	—	9.8	—	—	—	—
0.318	8.80	1910	0.09	10.2	1308	23.6	17	6.4
—	—	—	—	—	—	—	—	—
0.316	8.75	1880	0.09	10.4	1104	28	17	6.4
—	—	—	—	—	—	—	—	—
0.323	8.94	2260	0.09	9.0	204	225	22	8.3
0.100	2.77	1180	0.23	12.6	1340	21	10.6	3.75
—	—	—	—	—	840	35	10.6	4.0
0.099	2.73	1210	0.23	12.9	1340	21	10.0	3.75
—	—	—	—	—	1070	26	10.0	3.75
0.097	2.68	1200	0.23	13.2	960	30	10.2	3.75
—	—	—	—	—	960	30	10.2	3.75
0.098	2.70	1205	0.23	13.0	1190	23	10.0	3.75
—	—	—	—	—	1070	26	10.0	3.75
0.101	2.80	1175	0.23	12.9	—	—	10.4	3.9
—	—	—	—	—	840	35	10.4	3.9
0.096	2.65	1080	0.23	11.5	870	34	10.3	3.85

TABLE 10-11 REPRESENTATIVE METALS

Material	Nominal Composition (Essential Elements) (%)		Form and Condition	Typical Mechanical Properties			
				Yield Strength (0.2% offset) (1000 psi)	Tensile Strength (1000 psi)	Elongation in 2 in. (%)	Hardness, Brinell or Rockwell
Magnesium alloy AZ 31B	Mg	bal	Sheet annealed	22	37	21	56
Plate and sheet—ASTM	Al	3.0					
B90	Zn	1.0	Hard sheet	32	42	15	73
	Mn	0.2 min					
Magnesium alloy AZ 80A	Mg	bal	As forged	33	48	11	69
Forgings—ASTM B91	Al	8.5					
	Zn	0.5	Forged and aged	36	50	6	72
	Mn	0.15 min					
Magnesium alloys AZ 91A and AZ 91B	Mg	bal					
Castings—ASTM B94	Al	9.0	Die cast	22	33	3	63
	Zn	0.7					
	Mn	0.2 min					
Titanium Ti-35A	Ti	bal					
Forgings—ASTM B381	C	0.08 max					
Sheet, strip, and plate—ASTM B265	Fe	0.12 max	Sheet annealed	30	40	30	135
Pipe—ASTM B337	N_2	0.05 max					
Tubes—ASTM B338	H_2	0.015 max					
Bars—ASTM B348							
Ti-6 Al-4 V alloy	Ti	bal					
Sheet, strip, and plate—ASTM B265	Al	6.5	Sheet annealed	130	140	13	39 HRC
Bar—ASTM B348	V	4					
Forgings—ASTM B381	C	0.08 max	Heat treated	165	175	12	—
	Fe	0.25 max					
	N_2	0.05 max					
	H_2	0.015 max					
Nickel 211 ASTM F290	Ni	95.0					
	Mn	4.75	Annealed	35	75	40	140
	C	0.10					
Nickel (cast)	Ni	95.6					
	Cu	0.5					
	Fe	0.5	As cast	25[a]	57	22	110
	Mn	0.8					
	Si	1.5					
	C	0.8					
Duranickel alloy 301	Ni	bal					
	Al	4.5					
	Si	0.55	Hot rolled and aged	132	185	28	330
	Ti	0.5					
	Mn	0.25					
	Fe	0.15					
	C	0.15					

Typical Physical Properties								
Density (lb/cu in)	Specific Gravity	Melting Point (°F)	Specific Heat (32 to 212°F) (Btu/lb/°F)	Thermal-expansion Coefficient (32 to 212°F) (in 10^{-6} in./in./°F)	Thermal Conductivity (32 to 212°F) (Btu/sq ft/hr/°F/in.)	Electrical Resistivity (68°F) (ohms/cir mil ft)	Tensile Modulus of Elasticity (10^6 psi)	Torsional Modulus of Elasticity ($\times 10^6$ psi)
0.064	1.77	1170	0.245	14.5	657	55	6.5	2.4
—	—	—	—	—	—	—	—	—
0.065	1.80	1130	0.25	14.5	522	87	6.5	2.4
—	—	—	—	—	—	—	—	—
0.065	1.80	1105	0.25	14.5	493	102	6.5	2.4
0.163	4.50	3063	0.124	4.8	108	336	14.9	6.5
0.160	4.42	3000	0.135	4.9	50	1026	16.5	6.1
—	—	—	—	—	—	—	—	—
0.315	8.73	2600	0.11	7.4	306	102	30	11
0.301	8.34	2550	0.13	8.85	410	125	21.5	—
0.298	8.75	2620	0.104	7.2	165	255	30	11

TABLE 10-11 REPRESENTATIVE METALS

Material	Nominal Composition (Essential Elements) (%)		Form and Condition	Typical Mechanical Properties			
				Yield Strength (0.2% offset) (1000 psi)	Tensile Strength (1000 psi)	Elongation in 2 in. (%)	Hardness, Brinell or Rockwell
Monel alloy 400	Ni	bal					
Rod and Bar—ASTM B164	Cu	31.5					
Plate, sheet, and strip—ASTM B127	Fe	1.35	Rod hot-rolled annealed	30	79	48	125
	Mn	0.90					
Tube—ASTM B165	Si	0.15					
	C	0.12					
Inconel Alloy 600 plate	Ni	bal					
Sheet and strip—ASTM B168	Cr	15.8					
Rod and bar—ASTM B166	Fe	7.2	Rod hot-rolled	36	90	47	150
Pipe and tube—ASTM B163 and B167	Mn	0.2	annealed				
	Si	0.2					
	C	0.04					
	Cu	0.10					
Hastelloy Alloy W	Cr	5					
Wire—AMS 5786	Mo	24.5	Sheet annealed	53	123	55	—
Bar and forgings—AMS 5755A	Fe	5.5					
	Ni	bal					
Ingot iron			Hot rolled	29	45	26	90
	Fe	99.9 plus					
			Annealed	19	38	45	67
Wrought iron	Fe	bal	Hot rolled	30	48	30	100
Forgings—ASTM A73	Slag	2.5					
Carbon steel—SAE 1020	Fe	bal	Annealed	38	65	30	130
ASTM A285	C	0.20					
	Mn	0.45	Quenched and				
	Si	0.25	tempered at 1000° F	62	90	25	179
300 M alloy steel	Fe	bal					
Bar and forgings—AMS 6416	Mn	0.80					
	Si	1.6					
	Ni	1.85	Hardened	240	290	10	535
	Cr	0.85					
	Mo	0.38					
	V	0.08					
	C	0.43					
Cast gray iron	C	3.4					
ASTM A48 class 30	Si	1.8	As cast	—	32	—	190
	Mn	0.8					
	Fe	bal					
Malleable iron	C	2.5					
Castings—ASTM A47	Si	1	Annealed	33	52	12	130
	Mn	0.55 max					
	Fe	bal					

Typical physical properties								
Density (lb/cu in)	Specific Gravity	Melting Point (°F)	Specific Heat (32 to 212°F) (Btu/lb/°F)	Thermal-expansion Coefficient (32 to 212°F) (in 10^{-6} in./in./°F)	Thermal Conductivity (32 to 212°F) (Btu/sq ft/hr/°F/in.)	Electrical Resistivity (68°F) (ohms/cir mil ft)	Tensile Modulus of Elasticity (10^6 psi)	Torsional Modulus of Elasticity ($\times 10^6$ psi)
0.319	8.84	2460	0.102	7.7	151	307	26	9.5
0.304	8.43	2600	0.106	7.4	103	620	31	11
0.325	9.03	2400	—	6.3	—	—	—	—
0.284	7.86	2795	0.108	6.8	490	57	30.1	11.8
—	—	—	—	—	—	—	—	—
0.278	7.70	2750	0.11	6.35	418	70	29	—
0.284	7.86	2760	0.107	6.7	360	60	30	—
—	—	—	—	—	—	—	—	—
0.283	7.84	2740	0.107	6.7	360	60	30	11.6
0.260	7.20	2150	—	6.7	310	400	14	—
0.264	7.32	2250	0.122	6.6	—	180	25	—

TABLE 10-11 REPRESENTATIVE METALS

Material	Nominal Composition (Essential Elements) (%)	Form and Condition	Typical Mechanical Properties: Yield Strength (0.2% offset) (1000 psi)	Tensile Strength (1000 psi)	Elongation in 2 in. (%)	Hardness, Brinell or Rockwell
Ductile iron (Nickle containing) Grade 60-40-18 castings—ASTM A536	Fe bal C 3.6 Si 2.3 Mn 0.5 Ni 0.75	Annealed	47	65	24	160
Ductile iron (Nickel containing) Grade 120-90-02 castings—ASTM A536	Fe bal C 3.6 Si 2.3 Mn 0.5 Ni 0.75	Oil quenched and tempered	120	140	4	325
Type 201 stainless steel (UNS 20100) Plate, sheet, and strip—ASTM A412 Bar—ASTM A429	Fe bal Cr 17 Ni 4.5 Mn 6.5 N_2 0.25 max C 0.15 max	Strip annealed	55	115	60	90 HRB
Type 302 stainless steel (UNS 30200) Plate, sheet, and strip—ASTM A167 and A 240 Bar—ASTM A276 and A314 Wire—ASTM A313 Forgings—ASTM A473	Fe bal Cr 18 Ni 9 C 0.15 max	Sheet annealed Cold rolled	40 up to 165	90 up to 190	50 5	85 HRB up to 40 HRC
Type 303 and 303 Se stainless steel (UNS 30323) Bar—ASTM A276 and A314 Forgings—ASTM A473	Fe bal Cr 18 Ni 9 {S 0.15 min or {Se 0.15 min C 0.15 max	Bar annealed	35	90	50	160
Type 314 stainless steel (UNS 31400) Bar—ASTM A276 and A314	Fe bal Cr 25 Ni 20 Si 2.50 C 0.25 max	Bar annealed	50	100	45	180
Type 405 stainless steel (UNS 40500) Plate, sheet, and strip—ASTM A176 and A240 Tube—ASTM A268 Bar—ASTM A276 and A314	Fe bal Cr 12.5 C 0.08 max Al 0.20	Sheet annealed	40	65	25	75 HRB

Source: Courtesy of Inco.

Typical Physical Properties								
Density (lb/cu in)	Specific Gravity	Melting Point (°F)	Specific Heat (32 to 212°F) (Btu/lb/°F)	Thermal-expansion Coefficient (32 to 212°F) (in 10^{-6} in./in./°F)	Thermal Conductivity (32 to 212°F) (Btu/sq ft/hr/°F/in.)	Electrical Resistivity (68°F) (ohms/cir mil ft)	Tensile Modulus of Elasticity (10^6 psi)	Torsional Modulus of Elasticity (× 10^6 psi)
0.250	7.1	2150	—	6.2	276	399	24.5	9.3
0.252	7.2	2150	—	5.9	218	408	24.5	9.3
0.283	7.86	—	—	8.7	—	423	28.6	—
0.29	7.9	2590	0.12	9.6	113	432	28	12.5
—	—	—	—	—	—	—	—	—
0.29	—	2590	0.12	9.6	113	432	28	—
0.279	—	—	0.12	8.4	121	462	29	—
0.28	7.7	2790	0.11	6.0	—	360	29	—

TABLE 10-12 REPRESENTATIVE ULTRAHIGH-STRENGTH STEELS[a] (Courtesy Advanced Materials & Processes)

Steel and Composition[a] (%)	Tempering Temperature [°C (°F)]	Tensile Strength [MPa (10^3 psi)]	Yield Strength [MPa (10^3 psi)]	Elongation in 50 mm (2 in.)(%)	Reduction in Area (%)	Hardness[c]	Impact Energy[d] [J (ft · lbf)]
Medium-carbon low-alloy steels							
AISI/SAE 4130 (UNS G41300):[e]	205 (400)	1765 (256)	1520 (220)	10	33	475 HB	18 (13) I
0.28–0.33 C, 0.4–0.6 Mn,	425 (800)	1380 (200)	1170 (170)	16.5	49	375 HB	34 (25) I
0.2–0.35 Si, 0.8–1.1 Cr 0.15–0.25 Mo	650 (1200)	965 (140)	830 (120)	22	63	270 HB	135 (100) I
AISI/SAE 4140 (UNS G41400):[f]	205 (400)	1965 (285)	1740 (252)	11	42	578 HB	15 (11) I
0.38–0.43 C, 0.75–1 Mn,	425 (800)	1450 (210)	1340 (195)	15	50	429 HB	28 (21) I
0.2–0.35 Si, 0.8–1.1 Cr, 0.15–0.25 Mo	650 (1200)	900 (130)	790 (114)	21	61	277 HB	112 (83) I
AISI/SAE 4340 (UNS G43400):[h]	205 (400)	1980 (287)	1860 (270)	11	39	53 HRC	20 (15) I
0.38–0.43 C, 0.6–0.8 Mn,	425 (800)	1500 (217)	1365 (198)	14	48	46 HRC	16 (12) I
0.2–0.35 Si, 0.7–0.9 Cr, 1.65–2 Ni, 0.2–0.3 Mo	650 (1200)	1020 (148)	860 (125)	20	60	31 HRC	100 (74) I
Alloy 300M (UNS K44220):[i]	205 (400)	2140 (310)	1650 (240)	7	27	54.5 HRC	21.7 (16) CVN
0.4–0.46 C, 0.65–0.9 Mn, 1.45–1.8 Si, 0.7–0.95 Cr, 1.65–2 Ni, 0.3–0.45 Mo, 0.05 V min	425 (800)	1790 (260)	1480 (215)	8.5	23	45.5 HRC	13.6 (10) CVN

AISI/SAE 6150 (UNS G61500):[k]	205 (400)	2050 (298)	1810 (263)	1	5	610 HB	—
0.48–0.53 C, 0.7–0.9 Mn,	425 (800)	1585 (230)	1490 (216)	11	42	470 HB	14 (10) I
0.2–0.35 Si, 0.8–1.1 Cr, 0.15–0.25 V	595 (1100)	1150 (167)	1080 (157)	16	47	350 HB	28 (21) I
AISI/SAE 8640 (UNS G86400):[l]	205 (400)	1810 (263)	1670 (242)	8	25.8	55 HRC	11.5 (8.5) I
0.38–0.43 C, 0.75–1 Mn,	425 (800)	1380 (200)	1230 (179)	10.5	46.3	44 HRC	27.8 (20.5) I
0.2–0.35 Si, 0.4–0.6 Cr, 0.4–0.7 Ni, 0.15–0.25 Mo	650 (1200)	870 (126)	760 (110)	20.5	61	28 HRC	96.9 (71.5) I
Medium-alloy air-hardening steels							
H11 Mod (UNS T20811 Mod):[m]	540 (1000)	2010 (291)	1675 (243)	9.6	30.6	56 HRC	21 (15.5) CVN
0.37–0.43 C, 0.2–0.4 Mn, 0.8–1 Si, 4.75–5.25 Cr, 1.2–1.4 Mo, 0.4–0.6 V	650 (1200)	1060 (154)	850 (124)	14.1	41.2	33 HRC	40 (29.5) CVN
AISI/SAE H13 (UNS T20813):[n]	525 (980)	1960 (284)	1570 (228)	13[o]	46.2	52 HRC	16 (12) CVN
0.32–0.45 C, 0.2–0.5 Mn, 0.8–1.2 Si, 4.75–5.5 Cr, 1.1–1.75 Mo, 0.8–1.2 V	605 (1120)	1495 (217)	1290 (187)	15.4[o]	54	44 HRC	30 (22) CVN

TABLE 10-12 REPRESENTATIVE ULTRAHIGH-STRENGTH STEELS

Steel and Composition[a] (%)	Tempering Temperature [°C (°F)]	Tensile Strength [MPa (10^3 psi)]	Yield Strength [MPa (10^3 psi)]	Elongation in 50 mm (2 in.)(%)	Reduction in Area (%)	Hardness[c]	Impact Energy[d] [J (ft · lbf)]
						Fracture toughness, [K_k, MPa · $m^{1/2}$ (ksi · in.$^{1/2}$)]	
High-fracture-toughness steels							
HP 9-4-30 (UNS K91283):[p] 0.29–0.34 C, 0.10–0.35 Mn, 0.2 Si max, 0.9–1.1 Cr, 7–8 Ni, 0.9–1.1 Mn, 0.06–0.12 V, 4.25–4.75 Co	205 (400)[q]	1650–1790 (240–260)	1380–1450 (200–210)	8–12[o]	25–35	66–99 (60–90)	20–27 (15–20) CVN
	550 (1025)[r]	1520–1650 (220–240)	1310–1380 (190–200)	12–16[o]	35–50	99–115 (90–105)	24–34 (18–25) CVN
	Quenchant						
AF1410 (UNS K92571):[s] 0.13–0.17 C, 0.1 Mn max, 0.1 Si max, 1.8–2.2 Cr, 9.5–10.5 Ni, 0.9–1.1 Mo, 13.5–14.5 Co	Air	1680 (244)	1475 (214)	16	69	174 (158)	69 (51) CVN
	Oil	1750 (254)	1545 (224)	16	69	154 (140)	65 (48) CVN
	Water	1710 (248)	1570 (228)	16	70	160 (146)	65 (48) CVN

Source: Bruce A. Becherer and Thomas J. Witheford. "Heat Treating of Ultrahigh-Strength Steels," *ASM Handbook,* 10th ed., Vol. 4, *Heat Treating,* ASM International, Materials Park, Ohio, 1991, pp. 207–218.

[a]The ultrahigh-strength-steel category (1380 MPa, 200×10^3 psi, minimum yield strength) also includes the 18% Ni maraging steels and a variety of stainless-type steels.

[b]Phosphorus and sulfur contents may vary with steelmaking practice, but usually do not exceed 0.035% P and 0.04% S.

[c]HB, Brinell; HRC, Rockwell C.

[d]I, Izod; CVN, Charpy V-notch.

[e]Round bars, 25-mm (1 in.) diameter; water quenched from 845 to 870°C (1550 to 1600°F).

[f]Round bars, 13-mm (0.5 in.) diameter; oil quenched from 845°C (1550°F).

[g]EAF-VAR plate, 25-mm (1 in.) thick; oil quenched from 870°C (1600°F), tempered 2 + 2 hr.

[h]Oil quenched from 845°C (1550°F).

[i]Round bars, 25-mm (1 in.) diameter; oil quenched from 860°C (1575°F).

[j]Bars normalized at 900°C (1650°F), oil quenched from 845°C (1550°F), D-6A was developed by Ladish Co. Inc., Cudahy, Wis.

[k]Round bars, 14-mm (0.55 in.) diameter; normalized at 870°C (1600°F), oil quenched from 860°C (1575°F).

[l]Round bars, 13.5-mm (0.53 in.) diameter; oil quenched from 830°C (1525°F).

[m]Longitudinal properties. Heat treatment: air cooled from 1010°C (1850°F), double tempered 2 + 2 hr.

[n]Longitudinal properties of round bars. Heat treatment: oil quenched from 1010°C (1850°F) double tempered 2 + 2 hr.

[o]Elongation in 4D.

[p]Oil quenched from 845°C (1550°F), refrigerated to −73°C (−100°F), double tempered, HP 9-4-30 was developed by Republic Steel Corp. (now LTV Steel Co.), Cleveland, Ohio.

[q]Hardness, 49–53 HRC.

[r]Hardness, 44–48 HRC.

[s]VIM/VAR plate, 50-mm (2 in.) thick. Heat treatment: 675° C (1250°F) for 8 hr, air cooled; 900°C (1650°F) for 1 hr, quenched; 830°C (1525°F) for 1 hr, quenched; refrigerated at −73°C (−100°F) for 1 hr; 510°C (950°F) for 1 hr, air cooled.

TABLE 10-13 PROPERTIES OF SELECTED CERAMICS

	Density (lb/in^3) (kg/m^3)	Hardness (M, Mohs) (K, Knoop)	Tensile Strength (psi) (MPa)	Thermal Conductivity ($Btu \cdot in./hr \cdot ft^2 \cdot °F$) ($W/m \cdot K$)		Coefficient of Thermal Expansion ($10^{-6}\ F^{-1}$) ($10^{-6}\ K^{-1}$)
Alumina	0.14	M, 9	25,000	25°C	192–255	77–1830°F 4.3
(Al_2O_3)	3.8	K, 2500	172		27.7–36.7	298–1272 K 8.1
Beryllia	0.11	M, 9	23,000	25°C	1741	68–2550°F 5.28
(BeO)	2.92	K, 2000	159		250	293–1672 K 9.5
Boron carbide	0.087	M, 9	22,500	70°F	104–197	0–2550°F 1.73
(B_4C)	2.41	K, 2800	155		—	255–1672 K 3.1
Boron nitride	0.076	—	3,500	70°F	100–200	70–1800°F 4.17
(BN)	2.10		24.1			294–1255 K 7.5
Cordierite	0.065	M, 6.5	3,500	25°C	12–22	68–212°F 2.08
($2MgO \cdot 2Al_2O_3 \cdot 5SiO_2$)	1.8		24.1		1.7–3.2	293–373 K 3.7
Silicon carbide	0.11	M, 9	24,000	70°F	101	0–2552°F 2.4
(SiC)	3.17	K, 2500	165		—	225–1672 K 4.3
Steatite	0.09	M, 7.5	8,500	25°C	20–41	68–212°F 3.99
($MgO \cdot SiO_2$)	2.7	K, 1500	60		2.9–5.9	293–373 K 7.2
Zircon	0.13	M, 8	12,000	25°C	4.9–6.2	68–212°F 1.84
($ZrO_2 \cdot SiO_2$)	3.7		82.7			293–212 K 3.3

Source: Data adapted from *Machine Design,* '84 Materials Reference Issue 3M, and Ceramics Bulletin No. 757, *Materials Engineering,* '84 Materials Selection Issue.

Dielectric Constant at 10^6 Hz (Except as Noted)	Volume Resistivity	Compressive Strength (psi) (MPa)	Flexural Strength (psi) (MPa)		Impact Strength (in. · lb) (N · m)	Modulus of Elasticity (psi × 10^6) (GPa)	Safe Service Temperature (°F) (°C)
			70°F	2,250°F			
8.0–10.0	$>10^{20}$	340,000	48,000	31,000	6.5	50	3540
		2,344	311	214	0.73	379	1965
6.4–7.0	$>10^{14}$	260,000	33,000	—	—	47	4350
		1,793	228	—	—	324	2414
	—	420,000	44,000	—	—	65	1100
		2,896	303	—	—	448	611
4.1–4.8	—	45,000	—	—	—	7	3000
		310	—	—	—	48	1665
4.02–6.23	$>10^{20}$	40,000	8,000	—	2.5	7	2282
		276	55	—	0.28	48	1250
—	—	100,000	110,000	80,000	—	62	3200
		690	758	552	—	427	1776
5.9–6.3 6 × 10^1 cps (6 × 10^1 Hz)	$>10^{20}$	90,000	19,000	—	5.0	15	1832
		620	131	—	0.56	103	1016
8.0–10.0	$>10^{20}$	100,000	22,000	—	5.5	23	2012
		690	152	—	0.62	159	1117

TABLE 10-14 REPRESENTATIVE GLASSES

U.S. Customary																
				Corrosion Resistance			Thermal Expansion Multiply by 10^{-7} in./in./°F		Upper Working Temperatures (Mechanical Considerations Only)				Thermal Shock Resistance Plates 6 × 6 in.			
									Annealed		Tempered		Annealed			
Type	Color	Principal Use	Class	Weathering	Water	Acid	32 to 572°F	77°F to Setting Point	Normal Service (°F)	Extreme Service (°F)	Normal Service (°F)	Extreme Service (°F)	⅛ in. Thick (°F)	¼ in. Thick (°F)	½ in. Thick (°F)
(1) Soda lime	Clear	Lamp bulbs	I	3	2	2	52	58.3	230	860	428	482	149	122	95
(2) Potash soda lead	Clear	Lamp tubing	I	2	2	2	49.8	53.9	230	716	—	—	—	—	—
(3) Aluminosilicate ..	Clear	Electron tube	I	1	1	3	25.6	30	392	1202	752	842	257	212	158
(4) Borosilicate	Clear	Sealed-beam lamps	I	[3]1	[3]2	[3]2	20.4	21.2	446	860	500	500	320	266	294
(5) Borosilicate	Clear	General	I	2	2	2	18.9	20.6	446	842	482	482	320	266	194
(6) 96% Silica	Clear	High temp.	I	1	1	1	4.2	3.1*	1652	2192	—	—	—	—	—
(7) Fused silica	Clear	Optical	I	1	1	1	3.1	1.9*	1652	2012	—	—	—	—	—
(8) Glass-ceramic ...	White	Missle nose cones	II	—	1	4	31.7	—	1292	—	—	—	572	338	266

Source: Courtesy of Corning Glass Works.

TABLE 10-14 (continued)

	Thermal Stress Resistance °F	Viscosity Data: Strain Point (°F)	Annealing Point (°F)	Softening Point (°F)	Working Point (°F)	Knoop Hardness, KHN_{100}	Density (lb/ft^3)	Young's Modulus (Multiply by 10^6 psi)	Poisson's Ratio	Log_{10} of volume resistivity ($\Omega \cdot cm$): 77°F	482°F	662°F	Dielectric properties at 1 MHz, 68°F: Power Factor (%)	Dielectric Constant	Loss Factor (%)	Refractive Index
(1)	29	883	957	1285	1841	465	154	10.2	0.22	12.4	6.4	5.1	0.9	7.2	6.5	1.512
(2)	36	743	815	1166	1805	382	190.3	8.6	0.22	17+	10.1	8.0	0.12	6.7	0.8	1.560
(3)	47	1229	1310	1666	2134	514	164.7	12.5	0.24	17+	13.5	11.3	0.16	6.3	1.0	1.547
(4)	86	932	1011	1436	2133	—	140.4	9.3	0.19	18	8.1	6.6	0.45	4.85	2.18	1.476
(5)	94	892	973	1436	2188	442	139.8	9.0	0.20	17	9.4	7.7	0.18	4.5	0.79	1.473
(6)	396	1634	1868	2786	—	487	136	9.8	0.19	17+	9.7	8.1	0.04	3.8	0.15	1.458
(7)	515	1753	1983	2876	—	489	137.2	10.5	0.16	17+	11.8	10.2	0.001	3.8	0.0038	1.459
(8)	29	—	—	—	—	657	162.2	17.2	0.24	16.7	10.0	8.7	0.30	5.6	1.7	—

(Corning Glass Works)

TABLE 10-14 (continued)

SI															
				Corrosion Resistance			Thermal Expansion (Multiply by 10^{-7} cm/cm/°C)		Upper Working Temperatures (Mechanical Considerations Only)				Thermal Shock Resistance Plates 15 × 15 cm		
									Annealed		Tempered		Annealed		
Type	Color	Principal Use	Class	Weathering	Water	Acid	0–300°C	25°C to Setting Point	Normal Service (°C)	Extreme Service (°C)	Normal Service (°C)	Extreme Service (°C)	3.2 mm Thick (°C)	6.4 mm Thick (°C)	12.7 mm Thick (°C)
(1) Soda lime	Clear	Lamp bulbs	I	3	2	2	93.5	105	110	460	220	250	65	50	35
(2) Potash soda lead	Clear	Lamp tubing	I	2	2	2	89.5	97	110	380	—	—	65	50	35
(3) Aluminosilicate ..	Clear	Electron tube	I	1	1	3	46	54	200	650	400	450	125	100	70
(4) High lead	Clear	Solder sealing	II	1	1	4	84	92	100	300	—	—	—	—	—
(5) Borosilicate	Clear	General	I	2	2	2	34	37	230	450	250	250	160	130	90
(6) 96% Silica	Clear	High temp.	I	1	1	1	7.5	5.5*	900	1200	—	—	—	—	—
(7) Fused silica	Clear	Optical	I	1	1	1	5.5	3.5*	900	1100	—	—	—	—	—
(8) Glass-ceramic ...	White	Missle nose cones	II	—	1	4	57	—	700	—	—	—	200	170	130

TABLE 10-14 (continued)

	Thermal Stress Resistance °C	Knoop Hardness, KHN_{100}	Density (g/cm^3)	Young's Modulus (Multiply by 10^3 kg/mm^2)	Poisson's Ratio	Log_{10} of Volume Resistivity ($\Omega \cdot cm$)			Dielectric Properties at 1 MHz, 20°C			Refractive Index
						25°C	250°C	350°F	Power Factor (%)	Dielectric Constant	Loss Factor (%)	
(1)	16	465	2.47	7.1	0.22	12.4	6.4	5.1	0.9	7.2	6.5	1.512
(2)	20	382	3.05	6.0	0.22	17+	10.1	8.0	0.12	6.7	0.8	1.560
(3)	26	514	2.64	8.8	0.24	17+	13.5	11.3	0.16	6.3	1.0	1.547
(4)	21	—	5.42	5.6	0.28	17+	10.6	8.7	0.22	15	3.3	1.86
(5)	52	442	2.24	6.3	0.20	17	9.4	7.7	0.18	4.5	0.79	1.473
(6)	220	487	2.18	6.9	0.19	17+	9.7	8.1	0.04	3.8	0.15	1.458
(7)	286	489	2.20	7.4	0.16	17+	11.8	10.2	0.001	3.8	0.0038	1.459
(8)	16	657	2.6	12	0.24	16.7	10.0	8.7	0.30	5.6	1.7	—

HARDNESS/TENSILE-STRENGTH CONVERSION

Hardness conversion chart for hardenable carbon and alloy steels: approximate relationship between hardnesses and tensile-strength. Conversions from one scale to another are made at the intercepts with the curve crossing the chart. For example, follow the horizontal line representing 400 Diamond Pyramid Hardness to its intersection with the conversion curve. From this point follow vertically downward for equivalent Rockwell C hardness (41), horizontally to the right for Brinell hardness (379) and tensile strength (187,000 psi), and vertically upward for the equivalent Shore hardness (55).

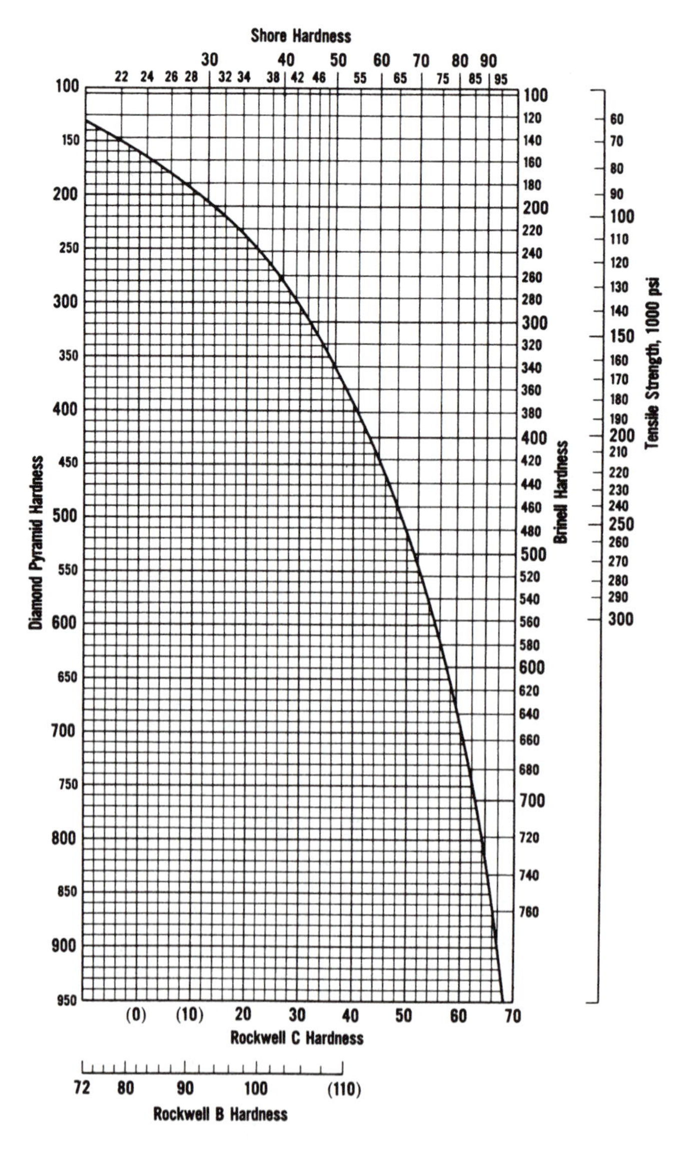

Source: Data from 1966 SAE handbook.

Index of Principles/Equations and Illustrative Problems

Index

Page numbers on which definitions appear are printed in **boldface.**